Advances in Intelligent Systems and Computing

191

For further volumes:
http://www.springer.com/series/11156

Zhenyu Du (Ed.)

Proceedings of the 2012 International Conference of Modern Computer Science and Applications

Editor
Dr. Zhenyu Du
Information Technology & Industrial
Engineering Research Center
Wuhan
China

ISSN 2194-5357 e-ISSN 2194-5365
ISBN 978-3-642-33029-2 e-ISBN 978-3-642-33030-8
DOI 10.1007/978-3-642-33030-8
Springer Heidelberg New York Dordrecht London

Library of Congress Control Number: 2012945427

Printed on acid-free paper

Springer is part of Springer Science+Business Media (www.springer.com)

Foreword

2012 International Conference of Modern Computer Science and Applications (MCSA 2012) was held on September 8, 2012 in Wuhan, China. This conference is sponsored by Information Technology & Industrial Engineering Research Center.

MCSA 2012 provides an excellent international forum for sharing knowledge and results in theory, methodology and applications of modern computer science and applications. The conference looks for significant contributions to all major fields of the modern computer science and applications in theoretical and practical aspects. The aim of the conference is to provide a platform to the researchers and practitioners from both academia as well as industry to meet and share cutting-edge development in the field.

This book covers significant recent developments in the computational methods, algorithms and scientific applications, computer networks, geometric modeling, graphics and visualization,computer science applications in education, computer science applications in business and social and modern and emerging applications.

We are glad this conference attracts your attentions, and thank your support to our conference. We will absorb remarkable suggestion, and make our conference more successful and perfect.

Contents

Computational Methods, Algorithms and Scientific Applications

A Study on the Distributed Real-time System Middleware Based on the DDS 1
Ren Haoli, Gao Yongming

An Iteration Method for Solving an Inverse Parabolic Equation in the Reproducing Kernel Space 7
Sufang Yue, Zongli Ma

Price Sensitive Support Vector Machine Based on Data Set Decomposition 13
Xia Guilin

Research on the Application Subdivision Pattern in Triangular Mesh Simplification 19
Chen Jingsong

Study on Reduction of Multi-Dimensional Qualitative Variables with Granular Computing 25
Liu Hong, Yang Shen, Li Haiyan

Isolated Distributed Architecture for Cloud-Storage Services Based on Structured Strategy 31
Ying Gao, Jianlin Zheng

A New Management Algorithm for Multiplex Networks 39
Yu Genjian

An Improved Min-Min Algorithm in Cloud Computing 47
Gang Liu, Jing Li, Jianchao Xu

A Novel Information Technology Based College Physical Education Model 53
Li Yan-xia, Wang Qiang, Zhang Ying-jian

The Improved Text Classification Method Based on Bayesian and k-NN 57
Wang Tao, Huo Liang, Yang Liu

The Basic Principle and Applications of the Search Engine Optimization . . . 63
Xie Xiaojie, Fang Yuan, Wu Jian

Semi-supervised Incremental Model for Lexical Category Acquisition 71
Bichuan Zhang, Xiaojie Wang

An Approach to TCM Syndrome Differentiation Based on Interval-Valued Intuitionistic Fuzzy Sets 77
Zhiling Hong, Meihong Wu

A Novel Artificial Neural Network Learning Algorithm 83
Tinggui Li, Qinhui Gong

Computer Networks

IPv6 Network Intrusion Detection Protocol Analysis Techniques 89
Zhang Zhi

Design of Wireless Industrial Network Based-On ZigBee and ARM11 95
Zhuguo Li, Bingwen Wang, Xiaoya Hu, Lizhu Feng

The Research on Building of a New Safe Campus Network 101
Ma Tao, Wei Shaoqian, Wu Baozhu, Qin Yihui

Research of High Definition Download Model Based on P2P&DRM 105
Zexun Wu

Resistance Test on Ethernet Switch against Dynamic VLAN Attack 111
Zhihong Liang, Baichao Li

Modeling Dynamic File Diffusion Behavior in P2P Networks 117
Baogang Chen, Jinlong Hu

Research on Earthquake Security Evaluating Using Cellular Neural Network 123
Zhou Chang-xian, Zheng Shao-peng

Research on Zigbee Based Resources Addressing Technology in Internet of Things 129
Zhang Chunzhi, Li Haoru, Hu Guangzhou, Bao Qingpeng, Geng Shu

A Scalable Model for Network Security Situation Based on Endsley Situation Model 137
ChenHao Deng

Information Diffusion Model Based on Social Network 145
Zhang Wei, Ye Yanqing, Tan Hanlin, Dai Qiwei, Li Taowei

Modeling and Application in Networked Control System Based on Predictive Functional Control with Incremented Weight 151
Zhang Jun, Luo Da-yong

The Research and Design of the Internet Interface Unit of the Remote Electromechanical Control 159
Zhu Hui-ling, Jiang Yan, Zhu Xin-yin, Xie Jing

Research of Security Router Technology on Internet Environment 167
Hu Liang, Zhang Qiansheng

The Firmware Upgrading Design of Embedded Equipment That Based on SNMP 173
Hu Liang, Zhang Qiansheng

The Design of Port Scanning Tool Based on TCP and UDP 179
Hu Liang, Zhang Qiansheng

Research on the System Model of Network Intrusion Detection 185
Yang Yunfeng

The Information System of the Radio and Television Network Based on SOA 191
Ran Zhai

Research of Communication System on Intelligent Electric Community Based on OPLC 197
Zengyou Sun, Peng Chen, Chuanhui Hao

Geometric Modeling, Graphics and Visualization

End-Orthodox Graphs Which Are the Join of N Split Graphs 203
Hailong Hou, Aifen Feng, Rui Gu

The Zero-Divisor Semigroups Determined by Graphs Gn(2,1) 209
Hailong Hou, Aifen Feng, Rui Gu

The Design of Fusion Semantics Automatic Labeling and Speech Recognition Image Retrieval System 215
Lu Weiyan, Wang Wenyan, Liu-Suqi

A Method to Automatic Detecting Coronal Mass Ejections in Coronagraph Based on Frequency Spectrum Analysis 223
Zeng Zhao-xian, Wei Ya-li, Liu Jin-sheng

Computer Science Applications in Education

The Study on Network Education Based on Java 2 Platform Enterprise Edition 229
Liu Chunli, Huang Linna

Study on a Learning Website of Self-directed English Study 235
Yan Pei

The Building of Network Virtual Laboratory for Physics Teaching 241
Baixin Zhang

Intelligent Computer Aided Second Language Learning Based on Constitutive Principle 247
Wanzhe Zhao

The Application of Computer - Based Training in Autism Treatment 253
Li Lin, Chunmei Li

Survey to CAT Mode Based on the Construction Theory 259
You Yangming, Wang Bingzhang, Chi ZhenFeng, Zhao BaoBin, Zhang Guoqing

Intelligent Distance Learning Model of Ethnic Minority 265
Jianru Zhang

Study on English Writing Based on Modern Information Technology 271
Wen Lai, Han Lai

The Study on College Sports Theory Teaching Based on Computer Intelligence Technology 277
Wenjie Zhu

Research on Multimedia English Teaching Model Based on Information Technology 283
Shuying Han, Xiuli Guo

The Study on University Library Information Service for Regional Characterized Economy Construction Based on Integrated Agent 289
Shuhua Han, Xuemei Su

The Research on Information Commons Based on Open Access in College Library 295
Xiaohui Li, Hongxia Wang, Hongyan Guo

The Development Trend of the 21st Century Computer Art 301
Weiming Deng

On Foundation Database System of Education MAN Based on Three-Layer Structure 307
Xiaohong Xiong

The Practice of Physical Training Based on Virtual Trainer Concept 313
Bin Tian

Development of Graduate Simulation Experiment Teaching System Based on Matlab Language 321
Chunhua Tu

Study on the Application of Fuzzy KNN to Chinese and English Recognition 327
Cuiping Zhang

Study on English Converting to Chinese of Database Fields 333
Cuiping Zhang

Cognitive Function Analysis on Computer English Metaphor 339
Cuiping Zhang

ASP.net-Based Boutique Web Design and Implementation of Curriculum 345
Wang Lifen

The Research on Objectives Design of Vocational CNC Machining Process 351
Shen Weihua

The Research on Online Examination System of PE Theory Courses 357
Xing-dong Yang

Computer Science Applications in Business and Social

Regression Method Based on SVM Classification and Its Application in Influence Prediction of a Liberalism Case in International Law 363
Gang Wu, Jinyan Chenggeng

The Establishment of the Database of Mathematics History 369
Xie Qiang

Research on Real-Time Monitoring and Test of Download Threads of the Literary Fiction Website —-Take "The Woman Warrior" as an Example . . . 375
Yanhua Xia

Design of Forest Fire Monitoring System in Guangxi Zhuang Autonomous Region Based on 3S Technology 381
Yuhong Li, Li He, Xin Yang

Analyze the Film Communication under Internet Circumstances 387
Yang Qi

Research on Data Mining in Remote Learning System of Party History Information 393
Ruifang Liu, Changcun Li, Qiwen Jin

The Study on Oil Prices' Effect on International Gas Prices Based on Using Wavelet Based Boltzmann Cooperative Neural Network 399
Xiazi Yi, Zhen Wang

Forecasting of Distribution of Tectonic Fracture by Principal Curvature Method Using Petrel Software 405
Li Zhijun, Wang Haiyin, Wang Hongfeng

Study on Bulk Terminal Berth Allocation Based on Heuristic Algorithm . . . 413
Hu Xiaona, Yan Wei, He Junliang, Bian Zhicheng

The Research of Intelligent Storage Space Allocation for Exported Containers Based on Rule Base 421
Yan Wei, Bao Xue, Zhao Ning, Bian Zhicheng

Study on Evaluation System of The Quayside Container Crane's Driver . . . 427
Liu Xi, Mi Weijian, Lu Houjun

Using a Hybrid Neural Network to Predict the NTD/USD Exchange Rate 433
Han-Chen Huang

Research on the Influential Factors of Customer Satisfaction for Hotels: The Artificial Neural Network Approach and Logistic Regression Analysis 441
Han-Chen Huang

Research on Analysis and Monitoring of Internet Public Opinion 449
Li Juan, Zhou Xueguang, Chen Bin

Nonlinear Water Price Model of Multi-Source for Urban Water User 455
Wang Li, Ligui Jie, Xiong Yan

A Collaborative Filtering Based Personalized TOP-K Recommender System for Housing 461
Lei Wang, Xiaowei Hu, Jingjing Wei, Xingyu Cui

An Express Transportation Model of Hub-and-Spoke Network with Distribution Center Group 467
Xiong Yan, Wang Jinghui, Zheng Liqun

Developing a Small Enterprise's Human Resource Management System -Based on the Theory of Software Engineering 473
Chen Yu

Research on Applying Folk Arts' Color in Computer Based Art Design 479
Hu Xiao-ying

Detecting Companies' Financial Distress by Multi-attribute Decision-Making Model 485
Li Na

Study on Advertisement Design Method Utilizing CAD Technology 491
Ru Cun-guang

A Solution Research on Remote Monitor for Power System of WEB Technology Based on B/S Mode 495
Guo Jun, Yu Jin-tao, Yang Yang, Wu Hao

The Research of Supply Chain Management Information Systems Based on Collaborative E-Commerce 501
Ruihui Mu

Analysis on Automation of Electric Power Systems Based on GIS 507
Jiang Chunmin, Yang Li

The Empirical Study on the Relationship of Enterprise Knowledge Source and Innovation Performance 511
Dan Zhu, Ailian Ren, Fenglei Wang

E-Business Diffusion: A Measure of E-Communication Effectiveness on the E-Enabled Interorganizational Collaboration 517
Wu Lu, Min He

The Application Research of Small Home Appliance Product Based on Computer Aided Ergonomics 523
Dong Junhua

Pro/E-Based Computer-Aided Design and Research of Small Household Electrical Appliances Mold 529
Li Baiqing

The Key Technology Research of Sport Technology Learning in the Platform Establishment That Based on ASP.NET 535
Tang Lvhua

Computer Application in the Statistical Work ... 541
Yang Liping

Study on Algorithm Design of Virtual Experiment System ... 547
Liu Jianjun, Liu Yanxia

The Application of GIS in the Real Estate Management System ... 553
Ru Qian

Analysis and Implementation of Computer System of College Students' PE Scores Management ... 559
Jianfeng Ma

The Study Based on IEC61850 Substation Information System Modeling ... 565
Yu Xian

Axis Transformation and Exponential Map Based on Human Motion Tracking ... 571
Yu Xue

Research on Characteristics of Workload and User Behaviors in P2P File-Sharing System ... 577
Baogang Chen, Jinlong Hu

Steady-State Value Computing Method to Improve Power System State Estimation Calculation Precision ... 583
Fang Liu, Hai Bao

Feasibility Study of P.E Contest Crowd Simulation ... 589
Bing Zhang

Design and Realization of Distributed Intelligent Monitoring Systems Using Power Plant ... 595
Zhemin Zhou

Modern and Emerging Applications

Output Feedback Control of Uncertain Chaotic Systems Using Hybrid Adaptive Wavelet Control ... 603
Chun-Sheng Chen

The Building of Oral Medical Diagnosis System Based on CSCW ... 609
Yonghua Xuan, Chun Li, Guoqing Cao, Ying Zhang

Zhichan Soup on PD after Transplantation of Neural Stem Cells in Rat Brain Content of DA and Its Metabolites Related Data Mining Research ... 615
Shi Huifen, Yang Xuming

Remote Monitoring and Control of Agriculture ... 623
Xiao Chu, Xianbin Cui, Dongdong Li

Research on Reform of Molding Materials and Technology 629
Pei Xuesheng

Research and Implementation on Temperature and Humidity Measurement System Based on FBG 633
Luo Yingxiang

Clinical Analysis of Non-ST-Segment Elevation Acute Myocardial Infarction 639
Huang Zhaohe, Liang Limei, Pan Xingshou, Lan Jingsheng, He Jinlong, Liu Yan

On the Batch Scheduling Problem in Steel Plants Based on Ant Colony Algorithm 645
Li Dawei, Zhang Ranran, Wang Li

A Multi-channel Real-Time Telemetry Data Acquisition Circuit Design 651
Wang Yue, Zhang Xiaolin, Li Huaizhou

Research on Optical Fiber Inspection Technology Based on Radio Frequency Identification 657
Yu Feng, Liu Wei, Liu Jifei

Usability Analyses of Finger Motion in Direct Touch Technology 663
Xiaofei Li, Feng Wang, Hui Deng, Jibin Yin

Design and Implementation of an Automatic Switching Chaotic System between Two Subsystems 669
Yao Sigai

Study on the Design and Application of Immersive Modeling System for Electronic Products with Multi-displayport Interface 675
Wang Xin-gong

An Efficient Approach for Computer Vision Based Human Motion Capture 681
Wang Yong-sheng

Design and Implementation of Instant Communication Systems Based on Ajax 687
Liu Tian, Wu Jun

Analysis of Characteristics in Japans Robot 695
Chen Beibei

The Development of Mobile Learning System Based on the Android Platform 701
Lingmei Kong

Deep Study of Computer Simulation of Human's Thinking Way and Philosophical Value 707
Peilu Yang

Water Quality Remote Monitoring System Based on Embedded Technology 713
Wang Xiaokai, Deng Xiuhua

Electrical Automation Technology in the Thermal Power 719
Chen Yijun

The Research of Electrical Automation and Control Technology 725
Yu Xian

The Military Strategy Threat of Cognitive System Based on Agent 731
Tiejian Yang, Minle Wang, Maoyan Fang

Based on Zigbee Technology Greenhouse Monitoring System 737
Mingtao Ma, Gang Feng

Author Index 741

A Study on the Distributed Real-time System Middleware Based on the DDS

Ren Haoli and Gao Yongming

Department of Information Equipment, Academy of Equipment, Beijing 101416, China

Abstract. This paper study on the distributed real-time system middleware. Presents a comprehensive overview of the Data Distribution Service standard (DDS) and describes its benefits for developing Distributed System applications. The standard is particularly designed for real-time systems that need to control timing and memory resources, have low latency and high robustness requirements. As such, DDS has the potential to provide the communication infrastructure for next generation precision assembly systems where a large number of independently controlled components need to communicate. To illustrate the benefits of DDS for precision assembly an example application was presented.

Keywords: Real-time System, DDS, middleware, Distributed system.

1 Introduction

In many distributed embedded real-time (DRE) applications have stringent deadlines by which the data must be delivered in order to process it on time to make critical decisions. Further, the data that is distributed must be valid when it arrives at its target. That is, if the data is too old when it is delivered, it could produce invalid results when used in computations [1]. This paper Presents a comprehensive overview of the Data Distribution Service standard (DDS) and describes its benefits for developing Distributed System applications. DDS is a platform-independent standard released by the Object Management Group (OMG) for data-centric publish-subscribe systems [2].

1.1 Distributed Applications

There are many distributed applications exist today, one requirement common to all distributed applications is the need to pass data between different threads of execution. These threads may be on the same processor, or spread across different nodes. You may also have a combination: multiple nodes, with multiple threads or processes on each one. Each of these nodes or processes is connected through a transport mechanism such as Ethernet, shared memory, VME bus backplane, or Infiniband. Basic protocols such as TCP/IP or higher level protocols such as HTTP can be used to provide standardized communication paths between each of the nodes. Shared memory (SM) access is typically used for processes running in the same node. It can also be used wherever common memory access is available. Figure 1 shows an example of a simple

Z. Du (Ed.): Proceedings of the 2012 International Conference of MCSA, AISC 191, pp. 1–6.
springerlink.com

distributed application. In this example, the embedded single board computer (SBC) is hardwired to a temperature sensor and connected to an Ethernet transport. It is responsible for gathering temperature sensor data at a specific rate. A workstation, also connected to the network, is responsible for displaying that data on a screen for an operator to view. One mechanism that can be used to facilitate this data communication path is the Data Distribution Service for Real Time Systems, known as DDS [3].

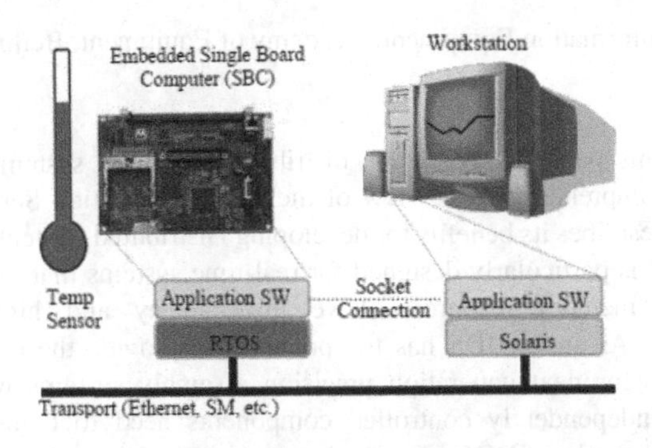

Fig. 1. Simple Distributed Application

1.2 Real-Time Publish-Subscribe Middleware DDS

Distributed systems connect devices, since devices are faster than people, these networks require performance well beyond the capabilities of traditional middleware. The Object Management Group's Data Distribution Service for Real Time Systems (DDS) Standard [4]. The OMG Data-Distribution Service (DDS) is a new specification for publish-subscribe data distribution systems. The purpose of the specification is to provide a common application-level interface that clearly defines the data-distribution service. DDS is sophisticated technology; It goes well beyond simple publishing and subscribing functions. It allows very high performance (tens of thousands of messages/sec) and fine control over many quality of service parameters so designers can carefully modulate information flow on the network [5].

DDS provides common application-level interfaces which allow processes to exchange information in form of topics. The latter are data flows which have an identifier and a data type. A typical architecture of a DDS application is illustrated in figure 2.

Applications that want to write data declare their intent to become "publishers" for a topic. Similarly, applications that want to read data from a topic declare their intent to become "subscribers". Underneath, the DDS middleware is responsible to distribute the information between a variable number of publishers and subscribers.

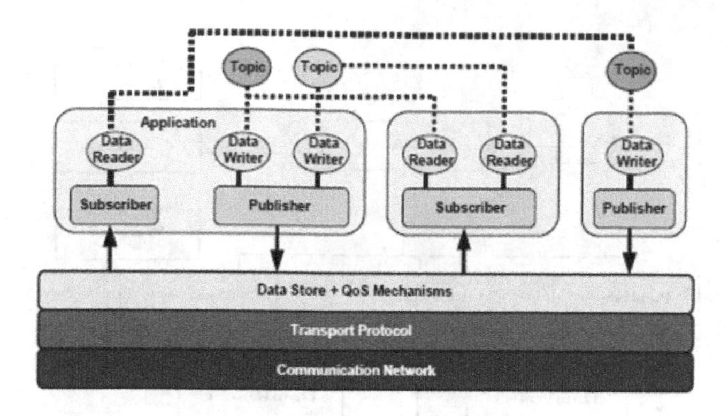

Fig. 2. Decoupling of publishers and subscribers with the DDS middleware

2 The Data Centric Publish/Subscribe Model

The DDS publish-subscribe model connects anonymous information producers (publishers) with information consumers (subscribers). The overall distributed application (the PS system) is composed of processes, each running in a separate address space possibly on different computers [6]. We will call each of these processes a "participant". A participant may simultaneously publish and subscribe to information.Figure3 illustrates the overall DCPS model, which consists of the following entities: DomainParticipant, DataWriter, DataReader, Publisher, Subscriber, and Topic. All these classes extend DCPSEntity, representing their ability to be configured through QoS policies, be notified of events via listener objects, and support conditions that can be waited upon by the application. Each specialization of the DCPSEntity base class has a corresponding specialized listener and a set of QoSPolicy values that are suitable to it. Publisher represents the objects responsible for data issuance. A Publisher may publish data of different data types. A DataWriter is a typed facade to a publisher; participants use DataWriter(s) to communicate the value of and changes to data of a given type. Once new data values have been communicated to the publisher, it is the Publisher's responsibility to determine when it is appropriate to issue the corresponding message and to actually perform the issuance.

A Subscriber receives published data and makes it available to the participant. A Subscriber may receive and dispatch data of different specified types. To access the received data, the participant must use a typed DataReader attached to the subscriber. The association of a DataWriter object with DataReader objects is done by means of the Topic. A Topic associates a name, a data type, and QoS related to the data itself. The type definition provides enough information for the service to manipulate the data. The definition can be done by means of a textual language or by means of an operational "plugin" that provides the necessary methods.

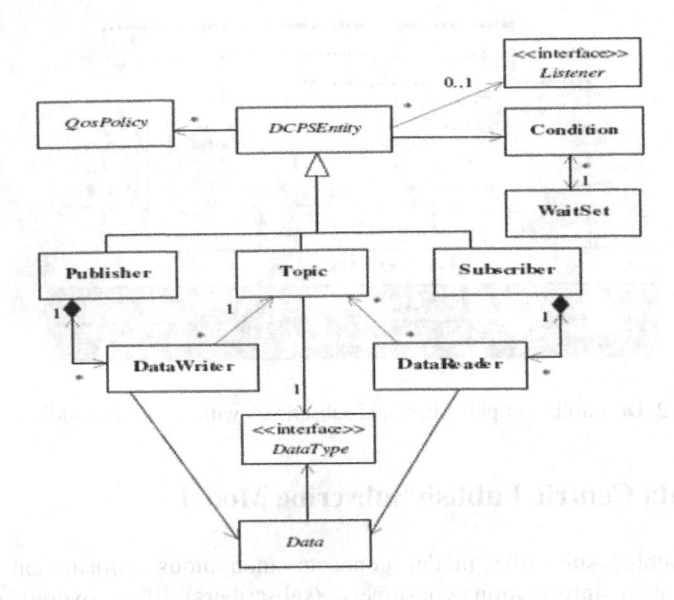

Fig. 3. UML diagram of the DCPS model

3 Example

This section presents the development of a software-based prototype for a modular active fixturing system. The prototype was developed using the commercially available DDS implementation OMG DDS 1.2 from Real-Time Innovations [7].

3.1 Design the System Framework

The system consists of a variable number of physical fixture modules, a fixture control software, a variable number of Human Machine Interfaces (HMI). For the sake of simplicity, each module consists of one linear actuator and three sensors. The former acts as the locating and clamping pin against the workpiece, while the sensors feedback reaction force, position and temperature of the contact point. The fixture modules are implemented as smart devices with local control routines for their embedded sensor/actuator devices. It is further assumed that each fixture module is configured with a unique numerical identifier and with meta information about its sensors and actuators. This way, the module is able to convert the signals coming from the sensors (e.g. a voltage) into meaningful information (e.g. reaction force in Newton) which is then published via DDS. The fixture control implements the global control routines of the fixture. It processes the data coming from the various modules and controls the movement of the actuators by publishing their desired status.

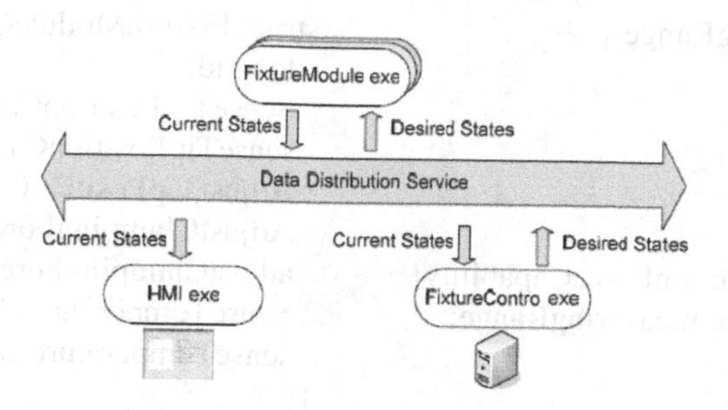

Fig. 4. Overview on the example application

3.2 Design the Data Structures

The first step of the data-centric application development is the definition of the data structures to be exchanged between the processes. In this context, there is a trade-off between efficient data transfer and flexible interpretation of data. On one hand, it shall be allowed to add fixture modules with varying capabilities, i.e. different hardware characteristics and representations of data. This makes it necessary that each module informs other systems about its capabilities which define how data has to be interpreted. On the other hand, it is not efficient to publish this meta-information with every sensor update. For this reason, it is proposed to separate actual state data from meta-information to interpret it. Therefore, two data structures are created for each capability of a fixture module. The first data structure is used for the transmission of the sensor readings or desired states for actuator devices during the manufacturing process. It is a very simple data structure that only consists of a field for the numeric module-ID and the data itself. Below an example is provided for the data structure for force sensor readings. Each attribute is defined with a data type, followed by a name.

```
struct Force {
      long module_id;
      double value;
};
```

Since this structure does not contain any information on how to use the data, an additional data structure is defined for each capability. It contains attributes describing the characteristics of the relevant capability like measuring range, resolution etc. In this prototype the meta-information only contains the measuring range for capabilities that result from the existence of sensor devices. This is further defined by attributes for the minimum and maximum measuring value, as well as the measuring unit. For the latter unique numerical constants have been defined. The following listing provides the data definitions for the capability that results from the existence of a force sensor. Similar structures have been defined for the other capabilities.

```
struct MeasuringRange {
  double min;
  double max;
  long unit;
};

struct SenseReactionForceCapability{
MeasuringRange measuringRange;
};
```

```
struct FixtureModuleCapabilityDef{
  long id;
  SenseTipPositionCapability
  senseTipPositionCapability;
  AdjustTipPositionCapability
  AdjustClampingForceCapability
  adjustClampingForceCapability;
  SenseTemperatureCapability
  senseTemperatureCapability;
};
```

Based on these data type definitions, the source code for all the subscribers, publishers, data readers and data writers are automatically generated in the specified programming language. These classes have to be used by the application programs that implement the fixture modules, the fixture control and any other participating system like monitoring applications.

4 Conclusions

In this paper, a new standard for data-centric publish-subscribe communication has been presented and put in context to the development of next-generation precision assembly platforms. The standard is called Data Distribution Service and is particularly targeting real-time applications which need to manage resource consumption and timeliness of the data transfer. DDS allows platform-independent many-to-many communication and alleviates a number of common problems which are of particular interest for the development of distributed assembly systems.

References

1. Cross, J.K., Schmidt, D.C.: Applying the Quality Connector Pattern to Optimize Distributed Real-time and Embedded Middleware. In: Rabhi, F., Gorlatch, S. (eds.) Patterns and Skeletons for Distributed and Parallel Computing. Springer Verlag (2002)
2. Object Management Group, Data Distribution Service for Real-Time Systems, Version 1.2 (June 2007), http://www.omg.org
3. Joshi, J.: A comparison and mapping of Data Distribution Service (DDS) and Java Messaging Service (JMS), Real-Time Innovations, Inc., Whitepaper (2006), http://www.rti.com
4. Real-Time CORBA Specification Version 1.2, Object Management Group (January 2005), http://www.omg.org/cgi-bin/apps/doc?formal/05-01-04.pdf
5. Joshi, J.: Data-Oriented Architecture, Real-Time Innovations, Inc., Whitepaper (2010), http://www.rti.com
6. Richards, M.: The Role of the Enterprise Service Bus (October 2010), http://www.infoq.com/presentations/Enterprise-Service-Bus
7. Data Distribution Service for Real-Time Systems Specification Version 1.1, Object Management Group (December 2005), http://www.omg.org/docs/formal/05-12-04.pdf

An Iteration Method for Solving an Inverse Parabolic Equation in the Reproducing Kernel Space

Sufang Yue* and Zongli Ma

School of Mathematics and Computing Science, Anqing Teacher College, Anqing, 246011, P.R. China

Abstract. In this paper, an iteration method is used to solve an inverse parabolic equation in the reproducing kernel space. The analytical solution u(t, x) is represented in form of series in the reproducing kernel space. In the mean time, two rapidly convergent sequences un and rn are produced which converge to the analytical solutions. Numerical experiments are presented to demonstrate the accuracy and the efficiency of the proposed method.

Keywords: Parabolic equations, Inverse problem, Iteration method, Reproducing kernel, analytical solution.

1 Introduction

Many physical phenomena can be described in terms of inverse parabolic equations. This type of equations have important application in various branches of pure and applied science, such as heat condition processes, chemical diffusion, thermoelasticity and control theory [1 – 6]. In general, these problems are ill-posed. Therefore, a variety of numerical techniques based on regularization, finite difference, finiteelement and finite volume methods are given to approximate solutions of the equations [7, 8].

In this paper, we present a new iteration method for the following inverse parabolic equation

$$w_t = w_{xx} + p(t)w(t, x) + f(t, x), \quad 0 \le x \le 1, \qquad 0 \le t \le T, \tag{1.1}$$

$$w(0, x) = \phi(x), 0 \le x \le 1, (1.1)$$

$$w(t, 0) = w(t, 1) = 0, \quad 0 \le t \le T,$$

$$w(t, x*) = E(t), 0 \le t \le T,$$

where f (t, x), $\phi(x)$ and E(t) are known functions and x*is a fixed prescribed interior point in (0, 1), while the functions w(t, x) and p(t) are unknown. If w(t, x) is a temperature, then Eq.(1.1)can be regarded as a control problem finding the control function p(t) such that the internal constraint is satisfied. The existence and uniqueness of the equations have been proved[9].

Employing a pair of transformations

$$r(t) = \exp(-\int_0^t p(s)ds), \tag{1.2}$$

* Corresponding author.

Z. Du (Ed.): Proceedings of the 2012 International Conference of MCSA, AISC 191, pp. 7–11.
springerlink.com

$$u(t, x) = r(t)w(t, x) - \phi(x). \tag{1.3}$$

Eq.(1.1) will be
ut= uxx+ r(t)f (t, x) + ϕ00(x) $0 \le x \le 1$, $0 \le t \le T$,
u(0, x) = 0, $0 \le x \le 1$, (1.4)
u(t, 0) = u(t, 1) = 0, $0 \le t \le T$,
u(t, x∗) =E(t)r(t) − ϕ(x∗), $0 \le t \le T$,
where $0 \le x \le 1, 0 \le t \le T$.

2 The Reproducing Kernel Spaces

In this section, several reproducing kernel spaces needed are introduced.

(1)The reproducing space W23[0, 1]Inner product spaceW23[0, 1] is defined as W23[0, 1] = {u(x) | u′, u', u'' are absolutely continuous real value functions, u, u0, u00, u(3)∈ L2[0, 1], u(0) = 0, u(1) = 0}. It is endowed with the inner product and the norm of the forms

$$(u(y), v(y))_{W_2^3[0,1]} = \sum_{i=0}^{2} u^{(i)}v^{(i)} + \int_0^1 u^{(3)}v^{(3)}dy, \quad u, v \in W_2^3[0,1]$$

$$\|u\|_{W_2^3[0,1]} = \sqrt{(u,u)_{W_2^3[0,1]}}. \tag{2.1}$$

Theorem 2.1. The spaceW23[0, 1] is a reproducing kernel space, that is, for any u(y) ∈ W23[0, 1] and each fixed x ∈ [0, 1], there existsRx(y) ∈ W23[0, 1], y∈ [0, 1], such that(u(y), Rx(y))W23[0,1]=u(x). The reproducing kernel Rx(y) can be denoted by

$$R_x(y) = \begin{cases} \sum_{i=0}^{5} c_i(x)y^{(i)}, & y \le x, \\ \sum_{i=0}^{5} d_i(x)y^{(i)}, & y > x. \end{cases} \tag{2.2}$$

Proof. Since Rx(y) ∈ W23[0, 1], it follows that

$$Rx(0) = 0, Rx(1) = 0. \tag{2.3}$$

Through several integrations by parts, we get

$$\int_0^1 u^{(3)} R_x^{(3)}(y)dy = \sum_{i=0}^{2} (-1)^i u^{(2-i)}(y) R_x^{(3+i)}(y)|_{y=0}^1 + (-1)^3 \int_0^1 u(y) R_x^{(6)}(y)dy$$

Also

$$\sum_{i=0}^{2} (-1)^i u^{(2-i)}(y) R_x^{(3+i)}(y) = \sum_{i=0}^{2} (-1)^{2-i} u^{(i)}(y) R_x^{(5-i)}(y)$$

then

$$\begin{aligned}(u(y), R_x(y)) = &\sum_{i=0}^{2} u^{(i)}(0)[R_x^{(i)}(0) - (-1)^{2-i} R_x^{(5-i)}(0)] \\ &+ \sum_{i=0}^{2} (-1)^{2-i} u^{(i)}(1) R_x^{(5-i)}(1) + (-1)^3 \int_0^1 u(y) R_x^{(6)}(y)dy\end{aligned} \tag{2.4}$$

Sinceu∈W23[0, 1], thus, u(0) =u(1) = 0.Suppose that Rx(y) satisfies the following generalized differential equations:

$$R_x^{(1)}(0)+R_x^{(4)}(0)=0, R_x^{(2)}(0)-R_x^{(3)}(0)=0, R_x^{(3)}(1)=0, R_x^{(4)}(1)=0. \quad (2.5)$$

Then (2.4) implies that

$$(u(y), R_x(y)) = (-1)^3 \int_0^1 u(y) R_x^{(6)}(y)dy.$$

If ∀y ∈ [0, 1],Rx(y) satisfy

$$(-1)^3 R_x^{(6)}(y) = \delta(y-x), \quad (2.6)$$

then Rx(y) is said to be a reproducing kernel of space W23[0, 1], that is(u(y), Rx(y)) W23 =u(x).

Characteristic equation of (2.6) is given by λ6= 0,then we can obtain characeristic values λ = 0(a 6multiple roots). So, let

$$R_x(y) = \begin{cases} \sum_{i=0}^5 c_i(x) y^{(i)}, & y \le x, \\ \sum_{i=0}^5 d_i(x) y^{(i)}, & y > x. \end{cases}$$

On the other hand, note that (2.6),let Rx(y) satisfy

Rx(k)(x + 0) = Rx(k) (x – 0), k = 0, 1, 2, 3, 4 (2.7)

Integrating (2.6) from x – ε to x + ε with respect to y and let ε → 0, we have the jump degree of R(5)x(y) at y = x

R(5)x(x – 0) – R(5)x(x + 0) = 1 (2.8)

Applying (2.3),(2.5),(2.7),(2.8),the unknown coefficients of (2.2) can be obtained(detail in Appendix).

(2)The reproducing kernel space W22[0, T]

Inner product space W22[0, T] is defined asW22[0, T] = {u(x) | u, u ’are absolutely continuous real value functions, u, u', u'' ∈ L2[0, T], u(0) = 0}. The inner product in W22[0, T]is given by

$$(u(y), v(y))_{W_2^2[0,T]} = \sum_{i=0}^{1} u^{(i)} v^{(i)} + \int_0^1 u''(y) v''(y)dy, \quad u, v \in W_2^2[0, T]$$

$$\|u\|_{W_2^2[0,T]} = \sqrt{(u, u)_{W_2^2[0,T]}}.$$

Similarly,W22[0, T]is a reproducing kernel space and the corresponding reproducing kernel is

$$R_x^{\{2\}}(y) = \begin{cases} xy + \frac{xy^2}{2} - \frac{y^3}{6}, & y \le x, \\ -\frac{x^3}{6} + \frac{1}{2}x(2+x)y, & y > x, \end{cases} \quad (2.9)$$

(3)The reproducing kernel space W21[0, 1]The inner product space W21[0, 1] is defined by W21[0, 1] = {u(x) | u is absolutely continuous real value function, u, u0∈ L2[0, 1]}. The inner product and norm in W21[0, 1] are given respectively by

$$(u(y), v(y))_{W_2^1[0,1]} = u(0)v(0) + \int_0^1 u'(y) v'(y)dy, \ \| u \|_{W_2^1[0,1]} = \sqrt{(u, u)_{W_2^1[0,1]}}.$$

where u(x), v(x) ∈ W21[0, 1]. The reproducing kernel is

$$\overline{R}_x(y) = \begin{cases} 1+y, & y \le x, \\ 1+x, & y > x, \end{cases}$$

3 The Exact Solution of Eq.(1.4)

If r(t) is known, then we can give the exact solution of Eq.(1.4). Note that Lu = ut−uxxin Eq.(1.4), it is clear that L : W(2,3)(Ω) → W(1,1)(Ω) is bounded linear operator. Put M= (t, x), Mi= (ti, xi), ϕi(M) =KMi(M) and ψi(M) = L∗ϕi(M) where K is the reproducing kernel of W(1,1)(Ω) and L∗is the conjugate operator of L. $\{\bar{\psi}_i(M)\}_{i=1}^{\infty}$ derives from Gram-Schmidt orthonormalization of $\{\psi_i(M)\}_{i=1}^{\infty}$,

$$\bar{\psi}_i(M) = \sum_{k=1}^{i} \beta_{ik}\psi_k(M), (\beta_{ii} > 0, i = 1, 2, \cdots). \tag{3.1}$$

Theorem 3.1. ForEq.(1.4),if $\{M_i\}_{i=1}^{\infty}$ is dense on Ω, then $\{\psi_i(M)\}_{i=1}^{\infty}$ is the complete system ofW(2,3)(Ω).

Proof. For each fixed u(M) ∈ W(2,3)(Ω), let (u(M), ψi(M)) = 0,(i = 1, 2, · · ·), that is,

(u(M), (L∗ϕi)(M)) = (Lu(·), ϕi(·)) = (Lu)(Mi) = 0 (3.2)

Note that $\{M_i\}_{i=1}^{\infty}$is dense on Ω, therefore (Lu)(M) = 0. It follows that u ≡ 0 from the existence of L−1. So the proof of the Theorem 3.1 is complete.

Theorem 3.2.If $\{M_i\}_{i=1}^{\infty}$ is dense on Ωand the solution of Eq.(1.4) is unique, then the solution of Eq.(1.4) satisfies the form

$$u(M) = \sum_{i=1}^{\infty}\sum_{k=1}^{i} \beta_{ik}F(M_k, r(t_k))\bar{\psi}_i(M), \tag{3.3}$$

where F(M, r(t)) = r(t)f (t, x) + ϕ00(x).

Proof. By Theorem 3.1, it is clear that $\{\bar{\psi}_i(M)\}_{i=1}^{\infty}$ is the complete orthonormal system of W(2,3)(Ω).Note that (v(M), ϕi(M)) = v(Mi) for each v(M) ∈ W(1,1)(Ω), then

$$\begin{aligned} u(M) &= \sum_{i=1}^{\infty}(u(M), \bar{\psi}_i(M))\bar{\psi}_i(M) \\ &= \sum_{i=1}^{\infty}\sum_{k=1}^{i} \beta_{ik}(u(M), L^*\varphi_k(M))\bar{\psi}_i(M) \\ &= \sum_{i=1}^{\infty}\sum_{k=1}^{i} \beta_{ik}(F(M, r(t)), \varphi_k(M))\bar{\psi}_i(M) \\ &= \sum_{i=1}^{\infty}\sum_{k=1}^{i} \beta_{ik}F(M_k, r(t_k))\bar{\psi}_i(M), \end{aligned} \tag{3.4}$$

which proves the theorem.

References

[1] Day, W.A.: Extension of a property of the heat equation to linear thermoelasticity and other theories. Quartly Appl. Math 40, 319–330 (1982)

[2] Macbain, J.A.: Inversion theory for parametrized diffusion problem. SIAMI. Appl. Math.18

[3] Cannon, J.R., Lin, Y.: An inverse problem of finding a parometer in a semilinear heat equation. J. Math. Anal. Appl. 145(2), 470–484 (1990)

[4] Cannon, J.R., Lin, Y., Xu, S.: Numerical procedures for the determination of an unknown coefficient in semi-linear parabolic differential equations. Inverse Probl. 10, 227–243 (1994)

[5] Cannon, J.R., Yin, H.M.: Numerical solutions of some parabolic inverse problems. Numer. Meth. Partial Differential Equations 2, 177–191 (1990)

[6] Cannon, J.R., Yin, H.M.: On a class of non-classical parabolic problems. J. Differential Equations 79(2), 266–288 (1989)

[7] Dehghan, M.: Numerical solution of one dimensional parabolic inverse problem. Applied Mathematics and Computation 136, 333–344 (2003)

[8] Fatullayev, A., Can, E.: Numerical procedures for determining unknown source parameter in parabolic equations. Mathematics and Computers in Simulation 54, 159–167 (2000)

[9] Cannon, J.R., Lin, Y., Wang, S.: Determination of source parameter in parabolic equations. Meccanica 27, 85–94 (1992)

[10] Cui, M., Geng, F.: A computational method for solving one-dimensional variable-coefficient Burgers equation. Applied Mathematics and Computation 188, 1389–1401, 3-399 (2007)

Price Sensitive Support Vector Machine Based on Data Set Decomposition

Xia Guilin

Computer Department of Chaohu College, Chaohu Anhui 238000

Abstract. This paper presents a training set decomposition of price-sensitive support vector machines for classification Firstly, decomposition of the training sample set, each subset of training a support vector machine can output a posteriori probability, get through the training of support vector machine on the training sample posterior probability using the learning process and the cost matrix, the true class label of the sample, in order to achieve the reconstruction of the sample, contains a sample misclassification cost information, cost-sensitive support vector machines, making the classification of unbalanced data sets, so that the smallest misclassification cost.

Keywords: Sensitive Support Vector Machine, Data set, Decomposition.

1 Theory and Inspiration of Bayesian Decision

Bayesian decision theory is an important part of the subjective Bayesian the Spirax inductive theory, which is founded by the British mathematician, Bayesian.

In the case of incomplete data, the Bayesian decision of the unknown part to the use of subjective probability to estimate the final Bayesian probability formula to fix the probability of occurrence final with the best decisions with expectations and revised probability made.

The basic idea of Bayesian decision theory can be described as: first, already know the class conditional probability density parameter expression and a priori probability of the premise, the use of Bayesian probability formula to convert it to a posteriori probability, the final utilization the size of the posterior probability to the optimal predictive decision-making.

Bayesian inverse probability and use it as a general reasoning method is a major contribution to the statistical inference. Bayesian formula is the expression of Bayes' theorem, a mathematical formula.

Assumption $B_1, B_2, ...$ is that a process may be a prerequisite, $P(B_i)$ is a priori probability of the prerequisite for the possibility of the size of the estimates. This process has been a result of A, then the Bayesian formula provides a prerequisite to make a new evaluation method based on the appearance of A, under the premise of A, $P(B_i \mid A)$, the probability of re-estimated as the posterior probability.

Z. Du (Ed.): Proceedings of the 2012 International Conference of MCSA, AISC 191, pp. 13–17.
springerlink.com

By the Bayesian formula and it is the basis to develop a set of methods and theories in real life has a very wide range of applications. Here is the expression of a mathematical formula of the Bayesian formula:

Bayesian formula: a partition of the set $D_1, D_2, \ldots, D_n$ is the sample space S, if $P(D_i)$ represents the probability of occurrence of event D_i and $P(D_i) > 0, i = 1,2,\ldots,n$. For any event x, and $P(x) > 0$, are:

$$P(D_j \mid x) = \frac{p(x \mid D_j) p(D_j)}{\sum_{i=1}^{n} p(x \mid D_i) p(D_i)}$$

2 Price Sensitive Support Vector Machine Based on Data Set Decomposition

Design ideas. Data using the sampling method to solve the problem of unbalanced data set classification, there may exist due to the increase of training samples on the sampling data set leads to learning, while down sampling because the training sample reduction in the number, the loss of sample classification. To avoid this situation, there is a way to divide the training set, the training set into a subset of a certain degree of balance in each sub-focus on the use of machine learning methods for training, and then integrated learning, this approach is neither increasing the number of training samples, will not be useful for the classification of information in the loss of sample.

Posterior probability cannot determine the types of samples to solve the problem of cost-sensitive data mining, when $i \neq j$, when $C(i, j) \neq C(j, i)$, if you rely only on x is great when a given sample misclassification cost can cost matrix to reconstruct the Bayeuxdecision theory the cost of sensitive issues for the realization of the embedding of different misclassification cost of Sri Lanka provides an implementation framework based on Bayesian decision theory to achieve cost-sensitive, making the misclassification cost of the global minimum.

Bayesian decision theory to deal with cost sensitive issues is the data set divided into sub-categories, the category is a subclass of minimum expected cost, relative to other subclasses, misuse categories of samples more costly if, then it will be the original does not belong to the subclass but in the last part of the sample allocated to the sub-class of the minimum expected cost, which is to modify the class mark of the samples to reconstruct the sample set of reasons.

Comprehensive analysis, propose a decomposition based on the data set the price-sensitive support vector machine (KCS-SVM).

Algorithm Description. There is a training set $L = \{(x_1, y_1), \ldots, (x_m, y_m), y_i \in \{-1, +1\}\}, 1 \le i \le m$, $L = L_+ \cup L_-$ of which: sample set of the positive class, negative class sample set $L_- = \{(x_{n+1}, -1), \ldots, (x_m, -1)\}$, represents a sample, $x_i \in X \subseteq R^n$, , n said the training set the number of positive samples, $m - n$ said the training set the number of negative samples. In this algorithm, first the negative class sample set arbitrary decomposition into $k = \left\lceil \frac{m-n}{n} \right\rceil$ Subset. And then decompose each negative category subset of L_{-i} and class sample collection L_{+} merge together, merging $L_i, 1 \le i \le k$ training set k。 The proportion of the number of positive and negative samples of training set can be controlled by adjusting the support vector machine is trained on each subset of the posterior probability of the output.

Take a sample training set $x_i \subseteq L$, Posterior probability in each sub-classifier, respectively $P_i(+1 \mid x), P_i(-1 \mid x)$, According to a set price matrix using meta-learning, the training sample misclassification cost $R_i(+1 \mid x) = \sum_i P_i(-1 \mid x) \cdot \cos t(-1,+1)$, $R_i(-1 \mid x) = \sum_i P_i(+1 \mid x) \cdot \cos t(+1,-1)$,

Take the minimum misclassification cost $\min(R)$, Conditions to judge the true class label of the sample, thus making the sample incorporates a misclassification cost, using the above-mentioned way to regain the class label of sample concentration of each sample, and then reconstruct the sample set, because the reconstruction of the sample set integrated misclassification cost, so you can use cost-sensitive support vector machine, decision-making with a misclassification cost function, making the classification of the overall misclassification minimal cost.

KCS-SVMSteps of the algorithm description: Input: set of sample L, a sigmoid function, the kernel function K, the cost matrix C, negative-class training set is divided into degrees

Output: cost-sensitive support vector machine (KCS-SVM),

Divided: the training set L_{-} were broken down into a subset of k-independent and equal size, which is

$\sum_{i=1}^{k} L_{-i} = L_{-}$ and $\bigcap_{i=1}^{k} L_{-i} = \phi$, ϕ Represents the empty set

Generate the training set: $L_i = L_{+} \cup L_{-i}$, $1 \le i \le k$

Training:1) $f_i(x) = wx + b \leftarrow g(SVM, L_i, k)$

2) for L In each sample x

{ for xIn each subset L_i In

{ $P_i(+1 \mid x) = 1/(1+\exp(Af(x)+B)$;

$P_i(-1 \mid x) = 1/(1-\exp(Af(x)+B)$;

$R_i(+1 \mid x) = \sum_i P_i(-1 \mid x) \cdot \cos t(-1,+1)$;

$R_i(-1 \mid x) = \sum_i P_i(+1 \mid x) \cdot \cos t(+1,-1)$; }

if $R_i(+1 \mid x) \le R_i(-1 \mid x)$

$\hat{y} = 1$

else

$\hat{y} = -1$

} Get set according to the sample after the reconstruction of the C and the posterior probability $\hat{L}$

$KCS - SVM \leftarrow g(CS - SVM, \hat{L}, K)$

3 Simulation Experiment

Because it is cost-sensitive support vector machine, and the known cost matrix, and evaluation criteria for cost-sensitive learning. Commonly used in cost-sensitive evaluation: classification error cost

($Total\ Costs$), the average misclassification cost ($Average$ Costs).

According to the two classification confusion matrix and cost matrix, $Total\ Costs$ And $Average\ Costs$ Were:

Total $Costs = FP \times \cos t(N,P) + FN \times \cos t(P,N)$

$$\text{Average } Costs = \frac{FP \times \cos t(N,P) + FN \times \cos t(P,N)}{TP + TN + FP + FN}$$

4 Conclusions

A cost-sensitive support vector machine based on the decomposition of the data set. The algorithm first sample set of decomposition rules in accordance with certain data sets broken down into several subsets, each subset of training support vector machine can output posterior probability mapping the output of the a posteriori probability to get samples, using the sigmoid function proposed by Platt , and then the use of learning theory and the cost of the minimum classification of the cost of the sample, the sample "true" class label for each training sample training is completed, the reconstruction of a data set, the new data set is the integration of the misclassification cost data sets, and then use the minimum cost-sensitive support vector machine on a new sample set to train a new classifier, the classifier can make misclassification cost. Simulation results show that the new cost-sensitive support vector function effectively reduce misclassification cost, to achieve good results.

References

[1] Han, H., Wang, W.-Y., Mao, B.-H.: Borderline-SMOTE: A New Over-Sampling Method in Imbalanced Data Sets Learning. In: Huang, D.-S., Zhang, X.-P., Huang, G.-B. (eds.) ICIC 2005. LNCS, vol. 3644, pp. 878–887. Springer, Heidelberg (2005)

[2] Tang, S., Chen, S.P.: The generation mechanism of synthetic minority class examples. In: International Conference on Information Technology and Applications in Biomedicine, pp. 444–447 (2008)

[3] Wang, J.J., Xu, M.T., Wang, H., Zhang, J.W.: Classification of imbalanced data by using the SMOTE algorithm and locally linear embedding. In: The 8th International Conference on Signal Processing (2007)
[4] Zhou, Z.H., Liu, X.Y.: Training cost-sensitive neural networks with methods addressing the class imbalance problem. IEEE Trans. Knowl. Data Eng. 18(1), 63–77 (2006)
[5] Akbani, R., Kwek, S.S., Japkowicz, N.: Applying Support Vector Machines to Imbalanced Datasets. In: Boulicaut, J.-F., Esposito, F., Giannotti, F., Pedreschi, D. (eds.) ECML 2004. LNCS (LNAI), vol. 3201, pp. 39–50. Springer, Heidelberg (2004)
[6] Weiss, G.M.: Mining with rarity: a unifying framework. ACM SIGKDD Explorations 6(1), 7–19 (2004)
[7] Kubat, M., Matwin, S.: Addressing the curse of imbalanced datasets.One-sided Sampling. In: Proceedings of the Fourteenth International Conference on Machine Learing, pp. 178–186. Tennessee, Nashville (1997)
[8] Weiss, G.M.: Mining with rarity: a unifying framework. ACM SIGKDD Explorations 6(1), 7–19 (2004)
[9] Drown, D.J., Khoshgoftaar, T.M., Narayanan, R.: Using evolutionary sampling to mine imbalanced data. In: The 6th International Conference on Machine Learning and Applications, pp. 363–368. IEEE Computer Society, Washington (2007)
[10] Chawla, N.V., Cieslak, D.A., Hall, L.O., et al.: Automatically Countering Imbalance and Empirical Relationship to Cost. Data Mining and Knowledge Discovery 17(2), 225–252 (2008)
[11] Garcia, S., Herrera, F.: Evolutionary Under-Sampling for Classification with Imbalanced DataSets: Proposal sand Taxonomy. Evolutionary Computation 17(3), 275–306 (2008)
[12] Han, H., Wang, W.-Y., Mao, B.-H.: Borderline-SMOTE: A New Over-Sampling Method in Imbalanced Data Sets Learning. In: Huang, D.-S., Zhang, X.-P., Huang, G.-B. (eds.) ICIC 2005. LNCS, vol. 3644, pp. 878–887. Springer, Heidelberg (2005)
[13] Hulse, J.V., Khoshgoftaar, T.M., Napolitano, A.: Experimental Perspectives on Learning from Imbalanced Data. In: Proceedings of the 24th International Conference on Machine Learning (ICML 2007), pp. 935–942. ACM, New York (2007)
[14] Wu, H., Xu, J.B., Zhang, S.F., Wen, H.: GPU Accelerated Dissipative Particle Dynamics with Parallel Cell-list Updating. IEIT Journal of Adaptive & Dynamic Computing 2011(2), 26–32 (2011) DOI=10.5813/www.ieit-web.org/IJADC/2011.2.4
[15] Zhou, J.J.: The Parallelization Design of Reservoir Numerical Simulator. IEIT Journal of Adaptive & Dynamic Computing 2011(2), 33–37 (2011), DOI=10.5813/www.ieit-web.org/IJADC/2011.2.5

Research on the Application Subdivision Pattern in Triangular Mesh Simplification

Chen Jingsong

Department of Mathematics and Information Science, Zhoukou Normal University, Henan Zhoukou, 466001

Abstract. This paper discusses the application of another type of subdivision pattern, namely, the subdivision surfaces work as a mesh surface approximation of the original data, thus the grid data is resampled by using sub-sampling points. Specifically, in a number of algorithms based on edge collapse or triangle, the triangle is removed to a new edge collapse or vertex, the algorithm presented with the segmentation method to calculate the new vertex location to replace the traditional quadratic surface fitting algorithm or direct averaging method of calculation. polyhedral mesh simplification, especially triangular mesh simplification has been for nearly a decade, during which a large number of proposed algorithm, also published a number of reviews, here for a comprehensive review will not only simplify the background needed for the grid and make a brief review of related work, the rest of the content of the proposed algorithm is described.

Keywords: segmentation model, triangular mesh, resolution.

1 Introduction

Polygon mesh triangle mesh is particularly easy to handle various graphics system With 3D scanners and CT-like popularity of three-dimensional data acquisition equipment, data grid generation is also more easily, so the computer graphics surface that has become a common way. With the data acquisition equipment to improve the accuracy, the generated mesh has a great amount of data, with the existing hardware is almost impossible for interactive processing. In addition, data on the network graph transmission will become a huge burden on all of these applications require mesh simplification, ie by calculating the surface to generate a mesh or less the number of vertices of the original shape but remained relatively coarse grid instead of the original grid. Since the early 1990s has made a lot of different needs ácording to mesh simplification algorithm [1-5].

Vertices according to the composition of the new grid, mesh simplification methods can be divided into three categories.

The first category, the new mesh vertices in the vertex of the original mesh subset is the method by removing the mesh vertices, edges, or triangles, and the resulting polygonal hole re-subdivision, these methods does not generate new vertices [6].

The second method is based on certain rules to regenerate all of the new mesh vertices and vertex sets a new generation of triangulation to generate a vertex fewer,

Z. Du (Ed.): Proceedings of the 2012 International Conference of MCSA, AISC 191, pp. 19–24.
springerlink.com

simpler topology to connect the approximation grid. Greg Turk proposed algorithm is the earliest of these methods; the Turk's method is called a "re-collage" The technology is used to generate new vertices and mesh topologies. Hoppe and others minimize the vertex associated with measure of the energy mesh simplification algorithm [7] also re-generate the mesh vertices. Matthias Eck re the original mesh grid, so that it has broken down connectivity, creating triangular mesh simplifies the link between multi-resolution analyses [7]. Zhou Quintilian a contract based on the triangle also belong to this class of algorithms [8].

The last category includes methods to generate a new mesh vertex is mixed, both the original mesh vertices, but also contains the newly generated vertex. Such algorithms usually folded edge or surface operations, when you want to delete a grid face or an edge when not directly face or edge and then delete the hole on the left to split, but replaced with a new vertex or edge of the surface is removed, make the new vertex "inheritance" of the original surface is removed or side of the topology [9].

This paper describes the algorithm is the third category, the simplified polygon mesh is triangular mesh the algorithm work includes two aspects, first, features expanded Zhou Xiaoyun vertex angle (Feature angle of vertex) [10] the concept of (called the "flatness") and to determine the flatness of the triangle mesh is deleted. Secondly, the segmentation method using a new vertex, after contraction as the location of the triangle, all triangles with adjacent vertices to be deleted are now connected with the new point, that is to remove the new vertex as the center of the hole produced by the star-triangle subdivision, as shown in Figure 1. This method is similar to the triangle removed very simple and easy to implement measure, and calculate the new vertex method can be used to simplify many existing algorithms [11].

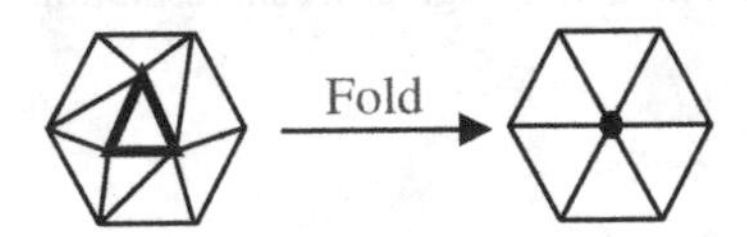

Fig. 1. Folded into a triangular point, all the triangles adjacent vertices are adjacent to the new vertex

2 Related Work

This section describes several simplified algorithm based on triangle collapse. Hamann was first proposed based on the triangular folded triangle mesh simplification algorithm [Hamann 1994]. He used the triangular shape of the curvature and weighted to give each factor the weight to generate a triangle determine the grid order of the triangle is deleted for the triangles to be deleted half-space test, if the resulting polygonal hole is not suitable for re-subdivision, then the re-election of a triangle repeat the above steps. Secondly, the triangle is removed set of adjacent vertices for quadratic polynomial surface fitting, the center of another set of neighbors along the triangle method to lead a straight line, so that a straight line and triangle collapse to the surface of the intersection. Gieng other people's work [12] Similarly, but with more detailed criteria for judging whether or not to delete the triangle, involving measurement, including the curvature, shape measurement, topology metrics (new vertices price), the error measure and a series of weight parameter of these methods is very time-consuming calculation of curvature, in addition, the second fitting surfaces must be constructed for solving linear equations, and no solution of the equation is likely to this time need to be fitted to increase the number of vertices repeat the above calculation the other hand, the second surface is constructed by the local vertex, obviously, different pieces fit quadratic surface and does not constitute a smooth surface between the even between the surface may not be continuous.

Kun Zhou collage in a re-curvature algorithm using an approximate measure of the triangle as an alternative to reduce the folding calculation [12] the algorithm from the original triangular surface mesh select a new vertex, which is equivalent to using a triangle where the plane to fit the surface, taking into account that the surface is generally curved surface, so choose the vertex folding is clearly not reasonable, the error is greater. Kun Zhou contraction in another triangle-based algorithm [13], such as the use of Garland human error matrix [14] as a measure of the triangle to delete or not, shall be required to produce the inverse of a matrix folded position of the vertex, when the inverse matrix does not exist to deal with them will encounter a similar surface fitting in touch the trouble.

Breakdown of the proposed algorithm to calculate the mode new folded triangle vertex position, do so at least two advantages, first, does not require the solution of linear equations or find the inverse matrix to avoid numerical instability may occur, followed by, although the calculations used only a small number of vertices of the triangle around, but the subdivision support the overall continuity, not the same as the quadratic surface fitting only support local continuous. Which overcomes the aforementioned types of methods exist the problem. In addition, the algorithm also gives the concept of flatness measure to simplify the calculation of the triangle fold.

3 Triangle 1 - Central and Flatness

Triangular mesh of a triangular face can be deleted and the deleted sequence from the adjacent triangles to determine the flatness of the region below the introduction of the triangle 1 - to describe the concept of triangular ring adjacent regions, and gives the definition of flatness of the template.

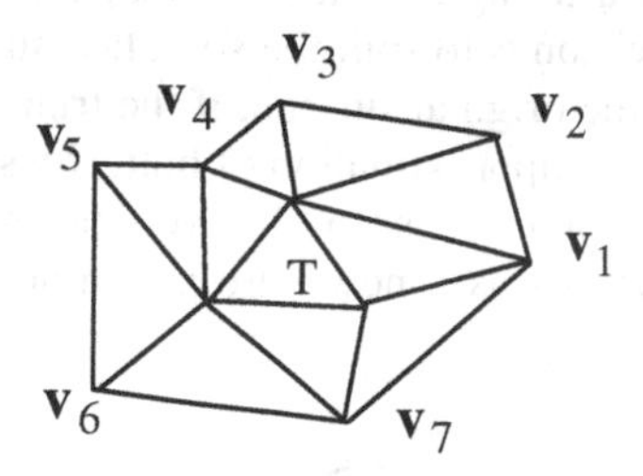

Fig. 2. 1 – ring of triangles T

3.1 Triangle 1 – Ring

Triangle 1 - The concept of vertex loop 1 - loop similar to the grid in a given triangle, and said one of the three vertices of all triangles adjacent to triangle subnet 1 - ring, denoted by shown in Figure 2, the triangular mesh is 1 - Ring 1 - Ring of the boundary polygon as a triangle of 1 - loop in Figure 2 in the 1 - loop is.

3.2 Flatness

In general, the more flat surface where you can use fewer triangles that remove these regions influence the shape of a triangle on the smaller. Flatness triangular mesh is used to describe the extent of the local close to the plane, the triangle degree of the vertex flat by the distance to the average plane [Schroeder 1992, Ma Xiaohu 1998, Li Jie 1998], curvature [Hamann 1994, Gieng 1998, Kun Zhou 1998], the fractal dimension [Jaggi 1993], and so to measure where the concept of vertex angle characteristics [Zhou Xiaoyun 1996] and extended to the triangle as a triangle on the flatness of the metric.

Note the unit normal to the triangle obviously triangle and triangle coplanar if and only if their inner product method, so if the grid of the law any two triangles are close to an inner product that the grid can be approximated is flat below the triangle look at the definition of flatness.

Definition 1 given a triangle, with its flatness is defined as 1 - ring all the inward plot the average triangle method:

$$P(T) = \frac{1}{|N_1(T) - \{T\}|} \sum_{T' \in N_T - \{T\}} (\mathbf{n}_T \bullet \mathbf{n}_{T'})$$ is clearly less than the absolute value of the real number 1, close to a description of the 1 - ring relatively flat, the smaller the more uneven if the number of adjacent triangles enough so a few of the triangle is not flat (ie smaller) of the flatness effect is small, so even set a threshold, when the 1 - loop inside a triangle made more than this threshold are not deleted.

3.3 Calculation of the New Vertex

Deleted after being folded triangle is not directly split the resulting multi-shaped hole, but to gradually shrink into a triangle vertex, so that the natural inheritance of the triangle topology, this subdivision is to split the so-called star. Of course, we hope that triangle shrink to a point on the original surface, if the triangle mesh is simplified as a result of the grid to determine an approximation of butterfly subdivision surfaces, select from the triangle "center" point leads to the vertical point of intersection with the butterfly subdivision surfaces as a new vertex is more appropriate, as shown in Figure 3.

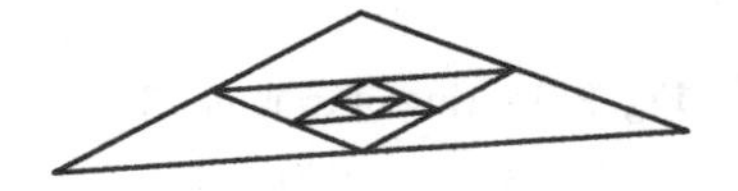

Fig. 3. Three E-vertices of a triangle formed converge at one point

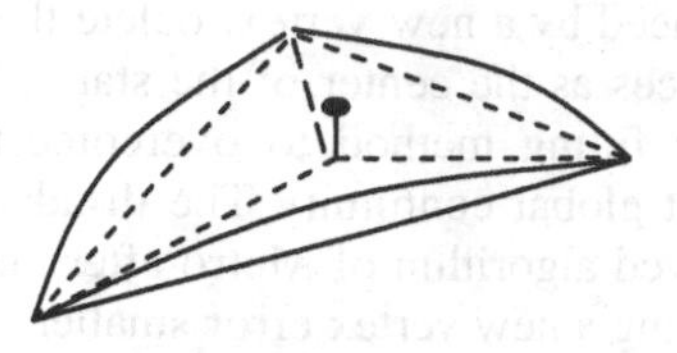

Fig. 4. Solid leads from the center point of vertical intersection with the surface

Of course, because there is no analytical expression for the surface, can not find the intersection by solving equations, the limits of our triangle center position instead. Taking into account through continuous segments, three E-vertices of the triangle will converge to determine the point (as shown in Figure 4 below), to this point as the limit position of the center triangle. The actual calculation can not go on infinitely subdivided, several segments can be obtained after taking the last three vertices of a triangle formed E centers (such as Figure 4 in the center of the innermost triangle.) approximation instead of the required new vertex. Specifically, given a triangle to be removed, find the 1 - and its a butterfly ring segments; and then find the three sides corresponding to the three vertices of the triangle E-1 - a butterfly ring segments, this process can be repeated several times to get the current after the E-center is the vertex of the triangle corresponding to the request.

4 Algorithm Description

Algorithm consists of four parts. First, calculate the flatness of the grid and all the triangles descending order, for the law to the triangle method with a neighbor to the inner triangle is less than the specified threshold are excluded. Second, the successive removed from the sequence of a triangle a triangle, a triangle with the butterfly pattern of contraction of the new vertex position. The third step is a new vertex as the central triangle is removed 1 - for the star polygon boundary loop is split Finally, re-calculation of the loop on all the vertices of a triangle and the adjacent newly created triangle contraction flatness inserted into the appropriate location, because these triangles 1 - ring changes, which flatness has changed the algorithm loop control triangle by the number of pre-determined limit.

The algorithm calculated the new vertex is only an approximation, and triangles for each segment need to extract the 1 - or the cost it takes to ring the larger Lee now people have been improved, improved algorithm based on edge folding, according to butterfly subdivision know the rules, the new vertex to the edge contraction can be expressed as a linear combination of the mesh vertices to avoid multiple segments.

5 Conclusions

This paper describes a folded triangle triangle-based mesh simplification algorithm, defined as a triangle collapse flatness measure, the greater flatness of the triangle shows the triangular mesh in the vicinity of the more flat, so the more the first is deleted, the deleted triangle by the replaced by a new vertex, delete the new vertices as a triangle hole after the polygon vertices as the center of the star split the new vertex with the butterfly pattern of surface fitting method to overcome traditional low efficiency, instability, does not support global continuity The disadvantage of Li Xian-min [Li Xian-min 2001] The improved algorithm of Metro after the error estimates show that the segmentation method using a new vertex error smaller.

Acknowledgment. The work is supported by the Natural Science project of Henan province Education Department of China under Grant Nos. 2010A520050.

References

[1] Ochiai, Y., Yasutomi, Z.: Improved method generating a free-form surface using integral equations. CAGD 17(3), 233–245 (2000)
[2] Peters, J., Reif, U.: Analysis of algorithms generalizing B-spline subdivision. SIAM Journal on Numerical Analysis 35(2), 728–748 (1998)
[3] Peters, J., Umlauf, G.: Gaussian and mean curvature of subdivision surfaces. TR 2000-01, CISE, University of Florida
[4] Popovic, J., Hoppe, H.: Progressive simplicial complexes. In: Computer Graphics Proceedings, Annual Conference Series, ACM SIGGRAPH, pp. 217–224 (1997)
[5] Prautzsch, H.: Analysis of-subdivision surfaces at extraordinary points. Oberwolfach (1995)
[6] Prautzsch, H., Umlauf, G.: A-subdivision algorithm. In: Farin, G., Bieri, H., Brunnet, G., DeRose, T. (eds.) Geometric Modeling, Computing Suppl., vol. 13, pp. 217–224. Springer-Verlag (1998)
[7] Prautzsch, H., Umlauf, G.: Improved triangular subdivision schemes. In: Wolter, F.-E., Patrikalakis, N.M. (eds.) Proceedings of the CGI 1998, Hannover, pp. 626–632 (1998)
[8] Prautzsch, H., Umlauf, G.: A and a subdivision scheme for triangular nets. Int. J. Shape Modelling 6(1), 21–35 (2000)
[9] Qin, H., Mandal, C., Vemuri, B.C.: Dynamic Catmull-Clark subdivision surfaces. IEEE Transactions on Visualization and Computer Graphics 4(3), 215–229 (1998)
[10] Reif, U.: A unified approach to subdivision algorithms near extraordinary vertices. CAGD 12(2), 153–174 (1995)
[11] Riesenfeld, R.F.: On Chaikin's algorithm. Computer Graphics and Image Processing 4, 153–174 (1975)
[12] Sanglikar, M.A., Koparkar, P., Joshi, V.N.: Modelling rolling ball blends for CAGD. CAGD 7(5), 399–414 (1990)
[13] Vani, M.P.: Computer Aided Interactive Process of Teaching Statistics Methodology – III Evaluation Questionnaire for LearnersThrough statistical display using Bar chart. IEIT Journal of Adaptive & Dynamic Computing 2011(4), 9–14 (2011), DOI=10.5813/www.ieit-web.org/IJADC/20114.2
[14] Ye, Y.H., Liu, W.P., Dao, B.: BIM-Based Durability Analysis for RC Structures. IEIT Journal of Adaptive & Dynamic Computing 2011(4), 15–24 (2011), DOI=10.5813/www.ieit-web.org/IJADC/20114.3

Study on Reduction of Multi-Dimensional Qualitative Variables with Granular Computing

Liu Hong, Yang Shen, and Li Haiyan

School of Science, University of Science and Technology Liaoning, China

Abstract. Based on the analysis of large complicated, multi-dimensional qualitative (MDQ) data is described by using the information system in rough set theory. Because there exists some deficiencies in reduction method of identification matrix, a model of MDQ variable reduction is given based on granular computing, and an example verifies its feasibility and advantages, which provides a new idea for the analysis of large complicated qualitative data.

Keywords: Rough set, Granular computing, MDQ variable.

1 Introduction

With the development of computer technology, the data obtained by humans is increasing rapidly, which is including a large amount of qualitative data. The distribution of qualitative data in statistics is generally described by using the contingency table, however, it is not convenient for the large data set. Therefore, the problem that needs to be solved firstly to analyze the high dimensional qualitative data is how to reduce the variable dimension.

The rough set is a mathematical tool of processing fuzzy and uncertain information, it has a consistency for the description essence of data and the data description of qualitative variables, and it is more suitable for the description of large data.

From the rough set theory, the variable reduction is to find the minimum variable set for object set with unchanged classification. According to this idea, the reduction method in rough set theory can be adopted by identification matrix. However, for the analysis of huge data, the limitation is that the computation is very complicated. Therefore, on the basis of this reduction idea, a MDQ variable reduction is given by applying the granular computer theory, and the attribute significance in information system is defined, finally, the example verifies its feasibility and effectiveness.

2 Variable Reduction of MDQ Data with Rough Set

2.1 Information System Description of MDQ Data

Assume that there are m observed objects, and n variables, when the research for qualitative data is carried out, an original data matrix can be obtained, $X = (x_{ij})_{m \times n}$,

Z. Du (Ed.): Proceedings of the 2012 International Conference of MCSA, AISC 191, pp. 25–30.
springerlink.com

where, x_{ij} $(i = 1,2,\cdots \in, m; j = 1,2,\cdots,n)$ denotes the value under the *j*th variable for the *i*th observed object. It is easy to see that this data matrix has the same form as the information system in rough set. Therefore, the MDQ data can be denoted by information system $S = (U, A, V, f)$, where U denotes the finite nonempty set of the object, which is called universe corresponding to the *m*th observed object in MDQ data, A denotes the finite nonempty set of the attribute corresponding to the *n*th variable, for example, a_1 denotes the sex, a_2 denotes the career etc. $V = \bigcup_{a \in A} V_a$, V_a denotes the range of attribute a corresponding to the value set in MDQ data, $f : U \times A \to V$ is the information function, it gives an information value for each attribute of each object, thus $\forall a \in A, x \in U, f(x,a) \in V_a$ is corresponding to the mapping relationship between each observed object and each variable value.

It is easy to see that the MDQ data that is described by information system not only overcomes the deficiencies existing in the distribution of a large amount of data, but also provides a theoretical basis for the following research.

2.2 Variable Reduction Model with Identification Matrix

After the original data is described by information system, the reduction is carried out for the variables. The variable reduction is to find a smaller variable subset P, and the classification of P to U is the same as the classification of A to U, the variable reduction in rough set is actually realized by identification matrix.

Assume that the observed object U can be classified into l classes by variable set A, which is:

$$U/A = \{[x_i]_A | x_i \in U\} = \{X_1, X_2, \cdots, X_l\}$$

and $D([x_i]_A, [x_j]_A) = \{a \in A, f_a(x_i) \neq f_a(x_j)\}$ is a division identification set, thus the division identification matrix is composed by all division identification sets.

$$D = D([x_i]_A, [x_j]_A) | [x_i]_A, [x_j]_A \in U/A$$

Firstly, find the l-dimensional subset of A, the intersection calculation is carried out with each element of identification set, if and only if $B \cap D([x_i]_A, [x_j]_A) = \varnothing$, this variable subset is contained in the simplest variable subset, and then calculate the 2, 3 and 4-dimensional subset of A, finally calculate all sets that satisfy the conditions, and the union set of them is taken as the simplest set B.

The identification matrix is an effective reduction method when the variable dimension is low. However, its calculation process is very complicated for the analysis of a large amount of data. Therefore, according to the relative theory of granular computing, the redundant variables are selected out by the significance of the variables to realize the reduction of the variables.

3 Reduction of MDQ Variables with Granular Computing

According to the granular computing theory, each variable in MDQ variables is the indiscernibility relation, not every variable can affect the classification of the object, the significance of the variables can be taken as the standard of the variable selection, the variable core is the starting point, and the redundant variables are selected out by the significance of the variables to realize the reduction of the variables. This method can process a large amount of data, and also improve the calculation efficiency.

3.1 Concepts and Symbol Description about Granular Computing

Granular is the initial concept of granular computer, it is the unit of research object. The granularity is used to measure the size of granular, the granularity obtained from different aspects is different, from the aspect of information theory, and the size of granular can be measured by information quantity.

Let $P \subseteq A$ be a variable subset, P on U can be divided into:

$$U/P = \{[x]_P \,|\, x \in U\} = \{X_1, X_2, \cdots, X_n\}$$

Define the information granularity $GD(P)$ of $P: GD(P) = \sum_{i=1}^{n} |x_i|^2 \Big/ |U|^2$

where $\sum_{i=1}^{n} |X_i|^2$ is the cardinal number of equivalent relation determined by variable subset P.

By using the definition of granularity of variable, the significance of each variable in MDQ variable can be analyzed. Assume that a is a variable in A, then if a is removed from A, the granularity of variable will be changed. The larger change of significance in A is, the more important a for A will be. $Sig_{A-\{a\}}(a)$ denotes the significance of a in A.

$$Sig_{A-\{a\}}(a) = GD(A)/GD(A-\{a\}) \tag{1}$$

If and only if $GD(A)/GD(A-\{a\}) \neq 1$, a is the core variable of A, then the core of A can be denoted as: $Core(A) = \{a \in A \,|\, Sig_{A-\{a\}}(a) < 1\}$.

The variable significance from above definition can be taken as the standard of variable selection, and the variable core is the starting point to calculate the reduction set of MDQ variables.

3.2 Variable Reduction Model with Granular Computing

Assume that there are m observed objects ($U = \{x_1, x_2, \cdots, x_m\}$), and n variables ($A=\{a_1,a_2,\cdots,a_n\}$), X_i ($i=1,2,\cdots,l$) is the division of A on U ($U/IND(A)=\{X_1,X_2,\cdots,X_l\}$).

The first problem that needs to be solved is to find a subset $Red(A)$ in A to satisfy the condition:

(1) Division of $Red(A)$ and A on U are the same;

(2) $Red(A)$ is the simplest subset of A.

According to the view of granular, each variable in MDQ variable has a certain resolution, hence each variable has its own granularity, the granularity will be smaller with the confluence of them, and a variable set with stronger resolution and smaller granularity is generated. Therefore, the variable significance from above definition can be taken as the standard of variable selection, and the variable core is the starting point to calculate the reduction set of MDQ variables by adding the numbers of variables. The steps of model calculation are listed below:

Step 1 Calculate variable core $Core(A)$.

Firstly, calculate the granularity $GD(A)$ of A, and then remove any variable a, observe whether the granularity formed by the rest variables change or not, if it changes, which means variable a is a core variable, otherwise it is not ($Core(A)$ may be an empty set), and let $Red(A) = Core(A)$.

Step 2 Calculate the division of A and $Red(A)$ on U respectively, and check whether they are equal or not.

If they are equal, $Red(A)$ is the reduction set, if not, then calculate $Sig_{Red(A)}(a)$ of all $a \in A - Red(A)$, select a_1 to satisfy $Sig_{Red(A)}(a_1) = \max_{a \in A - Red(A)} \{Sig_{Red(A)}(a)\}$, and a new reduction set $Red(A) = Core(A) \cup \{a_1\}$.

Step 3 Repeat step 2 until the reduction set is calculated.

3.3 Example Analysis

The experiment data is collected from the database in one commercial bank, and 2500 customer data are selected randomly, they are taken as the foundation of qualitative variable reduction. The basic information table is: customer number x_i, sex a_1, age a_2, salary a_3, marital status a_4, education level a_5, career a_6, house status a_7, health status a_8, total amount of account a_9, account balance a_{10}, telephone bank user a_{11}, network user a_{12}. All information is shown in table 1.

Table 1. Customer's basic information table in one commercial bank

U	a_1	a_2	a_3	a_4	a_5	a_6	a_7	a_8	a_9	a_{10}	a_{11}	a_{12}
x_1	1	2	2	2	3	2	1	1	3	3	1	1
x_2	1	3	1	2	1	4	2	1	2	2	1	1
...	...	...	...	...	...	...	...	...	...	...	...	...
x_{2500}	1	4	3	1	1	1	1	2	3	2	1	1

From table 1, $U = \{x_1, x_2, \cdots, x_{2500}\}$, $A = (a_1, a_2, \cdots, a_{12})$, $V = \{1,2,3,4\}$, where V_{a_1}: 1(Male), 2(Female); V_{a_2}: 1(under 25), 2(25-35), 3(35-45), 4(45 or over); V_{a_3}: 1(< \$10000), 2(\$10000-30000), 3(\$30000-50000), 4(> \$50000); V_{a_4}: 1(married), 2(unmarried); V_{a_5}: 1(college), 2(undergraduate), 3(postgraduate); V_{a_6}: 1(first industry), 2(second industry), 3(third industry); V_{a_7}, $V_{a_{11}}$, and $V_{a_{12}}$: 1(yes), 2(no); V_{a_8} : 1(bad), 2(normal), 3(good); V_{a_9} and $V_{a_{10}}$: 1(< \$50000), 2(\$50000-200000), 3(\$200000-500000), 4(> \$500000).

According to the mathematical model, a reduction is carried out for above 12 variables on the basis of unchanged customer's classification information by using Matlab, finally, the reduction set $Red(A)$ is determined, the results are:

From step 1, it is obtained that $U / IND(A) = \{X_1, X_2, \cdots, X_{48}\}$, then calculate the granularity of A, that is $GD(A) = 0.572$. Remove each variable in turn, the granularity of rest variables can be obtained, $Core(A) = \{a_3\}$ of A can be obtained, and let $Red(A) = Core(A)$.

From step 2, it is obtained that $U/IND(\mathrm{Re}\,d(A)) = \{Y_1, Y_2, Y_3\}$, thus the division of $Red(A)$ on U is not equal to that of A on U, then calculate the value $Sig_{Red(A)}(a)$ of all $a \in A - Red(A)$, the result is shown in Table 2.

Table 2. The significance of value

SIG	*A*										
	a_1	a_2	a_4	a_5	a_6	a_7	a_8	a_9	a_{10}	a_{11}	a_{12}
$Sig_{Red(A)}(a)$	0.162	0.237	0.318	0.337	0.342	0.626	0.242	0.564	0.577	0.298	0.206

From table 3, a_7 is obtained to satisfy $Sig_{Red(A)}(a_7) = \max_{a \in A-Red(A)} \{Sig_{Red(A)}(a)\}$, finally, let $Red(A) = Core(A) \cup \{a_7\}$.

Repeat step 2 until obtaining the reduction set $Red(A) = \{a_3, a_7, a_9, a_{10}\}$.

From above information, 2500 customer objects are classified into 48 classes, under the condition of keeping unchanged classification ability of information system, 12 variables are reduced to 4 variables a_3, a_7, a_9 and a_{10}, which means only salary, house status, total amount of account and account balance are considered when the classification is carried out, the classification result with 4 variables is the same as that with 12 variables. It verifies the feasibility of this method, improves the calculation efficiency and it is convenient for further research of MDQ data.

4 Conclusions

A variable reduction model with granular computing is given, and an example verifies its feasibility and advantages, which provides a new idea for the analysis of large complicated qualitative data. Because only the classification invariance is considered during the computing process, however the association of variables is not considered, therefore the data after reduction is suitable for further classification research.

References

[1] Lin, T.Y.: Granular computing. Announcement of the BISC Special Interest Group on Granular Computing (1997)

[2] Pawlak, Z.: Rough Sets: Theoretical Aspects of Reasoning about Date. Kluwer Academic Publishers, Dordrecht (1991)

[3] Zhao, M., He, L.: Research on knowledge reduction algorithm based on granular computing. Computer Science 36, 237–241 (2009)

[4] Wenxiu, Z., Guofang, Q.: Uncertain Decision Making Based on Rough Sets. Tsinghua University Press, Beijing (2005)

[5] Xiuhong, L., Kaiquan, S.: A attribute reduction algorithm based on knowledge granularity. Journal of Computer Applications 26, 76–78 (2006)

Isolated Distributed Architecture for Cloud-Storage Services Based on Structured Strategy

Ying Gao and Jianlin Zheng

College of Computer Science and Engineering, South China University of Technology, China
gaoying@scut.edu.cn, helinz@qq.com

Abstract. Cloud Security threats have greatly hindered the development of Cloud Computing and the promotion of cloud application, so how to protect the data stored in the cloud is not only the core issue of security, but also the biggest challenge of development. As these problems above, a trust-control architecture based on the concept of Cloud Storage Service was creatively proposed. Based on the structured strategy and isolated distributed storage-framework, the architecture of cloud services can eliminate the concerns about cloud security and achieve a high degree of safe, reliable and available cloud storage. It's rather a simple and efficient way proved by experiments.

Keywords: Trust Mechanism, Security, Structure, Storage, Cloud, Service.

1 Introduction

To promote Cloud Computing is very hard, for customers, trust is considered as the key issue of cloud computing. People will trust their money into banks, because banks are state-owned and behind the legal guarantee of the Government, but the security of cloud providers' data centers get no third-party of any credibility in the institutional guarantee [1]. Thus users are afraid of sharing their data in cloud centers. Cloud Security has become the biggest problem encountered in the development of Cloud Computing; therefore, how to provide a trusted mechanism to meet the security needs in the cloud environment is now an urgent problem to be solved.

2 Ideas of Structured Isolated Distributed Cloud-Storage Services

2.1 The Way to Solve Cloud Security

Relative to technology, cloud computing is essentially more of a model or a strategy on the innovation based on existing technologies. Cloud computing has changed the serving way, but hasn't overturned the traditional safe mode [2].

The difference is the change of safety measures, the location of deployment of safety equipment and the main body of the responsibility for security. Similar to the traditional way, the way to solve the issues of cloud computing security is also the combination of these three elements of the strategy, technology and people [3].

Z. Du (Ed.): Proceedings of the 2012 International Conference of MCSA, AISC 191, pp. 31–37.
springerlink.com

2.2 The Concept of Cloud Storage Service

The Cloud storage is a extension of cloud-computing concept in the infrastructure layer, however, it's currently mainly used as an attachment of cloud-computing platform, not as an independent standardized cloud service. In addition, cloud providers have built their own data centers; data just exists in respective centers.

Cloud computing means that the IT infrastructure runs as a service, and the service can be anything from the rental of the original hardware to using third-party APIs. Similarly, storage facilities should also be used as a basic resource of the cloud environment. The future of cloud storage should be cloud-storage services.

As the development of next-generation cloud storage, Storage As a Service is a change to existing way, it is a special form of architecture services. This cloud storage service is also transparent to users. It doesn't refer to a specific device or a cloud provider's cloud storage centers, it means the aggregates consisting of storage devices distributed in different physical geographical or various providers' storage-centers and users can use the services provided by a number of different cloud-storage centers.

The core of cloud storage service is the combination of cloud service agreements with cloud storage system and uses opening service interface standards to achieve the transformation of cloud storage to cloud-storage service. It's the core idea of the architecture proposed as the follow in this article.

2.3 The Solution of Isolated Distributed Architecture for Cloud-Storage Services Based on Structured Strategy

The isolated distributed cloud-storage services based on structured strategy have 3 aspects: structured, decentralized and isolated distributed storage. In essence, it breaks data into a couple of individually meaningless subsets of data basing on the scheme of data, and then use the isolated distributed cloud-storage services belonging to different cloud providers to store the subsets. For the cloud, it provides a secure available storage solution and can be trusted (Shown in Fig. 1).

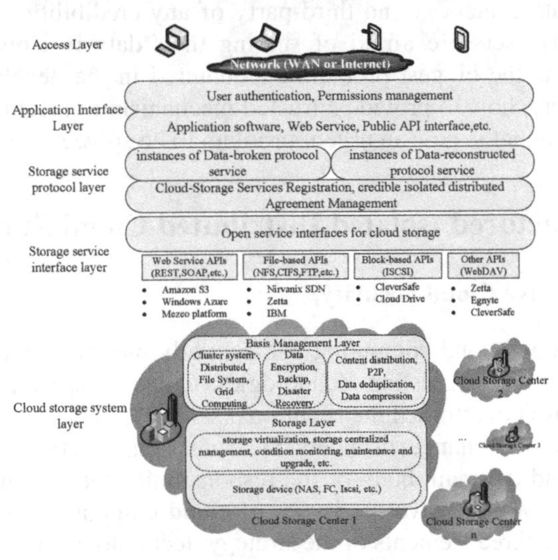

Fig. 1. Isolated distributed Services Architecture based on structured strategy

The architecture is a trust-mechanism. Data is stored in isolated distributed cloud-storage centers from different cloud-providers, in this way, users can recapture the control of data originally belonging to cloud providers, and then have a higher level of centralized control of the distributed control. Combined with tolerable intrusion approaching to data protection(data-broken based on data scheme),such separate meaningless data storage lets users no longer have to worry about the security of cloud storage. In addition, the method of isolated distributed storage disperses the overall risk, with the superposition of each cloud storage service provider's own security the system has a higher security, and the simple structured distribution makes it very efficient. In the architecture, cloud computing applications in accordance with the cloud-storage service agreement (data broken protocols, data reconstruction protocols and isolated distributed agreement) can create service instances in their own cloud computing environment to call the interfaces of cloud storage services for data storage, and also can directly use the third-party services which implement the structured isolated distributed interfaces above.

2.4 The Design of Structured Isolated Distributed Storage

The structured, isolated distributed storage-service layer involves three main service instances: data-broken, data-reconstruction and isolated distributed storage services.

The output of data-broken service instances is data segments based on the logical structure of data. Data submitted by cloud applications is divided into two categories: strong structured data and non-logical data. For the structured data, it's firstly segmented based on the structure of the data, secondly, each data segment continues to be broken and then forms several subsets which have the same scheme; For the second data, it will firstly be segmented based on a default rule-format.

After data subsets are formed, isolated distributed storage services will receive a notification. By its access to the user authentication and available cloud storage services provided by different cloud organizations, it forms the mapping of multi-subsets to cloud storage multi-services, then the mapping and all the subsets are stored in different isolated distributed cloud-storage centers. Finally, the stored path of mapping is stored in the cloud service registry and feedback results will be sent to users (Data write and modify operations both can get a result of feedback). (Fig. 2)

For the instance of data-reconstruction service, when the request of data-reading arrives, it gets the stored path of mapping from the cloud service registry, then requests required data pieces from the isolated distributed subsets according to the structured mapping and restore the data based on the structure. If the request is data-updating or data-deleting, the instance will proceed with corresponding operation according to the original mapping, and then update the mapping after deleting (it also mostly exists in the process of data-updating when the affected pieces count more) or updating the affected data pieces.

The isolated distributed storage-services run based on the open cloud-storage service interfaces in the lower layer. With the logical models for the Family Key-Value, Column, Document, the Graph and other data, this interface layer includes: REST, Thrift, and the Map Reduce, GET / PUT, language-specific API, the SQL sub-set.

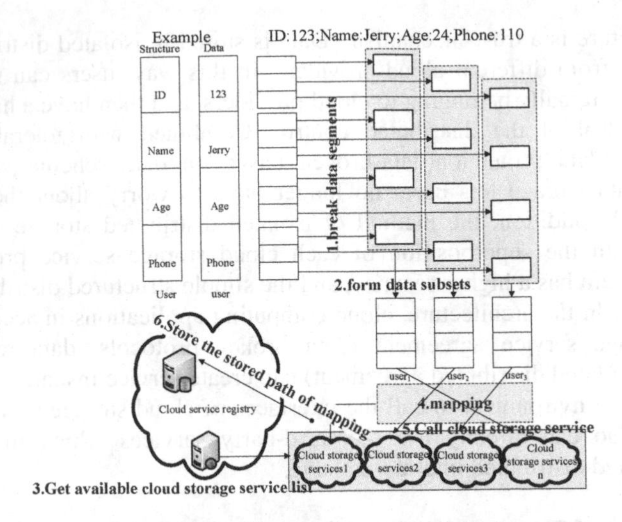

Fig. 2. The basic process for structured isolated distributed storage

2.5 The Method of Data Separation

As early as before the coming of cloud computing, Data separation and reconstruction in a distributed system have made great progress, currently, the most common and effective methods are rooted in information dispersal algorithms (IDA or RS), but in the cloud-storage services architecture proposed in this article, the separation and reconstruction of data are based on the information with its own schema. This simple and efficient data dispersion algorithm has a lower storage, computational and bandwidth cost than IDA and RS.

2.5.1 The Method of Data Separation

The basic idea of information dispersal algorithms based on redundant coding techniques is that a length L of the file F is split into n slices F_i ($1 \le i \le n$) in such a way that the file can be reconstructed by any m slices. In space, we can see that the sum of the n-slices F_i is as n/m times as original file, n/m is greater than or close to 1.

IDA equally divides the data volume D required to be distributed into fragments (size m).Assuming the size N of the volume is | D |, then the number of fragments N' is N/m. The data is encoded as n piece of data using stage independent n*m vector matrix multiplication, and its size is N'. The only difference between RS and IDA is that RS uses stage independent (n-m)*m Vander monde matrix to calculate the checksum data for each piece of data, then constitute the n fragments with the original data in accordance with the data left-justified way similar to RAID 6 parity. In the case of the same n, m and fault fragment data numbers, the time cost of IDA and RS is basically the same, but with the increase of n and m, the operating efficiency of the RS method will be slightly higher than the IDA [4].

2.5.2 The Information Dispersal Algorithm Based on Schema

The data dispersion algorithm based on the structure is rooted in the duality of information: form and content. The data of no schema does not exist, as long as the data exists in systems, whether it is stored or applied, it has its own schema. Data needs to be resolved and demonstrated its qualified sense, this is the content of data; how to construct or parse the data is the form of data that schema reflects. Thus, a complete and effective information object is a unity of form and content.

In the process of application, data all has its own form of organization (referred to here as a schema), the dispersion algorithm based on the structure of data is in fact decentralized using the schema that comes with the data, such as to a length L of information I, using the data field, table or file as the basic particle size, the form and content of the data are stripped out and data is dispersed into n data subsets S_i $(1 \leq i \leq n)$ with the same schema , L_i corresponds to the length of the S_i $(1 \leq i \leq n)$, where $\sum_{n}^{1} L_i = L$.

In space, the total length of data subsets in the algorithm based on schema equals to the original file's and the required storage space is less than the information dispersal algorithms' (IDA, RS); In time, because it is based on the data schema to simply segment and restructure with the integration of the structuring stage and the parsing stage about data, it is very efficient. It doesn't require coding with vector matrix multiplication, so the operating efficiency is better than the information dispersal algorithms. In addition, to avoid unnecessary duplication, disaster recovery and error correction processes are put on the cloud-storage services in the lower layer provided by various cloud providers, so it can highly reduce the consumption of time and space.

3 System Analysis and Verification

The architecture is different from the traditional encryption methods on ensuring information security; it uses the cryptography mechanisms in the organizational structure of information making the information content and form of separation, combined with the isolated distributed cloud-storage service solution to establish the Information Trust-System to ensure the reliability, availability and privacy of information.

Application data in the program after structured broken becomes separate meaningless subsets, and all the subsets are isolatedly distributedly stored in the cloud-storage mediums provided by different cloud providers with the pulled out schema and mapping file. Separate one or several data sub-bodies are not valid, only the schema mapping with all the required subsets can reconstruct the original message-body. Such content and form of isolated sharing, mutual checks and balances of the various cloud providers ensure the security of internal and external information, and users have no need to worry about the invasion of the cloud data from external hackers or internal staffs or users of cloud providers, theft and abuse etc. such a series of threats. To breakthrough the defense line of the architecture, at least need to pass the following checkposts: user security authentication, the key of schema mapping and cloud-storage service safety certification of various cloud providers. Therefore, this line of defense is very solid, and reliable cloud providers also protect the authority of the security of the system. The solution was through system simulation and verification in the LINUX environment, and the information dispersal algorithms were compared.

In contrasting schema-based dispersal algorithm with information dispersal algorithms, the inputs of the experiments are: data files with User structure (Fig. 2), the length is 1M, n = [5, 15]; the broken sub-unit of schema-based dispersal algorithm is file, and its basic granularity is table; m=n-1 in information dispersal algorithms. Experiments were repeated 10 times averaged to obtain the time-costs of dispersion and reconstruction (Fig. 3).Figures show that the efficiency of the schema-based dispersal algorithm is better than the information dispersal algorithms', and there is a linear relationship between its efficiency and the size of the original data file.

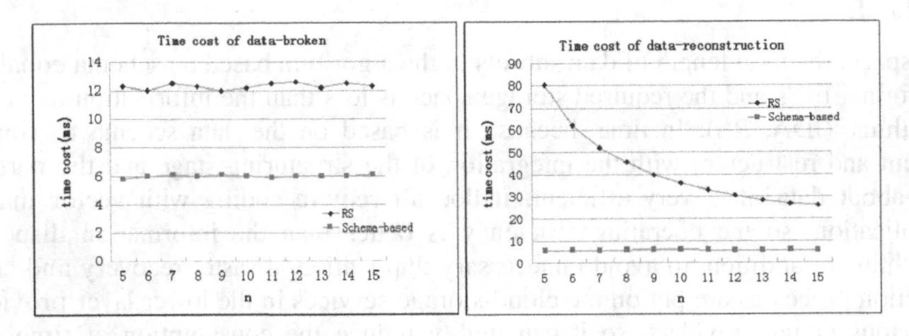

Fig. 3. Time cost of data-broken and data-reconstruction

4 Conclusion

Compared with distributed storage, this article creatively proposes a trust-control architecture based on the concept of cloud storage service. Users no longer have to worry about the security of cloud storage; it removes the biggest obstacle faced in the promotion of cloud, and has a great significance to the development of cloud.

Acknowledgement. The research is supported by:

- Science and Technology Planning Project of Guangdong Province (2009B010800047).
- Ministry of Education University-Industry-Research Project of Guangdong Province (2009B080702037).
- Ministry of Education University-Industry-Research Project of Guangdong Province (2010B090400535).
- The Fundamental Research Funds for the Central Universities (20112M0068).

References

[1] Yi, Y., Chun, L.Y.: Security Issues on the cloud computing environment. Computer Knowledge and Technology (5) (2009)
[2] From cloud computing to the cloud security. Information Systems Engineering (2009)
[3] Yi, C.S.: On the cloud computing security issues. Network Security Technology and Application (10), 22–23 (2009)
[4] Rabin, M.O.: Efficient Dispersal of Information for Security, Load Balancing, and Fault Tolerance. Journal of the Association for Computing Machinery (36), 2–4 (1989)

A New Management Algorithm for Multiplex Networks

Yu Genjian

Department of Computer Science, Minjiang University, Fuzhou, China

Abstract. The traditional network management architecture is not suitable for multiplex networks because of high complexity, heterogeneity and integrity of several different kinds of networks (internet, wireless mobile Ad Hoc network, satellite and other integrated network). This paper addressed a Novel Management Algorithm(NMA) for multiplex networks and verified it by using the Manager/Agent system. A special gateway was designed to connect two kinds of agents. The delivery time of message and the length of algorithm digital unit were simulated under two kinds of algorithms(NMA and SNMA). The results show the better performances of NMA.

Keywords: management protocol, multiplex network, Ad Hoc network, simulation.

1 Introduction

With the technology amalgamation of wireless mobile ad hoc network and other kinds of network, the development network technology trend in the future will definitely be the multiplex network, and more and more important in computer and wireless communications. The multiplex network consists of some different kinds of heterogeneous networks, such as Ad Hoc network, sensor network, internet and satellite systems. Its objective is to build an application system that can integrate the resource of complex system. Because of the high complexity, heterogeneity and integrity of complex network, the most challenging task is the network management. The network management algorithm is the core of the network management. The effective network management algorithms currently is either the SNMA[1](Simple Network Management Algorithm). The structure of the SNMA is shown in Figure 1, or the CMIP[2](Common Management Information Algorithm), the first one is extensively supported because of its simplicity, but its managed object is described by variable which make it lacking of the description of the overall situation, and the system management will be more and more huge and difficult to control with the evolvement of the network.

The CMIP is an object oriented algorithm[3-4] that can reflect the layer and successive relationship of the managed objects, but difficult to apply in realistic system because of the complicated algorithm definition, it is not adaptive to the multiplex network too. Based on the complex multiplex network, a novel network management algorithm was proposed in paper that superior to the CMIP, and the object oriented technology and monitor mechanism was adopted, and thus reduces the complexity of network management algorithm and improves the efficiency.

Z. Du (Ed.): Proceedings of the 2012 International Conference of MCSA, AISC 191, pp. 39–45.
springerlink.com

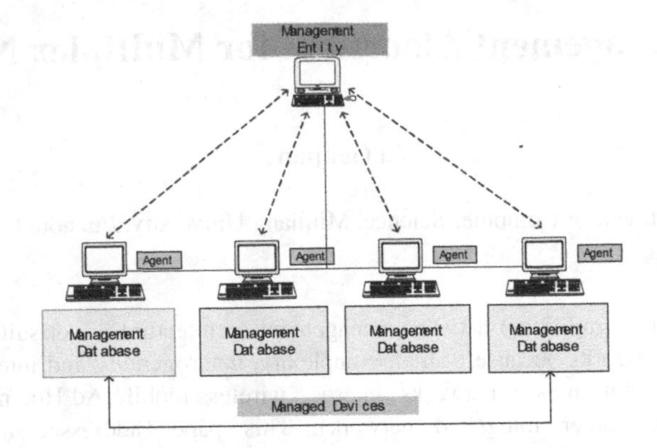

Fig. 1. The Structure of The Simple Network Management

2 The Algorithm Design

The algorithm belonged to the application layer, and the lower layer transmission mode can be decided during creating the connection by negotiation between the manager and the agent[5-7], so it will be not only able to avoid management information delay arising form connection-based service by the association control service element and the remote operations service element in CMIP, but also avoid the loss of management information in the SNMA.

In NMA design, five management primitives and two reply styles are designed to complete the network management of multiplex network. The management primitives are execute, get, set, inform and cancel. Get is used to provide the function of getting the value of NMA managed object attribute; set is used to modify the value of the managed object; inform is the important event-report send to manager which can be from manager or agent; cancel is used to cancel an operation of query.

The design entity of NMA is divided into manager and agent. Manager/Agent model is adopted between them. Agent responds to the operational request of Manager and returns the corresponding reply or confirming information. The communication model of NMA is as figure 2.

Message Dispatcher Object(MDO) is the main component in NMA mechanism which routes and delivers messages to the group members[8]. There is one MDO on each host in the system. They can be imagined as a part of agent servers that are running on every manager. MDO are created, set up and sent on all managers by the system administrator before the mechanism starts its work. This can easily be done using a MDO creator program. Agent monitors and operates the managed objects according the management information from Manager, including the following steps: receiving the management information, analyzing the management information, operating the managed objects and return the result back to Manager. Agent can be divided into three layers: communication layer, processing layer and system access control layer, which improve the ability of parallel processing. When an operational PDU (Algorithm Data Unit) comes, communication layer analyzes it and confirms its operational type and inserts it into the corresponding processing queue waiting for processing in processing layer. After being processed, result PDU is constructed and the sending module answers for sending it to Manager. Receiving, analyzing and sending is three parallel processes, and this asynchronous operation can improve the parallel ability of Agent. Inform sending module is in charge of sending the inform information in the inform queue to Manager. Communication layer sends the received PDU to processing layer, and processing layer will distinguish the type of managed object. Each MDO has the following components[9]: 1) MDO List; 2) Message Storage Queue; 3) List of the Local Group Members; 4) Maximum Message Transfer Time(MMTT); 5) Maximum Agent Migration Time(MAMT).

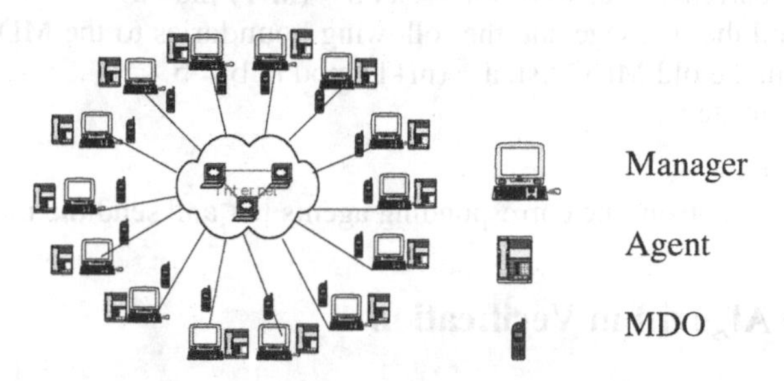

Fig. 2. Manager/Agent model with 14 hosts connected through the Internet

The description is a very important step in NMA design, which can eliminate the potential errors, there are 15 Transitions for Manager/Agent and two phases for MDO.

A. 15 Transitions for Manager/Agent

1. Manager sends connection request to Agent.
2. Manager receives connection-rejected reply from Agent.
3. Manager receives connection reply from Agent.
4. Manager sends operation PDU to Agent.
5. Manager receives operation PDU reply from Agent.
6. Manager receives disconnection reply request.
7. Manager sends disconnection request to Agent.
8. Agent receives connection request from Manager.
9. Agent sends connection-rejected reply from Manager.
10. Agent sends connection reply from Manager.
11. Agent receives disconnection request from Manager.
12. Agent receives operation PDU from Manager.
13. Agent sends operation PDU reply to Manager.
14. Agent releases occupied resources in the connection.
15. Manager release occupied resources in the connection.

Each MDO does the following 2 phases after receiving a message. Suppose the number of MDO is n and the boundaries of the Customized MDO list are (a, b).

B. Phase one

1. Get the message and the boundaries for the MDO list and calculate the customized MDO list.
2. If there is no boundaries.

1) If you are the first receiver MDO and your position in the MDO list is p set the boundaries as a = (p+1) mod n and b = (p-1) mod n.
2) Else go to Phase 2.
3) If b-a mod n < 2 send the message to the MDOs which are at indices a and b and go to Phase 2.
4) Find the median component of the customized list. Assume its position in the customized list is m then m = (a + ((b-a) mod n)/2) mod n.
5) Calculate the boundaries of the new customized list, which is the first half of the current list as follows: a = a , b = (m-1) mod n
6) Send the message and the following boundaries to the MDO, which is at index m in the old MDO list. a = (m+1) mod n, b = b
7) Go to step 1.

C. Phase two
Pick up agents from the corresponding agents list and send the message to them.

3 The Algorithm Verification

In the first place, using $A^T \cdot X = \omega_T$, here ω_T is a vector in which every element is 0, A is the incidence matrix of the Manager/Agent system used to describing NMA. By computing the matrix equation, we can get two invariants:

$$V_1 = (1,1,1,1,1,0,0,0,0,0,0,1,0,0,0)T$$
$$V_2 = (0,0,0,0,0,1,1,1,1,1,0,0,0,0,0)T \quad (1)$$

The above two invariant indicate: In the model of the Manager/Agent system which shown in Figure 2, the number of Token in all places in Manager is a constant, and so do the Agent, it means that in any time only one operation can be executed and the operation won't be sent again before it has been processed, which guarantees the reliability of NMA, and the Token number in Manager/Agent is unchanged, which shows the algorithm's conservation.

4 The Gateway Design

NMA is designed for multiplex information network system, so it must be compatible with the current network management system. Under this circumstance, the special gateway is needed to complete the communication between different algorithms.

The communications between NMA and the SNMA are supported and completed by the gateway which is shown in Figure 3. When Manager needs to control the equipment, it sends corresponding PDU to the agent of NMA, it will distinguish that the managed object is the equipment supported by the SNMA, then send the PDU to the gateway, and the gateway completes the transformation from NMA message to message of the SNMA. The communication between the gateway and the agent of the SNMA is based on algorithm itself. NMA is object-oriented one, and the attributes of managed objects, action and inform are all encapsulated in class. However, the managed object in the SNMA is described by tables or variables that only includes the static attribute, so the gateway must transform messages, that means transform primitives and identifier(see Figure 3).

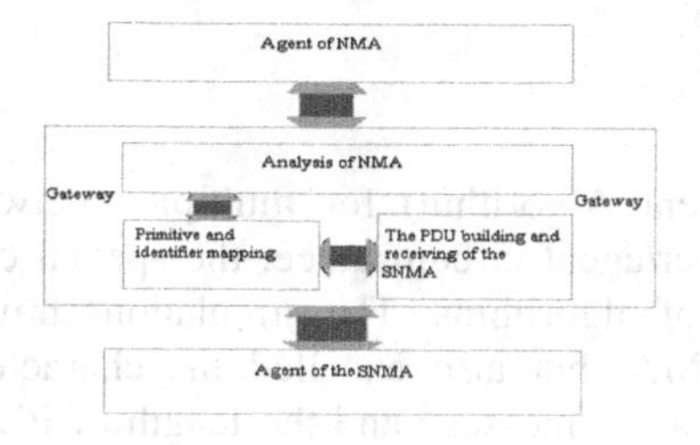

Fig. 3. Gateway connected two kinds of agents

5 Simulation

The test configuration was made of 16 managers/agents connected via a 100 Mbps Ethernet networks and with a 20 KB string message. There are manager, agent, gateway in the system respectively.

Each mechanism was executed 10 times and the average message delivery times were measured. The calculated time is the time between sending the message and receiving acknowledges from all group members. As it can be seen in figure 4-5, NMA mechanism is very scalable in comparison with Mobile Process Groups since it shows considerably less delay when the number of group members grows.

Transparent message delivery to a group and the length of PDU are two important properties of NMA mechanism. By transparent message delivery we mean the sender need not to know the group members and it just sends a message to the group and the mechanism will deliver the message to all group agents. This is a very important characteristic that makes implementation of the sender and the agents easy.

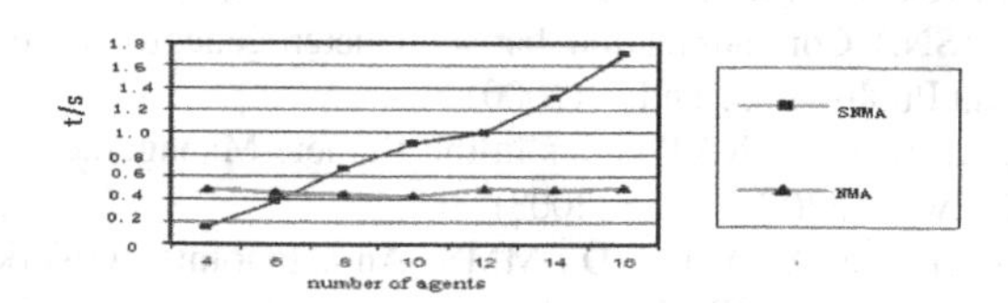

Fig. 4. Comparison of delivery time

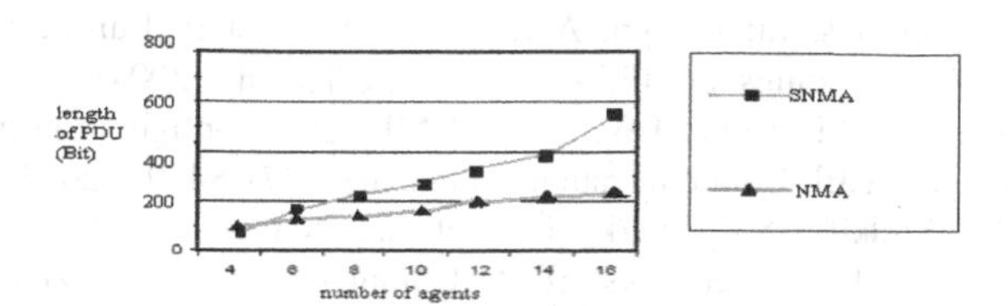

Fig. 5. Comparison of length of PDU

In Figure 4, the message delivery time of the SNMA will increase near linearly with the increment of the number of agents, but that's of NMA don't change obviously. In Figure 5, the PDU length of the SNMA will increase faster than that's of NMA the number of agents increase, longer PDU will be, result in the larger data stream, which will occupy the more bandwidth.

6 Conclusion

A NMA(New Management Algorithm) for multiplex network was proposed and verified under the manager/agent circumstance, the special connective gateway was designed for two kinds of algorithms. The simulation shows that NMA not only compatible with the SNMA, but also excelled the characteristic of two kinds of algorithms on delivery time of message and the length of PDU. We also try, in near future, on applying our method to other distributed multiplex systems such as the multiplex complex network that combine the wireless telecommunication network with satellite, internet system[10-14].

Acknowledgement. This work is supported by the special project funds of college of Fujian Province under Grant No. JK2010045 and the funds of Department of Education of Fujian Province under Grant. No. JA11199 and funds of Fuchun communications company.

References

[1] Wolberg, M., Chadha, R.: Using an adaptive management plane for policy-based network management traffic in MANETs. In: MILCOM, pp. 1133–1138 (2011)
[2] Hang, L., Lei, M.: Definition of managed objects for fault management of satellite network. In: IFITA 2009, vol. 1, pp. 107–110 (2009)
[3] Khan, A., Kellerer, W., Kozu, K.: Network sharing in the next mobile network: TCO reduction, management flexibility, and operational independence. IEEE Communications Magazine 49(10), 134–142 (2011)
[4] Dubuisson, O.: ASN.1-Communication between heterogeneous systems, pp. 156–178. Morgan Kaufmann Publishers, London (2000)
[5] Park, C., Jeong, K., Kim, S.: NAT Issues in the Remote Management of Home Network Devices. IEEE Network 22(5), 48–55 (2008)
[6] Gogineni, H., Greenberg, A., Maltz, D.: MMS: An autonomic network-layer foundation for network management. IEEE Journal on Selected Areas in Communications 28(1), 15–27 (2010)
[7] Laouamri, O., Aktouf, C.: Enhancing testability of system on chips using network management algorithms. In: Design, Automation and Test in Europe Conference and Exhibition, 2004 Proceedings, pp. 1370–1371. ACM, Toronto (2004)
[8] Choi, M.-J., Choi, H.-M., Hong, J.W., et al.: XML-based configuration management for IP network devices. IEEE Communications Magazine 42(7), 84–91 (2004)
[9] DaSilva, L.A., Midkiff, S.F., Park, J.S., et al.: Network mobility and algorithm interoperability in ad hoc networks. IEEE Communications Magazine 42(11), 88–96 (2004)

[10] Landry, R., Grace, K., Saidi, A.: On the design and management of heterogeneous networks: a predictability-based perspective. IEEE Communications Magazine 42(11), 80–87 (2004)

[11] Haddad, M., Elayoubi, S.E., Altman, E.: A Hybrid Approach for Radio Resource Management in Heterogeneous Cognitive Networks. IEEE Journal on Selected Areas in Communications 29(4), 831–842 (2011)

[12] Li, L., Buddhikot, M.M., Chekuri, C., et al.: Routing bandwidth guaranteed paths with local restoration in label switched networks. IEEE Selected Areas in Communications 23(2), 437–449 (2005)

[13] Yanfeng, Z., Qian, M., Bisdikian, C.: User-Centric Management of Wireless LANs. IEEE Transactions on Network and Service Management 8(3), 165–175 (2011)

[14] Akyildiz, I.F., Jiang, X., Mohanty, S.: A survey of mobility management in next-generation all-IP-based wireless systems. IEEE Wireless Communications 11(4), 16–28 (2004)

An Improved Min-Min Algorithm in Cloud Computing

Gang Liu, Jing Li, and Jianchao Xu

Department of Computer Science and Engineering,
Changchun University of Technology, China

Abstract. Scheduling of user tasks is a very complicated process in cloud computing environment. Min-min algorithm shortcoming is the long tasks may not be scheduled. Therefore, this paper puts forward an improved min-min algorithm that not only based on the min-min algorithm but also the three constraints. The three constraints are quality of service, the dynamic priority model and the cost of service. By using the software CloudSim to run the simulation experiment. Compared with the traditional min-min algorithm, the experimental results show it can make long tasks execute at reasonable time, increase resource utilization rate and meet users' requirements.

Keywords: Cloud Computing, Scheduling Algorithm, QOS, Dynamic Priority.

1 Introduction

Cloud computing is developed from the grid computing, distributed computing and parallel computing. It is a new business computing model and will distribute the tasks on the resources pool which is made up with the large number of computers, storage devices [2]. parallel and distributed system consisting of a collection of interconnected and virtual computers that are dynamically provisioned, and presented as one or more unified computing resources based on service-level agreements established through negotiation between the service provider and consumers [3]. Users can access to computing power, storage space and a variety of software service. At present, Google, IBM, Amazon, Microsoft and other IT vendors are studying and promoting cloud computing applications. Multi-tasks scheduling is the NP-Complete problems. Therefore, the cloud computing task scheduling has a great significance.

2 The Scheduling Model of Min-Min

The thinking of the min-min [4] algorithm is as quickly as possible to allocate each task to resources which can complete the task in the shortest possible time. Ease of description is defined as follows:

T_{exe}:represent the expected execution time; T_{start}: represent the starting time; T_{finish} :represent the finish time; T_{comp}: represent the completion time.

then,

$$T_{finish} = T_{exe} + T_{start}. \tag{1}$$

Z. Du (Ed.): Proceedings of the 2012 International Conference of MCSA, AISC 191, pp. 47–52.
springerlink.com

First, calculate the T_{exe} on each node in each machine. Then assign the earliest T_{exe} of the task to the corresponding machine node. Finally, the mapped task is deleted, and the process repeats until all tasks are mapped.

Completion time: $$T_{comp} = \max (T_{finish} = T_{exe} + T_{start}). \quad (2)$$

Algorithm procedure [5] is showed in figure 1.

(1) While there are tasks to schedule
(2) For all tasks j_i in the tasks set
(3) For all service resources r_j in the service resources set
(4) Get the value of ETC_{ij}
(5) End For
(6) End For
(7) Get *ETC* matrix that is composed of ETC_{ij}
(8) Get *Min_Time* from *ETC* matrix
(9) Find the task j_i with the minimum value of ETC_{ij} from *Min_Time*
(10) Assign task j_i to the resource r_j that provides it with the earliest completion time
(11) Delete task j_i form *J*
(12) Update the value of *ETC*
(13) Endwhile

Fig. 1. Algorithm procedure

Min-Min algorithm considers all which are not assigned tasks in each task mapping, Min-min algorithm will execute until the whole tasks set is empty. Min-min will execute short tasks in parallel and the long tasks will follow the short tasks. Min-min algorithm shortcoming is the short tasks scheduled first, until the machines are leisure to schedule and execute long tasks.Min-min can cause both the whole batch tasks executed time get longer and unbalanced load. Even long tasks can not be executed. Compared with the traditional Min-min algorithm, improved algorithm adds the three constraints strategy which can change this condition.

3 The Improved Algorithm

3.1 QOS Constraint

QOS indicates service performance of any combination of attributes. In cloud computing, QOS is the satisfaction standard of the users use cloud computing services. Cloud computing has the commercial characteristics, in order to make it has value, these attributes must be available, management, validation and billing, and when use them, they must be consistent and predictable, some attributes are even playing decisive roles. To classify and definition of the QOS purpose is to make agents manage and allocate resources, according to different types of QOS.

3.2 Priority Constraint

The priority of task is divided into static priority and dynamic priority. Static priority is defined before the tasks being executed and it remained the same priority in the scheduling stage. Static priority can reflect the willing of the user, but, the system is dynamic, static priority is difficult to reflect the changes of the system. If high priority tasks are scheduled constantly, it will cause low priority tasks not be scheduled. So, we introduced the concept of dynamic priority in this paper. It is based on the static priority and added the dynamic change factors, priority changes over time. Dynamic priority can consider a variety of factors [6], such as the initial quota of the tasks, resources consumed by the user (Run time, the actual CPU time, etc.), task waiting time and so on. Because the tasks have priorities, according to the level of priority to schedule the task, high priority scheduled firstly. This article uses a dynamic priority on the basis of the initial priority which is user-assigned.For illustrative, given the following definition [7]:

EP(Excepted Priority):a user-specified priority of each task, you can use as a static priority.

IP (Initial Priority): the shortest task has the highest priority. It is fit for min min algorithm.

DP(Dynamic Priority):dynamic priority, priority changes over time.

WTP(Waiting Time Priority):from task submission to the current time.

MP(Maximum Priority): the maximum value of priority and will not change over time.

Interval:a fixed time interval,user can specify them as their own.

Increment:waiting for the Interval time, the increasing size of priority .

for example: If the Interval = 10,Increment = 1, said that every 30s, the priority will gain 1.

The definition of dynamic priority :

$$DP = (WTP / Interval) * Increment. \tag{3}$$

The longer the waiting time, the higher the priority, the chance of the task "starve" will be smaller.

When the priority grows to a value (MP), it will no grow.

It is based on the user specifies the dynamic priority. In each scheduling cycle, according to the value of the priority to assign tasks.

3.3 The Cost of Service Model

Cloud computing is a market-oriented applications, user pays according to his service. The users can select the services which are cheap.

3.4 Improved Algorithm Procedure

Specific steps are showed in figure 2.

(1) While there are tasks to schedule
(2) For all tasks j_i in the tasks set
(3) For all service resource r_j in the service resources set
(4) Get the value of ETC_{ij}
(5) End For
(6) End For
(7) Get ETC matrix that is composed of ETC_{ij}
(8) Get Min_Time from ETC matrix
(9) Find the task j_i with the minimum value of ETC_{ij} from Min_Time
(10) Get the task j_i priority
(11) If j_i has the highest priority
(12) Assign task j_i to the resource r_j
(13) Delete task j_i from J
(14) Update the value of ETC
(15) Endwhile

Fig. 2. Improved algorithm procedure

Scheduling class needs to reduce the earliest completion time, the weight of initial priority, dynamic priority, and the priority of user expectation are W_1, W_2, W_3.Known by the definition of the front can get the $final\,Priority(i)$:

$$finalPriority(i) = w1 * initPriority(i) + w2 * \frac{waitTime(i)}{interval} * increment + w3 * except(i) \quad (4)$$

The task of the highest priority will be scheduled first, according to (4).User-defined weights, has a good self-adaptability.

3.5 Stimulate Experiment

Cloud computing simulation platform Cloudsim[1] was designed by Australia Melbourne university. It is established on the grid and Gridbus. Initial releases of CloudSim used SimJava as discrete event simulation engine [8].

Now using the first kind (completion time) and the second kind (Bandwidth) task scheduling simulation to validate the algorithm model. Create a group of tasks, its list of parameters, shown in the figure 3. Create a group of virtual machines which are performance differences and preferences. The virtual machine parameters are listed in table figure 4. Data sources from [9].

Cloudlet	Class Type	length	filesize	outpusizet	expectationtime	expectationBW
0	1	4000	2500	500	400	-
1	1	3000	2000	400	200	-
2	1	2000	800	300	150	-
3	1	5000	5000	2000	500	-
4	2	2000	800	300	-	2000
5	2	3000	2000	400	-	3000
6	2	800	300	300	-	1200
7	2	2500	1000	500	-	2000

Fig. 3. The tasks parameters

VMId	CPU	Memory	BandWith
0	4	2048	1200
1	2	1024	3000
2	2	1024	1000
3	1	512	1200

Fig. 4. The virtual machine parameters

Cloudlet	Class Type	STATUS	VMID	Time	ProcessingCost	VMBandWith	CUPs	Memory
6	2	SUCCESS	2	8	35	1000	2	1024
4	2	SUCCESS	0	20	71	1200	4	2048
2	1	SUCCESS	2	20	71	1000	2	1024
5	2	SUCCESS	1	30	114	3000	2	1024
0	1	SUCCESS	0	40	150	1200	4	2048
1	1	SUCCESS	1	50	200	3000	2	1024
7	2	SUCCESS	3	50	165	1200	1	512
3	1	SUCCESS	3	60	370	1200	1	512

Fig. 5. Simulation results of the traditional algorithm

Cloudlet	Class Type	STATUS	VMID	Time	ProcessingCost	VMBandWith	CUPs	Memory
6	2	SUCCESS	2	10	47	1000	2	1024
2	1	SUCCESS	0	20	71	1200	4	2048
4	2	SUCCESS	2	20	83	1000	2	1024
5	2	SUCCESS	1	28	223	3000	2	1024
0	1	SUCCESS	3	35	150	1200	1	512
3	1	SUCCESS	0	44	365	1200	4	2048
1	1	SUCCESS	1	56	200	3000	2	1024
7	2	SUCCESS	3	56	170	1200	1	512

Fig. 6. Simulation results of the improved algorithm

4 Conclusion

The purpose of the simulation is comparing the improved min-min algorithm with the min-min algorithm. Compared with the traditional min-min algorithm, the experimental results show it can increase resource utilization rate, long tasks can execute at reasonable time and meet users' requirements.

Acknowledgement. The work was supported by the Project of cloud computing resource planning strategy and Algorithm from the Research the Natural Science Foundation of Jilin Province (20101533) and the Project of the virtual resource scheduling algorithm of cloud computing environment from the Jilin Provincial Department of Education of "Twelfth Five-Year Plan" science and technology research projects (2011103).

References

[1] Calheiros, R.N., Rajiv, De Rose, C.A.F., Buyya, R.: Cloudsim: A Novel Framework for Modeling and Simulation of Cloud Computing Infrastructures and Services, http://www.gridbus.org/reports/CloudSim-ICPP2009.pdf
[2] China cloud computing. Peng Liu:cloud computing definition and characteristics, http://www.chinacloud.cn/.2009-2-25
[3] Buyya, R., Yeo, C.S., Venugopal, S., Broberg, J., Brandic, I.: Cloud Computing and Emerging IT Platforms: Vision, Hype, and Reality for Delivering Computing as the 5th Utility. Future Generation Computer Systems 25(6), 599–616 (2009)
[4] Braun, T.D., Siegenl, H.J., Beck, N.: A Comparison of Eleven Static Heuristics for Mapping a Class of Independent Tasks onto Heterogeneous Distributed Computing Systems. Journal of Parrallel and Distributed Computing, 801–837 (2001)
[5] Shan, H.X., He, S.X., Gregor, V.L.: QoS Guided Min-Min Heuristic for Grid Task Scheduling. Journal of Computer Science and Technology 18(4), 442–451 (2003)
[6] Deng, J.: User-oriented job fair scheduling algorithm of Cluster system. Beijing University of Posts and Telecommunications, Beijing (2008)
[7] Ding, M.: Improved Min-Min algorithm in grid computing. Northwestern University (2010) TP393.01
[8] Howell, F., Mcnab, R.: SimJava: A discrete event simulation library for java. In: Proceedings of the first International Conference on Web-Based Modeling and Simulation (1998)
[9] Zhao, C.: Job Scheduling Algorithm Research and Implementation of the cloud environment. Jiaotong University, Beijing (2009) TP393.01
[10] GRIDS Laboratory, http://www.gridbus.org.cloud
[11] Jasso, G.: The Theory of the Distributive-Justice Force in Human Affairs:Analyzing the Three Central Questions. In: Sociological Theories in Progress:New Formulation, pp. 354–387. Sage Pub., Newbury Park (1989)
[12] Braun, T.D., Siegenl, H.J., Beck, N.: A Comparison of Eleven Static Heuristics for Mapping a Class of Independent Tasks onto Heterogeneous Distributed Computing Systems. Journal of Parrallel and Distributed Computing, 801–837 (2001)
[13] Bharadwaj, V., Ghose, D., Robertazzi, T.G.: Divisible Load Theory: A New Paradigm for Load Scheduling in Distributed Systems. Cluster Comput. 6(1), 7–17 (2003)

A Novel Information Technology Based College Physical Education Model

Li Yan-xia, Wang Qiang, and Zhang Ying-jian

Dept.of Physical Education,
Lang Fang Normal University, China
xiaoxiali529@163.com

Abstract. In this paper, we propose a novel college physical education model using information technology. Computer technology is combined with communication network to make physical course contents more rich. Furthermore, multimedia data and online learning mode are introduced into physical course to make students understand the course more deeply. The proposed model is made up of three parts which are "Teachers section", "Multimedia data" and "Students section". Moreover, our college physical education model integrates teaching process with learning process using multimedia data. To validate the effectiveness of the proposed model, we design two experiments for performance evaluating. To test the performance of our model, 50 students are required to give feedbacks after attending the course based on our model. Experimental results show that our model is effective and interesting for physical course learning.

Keywords: Information technology, College physical education, Multimedia data, Online learning.

1 Introduction

The 21st century is a digital information age with fast development, and there are full of new technologies and new knowledge in this era. Meanwhile, physical teaching process should keep the pace with modern information technology. With the great development of modern science, more and more information techniques have been adopted in physical teaching, and the most important technology is computer aided technology, multimedia technology and so on.

In recent years, information technology has been one of the most favorite teaching tools in physical education process. The use of information technology makes the college physical education process more efficiently than the traditional approach. The reasons mainly lie in that information technology based physical teaching has many advantages compared with traditional teaching methods. Applying information technology in physical teaching, multimedia software and internet can be used in the whole process of physical teaching. This paper presents a novel model to reform traditional college physical teaching mode by applying information technology.

The rest of the paper is organized as follows. Section 2 introduces requirements of modern college physical education. Section 3 presents a novel information technology

Z. Du (Ed.): Proceedings of the 2012 International Conference of MCSA, AISC 191, pp. 53–56.
springerlink.com

based college physical education model. In section 4, we conduct experiments to show the effectiveness of the model we designed. In Section 5, we conclude the whole paper.

2 Requirements of Modern College Physical Education

By making full use of information technology in physical education, not only students but also teachers can benefit from this mode, and it is also helpful for students to grasp relevant knowledge and improve learning effect. Before designing the physical education model, we should analyze the requirements of modern college physical education in detail as follows.

2.1 Making Contents of Physical Course More Rich

Combining computer technology with communication technology, computer network has been widely used in modern education. Moreover, Web is the largest and most influential network in the world, which contains many online sports related websites. The information of these websites includes all aspects of sports, such as sports news, sports research and sports knowledge. Using information of the Web could help the students to learn physical course more effectively and more timely.

2.2 Making Students to Understand the Physical Course More Deeply

To make the physical course more interesting, multimedia data(such as video, image) could be used in physical teaching. Multimedia data can also help students understand the course more deeply and thus enhance teaching quality. For example, introducing image and video of Internet into physical course could explain the key teaching point and the difficult teaching point more clearly. On the other hand, online course is also an important method to enhance physical education. In this mode, students can learn the course from online multimedia resource repeatedly at any time, and teachers can also enrich and modify course contents more conveniently.

3 College Physical Education Model

Based on the above analysis, our college physical education model is made up of three parts, which are "Teachers section", "Multimedia data" and "Students section". As is shown in Fig.1, the proposed college physical education model combines teaching with learning process using multimedia data which includes image, video and animation.

For teachers, the whole teaching process contains four steps: 1) Collecting teaching contents which is related to the course, 2) Choosing teaching multimedia which could help students learn the course, 3) Making teaching plan according to course characteristics and learning ability of students; 4) Design teaching process to make full use of teaching materials and then make the teaching process more interesting.

For students, the learning process of our model includes four steps: 1) Making learning plan to generally know what are key problems in this course; 2) Classroom learning which is an important parts in learning process; 3) Online Learning which is an supplementary parts for classroom learning, and in this step students could use multimedia data to understand the course contents more deeply; 4) After online learning process, the students could give feedbacks of teaching quality to teachers.

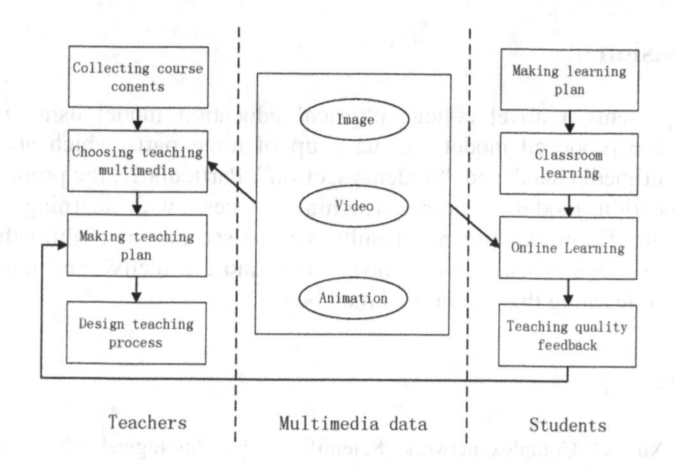

Fig. 1. Illustration of College Physical Education Model

4 Experiments and Analysis

To validate the effectiveness of the proposed model, we design two experiments for performance analyzing. To value the performance of our model, we arrange 50 students to value the proposed model after attending our course for a whole term. We ask each student to answer two questions by providing a rating score of between 1(very bad) and 10(excellent): 1) Question 1 is if the proposed model is more effective than traditional physical course? 2) Question 2 is if the proposed model is more interesting for students than traditional course?

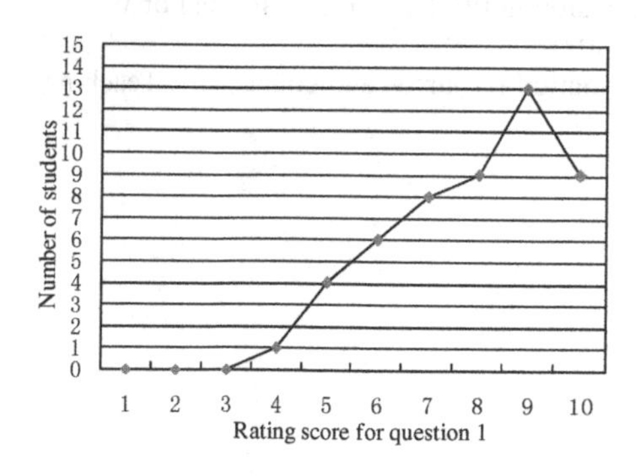

Fig. 2. Performance Evaluation on question 1

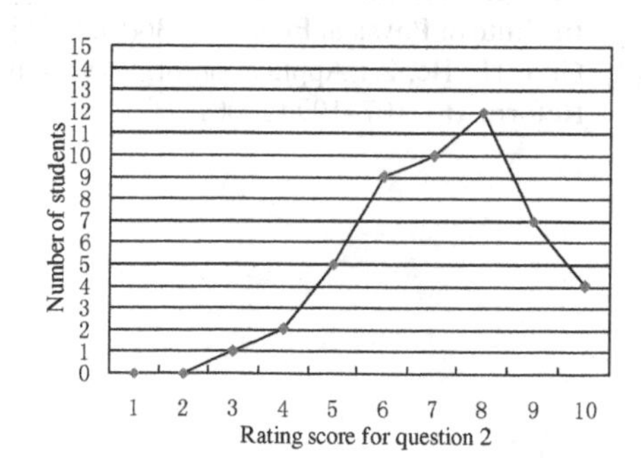

Fig. 3. Performance Evaluation on question 2

As is shown in Fig.2 and Fig.3, the average rating score of the above two questions are 7.9 and 7.18 respectively. Therefore, the conclusions can be drawn that the proposed model could enhance the effectiveness of physical teaching process and make students be interested in physical course learning.

5 Conclusion

This paper presents a novel college physical education model using information technology. The proposed model are made up of three parts which are "Teachers section", "Multimedia data" and "Students section". Particularly, the proposed college physical education model combines teaching process with learning process by multimedia data. From experimental results we can see that as multimedia data and online learning mode is used in our model, it is more effective and interesting for physical course learning than traditional methods.

References

[1] Guo, L., Xu, X.: Complex network. Scientific and technological education publishing house, Shanghai (2006)
[2] Sharma, G., Mazumdar, R.R.: A case for hybrid sensor networks. In: MobiHoc, pp. 366–377 (2006)
[3] Ye, X., Xu, L., Lin, L.: Topology optimization based on small-world phenomenon in wireless sensor networks. Journal of Fujian Normal University 9(5), 38–40 (2008)
[4] Du, H.: Technology-innovated multimedia courseware. Health Vocational Education 22(1), 124 (2004)
[5] Ye, Q., Wang, A.: Concept of quality education and college PE network course reform. Journal of Physical Education 6, 97–99 (2001)
[6] Geng, J.: Multimedia courseware's role in PE teaching reform. Journal of Harbin Institute of Physical Education (1) (2002)
[7] Lu, C.: Application of computer network in modern physical education. Journal of Shandong Physical Education Institute 3, 94–96 (2002)
[8] Cai, X.: Multimedia teaching of technology division in PE departments. Journal of Wuhan Institute of Physical Education 36(6), 87–89 (2002)
[9] Guo, H., He, Z.: Application of network information resources and college. PE Teaching Reform. (6), 107–109 (2000)

The Improved Text Classification Method Based on Bayesian and k-NN

Wang Tao, Huo Liang, and Yang Liu

Department of Information Management and Engineering Hebei Finance University China
Experimental Center of Economics and Management Hebei University China

Abstract. In order to improve the classification accuracy and speed, classification of the structure of this paper has been improved, is proposes a combination of Bayesian and k-nearest neighbor classifier model, which combines Bayesian classification method of classification rate fast and k-nearest neighbor method with higher classification accuracy advantages. Experimental results show that the method to ensure the classification rate under the premise of effectively improving the classification accuracy.

Keywords: Bayesian classification, k-NN, Text classification.

1 Introduction

Text classification is a classification algorithm using a known sample set to learn, to train a classifier, using the classification of unknown samples were automatically classified category. Commonly used classification algorithms are Bayesian methods, k-NN method, the center vector method, decision tree methods, support vector machine methods. The classification algorithm has advantages and disadvantages, so the use of a combination of strategy, these algorithms together, whichever benefits to its shortcomings, is a classification algorithm to optimize a variety of ways. Combination of strategies used in two forms: horizontal combinations and vertical combinations. Each horizontal layer which contains only the combination is a classifier, the classification level on the forecast with all properties as initial input to the next level, until the final layer gives the classification results. Vertical combination is the first layer classifier for the different categories of independent samples, the second layer of the results of these classifiers are combined according to some strategy[1] [2].

This paper presents a cross classification method using a central vector as the first layer classifier, Bayesian classifier as the second layer, k-NN classification as a third layer classifier. Experimental results show that the classification of the classification accuracy is higher than the single classifier accuracy.

2 Combination Classification Model

First, the training sample pretreatment. Text that is currently used mainly vector space model, the text that point to a vector space.

Z. Du (Ed.): Proceedings of the 2012 International Conference of MCSA, AISC 191, pp. 57–61.
springerlink.com

(1) First, the text word by word as a vector of these dimensions to represent text, which will match the text message said the problem with the vector into a vector space representation and matching.
(2) Then the text vector dimensionality reduction algorithm, scanning the document vector vocabulary, with synonyms, delete the word frequency is very small and the emergence of frequent words.
(3) The use of TF-IDF representation of lexical weights for processing.
(4) Use of mutual information (MI) method for extracting feature items, select a category in a high probability, select other categories of low-probability words as feature words.

2.1 The First Layer Classifier

The first layer classifier of this paper choose the simple algorithm, fast center vector method, the basic idea is, according to belong to a class of all training text vector calculation of the category of the center vector, and then calculate the text to be classified with each class centre vector similarity, and put it into the largest category similarity.

(1) The text of the test word, the formation of new test vectors.
(2) The characteristics of each type of query word contains the number of the vector.
(3) certain number of test vectors contain more text and description of the test in this category, the more similar, so the test text test vector containing the number assigned to a class of most, if category number up to more than one class, then the test text is not classified as an input to the next level of classification.

2.2 The Second Layer Classifier

The second layer of a Bayesian classifier selection method is faster but the classification accuracy rate is low, so the text in the text of the test, when added to forecast the probability of a threshold control, encountered a probability below this threshold value, will enter the next level, so you can improve the performance of Bayesian methods.

Specific steps are as follows:

(1) According to Equation 1 for each category feature words t_k is the probability vector w_k .

$$w(t_k \mid C_j) = \frac{1 + \sum_{i=1}^{|D|} tf(t_k, d_i)}{M + \sum_{s=1}^{M} \sum_{i=1}^{|D|} tf(t_s, d_i)}$$

Formula 1

Among them, the $w(t_k \mid c_j)$ is the probability of t_k appear in c_j , c_j is a category, d_i is an unknown type of text, t_k appear in the d_i feature items, $tf(t_k, d_i)$ as the number of occurrences in the d_i ,M is the general characteristics of the training set number of words, |D| is the number of such training text[3].

(2) In the new text arrives, the new text word, and then the text in accordance with the formula 2 to calculate the probability of belonging to the class:

$$P(C_j \mid d_i) = \frac{P(C_j)\prod_{k=1}^{M} P(t_k \mid C_j)^{tf(t_k,d_i)}}{\sum_{r=1}^{|C|} P(C_r)\prod_{k=1}^{M} P(t_k \mid C_r)^{tf(t_k,d_i)}}$$

Formula 2

Which $p(c_j) = \frac{\text{The number of training documents of } c_j}{\text{The totalnumber of training documents}}$

$P(C_r)$ is the similar meaning, $|C|$ is the total number of categories.

(3) To compare the new text belongs to all classes of probability, the probability of finding the maximum value, if the value is greater than the threshold value, the text assigned to this category, if less than the threshold, then the next layer of classification.

2.3 The Third Layer Classification

Consider these two main categories of classification rate, the third layer selected a better classification performance k-NN method.[4]

$$S_{ij} = \text{CosD}_{ij} = \frac{\overrightarrow{D_i}\,\overrightarrow{D_j}}{\left|\overrightarrow{D_i}\right|\left|\overrightarrow{D_j}\right|} = \frac{\sum_{k=1}^{M} w_{ik} w_{jk}}{\sqrt{\sum_{k=1}^{M}(w_{ik})^2 \cdot \sum_{k=1}^{M}(w_{jk})^2}}$$

Formula 3

(1) According to Formula 3 with the training text and the test text on each similarity;

(2) According to text similarity, text set in the training and testing of selected text in the k most similar texts;

(3) Test the k neighbors in the text, followed by calculating the weight of each class is calculated as Formula 4.

$$p(\vec{x}, C_j) = \begin{cases} 1 & \text{if } \sum_{\vec{d}_i \in KNN} Sim(\vec{x}, \vec{d}_i) y(\vec{d}_i, C_j) - b \geq 0 \\ 0 & \text{others} \end{cases}$$

Formula 4

Among them, $\vec{x}$ the eigenvector for the new text, $Sim(\vec{x}, \vec{d}_i)$ for the similarity calculation formula 4, b is the threshold, this value is a value chosen to be optimized, while the value of $y(\vec{d}_i, C_j)$ is 1 or 0, $\vec{d}_i$ Belongs to class C_j ,then the function value is 1, otherwise 0.

(4) Compare the weight class, the text assigned to the heaviest weight on that category.

3 Experimental Results and Analysis

In this paper, achieve the above types of Chinese text classification system on the Windows XP operating system, the database is SQL SERVER2000.

Corpus used in this article from the website "Chinese natural language processing open platform" provided by the Fudan University, Dr. Li Ronglu upload, repeat the text of which initial treatment and damage to documents, training documents 9586, test text 9044, is divided into 20 categories, including training and test set ratio of 1:1. First, preliminary processing, training text with the second scan of the sub-word method, feature selection using mutual information method is the test text has been over the word and feature selection process, and on this basis for classification. After statistical classification of the time contains only the time, does not include the initial processing time[5].

3.1 The Results

Table 1. The Accuracy of Classification

classification method	Center vector	Bayesian classification	k-NN (k=5)	Combined classification
Accuracy rate rate	65.72%	66.03%	71.51%	78.69%
Time(s) consuming(s)	437	492	3636	1311

The experimental results obtained, the center's fastest vector classification method, and only 437 seconds, and the Bayesian method than the classification accuracy increased by 0.5%. Bayesian methods and improved methods of combining k-nearest neighbor method of classification accuracy rate is highest, the classification accuracy rate higher than the Bayesian method 19.7%, higher than the k-nearest neighbor method 10%; it takes is a simple Bayesian 2.66 times approach, but only k-nearest neighbor method of 36.06%[6].

3.2 Experimental Results

Because K - nearest neighbor method each decision-making need to all training samples for comparison, calculation are deferred to the classification process, so the classification speed is slow. The Bayesian classifier, only needs to calculate the product to the test documents are classified, so the test time is short.

For improved Bayesian classification method, only calculating the containing characteristic word number, so than naive Bayesian classification method is faster, at the same time it with some small probability event, by naive Bayesian can classify text, can be correctly classified, so the classification accuracy rate than the naive Bayesian method is improved.

For improved Bayesian method and K - nearest neighbor method combining method, combines Bayesian classification speed and K - nearest neighbor classification accuracy rate high, the experimental results compared with the ideal.

4 Conclusion

This article discusses two is generally considered good classification, naive Bayesian method and K - nearest neighbor method, and the Bayesian method was improved, finally proposed combining Bayesian method and K - nearest neighbor method a new method. In this paper, to achieve the above four methods in the same corpus -- "Chinese natural language processing open platform", From the experimental results, the improved Bayesian classification method is faster, suitable for larger data sets and on-line real-time classification; Bayesian method and K - nearest neighbor method a new method combining classification accuracy was highest, applicable in high accuracy, at the same time its classification speed compared with the pure K - nearest neighbor method raise, can also handle a larger sample set of data or for real time processing.

References

[1] Langley, P., Iba, W., Thompson, K.: An analysis of Bayesian classifiers. In: Proceedings of the Tenth National Conference on Artificial Intelligence, pp. 223–228. AAAI Press, Menlo Park (1992)

[2] Geiger, F.N., Goldszmidt, D.: Bayesian network classifiers. Machine Learning 29(2/3), 131–163 (1997)

[3] Ramoni, M., Sebastiani, P.: Robust Bayes classifiers. Artificial Intelligence 125(122), 209–226 (2001)

[4] Cheng, J., Greiner, R.: Comparing Bayesian network classifiers. In: Laskey, K.B., Prade, H. (eds.) Proc. of the 15th Conf. on Uncertainty in Artificial Intelligence, pp. 101–108. Morgan Kaufmann Publishers, San Francisco (1999)

[5] Susumu, T.: A study on multi relation coefficient among variables. Proceedings of the School of Information Technology and Electronics of Tokai University 4(1), 67–72 (2004)

[6] Bocchieri, E., Mark, B.: Subspace Distribution clustering hidden Markov model. IEEE Transactions on Speech and Audio Processing 9(3), 264–275 (2001)

The Basic Principle and Applications of the Search Engine Optimization

Xie Xiaojie, Fang Yuan, and Wu Jian

Computer Science, Zhejiang University of Technology, China

Abstract. This paper analyzes the search engine algorithms and ranking principle, we discussed the site structure, keywords, single-page optimization, and search engine penalties. In addition, search engine optimization techniques developed in recent years, which have a very important role in the corporate website ranking. In this paper, we use some search engine optimization technology applications to discuss this increasing technology.

Keywords: Basic principle, Application, Search engine optimization.

1 Introduction

The search engines are commonly used tools for access to information resources through the network. Retrieved by the search engines, customers can easily find the relevant data and information in the sorted list, which listed by the degree of attention. In order to improve the site's traffic, the site owners always use many methods for occupying a good position in the list. The Search Engine User Attitudes report was released in April 2004 by the search engine marketing service provider iProspect, showed 81.7% of users do not browse the search results which located after the third page, and 52.2% of users will only focus on the first page of search results returned from the search engine. In other words, the user usually cares only about the top of pages provided by the search engines. Figure 1 shows the flow diagram of search engine.

2 The Basic Principles of the Search Engine Optimization

Figure 2 shows the search engine optimization strategy implement block diagram. And later we will discuss the basic principles of search engine optimization.

2.1 The Source of Search Engine Optimization and Some of Basic Definitions

In recent years, more than 90% of website traffic and 55 percent of online transactions are dependent on search engines. There are much higher return on investment (ROI) relative to other online or traditional marketing, search engine would take your potential customers to your side, and also would increase your sales. The world famous Google search engine provides the 150 million queries a day. If some e-commerce sites in the Google search engine ranking top 10, which will give the

Z. Du (Ed.): Proceedings of the 2012 International Conference of MCSA, AISC 191, pp. 63–69.
springerlink.com

companies a lot of orders. Based on these advantages, Web site creation and maintenance personnel specialize search engine optimization, it will be imperative refers to e-commerce sites. SEO (Search Engine Optimization, referred to as SEO), SEP (Search Engine Positioning), and SER (Search Engine Ranking) are the same kind of the work. Specifically, the work including: By studying the various search engines access to the Internet page, how to build an index, and how to determine the ranking for a particular keyword search results, how to optimize the site pages in a targeted manner and increase the ranking in search engines results, and finally, you can get much more website click-through rate, and ultimately enhance the sale of the site technical or publicity.

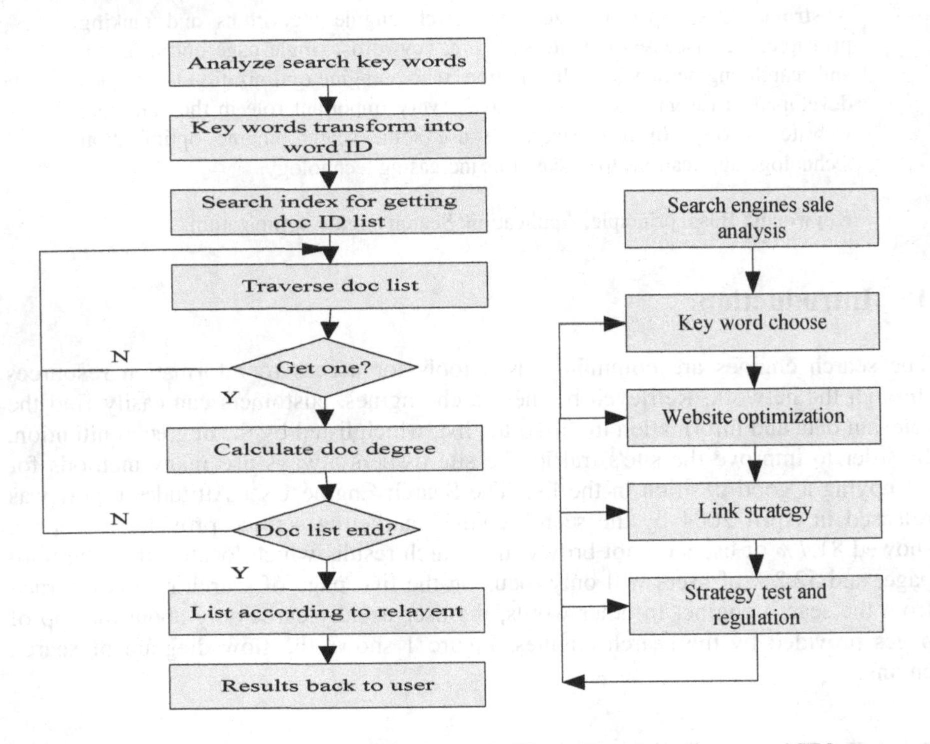

Fig. 1. The flow diagram of search engine **Fig. 2.** The block diagram of SEO strategy

2.2 Summary The Core of Search Engine Optimization - Keyword Strategy

There are input statements call key words used by the potential customers and target users when they want to find the corresponding website in the search engines.

The choose of Key words: First make sure core key words, and then around the core key words to the permutation and combination, so they can generate key phrases or short sentences. For enterprises and businesses, the core key words are their business scope, such as product or service name, industry location, as well as enterprise name or the brand name. In addition, people will be standing in the point of view of the customer to learn their ideas in the process of search. Let key words expansion into a series of phrase, and try to avoid using single words but a single vocabulary as the basis of expansion. Like this we may use the network marketing software to test for these key phrases and phrases, software function is to check your key words in other web pages for getting the frequency of using, and finally to choose the highest frequency one. In addition, it need to a multiple permutation and combination; Don't use the key word which meaning too broad; With their own brand as the key word; Can use the geographical position; As far as possible control the number of key words and so on, these all can effectively increase the retrieve probability, which has more important role to the optimization of the website.

The density of Key words: After determining the key word, the present frequency of key words in your web page is keyword density. That is, on one page, accounting for the proportion of the total text in the page, this index has played an important role in search engine optimization. Code label in HTML is excluded from the page content during calculating the keyword density. Usually, there will be thousands of words on the page, the search engine identify the important words to describe the site y the keyword density.The website will get good rankings if the key words have been set within the allowable range. Studies have shown that the keyword density within 2% -8% is considered to be useful to improve the website ranking. On the contrary, excessive use of keywords will result in a pile of keyword stuffing. Although the excessive using of keywords could trigger the search engines filter, the more serious problem is the potential loss caused by traffic due to difficult to read. A viable approach is to search optimized keyword in the search engines, and then with the help of keyword density query tools to get the keyword density in the top few sites, with reference to its value to set the critical density of your pages.

Density distribution: Keyword distribution principle is to place keywords in the appropriate position. According to keyword density may get the number of keywords, and then consider how the keywords are placed in a prominent position on the page. Highlight the keyword is one of the most important factor to attract search engine attention. The search engine spiders will focus on a specific part of the contents of Web pages, in the words of this concern part is much more important than other parts of the word. Is usually placed in the key: <Title>, and <meta> labels, titles (headings), hyperlink text, URL text and the top of the page of the texts. Table 1 shows an example of key words analysis.

Table 1. The example of key words analysis

Key words	density	Baidu Index	Google Index
A	0.26%	6010	7608
B	0.26%	4531	4690
C	0.39%	203	301

2.3 Search Engine Optimization Focus – Linking Strategy

Link is a connection from a page to point to another target, this target can be a web page, can be different locations of the same page, and can also be a picture and e-mail address or file, or even an application. User get rich website content through hyperlinks, and search engines is also tracking the layers along a website link to the page depth, to complete the site's information capture. When the search engines determine the ranking of a website, not only analyze the page content and structure but also analyze links around the site. Critical factors affecting website ranking is to get as much high-quality external links, also known as incoming links. Overall, access to hundreds of poor quality site links is less worth than a high-quality sites link.

2.4 Optimization of a Single Page

Name the Website: You can use keywords in the directory name and file name, if there are key phrases you need with a separator to separate them. We used a hyphen for separated, such as Welcome-To-Beijing.html.

Website title: Title length should be controlled within 40-60 letters, and highlight the theme by locating keywords in title. However, don't use irrelevant keywords as title.

Website description: This Meta value is only recognized by all search engines, within 40 words, each page should have a unique description.

Subject of the Website: To express the theme of the prominent pages of important sections in bold, large print, etc. Using keywords to describe HTML text, in the case does not affect the significance of Website theme.

External connections: The external connection is the most important factors that affect search engine rankings in ddition to the page title and subject, best able to enhance the ranking of connection from a higher PR value of site's home page connection. In addition, this text should be correlated to title and subject key words in subject website.

<H1> label: The search engines are interested in such information <h1> <h2>, and will increase the right weight so the most important information should be logo out with < h1> </ h1>, and minor important information should be logo out with <h2> </ h2>.

<b> <strong> : Label will also be well noted by the search engine. Although the weight is not better than <h1> <h2> can be used flexibly.

2.5 The Meaning of Search Engine Optimization

Through the search engines to find information and resources are the primary means of today's Internet users. There are many reasons to promote the implementation of search engine optimization, following the investigation report fully illustrates this point: First, the search engine marketing firm iCrossing done the survey found that: prior to online shopping, search engines are the most popular tools used to search for products and services, 74% of users search for products, while 54% of users to find the can shopping sites. Second, the Chinese search engine market research annual report noted that: the end of 2007, the market will reach 2.93 billion Yuan, has an increase of 76.5%, Q1 Chinese search engine market size reached 929 million Yuan in 2008, declined slightly to 4.7%, but the annual doubled year on year increase of 108.3%. For this reason, how to make your own website record by major search engines, and then get a higher rank to become the issue of Web site builders. Figure 2 shows the SEO process and techniques.

2.6 Search Engine Optimization Cheating and Punishment

Technology-based search engine site ranking process is completely done automatically by the spider, not human intervention, which provides the likelihood of success for those who deceive spiders means for ranking.

SEO cheating ways:

Keyword Superimposition and accumulation:

a) Keywords in any position of the page source code, deliberately by adding relevant keywords to the page content, and intentionally a lot of repetition of certain keywords.
b) Hidden text: The search engine can identify the web page but the user can not see the hidden text. Include using the text and background color, the small font text, text hidden layer, or abuse of the image ALT and other methods.
c) Cover up: The same Web site, search engine and user access to different content pages, including the use of redirection behavior.
d) Promotion of illegal operation of the site external

The illegal operation include: first, to repeatedly submit your site to search engines. Second, blog contamination. The comments of the blog site can be frequently posts, which buried connection to reach your site, this approach attempts to pollution blog to increase the number of connection to their site. Third, connect the farms. Refers to the pages of worthless information, this page addition to the list of a connection point to the websites of others, almost nothing else, the entire site is purely a switched connection, no matter is not related to the theme. Cheating is a complete departure from the original intention, the principle of favorable to the user, of the search engine ranking search results, which destroy the normal search results and impact on the quality and reputation of the search engines, and also clear out the loopholes in search engine technology, so once they are the detection of cheating will give different degrees of punishment depend on the seriousness of the case. For example, in March 2005, Google cleaning up some spam search results is the typical punishment cases.

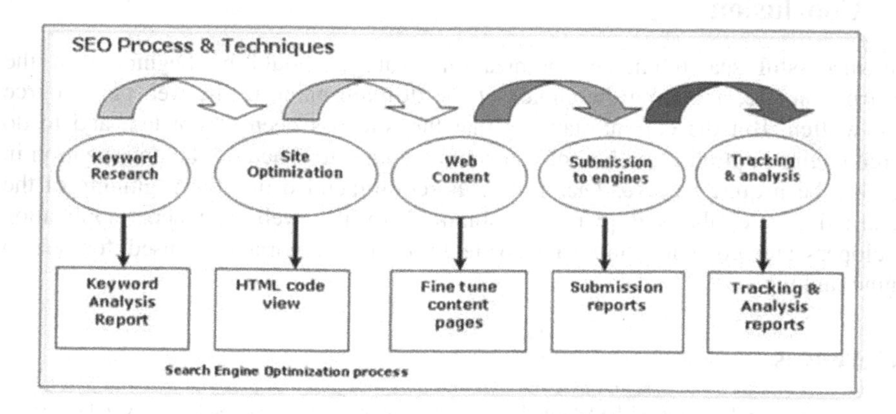

Fig. 2. The Search engine optimization process and techniques

3 Applications of the Search Engine Optimization

Dell China the site www.dell.com.cn Home titled "Dell China - computers, laptops, desktops, printers, workstations, servers, storage, electronic products and accessories,". Dell Home title not only covers the most important company information, but also includes the company's main products. This is the core keywords, with notebook computers, desktop computers. These keywords in Google search Dell's pages are ranked in the first few positions in the first screen. Google's ranking algorithm and organizational details are trade secrets, and well known is that the PageRank algorithm. It is an important content of the Google algorithm, which awarded as a U.S. patent in September 2001. The author is one of the Google founders Larry Page. Measure the value of the site based on the quantity and quality of the site's external and internal links. Behind the Page Rank concept is that each page to link to this page are the first ballot, is linked more, it means that other websites to vote more. This is the so-called "link popularity". Google's Page Rank score from 0 to 10; the Page Rank of 10 is the best, but very rare, similar to the Richter scale, the Page Rank level is not linear, but in accordance with an exponential scale. This is a strange mathematical terms, the mean PageRank4 not level and may be 6-7 times better than PageRank3. PageRank algorithm formula is very complicated. Simply, the value of Pagerank is subject to the following factors. The more number of pages crawled by Google, the more Pagerank higher the value. Google usually does not take the initiative to crawl all the pages, especially in the dynamic link to the URL with "?". The better approach is to produce a static HTML pages, often referred to as the "Site Map" or "Site Map", it contains all the URLs you want to add static pages and then submit to Google. Google's Pagerank system on the portal network directories such as DMOZ, Yahoo and Looksmart are particularly high regard. On Google's Pagerank, DMOZ links to a site, like a piece of gold as precious. PageRank is only part of the Google algorithm, so deliberately to pursue PageRank would not guarantee high ranking.

4 Conclusion

The successful search engine optimization strategy should be beginning of the construction site, and from the choice of the domain name to the web page source code written. But the current status is that the site has been completed, and to do search engines optimization before you submit your site. Then do the optimization in fact, has been quite passive. Therefore, it is recommended that the beginning of the construction site, the site planners submitted to the web designers, application developers and content editors a website building memorandum, used for search engine ranking.

References

[1] Lin, Y.Y.: Applications of the Search Engine Optimization. Software 8(11) (November 2009)

[2] Liu, L.: The discussion of the Search Engine Optimization technology. Software 8(8) (August 2009)

[3] Kent, P.: Search Engine Optimization For Dummie. Wiley Pbulishiing Inc. (2006)
[4] Xu, Z.: Website promotion of Search Engine Optimization. Science Information (11) (2008)
[5] Lawrence, S., Lee Giles, C.: Accessibility of information on the Web. Nature 400(8), 107–109 (1999)
[6] Sullivan, D.: Fifth Annual Search Engine Meeting Report, Boston, MA (April 2000)

Semi-supervised Incremental Model for Lexical Category Acquisition

Bichuan Zhang and Xiaojie Wang

Center for Intelligence Science and Technology,
Beijing University of Posts and Telecommunications, Beijing
bugnec@gmail.com, xjwang@bupt.edu.cn

Abstract. We present a novel semi-supervised incremental approach for discovering word categories, sets of words sharing a significant aspect of distributional context. We utilize high frequency words as seed in order to capture semantic information in form of symmetric similarity of word pair, lexical category is then created based on a new clustering algorithm proposed recently: affinity propagation (AP). Furthermore, we assess the performance using a new measure we proposed that meets three criteria: informativeness, diversity and purity. The quantitative and qualitative evaluation show that this semi-supervised incremental approach is plausible for induction of lexical categories from distributional data.

Keywords: computational model, language acquisition, lexical category, semi-supervised incremental model.

1 Introduction

In the first years of life, children have achieved the remarkable feat of how words are combined to form complex sentences. Recently researchers have begun to investigate the relevance of computational models of language acquisition. These investigations could have significance since an improved understanding of human language acquisition will not only benefit cognitive sciences in general but may also feed back to the Natural Language Processing (NLP) community. The development of computational models for language acquisition has been promoted by the availability of resources such as the CHILDES database (MacWhinney, 2000).

A number of computational models of category learning such as those of Clark (2001), Redington et al.(1998), Mintz(2002) have been developed, most of which regard the problem as one of grouping together words whose syntactic behavior is similar. Typically, the input for these models is taken from a corpus of child-directed speech, and these computational models have used distributional cues for category induction; these models confirm the ability of identifying categories, and show that distributional cues are informative for categorization.

Our main contributions are twofold. First, we propose a semi-supervised incremental model for lexical category acquisition by utilizing semantic information of seed words and distributional information about all words' surrounding context. We train our model on a corpus of child-directed speech (CDS) from CHILDES

Z. Du (Ed.): Proceedings of the 2012 International Conference of MCSA, AISC 191, pp. 71–76.
springerlink.com

(MacWhinney, 2000) and show that the model learns a fine-grained set of intuitive word categories. Second, we propose a novel evaluation, Cohesivity, approach by synthesizing many different kinds of criteria. We evaluate our model by applying our new measure. The results show that the categories induced by our model are informative about the properties of their members, rich diversity about the quantity of clusters, and pure about similarity among their members. Also, the evaluation metric can give more reasonable explanation for the clustering results and is efficient computationally.

2 Lexical Category Acquisition

In lexical category acquisition task, the gold-standard categories are formed according to "substitutability": if one word can be replaced by another and the resulting sentence is still grammatical and semantically correct, then there is a good chance that the two words belong to the same category. Children acquiring lexical category of their native language have access to a large amount of contextual information on the basis of speech they hear.

Much work has been done on lexical acquisition of all sorts. The three main distinguishing axes are (1) the type of corpus annotation and other human input used; (2) the type of lexical relationship targeted; and (3) the basic algorithmic approach (Davidov and Rappoport, 2006). Many of the papers cited below follow the same steps, and partition the vocabulary into a set of optimum clusters.

Semi-supervised learning is a good idea to reduce human labor and improve accuracy that is widely used in machine learning tasks. It is possible that understanding of the human cognitive model will lead to novel machine learning approaches (Mitchell, 2006). Do humans including children when learning language do semi-supervised learning? Maybe yes. Some evidence shows that we humans accumulate unlabeled input data, which we use to help doing cognitive tasks (Tomasello, 2006). Semantic information is often expensive to obtain; however, relatively little labeled data is guaranteed to improve the performance.

The process of learning word categories by children is necessarily incremental. Language acquisition is constrained by memory and processing limitations, and it is implausible that humans process large volumes of text at once and induce an optimum set of categories (Afra Alishahi and Grzegorz Chrupała, 2009).

3 General Method

In view of these considerations, we set out a semi-supervised incremental model to develop an approach to discovery of categories from large child-directed speech from CHILDES. We make use of both distribution information and supervised information to construct a word relationship graph, and then induce clusters in an incremental method. Fig. 1 illustrates our approach. It is composed of a number of modules.

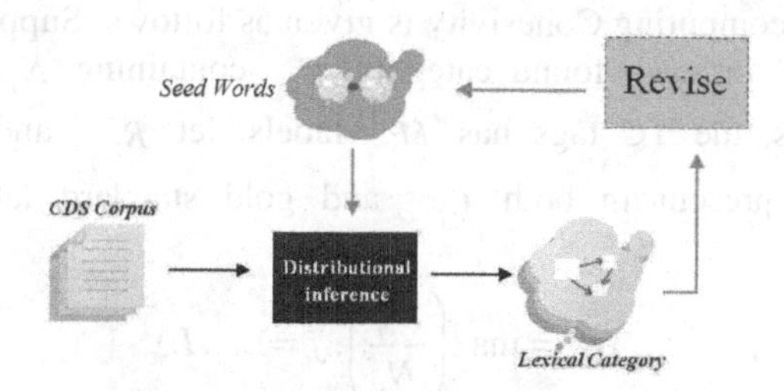

Fig. 1. The semi-supervised incremental model

Some clustering algorithms applied to lexical category acquisition task so far require the number of categories as input. This is problematic as we do not know how many categories exist in the data. Recently, a powerful algorithm called Affinity Propagation (AP) based on message passing was proposed by Frey and Dueck (2007). AP clustering method has been shown to be useful for many applications. We consider two similarity measures widely used in NLP: Euclidean distance and Cosine similarity.

Seed candidates are discovered using words in child-directed speech of high frequency. More specifically, we consider the following two types of pairwise constraints: must-link constraints, which specify that two samples should be assigned to the same cluster; and cannot-link constraints, which specify that two samples should be assigned to different clusters. In the later phase, there are already categories induced, we also choose 50 words as seed words manually, which are judged to be correctly acquired, then put them into the model.This process includes both semi-supervised issue and incremental ideal. The goal is clustering but there are some 'labeled data' in the form of must-links and cannot-links.

4 Evaluation Metric

While grouping word usages into categories, we attempt to trade off three conflicting criteria. First, the categories should be informative about the properties of their members. Second, the categories should be rich diversity about the quantity of clusters. Third, the categories should be pure about similarity among their members, which including syntactic and semantic similarity. POS only shows the syntactic similarity of words, besides POS tags, manually or semi-automatically annotated semantic lexicon (e.g. thesaurus) should be treated as a gold standard, which is a representation of the semantic similarity of all the word categories.

An appropriate evaluation created for formalizing all the three criteria and quantitative evaluation of analysis is called Cohesivity. In this paper, for Chinese lexical category acquisition task, the Peking University POS tagset (Yu et al. 2003) and TongYiCi CiLin (Mei et al., 1983), a Chinese semantic thesaurus, are used. TongYi Ci CiLin (TC) has classified Chinese lexicons in a 3-level hierarchy. Level 1 (Classes), consists of 12 classes, level 2 (Sections), consists of 94 classes. In our experiment, we have examined results up to the second level.

The formula for computing Cohesivity is given as follows. Suppose there are K categories induced, for each found category C_i containing N_i words, the POS tags has L labels, the TC tags has M labels, let R_{Pj} and R_{Tk} denote the number of words present in both C_i and gold standard label j or k respectively:

$$P_{Pi} = \max\left(\frac{R_{Pj}}{N_i}\right), j=1,...,L, \tag{5}$$

$$P_{Ti} = \max\left(\frac{R_{Tk}}{N_i}\right), k=1,...,M \tag{6}$$

Purity of one category is computed as the harmonic mean of distinct similarity of syntactic and semantic scores. Cohesivity can be weighted to favor the contributions of syntactic and semantic similarity.

$$Pu_i = \frac{(1+\beta)P_P \cdot P_T}{\beta \cdot P_P + P_T} \tag{7}$$

Similarly to the familiar F-measure, if β is greater than 1 syntactic similarity is weighted more strongly in the calculation, and vice versa.

Pu_i is the purity of category C_i ,then

$$Cohesivity = \sum_i \frac{N_i \log N_i}{Z} Pu_i \qquad \overline{N} = \frac{\sum N_i}{K}, \quad Z = \log \overline{N} \tag{9}$$

5 Experiment

We use Zhou corpus as input, Zhou corput offers CDS speeches in different stage at 14, 20, 26 and 32 months. In the experiment, for different stage, we show three different categorization results: categorization result without seed words (N in Table 5), result that model using the initial seed words (Yo) and that using the revised seed words (Yn). Since 14 month is the first stage, no acquired category exists, thus only results of the model with and without the initial seed are shown. The revision in this step is made by author manually, same as the setting for the initial seed words, we select 50 high frequency words form the result in each stage, group them into classes as the revised seed words. We choose Cosine similarity in this analysis, and set the feature dimension to be 300.

Table 1 shows the result of lexical categorization in semi-supervised incremental process. Scale and Word stand for the number of utterances and token type in the CDS corpus in different age stage respectively.

As seen in Table 1, the semi-supervised methods gains advantage in the comparison and at 14 month even no category formed without seed. The revised seed words help better for categorization compare with the initial seed words, but its performance approaches a plateau at 32 month of input. This suggests that early in the language development, the distributional information is not sufficient to achieve learning, while seed words as the forms of abstract knowledge representation bring in information from other cognitive resource that help "the Mind Get So Much from So Little" (Tenenbaum et al. 2011). When children grow up, the difference that fixed up by the abstract knowledge between mind and the date given become smaller, in other word, language acquisition become self-sufficiency.

Table 1. Categorization Result for different age stage

Age	Scale	Word	Seed	Number	Avg size	POS	TC	Cohesivity
14	1472	558	Y	7	8.9	68.2%	33.4%	23.28
			N	-	-	-	-	-
20	4474	1031	Y o	72	10.9	63.8	35.7	311.7
			Y n	73	10.8	63.6	37.1	315.0
			N	60	10	60.9%	34.5%	230.9
26	7003	1359	Y o	85	10.5	61.7	36.3	345.2
			Y n	87	10.2	64.4	36.2	349.1
			N	85	10.5	65.0	32.5	337.2
32	8622	1540	Y o	99	12.1	64.3%	34.9%	474.4
			Y n	96	12.5	64.6%	34.9%	474.7
			N	94	12.7	64.8	35.8	468.7

Although the results of the previous analyses provide a general concept of the lexical category induced, it is necessary to analyses the concrete clusters deeply. This requires considerable time and efforts, even if only the results of one stage are considered. In the next analysis, we give qualitative analysis of the acquisition result and concentrate to investigate the category of two words varying with time.

6 Conclusions

We have presented a semi-supervised incremental model for lexical category acquisition that aims to optimize the category assignments by utilizing semantic information of seed words and distributional information about all words' surrounding context.

We have also presented a novel evaluation,Cohesivity, approach by synthesizing many different kinds of criteria. It is therefore highly desirable to develop an evaluation measure that makes reference to gold standard; the ideal measure needs to be applicable to a wide range of different acquisition models. We evaluate our model by applying new measure. The results show that the evaluation metric can give more reasonable explanation for the clustering results and is efficient computationally. The results suggest that a fine-grained set of categories which combine similarity of syntactic and semantic are more appropriate than the part of speech categories.

Acknowledgments. This research has been supported by NSFC90920006 and RFDP 20090005110005.

References

[1] Redington, M., Crater, N., Finch, S.: Distributional information: A powerful cue for acquiring syntactic categories. Cognitive Science: A Multidisciplinary Journal (1998)
[2] Mintz, T.: Category induction from distributional cues in an artificial language. Memory and Cognition 30(5) (2002)
[3] Davidov, D., Rappoport, A.: Efficient Unsupervised Discovery of Word Categories using Symmetric Patterns and High Frequency Words. In: COLING-ACL 2006 (2006)
[4] Frey, J., Dueck, D.: Clustering by Passing Messages Between Data Points. Science 315, 972–976 (2007)
[5] Parisien, C., Fazly, A., Stevenson, S.: An incremental bayesian model for learning syntactic categories. In: Proceedings of the Twelfth Conference on Computational Natural Language Learning (2008)
[6] Frank, S., Goldwater, S., Keller, F.: Evaluating models of syntactic category acquisition without using a gold standard. In: Proceedings of the 31st Annual Meeting of the Cognitive Science Society (2009)
[7] Chrupala, G., Alishahi, A.: Online Entropy-based Model of Lexical Category Acquisition. In: Proceedings of the Fourteenth Conference on Computational Natural Language Learning (2010)
[8] Brown, P., Mercer, R., Della Pietra, V., Lai, J.: Class-based n-gram models of natural language. Computational Linguistics 18(4), 467–479 (1992)
[9] Clark, A.: Inducing syntactic categories by context distribution clustering. In: Proceedings of the 2nd Workshop on Learning Language in Logic and the 4th Conference on Computational Natural Language Learning, pp. 91–94 (2000)
[10] Tomasello, M.: Acquiring linguistic constructions. In: Siegler, R., Kuhn, D. (eds.) Handbook of Child Psychology: Cognition, Perception and Language. Wiley Publishers (2006) (in Press)
[11] Yu, S., Duan, H., Zhu, S., Swen, B., Chang, B.: Specification for corpus processing at Peking University: Word segmentation, POS tagging and phonetic notation. Journal of Chinese Language and Computing 13(2), 121–158 (2003)
[12] Mei, J., Zhu, Y., Gao, Y., Yin, H.: TongYiCi CiLin, ShangHai DianShu ChuBanShe (1983)
[13] MacWhinney, B.: The CHILDES project: Tools for analyzing talk. Lawrence Erlbaum Associates Inc., US (2000)
[14] Mitchell, T.: The discipline of machine learning (Technical Report CMUML-06-108). Carnegie Mellon University (2006)
[15] Tenenbaum, J.B., Kemp, C., Griffiths, T.L., Goodman, N.D.: How to Grow a Mind: Statistics, Structure, and Abstraction. Science 331(6022), 1279–1285 (2011)
[16] Davidov, D., Rappoport, A.: Efficient Unsupervised Discovery of Word Categories using Symmetric Patterns and High Frequency Words. In: COLING-ACL 2006 (2006)
[17] Alishahi, A., Chrupała, G.: Lexical category acquisition as an incremental process. In: Proceedings of the CogSci 2009 Workshop on Psycho Computational Models of Human Language Acquisition, Amsterdam (2009)

An Approach to TCM Syndrome Differentiation Based on Interval-Valued Intuitionistic Fuzzy Sets

Zhiling Hong[1] and Meihong Wu[2,*]

[1] Department of Psychology, Peking University
[2] Department of Computer Science, Xiamen University
zhiling.hong@pku.edu.cn, wmh@xmu.edu.cn

Abstract. Correct diagnosis is prerequisite to the treatment, prognosis and prevention of disease. Traditional Chinese Medicine theory provides a framework for assessing any particular manifestation of symptoms and signs, in which syndrome match is especially useful in analyzing clinical cases. Unfortunately, the two-value method cannot exactly represent the knowledge about syndrome differentiation in TCM diagnosis system. In this paper, we present a new method for syndrome differentiation in the Traditional Chinese Medical clinical analysis based on interval-valued intuitionistic fuzzy sets, furthermore we point out the proper method for decision-making in the Traditional Chinese Medical diagnosis process via intuitionistic fuzzy cognitive match between syndromes.

Keywords: interval-valued intuitionistic fuzzy sets, syndrome differentiation, symptoms, TCM.

1 Introduction

According to the characteristics of Traditional Chinese Medicine, the process of diagnosis is the combination of disease differentiation and syndrome differentiation. Syndrome is a summarization of the development of a disease at a certain stage, including cause, location, nature, pathogenesis and the relevant symptoms and signs. The differentiation of disease and the differentiation of syndrome refer to the understanding of the nature of a disease from different angles, therefore the syndrome match is especially useful in analyzing clinical cases. As we all know, correct diagnosis is prerequisite to the treatment, prognosis and prevention of disease. Unlike western medicine, in which a clear diagnosis of specific disease is necessary to determine a clinical strategy, TCM theory provides a framework for assessing any particular manifestation of symptoms and signs, classifying the individual's condition within some region of an *N*-dimensional vector space that defines the total set of possibilities for systemic-metabolic characteristics.

In a recent work by Wu et al., 2009 [8], they proposed an intelligent Multi-Agent TCM diagnosis system, which focuses on an elucidation of the theory and methods of TCM in examining pathological conditions as well as analyzing and differentiating

---

* Corresponding author.

Z. Du (Ed.): Proceedings of the 2012 International Conference of MCSA, AISC 191, pp. 77–81.
springerlink.com

syndromes. Unfortunately the two-value method cannot exactly represent the knowledge about syndrome differentiation which is the crucial part of TCM diagnostic process, since patients' symptoms and doctors' diagnosis are matters with strong vagueness and uncertainty. With regard to this characteristic, we present a new method for syndrome differentiation in the TCM decision-making process on the basis of system proposed by Wu et al., 2009. In contrast to many traditional methods of analysis, which are oriented toward the use of numerical techniques, this diagnostic model involves the use of variables whose values are fuzzy sets [2]. Interval-valued fuzzy set theory[1,3,4,5] is an increasingly popular extension of fuzzy set theory where traditional [0, 1]-valued membership degrees are replaced by intervals in [0, 1] that approximate the (partially unknown) exact degrees. The approaches of Interval-valued fuzzy set theory have the virtue of complementing fuzzy sets that is able to model vagueness, with an ability to model uncertainty as well, and interval-valued intuitionistic fuzzy set theory reflect this uncertainty by the length of the interval membership degree for every membership degree. Interval-valued fuzzy set theory emerged from the observation that in a lot of cases, no objective procedure is available to select the crisp membership degrees of elements in a fuzzy set. It was suggested to alleviate that problem by allowing specifying only an interval to which the actual membership degree is assumed to belong. Therefore, in this paper we present interval-valued intuitionistic fuzzy sets as a tool for reasoning in the TCM medical diagnosis process.

This paper is organized as follows: In Section 2, we propose a new approach for clinical analysis based on interval-valued intuitionistic fuzzy sets, which is an approach to gain intuitionistic medical knowledge in TCM fuzzy model; In Section 3 we propose a new method for syndrome differentiation in the TCM decision-making process by largest degree of intuitionistic cognitive fuzzy match, the method can distinguish syndromes that share same symptoms and signs and difficult to distinguish correctly; and in the last section, a brief summary is presented.

2 TCM Fuzzy Model in the Basis of Interval-Valued Intuitionistic Fuzzy Set Theory

Definition 1. An interval-valued intuitionistic fuzzy set[1], A in X, is given by:

$$A=\{\langle x, M_A(x), N_A(x)\rangle \mid x \in X\}$$

where: $M_A(x): X \to D[0,1]$, $N_A(x): X \to D[0,1]$, with condition that $0 \le \sup_x(M_A(x)) + \sup_x(N_A(x)) \le 1, \forall x \in X$.

The interval $M_A(x)$ *and* $N_A(x)$ denote the degree of membership and the degree of non-membership of the element x to the set A , respectively. And let $D[0,1]$ be the set of all closed subintervals of the interval[0,1] and $X(\neq \varnothing)$ be a given set. $M_A(x)$ *and* $N_A(x)$ are closed intervals whose lower fuzzy set and upper fuzzy set are, respectively, denoted by $M_{AL}(x)$, $M_{AU}(x)$ and $N_{AL}(x)$, $N_{AU}(x)$.

Definition 2. If *A, B* are two interval-valued intuitionistic fuzzy sets of *X* then the subset relation, equality ,complement, union and intersection are defined as follows:

$$A \subset B \text{ iff } \forall x \in X, [M_{AL}(x) \le M_{BL}(x), M_{AU}(x) \le M_{BU}(x) \text{ and } N_{AL}(x) \ge N_{BL}(x), N_{AL}(x) \ge N_{BL}(x)]$$

$$A = B \text{ iff } \forall x \in X, [A \subset B \text{ and } B \subset A]$$

$$A_c = \{\langle x, N_A(x), M_A(x) \rangle | x \in X\}$$

$$A \cup B = \{\langle x, \max(M_A(x), M_B(x)), \min(N_A(x), N_B(x)) \rangle | x \in X\}$$

$$A \cap B = \{\langle x, \min(M_A(x), M_B(x)), \max(N_A(x), N_B(x)) \rangle | x \in X\}$$

Definition 3. For a family of interval-valued intuitionistic fuzzy sets of X , $\{A_i, i \in I\}$ then the union and the intersection are defined as follows:

$$\bigcup_i A_i = \{\langle x, [\max M_{A_iL}(x), \max M_{A_iU}(x)], [\min N_{A_iL}(x), \min N_{A_iU}(x)] \rangle | x \in X\}$$

$$\bigcap_i A_i = \{\langle x, [\min M_{A_iL}(x), \min M_{A_iU}(x)], [\max N_{A_iL}(x), \max N_{A_iU}(x)] \rangle | x \in X\}$$

Definitions of each TCM syndrome generally includes a set of key symptoms and signs, of which only a subset may be present in specific cases and yet still qualify as being characterized by that syndrome. As present in the model we proposed, all the knowledge base is formulated in terms of interval-valued intuitionistic fuzzy sets.

Definition 4. Let *X* and *Y* be two sets. An intuitionistic fuzzy relation[9] *R* from *X* to *Y* is an intuitionistic fuzzy set of $X \times Y$ characterized by the membership function M_R and non-membership function N_R .An intuitionistic fuzzy relation *R* from *X* to *Y* will be denoted by $R(X \to Y)$.

We also define an intuitionistic fuzzy relation *R* from the set of syndromes *S* to the set of diagnoses *D*, that's on $S \times D$, which reveals the degree of association and the degree of non-association between syndromes and diagnosis, which is more in accordance with the thinking processes of an excellent Doctor of traditional Chinese medicine in diagnosing difficult and complicated cases. The diagnosis criteria can be represented by logical operations over groups of fuzzy sets by modeling rules through fuzzy logic.

3 The Intuitionistic Fuzzy Cognitive Match in TCM Diagnosis Decision-Making

Correctly identifying subsets of symptoms as being more likely associated with one or the other syndrome is a challenging task and is a fundamental skill of clinical health assessment or diagnosis.

In this section, we point out the final proper diagnosis for syndrome differentiation by largest degree of intuitionistic cognitive fuzzy match, the method can distinguish syndromes that share same symptoms and signs and difficult to distinguish correctly.

Let $X=\{x_1,x_2,\cdots x_n\}$ be the universe of discourse, A, B be two interval-valued intuitionistic fuzzy sets in X, where $A=\{\langle x_i, M_A(x_i), N_A(x_i)\rangle \mid x_i\in X\}$, $B=\{\langle x_i, M_B(x_i), N_B(x_i)\rangle \mid x_i\in X\}$, that is:

$$A=\{\langle x_i, [M_{AL}(x_i), M_{AU}(x_i)], [N_{AL}(x_i), N_{AU}(x_i)]\rangle \mid x_i\in X\}$$

$$B=\{\langle x_i, [M_{BL}(x_i), M_{BU}(x_i)], [N_{BL}(x_i), N_{BU}(x_i)]\rangle \mid x_i\in X\}.$$

The Euclidean distance between interval-valued intuitionistic fuzzy sets A and B is defined as follows[6]:

$$d(A,B)=\sqrt{\frac{1}{4}\sum_{i=1}^{n}\left[(M_{AL}(x_i)-M_{BL}(x_i))^2+(M_{AU}(x_i)-M_{BU}(x_i))^2+(N_{AL}(x_i)-N_{BL}(x_i))^2+(N_{AU}(x_i)-N_{BU}(x_i))^2\right]}$$

And the following is weighted Euclidean distance between interval-valued intuitionistic fuzzy sets A and B [6]:

$$W_d(A,B)=\sqrt{\frac{1}{4}\sum_{i=1}^{n}w_i^*\left[(M_{AL}(x_i)-M_{BL}(x_i))^2+(M_{AU}(x_i)-M_{BU}(x_i))^2+(N_{AL}(x_i)-N_{BL}(x_i))^2+(N_{AU}(x_i)-N_{BU}(x_i))^2\right]}$$

then we propose the intuitionistic fuzzy cognitive match:

$$W_{match}(A,B)=1-\sqrt{\frac{1}{4}\sum_{i=1}^{n}w_i^*\left[(M_{AL}(x_i)-M_{BL}(x_i))^2+(M_{AU}(x_i)-M_{BU}(x_i))^2+(N_{AL}(x_i)-N_{BL}(x_i))^2+(N_{AU}(x_i)-N_{BU}(x_i))^2\right]}$$

4 Summary

Extending the concept of fuzzy set, many scholars introduced various notions of higher-order fuzzy sets. Among them, interval-valued intuitionistic fuzzy sets, provides us with a flexible mathematical framework to cope with imperfect and/or imprecise information. On the basis of interval-valued intuitionistic fuzzy sets theory, we introduce an intelligent fuzzy diagnostic model according to the characteristics of Traditional Chinese Medicine, which realize intelligent diagnosis by modeling medical diagnosis rules via intuitionistic fuzzy relations, then we propose a new approach for clinical analysis based on interval-valued intuitionistic fuzzy sets, which is an approach to gain intuitionistic medical knowledge in TCM fuzzy model. At last we point out the final proper diagnosis for syndrome differentiation by largest degree of intuitionistic cognitive fuzzy match. The method we introduced can distinguish syndromes that share same symptoms and signs and difficult to distinguish correctly, which intensifies the intelligence of the whole TCM diagnostic process.

References

[1] Atanassov, K., Gargov, G.: Interval valued intuitionistic fuzzy sets. Fuzzy Sets and Systems 31, 343–349 (1989)

[2] Zadeh, L.: The concept of a linguistic variable and its application to approximate reasoning-I. Inform. Sci. 8, 199–249 (1975)

[3] Biswas, R.: Intuitionistic fuzzy relations. Bull. Sous. Ens.Flous. Appl. (BUSEFAL) 70, 22–29 (1997)

[4] Atanassov Krassimir, T., Janusz, K., Eulalia, S., et al.: On Separability of Intuitionisitc Fuzzy Sets. LNCS (LNAI), vol. 27, pp. 285–292 (2003)

[5] Mondai, T.K., Samanta, S.K.: Topology of interval-valued intuitionistic fuzzy sets. Fuzzy Sets and Systems 119(3), 483–494 (2001)

[6] Zeshui, X.: On similarity measures of interval-valued intuitionistic fuzzy sets and their application to pattern recognitions. Journal of Southeast University 01 (2007)

[7] Feng, C.: Fuzzy multicriteria decision-making in distribution of factories: an application of approximate reasoning. Fuzzy Sets and Systems 71, 197–205 (1995)

[8] Wu, M., Zhou, C.: Syndrome Differentiation in Intelligent TCM Diagnosis System. In: 2009 World Congress on Computer Science and Information Engineering, USA, pp. 387–391 (2009)

A Novel Artificial Neural Network Learning Algorithm

Tinggui Li and Qinhui Gong

Luzhou Vocational and Technical College

Abstract. The unit feedback recursive neural network model which is widely used at present has been analyzed. It makes the unit feedback recursive neural network have the same dynamic process and time delay characteristic. The applications of the unit recursive neural networks are limited. For its shortcomings, we proposed another state feedback recursive neuron model, and their state feedback recursive neural network model. In this neural network model, the static weight of the neural network explained the static transmission performance, and the state feedback recursive factor indicated the dynamic performance of neural networks, the different state feedback recursion factor indicated the dynamic process time of the different systems.

Keywords: artificial neural network, learning algorithm, state feedback.

1 Introduction

The artificial neural network theory mainly researches the structure of artificial neural networks, learning algorithm and convergence which is based on the biological neural networks working mechanism. The forward neural network structure is concise, the learning algorithm is simple, and its significance was clear. It has been widely studied in-depth. The forward neural network established relationships, which is usually static, between the input and output, but in practical, the application of all controlled object was usually dynamic. Therefore, the static neural network modeling is not able to describe system dynamic performance accurately [1]. The neural networks which describe the dynamic performance should contain the dynamic characteristics and the ability to store dynamic information. To accomplish this function there is generally delayed feedback or information feedback in the network, such networks are known as the recursive neural networks or state feedback neural network. The recursive neural network has already become a widely researched topic. [2].

The quantificational research for the evolutionary process of the recursive neural network is a significant topic. In recent years, many scholars have used recursive neural networks for nonlinear dynamic systems to establish mathematical models [3].The output of the system is only determined by the dynamic system condition and exterior input, the recursive neural network itself will be a non-linear dynamic system, therefore, studying its state evolution is a necessity [4].

2 State Feedback Dynamic Neural Model

Neuron model is related to the performance of neural networks directly. The neuron receives input signals from all directions, after weighted sum of space and time it becomes into the neuron state s(t). Finally it is transformed with the non-linear to

Z. Du (Ed.): Proceedings of the 2012 International Conference of MCSA, AISC 191, pp. 83–88.
springerlink.com

achieve non-linear mapping. Neuron model plays an important role in the performance of the neural network, and it will directly indicate the dynamic performance of neural networks, so a reasonable model of neurons is essential. The dynamic neuron model with state feedback is shown in Figure 1.

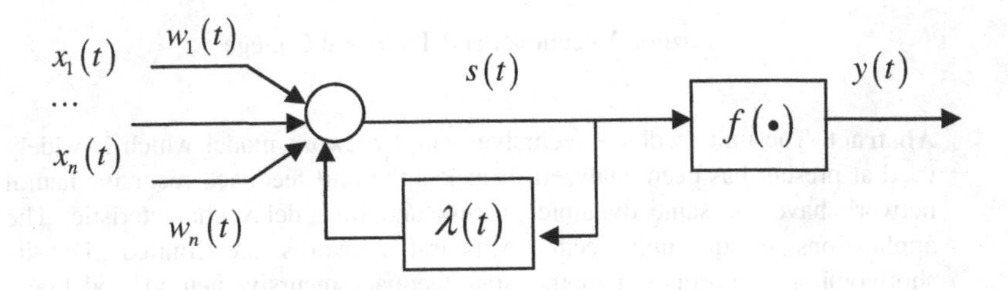

Fig. 1. The state feedback recursive dynamic neuron model

According to spatial and temporal characteristics of neurons, the state s(t) is as

$$s(t)=\sum_{i=1}^{n} w_i(t)x_i(t)+\lambda(t)s(t-1)-b(t) \tag{1}$$

Where $x_i(t)$ is the ith input of the neuron, $w_i(t)$ is the ith input weighted, $\lambda(t)$ is the state feedback, and then neuron output is

$$y(t)=f(s(t)) \tag{2}$$

In order to write easily, the neuron threshold b (t) may be rewritten as weight $w_0(t)$, its input $x_0(t)$ is a constant -1. Then the equation (1) can be written as

$$s(t)=\sum_{i=0}^{n} w_i(t)x_i(t)+\lambda(t)s(t-1) \tag{3}$$

From this we can see that the weight $w_i(t)$ implies the static characteristic of neurons, which is called the static weight. The state feedback λ(t) implies the dynamic factor of neurons, which is called the dynamic feedback coefficient. This neural model not only expressed the static memory characteristics of neurons, but also the dynamic evolution properties. This model is called the state feedback neuron model.

3 State Feedback Neural Network Model

The ability of neural networks to process information is not only related to the neurons performance, but also the structure of the neural network, which directly implies the performance of neural networks. Therefore, the study for the neural network structure is very necessary. The state-feedback neural model will be combined to network in accordance with the direction of information forward flowing. For easy calculation, assuming that there is no information transmitted between layers, the feed forward neural network model with the different state feedback coefficients can be constituted, as shown in Fig.2.

Supposes the input pattern vector of the state feedback neural network model is $u(t)=\left(u_1(t),u_2(t),\ldots,u_n(t)\right)^T$, the active function state vector is $s(t)=\left(s_1(t),s_2(t),\ldots,s_p(t)\right)^T$, the state feedback factor vector is $\lambda(t)=\left(\lambda_1(t),\lambda_2(t),\ldots,\lambda_p(t)\right)^T$, the hidden output pattern vector is $x(t)=\left(x_1(t),x_2(t),\ldots,x_p(t)\right)^T$; the output pattern vector is $y(t)=\left(y_1(t),y_2(t),\ldots,y_m(t)\right)^T$; The weighted matrix from the input to hidden is $\{w_{ji}(t)\}, i=1,2,\ldots,n, j=1,2,\ldots,p$, the weighted matrix from the hidden to output is $\{w_{kj}(t)\}, j=1,2,\ldots,p, k=1,2,\ldots,m$, the neuron threshold in the hidden is $\{b_j(t)\}, j=1,2,\ldots,p$; The threshold in the output is $\{b_k(t)\}, k=1,2,\ldots,m$

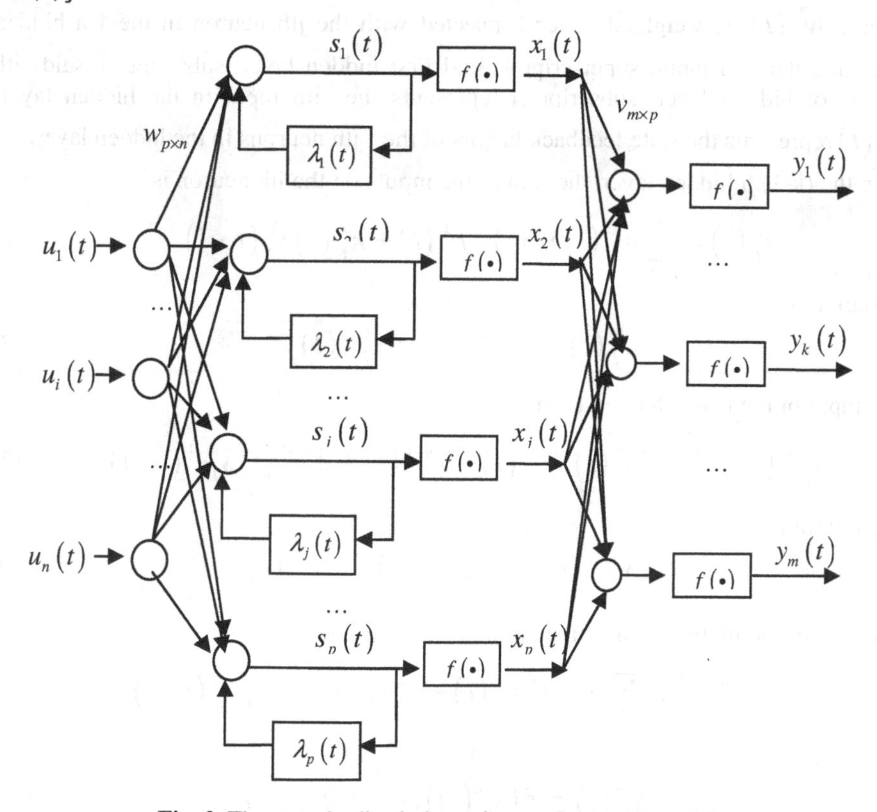

Fig. 2. The state feedback dynamic neural network model

4 Learning Algorithm of the State Feedback Neural Network

The processing information performance of the artificial neural network is not only related to the structure, but also to the weighted parameters of neuron and the neuron active function and so on. If the neural network structure is the same but the parameters are different, it can also present the different characteristic. Generally, once the neural network structure and active function are determined, the neural network performance will mainly depend on the neuron connection parameters. Neuron parameters can be obtained by learning the external environment, and it can be changed to meet the application. This needs to study the learning algorithm of the neural network.

When the input layer neurons in the state feedback neural network accept input signals, the input signal passes through the input layer, layer by layer to the output layer. When the input signal is transmitted to the first hidden layer, the sum of all the input on the jth neuron in the 1th hidden layer is

$$s_j^1(t)=\sum_{i=1}^{n} w_{ji}^1(t)x_i^1(t)-b_j^1(t)+\lambda_j^1(t)s_j^1(t-1) \tag{4}$$

The response of the jth neuron in the 1th hidden layer is

$$y_j^1(t)=f\left(s_j^1(t)\right)=x_j^2(t) \tag{5}$$

where, $w_{ji}^1(t)$ is weighted factor connected with the jth neuron in the 1th hidden layer and the ith input, superscript 1, said first hidden layer, subscript j, said jth neuron of hidden layer, subscript i represents the ith input in the hidden layer, $\lambda_j(t)$ represents the state feedback factors of the jth neurons in the hidden layer.

In the (k-1)th hidden layer, the sum of the inputs on the jth neuron is

$$s_j^k(t)=\sum_{i=1}^{n} w_{ji}^k(t)x_i^k(t)-b_j^k(t)+\lambda_j^k(t)s_j^k(t-1) \tag{6}$$

Its output is

$$y_j^k(t)=f\left(s_j^k(t)\right)=x_j^{k+1}(t) \tag{7}$$

The input of the (m-1)-hidden layer is

$$s_j^{m-1}(t)=\sum_{i=1}^{n} w_{ji}^{m-1}(t)x_i^{m-1}(t)-b_j^{m-1}(t)+\lambda_j^{m-1}(t)s_j^{m-1}(t-1) \tag{8}$$

The output is

$$y_j^{m-1}(t)=f\left(s_j^{m-1}(t)\right)=x_j^{m-2}(t) \tag{9}$$

The sum of the inputs on the kth neuron in the output layer is

$$s_k^m(t)=\sum_{j=1}^{n} w_{kj}^m(t)x_k^m(t)-b_k^m(t)+\lambda_k^m(t)s_k^m(t-1) \tag{10}$$

Its output is

$$y_k^m(t)=f\left(s_k^m(t)\right),k=1,2,\ldots,m \tag{11}$$

The input signal transmission is the process that the input signal passes through each neuron layer, and finally reaches the output layer, and generates response in the neurons.

From the neural network signal transfer process it can be seen that the neural network output is a function based on the input signal $u_i(t)$, weighted $w_{ji}(t)$ and state feedback $\lambda_j(t)$. If the input signal remains unchanged, the weights $w_{ji}(t)$ and state feedback $\lambda_j(t)$ of the neural network is adjusted, and then the neural network's output will be changed with it. Therefore, the neural network weight $w_{ji}(t)$ and state feedback $\lambda_j(t)$ may be adjusted in order to improve the performance of neural networks, which would make the performance of neural networks approaching the desired requirements. Their difference is only the weight, which is adjusted to change the static performance of neural networks, while the state feedback coefficients are adjusted to change the neural network dynamic performance.

Learning algorithm is essential in search a minimum error performance function. In this algorithm, the method of steepest descent in the nonlinear programming can be used, that is, the weight may be corrected by the negative gradient of error function. To illustrate the error gradient descent algorithm, first, the quadratic sum of the difference between the desired output $d_k^m(t)$ and the actual output $y_k^m(t)$ is defined as the error performance function E(t), that is

$$E(t)=\frac{1}{2}\sum_{k=1}^{m}\left(d_k^m(t)-y_k^m(t)\right)^2=\frac{1}{2}\left\|d(t)-y(t)\right\|^2 \tag{12}$$

Where, $d_k^m(t)$ is the kth neuron expected output in the output layer m, whose role is a teacher signal; $y_k^m(t)$ is the k-neuron output. The error performance function E (t) is changed with the connection weights by the gradient descent, and then the partial derivative of the error E (t) to the output $y_k^m(t)$ is required to change with the error gradient descent. The definition of the error E (t) is expanded to the hidden layer, there are.

$$\begin{aligned}E(t)&=\frac{1}{2}\sum_{k=1}^{m}\left(d_k^m(t)-y_k^m(t)\right)^2\\&=\frac{1}{2}\sum_{k=1}^{m}\left\{d_k^m(t)-f\Big(\sum_{j=0}^{p}w_{kj}^m(t)\cdots\right.\\&\left.f\left(\sum_{i=0}^{n}w_{ji}^k(t)x_i^k(t)+\lambda_j^k(t)s_j^k(t-1)\right)+\cdots+\lambda_k^m(t)s_k^m(t-1)\right\}^2\end{aligned} \tag{13}$$

From this we can see that the neural network output error is a function of weights $w_{kj}^k(t)$ and state feedbacks $\lambda^k(t)$, so the weights $w_{kj}^k(t)$ and state feedbacks $\lambda_i^k(t)$ are adjusted which is how to change the system error E(t).

Clearly, the static weight and the state feedback factor can be adjusted, and its aim is to reduce the error between the neuron output and the desired output. Therefore, the static weight and the state feedback factors can be adjusted so that its changes should be proportional to the negative gradient of the error, that is.

$$\Delta w_{ji}^{k}(t) = -\eta(t)\frac{\partial E(t)}{\partial w_{ji}^{k}(t)} \tag{14}$$

$$\Delta \lambda_{j}^{k}(t) = -\beta(t)\frac{\partial E(t)}{\partial \lambda_{j}^{k}(t)} \tag{15}$$

Where, the minus indicates the gradient descent, the factor $\eta(t) \in (0,1)$ and $\beta(t) \in (0,1)$ that is the proportion coefficients, which is the learning rate in the training. From this we can see that state feedback neural network learning algorithm belongs to Delta learning rule, these algorithms are often referred to as errors gradient descent algorithm.

5 Conclusion

Therefore, the state-feedback neural network dynamic characteristics and learning strategies, which have great theoretical and application significance, are studied. In theory, the learning algorithm for the state feedback neural network is derived and proved, and the state feedback neural network learning algorithm is summed up in the form of theorems.

References

[1] Cong, S., Dai, Y.: Structure of recurrent neural networks. Computer Application 24(8), 18–20 (2010)

[2] Yilei, W., Qing, S., Sheng, L.: A Normalized Adaptive Training of Recurrent Neural Networks With Augmented Error Gradient. IEEE Transactions on Neural Networks 19(2), 351–356 (2009)

[3] Song, Q., Wu, Y., Soh, Y.C.: Robust Adaptive Gradient-Descent Training Algorithm for Recurrent Neural Networks in Discrete Time Domain. IEEE Transactions on Neural Networks 19(11), 1841–1853 (2010)

[4] Han, M., Shi, Z.-W., Xi, J.-H.: Learning the trajectories of periodic attractor using recurrent neural network. Control Theory& Application 23(4), 497–502 (2006)

[5] Yang, S.-S., Siu, S., Ho, C.-L.: Analysis of the Initial Values in Split-Complex Backpropagation Algorithm. IEEE Transactions on Neural Networks 19(9), 1564–1573 (2008)

[6] Cong, S., Liang, Y.-Y., Li, G.-D.: Multivariable Adaptive PID-like Neural Network Controller and Its Design Method. Information and Control 35(5), 568–573 (2006)

IPv6 Network Intrusion Detection Protocol Analysis Techniques

Zhang Zhi

Jiangxi Science & Technology Normal University,
Nanchang, Jiangxi, 330013 China
zhangzhi@cssci.info

Abstract. Network intrusion detection problem, the use of protocol analysis techniques, under the IPv6 network protocol analysis test, the context of network protocols, network application layer protocol design, gives a new protocol analysis based network intrusion detection design ideas. And the traditional pattern matching algorithms, designed for the detection engine IPv4/IPv6 network also provides input to improve the detection efficiency and detection efficiency.

Keywords: IPv6, intrusion detection, protocol analysis, network security.

1 Introduction

The early 1990s, intrusion detection technology began to be widely used to protect the user's information security, and the recent emergence of protocol analysis techniques to intrusion detection technology to add new vitality. Protocol analysis is based on the traditional pattern matching techniques to develop a new security technology. Protocol analysis is the use of network data specifications, formatting features, combined with the high-speed packet capture, protocol analysis, session recombinant technology, the information from the protocol data communication both in the analysis of interaction processes, and technologies to accomplish a specific function.

As the existing IPv4 network almost no consideration of network information security, network application layer only in some security tools, such as E-mail encryption, SSL and so on. To enhance the security of the Internet, since 1995, IETF IPSec protocol developed to protect the security of IP communications, security measures will be placed in the network layer, into a standard service. A new generation of protocol analysis techniques uses the IPv6 network protocol rules of the height; speed and accuracy in the detection of other aspects have been greatly improved.

2 IPv6 Network Protocol Analysis Test

2.1 Identification of Data

If the LAN transmission of IPv6 packets, and its basic structure as shown in Table 1. Of which: Ethernet header: include destination MAC address (6 bytes), the source MAC address (6 bytes), protocol type (or length, 2 bytes).

Z. Du (Ed.): Proceedings of the 2012 International Conference of MCSA, AISC 191, pp. 89–94.
springerlink.com

Table 1. Ethernet frame preload IPv6 packet

14 bytes	≥ 40 bytes	Variable-length	Variable-length	4 bytes
Ethernet header	IPv6 header and extension headers may be	Other protocols in the head	Load	The end of the Ethernet frame

IPv6 head: the head and extension headers can be included. Among them, the head of each group must IPV6, and expansion of the head is based on the specific circumstances of the options. In other words, IPv6 is a common pattern: a fixed length of head + a set of basic options change the number of extension headers.

The head of the other protocols: IPv6 protocol capable of delivering data in many different formats and protocols of the length of the head is different. Common protocols are: TCP, UDP, ICMPv6, etc. TCP data segment length of 20 bytes of header, UDP header data segment length of 8 bytes.

2.2 Protocol Analysis Test

Combination of the IPv6 packet header format and extended format head to head a group containing only basic, protocol analysis techniques to study the process. Frame fragment shown in Figure 1, which features used protocol byte of analytic functions has been flagged with an underscore.

Protocol format that in the first 13 bytes of Ethernet frames at the beginning, contains 2 bytes of the network layer protocol identifier. Agreement to use this information for analytic functions the first step of testing work, which ignores the first 12 bytes, skip the first 13 bytes of position, and read 2-byte protocol identification.

The first step, skips to 13 bytes, and read 2-byte protocol identifier (for example, 86DD). Read value is 86DD, it indicates that the Ethernet frames to carry the IPv6 packet, protocol parsing function uses this information to the second step of the testing work, the format of the IPv6 protocol, IPv6 packets at the beginning of the first 21 bytes , contains a 1-byte transport-layer protocol identifier. Therefore, protocol analysis function will be ignored from the first 15 to 20 bytes, skip the first 21 bytes of read 1 byte at the transport layer protocol identifier.

The second step, skip to 21 bytes to read 1 byte of the fourth layer protocol identification (for example 06). Read value is 06, indicating that the IPv6 packet carrying the TCP segment, protocol analysis functions use this information to the third step of the testing work, according to the TCP protocol format, that in the first 55 bytes of TCP header at the beginning, contains 4 bytes of application-layer protocol identification, two each 2-byte application layer identity is the source and destination port number. Therefore, protocol analysis function will be ignored from the first 22 to 54 bytes, skip the first 55 bytes to read two port numbers.

Another possibility is that in the IPv6 header, the next header field value is determined with the expansion in the back of the head or the head of some other protocols (such as TCP, UDP, ICMPv6, etc.). Which, under a common head of the recommended value (decimal) are: HOP-BY-HOP option to head 0; TCP 06; UDP 17; SIP-SR (routing header) 43; SIP-FRAG (sub- head) 44; ESP (encryption security payload) 50; AH (authentication header) 51; ICMPv6 (Internet Control Message Protocol) 58; the last head to head, without a head under 59; destination option header 60 and so on, these values are in RFC 1700 in detail. Agreement will be based on an analytic function of the value of the next header field to decide what to do next, in this case assumed that there is no extension headers.

The third step, skip 55 bytes read port number. If a port number is 0080, indicates that the TCP segment is carried in the HTTP packet, protocol parsing function uses this information to the next step of the analytical work.

```
0000 ** ** ** ** ** ** ** ** ** ** ** ** 86 DD ** **
0010 ** ** ** ** 06 ** ** ** ** ** ** ** ** ** ** **
0020 ** ** ** ** ** ** ** ** ** ** ** ** ** ** ** **
0030 ** ** ** ** ** ** 00 80 ** ** ** ** ** ** ** **
0040 ** ** ** ** ** ** ** ** ** ** ** ** ** ** ** **
0050 ** ** ** ** ** ** ** ** ** ** ** ** ** ** ** **
```

Fig. 1. Frame Fragment

The fourth step, the protocol parsing functions starting from the first 75 bytes to read at the URL string.

At this point, based on protocol analysis, intrusion detection systems analysis has found the location of the application layer data, and is aware of the application layer protocol is HTTP, TCP packet from 75 bytes at the beginning of the URL information. The URL string will be submitted to the HTTP parsing function for further processing.

As mentioned above, URL location of the first byte of the string to be submitted to the analytic function. Analytic function is a command parser, intrusion detection engines typically include a variety of command syntax parser, so that it can target different high-level protocols, including Telnet, HTTP, FTP, SMTP, SNMP, DNS, and so the user command, the data will be decoded to the different analytic functions for further detailed analysis.

Command parsing function with the command string and reading all the possible attack of the deformation, and the ability to find the essential meaning. For example, analytic functions can be found in "/. / Phf" and "/ phf", or with other "/." By the deformation, is the same kind of aggressive behavior, which is "/ phf attack." Thus, the characteristics of attack requires only a feature of the library, you can detect all possible variants of this attack. Therefore, based on protocol analysis and command parsing intrusion detection system, with the usual pattern matching systems, generally have a smaller attack signature database.

Protocol decoding while offering improved the detection efficiency, because the system at every level agreement are resolved along the protocol stack up, so you can use all the current information known to the agreements, to exclude all types of attacks fall into this agreement . For example, if the fourth-layer protocol type is TCP, then there is no need to search for the UDP protocol signatures. If the high-level protocol is SNMP, there is no need to search for Telnet and HTTP attacks.

3 IPv6 Network Protocol Analysis of the Context

Based on protocol analysis of intrusion detection system, protocol data packets will be completely decoded, if you set the IP fragmentation flags, it will first be re-grouped, and then analyze whether the attack occurred. By re-grouping, the system can detect, such as data fragmentation, TCP or RPC section of the border to avoid detection of fraud and other attacks. At the same time, the system during the protocol decoding, will conduct a thorough verification protocol, which means checking all the protocol field, the existence of illegal or suspicious values, including whether to use a reserved field, the default value is abnormal , improper options, serial number out of order, serial number jump number, serial number overlap, calibration errors.

3.1 Sub-merged

Segment of the merger on network intrusion detection system is important. First, there are some ways to attack the use of the operating system protocol stack sub-merged to achieve the loophole, such as the famous teardrop attack is to send in a short time a number of overlapping pairs of offset IP fragmentation, the target receives this When will the merger of sub-section, due to its overlapping offset memory errors, and even lead to the collapse of the protocol stack. The attack from a single group is unable to identify the need to simulate the operating system, protocol analysis of sub-merged, in order to find illegal segmentation.

In addition, the fragments used to bypass the firewall attack a complete TCP header mounted separately within multiple IP segments, resulting in some packet filtering firewall can not be checked under the header information. This attack also needs to deal with intrusion detection software combined fragments to restore the true face of the packet.

3.2 Restructuring the Session

TCP connection state is communication important information; connection status can make full use of more accurate and efficient detection, but also to avoid the IDS itself against denial of service attacks. For example, Snort Intrusion defines the characteristics of an HTTP header for the TCP ACK flag is 1 and the data payload contains a string. But ignores an important fact: ACK packet position 1 is not necessarily a connection. An attacker could send a random source address of a tool for data packets, the ACK bit set to 1, making the continuous alarm IDS, IDS itself cause a denial of service. This is for the IDS of the "flood alarm (Alert flooding)" attack. If the connection status as a condition in itself, would avoid this situation. Because to achieve the "connection has been established," the state must pass three-way handshake, so an attacker can not use random source address to launch a "flood warning" attack.

In addition, some protocols such as Telnet, a character may be a way to transfer data. In this manner, a packet can not be obtained from sufficient information. Session re-use can be a session from the session during the party stitching up all the data, reconstruct the original information, and then carry out the inspection.

4 IPv6 Network Application Layer Protocol Analyses

There are many kinds of application layer protocols, each protocol format are not the same and more complex. Application layer protocol analysis module has been known to use protocol format, accurate positioning of a specific command or valid data to narrow the range of string matching, string matching to reduce the blindness, in order to achieve improved detection efficiency. In this case to common HTTP and e-mail as an example to illustrate.

4.1 HTTP Protocol

A complete HTTP request message contains a series of fields, such as command, URL, HTTP protocol version number, encoding, content type, etc.

Example of an HTTP request:

GET http://www.server.com/example.html HTTP/1.0

Sometimes the attacker will use the encoding tool of the requested content into hexadecimal format, such as% 255c, so as to avoid the server's security checks. In analyzing the HTTP request to the first of these codes into the original character, to see whether the use of illegal request code camouflage.

Since the majority of attacks are to use a specific command (such as GET), and therefore the request command is part of a signature.

Most attacks on the performance of a specially crafted URL, for example "/ scripts / ..% 1 c% 1 c.. / winnt /", or loopholes in the script and the script has access to the directory, so the detection engine needs to be done is to check if the URL string contains a signature. Therefore only need to locate the beginning of the URL field and the length of the detection engine, the pattern matching algorithm can work.

HTTP request URL string can still be a more detailed analysis. The CGI program, for example, CGI programs to be placed in the specified cgi-bin directory, usually GET or POST to receive data. Here is the general format of a CGI request:

http://www.server.com/cgi-bin/test-cgi?arg 1 = vall & args = val2

Before CGI scripts are usually the path and file name, CGI parameters part of the "?" Start parameters with "&" separated each parameter entry form: parameter name = parameter value. Signatures may be contained in the path and file name parts, such as a flaw in the script; it may be included in the parameters section, the performance of a particular parameter name or value. Path name of the CGI request and parameter parsing, you can further narrow the search to improve the detection efficiency of the system.

4.2 E-mail

E-mail message with a clear format. Specified in RFC 822, the message consists of a basic envelope, a number of header field, a blank line and message body composition. Each header field consists of a line of ASCII text, comprising the field name, colon, and most fields have a value. These fields include sender and recipient e-mail address, Cc address. RFC 1341 and RFC 1521 for such an extended format of the program, but still readable message header structure.

Most programs contain viruses and malicious code has clear e-mail features, such as the theme of "I love you", or the attachment name suffix is ". Vbs", some legal, military or political-related messages such as "bribery", "missile" or "attack" and other such e-mail features are usually included in the message's subject, content, attachment file name, content and e-mail attachments the sender and recipient addresses. This information can be obtained directly from the message header, message body without the need to do the whole pattern matching.

5 Conclusion

In this paper, IPv6 network environment, for the current problem of intrusion detection technology, while maintaining the original IPv4 IDS detection engine based on the merits, mining IPv6 features, in-depth study of next-generation network architecture, presents a new IPv6-based protocol analysis of network intrusion detection system design ideas to improve the effectiveness of intrusion detection and detection efficiency. With the deployment of IPv6 technology and network security, intrusion detection technology to adapt to the new IPv6 network environments, how to deal with new attacks, the focus of future research will be after.

References

1. Lee, W.: A Data Mining Framework for Building Intrusion Detection Model. In: IEEE Symposium on Security and Privacy, pp. 120–132 (1999)
2. Madi, T.: TCP/1P Professional Guide. Beijing Hope Electronic Press (2000)
3. Tidwell, T., Larson, R., Fitch, K., et al.: Modeling Internet Attacks. In: Proceedings of the 2001 IEEE Workshop on Information Assurance and Security, pp. 54-59 (2001)
4. Mykerjee, B., et al.: Network Intrusion Detection. IEEE Network 8(3), 26–41 (1994)

Design of Wireless Industrial Network Based-On ZigBee and ARM11

Zhuguo Li, Bingwen Wang, Xiaoya Hu, and Lizhu Feng

Huazhong University of Science and Technology
Wuhan, Hubei, China
lzg440@163.com

Abstract. With the rapid development of wireless technologies, many advanced wireless devices will be blended into industry control family. Three hot short-range wireless technologies are compared together and then ZigBee is selected out of them as implementation supportive technology of wireless industrial network. The overall system constitute of wireless sensor network is summarized, and under ARM11 hardware platform, which is with low price and easy to be developed on, hardware and software designs of sensor nodes and gateways are analyzed according to the design requirements.

Keywords: Wireless Sensor Network (WSN), ZigBee, ARM11, Sensor Node, Sensor Design.

1 Introduction

With the rapid development of industrial automation, the demand of real-time data transmissions and the openness of data link become more stringent and the limits of wired control networks grow more prominent [1]. In large enterprises, characterized by scattered districts, complex business division of labor, numerous facilities, high price assets and hazard manufacture circumstances, the status of equipments must be monitored and controlled at all times [2]. In those harzard industrial surroundings, reliable data transmission networks are urgent to be established.

Sensor networks are combined with technologies of sensors, computation embedded, distributed information disposal and commutations. As an industrial communication way, wireless technologies are improved rapidly and become growth points in industrial commutation market. With the birth of various fresh wireless technologies, wireless communication shows good prospects in findustrial automation field.Short-range wireless network, of safe and reliable data transmission, easy and flexible deployment, low equipment cost and long battery life, reveal profound and potential development in industrial control territory, therefore it becomes one of research hot spots. Now, three promising short-range wireless network technologies are wireless local area network, Blue Tooth and ZigBee, as compared in table 1. In the view of practice, ZigBee can be the best choice in industrial control network.

Z. Du (Ed.): Proceedings of the 2012 International Conference of MCSA, AISC 191, pp. 95–100.
springerlink.com

Table 1. Comparison of WLAN, Blue Tooth and ZigBee

Feature	WLAN	Blue Tooth	ZigBee
Data Rate	11Mbps	1Mbps	250kbps
Nodes per master	32	7	64,000
Slave enumeration latency	Up to 3s	Up to 10s	30ms
Range	100m	10m	70m
Battery life	Hours	A week	1-year or longer
Bill of material	$9	$6	$3
Complexity	Complex	Very Complex	Simple

Among those wireless transmission protocols, ZigBee is of lowest energy consumption and cost, enough data rate and range, large node capacity and good security, so it becomes the first choice when deploying wireless sensor networks [3]. ZigBee protocol, which is a complete protocol standard and becomes an international standard protocol, based on IEEE 802.15.4 standard is added with network layer, application layer and security layer. ZigBee is mainly used between electronic devices with low data rate, which are not far from each other. Typical data types include cyclic data, interval data and low response timed repeated data. It mainly face towards personal computer peripheral, consumer electronic devices, home automation, toys, medical care, industrial control, etc.

With the rapid development of wireless technologies, many advanced wireless devices will blended into industry control family. In this paper, three hot short-range wireless technologies are compared together and then ZigBee is selected as implementation support technology of wireless industrial network. The system overall constitute of WSN is summarized. Under ARM11 hardware platform, which is with low price and easy to develop, hardware and software designs of sensor nodes and gateways are introduced.

2 System Overall Constitute

WSNs system discussed here are made up of sensor nodes, gateways and supervisor nodes [4]. Gateway concerns themselves with collection and disposal of data from or to sensor nodes, as well communication with outside networks. As a sensor node, it responses requests of gateways, gathers information around itself, e.g. temperature or pressure signals and still acts as a simple gateway, forwarding data through route protocol direct or as a relay in multihop packet delivery, then transmitting data to or from remote center by means of the temporary convergence link. Supervisor nodes receive data from sensor nodes, analyze and handle with them properly. In occasions with no tight real-time requirements, sensor nodes can also play the role of actuator nodes, which carry out commands by customers or software and manipulate actuators in factory field. Those nodes can be scattered at will or embedded manually in field to be monitored. WSNs have the abilities to self-organize and self-diagnose and self-heal, if designed optimally.

After ZigBee showed up, many researchers are exploring the application. Some addressed technologies through WSNs, but failed to offer a concrete implementation. Although off-the-shelf products have shown up, they are expensive and not easy to self-program by users. Some designed a sensor node with an MCU and a ZigBee transceiver, in order to fulfill simple system design, such as CC2430 as a sensor node [5] and ATMega128L or MSP430 as a controller node [4]. In these methods, hardware design was complex, with big size and high cost.

WSN designed here uses CC2530 as a sensor or actuator node, which collects data and execute commands by customers [6], S3C6410, a high-speed MCU, as a gateway and communicates with operators in Internet or Intranet.

3 Hardware Design

3.1 Sensor Node

CC2530 made by TI Corporation is an upgrade alternative chip for CC2430.

Hardware architecture of a sensor node[6] shown in fig. 1. Connected with an RF antenna, simply design of peripheral and power circuit according to the hardware manual, equipped with a proper sensing chip, CC2530 can act as a sensor node.

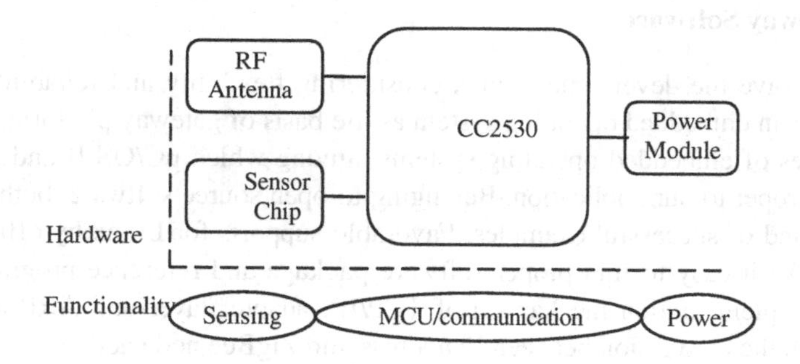

Fig. 1. Hardware and functionality of a sensor node

3.2 Gateway

Considering probable complex communication problems in WSNs, we decided to use S3C6410, made by SAMSUNG, a common product of ARM11 serials, to play the role of a gateway. Its simple and stable design is very suitable for products with tight request on power.

As to S3C6410, peripheral circuit design must be carried out, in order to from proper hardware platform. Towards communication with sensor nodes, CC2530 connected via SPI. Hardware block diagram of a gateway shown in Fig. 2.

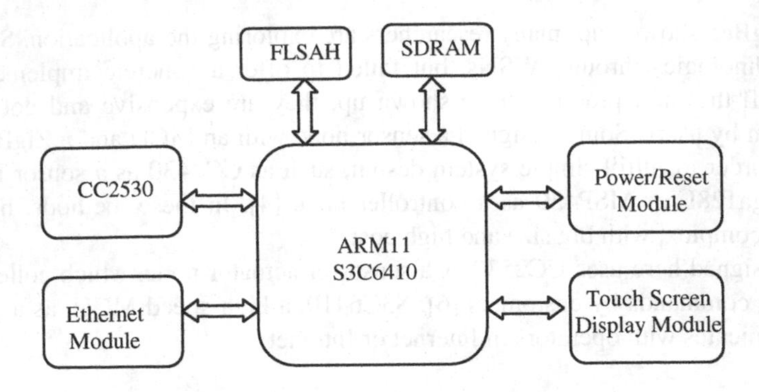

Fig. 2. Gateway hardware block diagram

4 Software Design

In the whole WSN, data between nodes are superframes [7] in the mode of IEEE 802.15.4. Such frames must be transmitted by guaranteed real-time and stability.Software for CC2530 sensor nodes may be developed under IAR.

4.1 Gateway Software

In order to save the development time, considering flexibility and reliability, we decided to use an embedded operating system as the basis of gateway platform. There are several types of embedded operating systems, among which μC/OS II and embedded Linux are proper to our application. Belonging to open source software, both are widely applied and of successful examples. Favorable supports for Linux by official ARM website make it easy to find proper software package and reference programs. Referenced to implementation mechanism of TCP/IP, adapter layer and ARP are carried out, to finish the conversion between IP address and ZigBee node address.

4.2 Supervisory Software Design

In some micro or tiny systems, supervisory software may be integrated within ARM11, to implement basic data acquisition and controls by commands. However, in medium or large systems with large screens, in which many functions, such as large-scale historical data queries, or advanced scheduling algorithms, must be integrated within supervisory nodes, complex functionalities of supervisory software are required. Functionalities of supervisory software are shown in Fig. 3. Among the modules, communication, data acquisition and database are basic.

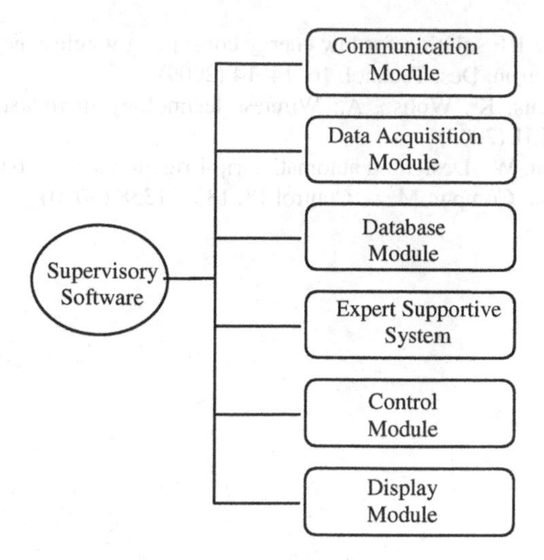

Fig. 3. Functionalities of supervisory software

These program modules may be carried out by using C-like programming languages by engineering methods. When needed, an interface program for commercial supervisory software may be developed as an open bridge like industrial common standard, such as OPC or ODBC.

5 Conclusion

WSNs based on ZigBee are simple in network deployment, extensible in scale and of low manufacturing cost and stable performance. Conformed to automation, intelligence and networking in industrial supervisory environment, we proposed a system solution based on WSNs for industrial monitoring and control networks, which is of good application perspective and promotion value[8].

References

[1] Jingyun, C., Xiangping, Z.: ZigBee wireless communication technology in industrial controls. Radio Eng. 36, 61–64 (2006)
[2] Caijun, H., Houhui, F., Yuqiu, G., Li, H.: Application of ZigBee technology in industrial monitoring network. Applied Comput. Syst. 19, 179–182 (2010)
[3] Guangrong, W., Jianxiong, Z.: Realization of wireless sensor network based on CC2430. Mod. Electronic Technol. 31, 121–124 (2008)
[4] Chenglong, L., Zhihao, L.: Design of multifunctional ZigBee gateway based on AT91SAM9260. Process Autom. Instrum. 30, 30–33 (2009)
[5] Xinchun, L., Yongxi, L., Cuiyan, F.: Design of energy-saving system for academic based APM9 and CC2430. Applied Computer Systems 19, 230–232, 235 (2010)

[6] Qiubo, C.: 2.4GHz RF solution for low energy consuming wireless network system development. EDN Electron. Des. Technol. 16, 14–14 (2009)
[7] Willig, A., Matheus, K., Wolisz, A.: Wireless technology in industrial networks. Proc. IEEE 93, 1130–1151 (2005)
[8] Ting, Y., Xiaochan, W.: Design on automatic drip irrigation system based on ZigBee wireless sensor network. Comput. Meas. Control 18, 1332–1338 (2010)

The Research on Building of a New Safe Campus Network

Ma Tao, Wei Shaoqian, Wu Baozhu, and Qin Yihui

Teachers' College of Beijing Union University

Abstract. With the continuous development and application of the campus network, network management and security issues are also increasingly complex. In this paper, combined with the practical work of the campus network management, an analysis of the factors threatening the campus network security, and maintain network security strategy two, the safety of the campus network management, to protect the campus network security strategy.

Keywords: network, security policy, data, access.

1 Introduction

The campus network infrastructure as schools, play many roles of teaching, research, management and foreign exchange. The campus network security has a direct impact on classroom teaching. However, many of the campus network security construction lagged far behind, making it planted early in the construction safety hazards; put into operation, but also not established strictly enforce the necessary rules and regulations, staff safety consciousness, neglect of safety management. These phenomena will lead to the campus network security crisis, seriously affecting the normal operation of the campus network data and network equipment. Thus enhancing the security building is built, with the important safeguard of the campus network. I combine a few years to gradually explore the construction of the campus network security management, accumulated some experience and lessons learned, and talk about how to strengthen the campus network security management measures.

2 The Campus Network Security Policy

Security policy refers to a specific environment, to guarantee a certain level of security protection rules to be observed. Security policy, including strict management, advanced technology and related laws. The security policy decision to adopt the ways and means to ensure the security of network systems. First, be aware of their need, to formulate appropriate strategy to meet the needs of the program, before considering the technical implementation.

2.1 Physical Security Strategy

The purpose of physical security to protect routers, switches, workstations, a variety of network servers, printers and other hardware entities and communication links

Z. Du (Ed.): Proceedings of the 2012 International Conference of MCSA, AISC 191, pp. 101–104.
springerlink.com

from natural disasters, man-made destruction and wiretapping attacks; to ensure that the network equipment has a good electromagnetic compatibility environment; safekeeping of backup tapes and documentation; to prevent illegal entry into the engine room to carry out sabotage activities.

Physical security is to protect the computer network equipment, facilities, and other media from earthquakes, floods, fires and other environmental accidents, as well as the process of damage caused by the human operator mistakes or errors, and a variety of computer crime. Its purpose is to protect computer systems, web servers, printers and other hardware entities and communication link-layer network equipment from natural disasters, the destruction and take the line of attack.

2.2 Access Control Policy

The main task of access control is to ensure that network resources against unauthorized use and access, it is to ensure that network security is the most important one of the core strategy. Control network access control, network permissions, directory-level security controls, property security controls, network server security control, network monitoring and lock control, network port and node security control various forms of access control.

2.3 Information Encryption Strategy

"Encrypted", especially in the protection of network security data security is irreplaceable, incomparable role. Message encryption algorithms, protocols, management of the huge system. Based on the encryption algorithm, cryptographic protocol is the key, key management is to protect. As the core of the confidentiality, encryption and decryption of the core program and related procedures, such as the login system, user management system should be developed in the possible case.

2.4 Firewall Control Strategy

A firewall is a recent development of a protective computer network security and technical measures; it is a barrier to prevent network hackers to access a network of institutions. It is located between the two networks the implementation of the system control strategy (which may be software or hardware or a combination of both), used to limit the external illegal (unauthorized) users to access internal network resources, set up the corresponding network traffic monitoring system to isolate the internal and external networks, to block the intrusion of the external network, to prevent theft or from the damaging effects of malicious attacks.

2.5 Backup and Mirroring Technology

Improve the integrity of backup and mirroring technology. Backup technology is the most commonly used measures to improve data integrity; it is to make a backup of the need to protect the data in another place, and lose the original can also use the data backup. Mirroring technology, two devices perform exactly the same work, if one fails, another can continue to work.

2.6 Network Intrusion Detection Technology

Trying to undermine the integrity of information systems, confidentiality, the credibility of any network activity, is known as network intrusion. Intrusion detection is defined as: to identify the malicious intention and act against the computer or network resources and this made the reaction process. It not only detects intrusion from outside, but also to detect unauthorized activities from internal users. Intrusion detection applied to the offensive strategy, the data it provides is not only possible to find legitimate users from abusing the privilege, but also may provide valid evidence to be investigated for the intruder legal responsibility to a certain extent.

2.7 Filter Harmful Information

Campus network, due to the use of population-specific, must be the harmful information of the network filtering to prevent pornography, violence and reactionary information harm the students' physical and mental health must be the combination of a complete set of network management and information filtering system. Harmful information filtering management of the campus computer to access the Internet.

2.8 Network Security Management Practices

In network security, in addition to the use of technical measures, and formulate relevant rules and regulations, and will play a very effective role for ensuring network security and reliable operation. As a core element, the rules and regulations should always be run through the safety of the system life cycle. Network security management system should include: to determine the safety level and security management scope; formulate relevant network operating procedures and personnel access to the engine room management system; the development of network systems, maintenance systems and emergency measures.

3 Conclusions

The campus network security issues, equipment, technology, and management problems. In terms of the management staff for the campus network, be sure to raise the awareness of network security, enhance network security technology to master, focusing on the knowledge of network security training of students to faculty, but also need to develop a complete set of rules and regulations to regulate the behavior of Internet.

References

[1] Data communication and computer networks, high pass good Higher Education Press (July 2001)

[2] Xie, M.D.: The Survey of Latest Researches on Online Code Dissemination in Wireless Sensor Networks. IEIT Journal of Adaptive & Dynamic Computing (1), 23–28 (2011), DOI=10.5813/www.ieit-web.org/IJADC/2011.1.4

[3] Zhao, L., Mao, Y.X.: GOBO: a Sub-Ontology API for Gene Ontology. IEIT Journal of Adaptive & Dynamic Computing (1), 29–32 (2011), DOI=10.5813/www.ieit-web.org/IJADC/2011.1.5
[4] Wu, H., Xu, J.B., Zhang, S.F., Wen, H.: GPU Accelerated Dissipative Particle Dynamics with Parallel Cell-list Updating. IEIT Journal of Adaptive & Dynamic Computing (2), 26–32 (2011), DOI=10.5813/www.ieit-web.org/IJADC/2011.2.4
[5] Zhou, J.J.: The Parallelization Design of Reservoir Numerical Simulator. IEIT Journal of Adaptive & Dynamic Computing (2), 33–37 (2011), DOI=10.5813/www.ieit-web.org/IJADC/2011.2.5

Research of High Definition Download Model Based on P2P&DRM

Zexun Wu

Fujian Radio and Television University, Longyan Branch, 364000 China

Abstract. In this model, the .Net framework is regarded as service management point, applying the WEC technique, the P2P and DRM technique is introduced to serve for the requirement and implemented the combine with the existing platform. In the model, the point is the research about the implementation which from the video release to the user download and monitoring of the resources using process. The advantages of it are the unified management and accelerate download and the resources protection. This model offers a better billing mode for operators and SP.

Keywords: High Definition Download, WEB, .Net, P2P, DRM.

1 Introduction

With the growing popularity of broadband, video and audio content on the Internet has become increasingly diverse. More and more people develop a way through the network to access and watch video content consumption habits, and operators to promote products through the years of broadband star world occupy the strategic objectives of the user's living room has taken an important step . With in-depth promotion of business, has achieved initial results. And nurture a mature star wide user group of consumers. With the wide-Star users continue to increase, the market for the business such as support for HD offline play, pay-per-view, a number of new demands. If you can successfully meet these requirements, great benefits will follow-up business to promote and build a more flexible business model. It is in this context, the operator of the project requirements of broadband high-definition download platform. Telecom operators access to terminal equipment through the integration of audio and video content providers and video content, providing a convenient platform for free viewing and download audio and video content, creating a broad and huge potential of audio-visual consumer market.

2 System Model

2.1 Overall Description

Download high-definition model with broadband Star world business management system compatible with the introduction of DRM and HTTP/P2P, service system, upgrade the client software on the broadband star world existing terminal P2P

Z. Du (Ed.): Proceedings of the 2012 International Conference of MCSA, AISC 191, pp. 105–110.
springerlink.com

download capabilities of HD sources for download through PC for playback. System using the MS SQLServer2008 for database management systems, in C # for mainstream development language, C, + + complete DRM dynamic link library, using the WebService as each subsystem interface to achieve, based on these technologies, to build the one kind of the P2P + the DRM-based high-definition download platform model and download acceleration and on-chip source protection.

2.2 P2P Applications

P2P networks, also known as peer-to-peer network to break the traditional network of C / S mode - set it in the transmission mode. C / S architecture, the client to the server to issue a service request, the server response to client requests and provide the necessary services, the status of the client and server is based on the resources not. Therefore, when a user makes too many requests, the server is overloaded, the reaction, however, resulting in a denial of service, resulting in server crashes, the link bandwidth is idle, has been wasted. P2P mode network service, just to fully tap the idle resources of the network, each node in the P2P network is both a server and client resources, load balancing assigned to each node in the network, so that each a node will have to bear part of the computing tasks and storage tasks. Therefore, the amount of resources in the network with the number of nodes join the network is proportional to the number of nodes, the more contribution of more resources, and its quality of service higher. [1-2]

2.3 Streaming Media

Streaming media refers to a series of media data is compressed, segmented transmission of data through the network, real-time on the network transmit audio and video for the viewing of a technology and process, this technology allows data packets can be sent like water; If you do not use this technology, you must download the entire media file before using. Transmission site of streaming audio and video stored on the server of the film when the viewer to watch these video files, audio and video data immediately after delivery to the viewer's computer by particular player software to play. In the P2P streaming system, each user is a node with multiple users according to their respective load conditions and equipment performance to establish a connection in order to share each other's streaming media data, this approach is both balanced server load, but also can improve the quality of the server for each user, which can overcome the traditional streaming media bandwidth capacity is not enough.

2.4 Trust Management

This model is based on a multi-system environment, we use PKI erected in strict confidence mechanism both in the communications, and trust management system allows the unknown to each other information the Union to establish a trust relationship based on the certificate. , Unified authentication is a core part of our design model. We rely on the certificate trust list; this list is by a trusted unified certification. The list itself is signed by electronic means, to ensure its integrity.

It replaces the need for cross-validation process. Registered in the certificate trust table cooperation Unified Certification Services are well known and trusted each other.

All information is signed and encrypted using PKI technology, the user agent can not have their PIN to ensure reliable user agent services due to their high number and dynamic, the other hand, the other instances have their own The private key [4-7].

3 Model Constructions

3.1 Overall Technology Route of Model

Select Net technology system as an application integration platform for the entire system. Net framework is a technologically advanced, fully functional, reliable, fast and safe framework. Net can quickly build distribution, scalable, portable, safe and reliable the server-side configuration.

In order to achieve customization and unified authentication, unified content management purposes, we use the portal. The characteristics of the portal are a unique, integrated, personalized and holistic. Among them: the only requirements of the enterprise, but also the significance of the portal; integration limit of reality, and reflects the continuity of the business; personalized customer preferences, but also the vitality of the portal; and integrity is the high level of information requirements.

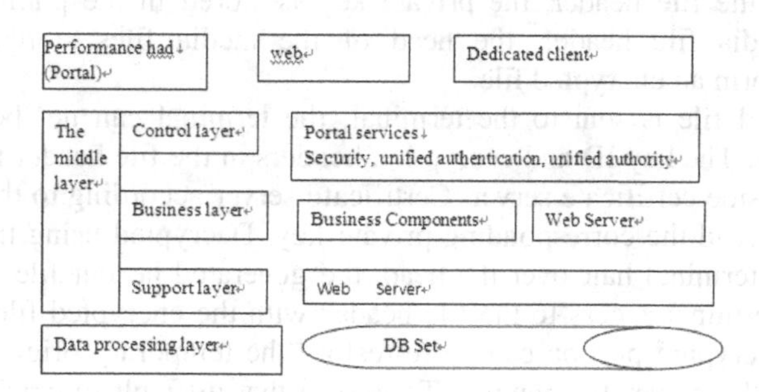

Fig. 1. The Route of Technology

The entire system for three-tier structure, namely, the presentation layer, middle layer, data layer is completely separate. The technical architecture can best guarantee of efficiency and stability of the system. Added to the interface layer, making the content management and application system separation, not only to make the system easy to expand and upgrade, and improve the integration of system security.

3.2 Model of the Target and the Network Topology

This model, when completed, will achieve the following objectives.

First, the definition film source (10-20Mbps bit rate) can not be online real-time viewing. To a local or external hard drive, you can watch by downloading new sources.

Second, to solve the current on-demand process due to occasional network congestion, resulting in plays halfway cards, to make the playback smoother.

Third, to support the pre-set-top box device customization and download audio and video programs remotely via a PC, thus avoiding the bad experience of the buffer in front of the TV, etc. Download. Fourth, we can easily build a business model of broadband audio and video downloads, and more convenient to introduce the SP with the program source, to achieve a win-win, win.

Fifth, we can build a media advertising to download the distribution network business model, facilitate the advertising and media through the platform, and advertising.

3.3 DRM Design

DRM implementation mechanisms, including part of the back-end business systems and terminals. The back-end business system is mainly responsible for SP providers to upload media files encrypted. The terminal part is mainly responsible for the decryption play the video.

The platform side consists of two main systems: the media encryption system and the certificate issuing system. Need to download the release of media content, the platform side can produce the key ID and the random seed. The key ID and the random seed to generate the encryption key of the media file header, and asymmetric (RSA) encryption mechanism, and to generate public and private key. Use the public key to encrypt the file header, the private key is stored in the platform side. And change the media file header, the head of the media files stored key ID and encryption, to form an encrypted file.

The encrypted file is sent to the terminal, the terminal can not be read directly decode and play. The key ID and encryption headers in the file header need to be sent to the platform-side certificate server. Certificate server according to the key ID from the database to read the corresponding private key. Decrypted using the read private key encryption terminal hair over the head, the generated header file is sent back to the terminal. Terminal access to the file header with the encrypted file to extract the media is not encrypted portion can be re-restore the temporary original media files, use this file to the player for playing. To ensure that difficult to crack, need special treatment, in the end the header and the restore process to ensure document confidentiality.

In order to guarantee the security of the system, the transmission of confidential messages is transmitted through an encrypted message, if necessary, to sign and encrypt the user through the web server to complete registration and browser want to download movies. After successful registration, the certificate server to generate the user public and private key pair, and the private key in a safe manner transmitted to the user;, web server load to determine when the user selected they will need to download the movie and then decide from which file server to obtain the specified film and sent to the user. Calculated for the key in order to ensure that only legitimate users can watch their encryption to encrypt media files need to be the connection of the Random Number Generator to generate random numbers with the key ID, One-way Function, and dense key and key ID is stored in the database, to decrypt need. Find the information needed to decrypt the key stored in the file header.

In order to guarantee the security of the system, in addition to the movie file is encrypted, but also the need for various types of keys for protection, otherwise the security of the media file is empty talk. The user's private key will be the user password to transform the generated encryption key, the file header contains the decryption information to prevent identity impersonate, will also use signature and encryption are two means at the same time, will be filled in order to prevent replay attacks the key needed to decrypt the media file, similar to the protection of the header, will also be at the same time the use of signatures, encryption, and fill the three means.

Finally, in order to protect media files after decryption, each player confusion, play a certain number of times, it will be completely distortion can not be played, thus avoiding illegal copy.

3.4 P2P Model and Terminal Access to the Main Flow

PC or terminal to download P2P/HTTP/FTP a variety of ways to download, so not only improve the HD download speeds, but also to save some bandwidth for the server, reduce operating costs. Terminal to download and play process as shown below.

4 Conclusions

In this paper, a three-tier technology, the use of popular Net development framework and the introduction of P2P + DRM technology, designed a high-definition download model, the model of HD sources for the preceding management, and set the accounting principles, the terminal downloaded through P2P/HTTP manner specified new sources in the terminal for playback, the introduction of DRM monitoring new sources, and to prevent the spread of new sources of copy. Data exchange in the design of the model are based on XML, because XML has a good scalability, platform-independent structured data description capability, so that the model has good flexibility, scalability and cross-platform for the model further optimization and integration and provide a basis for.

References

[1] Wu, J.-Q., Liu, F., Peng, Y.-X.: EPSS:An Extensible Peer-to-Peer Streaming Simulator. Computer Engineering & Science (7), 101–105 (2011)

[2] Jin, S.: The Study of Copyright Protection in P2P Network Mode. Journal of the Postgraduate of Zhongnan University of Economics and Law (1), 58–61, 98 (2011)

[3] Lin, K., Yang, M., Mao, D.-L.: Scheme for Performance Improvement of P2P Live Video Streaming on LANs. Journal of Chinese Computer Systems. Journal of Chinese Computer Systems (7) (2011)

[4] Luo, C., Ouyang, J., Zhang, W.: Design of Trust and Authorization Service Platform Based on SAML. Computer Engineering 31(13), 118–120 (2005)

[5] Cheng, A., Yu, Q., Yao, X., Ye, Y.: Application Research on Authorization Management Model of PMI Based on Attribute Certificate. Computer Engineering (4), 162–164 (2006)
[6] Zhao, Z.L., Liu, B., Li, W.: Image Clustering Based on Extreme K-means Algorithm. IEIT Journal of Adaptive & Dynamic Computing (1), 12–16 (2012), DOI=10.5813/www.ieit-web.org/IJADC/2012.1.3
[7] Zheng, L.P., Hu, X.M., Guo, M.: On the q-Szasz Operators on Two Variables. IEIT Journal of Adaptive & Dynamic Computing (1), 17–21 (2012), DOI=10.5813/www.ieit-web.org/IJADC/2012.1.4

Resistance Test on Ethernet Switch against Dynamic VLAN Attack

Zhihong Liang[1] and Baichao Li[2]

[1] Electric Power Research Institute of Guangdong Power Gird Corporation, Guangzhou, China
[2] School of Information Science and Technology, Sun Yet-Sen University, Guangzhou, China
liangzhihong.gz@gmail.com

Abstract. In this paper, we describe the threat of DoS attack exploit dynamic VLAN attributes against Ethernet switches and propound a DoS attack resistance testing to evaluate the performance of the switch under attack. The testing scenario including network topology and testing procedure are introduced in detail. We also put forward a cast study on Ruijie RG-S2760-24 and then take deep analysis of the results about throughput and response time. At last, we develop a score function to describe the attack resistance ability of the switch, which helps network administrators to assess the switch's performance and take comparisons among different switches.

Keywords: Ethernet switch testing, dynamic VLAN, DoS attack resistance.

1 Introductions

The commonest approach to configuring VLANs is port-based VLANs, also referred to static VLANs, which means assigning each port of the switch to a particular VLAN. The creation and the change of the port-to-VLAN assignment have to be done manual, which is a time-consuming and tedious task. In order to make VLAN assignment more scalable, Dynamic VLANs are introduced. Administrators can assign switch ports to VLAN dynamically based on information such as the source MAC address of the device connected to the switch or the username used to log onto that device. When a device enters the network, the access switch launches a query to a database for the device's VLAN membership. The use of dynamic VLANs liberates webmasters from labor and mistakes, especially when the VLAN assignment has to be changed frequently. But, dynamic VLAN also brings network vulnerability. The vicious user can change his VLAN group frequently by sending packets with a forged MAC address or username to launch a denial of service attack (DoS attack) against the switch. The switch will be snowed under with requests of changing its VLAN assignment; then come in for a drop in throughput, and can't response normal users' requests. In this sense, the switch under attack is out of service. Since VLAN is widely used in nowadays network; it is foreseen that DoS attacks exploiting dynamic VLAN attribute against switches will become a big problem before long. So a method to estimate the influence of attacks to the switch is in urgent need.

Z. Du (Ed.): Proceedings of the 2012 International Conference of MCSA, AISC 191, pp. 111–116.
springerlink.com

In this paper, we propose a DoS attack resistance test to evaluate the performance of the switch. Some related work is shown in section 2. The testing scenario is described in section 3; so is a case study on Ruijie RG-S2760-24 which is a layer 3 switch supporting dynamic VLAN assignment. In section 4, we discuss the test results, and put forward a general score function to assess a switch's surviving ability against a DoS attack. At last, we draw a conclusion in section 5.

2 Related Works

Switch, commonly knew as Ethernet Bridge, is a very important kind of network devices. Its performance requirements are defined in RFC 2544, including maximum frame rate, throughput, latency, frame loss rate, system recovery and so on. Extensive works have been done on testing the performance of network switch. Most of them focus on the most important issues: throughput, latency and frame loss rate[1]. Furthermore, a lot of improved testing scenarios[2] and testing method[3] are brought forward. They provide a standard on network benchmark testing and performance measurements of a switch.

Besides switches' performance, their security issue catches a lot of attention as well. There are many kinds of attack specifically point against switches, such as MAC address flooding attack, STP (spanning tree protocol) attack, ARP treating attack and VTP (VLAN Trunk Protocol) attack. In order to protect the switch, plenty of solutions have been put forward, most of which are aimed at protecting VLAN assignment and related information[4]. The introduction of authentication is a great help to network security as well. To our surprise, DoS attack against switches does not catch adequate attention. The vast majority of DoS attacks aim at web servers and web applications presently, so does the analysis and protection of DoS attacks[5]. As dynamic VLAN is used more and more widely, the study on it is increasing. But most of them only talk about how to deploy dynamic VLAN; on the contrary, little work has been done on researching the risk brought by dynamic VLAN assignment.

3 Testing Scenario

In this section, we will describe the DOS attack resistance testing scenario in detail. First of all, we will describe how to deploying dynamic VLAN system. Then the testing network topology and the steps of the testing are propounded. At last, a deep observation of the testing result is carried out so as to evaluate the switch's ability of resisting DOS attack.

3.1 Deploying Dynamic VLAN

802.1X authentication involves three parties; a supplicant, an authenticator, and an authentication server, which refer to a client device (such as a laptop), a network device (such as an Ethernet switch or wireless access point), and a host running software supporting the RADIUS respectively. In this testing scenario, we use Ruijie RG-S2760-24 as an authenticator, and FreeRADIUS as the authentication server. Xsupplicant is an open source a supplicant. We modify the source codes of

Xsupplicant based on version 1.2.8 in order to let it adapt to the specified requirements such as marking the time cost by one successful login and trying to continuously login and logoff.

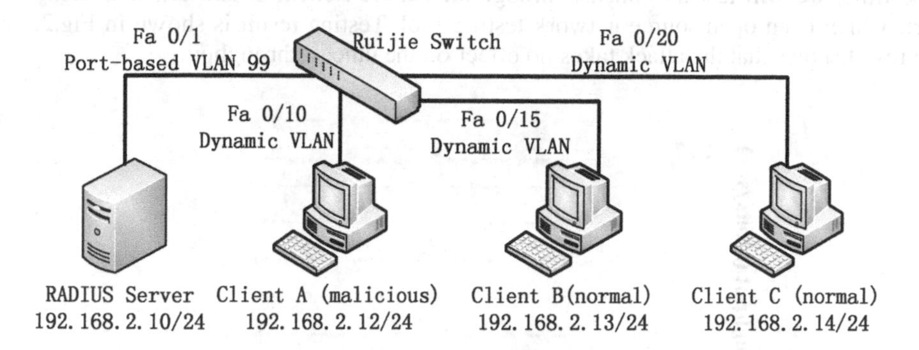

Fig. 1. Network topology of the testing scenario

Network topology of the testing scenario is shown in Fig.1. Four hosts are involved, one is the RADIUS Server deployed with FreeRADIUS and Mysql database, the other are clients deployed with modified Xsupplicant. In order to support dynamic VLAN assignment, the switch needs to be configured appropriately. We need to turn on the AAA (Authentication, Authorization, and Accounting) module; then tell the switch the IP address of a RADIUS server and the key shared between them. The last step is to enable the dox1x authentication on the ports which connected to the client, and configure the ports get their VLAN assignment dynamically.

The users' information including the username, password and VLAN assignment is stored in the RADIUS server's database. It is worth mentioning that the VLAN which the port is assigned to should be already created in the switch manually or by means of VTP.

3.2 Testing Methodology and Result Analysis

In this section, we demonstrate two effects of the DoS attack exploiting dynamic VLAN attribute. To aim at assessing the ability of the switch precisely, we don't complete the whole procedure of 802.1x authentication when a high speed login and logout is going to be executed (for example, 10 times per second). We only perform the first step, i.e. send EAPOL-Request from the supplicant to the switch. Consequently, the switch has to handle this request and reserve some resource for further operations. On the contrary, that is none of the server's business. In another word, the handling capacity of the server do not affects our testing.

3.2.1 Effect on Throughput

In this testing, the malicious client A keeps trying login and logoff over and over again. Each time client A logs in, the FastEthernet port 0/10 will be assigned to VLAN 10, i.e. VLAN client, which client B and client C are also assigned to.

Each time client A logs off, the FastEthernet port 0/10 will be assigned to VLAN 1, i.e. the default VLAN that every port of the switch is assigned to if the port is not configured specially. The frequency of client A's login will enhance gradually. At the same time, we will test the unicast throughput between client B and client C using iperf, which is an open source network testing tool. Testing result is shown in Fig.2. The result shows that the attack takes no effect on the unicast throughput.

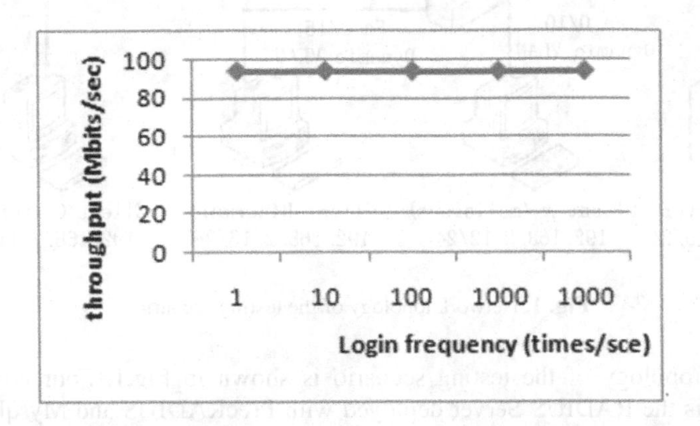

Fig. 2. The attack effect on unicast throughput

To our surprise, the attack doesn't affect the unicast throughput at all. It implies that packets switching is done by specific chip individually and doesn't cost any computation capacity of the CPU on the switch.

Since only performing the first step of the authentication procedure doesn't change the VLAN assignment, in other words, it doesn't change the broadcast domain; the attack effect has nothing different between unicast and broadcast. But completing the whole procedure would cost at least 2 seconds. It is far insufficient from launching an attack. So we adopt another strategy in testing the attack effect on broadcast throughput: increase the numbers of malicious clients. Testing result is shown in Fig.3. It is obvious that the rapid variation of broadcast domains confuse the switch.

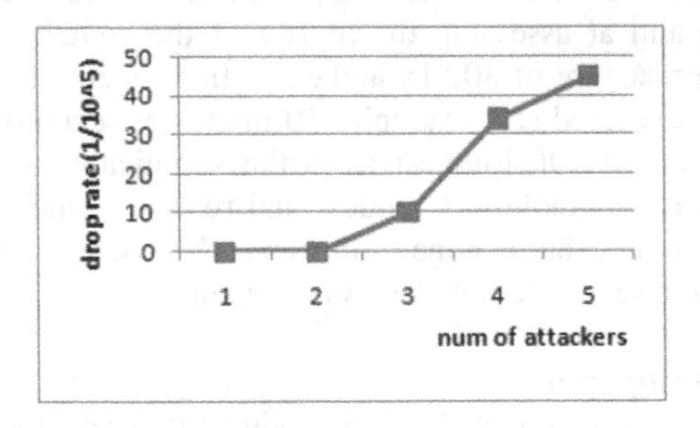

Fig. 3. The attack effect on broadcast drop rate

3.2.2 Effect on Response Time

The malicious client A does the same thing as foregoing, i.e. client A sends EAPOL-Request to the switch at a high speed. We mark the switch's response time of client B's supplication during A's attack. The result is shown in Fig.4. We can see that if the attack frequency is less than 1000 per time, the response time is almost constant. When the attack strength grows up to 1000 times per second, the response time increase to 2.15s and become unsteady, 15% of the login attempt will fail. When the attack strength grows up to 2000 times per second, the service is totally unavailable, that is, all the login attempts will fail.

The attack effect on authentication requests success rate is shown in Fig.5. It is obvious that the critical attack strength is 1000 times per second. It is noteworthy that the switch's performance doesn't drop gradually; on the contrary, it drops steeply. We can infer that the switch's ability of handling the requests is limited by memory capacity, rather than computation capacity.

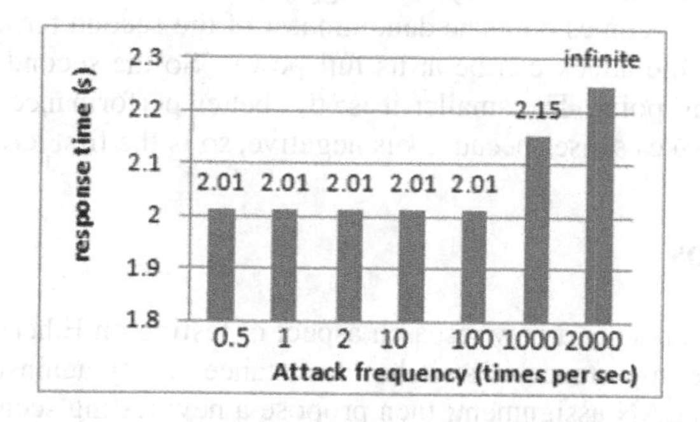

Fig. 4. The attack effect on response time

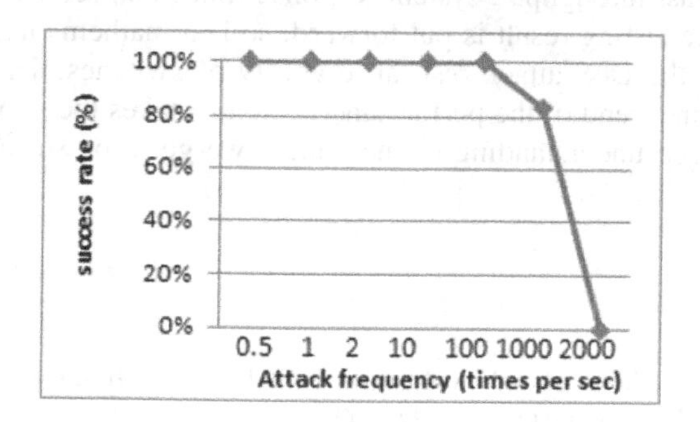

Fig. 5. The attack effect on authentication requests success rate

4 Modeling

We propose a score function to introduce comparison among different kinds of switches. Two things should be taken into account: the decline trend of the performance and the survival ability. The former is demonstrated by the slope of the performance-attack strength curve, the latter is demonstrated by the threshold of crash-down point. They stand for the computation capacity and the storage capacity respectively.

$$S_{resis} = \frac{f_{crash} \times k}{P_0} + \frac{f_{crash}}{B_{link} / L_{packet}}$$

S_{resis} denotes the DoS attack resistance ability of the switch, i.e. the score that the switch gets in the testing. f_{crash} refers to the critical frequency of the attack. As soon as the attack frequency climb up to this threshold, the switch's performance will become totally unacceptable. k is the slope of the performance vs. attack-strength curve. P_0 is the initial performance when the switch isn't under attack. B_{link} denotes the bandwidth of the physical link. L_{packet} denotes the EAPOL-Start packet's size.

The first term of the equation describes the normalized drop before the switch suffers the crash-down attack frequency. The bigger its absolute value is; the worse the switch's performance comes out. The denominator of the second term of the equation describes how fast the attack can be at its full power. So the second term means the normalized crashing point. The smaller it is; the better performance the switch gets. The whole score makes senses because k is negative, so is the first term.

5 Conclusions

This paper aims at opening a new research aspect of testing on Ethernet switches. We first emphasize the importance of switches' resistance ability against DoS attack exploiting dynamic VLAN assignment; then propose a new testing scenario and carry it out on Ruijie RG-S2760-24 and focus on how much the attack will affect the unicast throughput, broadcast throughput, system response times and request success rate. A deep analysis of the testing result is put forward, so is a mathematical score function that help to assess the DoS attack resistance ability of switches. The score not only represents the decline trend of the performance, but also takes the survival ability into account. With a deep understanding of the attack, we get a powerful tool to defend against it.

References

1. He, X.: Testing Methods of a Few Important Performance Index for Ethernet Switch. Communications Technology (11), 97–99 (2003)
2. Yuan, T.: Integra ted test environment for layer 3 switches. Computer Applications 25(10), 2267–2668 (2005)
3. Tao, S.: Research on performance testing method of layer 3 switch, Master's Thesis, Xidian University
4. Jinbo, H., Lichun, G.: Attacks and Security Guards Based on VLAN System. Network Security (145), 32–33 (2008)
5. Abu-Nimeh, S., Nair, S., Marchetti, M.: Avoiding Denial of Service via Stress Testing. In: Proceedings of the IEEE International Conference on Computer Systems and Applications (2006)

Modeling Dynamic File Diffusion Behavior in P2P Networks

Baogang Chen[1] and Jinlong Hu[2]

[1] College of Information and Management Science,
Henan Agriculture University, Zhengzhou, China, 450002
[2] Communication and Computer Network Laboratory of Guangdong Province,
South China University of Technology, Guangzhou, China, 510641

Abstract. In this paper, according to the characteristics of popular files downloading and transmission, the various states of nodes in P2P file-sharing system are researched. Based on these various states of nodes, through a multi-species epidemic model with spatial dynamic, file diffusion model of P2P file-sharing system is proposed. By experiments, P2P file diffusion model is proved to accords with the actual situation and to have the ability simulating the behaviors of the peers in P2P network.

Keywords: P2P Networks, file diffusion, SEIR epidemic model, spatial dynamics.

1 Introduction

The large number of popular file downloading behavior is similar to the process of the infectious disease spread in P2P file-sharing system, which can be described with infectious diseases dynamics. In the field of medicine, Many infectious diseases propagation model have been investigated for a long history, which are the effective ways to research process of infectious diseases propagation and to predict the outbreak of infectious diseases.

Most of existing researches on the characteristics of the files replication and diffusion only consider the steady state performance of P2P networks, and does not consider the unstable states in the file diffusion process [1,2], or not fully reflect the status of nodes in the system [3,4]. In this paper, the study of infectious diseases dynamics theories are referenced, and the various states of users nodes are comprehensively examined in P2P file-sharing system, hereafter a new model is proposed for file diffusion in P2P file sharing system.

2 Dynamic Model of Infectious Diseases

In 1927, Kermack and McKendrick provided famous SIR compartment model when they investigated the law of epidemics [5]. And later, a class of SEIR Epidemic Model with Latent period was made on this basis. In the SEIR model, the population is divided

Z. Du (Ed.): Proceedings of the 2012 International Conference of MCSA, AISC 191, pp. 117–122.
springerlink.com

into four groups. susceptible to infection denoted by type S; the infected denoted by type I; if the infected have a period of incubation before ill, and the period of infection is not contagious, then these people in the incubation period, denoted by E; recovered class denoted by type R.

In 2005, Julien Arino et al proposed a disease transmission model with spatial dynamics [6]. An SEIR epidemic model with spatial dynamics is considered for a population consisting of S species and occupying N spatial patches. The total population for species i in patch p is N_{ip} and the population for species i is $N_i^0 > 0$, a fixed constant. At time t, the numbers of susceptible, exposed, infectious and recovered individuals of species i in patch p at time t are denoted by Sip, Eip, Iip and Rip, respectively. $1/d_{ip} > 0$, $1/\omega_{ip} > 0$, $1/\gamma_{ip} > 0$ are the average lifetime, latent period and infectious period for species i in patch p, respectively. The disease is assumed to be horizontally transmitted within and between species according to standard incidence with $\beta_{jip} \geq 0$, the rate of disease transfer from species j to species i in patch p. The dynamics for species i = 1, . . . , s in patch p = 1, . . . , n is given by the following system of 4sn equations:

$$\frac{dS_{ip}}{dt} = d_{ip}(N_{ip} - S_{ip}) - \sum_{j=1}^{s} \beta_{ijp} S_{ip} \frac{I_{jp}}{N_{jp}} + \sum_{q=1}^{n} m_{ipq} S_{iq} - \sum_{q=1}^{n} m_{iqp} S_{ip} \tag{1}$$

$$\frac{dE_{ip}}{dt} = \sum_{j=1}^{s} \beta_{ijp} S_{ip} \frac{I_{jp}}{N_{jp}} - (d_{ip} + \omega_{ip}) E_{ip} + \sum_{q=1}^{n} m_{ipq} E_{iq} - \sum_{q=1}^{n} m_{iqp} E_{ip} \tag{2}$$

$$\frac{dI_{ip}}{dt} = \omega_{ip} E_{ip} - (d_{ip} + \gamma_{ip}) I_{ip} + \sum_{q=1}^{n} m_{ipq} I_{iq} - \sum_{q=1}^{n} m_{iqp} I_{ip} \tag{3}$$

$$\frac{dR_{ip}}{dt} = \gamma_{ip} I_{ip} - d_{ip} R_{ip} + \sum_{q=1}^{n} m_{ipq} R_{iq} - \sum_{q=1}^{n} m_{iqp} R_{ip} \tag{4}$$

3 Conformation of P2P File Diffusion Model

In P2P file-sharing system ,the state of user node is called susceptibility before they search for files and propose their downloading request, denoted by class W; When they enter into downloading queue after they propose downloading request, the state of user node is called latent period, denoted by class D; When they share file for period of time after accomplishing their downloading task, the state of user node is called infection, denoted by class S; When users are no longer interested in the file and delete it, or users nodes does not share the file after they downloading the file, this status is called restoration, denoted by Class I.

Let X be any particular type of nodes, and P_X denotes the number of nodes in X class. In system, the user node has two states: on-line and off-line, and between them can be transformed into each other. Therefore, all nodes within the system are divided into four different compartments, and each node has both online and offline status, such as W_{on} and W_{off} .

In the following, through analysis of user nodes number changes in various types, we get the process model of the file diffusion behavior.

(1) Change rate of the node class W_{on}

Firstly, a node of class W_{on} changes its state from W_{on} to D_{on} because it searches the download file and put forward download request. Suppose the rate of a file being inquired, the ratio of the current file shared and the total number of nodes in the system are proportional. Suppose the average rate of the user sending out file queries and file download request is λ . According to standard incidence rate of infectious diseases $\beta SI / N$, then the class node W_{on} will be converted to class D_{on} with rate $\lambda P_{Won} P_{Son} / N_P$. Meanwhile, when some nodes enter into offline, the number of nodes online will be reduced. If rate of offline is set at λ_{on-off} , the nodes of class W_{on} will transfer into class of W_{off} at rate $\lambda_{on-off} P_{Won}$ .As downloading nodes will lose downloading sources, so they will be forced to seek new downloading source. Therefore, these nodes will be reentering the W_{on} class from D_{on} class, and then set the transfer rate of occurrence as r_1 . When the node state transition from offline to online, the number of W_{on} class nodes will increase. Assuming that this conversion occurs with rate λ_{off-on} , then the change rate of class W_{on} can be expressed as:

$$\frac{dP_{Won}}{dt} = -\lambda P_{Won} P_{Son} / N_P - \lambda_{on-off} P_{Won} + r_1 P_{Don} + \lambda_{off-on} P_{Woff} \tag{5}$$

(2) Change rate of the node class D_{on}

There are four types of situations which cause the nodes out of the Class D_{on} : nodes of class D_{on} terminate download back to Class W_{on} because the source nodes are no longer share files; nodes of class D_{on} enter into Class S_{on} due to share the file after downloading it; nodes of class D_{on} enter into class I_{on} due to does not share the files after downloading the file; nodes of class D_{on} occur state transition from online to offline. Set the rate from class D_{on} into W_{on} as r1; download rate as μ ; file sharing probability as p_{share} ; the rate that class D_{on} enter into S_{on} is μp_{share} ;and the rate that class D_{on} change into I_{on} is $\mu(1-p_{share})$. At the same time, the number of class D_{on} is increased, caused by two cases: the nodes from class W_{on} to class D_{on} and from offline to online. Set the rate from W_{on} to D_{on} is $\lambda P_{Won} P_{Son} / N_P$ and the rate from offline to online is $\lambda_{off-on} P_{Doff}$. Then the change rate of class D_{on} can be expressed as:

$$\frac{dP_{Don}}{dt} = \lambda P_{Won} P_{Son} / N_P - \mu P_{Don} - r_1 P_{Don} - \lambda_{on-off} P_{Don} + \lambda_{off-on} P_{Doff} \tag{6}$$

(3) Change rate of the node class S_{on}

Nodes leave the class S_{on} in two situations: don't share files and change state from online to offline. Let the average time for each node to share the file is $1/\delta$, so the rate of class S_{on} enter into class I_{on} is δ.Meanwhile, there are two cases to increase the number of class S_{on} nodes: nodes of class D_{on} come into class S_{on} and nodes of class D_{on} change state from offline to online. Suppose the rate at which the transition D_{on} class into S_{on} class is μp_{share}, and the transition from class S_{off} to S_{on} occurs at rate $\lambda_{off-on} P_{Soff}$. Then the change rate of class S_{on} is given as:

$$\frac{dP_{Son}}{dt} = \mu p_{share} P_{Don} - \delta P_{Son} - \lambda_{on-off} P_{Son} + \lambda_{off-on} P_{Soff} \tag{7}$$

(4) Change rate of the node class I_{on}

When the nodes of class D_{on} give up to share the download file, they will directly come into the class I_{on} at rate of $\mu(1-p_{share})$; Meanwhile, nodes of class S_{on} end to share file at rate δ.Then the total transformation rate of class I_{on} nodes can be expressed as:

$$\frac{dP_{Ion}}{dt} = \delta P_{Son} + \mu(1-p_{share}) P_{Don} - \lambda_{on-off} P_{Ion} + \lambda_{off-on} P_{Ioff} \tag{8}$$

(5) Change rate of offline nodes

Offline nodes have four classes W_{off}、D_{off}、S_{off} and I_{off}. Set all kinds of nodes transition state rate from online to offline and in turn are the same. So, we have

$$\frac{dP_{Woff}}{dt} = \lambda_{on-off} P_{Won} - \lambda_{off-on} P_{Woff} \tag{9}$$

$$\frac{dP_{Doff}}{dt} = \lambda_{on-off} P_{Don} - \lambda_{off-on} P_{Doff} \tag{10}$$

$$\frac{dP_{Soff}}{dt} = \lambda_{on-off} P_{Son} - \lambda_{off-on} P_{Soff} \tag{11}$$

$$\frac{dP_{Ioff}}{dt} = \lambda_{on-off} P_{Ion} - \lambda_{off-on} P_{Ioff} \tag{12}$$

4 Experiment and Analysis

The model assumes that all user nodes are initially interested in a particular file, then we select RMVB type files downloaded rank in the top 7 in MAZE log, and file name called A, B,..., G respectively.

Due to the users are interested in download files initially, then download request interval is equivalent to the average time interval of all users' file download request. As users of MAZE system have obvious periodicity and "day mode", so online and offline times of nodes are set up at 12 hours. Taking into account the user log lasting for one week, the average time of file shared can be treated as half week (84 hours). Assuming that the number of user nodes are distributed in download queue evenly, when the user doesn't share file, then the corresponding proportion of users are forced to choose download source again. Therefore, the average leaving rate of downloading node r_1 equals to the average time of user nodes to share files. Parameter values in Table 1.

Table 1. Experimental parameter values

r_1	δ	p_{share}	λ	λ_{on-off}	λ_{off-on}	N_P
1. 98E-4	1. 98E-4	0. 122	0. 00612	0. 00138	0. 00138	100068

In order to obtain the initial number of nodes that share files, the log data is divided into two parts: the first 12 hours log data and the remaining time of the data. The number of nodes that share files initially are nodes that finish downloading file within 12 hours multiplied by the factor p_{share} and then plus number of nodes available to share file in the beginning. The initial value of P_{Son} and P_{Soff} are set to the half number of nodes sharing file initially. The initial values of P_{Don} , P_{Doff} , P_{Ion} , P_{Ioff} are set to 0; P_{Won} and P_{Woff} are set to the half value that N_P minus P_{Son} and P_{Soff} .The average download rate is defined as all users' download traffic divided by time between all nodes entering into download queue and end of download. The average download rate divided by the file size is the rate of download accomplished per unit time.

With time granularity in minutes, and using parameter values acquired within 12 hours first, we calculate the number of file downloaded completely in latter 156 hours. The result is shown in Table 2. The comparison of user log data and model results are shown in Figure 1, 2. As can be seen from Table 2, the differences between the top four RMVB files of user log data and the model results are small. But at the back rank of the files in table, the differences seem obvious. The experimental results and assumptions are related. Suppose that in addition to users sharing file, all other users are initially interested in and will download the file, but the reality is that files on the back list are not very popular, and not all users want to download. So model results biased.

Table 2. Experimental result and real data

rank	name	size (MB)	Download nodes	sharing node	number of nodes downloaded fully	
					log	model
1	A	386. 52	5838	157	2849	3106
2	B	383. 33	5304	146	2678	2905
3	C	441. 87	4368	138	2455	2791
4	D	465. 76	3313	114	2021	2269
5	E	158. 65	2629	96	1690	1997
6	F	383. 33	1722	33	586	720
7	G	734. 61	1399	66	874	1263

5 Conclusion

This paper analyzes the user node state changes based on dynamic model theory of infectious diseases, and describe file diffusion model of P2P file sharing system. Because of considering the online and offline status in user node, therefore this model can be more close to the actual situation. Experimental analysis indicates that the model can describe the diffusion of most popular file in P2P file sharing system very well. How to use the model in depth analysis, and to research more P2P file sharing system behavior and performance characteristics, is one of the main works in the future.

References

1. Lo, P.F., Giovannil, N., Giuseppe, B.: The effect of heterogeneous link capacities in BitTorrent-like file sharing systems. In: Proceedings of the First International workshop on Hot Topics in Peer-to-Peer Systems, Volendam, The Netherlands, pp. 40–47 (2004)
2. Qiu, D.Y., Srikant, R.: Modeling and performance analysis of BitTorrent_like peer-to-peer networks. In: Proceedings of the ACM SIGCOMM 2004: Conference on Computer Communications, New York, USA, pp. 367–377 (2004)
3. Leibnitz, K., Hossfeld, T., Wakamiya, N., et al.: Modeling of epidemic diffusion in Peer-to-Peer file-sharing networks. In: Proceedings of the 2nd International Workshop on Biologically Inspired Approaches for Advanced Information Technology, Osaka, Japan, pp. 322–329 (2006)
4. Ni, J., Lin, J., Harrington, S.J., et al.: Designing File Replication Schemes for Peer-to-Peer File Sharing Systems. In: IEEE International Conference on Communications, Beijing, China, pp. 5609–5612 (2008)
5. Kermack, W.O., McKendrick, A.G.: Contributions to the mathematical theory of epidemics. In: Proceedings of the Royal Society. Series A, vol. 115, pp. 700–721 (1927)
6. Arnio, J., Davis, J., Hartley, D., et al.: A multi-species epidemic model with spatial dynamics. Mathematical Medicine and Biology 22(2), 129–142 (2005)
7. Arnio, J., Jordan, R., van den Driessche, P.: Quarantine in a multi-species epidemic model with spatial dynamics. Mathematical Biosciences 206(1), 46–60 (2007)

Research on Earthquake Security Evaluating Using Cellular Neural Network

Zhou Chang-xian and Zheng Shao-peng

Xiamen Seismic Survey Research Center
zhoungxiang@163.com

Abstract. Earthquake security evaluating is very important to forecast earthquake, and minimize earthquake damage. In this paper, we propose a novel earthquake security evaluating algorithm by cellular neural network. We introduce nonlinear weight functions based cellular neural network algorithm to evaluate earthquake security by minimizing relative mean square error of a given sample. To validate the effectiveness of our algorithm, we conduct a experiment on 20 samples by seven security evaluating factors. Experimental results show the effectiveness of the proposed algorithm.

Keywords: Cellular Neural Network, Earthquake, Security Evaluating, Earthquake Magnitude.

1 Introduction

As earthquake occurrence is one of the significant events in nature which leads to both great financial and physical harms, it is of great importance to evaluating earthquake security of a given region effectively before earthquake happens. Therefore, accurate earthquake security evaluating is urgently required to minimize earthquake damage. Unfortunately, most of the phenomena related to earthquakes have not been cleared.

The destination of earthquake security evaluating is to forecast an earthquake of a specific magnitude will occur in a given place at a given time. As we all know that the mechanism of earthquake is very complex. In addition, some factors can be used in earthquake security evaluating, which are water level changes of wells, temperature changes of spring waters, Radon emission changes, earthquake vapor, changes in earth magnetic field and so on. Hence, with the factors been input into the a Cellular Neural Network based earthquake security evaluating model, the earthquake risks of these factors can be evaluated by calculating the output of the model.

The rest of the paper is organized as follows. Section 2 introduces the earthquake security evaluating problem. Section 3 presents our approach to evaluate earthquake security by cellular neural networks. In Section 4, we conduct experiments to show the effectiveness of the proposed approach. Section 5 concludes the whole paper.

Z. Du (Ed.): Proceedings of the 2012 International Conference of MCSA, AISC 191, pp. 123–127.
springerlink.com

2 Overview of Earthquake Security Evaluating

As earthquake security evaluating problem is very important, many researchers have done many works on it. In this section, we will show several methods to solve earthquake security evaluating problem.

One approach to evaluate earthquake security is to explore what causes earthquakes and analyze large amounts of historical data to create a mathematical model of the earth. This approach works well for weather forecasting and is called Numerical weather prediction. The theory of plate tectonics was born in the mid 1960s, and confirmed in the early 1970s, However, measuring underground pressures and underground rock movement is more difficult and more expensive than measuring atmospheric pressure. Therefore, this method is not satisfied by researchers.

Another approach it to identify, detect and then measure some kind of phenomena on the earth. Possible precursors of earthquakes under investigation are seismicity, changes in the ionosphere, various types of EM precursors including infrared and radio waves, radon emissions, and even unusual animal behavior. It has been prove that this method is effective, therefore, we adopt the main idea of this method in our earthquake security evaluating research.

3 Evaluating Earthquake Security by Cellular Neural Networks

The Cellular Neural Networks(CNN) model was first proposed by Chua and Yang in 1988 [1]. In recent years, CNN has been widely used both in research field and applications [2][3]. Stability analysis about the standard CNN has been proposed in [4]. Because popular nonlinear qualitative analysis methods always require the activation function to be differentiable, and the CNN neuron activation function is continuous but non-differentiable, the qualitative analysis of CNN turns out to be difficult. Therefore, we determine to use CNN in our earth security evaluating model.

In this section, we will give the formal description of CNN as follows.

$$x_i'(t) = -\alpha_i x_i(t) + \sum_{j=1}^{n} b_{ij} f_j(x_j(t)) + \sum_{j=1}^{n} c_{ij} g_j(x_j \beta(t)) + d_i,\ \alpha_i > 0, i \in [1, n] \tag{1}$$

where $\beta(t) = \theta_i$ and $t \in [\theta_i, \theta_{i+1})$ is satisfied. The main idea of our approach lies in that we use nonlinear weight functions based CNN algorithm to evaluate earthquake security as follows.

$$x_i(t_{n+1}) = -x_i(t_n) + \sum_{\lambda=0}^{T} \sum_{k=1}^{K} \sum_{j \in S_i(r)} \alpha_j^{(k)(\lambda)} \cdot x_j^k(t_{n-\lambda}) \tag{2}$$

where $S_i(r)$ represents cell S_i 's sphere influence. Considering a direct neighborhood interaction, the 1th layer can be regarded as a predictor by Eq.3.

$$\overline{x}_i^{l}(t_{n+1}) = -x_i^l(t_n) + \sum_{l^*=1}^{2} \sum_{j=-1}^{i+1} \alpha_j^{l^* l}(x_{i+j}^{l^*} \cdot (t_n)) + \sum_{l^*=1}^{2} \sum_{j=-1}^{i+1} \hat{\alpha}_j^{l^* l}(x_{i+j}^{l^*} \cdot (t_{n-1})) \tag{3}$$

where $\alpha_j^{l^* l}$ and $\hat{\alpha}_j^{l^* l}$ represent the couplings between layer l^* and l. The dynamics of CNN algorithm require a gene with 72 parameters. Afterwards, in the optimization procedure , relative mean square error for sample h is defined as follows.

$$err(h) = \sum_l \sum_n \sum_i \frac{(x_i^l(t_n) - \hat{x}_i^l(t_n))^2}{x_i^{l2}(t_n)} \tag{4}$$

In our earthquake security evaluating method, the final destination is to minimize $err(h)$ in Eq.4.

4 Experiments

In order to test the method proposed in this paper, general BP ANN[5] and GA-BP ANN[6] are respectively adopted to predict earthquake comparing the performance of our method, and experimental results are also analyzed in this section. We choose 20 samples to make performance evaluating, and seven security evaluating factors are adopted as follows(Shown in Table.1).

Table 1. Factors used for earthquake security evaluating

Factor 1	Earthquake accumulation frequency about magnitude larger than 3 in six months
Factor 2	Energy accumulation in six months
Factor 3	b value
Factor 4	Unusual earth swarms number
Factor 5	Earthquake banding number
Factor 6	Current state is in activity cycle or not
Factor 7	Magnitude in relative area

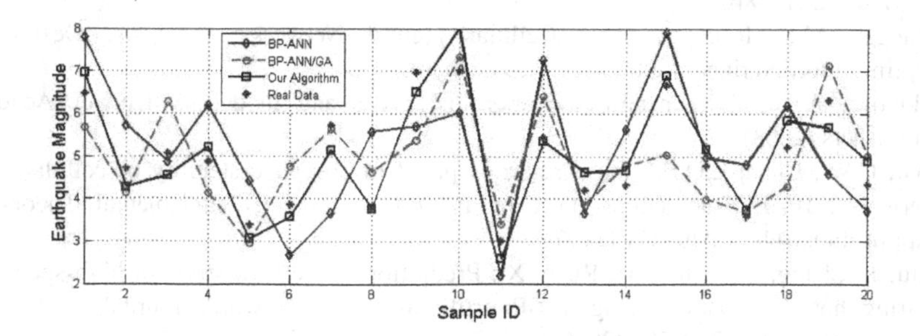

Fig. 1. Earthquake magnitude evaluating results for different methods

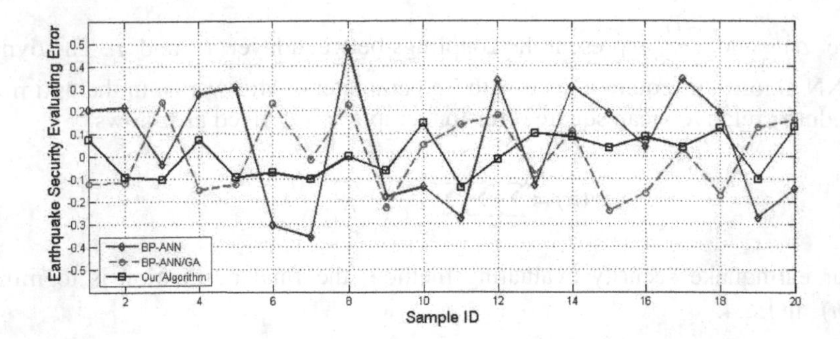

Fig. 2. Earthquake security evaluating results for different methods

From Fig.1 and Fig.2, we can see that the performance of earthquake magnitude of our algorithm is superior to BP-ANN and BP-ANN/GA based approach. The average error rate for Our algorithm, BP-ANN based approach and BP-ANN/GA based approach are 0.084, 0.236 and 0.146 respectively.

5 Conclusions

This paper present a new earthquake security evaluating algorithm by cellular neural network. We adopt nonlinear weight functions based CNN algorithm to evaluate earthquake security by minimizing relative mean square error of a given sample. We design a experiment on 20 samples by seven security evaluating factors and experimental results show that the performance of our algorithm is better than other two traditional methods.

References

[1] Chua, L.O., Yang, L.: Cellular neural networks: Theory. IEEE Trans. Circuits Syst. 35, 1257–1272 (1988b)

[2] Hänggi, M., Moschytz, G.S.: Cellular Neural Networks: Analysis, Design and Optimization. Springer (2000)

[3] Slavova, A.: Cellular neural networks: dynamics and modeling. Kluwer Academic Publishers (2003)

[4] Wu, C.W., Chua, L.O.: A more rigorous proof of complete stability of cellular neural networks. IEEE Transactions on Circuits and Systems I: Fundamental Theory and Applications 44(4), 370–371 (1997)

[5] Liu, J., Chang, H., Hsu, T.Y., Ruan, X.: Prediction of the flow stress of high-speed steel during hot deformation using a BP artificial neural network. Journal of Materials Processing 103(2), 200–205 (2000)

[6] Wong, M.L.D., Nandi, A.K.: Automatic digital modulation recognition using artificial neural network and genetic algorithm, vol. 84(2), pp. 351–365 (2004)

[7] Botev, E.A., Glavcheva, R.P.: Modern earthquake monitoring in central Balkan region. Recent Advances in Space Technologies, 208–213 (2003)

[8] Niwa, S., Takumi, I., Hata, M.: The precursor signal detection from electromagnetic waves for predicting great earthquakes using Kalman filter. In: IEEE International Geosciences and Remote Sensing Symposium, pp. 3620–3622 (2003)
[9] Leach Jr., R.R., Dowla, F.U.: Earthquake early warning system using real-time signal processing, Neural Networks for Signal processing. In: Proceedings of the 1996 IEEE Signal Processing Society Workshop, pp. 463–472 (1996)
[10] Lakkos, S., Hadjiprocopis, A., Comley, R., Smith, P.: A neural network scheme for Earthquake prediction based on theseismic electric signals. In: Proceedings of the 1994 IEEE Workshop, pp. 681–689 (1994)

Research on Zigbee Based Resources Addressing Technology in Internet of Things

Zhang Chunzhi, Li Haoru, Hu Guangzhou, Bao Qingpeng, and Geng Shu

Department of Information Engineering and Computer Technology,
Harbin Institute of Petroleum, China

Abstract. Internet of things is becoming more and more important in both industrial and consumer sides according to development of network technologies, especially the high development of mobile internet. The resources size is also becoming larger and larger. Since all resources are allocated by addressing system, it would cost a long time in current solutions with a higher repletion frequency. Therefore, this paper researched on Zigbee based resources address system and designed its addressing procedures, network establishment strategy and data process system. After stimulate with conventional distributed addressing system and centralized system, the result shows this paper designed Zigbee based resources addressing system can work well with a lower process time and repletion frequency.

Keywords: Internet of things, Zigbee, Resource addressing, Mobile internet.

1 Introduction

Internet of Things is an important part of new generation of information technology. It is an internet connecting with objects. This is two meanings: the first, internet of things extends and expends contents based on internet; the second, its user side connects all things to information exchange and communicate. The definition of internet of things is an internet that connects all things according to contract agreement to information exchange and communicates to realize intelligent identification, orientation, supervisory control and management of things through information sensing equipment such like radio frequency identification (RFID), smart sense, global positioning system and laser scanner. Internet of things widely applies all kinds of perception technologies. There are mass of types of sensor which can capture different information contents and forms in the internet of things. Besides, data is real-time captured by sensor, because sensor can capture information data according to certain frequency cycle to update data. The core of internet of things is still based on internet. It just connects kinds of wire and wireless network with internet and sends things' information out accurately. The way of sending information which captured by sensor in internet of things is network transmission generally. In the process of transmission, in order to guarantee the safety and timeliness, internet of things must sets and adapts types of heterogeneous networks and agreements.

Internet of things not only provides sensor connection, but also intelligent controls things, because it has the ability of intelligent processing. Internet of things extends its

Z. Du (Ed.): Proceedings of the 2012 International Conference of MCSA, AISC 191, pp. 129–135.
springerlink.com

application fields through connecting sensor with intelligent processing and using mess of intelligent technologies such like cloud computing and pattern recognition. Another way to discover new application fields and patterns is to adapt different users' needs by handling valuable data captured by sensor.

2 Zigbee Technology Synopsis

2.1 Current Zigbee Technology

Wireless communication technology works for improving wireless transmission rate. In the application field of wireless communication, data transmission quantity captured by sensor is small so that present wireless communication technology increases the complexity of the design and the cost. Therefore, Zigbee protocol is proposed to meet the communication requirement of sense and control equipment in wireless network. It successfully solves the communication problem between sense equipment and control equipment. Zigbee protocol based on IEEE 802.15.4 standard, but expands IEEE and standardizes network layer protocol and API, because IEEE only process low-level physical layer protocol and MAC protocol. Every Zigbee network node can support 31 sensors or controlled devices at most, and every sensor or controlled device has 8 different interface modes to collect and transfer digital quantity and analog quantity. The main technological character of Zigbee is showed by chart 1.

Chart 1. The main technological character of Zigbee

Character	Value
Data Rate	868 M Hz : 20 kbps 915 M Hz: 40 kbps 2.4 G Hz : 250 kbps
Communication Range	20 – 100 m
Communication Delay	> 15 ms
Number of Channel	868 / 915 M Hz : 11 2.4 G Hz :16
Addressing System	46 bit IEEE address, 8 bit network address
Channel Access	CSMA / CA, Time CSMA / CA
Temperature	-40°C - 85°C

2.2 Zigbee Instance in Internet of Things

Zigbee protocol defines two types of physical equipment which are used peer to peer. Those are full function equipment and reduced function equipment. Full function equipment can support any topological structures as router, network coordinator or terminal node and communicate with any equipment as controller. Reduced function equipment just supports star topology structure and can't be router or network coordinator but communicate. Therefore, Zigbee can constitute three network topological structures: star network, tree network and mesh network.

2.3 Disadvanteges of Current Solutions

Since the high development of network, especially mobile internet devices, the demand on Zigbee and internet of things is becoming higher and higher, especially the resource addressing technology. Current solutions cannot meet higher demands in the following areas: On the one hand, Internet workload is becoming heavier and heavier, and internet of things is becoming more and more lager, how to address resources quickly is a challenge for current solution. On the other hand, Current solution cannot handle large amount of things effectively, and differentiate things according to their category and resources.

3 Research on Zigbee Based Resource Addressing Technology

3.1 Resources Address Technology

Resource addressing models of internet is based on the essential rule and relationship theory of internet resource addressing technology. Therefore, the resource addressing model can divided into two parts, one is level iteration model and the other is application structure model. The level iteration model represent for the network resource addressing technology essential rule theory.

In level iteration model, iteration means resources get the address by resource name and then use this address as resource name to get the new resource address. The output of iteration can be used as the resource name of the next level iteration and the end of iteration is the direct address of network. Level means each iteration strut a level in resource addressing system and also each iteration is also an independent resource address sub system. Each level of resource addressing system didn't need to know information of every level, just need to know adjacent level interfaces. Therefore, this mechanism reduce the system coupling factor and it makes each level can handle and process information belongs to them and take little effort on the system iteration among different levels. The instance method in each level of resources addressing system can also be different; it can also be same for different levels. A sub resource addressing system can be reused by a set of bottom level sub system and also be recalled by a set of up level sub system. The system architecture of level iteration model is showed in figure 1.

3.2 Resources Addressing Procedures

In the level iteration model, the name of resource in layer N represented as F_N, resources address represented as A_N. Then, the resources address and resource name can be expressed in namespace as:

$$NameSpace_{F_N} = \{F_1, F_2, F_3, \cdots, F_k\} \tag{1}$$

$$NameSpace_{A_N} = \{A_1, A_2, A_3, \cdots, A_k\} \tag{2}$$

We used G_N as the network resource addressing function. These two parameters are qualified by the following formula:

$$NameSpace_{A_N} = G_N \bullet NameSpace_{F_N} \tag{3}$$

In this formula, $F_i = F_j \Leftrightarrow G_N(F_i) = G_N(F_j)$.

At last, if the number of levels is M, the level iteration model can be expressed as:

$$NameSpace_{A_N} = G_N(G_{N-1}(\cdots(G_K(NameSpace_{F_K})))) \quad (4)$$

In this formula, $1 \le K < N \le M$. When the number of N equals to M, resources addresses in this level is the MAC address of resources.

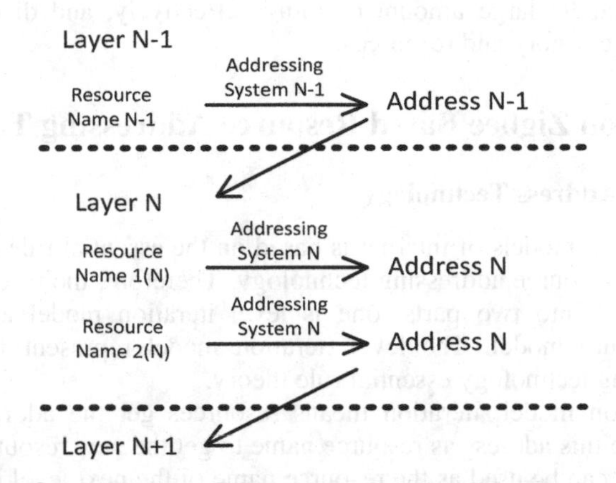

Fig. 1. System architecture of level iteration model

3.3 Network Establishment Strategy

The main process of network establishment strategy is the establish process of network coordinator. Coordinator is the key component to establish network. The process of network establishment is initialized by a primitive in application level called NetworkFormation.request. Then, establish the network by function NetworkFormationRequest in application level. The message sequence chart is showed in figure 2.

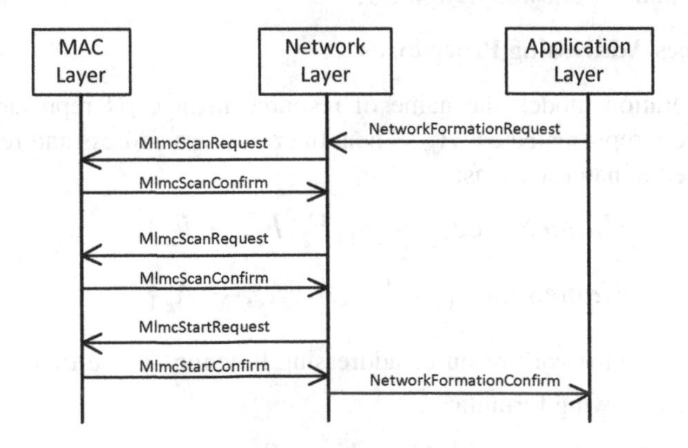

Fig. 2. The message sequence chart of network establish process

Application layer initialize a network formation request to network layer, and then the network layer would lunch an addressing scan request to MAC layer, after process energy detect scan, the MAC layer send addressing scan confirm message to MAC layer. The Network work layer send another address scan request to MAC layer for initiative scan, and then the MAC layer send a scan response back after successful. And then, the network layer send scan request for addressing selection and MAC layer send a confirm message after successful allocated. Finally, the network layer send confirm message named NetworkFormationConfirm to application layer and the network establishment process successfully.

3.4 Data Sending and Receiving

Data sending and receiving mission are realized by Process Event. Users can call function send data function to send data to Zigbee network. The input parameter destination is the 16 bit network address of target devices. If the setting is null, and the real target address needs to be searched in table. The parameter length is the length of sending data. There also need a point to the sending address and sending option. If the sending radius is zero, they are sending in random. And then call function data request to packaging the data.

4 Stimulate Experiment and Validation

In order validate the efficacy and ability of this paper designed Zigbee based resource addressing technology; we designed a set of experiment to test it. We use the OID code as items standard identification number, the mini - mCode as items numbers. The hardware environment is on a personal computer which installed Linux operate system and with an Intel core processor. After configure the environment successfully, we test the resource addressing system by different size of sample with its number from 500 to 5000. We then compared Zigbee based addressing system with distributed addressing system and centralized system. We had tested these three kinds of method three times and choose the average value of each method which is tested in three times. The test result is showed in figure 3. We also test the

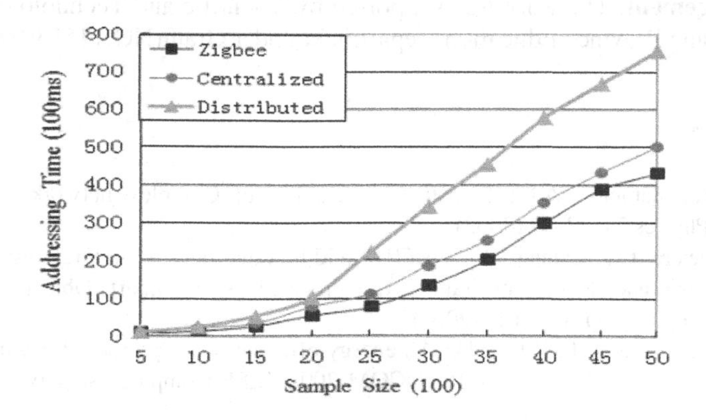

Fig. 4. Address times of Zigbee based system

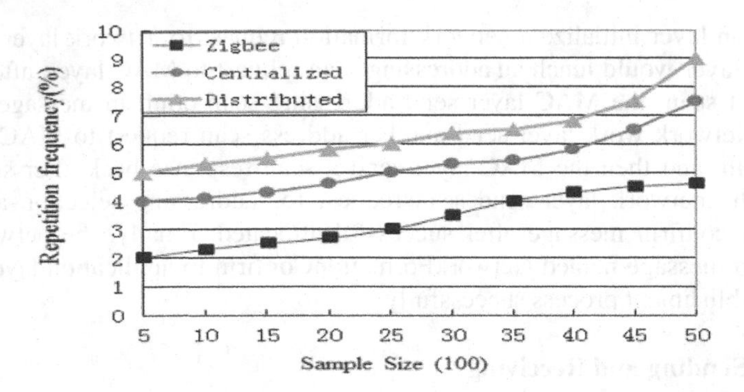

Fig. 5. Repetition frequency of Zigbee based system

repetition frequency of this paper designed Zigbee based resource addressing model. The result is showed in figure 4, according to the size of sample increasing, the repetition frequency reduced aligns. Zigbee based addressing model keep low repletion frequency compared with distributed addressing system and centralized system.

5 Conclusions

According to the high development of internet technology, things of internet is becoming more and more important in daily life and has become a common way in industrial and consumer electronics. But conventional addressing system can't allocate large amount of things effectively. Therefore, these papers research on Zigbee based resource address system, designed its addressing process, resource addressing procedures and data process. After experiment in emulate environment, it shows this paper designed Zigbee based resource address system works effectively with low repletion frequency. The result gave a reference for researching on resource address system.

Acknowledgement. The work was supported by Scientific and Technological Project of Heilongjiang Province Education Department under Grant(NO.11553018).

References

[1] Albert, R., Barabasi, A.L.: Statistical Mechanics of Complex networks. Reviews of Modern Physics 74, 47–97 (2002)

[2] Floerkemeier, C., Lampe, M.: RFID middleware design -addressing application requirements and RFID constraints. In: Proceedings of Smart Objects Conference, Grenoble, France, pp. 118–121 (2003)

[3] Ye, W., Heidemann, J., Estrin, D.: An energy-efficient MAC protocol for wireless sensor network. In: Proceedings of the INFOCOM 2002. IEEE Computer Society, SanFrancisco (2002)

[4] Ondrej, S., Zdenek, B., Fetal, P.: Zigbee Technology and Device Design. Networking. In: 2006 International Conference on International Conference on Systems and Conference on Mobile Communications and Learning Technologies (2006)
[5] Arici, T., Altunbasak, Y.: Adaptive Sensing for Environment Monitoring Using Wireless Sensor Networks. In: IEEE Wireless Communications and Networking Conference (WCNC), Atlanta, GA (2004)
[6] Polastre, J., Hill, J., Culler, R.: Versatile Low Power Media Access for Wireless Sensor Networks. In: Proceedings of 2nd ACM Conference on Enbedded Network Sensor System, Baltimore, MD, USA, pp. 95–1070 (2004)

A Scalable Model for Network Security Situation Based on Endsley Situation Model

ChenHao Deng

Institute of Electrical Engineering & Automation,
Tianjin University, China

Abstract. Network Security Situation (NSS) awareness provides a macroscopical safety management, yet situation awareness is lacking in unified standard and norm. Base on the research by Endsley, this paper proposes a scalable model for network security situation and introduces the concept into the field of network security. In situation data processing, this model introduces the space/time knowledge base to regulate the situation extraction process, and takes the situation as entity object for modeling. The model includes attack frequency, attack time and space information to simplify the situation extraction process. This model is evaluated using real data to prove its usefulness and efficiency.

Keywords: network security situation, Endsley situation model.

1 Introduction

As a hot research topic, the concept of situational awareness was originated from human factors in automation, and was introduced into network security to form the Cyberspace Situational Awareness (CSA) to become the development direction of network security management and make more and more industry experts recognize the macroscopical network security management[1]. Currently, the network security situation awareness is still lacking in uniform standards. Also, the researches on the understanding of security situation are much the different, resulting in the diversified ways to achieve network situation awareness, such as quantifying network risk to take risk index as an indicator for situation assessment[2]; and multi-behavior information fusion framework together with quantitative and qualitative analysis also belongs to this category[3]. In addition, the qualitative assessment scheme based on attack scenarios [4], and the visual flow situation analysis using NetFlows, have enriched the implementation forms of CSA[5]. The differences among the implementation of the above schemes are a little great, and the standard situation awareness process is lacking. Due to the variation characteristics of NSS, the NSS prediction system shows more like a grey system. Therefore, the grey model is usually utilized to forecast the security situation. Aiming at the shortages existing in the prediction model based on conventional grey model, this paper introduces the traditional and mature Endsley Situation Model (ESM)[6] into the field of network security, and improves it to form the more practical situation extraction solution to network security situational awareness, and adopts adaptive parameters and filling method to enhance the precision.

Z. Du (Ed.): Proceedings of the 2012 International Conference of MCSA, AISC 191, pp. 137–143.
springerlink.com

2 Proposed Situation Model Analysis and Framework Design

2.1 Situation Model and Process Framework

The events transacted by ESM are different from the underlying events in network security, but its data processing can be used by NSS analysis for reference. The objects in the environment are defined by ESM as threat units, all units comprising a multi-unit group includes the parameters information of interest. Certain threat units are as the input of situation extraction module to compare with the historical situation knowledge and ultimately obtain the standard situation information. According to ESM processing, this paper proposes the framework of NSS extraction, as shown in Fig. 1.

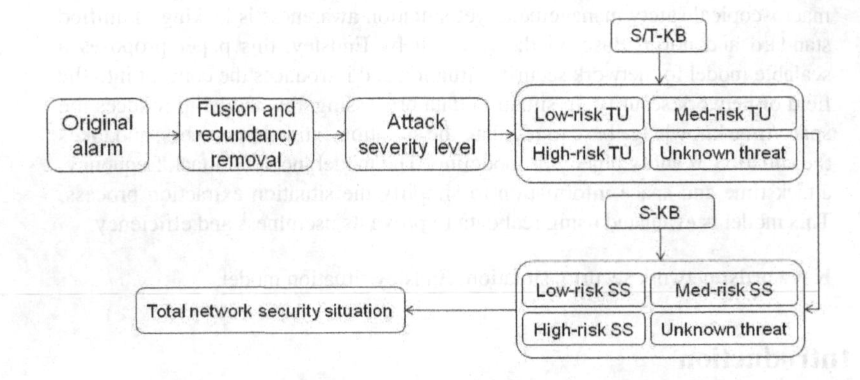

Fig. 1. Framework of network security situation extraction

In situation analysis process, network attacks are divided into high-risk, mid-risk, low-risk and unknown threat according to attack severity, and categorized into four threat units of $P_{high}(t)$, $P_{medium}(t)$, $P_{low}(t)$ and $P_{unknown}(t)$ to construct the total network security situation, that is:

$$S(t)=\{P_{high}(t),P_{medium}(t),P_{low}(t),P_{unknown}(t)\} \tag{1}$$

Where $$P_X(t)=\{Count,TimeIndex,SpaceIndex\} \tag{2}$$

In Equations (1) and (2), t is evaluating time duration, Count is event statistical value in the evaluating time duration, TimeIndex and SpaceIndex are time factor and space factor of this attack, respectively. The situation of one threat unit in one evaluating time duration can be denoted as:

$$\begin{aligned}&P_X(t)\times S/T-KB=\\&\{Count,TimeIndex,SpaceIndex\}\times S/T-KB\rightarrow\\&Count\times W_T(TimeIndex)\times W_S(SpaceIndex)\end{aligned} \tag{3}$$

where $W_T(TimeIndex)$ and $W_S(SpaceIndex)$ show the time result and space result of weighted coefficients assignment function, respectively.

As above mentioned In a word, the total situation extraction process can be expressed as:

$$S(t)\times S-KB=$$
$$\{P_{high}(t),P_{medium}(t),P_{low}(t),P_{unknown}(t)\}\times S-KB\rightarrow$$
$$\left[P_{high}(t)\ P_{medium}(t)\ P_{low}(t)\ P_{unknown}(t)\right]\times$$
$$\left[10^{w_{high}}\ 10^{w_{medium}}\ 10^{w_{low}}\ 0\right]^T=$$
$$P_{high}(t)\times 10^{w_{high}}+P_{medium}(t)\times 10^{w_{medium}}+P_{low}(t)\times 10^{w_{low}} \tag{4}$$

where S-KB is $\left[w_{high}\ w_{medium}\ w_{low}\ 0\right]^T$. In view of the damage of one high-risk network attack being much more than that of three low-risk attacks, this paper modifies Equations (4) to be $\left[10^{w_{high}}\ 10^{w_{medium}}\ 10^{w_{low}}\ 0\right]^T$ to make the situation extraction process suitable for the actuality.

2.2 Security Domain Partitioning and the Assignment of SpaceIndex

Network system often partitions the security domains to construct different trust levels. When different security domains suffer the same attack, the hazards faced by the network system are different. SpaceIndex in threat unit can represent different security domains, which can be as input of WCAF to obtain the importance indicators of different security domains. Security domain partitioning rule is shown in Tab. 1.

Table 1. Security Domain Level Partitioning and Indicators Assinment

Security domain level	Equipment and services	Security level indicators
High	Web Server, DNS, etc	3
Medium	Proxy, Non-important DB, etc	2
Low	Web Browser, Console, etc	1

2.3 Alarm Fusion Module

Traditional Intrusion Detection System (IDS) technology is based on misuse detection, adopting signature) match. Its advantages are high accuracy and relatively simple implementation (such as Snort, etc.); however, it can't detect a new attack, which causes great risk of network security. As one of IDS technology, anomaly detection can detect the intrusion which has never occurred, yet with the problem of high false alarm rate. Network Anomaly detection is a much difficult problem with more complex high-level application protocol, resulting in the difficulties of modeling and simulation; the real-time requirements is relatively high, also resulting in the bottleneck in detection performance. In order to solve the bottleneck problem, a lot of related work has been carried out in the modeling process of detection model, for example, the signatures are graded based on the costs and establish the resource-sensitive intrusion detection model in signature extraction process. The information extracted from original audit data that can describe the records outline are called as signatures. Signature extraction is an important step in constructing ID models[7]. IDS will generate a lot of redundant low-risk alarms. If without redundancy removal, the flood of low-risk alarms will affect the identification of high-risk attacks by situation analysis. This paper chooses the sliding time window to filter redundancy and takes different lengths of time windows to compare the effects of redundancy removal, so that retain the useful information as far as possible while eliminating the excessive redundancy.

2.4 One Application in RFID Service

In order to bring forth the visitor-oriented service in the mega-event, the scenario of showing the way to blind men is proposed for IDS application in RFID service. We can imagine that the visitor's RFID tag has stored his personal information. The RFID readers throughout the venue can scan his tag and record the pavilion he has visited and the exhibits he has seen, which are his new private information. Blind paths are paved in the venue, and RFID tags are placed beneath at those main nodes. Blind men visitors need borrow the special earphones and crutches at the entrance, and a mobile RFID reader is inside the crutch. When the blind man walks along the blind patch, the reader inside his crutch reads out the nearby tag information, and delivers the important place information to his earphone, such as current position, road introduction, attentions, surrounding environment. For example, when he arrives at a corner at the blind path, his crutch can read the RFID tag beneath and the information is, e.g., "China pavilion, 30 meters ahead". Another example, when he arrives at an intersection at the blind path, the tag beneath gives information "Please turn left to a café, go ahead to Singapore pavilion, and turn right to a shutter bus stop". According to the private information in blind man's RFID ticket, the above information should be provided in the language he is familiar with.

3 Optimized Evaluation for Scalable Modela

3.1 Signature Extraction

When using data mining techniques to analyze the raw network data, the raw data should firstly be pre-processed. This process is to cut and normalize the data according to algorithm needs. Audit data has several attributes, so that Extracting and dealing with the certain attributes suitable for application can improve the detection accuracy and efficiency while reducing the complexity of the process, thus providing the theoretical basis and data support for future intrusion detection modelling. Signature extraction is an optimization problem, not a simple quantity problem. The details can be found in reference. Data set contains 24 kinds of common attacks which can be divided into four categories: DoS attack (Denial of Service), R2L attack (unauthorized remote access), U2R attack (local unauthorized access) and Probe (port scan). Keep the number of total records changeless, dynamically adjust the signature species and observe the changes of detection rate and false detection rate; paint the assessment maps. In order to obtain accurate evaluation statistical results, arbitrarily extract data from KDD data set to obtain four data sets and evaluate them, as shown in Tab. 2. The assessment tool is K-Means algorithm developed using VC++.

Table 2. Example Numbers in Ecord Set and Intrusion Example Numbers

	Example numbers in record set	Intrusion example numbers
Data set 1	15,236	1,256
Data set 2	15,809	976
Data set 3	15,024	56
Data set 4	16,466	1,832

Therefore, in the process of network intrusion detection modelling, as long as the modelling is according to the time windows statistical data signatures, the accuracy and efficiency can just be ensured. Similarly, in evaluating a new algorithm, the signatures based on time windows statistic can be extracted for verification. In the optimization assessment for feature extraction, it is just to simply select the signatures from the aspect of category, and use algorithm comparison to select the different feature categories and test the impacts on detection results, while paying less attention to the deeper considerations. For example, when the algorithm calculates the Individual signatures, different signatures has different impacts on algorithm results and detection efficiency. Each record in the record set has concentrated the signature information of 3 categories of information, without considering the interaction and constraints among various categories. IDS development will require higher efficiency, ease of use and expansion ability. The "plug" form will correspond to the new intrusion detection device attached to the Inherent frame, and this calculation is significantly less than the re-fusion to generate a new unified model. This paper about signature extraction can provide basis to new detection model, and use as little signature and category as possible to finally identify the new Intrusion and reduce the computational complexity.

3.2 Model Simulation and Experimental Evaluation

In the simulation environment of campus network, the LAN is connected directly Internet, with the deployment of Web, OA and Proxy services. According to the deployed services and host resources Importance, the network segment is divided into two security domains, namely SD-1 and SD-2.

Data acquisition continues for 5 days, obtaining a total of about 305,000 data. Randomly choose 3 days of data, which are labelled Day1, Day2 and Day3, respectively, and carry out redundancy removal of the raw data. When 20s and 30s are selected as the length of time window, the effects of data reduction are similar, so this paper selects 20s as the window length. Refer to Snort User Manual for attack severity classes to access to the disaggregated data daily and statistical value of each class.

TimeIndex and SpaceIndex in situation analysis depend on the based on the evaluation intervals and security domain partitioning. This paper makes one day as the assessment period, and assigns TimeIndex in the range from 1 to 24 according to the allocation of hours. Importance of evaluation intervals is categorized into 3 levels based on network flow. Tab. 3 gives the partitioning of evaluation interval importance levels and normalized weighted coefficients.

Table 3. Evaluation Interval Partitioning and Weighted Coefficients

Evaluation intervals	Importance levels	Weighted coefficients
0~1, 1~2, 2~3, 3~4, 21~22, 22~23, 23~0	1	0.0192
4~5, 5~6, 6~7, 7~8, 19~20, 20~21	2	0.0385
8~9, 9~10, 10~11, 11~12, 12~13, 13~14, 14~15, 15~16, 16~17, 17~18, 18~19	3	0.0577

4 Conclusion

The experimental results accord with the original intentions of design model and framework, and the obtained visual effects can help users identify potential risks. Despite the resulting a large number of low-risk alarms, the high-risk alarms still decide the situation changes. Although a large number of alarms occur in low-level security domain, the situation in high-level security domain still plays a decisive role on the situation changes. While ensuring that the high-risk attacks can be highlighted, this paper does not ignore the impact of low-risk attacks, which is much helpful for the users to detect the potential problems in the network in time and then make decisions. In this paper, we have discussed the scalable network situation awareness, and improved the model. But we need to notice that the unknown threat situation awareness is more complex. The issue about the decomposition of the united situation awareness and forecasting of the complicated situation awareness should be researched in the future.

References

[1] Bass, T.: Intrusion detection systems and multi-sensor data fusion: creating cyberspace situation awareness. Communications of the ACM 43(4), 99–105 (2000)

[2] Ren, W., Jiang, X.H., Sun, T.F.: RBFNN-based prediction of networks security situation. Computer Engineering and Applications 42(31), 136–139 (2006)

[3] Guo, Z.J., Song, X.Q., Ye, J.: A Verhulst model on time series error corrected for port throughput forecasting. Journal of the Eastern Asia Society for Transportation Studies 6, 881–891 (2005)

[4] Lai, J.B., Wang, H.Q., Zhao, L.: Study of network security situation awareness model based on simple additive weight and grey theory. In: Proceedings of 2006 International Conference on Computational Intelligence and Security, pp. 1545–1548. IEEE Press, Hangzhou (2006)

[5] Yin, X.X., Yurcik, W., Slagell, A.: The design of VisFlowConnect-IP: a link analysis system for IP security situational awareness. In: Proceedings of the Third IEEE International Workshop on Information Assurance, pp. 141–153 (2005)
[6] Endsley, M.R.: Toward a theory of situation awareness in dynamic systems. Human Factors 37(1), 32–64 (1995)
[7] Sengupta, S., Andriamanalimanana, B.: Towards data mining temporal patterns for anomaly intrusion detection systems: technology and applications. In: IEEE International Conference on Intelligence Data Acquisition and Advanced Computing Systems, pp. 45–51 (2003)

Information Diffusion Model Based on Social Network

Zhang Wei, Ye Yanqing, Tan Hanlin, Dai Qiwei, and Li Taowei

National University Defense Technology, Hunan Changsha 410072

Abstract. Analyzing the process of information dissemination on online social network is a complex work because of large number of users and their complex topology relationship. Inspired by epidemic dynamics, we proposed two models that focus on the two different aspects of complicacy of the problem. The Improved SI Model pays more attention to topology relationship. In this model, we classify users into $M-m+1$ categories by their degrees and focus on influence of degree distribution of the online social network. From this model we found that more friends have greater influence on receiving message rather than spreading message. In Improved SIR Model, we do not focus on the specific network topology and utilize theory of probability and differential equations to describe the information dissemination process. From the model, we found that the total number of online social network users does not have great impact on the information spreading speed.

Keywords: social network, information diffusion model, SI, SIR.

1 Introduction

A social networking service is an online service, platform, or site that focuses on building and reflecting of social networks or social relations among people [1].However, with the social networking service providing us with a great convenience, it also brings us a great problem. In this paper, we have utilized the SIR model, which is a good and simple model for many infectious epidemics including measles, mumps and rubella [2]. Social networks are just similar to SIR model. The susceptible people are those users who can receive the criminal message. The infectious are those who receive and transmit the message. And the recovered are those who know the covered truths. If they receive the message, they would not transmit it.

2 Improved SI Model for Information Dissemination

Here we use graph theory to describe the network, where nodes denote hot members of the online social network and edges denotes friendship in the online social network. To further describe the features of hot members with different numbers of friends,

Z. Du (Ed.): Proceedings of the 2012 International Conference of MCSA, AISC 191, pp. 145–150.
springerlink.com

we classify users into $M-m+1$ categories by their degrees (number of friends), where m is the smallest degree and M is the maximum degree among all nodes in the network [3].Finally we combine complex network theory and epidemic dynamics theory to propose our model for information dissemination process.

SI (Susceptible Infectious) Model [4] is a classic model to describe epidemic transmission process. Yet epidemic transmission among population is similar to information dissemination in online social network. Here we improved SI model to describe information dissemination.

First we divide all users, i.e. nodes in the network, into two categories: S for susceptible users and I for infectious users. Susceptible users are those that haven't received and believed the crime message and infectious users are those that have.

Then we make a few rules to clarify our model:

- If a susceptible user believes and transfers the crime message, we say he or she becomes infectious.
- If one of a susceptible user's friends becomes infectious, the probability of that user to become infectious within time span Δt is p_1.

We define $S(k,t)$, $I(k,t)$ and $N(k,t)$ to denote the number of susceptible, infectious and all k-degree nodes at time t. Here k-degree means that the node has k adjacent nodes, i.e. the user has k friends. And we define infectious density $\rho^i(k,t)$ and susceptible density $\rho^s(k,t)$. Here

$$\rho^i(k,t) = \frac{I(k,t)}{N(k,t)}, \rho^s(k,t) = \frac{S(k,t)}{N(k,t)}$$

Suppose node j is susceptible at time t, p_{ss}^j donates the probability that j remains susceptible within time span $[t, t+\Delta t]$. Then we have

$$p_{ss}^j = (1-\Delta t p_1)^g \tag{2-1}$$

Where g denotes the number of infectious nodes that are adjacent to node j.

Then we define $w(k,t)$ to denote the probability of a k-degree node connected to an infectious node at time t.Then we have

$$\begin{aligned} w(k,t) &= \sum_{k'=m}^{M} p(k' \mid k) p(i_{k'} \mid s_k) \\ &\approx \sum_{k'=m}^{M} p(k' \mid k) \rho^i(k',t) \end{aligned} \tag{2-2}$$

Here $p(k' \mid k)$ denotes the probability of a k-degree node connected to k'-degree node. And $p(s_{k'} \mid i_k)$ denotes the conditional probability of a k'-degree node to be infectious on condition that it is connected to a k-degree susceptible node. It can be estimated by the infectious density $\rho^i(k',t)$.Recall that m is the smallest degree and M is the maximum degree among all nodes in the network. And the average probability of a k-degree susceptible node to remain susceptible is

$$\overline{p_{ss}}(k,t) = (1 - \Delta t p_1 w(k,t))^k \tag{2-3}$$

Combine Eq. (2-2) and Eq. (2-3) we have

$$\overline{p_{ss}}(k,t) = (1 - \Delta t p_1 \sum_{k'=m}^{M} p(k'|k)\rho^i(k',t))^k \tag{2-4}$$

Alike SI model, we establish our model equations as Eq. (2-5).

$$\begin{aligned} S(k,t+\Delta t) &= S(k,t) - S(k,t)(1 - \overline{p_{ss}}(k,t)) \\ I(k,t+\Delta t) &= I(k,t) + S(k,t)(1 - \overline{p_{ss}}(k,t)) \end{aligned} \tag{2-5}$$

$$\begin{aligned} \frac{I(k,t+\Delta t) - I(k,t)}{N(k,t)\Delta t} &= \frac{S(k,t)}{N(k,t)\Delta t}(1 - \overline{p_{ss}}(k,t)) \\ &= \frac{S(k,t)}{N(k,t)\Delta t}(1 - (1 - \Delta t p_1 \sum_{k'=m}^{M} p(k'|k)\rho^i(k',t))^k) \end{aligned}$$

Let $\Delta t \to 0$, and Taylor expansion the right side we have

$$\frac{\partial \rho^i(k,t)}{\partial t} = p_1 k \rho^s(k,t) \sum_{k'=m}^{M} p(k'|k)\rho^i(k',t) \tag{2-6}$$

Note that $\rho^i(k,t) = 1 - \rho^s(k,t)$, we have our model

$$\frac{\partial \rho^i(k,t)}{\partial t} = p_1 k (1 - \rho^i(k,t)) \sum_{k'=m}^{M} p(k'|k)\rho^i(k',t) \tag{2-7}$$

$$Here\ k = m, m+1, \ldots, M$$

According to Reference [2-2], we have

$$p(k'|k) = \frac{k' p(k')}{\bar{k}} \tag{2-8}$$

Where $\bar{k}$ is the average degree of the social network and $p(k')$ is the degree distribution.

According to Reference [2-3] and our assumption that the degree of nodes subject to power law distributions, we have

$$p(k') = 2m^2 k'^{-3} \tag{2-9}$$

Where m is the smallest degree among the network as stated before .

Combine Eq. (2-7), Eq. (2-8) and Eq. (2-9), we have

$$\frac{\partial \rho^i(k,t)}{\partial t} = \frac{2m^2 p_1 k}{\bar{k}}(1 - \rho^i(k,t)) \sum_{k'=m}^{M} \frac{\rho^i(k',t)}{k'^2} \tag{2-10}$$

$$Here\ k = m, m+1, \ldots, M$$

Eq. (2-10) is the final equation set we have to describe the information dissemination process among the social network.

3 Improved SIR Model for Information Dissemination

SIR Model is also a classic model for epidemic transmission [5]. Here we simulate the principle of SIR Model to establish our improved SIR Model. To simplify the problem, we do not focus on the specific network topology and utilize theory of percentage and differential equations to describe the information dissemination process.

3.1 Variables for Model

We define variables s, i, r and s_2 in Table 3-1.

Table 3-1. Definitions of variables and their initial conditions

Denotes	Description
$s(t)$	s denotes the percentage of susceptible nodes.
$i(t)$	i denotes the percentage of infected nodes
$r(t)$	r denotes the percentage of recovered nodes
$s_2(t)$	s_2 denotes the percentage of susceptible nodes that have seen the refuting message from *Those Who Knew* but do not believe it.

3.2 Model Design

We analyzed the transformation relationship among s, i, r and s_2 in Figure 3-1.

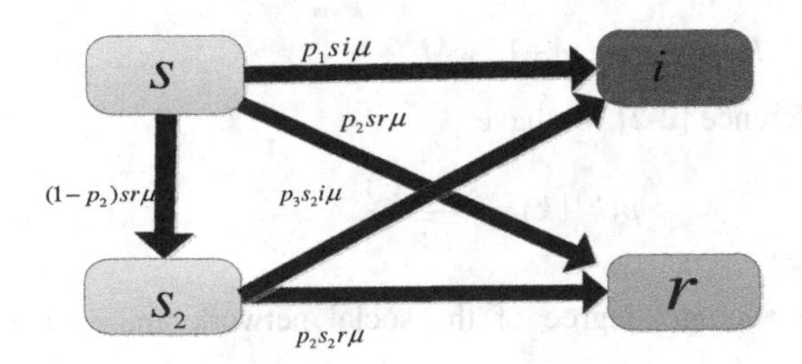

Fig. 3-1. Transformation relationship among s, i, r and s_2

We propose our improved SIR model according to transformation relationship shown in Eq. (3-1).

$$\begin{aligned}
\frac{ds}{dt} &= (-p_1 si - p_2 sr - (1-p_2)sr)\mu, \\
\frac{di}{dt} &= (p_1 si + p_3 s_2 i - \lambda si)\mu, \\
\frac{dr}{dt} &= (p_2 sr + p_2 s_2 i + \lambda sr)\mu, \\
\frac{ds_2}{dt} &= ((1-p_2)sr - p_3 s_2 i - p_2 s_2 r)\mu.
\end{aligned} \tag{3-1}$$

Definitions of parameters are given in Table 3-2.

Table 3-2. Definitions of parameters and their units

Parameter	Description
u	The probability for two logged in friend users to communicate.
λ	The probability for an infectious node to meet a *Those Who Knew* user.
p_1	The probability for a susceptible user to become infectious when he or she meets an infectious user.
p_2	The probability for a susceptible user to believe in *Those Who Knew* when he or she meets a *Those Who Knew* user.
p_3	The probability for a susceptible user who has met but doesn't believe in *Those Who Knew* to believe the crime message when he or she sees a crime message.

4 Conclusion

The information dissemination process in online social network is a complex problem. We proposed two models: Improved SI Model and Improved SIR Model to try to solve the problem.

In Improved SI Model, we classify users into $M-m+1$ categories by their degrees and focus on influence of degree distribution of the online social network. From our model we found that more friends have greater influence on receiving message rather than spreading message.

In Improved SIR Model, we do not focus on the specific network topology and utilize theory of probability and differential equations to describe the information dissemination process. From our model we found that with Those Who Knew in the network, the time required for infectious node to reach 80% is minutes. It can be concluded that with normal distributed age the information spreads slower. When the number of users reaches 10,000,000, the time is also minutes, which indicates that the total number of online social network users does not have great impact on the information spreading speed.

The two models focus on different complicacies of the problem. So their results lack comparability. The model suffers from considerable uncertainty. Because equation set of Improved SR Model is complex and difficult to get analytical or numerical solution, approximate solution may not provide accurate predictions. And Improved SIR Model does not focus on the topology of the online social network. However, it is believed that the behavior of both the Improved SI Model and the Improved SIR model can represent realistic trends.

References

[1] Xu, J., Chen, H.: Criminal Network Analysis and Visualization. Common. ACM 48, 100–107 (2005)

[2] Matthew, J., Marc, M., David, J.: Knowledge Discovery Laboratory Graph Clustering with Network Structure Indices

[3] Acta Phys. Sin. 60 (5), 050501 (2011)

[4] Zhang, Y.-C., Liu, Y., Zhang, H., Cheng, H.: The research of information dissemination model on online social network. Acta Phys. Sin. 60(5), 050501 (2011)
[5] Research on fitting of SIR model on prevalence of SARS in Beijing city. 1. Department of Mathematics, China Medical University, Shenyang 110001, China; 2. Department of Epidemiology, China Medical University, Shenyang 110001
[6] Vazquez, A., Weight, M.: Phys. Rev. E 67, 027101 (2003)
[7] Hu, H., Wang, L.: A brief research history of power law
[8] Hu, H.B., Han, D.Y., Wang, X.F.: Physical A 389, 1065 (2010)

Modeling and Application in Networked Control System Based on Predictive Functional Control with Incremented Weight

Zhang Jun[1,2] and Luo Da-yong[1]

[1] School of Information Science and Engineering,
Central South University,
Changsha 410075, China
[2] Department of Logistics Engineering,
Hu'nan Modern Logistics Occupation Technical College,
Changsha 410100, China
Linecon23@163.com

Abstract. Sensor, network and actuator and etc consist of Networked Control Systems(NCS) that make system flexible but also bring networked delay to system. In real application, networked delay usually is longer than a sampling time of system and affects the whole system's stability. In this paper, it changes the random long delay into certain long delay, provides a method based on Predictive Functional Control(PFC) with incremented weight concerning about the effect of inputs and designs a reasonable controller that make system stable. Finally, it compares with a traditional PID control through simulation and the result proves that takes the method mentioned in this paper can better make system's output stable.

Keywords: Networked Control Systems(NCS), long delay, incremented weight, Predictive Functional Control (PFC).

1 Introduction

Networked control system (NCS) consists of control system sensors, controllers and actuators, etc, and is a closed loop control system through communication network. Compared to the traditional point-to-point control systems, NCS is easy for resource sharing and remote monitoring, is easy to maintain and expand so as to increase the flexibility and reliability of system[1-2]. However, because of the introduction of communication network, network data transmission causes network delay, which affects the control performance of system in some degree and even causes the instability of system. So we can use stochastic control and deterministic control methods to implement. Stochastic control method takes delay as random variable, and use stochastic control principle to design network controller; but deterministic control method sets buffer through the receiver of controller and actuator, and makes network delay to be determination of parameter, changes random delay closed-loop network control system into determinate closed-loop network control system[3-4].

Z. Du (Ed.): Proceedings of the 2012 International Conference of MCSA, AISC 191, pp. 151–158.
springerlink.com

There are many research results about network delay-how to reduce or remove the effect of network delay for the whole system. Reference [5-6] showed that the network delay of network control system was reconstructed to be deterministic delay; reference [7] designed delay predictive compensator of NCS according to discrete model of NCS; reference [8] used a future control sequence to compensate for the forward communication time delay and a model predictor to compensate the delay in the backward channel of NCS; reference [9] designed a novel predictive controller to compensate the effect of network induced delay and data packet dropout, and analyzed the stability of closed-loop system.

In the 1980s, Richalet and Kuntsc proposed Predictive Functional Control (PFC). PFC is better applied to large inertia system, time varying system and delay system. And PFC is also an algorithm based on model reference and uses multi-step prediction, dynamic optimization and feedback correction to ensure good control performance. Itself also has own characteristics which is to make control principle structure so that when the change rate of set value is less than or equal to the threshold value in prediction horizon, control input is preselected linear combination of basis functions, and predicted output is output response linear combination of every basis function, that is:

$$u(k+i)=\sum_{n=0}^{N}\mu_n \bullet f_n(i) \ , \quad y_o(k+i)=\sum_{n=0}^{N}\mu_n \bullet y_n(i) \qquad (1),$$

where, $\mu_n(n=0,...,N, N=p-1)$ is the weighted coefficient of the linear combination of basis functions; $f_n(i)(n=0,...,N)$ is the value of $k+i$ sampling period of basis functions; N is the number of basis functions, and is determined by control model, in this work, $N=p-1$; p is the length of prediction horizon; $y_n(i)$ is the output of every basis function, respectively. The advantages of the method are simper calculation and smaller computational complexity of real-time control, and overcome the irregular control input which is caused by other model predictive control. At the same time this method has better tracking ability and anti-interference capability, etc.

When analyzing NCS of uncertain delay, this work firstly changed random delay into certain delay, secondly used the characteristic of predictive functional control to control NCS of certain delay and took into the error of system output by the performance index of universal predictive functional control, but ignored the impact of control input on control performance. So this work introduced the incremental weighted control method of system input in system performance index function to ensure the stability of system.

2 The Modeling of Certain Long Delay of NCS

The following described that the network delay of NCS mainly contained sensor-controller delay τ_{sc} and actuator delay τ_{ca} and controller computational delay τ_c. With the fast development of modern electronic and computer, controller becomes more and more intelligent, the computing capability of controller becomes stronger and stronger, computing time becomes smaller and smaller, so τ_c is not considered in the following network delay, and whole NCS delay τ_k is $\tau_k = \tau_{sc} + \tau_{ca}$. If $\tau_k > T_s$ (T_s, sampling period), then the whole network delay is named as long delay.

Therefore this work defined the following assumption for NCS: sensor uses time-driven and collects data by sampling period T; actuator uses event-driven and changes the uncertainty of τ_{ca} into certainty by setting buffer once actuator receives the order from controller; controller uses event-driven to receive the data information from sensor and places the information in buffer and makes relevant computing to transfer uncertain network delay into certain delay; if there is not data loss in the working process of the system, we think that the sampling period of the system is a fixed value h. Certain NCS structure diagram Figure 1 is shown:

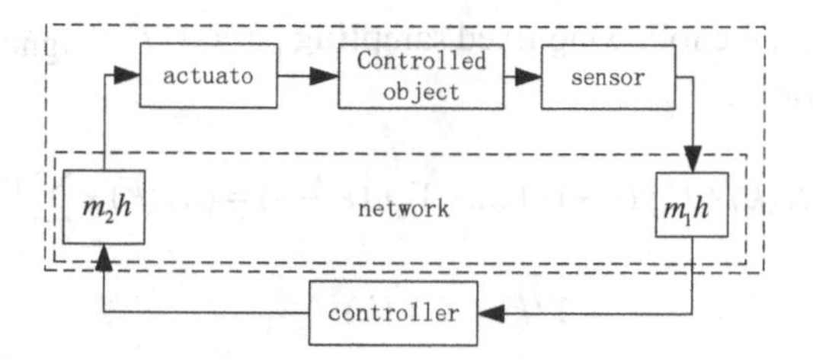

Fig. 1. Structure diagram of certain NCS

Figure 1 took actuator, collected object G_p, sensor and network as a generalized controlled object. The assumed continuous state equations are:

$$\dot{x}(t) = Ax(t) + Bu(t), y(t) = Cx(t), u(t) = -Kx(t-\tau) \qquad (2)$$

where $x(t)$ is the state of controlled object, $u(t)$ is the input of controlled object, $y(t)$ is the output of controlled object, A, B, C are the matrix of appropriate dimension, and $\tau = (m_1 + m_2)h$, h is buffer time which is artificially set, τ is the following new network delay and $\tau > \tau_k$.

The following is to discuss NCS network delay. If NCS network delay $T < \tau < nT$, $n > 1$, a period will appear n piecewise continuous control signals, showed in Figure 2.

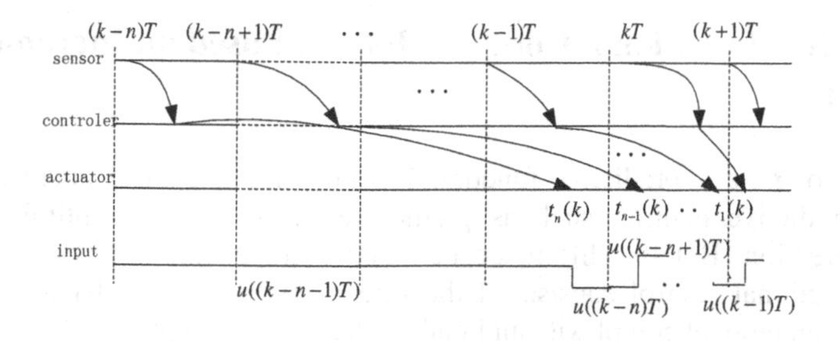

Fig. 2. Signal Sequence of NCS with long delay

On the basis of the above analysis, the control signals of k sampling period appears are $u((k-1)T),....,u((k-n+1)T),u((k-n)T)$, respectively, and the corresponding time of control variable are $kT+t_1(k),....,kT+t_{n-1}(k),kT+t_n(k)$, so Eq.(2) is discrete as:

$$X((k+1)T)=\phi X(kT)+\Gamma_1 u((k-1)T)+....+\Gamma_{n-1}u((k-n+1)T)+\Gamma_n u((k-n)T) \quad (3)$$
$$Y(kT)=CX(kT)$$

where, $\phi=e^{AT},\Gamma_1=\int_{kT+t_1(k)}^{(k+1)T} e^{A((k+1)T-\tau)}d\tau B$,...., $\Gamma_n=\int_{kT}^{kT+t_{n-1}(k)} e^{A((k+1)T-\tau)}d\tau B$.

In Eq (3), when k is different sampling period, there is 0 in the discrete $\Gamma_1,....,\Gamma_n$.

In order to discuss, the following fixed sampling period T is ignored in Eq. (3), the simplified expression is:

$$X(k+1)=\phi X(k)+\Gamma_1 u(k-1)+....+\Gamma_n u(k-n)=\phi X(k)+\sum_{i=1}^{n}\Gamma_i u(k-i) \quad (4)$$

$$Y(k)=CX(k) \quad (5)$$

According to the recursive and inductive of Eq. (4)and Eq.(5), the kth sampling period versus the future $k+p$ discrete state expression was derived from,

$$X(k+p)=\phi^p X(k)+\sum_{i=0}^{n-1}\sum_{j=0}^{p-1}\phi^{p-j-1}\Gamma_{1+j+i}u(k-i)+\sum_{i=1}^{p-1}\sum_{j=0}^{p-1-i}\phi^j\Gamma_{p-j-i}u(k+i) \quad (6)$$

$$Y(k+p)=CX(k+p)=C\phi^p X(k)+\sum_{i=0}^{n}H(i)u(k-i)+\sum_{i=1}^{p-1}G(i)u(k+i) \quad (7)$$

where, let $F=\phi^p$, $G(i)=C\sum_{j=0}^{p-1-i}\phi^j\Gamma_{p-i-j}$, $H(i)=C\sum_{j=0}^{p-1}\phi^{p-j-1}\Gamma_{1+j+i}$. And, $1+j+i>n$, $\Gamma_{1+j+i}=0$.

It was seen from Eq. (6) and Eq. (7), the first two of the right side of equation were all the direct recursive parameter in the kth sampling time, the third was the system input and output of the kth sampling period versus the future p system control variable.

3 Predictive Functional Control Method Based on Incremented Weight

On the basis of part 1, predictive functional control has the three characteristic of traditional predictive control, such as predictive model, rolling optimization and feedback correction. But the difference between them is that the control input of predictive functional control consists of the linear combination of basis function. It simplifies computational complexity and makes the control simple.

According to the definition of PFC model output, the system output is divided into two parts. One part is the free output of system, $y_1(k+p) = C\phi^p X(k) + \sum_{i=0}^{n} H(i)u(k-i)$, the other is the passive output of system, $y_2(k+p) = \sum_{i=0}^{p-1} G(i)u(k+i)$, so:

$$Y(k+p) = y_1(k+p) + y_2(k+p) \tag{8}$$

In Eq. (8), $Y(k+p)$ is the output which is derived in the kth time. But the actual output of the future $k+p$ th time, $\tilde{Y}(k+p)$ is unknown in kth time. There is error between model output and process output when the system is coupled with the effect of model mismatch, two-input and noise, etc. so error compensate is the necessary to system. The modified expression is:

$$\tilde{Y}(k+p) = Y(k+p) + e(k+p), e(k+p) = \tilde{Y}(k) - Y(k) \tag{9}$$

where, $\tilde{Y}(k)$ is the system actual output in the kth time, $Y(k)$ is the model predicted output in the kth time, $e(k+p)$ is the future error.

In NCS, the output of predictive horizon p can be derived from Eq. (9), and we hope that predicted output achieve asymptotic stability along reference path and eventually reach to reference value. In this paper, setting the first order exponential function was:

$$y_r(k+p) = y_c(k+p) - \alpha^p (y_c(k) - y(k)) \tag{10}$$

Where, $\alpha = e^{(-Ts/Tr)}$, Ts is sampling period, Tr is the response time of reference trajectory, y_r is reference value, y is process output.

The following performance index function not only considers system actual output and reference output, but also considers the increment of system control input, so the performance index function introduces incremental weighted, the modified form is:

$$\begin{aligned}
J_p &= \sum_{i=1}^{p} [\tilde{Y}(k+i) - y_r(k+i)]^2 + \lambda^2 \sum_{i=1}^{p} \Delta u(k+i-1)^2 \\
&= \sum_{i=1}^{p} [Y(k+i) + \tilde{Y}(k) - Y(k) - y_r(k+i)]^2 + \lambda^2 \sum_{i=1}^{p} \Delta u(k+i-1)^2 \\
&= \sum_{i=1}^{p} \begin{matrix} [y_1(k+i) + \sum_{j=0}^{i-1} G(j)u(k+j) + \tilde{Y}(k) - Y(k) - y_c(k+i) \\ + \alpha^i (y_c(k) - y(k))]^2 \end{matrix} + \lambda^2 \sum_{i=1}^{p} \Delta u(k+i-1)^2 \\
&= \sum_{i=1}^{p} [\sum_{j=0}^{i-1} G(j)u(k+j) - w(i)]^2 + \lambda^2 \sum_{i=1}^{p} [u(k+i-1) - u(k+i-2)]^2
\end{aligned} \tag{11}$$

where, $w(i) = -y_1(k+i) - \tilde{Y}(k) + Y(k) + y_c(k+i) - \alpha^i (y_c(k) - y(k))$, λ^2 is the weighted coefficient of system out, $\Delta u(k+i-1) = u(k+i-1) - u(k+i-2)$,

$$(i = 1, 2, ..., p) \tag{12}$$

On the basis of the above, in order to make sure system output to reach to reference output, let Eq. (1) into Eq. (13), and the weighted coefficient μ vector $\mu = [\mu_0, \mu_1, ..., \mu_N]$ derivation of the performance index function J_p is:

$$\frac{\partial J_p}{\partial \mu} = \frac{\partial \{\sum_{i=1}^{p}[\sum_{j=0}^{i-1} G(j)u(k+j) - w(i)]^2 + \lambda^2 \sum_{i=1}^{p}[u(k+i-1) - u(k+i-2)]^2\}}{\partial \mu}$$
$$= \frac{\partial \{\sum_{i=1}^{p}[\mu g(i) - w(i)]^2 + \sum_{i=1}^{p}[\mu \lambda l(i)]^2\}}{\partial \mu} = 2[\mu \cdot (R \cdot R^T + \lambda L \cdot L^T) - W \cdot R^T] \tag{13}$$

Where, $N = p-1$, $\mu = [\mu_0, \mu_1, ..., \mu_N]$, $g(i) = [g_0(k+i), g_1(k+i), ..., g_N(k+i)]^T$, $l(i) = [l_0(k+i-1), l_1(k+i-1), ..., l_N(k+i-1)]^T$, $g_n(k+i) = \sum_{j=0}^{i-1} G(j) f_n(j)$, $l_n(k+i-1) = f_n(i-1) - f_n(i-2)$, $W = [w(1), w(2), ..., w(p)]$, $R = [g(1), g(2), ..., g(p)]$, $L = [l(1), l(2), ..., l(p)]$, $n = 0,1, ..., N$, $i = 1, ..., p$.

Set Eq. (13) as 0, then $\frac{\partial J_p}{\partial \mu} = 0$, so:

$$\mu = W \cdot R^T \cdot (R \cdot R^T + \lambda L \cdot L^T)^{-1} \tag{14}$$

Bring Eq. (14) into Eq. (1), then:

$$\begin{bmatrix} u(k) \\ ... \\ u(k+N) \end{bmatrix} = \begin{bmatrix} \sum_{n=0}^{N} \mu_n f_n(0) \\ ..... \\ \sum_{n=0}^{N} \mu_n f_n(N) \end{bmatrix} = \begin{bmatrix} W \cdot R^T \cdot (R \cdot R^T + \lambda L \cdot L^T)^{-1} \cdot f_n(0) \\ ..... \\ W \cdot R^T \cdot (R \cdot R^T + \lambda L \cdot L^T)^{-1} \cdot f_n(N) \end{bmatrix} \tag{15}$$

To take the first term of Eq. (15), that is the system input control quantity $u(k)$ in the kth time.

4 Simulation Experiment

If it is assumed that the system is single input and single output system, and let the continuity time equation as $y(t)|_{1\times1} = \frac{14s-10}{-13s^2-16s+18} u(t)|_{1\times1}$, and the parameters of network delay as $\tau_{sc} = 1.6s$, $\tau_{ca} = 2.2s$, and sampling period as $T_s = 0.5s$. On the basis of the above, $\tau = 4s$ after the contrived fixed network delay, the length of predictive domain is $p = 8$, or the maximum number of control quantity in a sampling period is 8, such as $u(k+i), i = 0,1, ..., 7$.

The following is the simulation of control quantity which equal to 8 in a period, the emersion time is $t_7(k)=0.4$, $t_6(k)=0.8$, $t_5(k)=1.1$, $t_4(k)=2.3$, $t_3(k)=3.2$, $t_2(k)=3.5$, $t_1(k)=3.7$, respectively.

In order to verify the simple and useful control model based on basis function combination which was used in control input of the predictive function. In this work, the form of basis function used step function in common. The value of basis function is: $F=\begin{bmatrix} 1 & 0 & 0 & .. & 0 \\ 1 & 1^1 & 1^2 & .. & 1^7 \\ 1 & .. & .. & .. & .. \\ 1 & 7^1 & 7^2 & .. & 7^7 \end{bmatrix}_{8X8}$, System output gradually reached to 1 along reference.

And the same system used the incremental PID control model in common. Compared to the above PFC control model, the results showed Figure 3 through Matlab simulation.

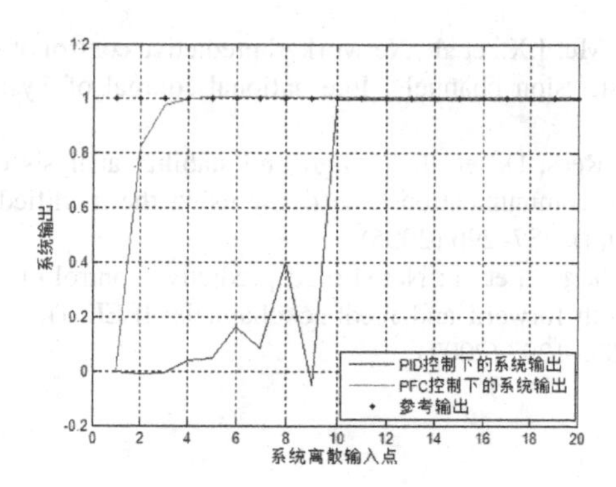

Fig. 3. Outputs of system base on PFC or PID

The simulation results showed that the same time used PFC and usual PID control model respectively. Figure 3 showed that 1) PFC control model, the system output gradually reached to 1, the output curve was smooth and quickly reached to reference output; 2) PID control model, the system out reached to stable value after oscillation, the oscillation time was longer than that in PFC. The PFC control model was doable according to the comparation of simulation results.

5 Conclusion

Previous parts discussed the basis theory of PFC and the control effect of PFC on NCS. PFC used basis function combination in system input, which simplified the system computational complexity. That is a advantage of PFC. But the selection of basis function determined the control precision of system, the length of predictive horizon also effected the stability of system. These parameters have different impact on the design of controller, so they are differently considered in the design of controller so as to make system stable.

References

[1] Walsh, G.C., Beld Iman, O., Bushnell, G.L.: Errorencoding algorithms for networked control systems. In: Proc. of the 38th Conference on Decision & Control Phoenix, Arizona, USA, vol. 5(5), pp. 4932–4938 (1999)

[2] Zhang, W., Bran Ickym, S., Phillops, S.M.: Stability of networked control systems. IEEE Control Systems Magazine 21(1), 84299 (2001)

[3] Hu, S.-S., Zhu, Q.-X.: Stochastic optimal control and analysis of stability of networked control systems with long delay. Automatica 39(1), 1877–1884 (2003)

[4] Wei, Z., Michael, S.B., Philips, S.M.: Stability of networked control systems. IEEE Control Systems Magazine 121(1), 84–99 (2001)

[5] Luck, R., Ray, A.: An observer-based compensator for distributed delays. Automatica 26(5), 903–908 (1990)

[6] Luck, R., Ray, A.: Experimental verification of a delay compensation algorithm for integrated communication and control systems. International Journal of Control 59(4), 1357–1372 (1994)

[7] Liu, G.P., Chai, S.C., Mu, J.X., et al.: Networked predictive control of systems with random delay in signal transmission channels. International Journal of Systems Science 39(11), 1055–1064 (2008)

[8] Liu, G.P., Mu, J.X., Rees, D., et al.: Design and stability analysis of networked control systems with random communication time delay using the modified MPC. International Journal of Control 79(4), 287–296 (2006)

[9] Liu, G.P., Xia, Y., Chert, J., et al.: Networked predictive control of systems with random network delays in both forward and feedbackchannels. IEEE Transactions on Industrial Electronics 54(3), 1282–1297 (2007)

The Research and Design of the Internet Interface Unit of the Remote Electromechanical Control

Zhu Hui-ling[1], Jiang Yan[1], Zhu Xin-yin[2], and Xie Jing[1]

[1] School of Mechanical Engineering,
University of South China, China

[2] The Affiliated Nanhua Hospital of University of South China, China

Abstract. Internet is one kind of open network and the equipments that accord with protocols linking to Internet can realize the cheap, effective remote control. The remote control system based on Internet is an important research direction of the remote control field. How to link the mechanical and electrical equipments to Internet is the key technique of the remote control system. The remote electromechanical control system based on Internet needs one kind of network interface device which is highly reliable and cheap and has good compatibility. After the thorough analysis and study of access theory and control theory of the remote control system, this paper has analyzed the research and design of the Internet interface unit of the remote electromechanical control around the remote control theory based on Internet combining with the requirement of the system design and choosing DSP as the main processor.

Keywords: Remote control. Internet. Interface unit.

1 Introduction

With the continuous development of industrial automation technology, the electromechanical control system structure also development to distribution type control (DCS) and the total line type control (FCS) structure from the original direct numerical control (DDC) and supervision and control (SCC) structure, and then the special control network for the industry control system began to form [1]. However, with the continuous application of FCS, its faults are constantly reflected. The limitation of the fieldbus technology itself and the coexistence situation of a variety of bus standard segmentation limit the further applications of the fieldbus technology in the field of industrial automation. In addition, in the enterprise automation system, due to adopting control network based on the fieldbus being isolated outside the Internet, the whole enterprise informatization integrating is unable to realize [2]. This limitation prompted the emergence of the remote control based on Internet. The remote control is based on Internet refers to use the computer through the Internet network system to realize the monitoring and controlling on running process of the remote device control system.

Z. Du (Ed.): Proceedings of the 2012 International Conference of MCSA, AISC 191, pp. 159–165.
springerlink.com

The computer software and hardware system which can realize the remote control is called remote control system [3-5]. In the remote control systems based on Internet, the controllers can through the Internet monitor and control the production system and the equipment running status and various parameters, the controllers don't have on the scene, which can save a lot of manpower and resources. Management personnel can monitor the remote production operation and send out a scheduling instruction according to the business needs; Research institutions can easily use local rich software and hardware resources to carry out the advanced process control of the remote objects.

How to connect the existing electromechanical equipment with Internet, so that people can get these equipment running status information and control their operation remotely, has become the focus of attention of the world today [6]. The electromechanical equipment Internet access technology is one of the key technologies realizing the remote electromechanical control, remote control system needs to be a high reliability, and high speed, compatibility and reliability are very good, and low cost of a network interface, so that the mechanical and electrical equipment, high efficient Internet convenient connected. The remote electromechanical control system based on Internet needs one kind of network interface device which is highly reliable and cheap and has good compatibility.

2 The Remote Control Based on Internet

Internet is the world's largest computer network communication system. It is a global public information resource and has widespread use for the user. Internet has become the floorboard of countless information resources, it is a non-polar network, not being controlled by one person or a certain organization, everyone can participate in Internet, and everyone can exchange information and share online resources. Internet provides important method of information sharing for people to engage in scientific research, business activities and so on, which has become a sea of human wisdom and the treasure of knowledge and it can make people get countless benefit from the all-embracing global information world.

Now Internet is transiting to the next generation of the Internet (NGI: Next Generation Internet), and Japan had completed the transition to ipv6 in 2005. The U.S. defense department also has finished the deployment of ipv6 in June 2005, and in the 2008 fiscal year, the transitions of ipv6 had competed in the department of defense network and internal network with a budget of $30 billion. Be in China at present about the next generation of the Internet construction has also in full swing. China's first the next generation of the Internet CERNET2 has been opened to the public. Compared with the first generation of the Internet, the next generation of the Internet will be faster, bigger, more and more security, timelier, more convenient. Faster is that the transmission speed of the next generation of the Internet will increase from 1000 to 10000 times comparing with the current network. Bigger is that the next generation of the Internet will gradually give up IPV4 and launching the IPV6 address agreement, therefore, original limited IP address become infinite rich, and big enough to equip every single sand in the earth with a IP address, that is, each appliance of the family, every instrument and machinery of the factory can be assigned a IP address, entering the network world, all of which can be controlled through the network. What it brings the human, is not only a change, but also a qualitative change. More safety is that a lot of security danger troubling current computer network will obtain the effective control in the next generation of the Internet. Our future will be a comprehensive Internet era, which makes the modern production life become more convenient and quick. We can feel Internet technology of the mechanical and electronic equipment will work into the People's Daily life and work, and unstoppable [7].

3 The Design of the Data Processing Module and Network Access Module of the Interface Unit

The main functions of Internet network interface unit contain: receiving the control command from the control terminal, according to the established agreements in the control terminal the processor of the network interface unit analysis the control command, and sends the control command to the field equipment to carry out, the network interface unit can collect the operation state of the field equipment and transfer it to control terminals. According to the existing shortage of the existing related products and technology, we design a intelligent network interface unit for remote control, it can use the lowest cost, in the most simple forms, provide network capacity for equipments. The general structure of the network interface unit is shown in Fig. 1. The whole network interface unit according to the function is divided into two modules: data processing module and network access module. The data processing module is the control center of the entire unit, including processors DSP, FLASH, SRAM and the expansion interface connected with other equipments. The network access module is used to realize the network access.

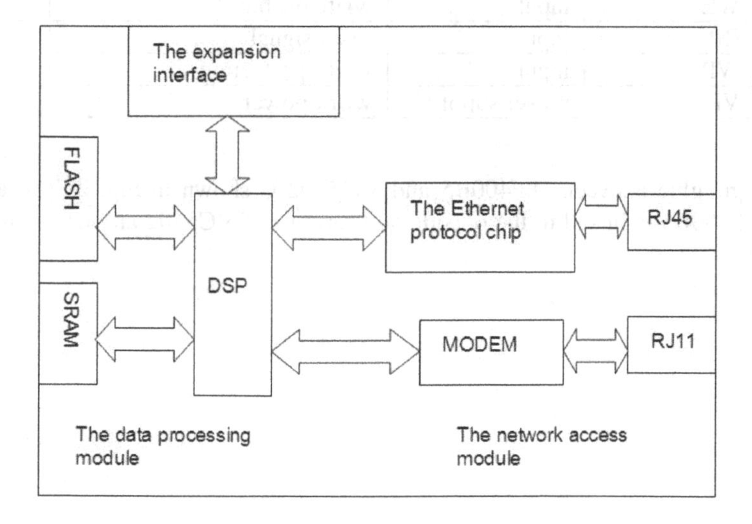

Fig. 1. The general structure diagram of the network interface unit

In embedded equipment, the speed of processors and its related software technology is an important sign to measure the performance of the whole embedded system. Digital signal processor (DSP) adopts the improved Harvard structure, programs and data having independent storage space, having their own independent program bus and the data bus, due to addressing for data and programs simultaneously, which greatly enhance the ability of handling data, and is very suitable for real-time digital signal processing. In our system, we consider TMS320VC5402 produced in TI. The TMS320VC5402 pieces ROM F800h~FBFFh is the self-guidance program area within the slice, and the guidance within the slice is to use the guidance programs within the ROM to load the programs outside the FLASH to programs memory running. Bootloader of ROM within TMS32OVC5402 is used to guide the user programs to RAM at time of power-on-reset to ensure its run at full speed. The system adopts 16 Flash parallel loading methods and the whole guidance process is: after the power-on-reset of VC5402, firstly inquires the MP/MC pin level and judge the work means of DSP, if it is low level, which means DSP being set as the micro computer work mean and alternatively the microprocessor work mean.

The bootstrap process in this system is completed by 28F400B5-T80 Flash memory device and VC5402 jointly. 28F400B5-T80 is four million Flash memory devices with 44 pins and adopts Plastic Small Outline Package (PSOP), whose control logic signals are droved by 6 pins:

Table 1. The function of partial pins in 28F400B5

Pin	Type	Functional description
CE	input	chip select
OE	input	output enable
WE	input	write enable
RP	input	reset signal
WP	input	write-protection
VPP	power supply	work power

The logic glue between 28F400B5 and VC5402 is shown in Fig. 2. The address bus of 28F400B5 is linked to the outside address bus of VC5402 and CE is linked to the pin DS of DSP.

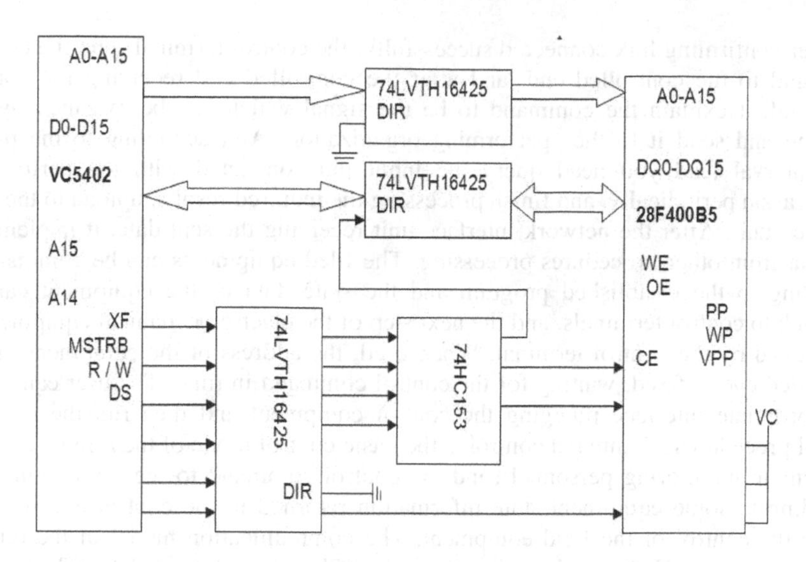

Fig. 2. The logic glue between 28F400B5 and VC5402

4 The Implementation and Application Research

The architecture diagram based on Internet of the whole control system is shown in Fig. 3. Because the control system is divided into the controlled end and the control terminal, before access to the network we must write the address of the controlled end and the control terminal to the corresponding program respectively. Usually, the controlled end is always connected to the network and ready to connect with the control terminal at any time, receiving the control commands from the control terminal, and the control terminal access to network only when needing control the controlled end.

This system determines the communication type of the remote control via inquiring. After having accessed the network, the connected network type and outgoing connection settings will be determined via inquiring the port level. If directly access Internet, then use the IP address stored in the storage area to connect with the controlled terminals and perform network control program; if access telephone network, then use automatic dial-up to connect with the controlled terminals, executive the remote control procedures under the dial-up way.

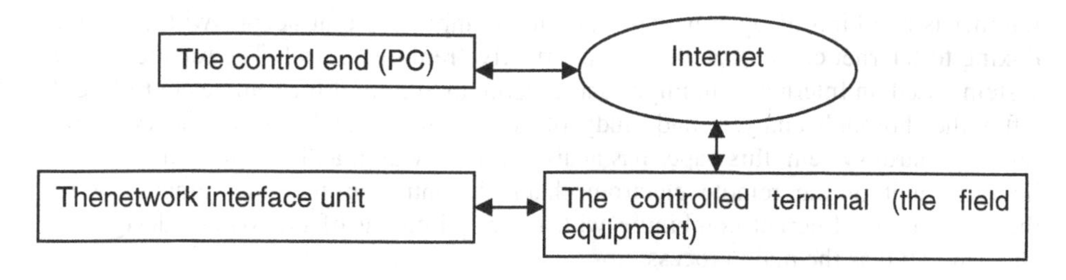

Fig. 3. The architecture diagram of the whole control system

After confirming link connected successfully, the control terminal send the control command to the controlled end, and after the controlled end receiving the control command, it explain the command to be the signal which can be executed by the machine and send it to the performing organization. And according to the preset time interval lead you need query the input pin connected with the performing organization periodically, and finish processing the inquired results, upload to the web interface unit. After the network interface unit receiving the sent data, it implements the data from other procedures processing. The filed equipments can be commanded according to the established program and the state data of the equipment can be feedback to control terminals, and the next step of the electromechanical equipment is determined by the control terminal. When used, the address of the equipment of the controlled end is fixed, waiting for the control command in time. The user can select an appropriate interface plugging the control equipment and then run the relevant control procedures. Command control is the basic control mode of the remote control, and remote monitoring personnel send the control command to the field equipment according to some equipment state information returned to the control end so as to realize the control of the field equipment. The communication model of the remote control based on Internet includes the C/S (Client/Server) model and the B/S (Browser/Server) model. The control data is via the control terminal sent to the shop floor control system and used for control command and other control data. Control data generally includes: the control command, the parameters of the shop floor control system itself needing adjustment and the control application program, etc. The transmission of the control data need to ensure the accuracy, the stability and rapidity.

In the actual remote control, the transmission of the control data can be divided into three parts: one part of it is the real-time requirement short control commands, and the second part is generally no real time requirement of the parameters of the computer control system, and the third is a lot of almost no real-time demand program files. Different transmission modes can be utilized on the three data. Control command adopts the flow socket or datagram socket because of short and relatively high transmission frequency. Due to the large amount of data and requesting transmission accuracy, the program files adopt application layer transfer agreement such as file transfer protocol TFP. The parameters transmission of the control system can adopt the first or the second one according to the actual situation of the computer control system.

5 Conclusions

Internet is one kind of open network and the equipments that accord with protocols linking to Internet can realize the cheap, effective remote control. The remote control system based on Internet is an important research direction of the remote control field. After the thorough analysis and study of access theory and control theory of the remote control system, this paper has analyzed the research and design of the Internet interface unit of the remote electromechanical control around the remote control theory based on Internet combining with the requirement of the system design and choosing DSP as the main processor.

References

[1] Shen, L.: The Current situation and development trend of the electromechanical control technology. The Southern Metallurgy College Journals 21, 125–129 (2004)
[2] Bai, Y., Wu, H., Yang, G.: The Distributed control system and filedbus control system, pp. 9–13. China Electric Power Press (2000)
[3] Overstreet, J.W., Tzes, A.: An Internet-Based Real-Time Control Engineering Laboratory. IEEE Control Systems 10, 19–34 (2000)
[4] Xiong, C., Wang, X.: The Remote Control Network based on Internet. Automation Instrument 22, 4–7 (2001)
[5] Forouzan, B.A.: Data Communication and network, pp. 32–54. Mechanical Industry Press (2002)
[6] Liu, J., Peng, Y.: The Application of the SCM PI Technology in Intelligent village. The Worldwide Electronic Components 12, 40–42 (2003)

Research of Security Router Technology on Internet Environment

Hu Liang[1] and Zhang Qiansheng[2]

[1] Jiangxi Science&Technology Normal University, Nanchang, Jiangxi, 330038
[2] Wenzhou College of Science &Technology, Wenzhou, Zhejiang, 325006
hliang1972@126.com, 328000010@qq.com

Abstract. With the society constantly information-making and networking, the demand of network equipments also increase. Against the security problems, this paper realizes IPSec protocol by using network processor, and uses the FPGA chip to do data encryption, and then design the hardware and software structure of the security router. All in all, the system owns advantages of low price, strong performance, high universality and good extensibility.

Keywords: Internet, security router, network processor, IPSec.

1 Introduction

As communication technology continues to be mature and develops rapidly, we have completely entered the information age and the Internet age. The interoperability and interconnection of network provides a convenient means for people to share information, thus becoming a huge repository of information. Currently, the network has become an indispensible part of many people's work and life. In economic life, due to the widespread use of network technology, the working efficiency of the information resource dependent industries (such as banks, railways, logistics, etc.) increases, the cost enjoys a sharp decline and the economic benefit increases day by day.

With the irresistible trend of the global informatization, governments around the world, the military and enterprises are constantly increasing the pace of information construction. Government departments of many countries, all kinds of scientific research institutions, and enterprisers have connected to the Internet via such gate as the routers, providing a convenient way to fully share and use of information resources on the network. With the accelerated process of modernization and information construction ,the internet has been playing a more and more prominent role in the access, transmission and processing of information and has become the object and even the second front for which the countries compete . Especially in the era of the 21st century when the information is the lifeblood, the Internet plays the key position that cannot be replaced.

Nevertheless, with the further development of the Internet, there comes all kinds of problems no matter in technology or social ethics, among which the safety of the information affects a lot. Being the transportation hub of the transmission of Internet information, the router is the core equipment in implementing the interconnection of

Z. Du (Ed.): Proceedings of the 2012 International Conference of MCSA, AISC 191, pp. 167–172.
springerlink.com

the entire network. Therefore, to ensure the safety as well as reliability of the information transmission on the Internet, we must consider the problems of the security of the routers in the interconnection and terminal on the Internet. The security of the Internet information must be guaranteed to the greatest extent in the premise condition that the data forwarding performance works well.

2 Introduction to the Security Routers

Up to now, the security industry has not yet been able to verify and come up with comprehensive norms and standards in testing the security features of routers. And the definition of security router has not a uniform standard. In general, the so-called security router is made by upgrading and transforming the hardware or software of ordinary router. Meanwhile the ordinary router is added by the appropriate security measures which can improve the reliability of the router itself and effectively prevent such attacks as the intrusion of the net through the route. Common security measures are as follows: increasing the module of security control, use of the Access Control List, the protocol stack to add a safety routing protocol, the authentication at the entrance and other measures to ensure the security in software level. In the narrow sense, the security router not only solves the security problem of enterprises' network from such two levels as network management and data communications, but ensures their own safety and reliability as well. Moreover, it provides secure and reliable communications services for communications networks on the basis that the data forwarding performance is ensured[1]. From the technical point of view, the security router is the dedicated accessing equipment for network security. In addition to the mandatory security features of ordinary routers, it must also support IPSec security protocol, and even integrate firewall and intrusion detection. For the application environment which calls for higher security requirements, it must also ensure that the operation of the device itself is credible.

On the design and implementation of security router, the edge router or the hardware and software architecture with the access of the security function of network and hardware security features are used due to the requirement that the data security and data forwarding efficiency are to the maximum extent; From the perspective of function, it is mainly used to ensure that the hardware and the system software of the security router work safely. And it also ensures the privacy and integrity of the connected network and network transmission of information. Generally speaking, the security router mainly includes the following features[2]:

- Convenient and effective network management which offers performance management, information flow management, configuration management, security and other management functions;
- Spare no efforts to ensure that the data forwarded is accurate, timely and safe to its destination;
- To reduce the influence on the forwarding and processing efficiency of the normal data packet as small as possible;
- To accelerate the speed of encrypting the network transmission encryption, integrity check, the authentication of source address and so on.
- To transfer the network address in accordance with the agreement

- Being able to control every access right of the link;
- Being able to transfer the authentication information to the internal LAN or external WAN upon request to the requirements, and ensure the privacy of the information

3 Design of the Security Router Based on the Network Processor

3.1 The Overview of Network Processor

Network Processor is a kind of programmable electronic chip, mainly aimed at various tasks dealt by the network data in the field of communication, for example, protocol analysis, packet processing, Fire Wall, gathering of voice and data, routing search, quality assurance and so on. The internal network processor usually consists of several microcode processors and several hardware coprocessors, which could realize high performance of business transformation and data processing through the way of programming. The core of network processor is microcode processor unit, which can process the data package intelligently in the way of high speed and capacity, such as unpack, classification and retransmission of the protocol data. Therefore, microcode processor unit is often called the processing engine of data package. According to its different types Hardware coprocessor can realize the functions as the restructuring of data packet, accelerating of table look-up, the message queue and buffer management, management of retransmission order, AC Access Control, and multicast communication, etc.[3]

As the network processor could own different functions through the way of the soft programming, it possesses both of the two advantages of flexibility and property. Therefore, the network processor owes much more incomparable advantages compared with ASIC in technique, which mainly include: (1) programmability; (2) high-speed data processing; (3) parallel processing; (4) deep data processing; (4) scalability; (5) modular design;

3.2 The Designing Principles of the Security Router

In the process of designing the router, such aspects below should be considered:

- The reliability and circuit safety measures of the router
- The privacy of identity authentication
- Data encryption
- Intrusion detection and prevention measures
- Security strategy management

The reliability and line security of the router itself must be supported by redundant backup when designing the interface of the routers. When the main interface goes wrong due to the the accidental reasons, another backup interface can temporarily replace the host interface to ensure that the router works well in the network; When the network flow is over loaded, the backup interface can also share the load of the task of the host interface.

In general, there are two types of identity authentications: the first is to authenticate the identity according to the access mode and the routing information of the router;

the other is to authenticate the identity based on the user's information, i.e. the access control. Visiting the router firstly needs the cascade protection of owner password, which can realize source access control strategy based on IP address through the filtering of agreement data packet. And the security router should be able to provide access control service functions for the users' access control and realize network access control for the access subscribers through setting up the filter conditions of IP and MAC address. Besides, through the transformation of the internal and external address, the private IP address in the local area network can be hidden when using the router communication, and accessing external network only with WAN IP address.

With regard to data encryption, the messages transmitted by the router can all be encrypted. That is to say, even though the data is transmitted on the Internet, it can also ensure the privacy and integrity of the agreement data packet, and the reliability of the data newspaper. For constructing a virtual private network VPN on the internet, encryption for agreement data can guarantee the safe arrival of the tunnel of data transmission. Generally, it can be realized by the hardware for data encryption, designed in the embedded type, and then the encrypted hardware module will be directly integrated with the circuit board of the router.

In terms of network attack, a security router needs to have a good function of intrusion detection and thorough protection measures. It should have various attack tests to make sure whether the router can deal with the current network environment. Likewise, the security router needs to include all kinds of security strategies to ensure that whether the security router can cope with many kinds of processing conditions, so it must have the management function of security strategy.

3.3 The Designing of Hardware System Structure

The CPU of the security router picked the S3C2440 ARM chip, which includes the CMOS macro processor unit and storage unit using 0.18 um electronic technology and with the internal Bus using Advanced Microcontroller Bus Architecture (AMBA). The new Bus structure[4], is provided with a wholly static design, using some technologies of low power consumption. It also has a compact system structure, especially for the price and the application of the power consumption. Network processing chip is adopted a chip BCM1480 made by Broadcom based on MIPS64 nuclear on-chip system[5], including four SB-1 MIPS64 nuclear. The chip has 4 sequence launch line structures and every clock cycle can launch two storage instructions and two ALU instructions. Each nuclear need 32 KB to cache and 32 KB data to buffer; one DDR memory controller, three high-speed HT ports are used to connect to other BCM1480 chip or HT to bridge the I/O chip connection. Each port can also be configured to a 10 Gbps network equipment of SPI-4.2 interface; four 10 / 100/1000 Ethernet MAC, which can be easily connected to the LAN or control backplane; one PCI-X local bus, used to directly connected to the I/O devices; General bus and Flash I/O; two system management bus SMBus; four serial ports can be used for control port. S3C2440 can connect with other peripherals through the PCI bus. MII interface connects to the 10/100 M Ethernet card, and uses 60 x bus and network processing chip in order to communicate. FLASH and SDRAM are used to exchange and control the data. The system is equipped with a special FPGA, as a special hardware encryption chip, based on good FPGA programmable and extensible, and the security router can in time change encryption algorithm and deal with logic.

3.4 Design of System Software

The system provides the two following interfaces to the end users: the web interface based on the PHP and the CLI command interface based on the serial port. Following the interface layer is the general command parsing layer which is in charge of the interpretation of the two interfaces to send user's message, change it into the system command and select the corresponding function modules. Processed by the corresponding program modules, it will display the implementation interface in the previous interface. The module supported by the system is mainly responsible for providing a variety of basic system API s and configuration features. The system management module is mainly for the API encapsulation of the Linux system management interface. At the same time, it will leave an interface in advance for possible future use, thus promoting the follow-up function expansion and code updates.

Network processing module is the core of the system software which is mainly responsible for the management and control of FPGA, NIC and network processing chip as well as the equipment maintenance and management of the log. Network processing module achieves the hardware support of the IPSec through the network processing chip. And it can deal with the IPSec agreement head; decry the content of the agreement with the coordination of the encryption circuit of the FPGA, and check the integrity and privacy of the content. In addition, the network processing module is also responsible for the decryption of the sensitive information which the users point out. It supports the detection and unpacking of the application protocol, such as HTTP,TP and P2P.

4 Access Control and Its Applicability Analysis

Access control also named visit control, which realizes the internal security of the network system to some extent. Security access strategies is a kind of rules, which is used to determine whether the main body have access ability to the subject object. The visit includes three levels as follows:

- System level control. Its task is implementing safety management from the whole system in order to prevent the users not permitted accessing to host or network and obtaining system resources and services. So it has various control measures of user registration, user login, system security access.
- User level control. It classifies its users in the network system, and distributes them different access permission. There are super users, network operators, average consumers, general customers.
- Resource level control. All kinds of resources have their own visited access permission in the network system, and have none relationship with the users in order to protect the safety. There are directory security set and file security set.

5 Conclusion

Nowadays information technology is developing at an unprecedented speed and the Internet has already walked into thousands of families. More and more people are in demand of the network terminal, and with the popularity of the Internet, the security problem becomes the key link that needs to be urgently addressed. This paper introduces a design program of the security router based on the network exchange chip, which can achieve the hardware of IPSec agreement and decrypt with the FPGA. The designing program is simple to develop, short in cycle and easy to extend.

References

[1] Seifert, R.: Gigabit Ethernet Technology and Application. Mechanical Industry Press, Bejing (2000)
[2] Ling, C., Jun, H.: Test Design and Implementation of Gigabit Ethernet. Communication Technology 40(11), 214–215, 38 (2007)
[3] Feng, C., Yan, H., Bin, Y.: The Ethernet Interface Design Based on Embedded Technology. Communication Technology 50(05), 127–129 (2010)
[4] Xiaozhe, Z., Zaifang, W., Qiong, L., Guiming, F.: The Design and Implementation of the Integration between Switches and Routers. Computer Engineering and Application (2004)
[5] Comer, D.E.: Network Processor and Network System Design, vol. 01. Publishing House of Electronics Industry (2004)

The Firmware Upgrading Design of Embedded Equipment That Based on SNMP

Hu Liang[1], and Zhang Qiansheng[2]

[1]Jiangxi Science&Technology Normal University, Nanchang, Jiangxi, 330038
[2] Wenzhou College of Science &Technology, Wenzhou, Zhejiang, 325006
hliang1972@126.com, 32800010@qq.com

Abstract. With the development of the network size, the network is more and more complex. It is more and more difficult to manage the network devices. In this paper, the embedded equipment upgrading soft based on SNMP is provided. It judges whether the network device is needed to upgrade by analyzing the state data. Moreover, it can report the abnormal by sound and light. The soft has the advantage of simple development and upgrading the device's firmware quickly and accurately.

Keywords: SNMP, Embedded Equipment, Firmware Upgrading.

1 Introduction

In recent years, with the ceaseless development of network scale, the network equipment management and maintenance becomes more and more difficult. Therefore, the network management software increases the importance. The good network management software can correctly influence the network connection, monitor the network condition, find out the stoppages and problems rapidly, and maintain each of the node equipment in the network. Especially in these years, the large amount of embedded equipment in the network brings much trouble to maintain and upgrade the firmware. However, the SNMP protocol makes everything easier. It is still under the widely application now.

2 SNMP Overview

The SNMP is the set of protocol that defined by IETF (Internet Engineering Task Force). [1] This protocol is based on the SGMP (Simple Gateway Monitor Protocol). The SNMP network management frame includes SMI (Structure of Management Information), MIB (Management Information Basement), and the SNMP. SMI defines the information organization, formation, and symbol that used to form the SNMP network management frame. It is the basement to describe the information exchange of MIB object and the SNMP protocol. MIB defines the object collection of SNMP management, and points out the variable quantity of network element maintain (the information that can manage the process research and setting).

Z. Du (Ed.): Proceedings of the 2012 International Conference of MCSA, AISC 191, pp. 173–178.
springerlink.com

2.1 SNMP Basic Principle

SNMP uses the special form of the Client / Server model [2]: Agency/ Management station model. This mode will finish the network management and maintenance through workstation management and the exchange work among SNMP agencies. SNMP management station will send search requests to each SNMP agency and obtains various types of information that defined in the MIB.

SNMP agency and management station will use the standard defined message format in SNMP to precede the communication. Each kind of information is the independent data report. SNMP uses the data report of UDP user as the transmission layer protocol [3] to do the non-connection network transport. The standard SNMP information report includes SNMP master head and protocol data unit (PDU). The SNMP masterhead divides into SNMP version mark and community name.

The data report structure of standard SNMP is figure 1.

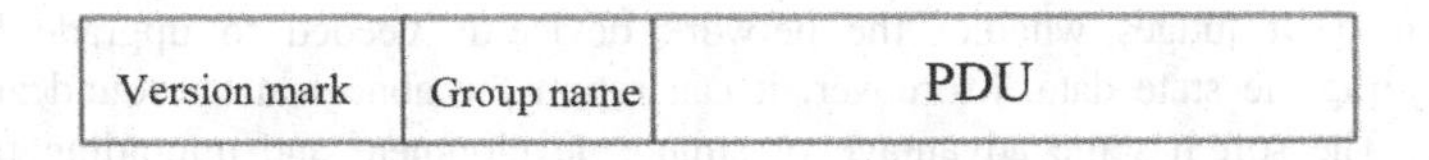

Fig. 1. Data report structure of standard SNMP

- Version identifier: ensure each SNMP agency will use the same protocol and each SNMP agency will directly abandon the different SNMP data report that different from the itself version mark.
- Community Name: use for the identification of the SNMP management station. If the network SNMP agency sets the manager identification, SNMP agency will test the group name and the management station IP. If the test fails, SNMP agency will send the fail Trap information to SNMP management station.
- SNMP protocol data unit (PUD) :PDU declares the information type and parameters of the SNMP data report.

2.2 The Management of Information Basement (MIB)

MIB is the core center of SNMP protocol and records the management objects attribute set. All the defined objects in MIB form the tree structure and each node can represent by OID (Object Identity). OID expresses the MIB tree route from the root to the relative node.

At present, the MIB tree has two sub-trees. No. 1 sub-tree is ISO, which maintain by the International Organization for Standardization. In the ISO sub-tree, 1.3.6.1.2.1 sub-tree is the famous MIB2 sub-tree. It includes 8 object groups such as System、Interface、Address、and Translation. Almost all the equipment that can support the SNMP equipment will support MIB2 sub-tree. 1.3.6.1.4 is the other important sub-tree as well as the private sub-tree. It is used to expand the private MIB basement in the enterprise. This sub-tree has only one sub-node, which name is enterprises. In the enterprises sub-tree, each manufacturer that registered object identifier will have one branch. This article will use the OID: 1.3.6.1.4.1.5651 from Beijing Taikang

Enterprise. MIB2 achieves the equipment basement and universal administration and the MIB in the private phase describes the equipment characteristic. We can use these two trees to manage the relative equipments. [4]

The practical work of the MIB is planning the private MIB tree. The process is classified the equipment control node, use ANS.1 to describe the regulation. This article uses the Mib Editor that provided by AdventNet Agent Toolkit C Edition. This tool can use imaging interface to design MIB tree. We can based on the management requirement to plan the private MIB tree structure and define the node name, location, type, access limitation and simple annotation. Mib Editor tool can produce .mib file of ANS.1 program.

2.3 Five Information Types of SNMP

The SNMP defines five information types [5]: Get-Request, Get-Response, Get-Next-Request, Set-Request, and Trap. Get-request and Get-Next-Request will send to the SNMP agency by the network management workstation that install by SNMP supervisor. It will request the searching equipment information, agency the Get-Response. SNMP supervisor uses Set-Request to agency monitor setting parameters. These can through thc achievement of read and write operation in the management information database. Only Trap is the unsolicited message form the agency, it can use to report how equipment starts, off, and other error conditions to the SNMP supervisor.

The agency process port uses UDP port 161 to receive Get or Set report. Moreover, the management end uses UDP port 162 to receive Trap report.

3 The Design of Embedded Firmware Upgrading

3.1 Software Frame

This system is based on the Linux+ Qt, it uses C/C++ language. The monitored system is the network system that connected to the core transformer. The main program function is upgrading the embedded equipment in the monitored system while developing the new firmware equipment.

The whole system can divided into three layers: network information processing, firmware upgrading and display interface.

The lowest layer is the network information processing. The model uses SNMP++ open source to obtain the SNMP information that is based on the monitored network equipment (router, switchboard, and so on). Moreover, package the SNMP information into a different single data structure. Then, process the packaged SNMP information, which includes form the network connection topology and forme the equipment condition information. The network topology achievement is mainly based on the SNMP system, Interface, CDP table and the TpFdb of SNMP (port forwarding table). Based on the CDP table, we can find the neighbor equipment such as router, switch, and so on. The equipment that cannot support CDP will obtain the equipment name and supported protocol that is based on the system. It is possible to estimate router and switch that in line with the supported protocol. The terminal MAC through a switch port can be obtained from the port-forwarding table. This information can form the network topology. The equipment status information follows Interface, CUP, MEM, Hsrp table can obtain the CUP and internal memory condition, port condition, port speed of receiving and dispatch the data package, port read speed, port packet lost and packet error rate and other information. Through the evaluation of this information we can form the unusual condition in the network and alarm the error state.

The firmware-upgrading model is based on the information processing model select the network information and equipment information. Based on the present network firmware basement of the different edition, evaluate them. If we judge, the equipment needs to be upgraded, make it red in the network topology, provide the firmware edition information, and upgrade the new brief introduction at the same time.

The operation interface provides the display of network topology, the terminal work condition, and provides the man-machine interface to send firmware upgrading order. Moreover, use light and sound alarm to warm the error and unusual condition in the network that convenient for the manager to research and monitor,

3.2 Network Information Process

Before Selecting the network and equipment information, we need to make the network topology discovery. Only the topology discovery can reach the equipment information selection and firmware upgrading. During the topology discovery process, we write network equipment information into the database can provide information to the firmware upgrading. Therefore, the first step to do the firmware upgrading is selecting network information. Topology discovery is the basement of network information selection. The basic thinking is through one or some given seed, use the thought provoking to gradually expand the equipment collection.

The classification will base on the network protocol layer. The topology discovery can divided into three layer topology discovery and two layer topology discovery. The primary is used in finding out the relationship such as routers in the third layer network equipment of the OSI model. The latter is finding out the topology interconnection of second layer network equipment in the OSI model such as network fridge, switchboard.

This article will use three layer topology discovery. The algorithm thinking is: when make the topology discovery, the first is select one or more known the IP of three layer router. Then start at these seed points, use the wide first research to search the network node until to reach the max hop that under the settlement. This will find out all the three layers equipment, sub- network and the topology relation. The detailed algorithm is in the following.

- Set the relative topology discovery start date: the topology discovery needs the following parameter equipment- equipment table Tdevice, seed nod table Tseed, Router table Troute, Port address table Tadd, relation table Tlink (includes a router and sub-network relation)
- Prepare the data structure that needed by topology discovery: equipment sequence QDevice, each object includes one IP address and the relative hop count. We can mark with (IP, hop)

- Before topology discovery, if the seed node table Tseed war put into the known IP address, blank the table of Tdevice、Tadd、Troute、Tlink. If the QDevice sequence is blank, the maximum hop count of topology discovery will be Max HOP.

3.3 Firmware Upgrading

The function of the firmware-upgrading model establishes up of the network information-processing model. It is mainly processing the network and equipment information from the lower layer. Through the compare between the equipment firmware edition and the present edition, judge whether it is past the time or need to upgrade. If the equipment needs to upgrade, the firmware upgrading model will through the network information processing model to integrate the information of Get-Request、Get-Next-Request and Set-Request in the SNMP protocol which will send to SNMP protocol. In the last, the model based on the network and equipment condition, it will select the right time, and send the Set-Request information to set the upgrading order. After receiving the information, it will finish the upgrade by automatic reboot. The process is figure 2.

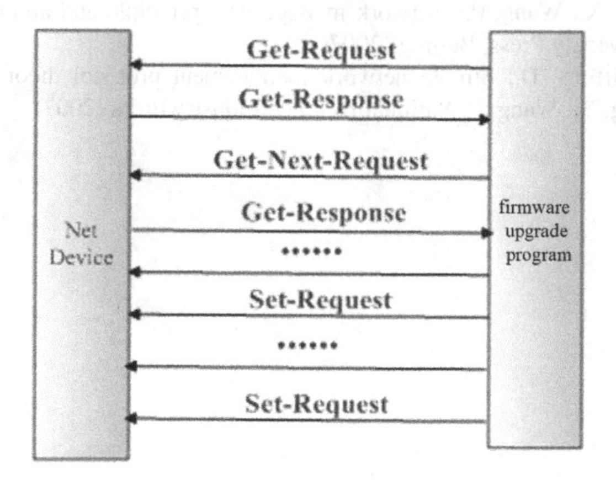

Fig. 2. SNMP interactive Message while firmware upgrading

In figure 3, when the equipment needs the firmware upgrading, it needs to send Get-Request information to obtain the newest edition information, and then send the received information to send Get-Response information. After that, it will send the received information by packaging. If the edition information cannot judge the required firmware information and integration information, there needs to send the Get-Next-Request in order to obtain the information until enough. And then, the firmware upgrading process will send N Set-Request information to the target equipment for the pre-upgrade setting. N value will differ from the various equipment types and firmware editions. After setting, the firmware-upgrading program will send the last Set-Request information, set the upgrade order, and notice the equipment to do the upgrading. In the last, automatically reboot the equipment, the firmware upgrading begins.

4 Summary

This system through the real time network condition achievement, through the condition evaluation can rapidly search the required firmware equipment and unusual events. It can provide great help for the network manager to manage the network equipment and fixer network error. Moreover, the easy operation of SNMP reduces the development difficulty, decrease the development period. SNMP places more and more important function in the network management.

References

[1] Yang, J., Ren, X., Wang, P.: Network management principle and application technology. Tsinghua University Press, Beijing (2007)
[2] RFC 1157, SNMP: Simple Network Management Protocol (May 1990)
[3] Li, M.: SNMP Simple Network Management Protocol. Electronic Industry Press, Beijing (2007)
[4] Stallings, W., Hu, C., Wang, K.: SNMP Network Management. China Electronic Power Press (2009)
[5] Yang, J., Ren, X., Wang, P.: Network management principle and application technology. Tsinghua University Press, Beijing (2007)
[6] Hein, M., Griffiths, D.: Simple network management protocol theory and practice. In: Xing G., Yang, Y., Wang, P. National Defence Industry Press (2007)

The Design of Port Scanning Tool Based on TCP and UDP

Hu Liang[1] and Zhang Qiansheng[2]

[1]Jiangxi Science &Technology Normal University, Nanchang, Jiangxi, 330038
[2] Wenzhou College of Science &Technology, Wenzhou, Zhejiang, 325006
hliang1972@126.com, 32800010@qq.com

Abstract. With the society constantly information-making and networking, network security is becoming the increasingly service. Port scanning is a common vulnerability detection tool. This article mainly introduces the principles of port scanning, analyzes the common port scanning based on TCP and UDP, and designs the port scanning tool. It has certain significance for network security.

Keywords: TCP, UDP, Port Scanning.

1 Introduction

With the rapid development of the Internet, the information has greatly promoted the economy development and social improvement and it has been the necessary part of people's life. [1] At the same time, the network security expresses the importance. The increase of relative application and system requirement provides some safeguard for the daily net play. However, the network security of the host cannot obtain the perfect security. It is very common that hacker attacks the online security and brings shadow for network living.

The function of port scanning tool is sequencing all the service port of openness network in the network service host. After that, utilize the relative port security leak to protect the network host. [2] The port scanning tool is the security software to reduce the security-hidden danger that can search the openness port of host. Moreover, it is also can find out the security leak, and apply the hacker attack. This article introduces the working function of port scanning that based on the TCP, and UDP protocol and designs the relative port scanning tool.

2 Brief Introduction of Port Scanning Technology

2.1 The Basic Principle of Port Scanning

The port scanning technology can through one exploring method and technology to detect the hidden trouble of the system or the network. Compare with other host in the network, each computer is the closed space. Moreover, the channel to communicate with the outside world is the port. In the OSI model, the port belongs to the transmission layer. The transmission layer will identify each service through one port

Z. Du (Ed.): Proceedings of the 2012 International Conference of MCSA, AISC 191, pp. 179–183.
springerlink.com

number. The port is one 16-bit address. It can divide into two types: (1) the universally port number: distributed by ICANN, the value is 0-1023. (2) The ordinary port number: can random distribute the customer process to the requested service. The scanning port target is exploring the openness service port of the target host. The other one is adjust the operation system based on the openness service and other network information.

In the transmission layer, there has two important transmission protocol—transmission control protocol (TCP) and the user data report protocol (UDP). The so-called port scanning has the very simple principle. It uses the socket import API that provided by the operation system and does the connection with TCP or UDP. Moreover, utilizes the protocol character to connect each port of the network target host, based on the backward result and connection to judge the port open. [3] The port scanning will try to connect the various services of the different ports by the long-distance TCP/IP protocol. At the same time, record the port response of network target host, through evaluation to obtain the useful information about the target host. If attacker uses the port scanning, he can hide when obtaining the port scanning result in order to avoid the reverse trace. Unless the target host is totally in idle (as well as no listen port, there has no such kind of system), the long interval port scanning can hardly to find.

The target of port scanning has four:

1) Explore the openness port of target host;
2) Judge the operation system of target host (Windows, Linux or UNIX etc);
3) Identify the special application program or service edition number;
4) Explore the system leak of target host.

2.2 The Common Technology of Port Scanning

At present, the technology of port scanning is much mature and we will have some examples in the following.

(1) TCP-SYN scanning

This is kind of half openness scanning technology based on the TCP protocol. The advantage to use this kind of scanning technology, there do not need to proceed the completely TCP handshaking process. During the scanning process, the scanning program will send one SYN data package to the target host for disguising the practical connection and wait the target host response. If there has the return SYN/ACK data package, it means the target port is under the monitor. If there has the return RST data package, it means the target port is not under the monitor. This scanning technology has the advantage of not live any access record in the target host. The disadvantage is only the highest operation permission can establish the self-defined SYN data package. [4]

(2) TCP connect() scanning

Until the present, the simplest basic TCP scanning technology is the connect system API that supported by the operation system and try to connect with every target computer port. The successful connect means the port is under the monitor condition. Otherwise, the port supports no service. The biggest advantage of this scanning technology has no special permission with rapid scanning speed.

(3) IP fractional scanning

This is not the new method and it is just the transformation of the scanning technology. During the scanning process, do not directly send TCP probe package. It will divide the package into several small IP data packages. Through the process, the TCP data head is divided to the different packages and hardly find the scanning during the system protecting. However, the disadvantages are obviously. The data package formation is complicated and will have some unpredictable problems and troubles while processing the large amount of small data packages.

(4) TCP-Fin scanning

This kind of scanning technology uses the FIN data package can easily avoid the malpractice of the SYN scanning through the characteristic of firewall and data package filter. The principle is: the closed target host port will use RST information to respond the received FIN data package. However, the openness host port will waiver to response the FIND data package. Some operation systems will response the RST information with ignoring the openness of port. The scanning under this condition becomes invalid. But, this method can use for distinguishing the operation system Unix and NT.

(5) TCP reverse and Ident scanning

This scanning uses the characteristic of Ident protocol can display the processing user name. Through Ident protocol can connect to the special port of http protocol, then use Identd information to search the highest user permission of the server. The biggest disadvantage of this method is only using under the perfect TCP connection of the target port.

(6) UDP recvfrom and write scanning

This kind of scanning is commonly using in the normal users that cannot directly connect to the operation host port. At this time, under the Linux system can tentative and indirectly obtain the port condition. Rewrite the closed port will not success. We invoke recvfrom operation on the no blocking UDP socket, if we obtain the wrong ICMP message, back to "EAGAIN" and try again. Otherwise, return ECONNREFUSED. If we receive the normal ICMP message, it means the target host directly refused. We can explore out whether the port is open from this.

(7) UDP ICMP port and unarrival scanning

This scanning technology uses the characteristic of UDP message. In the UDP protocol, it will not send the identification signal or feedback the error data package for the port openness and close. The large amount of hosts send UDP messages to the closed UDP port, there will return ICMP_PORT_UNREACH. From this, we can determine which port is closed. This scanning has the shortage of slow speed, because RFC has strict regulation for the rate of ICMP error information and there need the highest user permission.

(8) The high-level ICMP scannign

This scanning achieves on the basement of ICMP protocol. It is mainly use the basic usage of ICMP protocol—beep on error. Based on the ICMP protocol, if there has something wrong with the received network package protocol, the receiving terminal will create one ICMP error message of "Destination Unreachable". [5]

If there has edition error while transmitting the IP data package, the target host will directly abandon this data package. If the verification expresses the mistake, the router will abandon this data package directly. However, the system of AIX, HP/UX will not send ICMP unreachable information.

3 The Design of Port Scanning Tool

3.1 Total Software Structure

The port scanning tool is the software that specially scans the port and record the result detail by detail. The different user requires various port scanning degrees. However, most of them hope to obtain the service information on the operation of target host.

Generally, the scanning tool need to has the arbitrarily combination of the following functions.

- Explore one or more port condition of the target host
- Explore all the host ports on the special network
- Explore the operated service program on the special target host
- Explore the operation system and edition on the special host and try to find out the system leak
- Explore the user information on the special host

Form the above discussion, the scanning tool that we designed has the total structure in figure 1.

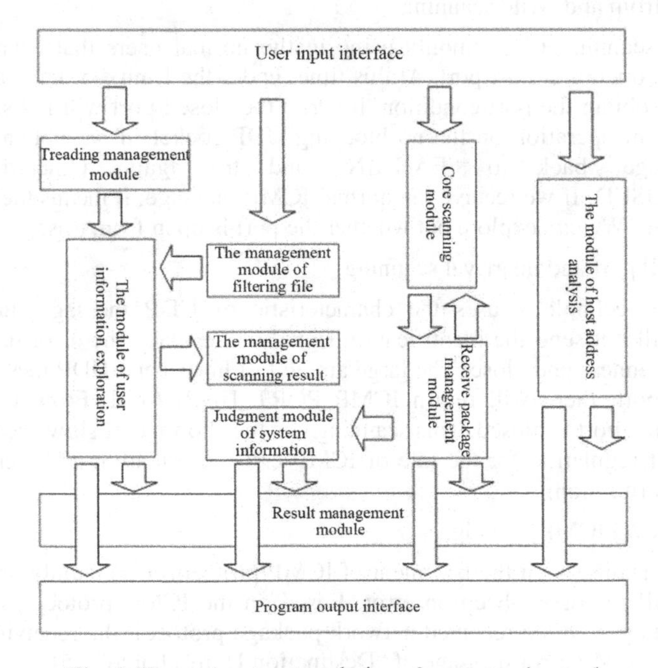

Fig. 1. The entire structure design of port scanning tool

In figure1. the port scanning tool can divide into 8 modules: treading management module, the module of user information exploration, the management module of filtering file, the management module of scanning result, judgment module of system information, core scanning module, receive package management module, the module of host address analysis and the result management module. The completed cooperation among modules is when inputting the parameters such as host address, the port number range and the extra obtained user information that need to scanned on the input interface, the treading management module will create the relative scanning treading. There will start the core-scanning module and module of host address evaluation. When the core-scanning module searching for the information, there needs to receive the receive package module to coordinate the data proceed and evaluation. At the same, there needs the module of user information exploration to explore the user code and catalog based on the information from the management module of filter file. The management module of result management will record the last result. The module of information judgment will judge the host operation system and edition from the comprehensive information. At last, all the information will output to the module of result management directly or indirectly. After the formatting, the information will send to the program output interface.

3.2 Target Process Design

Similar with the socket connection and communication process, scanning process is the exploring communication process between scanning tool and target host. The TCP scanning establishes TCP connection and communication after the process of Bind, then close the connection and socket. UDP is directly receive and dispute the data. When the scanning tool communicates with target host by TCP and UDP communication, we can use various scanning technologies in the article of 1.2. We can based on the return result to finish the port condition and the exploration of host information.

4 Summary

With the rapid development of the Internet technology, the port scanning technology is under the improvement. Port scanning is the important method that can secure the system to avoid the attack. There needs the further research about scan the port in higher efficiency and avoid being scanned.

References

[1] Guo, X., Zhang, J., et al.: Network program design. Wuhan University Press (2004)
[2] Ji, Z., Cai, L.: The vague detection strategy based on the port scanning. Journal of Computer Applications 23(10), 87–92 (2003)
[3] Liu, H., Wang, L., Wang, J.: The computer application of the vulnerability scanning system based on the network, vol. 23(98-99), p. 102 (2003)
[4] Liu, Y.: The information network security of vulnerability scanning based on plugni, vol. (12), pp. 4–50 (2003)
[5] Cai, Q.: The research of TCPSYN port scanning. Journal of Guangxi University of Technology 13(1), 25–27 (2002)

Research on the System Model of Network Intrusion Detection

Yang Yunfeng

Department of Computer and Information Science,
Hechi University, Yizhou 546300, China
yangyunfeng1233@163.com

Abstract. With the rapid development of the network, network security has become one of the most important issues, as it directly affects the interests of the state, enterprises and individuals. Further more, with the increasingly mature of the attack knowledge, complex and diverse attack tools and techniques, the current simple firewall technology has been unable to meet the needs of the people. Facing so many challenges and threats, intrusion detection and prevention technology is bound to become one of the core technologies in the security audit. The iintrusion detection, playing a role of active defense, is an effective complement to the firewall, and is an important part of network security. This paper mainly analyzes the decision tree algorithm and improved Naive Bayes algorithm, proving the effectiveness of the improved Naive Bayes algorithm.

Keywords: Network, Intrusion detection, System model.

1 Introduction

In the terms of usage, the main hazards of computer viruses are: destruction of computer data; Preemption of system resources; impact to computer speed; damage to computer hardware; destruction of the network and tampering with the site content, causing the network in a state of paralysis. From an economic point of view, the losses caused by computer viruses to the global can be calculated at trillion US dollars. From national security point of view, in recent years, the government websites have become attack objectives by hackers. Information security represents a country's comprehensive national strength, so it has increasingly been concerned by all the countries around the world. Traditionally, the firewall, a security and effective prevention technology, is the most safety precautions. It can be effectively monitoring the activities between intranet and network and effectively prevent the information transferred by Firewall. However, it could do nothing to the attacks within the network, because it stays between the protected network and external network, and the internal network information couldn't be protected. This means that the firewall is unable to stop those attacks do not go through itself. In addition, Firewall is difficult to be managed and its own has security vulnerabilities. Thus, the firewall can not guarantee the security of computers and networks.

Z. Du (Ed.): Proceedings of the 2012 International Conference of MCSA, AISC 191, pp. 185–190.
springerlink.com

With the increasing development of attack tools and techniques, a simple firewall technology has been unable to meet the needs of the people. Therefore, to find a new way has become an important way to guarantee the insecurity of the network. The intrusion detection technology, playing as an effective complement way to the firewall, came into being. It is the second line of defense and is an important part of network security. Tracing back to the intrusion detection, we find that it valued a lot in the field of network security, to which it made important contributions, so research on network intrusion detection measurement technology has important practical significance.

2 Host-Based Intrusion Detection Systems Based on Host and Network

Based on different sources of data, the intrusion detection can be divided into host-based intrusion detection system and network-based intrusion detection system. As the name suggests, the host-based intrusion detection system gets information mainly from the host. The information includes the host system files, logs and audit files etc. Through analysis of this information, it detects attacked behavior. Such intrusion detection system has the following advantages: able to accurately detect attacks with high detection rate; able to detect a specific system activity, file access behavior, and unauthorized access behavior; has a wide range of applications that it can be installed to different host; with fast and real-time response. At the same time, it also has many deficiencies: poor portability that it generally only be used for the same system platform; affect the performance of the host; difficult to detect such behavior that if the intruder is familiar with the system and it changes the system log to hide its intrusion; unable to detect some corresponding attacks because the information provided by the host is quite limited and some aggressive behavior will not be reflected in the host system file.

The network-based intrusion detection system obtains data through monitoring the packets in the network, and then through analysis, statistics and match to find aggressive behavior. Such intrusion detection system has the following advantages: it is more flexibility and portability, because it does not depend on the operating system platform and the network protocol is standard; it is able to obtain more information on the premise of not affecting host performance; can detect the behaviors taking place without success or malicious intent to protect all the host of the segment; with fast respond and countermeasures when it detects attacks. In contrast, it also has some shortcomings: it may omission some attacked behaviors to some encrypted packets, as it has no right to process such information; it can only monitor the same data segment, so the scope of monitoring is small. If to expand the scope of monitoring, it will cause cost increasing and the huge amount of information reduces its processing capability.

Expert system-based misuse detection is a kind of inference rules which is based on the experience of the experts on the analysis of aggressive behavior, that is, if - then rules. If is a rule base of specific attack described, then is appropriate measures based on the judgment in the current behavior according to the rules. From this, it can be seen that the expert system needs to extract rule base from a large amounts of data, which itself is very difficult, and can't meet the needs of real-time, so this system is generally not used in commercial products, but more applied in research.

3 Application of the Decision Tree to Intrusion Detection System

Decision tree, also known as the decision tree is a binary tree-like structure. The top node of the tree is root node, the middle node of that represents a test of certain of attribute. Different attributes constitute different branches, each one of which represents a test output and the leaf nodes represent the results of the classification. The construction process of the decision tree is a top-down process.

ID3 is a decision tree learning algorithm based on information entropy proposed by Quinlan and it is on the basis of information theory. Its classification is a top-down process. Its root node has the greatest information gain, and then selects the greatest information gain attributes in pending ones as a branch node. This growth algorithm is continued until there are no other pending branch variables so far. The standard ID3 algorithm is without pruning operation.

The information gain is calculated as follows:

$$I\left(s_1, s_2, \ldots\ldots, s_m\right) = -\sum_{i=1}^{m} P_i \log_2 P_i. \tag{1}$$

Of which $P_i = \frac{s_i}{s}$ represents the probability of each record of Ci.

Assume attribute A has n different value, so the training set D can be divided into n different subsets. If assume A is the optimum splitting attribute, the entropy of the subsets divided by A is:

$$E\left(A\right) = \sum_{i=1}^{n} \frac{s_{1i} + \ldots\ldots + s_{mi}}{s} I\left(s_{1i} + \ldots\ldots + s_{mi}\right). \tag{2}$$

Thus, the information gain obtained by a branch generated by A is:

$$Gain(A) = I\left(s_{1,\ldots\ldots,} s_m\right) - E\left(A\right). \tag{3}$$

ID3 algorithm is described as follows:

1) To initialize the root node of the decision tree;
2) The algorithm stops if all nodes belong to the same class;
3) Otherwise, calculate the information gain of the pending attributes and selects the greatest information gain attributes as a branch node.
4) obtain the subset of this branch's attribute, and generate the corresponding leaf node;
5) Go to (2);

ID3 algorithm has the advantages of less calculation and simple generated rules, but its information gain calculation method require the attribute with more values, thus, the C4.5 algorithm comes being. C4.5 algorithm can discretize continuous attributes and process incomplete data, achieving pruning. In order to solve the issue of lack of information gain caused by attribute selection, C4.5 uses the information gain ratio to determine the branch node attribute. Its information gain ratio formula is:

$$GainRatio(A) = \frac{Gain(A)}{SplitInfo(A)}. \tag{4}$$

Of which

$$SplitInfo(A) = -\sum_{i=1}^{n} \frac{s_{1i} + \ldots\ldots s_{mi}}{s} \log_2 \left(\frac{s_{1i} + \ldots\ldots s_{mi}}{s} \right). \tag{5}$$

C4.5 algorithm will be pruning after the decision tree generated. If the classification error rate of a node exceeds a certain thresholds, we cut off all branches of this node, changing them to leaf nodes, and specify category for them. Compared to the ID3 algorithm, the accuracy of this algorithm is greatly improved and the generated rules are easier to understand. However, the data used by the C4.5 algorithm have to be stored in memory, so when the training set is particularly large, the program could not run.

4 Improved Naïve Bayes Algorithm and Its Application

Traditional Naive Bayes algorithm uses normal distribution when calculating the probability of continuous attributes, while calculating in the probability of discrete attributes, it uses simple probability calculation. For this, we made the following improvement for Bayesian method:

The probability $P(x_{1,}|C_i), P(x_{2,}|C_i), \ldots\ldots, P(x_{n,}|C_i)$ can be calculated from training set, the method is as follows:

For discrete attribute

$$P(x_{1,}|C_i) = \frac{N(x_1|C_i)}{N(C_i)}. \tag{6}$$

Of which $N(C_i)$ is the number recorded in the catalogue C_i and $N(x_1|C_i)$ is the number of the attributes x_1 recorded in the category C_i. If there are no records of x_1 in $P(x_{1,}|C_i) = 0.0001$, assume $P(x_i|C_i) = 0.0001$ to prevent posterior probability become 0. In fact, a certain attribute recorded by test should not be regarded to be out of the category when it never appears in some sort of training sets. If so, the classifier obtained will be lack of predictive ability in detecting those new record not appeared in the training set. To solve this problem and improve the detection rate of the classifier, we propose this method, which is simple to use compared to other existing methods. The reason why we set $P(x_i|C_i) = 0.0001$ is because the probability of such case is quite small, which means the probability will never be great even if happening.

For continuous attributes, the experiment uses Gaussian Kernel density to conduct estimation for the continuous attributes:

$$P\left(x_i \middle| C_i\right) = \left(n_{Ci}\sigma\right) \sum_{x_k : C(x_k) = C_i} K\left(\frac{x_k - \mu_i}{\sigma}\right). \tag{7}$$

Of which

$$K\left(x\right) = g\left(x, 0, 1\right) \tag{8}$$

$$g\left(x, \mu, \sigma\right) = \frac{1}{\sqrt{2\pi}\sigma} e^{\frac{-(x-\mu)^2}{2\sigma^2}}. \tag{9}$$

If a posterior probability is 0, the $P\left(x_i \middle| C_i\right) = 0.0001$, which can prevent the probability of measured record all be 0, because in such case, it has no way to judge which category the measured record belongs to. The reason why use Gaussian Kernel density method is the different recorded attributes have big differences and not all attributes values are meet normal distribution. The value distributions of different attributes have different characteristics. If all of them use the uniform normal distribution, the model can't maximum describe the authenticity of the data, and the Gaussian Kernel density method could make different attributes select kernel functions based on their own characteristics. It from the maximum extent reflects the authenticity of the data and improves the performance of the system.

The typical Naive Bayes algorithm is that all the models choose the same attributes. However, in fact, although a thing can be described by several attributes, some attributes among them are quite important to the thing. These attributes can generally distinct the thing and other attributes play an auxiliary role, which make the things more specific. To judge which category a thing belongs to is not to require that all the attributes meet the standards, but depends on the main characteristics. Actually, we have to use same attributes to describe different records to distinguish data. As the same attributes affect differently to the different data, different models can choose different attributes when establishing model, which will prevent interference information, reduce error and increase detection rate.

5 Summary

With the rapid development of the network, network security has become one of the most important issues, as it directly affects the interests of the state, enterprises and individuals. Further more, with the increasingly mature of the attack knowledge, complex and diverse attack tools and techniques, the current simple firewall technology has been unable to meet the needs of the people. Facing so many challenges and threats, intrusion detection and prevention technology is bound to become one of the core technologies in the security audit. The intrusion detection, playing a role of active defense, is an effective complement to the firewall, and is an important part of network security. This paper mainly analyzes the decision tree algorithm and improved Naive Bayes algorithm, proving the effectiveness of the improved Naive Bayes algorithm.

Acknowledgement. The work was supported by the Projects of Guangxi Provincial (201010LX493) and Hechi University(2011A--N006).

References

[1] Peddabachigari, S., Abraham, A., Grosan, C., Thomas, J.: Modelling Intrusion Detection System Using Hybrid Systems. J. Network Comput. Appl. 30, 114–132 (2007)

[2] Xiang, C., Chong, M.Y., Zhu, H.L.: Design of Multiple-level Tree Classifiers for Intrusion Detection System. In: Proc.2004 IEEE Conf. on Cybernetics and Intelligent Systems, Singapore, pp. 872–877 (December 2007)

[3] Xiang, C., Yong, P.C., Meng, L.S.: Design of Multiple-level Hybrid Classifier for Intrusion Detection System Using Bayesian Clustering and Decision Trees. Pattern Recognition Letters 29, 918–924 (2008)

[4] Kumar, S., Spafford, E.H.: Software Architecture to Support Misuse Intrusion Detection. In: Proceedings of the 18th National Information Security Conference, pp. 194–204 (2005)

[5] Smaha, S.E.: Haystack: An Intrusion Detection System. In: Proceedings of the IEEE Fourth Aerospace Computer Security Applications Conference, Orlando, FL, pp. 37–44 (2008)

[6] Lee, W.: A Data Mining for Constructing Features and Models for Intrusion Detection System, Ph. D. Dissertation, Columbia University (2009)

[7] Lee, W., Stolfo, S.J.: Data Mining Approaches for Intrusion Detection. In: Proc. the 7th USENIX Security Symposium, San Antonio, TX (2008)

[8] Pfahringer, B.: Winning the KDD99 Classification Cup: Bagged Boosting. In: SIGKDD Explorations, 2000 ACM SIGKDD, vol. 1(2), pp. 65–66 (January 2000)

[9] Kruegel, C., Mutz, D., Robertson, W., Valeur, F.: Bayesian Event Classification for Intrusion Detection. In: Proceedings of the 19th Annual Computer Security Applications Conference, Las Vegas, NV (2007)

[10] Valdes, A., Skinner, K.: Adaptive Model-based Monitoring for Cyber Attack Detection. In: Recent Advances in Intrusion Detection, Toulouse, France, pp. 80–92 (2008)

[11] Ye, N., Xu, M., Emran, S.M.: Probabilistic Networks with Undirected Links for Anomaly Detection. In: Proceedings of the IEEE Systems, Man and Cybernetics Information Assurance and Security Workshop, West Point, NY (2009)

[12] Portnoy, L., Eskin, E., Stolfo, S.J.: Intrusion Detection with Unlabeled Data Using Clustering. In: Proceedings of the ACM Workshop on Data Mining Applied to Security, Philadelphia, PA (2007)

The Information System of the Radio and Television Network Based on SOA

Ran Zhai

Information Engineer College,
Communication University of China,
Beijing 100024,
China
feng56yun@yahoo.com.cn

Abstract. SOA is a software architecture in which services are wrapped as independent unit and loosely coupled by the defined interfaces. It has been used in many enterprise-level systems. In this paper, we describe the realization mode of SOA firstly. Then an architectrure based on SOA used in the information system of the radio and television network is proposed. The structure of the architecture is divided into four layers: SODA, the data source service, ESB and the Applications. .

Keywords: SOA, Radio and television network, Software engineering.

1 The SOA Architecture

In SOA, the implementation of the services is transparent to the users, and they don't need to care about how to realize the services. The users can pay more attention to the result . In a SOA system, The information is exchanged using the efficient protocols. As the protocols are usually neutral and open, the realization of the services is ignored.

The newest SOA is based on XML language(it is also called Web Services Definition Language, WSDL) to difine the interfaces. It's more effiencient and flexible than the Interface Definition Language(IDL) in CORBA.

In this paper we describe an SOA-based distributed software architecture and present its protocols between the services and the users. Then we'll show how the archtitecture is used in Radio and Television System.

To complete an SOA-based architecture, we should not only describe the services, but also define the workflow between different unit in the system. The workflow should be reliable, safety and worthy of trust. The key point is to find the transform point between the service request and the services. Fig. 1 below shows the typical SOA structure.

Z. Du (Ed.): Proceedings of the 2012 International Conference of MCSA, AISC 191, pp. 191–196.
springerlink.com

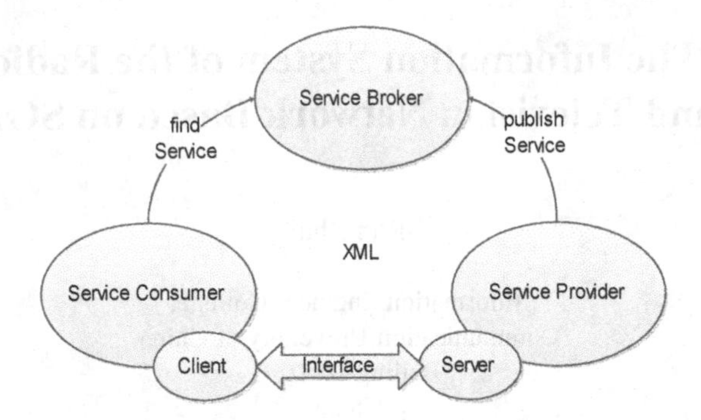

Fig. 1. The typical SOA structure

From the paragraph above, we can draw the following conclusions

(1) The workflow is easy and efficient: Firstly, the Service Provider register the name and description of its service in the Service Broker. The relevant Secret Key and protocols will also be delivered. The duty of the service Broker is to collect and store the description, protocols and Secret key of the service , thus providing them to the service users. With the protocols finding in the Service Broker, the user can begin to interact with the Provider. Finally, the architecture is like the Client/Server architecture.
(2) The key of the architecture is Service. It must be well-encapsulated. Howerver, the realization is Insignificant
(3) In the architecture , the information flow is based on precisely-defined neutral and open protocols. The appearance of XML language solves the problem perfectly. It's more general, flexibility and easy than the traditional SGML language. The Process of the regeister and the description of the Service usually use the UDDI, WSDL and so on.
(4) Undoubtedly, the safety of the Service Broker is important. Located in the central part of the architecture, it is the link between services and users. In essence, distributed SOA is to realize service reuse in loose coupling mode.

2 The Information System in Radio and Television Network Based on SOA

Having the weak points of high-cost, tightly coupled and being independent with othes, tradition applications , such as OA or EMP, in the information system of Radio and Television System is usually one-off. They are hardly accustomed to the Radio and Television which is integrating with telecommunication network and Internet. The SOA architecture has the characteristics of reusable and loosely coupled. The SOA architecture aims to build a platform-neutral, highly-flexible and standard-based IT environment to better cope with the ever-changing technical and service environments. Combining with the feathers of the radio and television network, the system is divided into four layers: the SODA, data source service, the ESB(enterprise service bus) and the applications.

The security systems including physical security, network security application security, data security and the standards and norms system including technical standards, management standards, service standards run through these four layers, making the architecture more perfect and specification. Description of the four layers are as follows.

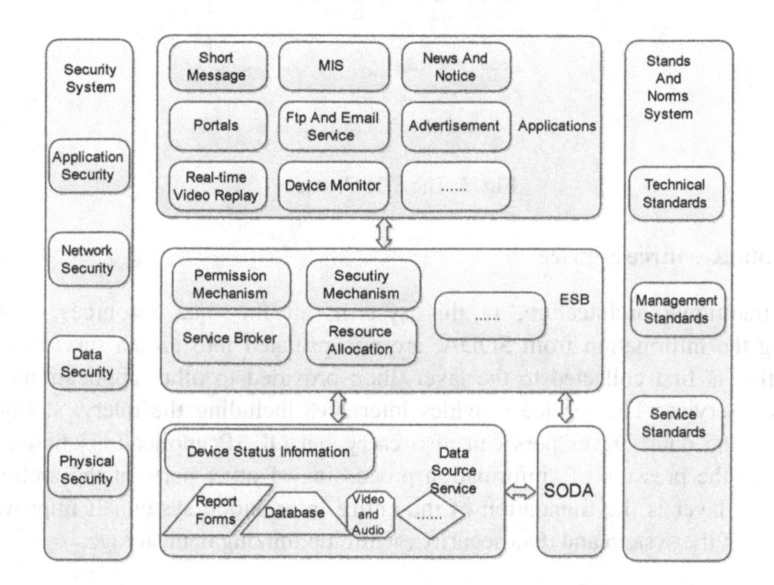

Fig. 2. The structure of the information system

2.1 The SODA

SODA means the service-oriented device architecture. It was firstly proposed by IBM and the University of Florida. SODA is committed to take full advantage of the existing standards of the field of IT and embedded system, to providing a standard platform for the integration of the embedded devices and SOA technology. Given the special characteristic that there are many embedded devices in the radio and television system, SODA is introducted to the architecture. SODA provides standard interfaces, converting the function of the devices to the software services which are unrelated with the hardware services, archiving the following objectives:

(1) makeing seamless connection between software and embedded devices.
(2) Integrate once , deploy everywhere, allowing the users to focus on the overall program rather than caught in the connection of the embedded devices
(3) By establishing a common inferface and DDL between the software applications and different hardware protocols , forming a unified data exchange standards.

Fig. 2 is a schematic of the SODA.

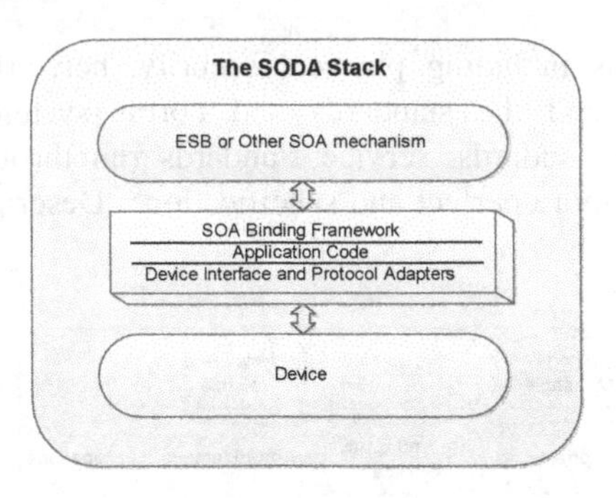

Fig. 3. The SODA stack

2.2 Data Resource Service

Unlike traditional architecture, in this system, all the data resources, including including the information from SODA, are encapsulated into a data service. All the information is first collected to the layer, then provided to other applications or the users as a service. The service provides interfaces including the query, storage and backup of the data. Developers can also carry out OLAP applications based on it. Bearing all the pressure of information processing of other parts of the architecture data source layer is the foundation of the entire information system. It improves the scalability of the system and data security , while optimizing data storage.

2.3 ESB (Enterprise Service Bus)

ESB connects users, applications and data source serveice. In the architecture, it is responsible for the communication and the resource allocation. The service broker is the core of this layer. Permission and security mechanism ensures the whole architecture works well.

2.4 Applications

Application layer is the collction of the services. The advantage of the arichitecture is mainl shown in this layer. Developers can easily develop new application and integrate it to this layer with defined interfaces and protocols. The more important point is that, due to the existence of XML, the concrete realization of these services is insignificant. It may be J2EE, LAMP or others.

Fig. 4 is the Network topology of the system.

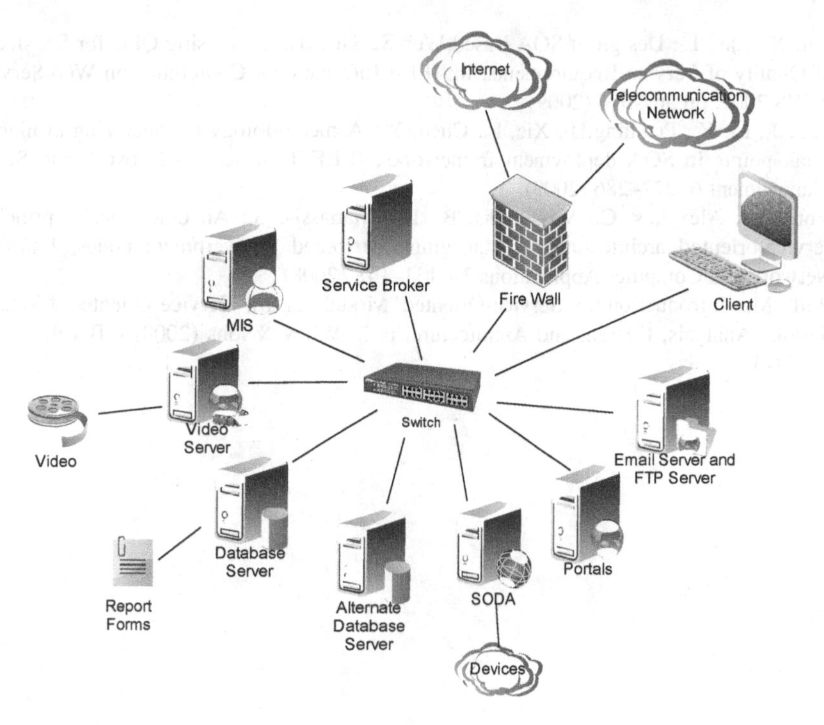

Fig. 4. The Network topology of the system

3 Conclusions

Comparing with traditional architecture. SOA is loosely coupled. highly efficient and more flexible. It is more suitable for the enterprise-level system. The information system in radio and television network take full advantage of the private network of the radio and television system. making the integrations of the applications more easily. SOA must be the mainstream architecture.

References

[1] Fang, Y.-M., Lee, B.-J., Chou, T.-Y., Lin, Y.-I., Lien, J.-C.: The implementation of SOA within grid structure for disaster monitoring. Expert Systems with Applications. Uncorrected Proof (June 2008) (in Press)

[2] Li, H., Wu, Z.: Research on Distributed Architecture Based on SOA. In: International Conference on Communication Software and Networks, ICCSN 2009, pp. 670–674 (2009)

[3] Peng, K.-Y., Lui, S.-C., Chen, M.-T.: A Study of Design and Implementation on SOAGovernance: A Service Oriented Monitoring and Alarming Perspective. In: IEEE International Symposium on Service-Oriented System Engineering, SOSE 2008, pp. 215–220 (2008)

[4] Liu, X., Zhu, L.: Design of SOA Based Web Service Systems Using QFD for Satisfaction of Quality of Service Requirements. In: IEEE International Conference on Web Services, ICWS 2009, pp.567-574 (2009)
[5] Luo, J., Li, Y., Pershing, J., Xie, L., Chen, Y.: A methodology for analyzing availability weak points in SOA deployment frameworks. IEEE Transactions Network and Service Management 6, 277–286 (2009)
[6] Votis, K., Alexakos, C., Vassiliadis, B., Likothanassis, S.: An ontologically principled service-oriented architecture for managing distributed e-government nodes. Journal of Network and Computer Applications 31, 131–148 (2008)
[7] Bell, M.: Introduction to Service-Oriented Modeling. In: Service-Oriented Modeling: Service Analysis, Design, and Architecture, p. 3. Wiley &Sons (2008) ISBN 978-0-470-14111-3

Research of Communication System on Intelligent Electric Community Based on OPLC

Zengyou Sun, Peng Chen, and Chuanhui Hao

School of Information Engineering,
Northeast Dianli University, Jilin City, Jilin, China
{sunzengyou,hchh518666}@163.com

Abstract. Article illustrates working principle of OPLC construction for the communication network in the intelligent community and EPON technology , as well as utilizes service agreement(SLA) queue priority of dynamic bandwidth allocation algorithm for EPON uplink access, not only ensures the operational bandwidth allocation fairness of the queue ,but also increases throughout of bandwidth. At last it exhibits construction scheme of communication network.

Keywords: OPLC, Communication network, Network convergence, EPON.

1 Instruction

Along with the development of modern communications, computers networking, automation, and other technologies, intelligent device of family sprawled from inexistence, intelligent systems community gradually developed from a simple building intercom system to the premises distribution system, computer network, the security systems of the public district, automation systems of device management, as well as property management system for equipment management in the community. Intelligent electricity community is utilizing technology such as smart metering, the terminal for smart user interaction, established connectivity of real time between users and electric network, and open digital networks ,achieves the entire collection, whole coverage. It reduces labor costs, reduces electricity losses, improves measurement accuracy, reduces human error, promotes mechanism innovation of marketing management, safe and reliable power supply, Electrovalency of a flight of stairs, spurs energy-saving and pollution reduction, and guidance the reasonable electric-consumption. Its service covers information of power-consumption and automation in the district power supply, accomplishes fiber to the family client and triple nets(radio and television network ,telecommunication network ,Internet)fusion, platform of intelligent service , photovoltaic power generation system, smart home services, such as the automation of water, gas and electric metering.

Construction of communication network in intelligent district aimed at achieving power fiber to the client, that is used in low voltage communication access network OPLC, laying optical fiber low-voltage power line and realization to the table for each household, combined with passive optical network technology, bearing power collection, the smart electric interactive, "triple nets fusion" business, and for the

Z. Du (Ed.): Proceedings of the 2012 International Conference of MCSA, AISC 191, pp. 197–202.
springerlink.com

protection of distribution network automation. Intelligent communication network construction form logic power private communication subnet on to business and public communication network independent network.

2 Communication Network and Its Key Technology

Intelligent electricity business requirements on the basis of analysis, to analyze the structure of communication systems, to expand the core functionality and features into the individual subsystems, communication network is divided into unit from the logic layer, layer and business layer. Equipment includes electric power optical fiber to the home community for all access electric power optical network terminal equipment, such as: smart energy, intelligent home appliances and so on. Communication layer is based on electric power optical fiber network of EPON technology, EPON structure is a point to of single fiber bi-directional optical access network by routing the network side optical line terminal(OLT), a user-side optical network unit(ONU) and the optical distribution network(ODN).Business is hosting all of the business system of intelligent communication. Figure 1 power fiber to the home is a logical architecture diagram.

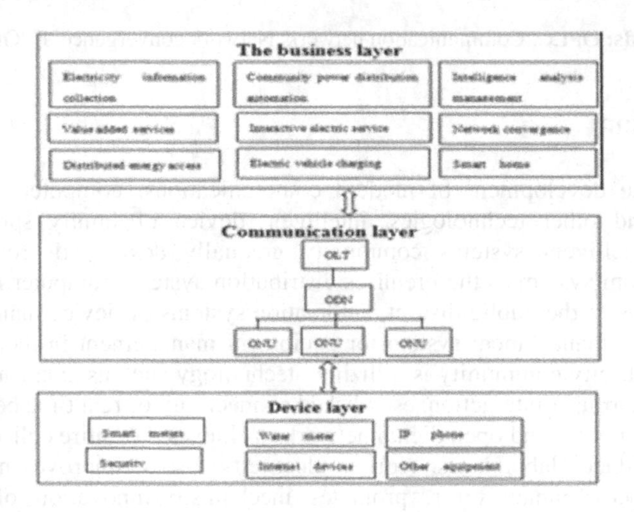

Fig. 1. Power fiber to the home is a logical architecture diagram

Optical fiber composite low-voltage cables(OPLC) product is a fusion of traditional low-voltage cable light unit, is a kind of low-voltage electrical and optical communication of double transmission capacity of composite cable. OPLC adapted to the 0.6/1 KV and following main voltage levels, the solution to the low-voltage distribution networks, customer network required advanced, reliable communications medium. Fiber composite low-voltage cables integrated power network resources, network infrastructure of "sharing", to solve the wire, cable, telephone lines, cable lines and other communication in Nan tong, avoid repeated construction due to interfere with the normal life of the masses.

Electric power optical fiber composites cable than traditional low-voltage cable costs less than 10% per cent, OPLC smart power integrated communications networks in the community-building cost around 40%.

3 EPON Technology

Typical of EPON system[3] as shown in figure 2, by the OLT,ONU composition ,OLT by ODN are connected to each ONU,OLT in the root node, place in the center office side, it is the main co-ordinating control center, control channel connection. ONU is placed in the user field, access the user terminal. ODN complete downlink and uplink signal of optical signal power distribution set, are passive devices connected OLT and ONU, their light between 1:16 – 1:128.EPON uses IEEE802.3 compliant Ethernet frame carrying business information agreement. EPON systems in downlink link, OLT to broadcast to send Ethernet frames.ONU on the uplink data frame in a sudden manner through joint transfer of passive coax distribution network to the OLT, uplink channel access is a multiple access method, generally uses TDMA[4] time division multiple access. Downward direction using broadcast mode.More than data from OLT to ONU according to different time slices to broadcast the following line(TDM time division multiplexing),through 1:N in the ODN (typically 1:32)passive, allocated to all PON ONU optical network unit. When OLT started, it will periodic to in all port broadcast allows access of gap allows access information, when ONU electric under allows access information, launched registered requests, implementation OLT on ONU of certification, allows requests registered of legitimate of ONU access, and to ONU distribution a only of logic chain road identifies(LLID), when data signal arrived the ONU, ONU under LLID in physical layer made judge, received to it of data frame, abandoned not to themselves of data frame.

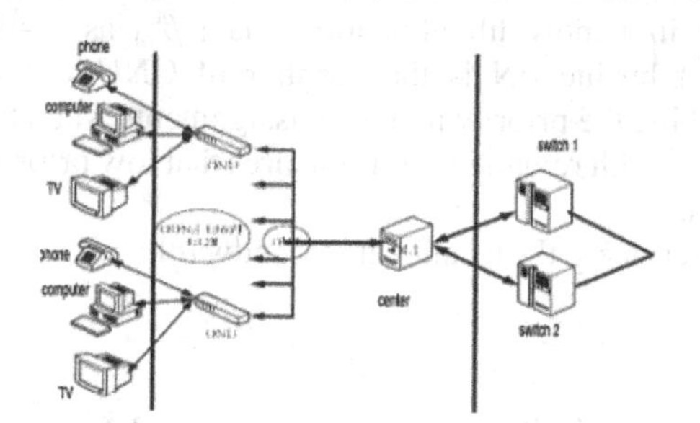

Fig. 2. Structure of EPON system

4 Dynamic Bandwidth Allocation Algorithm for EOPN Upstream Access

This is a support service level agreement (SLA)of the queue priority assignment algorithm. Under the current user ONU ends the connection there are three main types: voice, video, data EPON services level is divided into 3 priority level[5-6]: high priority voice traffic, video business priority and low-priority data service. According to the SLA 's different, assuming the minium bandwidth for each ONU ensure to Bi^{min} total bandwidth occupied by high-priority for B_C it by controlling information bandwidth and voice B_{voice} composition in real time. Amount of bandwidth, low priority in B_M and B_L respectively.ONU has registered the total bandwidth to B_{use} .

High-priority business on-demand bandwidth allocated to it, namely:

$$B_{voice} = \sum_{i=1}^{N} G^{i}_{voice} \tag{1}$$

Where G^{i}_{voice} is i - i of ONU, real-time voice traffic bandwidth. If there are no higher priority of business, only B_i^{min} size for the transfer control information.

Priority bandwidth allocation:

To ONU i - i priority video traffic in bandwidth allocation policy with maximum and minimum window policy: i G_M maximum transfer window as $TMW^{[i]}_{max}$.Its allocated bandwidth will not exceed the allocated bandwidth of the system, at the same time the minimum bandwidth guarantee TMW_{min} ,priority video traffic in light load and heavy load conditions may be allocated under the authorization window of maximum and minimum bandwidth calculation formula is as follow:

$$TMW^{i}_{max} = \frac{(B_{use} - NG^{i}_{C} - \sum_{i=1}^{N} G^{i}_{voice})\beta^{i}_{M}\sigma^{i}}{N} \tag{2}$$

$$TMW^{[i]}_{min} = \frac{(B_{use} - NG^{i}_{C} - \sum_{i=1}^{N} G^{i}_{voice}) \times \alpha^{i}_{M}}{N} \tag{3}$$

When $TMW^{[i]}_{min} < \beta^{i}_{M} < TMW^{i}_{max}$, $G^{i}_{M} = \beta^{i}_{M}$; when $TMW^{[i]}_{min} > \beta^{i}_{M}$, $G^{i}_{M} = TMW^{[i]}_{min}$; when $TMW^{i}_{max} < \beta^{i}_{M}$, $G^{i}_{M} = < TMW^{i}_{ma}$。 Where α^{i}_{M} coefficient of secondary priority traffic bandwidth guarantee for the system, which is ensuring stability at low load priority video traffic in bandwidth allocation; data β^{i}_{M} as i - i in ONU priority bandwidth requests of business;N is the number of ONU,$\sigma^{i}(0 < \sigma^{i} < 1)$for equitable adjustment factor, making the priority in queue assignments when the bandwidth is still a small amount of bandwidth remaining, this ensures that low priority queue is assigned to a certain bandwidth.

Low-priority data service g iL, its allocated bandwidth:

$$G^{i}_{L} = \frac{\left(B_{use} - NG^{i}_{C} - \sum_{i=1}^{N} G^{i}_{voice} - \sum_{i=1}^{N} G^{i}_{M}\right) \cdot \delta^{i}_{L}}{\sum_{i=1}^{N} \delta^{i}_{L}} \tag{4}$$

Where δ^{i}_{L} a i-ONU I lower priority bandwidth requests of data services. From the above you can get I have allocated bandwidth to ONU I registered:

$$G^{i}_{total} = G^{i}_{M} + G^{i}_{L} + G^{i}_{C} + G^{i}_{voice} \tag{5}$$

Where ONU I registered for part I of the allocation of the total bandwidth.

5 Construction Plan

Communication network construction divides into the electric power communication network communication network and the public network communication subnet with two parts. Power dedicated to communications networks provide communications support for electricity information collection network, the collection points are a fiber access, broadband communications network. Public communications network bring IPTV to community users, such as Internet, phone service provides communications network platform, each of the tow types of network to establish a fiber-optic network, using the same cable core construction of tow different sets of EPON system so that the tow kinds of business to achieve physical isolation, ensuring data transmission security, stability. Blow shows a typical intelligent electro-communications networks in the community-building programme, which physically divides into three parts, trunk, feeder lines, wiring,main part of the them using ordinary optical cable, by means of aerial, buried pipelines and other connection center to junction points(light boxes); Feeder parts feeder cable(or the introduction of optical cable), through aerial, shallow buried pipelines, the road connecting the junction point to branch point(of a community or the focal point of the building), wiring parts wiring parts wire cable, connecting branch points to a user; And users within the family home cable(also known as the indoor optical fiber cables). You can access using EPON/GPON technology equipment to achieve development of broadband services. As shown Figure 3.

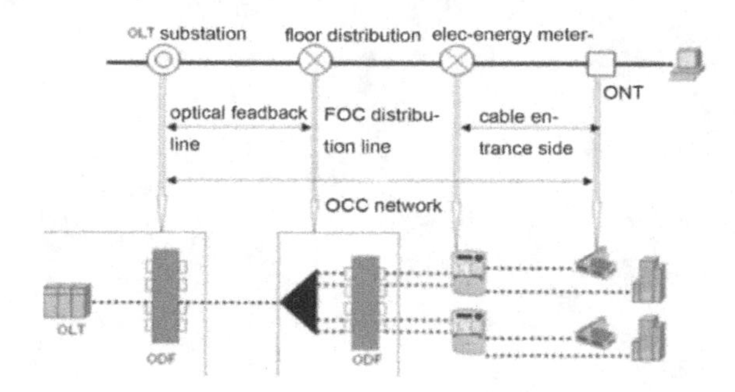

Fig. 3. Communication network construction scheme

6 Conclusion

As intelligent grid in distribution side and electricity side related business of rapid development, OPLC cable for three network fusion provides has reliable protection, this high price broadband fiber products can implementation end user of access, EPON system to its powerful of exchange function and rich of business policy, unified of more business access will gradually become intelligent electricity community construction of mainstream technology.

References

[1] Deng, G., Chen, J.: Intelligent power communication and its key technology. Hu Bei Electric Power (s1) (2010)

[2] Li, X., Yang, J., Qi, X.: Analysis and discussion on application of optical fiber composite cable. Power System Communication 7, 24–27 (2011)

[3] Jang, S.-H., Kim, J.-M., Jang, J.-W.: A new DBA algorithm supporting priority queues and fairness for EPON. In: Proceedings of Asia-pacific Optical Communications, vol. 12(07), pp. 56–63 (2004)
[4] Wei, Q.: A high efficient dynamic bandwidth scheme for QoS over EPON system. In: Proceedings of Asia-pacific Optical Communications, vol. 15(07), pp. 151–157 (2005)
[5] Kramer, G., Mukherjee, B., Maislos, A.: Ethernet passive optical networks. In: Dixit, S. (ed.) IP Over WDM: Building the Next Generation Optical Internet, pp. 234–239. John Wiley & Sons, Inc. (2003)
[6] An, F.T., Bae, H., Hsueh, Y.L., et al.: A new media access control protocol guaranteeing fairness among users in Ethernet-based passive optical networks. IEEE OFC 1(12), 134–135 (2003)

End-Orthodox Graphs Which Are the Join of N Split Graphs

Hailong Hou, Aifen Feng, and Rui Gu

School of Mathematics and Statistics,
Henan University of Science and Technology,
Luoyang, 471003, P.R. China

Abstract. A graph X is said to be End-orthodox if its endomorphism monoid End(X) is an orthodox semigroup. In this paper, we characterize the endomorphism monoid of the join of n split graphs. We give the conditions under which the endomorphism monoid of a join of n split graphs is orthodox.

Keywords: Monoid, orthodox semigroup, join.

1 Introduction

Endomorphism monoids of graphs are generalizations of automorphism groups of graphs. In recent years much attention has been paid to endomorphism monoids of graphs and many interesting results concerning graphs and their endomorphism monoids have been obtained. The aim of this research is try to establish the relationship between graph theory and algebraic theory of semigroups and to apply the theory of semigroups to graph theory.

Let X and Y be two graphs. A mapping from $V(X)$ to $V(Y)$ is called a homomorphism if $\{a,b\}\in E(X)$ implies that $\{f(a),f(b)\}\in E(Y)$ .A homomorphism from X to itself is called an endomorphism of X . An endomorphism f of X is said to be half-strong if $\{f(a),f(b)\}\in E(X)$ implies that there exist $c\in f^{-1}(a)$ and $d\in f^{-1}(b)$ such that $\{c,d\}\in E(X)$. Denote by $End(X)$ and $hEnd(X)$ the set of endomorphisms and half-strong endomorphisms of X . It is known that $End(X)$ forms a monoid with respect to the composition of mappings and is called the endomorphism monoid (or briefly monoid) of X . The reader is referred to [1] and [2] for all the notation and terminology not defined here. We list some known results which will be used frequently in the sequel to end this section.

Lemma 1.1([5]). Let X and Y be two graphs. If $X+Y$ is End-regular, then both X and Y are End-regular

Lemma 1.2([6]). Let X be a connected split graph with $V(X)=K\cup S$, where S is an independent set and K is a maximal complete set, $|K|=n$. Then X is End-regular if and only if there exists $r\in\{1,2,\cdots,n\}$ such that $d(x)=r$ for any $x\in S$.

Z. Du (Ed.): Proceedings of the 2012 International Conference of MCSA, AISC 191, pp. 203–207.
springerlink.com

Lemma 1.3([4]). Let $X_1, X_2, \cdots, X_n$ be n split graphs. Then $X_1, X_2, \cdots, X_n$ is End-regular if and only if

(1) X_i is End-regular for any $1 \le i \le n$,

(2) $q_i - d_i = q_j - d_j$ for any $1 \le i \le n$ and $1 \le j \le n$,

(3) There are no two vertices $x_1, x_2 \in S_t$ such that $N_{X_t}(x_1) \cup N_{X_t}(x_2) = V(K_t)$ for any $1 \le t \le n$.

Lemma 1.4([3]). Let X be a split graph with $V(X) = K \cup S$, where S is an independent set and K is a maximal complete set, If for some $y_i \ne y_j$, $N(y_i) \subseteq N(y_j)$, then X is not End-orthodox.

2 Main Results

The join of n split graphs whose endomorphism monoids is regular was characterized in [6]. In this section, we will give the conditions under which the endomorphism monoid of the join of n split graph is orthodox.

Let X_i ($i = 1, 2, \cdots, n$) be a split graph with $V(X_i) = V(K_i) \cup S_i$, where $S_i = \{x_{i1}, \cdots, x_{ip_i}\}$ is an independent set and $V(K_i) = \{k_{i1}, k_{i2}, \cdots, k_{iq_i}\}$ is a maximal complete set. Let X_i be a split graph withvertex set $V(X_i) = V(K_i) \cup S_i$, where $S_i = \{x_{i1}, \cdots, x_{ip_i}\}$ is an independent set and $V(K_i) = \{k_{i1}, k_{i2}, \cdots, k_{iq_i}\}$ is a maximal complete set. Then the vertex set $V(X_1 + X_2 + \cdots + X_n)$ can be partitioned into $n+1$ parts $K, S_1, S_2, \cdots, S_n$, where $V(K) = V(K_1) \cup V(K_2) \cup \cdots \cup V(K_n)$ is a complete set, $S_1, S_2, \cdots, S_n$ are independent sets. Obviously the subgraph of $X_1 + X_2 + \cdots + X_n$ induced by K is a complete graph and the subgraph of $X_1 + X_2 + \cdots + X_n$ induced by $S_1 \cup S_2 \cup \cdots \cup S_n$ is a complete n partite graph. By Lemma 1.3, we know if $X + Y$ is End-regular, then both of X and Y are End-regular. Clearly, If $X_1 + X_2 + \cdots + X_n$ is End-regular, then X_i is End-regular for any $1 \le i \le n$. So we always assume that X_i are End-regular sprit graphs in the sequel unless otherwise stated. Moreover, let d_i be the valency of the vertices of S_i in X_i. Clearly, if X_i is connected, then $1 \le d_i \le n-1$; if X_i is non-connected, then $d_i = 0$.

Lemma 2.1. Let $G_1, G_2, \cdots, G_n$ be n graphs. If $G_1 + G_2 + \cdots + G_n$ is End-orthodox, then G_i is End- orthodox for any $1 \le i \le n$.

Proof. Since $G_1 + G_2 + \cdots + G_n$ is End-orthodox, $G_1 + G_2 + \cdots + G_n$ is End-regular. By Lemma 1.3, G_i is End-regular for any $1 \le i \le n$. To show G_i is End-orthodox, we only need to prove that the composition of any two idempotent endomorphisms of G_i is also an idempotent.

Let f_1 and f_2 be two idempotents in $End(G_i)$. Define two mappings g_1 and g_2 from $V(G_1+G_2+\cdots+G_n)$ to itself by

$$g_1(x)=\begin{cases} f_1(x), & \text{if } x\in V(G_i) \\ x, & \text{if } x\in V(G_1+G_2+\cdots+G_n)\setminus V(G_i) \end{cases}.$$

$$g_2(x)=\begin{cases} f_2(x), & \text{if } x\in V(G_i) \\ x, & \text{if } x\in V(G_1+G_2+\cdots+G_n)\setminus V(G_i) \end{cases}.$$

Then g_1 and g_2 are two idempotents of $End(G_1+G_2+\cdots+G_n)$ and so g_1g_2 is also an idempotent of $End(G_1+G_2+\cdots+G_n)$ since $G_1+G_2+\cdots+G_n$ is End-orthodox. Clearly, $f_1f_2=(g_1g_2)|_{G_i}$, the restriction of g_1g_2 to G_i. Hence f_1f_2 is an idempotent of $End(G_i)$ as required.

Lemma 2.2. Let G be a graph. Then G is End-orthodox if and only if $G+K_n$ is End-orthodox for any positive integer n.

Proof. If $G+K_n$ is End-orthodox, then by Lemma 4.1, G is End- orthodox.

Conversely, for any positive integer n, by Lemma 1.4, if X is End- regular, then $X+K_n$ is End-regular. Let f be an idempotent of $End(G+K_n)$. Note that $\varpi(G+K_n)=\varpi(G)+n$, $V(K_n)\subset I_f$ and $f|_{K_n}=1|_{K_n}$, the identity mapping on K_n. Hence $f(V(G))\subseteq V(G)$ and $f|_G\in Idpt(G)$.

If f_1 and f_2 are two idempotents of $End(G+K_n)$, let $g_1 = f_1|_G$ and $g_2=f_2|_G$. Then $g_1, g_2\in Idpt(G)$ and so $g_1g_2\in Idpt(G)$. Now $(f_1f_2)|_{K_n}=1|_{K_n}$ and $(f_1f_2)|_G=g_1g_2$ imply that f_1f_2 is an idempotent of $End(G+K_n)$. Consequently $G+K_n$ is End- orthodox.

Let X_i $(i=1,2,\cdots,n)$ be two split graphs. If $X_1+X_2+\cdots+X_n$ is End-orthodox, then $X_1+X_2+\cdots+X_n$ is End-regular and X_i is End- orthodox for any $1\le i\le n$. The following lemma describes the idempotent endomorphisms of certain End-regular graphs $X_1+X_2+\cdots+X_n$.

Lemma 2.3. Let $X_1, X_2,\cdots, X_n$ be n split graphs with $d_i\le q_i-2$ for any $1\le i\le n$. If $N_X(x_1)\ne N_X(x_2)$ for any two vertices $x_1, x_2 \in S_1\cup S_2\cup\cdots\cup S_n$, then $f\in End(X_1+X_2+\cdots+X_n)$ is a retraction (idempotents) if and only if

(1) $f(x)=x$ for any $x\in V(K)$.

(2) For any $y\in S_1\cup S_2\cup\cdots\cup S_n$, either $f(y)\in V(K)\setminus N(y)$, or $f(y)=y$.

Proof. Note that under the hypothesis of lemma, $X_1+X_2+\cdots+X_n$ has an unique maximum clique K.

Lemma 2.4. Let $X_1, X_2, \cdots, X_n$ be n split graphs with $d_i \le q_i - 2$ for any $1 \le i \le n$. Then $X_1 + X_2 + \cdots + X_n$ is End-orthodox if and only if

(1) $X_1 + X_2 + \cdots + X_n$ is End-regular,

(2) $N_X(x_1) \ne N_X(x_2)$ for any two vertices $x_1, x_2 \in S_1 \cup S_2 \cup \cdots \cup S_n$.

Proof. Necessity is obvious.

Conversely, since $X_1 + X_2 + \cdots + X_n$ is End-regular, we only need to prove that the composition of two idempotent endomorphisms is also an idempotent. Let f be an arbitrary idempotent of $End(X_1 + X_2 \cdots + X_n)$ .Then $f|_{V(K)} = V(K)$ and either $f(x) = x$ or $f(x) = k_x$ for any $x \in S_1 \cup S_2 \cup \cdots \cup S_n$, where k_x is a vertex in $V(K)$ such that $\{x, k_x\} \notin E(X_1 + X_2 + \cdots + X_n)$. Now the assertion follows immediately.

Lemma 2.5. Let $X_1, X_2, \cdots, X_n$ be n split graphs with $d_i \le q_i - 1$ for any $1 \le i \le n$. Then $X_1 + X_2 + \cdots + X_n$ is End-orthodox if and only if $|S_1| = |S_2| = \cdots = |S_n|$.

Proof. Necessity is obvious.

Conversely, $X_1 + X_2 + \cdots + X_n$ is a join of a complete graph and a complete n partite graph. Since any complete n partite graph is End- orthodox, it follows from Lemma 2.2 that $X_1 + X_2 + \cdots + X_n$ is End- orthodox.

Theorem 2.6. Let $X_1, X_2, \cdots, X_n$ be n split graphs. Then $X_1 + X_2 + \cdots + X_n$ is End-orthodox if and only if

(1) X_i is End-regular for any $1 \le i \le n$,

(2) $q_i - d_i = q_j - d_j$ for any $1 \le i \le n$ and $1 \le j \le n$,

(3) There are no two vertices $x_1, x_2 \in S_t$ such that $N_{X_t}(x_1) \cup N_{X_t}(x_2) = V(K_t)$ for any $1 \le t \le n$,

(4) $N_X(s_1) \ne N_X(s_2)$ for any two vertices $s_1, s_2 \in S_1 \cup S_2 \cup \cdots \cup S_n$.

Proof. If $X_1 + X_2 + \cdots + X_n$ is orthodox, then $X_1 + X_2 + \cdots + X_n$ is regular and so both of X_i is regular for any $1 \le i \le n$. Now it follows immediately from Lemma 1.3, Lemma 2.4 and 2.5.

In conjunction with Lemma 1.3, we obtain another version of the previous theorem as follows:

Theorem 2.6*. Let $X_1, X_2, \cdots, X_n$ be n split graphs. Then $X_1 + X_2 + \cdots + X_n$ is End-orthodox if and only if

(1) $X_1 + X_2 + \cdots + X_n$ is End-regular;

(2) X_i is End-orthodox for any $1 \le i \le n$.

Acknowledgement. The authors want to express their gratitude to the referees for their helpful suggestions and comments.

This research was partially supported by the National Natural Science Foundation of China (No.10971053), the Natural Science Foundation of Shandong province (No. 2010ZRE09006) and Youth Foundation of Henan University of Science and Technology (No. 2010QN0036).

References

[1] Godsil, C., Royle, G.: Algebraic Graph Theory. Springer, New York (2000)
[2] Howie, J.M.: Fundamentals of Semigroup Theory. Clarendon Press, Uxford (1995)
[3] Fan, S.: Retractions of Split Graphs and End-orthodox Split Graphs. Discrete Mathematics 257, 161–164 (2002)
[4] Hou, H., Feng, A.: End-regularity of the join of n split graphs (to appear)
[5] Li, W.: Graphs with Regular Monoid. Discrete Mathematics 265, 105–118 (2003)
[6] Li, W., Chen, J.: Endomorphism-regularity of Split Graphs. European Journal of Combinatorics 22, 207–216 (2001)

The Zero-Divisor Semigroups Determined by Graphs Gn(2,1)

Hailong Hou, Aifen Feng, and Rui Gu

School of Mathematics and Statistics,
Henan University of Science and Technology, Luoyang, 471003, P.R. China

Abstract. In this paper, we characterize the structures of the commutative zero-divisor semigroups determined by graphs Gn(2,1). We also give a formula to calculate the number of mutually non-isomorphic commutative zero-divisor semigroups of these graphs.

Keywords: Zero-divisor Graph, Zero-divisor Semigroup, Complete graph.

1 Introduction

For any commutative semigroup S with zero element 0, the zero-divisor graph G(S) corresponding to S is a graph whose vertex set is the set of all zero divisors of S, and two vertices x and y are joined by an edge in case xy = 0. In [1,2], some fundamental properties and proper structures of these graphs were given.

Let G be a graph. The vertex set of G is denoted by V (G) and the edge set of G is denoted by E(G). If two vertices a and b are adjacent in graph G, then the edge connecting a and b is denoted by a-b. The reader is referred to [3,4,5] for all the notation and terminology not defined here.

2 Commutative Zero-Divisor Semigroups Determined by Gn(2,1)

In this section, we characterize the structures of zero-divisor semigroups determined by a class of graphs, denoted by Gn(2; 1). For every n ≥ 3, Gn(2; 1) is a graph with vertex set $M_n \cup \{x_1, x_2\}$, where the subgraph of Gn(2; 1) induced by vertex set $M_n = \{a_1, a_2, ..., a_n\}$ is a complete graph Kn and $x_1 - a_1$, $x_2 - a_2$, $x_2 - a_3$. We denote by $M_n(2,1) = M_n \cup \{0, x_1, x_2\}$ the possible zero-divisor semigroups determined by Gn(2; 1).

Theorem 2.1. There are four commutative zero-divisor semigroups determined by G3(2; 1) up to isomorphism. The multiplication tables of them are Table1-Table4.

Proof. We suppose $M_3(2,1)$ is a commutative zero-divisor semigroup such that G($M_3(2,1)$) = G3(2; 1). By the definition of zero-divisor graph, we immediately have $a_1 x_1 = 0$, $a_i x_2 = 0$ (i=2,3) and $a_j a_k = 0$ for any $j \neq k$. Note that $\{0, a_1\}$ and

Z. Du (Ed.): Proceedings of the 2012 International Conference of MCSA, AISC 191, pp. 209–213.
springerlink.com

$\{0,a_2,a_3\}$ are ideals in $M_3(2,1)$. Therefore we have $a_1x_2=a_1$, $a_2x_1\in\{a_2,a_3\}$ and $a_3x_1\in\{a_2,a_3\}$.

Since $a_2x_1^2\neq 0$ and $a_3x_1^2\neq 0$, then we have $x_1^2\neq 0$ and x_1^2 is not adjacent to both of a_2 and a_3. Therefore we have $x_1^2=x_1$. Now $a_1x_2^2=a_1x_2=a_1$. Thus $x_2^2\neq 0$ and x_2 is not adjacent to a_1. Therefore we have $x_2^2\in\{a_1,x_2\}$. In the following, we divide it into two cases to discuss:

Case 1. Assume $x_2^2=x_2$. We consider x_1x_2. Since x_1 is not adjacent to x_2 in graph G3(2; 1), $x_1x_2\neq 0$. Now from $a_1x_1x_2=0$ and $a_2x_1x_2=0$, we obtain that $x_1x_2\notin\{x_1,x_2\}$ and therefore we have $x_1x_2\in M_3$. If $a_2x_1x_2=a_1$, then we have $x_1x_2=x_1^2x_2=a_1x_1=0$. A contradiction. If $x_1x_2=a_i$ (i = 2, 3), then we have $x_1x_2=x_1x_2^2=a_ix_2=0$. Also a contradiction. Hence there is no semigroup determined by G3(2; 1) in this case.

Case 2. Assume $x_1^2=a_1$. Then $a_1^2=a_1x_2^2=a_1$. With a similar argument of case 1, we have $\{x_1,x_2\}\notin\{0,x_1,x_2,a_1\}$. Therefore $\{x_1,x_2\}\in\{a_2,a_3\}$. By symmetry, we may suppose $x_1x_2=a_2$. Now, we have $a_2^2=a_2x_1x_2=0$ and $a_2x_1=x_1x_2x_1=x_1x_2=a_2$. If $a_3x_1=a_2$, then $a_3^2x_1^2=a_3^2x_1=a_2^2=0$. Thus $a_3^2\in\{0,a_1\}\cap\{0,a_2,a_3\}$ and hence $a_3^2=0$; if $a_3x_1=a_3$, then $a_3^2\in\{0,a_2,a_3\}$ since $\{0,a_2,a_3\}$ is an ideal in $M_3(2,1)$. We can list all the possible multiplication tables as Table 1-Table 4.

The final work is to verify each table defines an associative binary operation. Fortunately, these multiplication tables are associative. In fact, the associativity of thetables can be verified by using a program provided by E.W.H.Lee. This complete the whole proof.

Theorem 2.2. There are ten commutative zero-divisor semigroups determined byG4(2; 1) up to isomorphism. The multiplication tables of them are Table 5-Table 14.

Proof. We suppose $M_4(2,1)$ is a commutative zero-divisor semigroup such that $G(M_4(2,1))=G_4(2,1)$. A similar argument of Theorem 2.1 will show that the possible multiplication tables up to isomorphism are Table5-Table 14.

	a_1	a_2	a_3	x_1	x_2
a_1	a_1	0	0	0	a_1
a_2		0	0	a_2	0
a_3			0	a_2	0
x_1				x_1	a_2
x_2					a_1

Table 1

	a_1	a_2	a_3	x_1	x_2
a_1	a_1	0	0	0	a_1
a_2		0	0	a_2	0
a_3			$0, a_2, a_3$	a_3	0
x_1				x_1	a_2
x_2					a_1

Table 2, 3, 4

	a_1	a_2	a_3	a_4	x_1	x_2
a_1	$0, a_1$	0	0	0	0	a_1
a_2		0	0	0	a_2	0
a_3			0	0	a_2	0
a_4				a_4	a_4	a_4
x_1					x_1	a_4
x_2						x_2

Table 5, 6

	a_1	a_2	a_3	a_4	x_1	x_2
a_1	$0, a_1$	0	0	0	0	a_1
a_2		a_2	0	0	a_2	0
a_3			a_3	0	a_3	0
a_4				a_4	a_4	a_4
x_1					x_1	a_4
x_2						x_2

Table 7, 8

	a_1	a_2	a_3	a_4	x_1	x_2
a_1	$0, a_1$	0	0	0	0	a_1
a_2		0	0	0	a_2	0
a_3			$0, a_2, a_3$	0	a_3	0
a_4				a_4	a_4	a_4
x_1					x_1	a_4
x_2						x_2

Table 9 – 14

The final work is also to verify each table de_nes an associative binary operation. This work was done by the program provided by E.W.H.Lee. This complete the whole proof.

Theorem 2.3. A semigroup $M_n(2,1)$ is a zero-divisor semigroup of Gn(2,1) if and only if the following conditions holds:

(1) $a_1x_1 = 0$, $a_2x_2 = 0$, $a_3x_2 = 0$, $a_1x_2 = a_1$, $x_1^2 = x_1$, $x_2^2 = x_2$, $a_2^2 \in \{0, a_2, a_3\}$, $a_3^2 \in \{0, a_2, a_3\}$ and $a_ia_j = 0$ for any $i \neq j$.

(2) $x_1x_2 \in \{a_4, ..., a_n\}$. If $x_1x_2 = a_t$ for some $t \geq 4$, then we have $a_tx_k = a_t$ ($k = 1, 2$), $a_t^2 = a_t$, $a_rx_2 = a_1$ for any $r \geq 4$ and $r \neq t$, $a_j^2 = 0$ for any $j \neq 2, 3, t$.

(3) $a_rx_1 \in \{a_2, a_3\}$ for any $r \neq 1, t$. If $a_rx_1 = a_2(a_3)$ for $r \neq 2(3)$, then $a_2x_1 = a_2$ ($a_3x_1 = a_3$) and $a_2^2 = 0$ ($a_3^2 = 0$). In particular, if $a_3x_1 = a_2$ ($a_2x_1 = a_3$), then $a_3^2 = 0$ ($a_2^2 = 0$).

Moreover, we denote by P_n the number of non-isomorphism zero-divisor semigroups determined by Gn(2,1). If n is odd, then $P_n = \frac{n+3}{2}$; If n is even, then $P_n = \frac{n+4}{2}$.

Proof. We suppose $M_n(2,1)$ is a commutative zero-divisor semigroup such that $G(M_n(2,1)) = G_n(2,1)$. With a similar argument of Theorem 2.1, we have $a_1x_1 = 0$,

$a_k x_2 = 0$ (k=2,3), $a_i a_j = 0$ for any $i \neq j$, $a_1 x_2 = a_1$, $a_2 x_1 \in \{a_2, a_3\}$. We divide it into two cases to discuss:

Case 1. Assume $x_2^2 = a_1$. A similar argument of Theorem 2.1 will show that $x_1 x_2 \notin \{0, x_1, x_2, a_1\}$. Now, if $x_1 x_2 = a_t$ for some $t \geq 4$, then $a_t x_2 = x_1 x_2^2 = x_1 a_1 = 0$. It is a contradiction since a_t is not adjacent to x_2. Therefore we have $x_1 x_2 \in \{a_2, a_3\}$. By symmetry, we may suppose $x_1 x_2 = a_2$. Note that $a_r x_1 x_2 = 0$ for any $r \geq 4$, then we have $a_r x_2 = a_1$. Thus $a_1^2 = a_1 a_r x_2 = a_1$. But now we have $a_1^2 = a_1 x_2^2 = a_1$. A contradiction. Hence there is no semigroup determined by Gn(2; 1) in this case.

Case 2. Assume $x_2^2 = x_2$. We consider $x_1 x_2$. A similar argument of Theorem 2.1 will show that $x_1 x_2 \notin \{0, x_1, x_2, a_1, a_2, a_3\}$. Therefore we have $x_1 x_2 \in \{a_4, ..., a_n\}$. If $x_1 x_2 = a_t$ for some $t \geq 4$. Then we have $a_t x_1 = x_1^2 x_2 = x_1 x_2 = a_t$, $a_t x_2 = x_1 x_2^2 = x_1 x_2 = a_t$, $a_t^2 = x_1^2 x_2^2 = a_t$. Note that a $a_r x_1 x_2 = a_r a_4 = 0$ for any $r \geq 4$ and $r \neq t$, then we have $a_r x_2 = a_1$ and $a_r x_1 \in \{a_2, a_3\}$. Now, $a_1^2 = a_1 a_r x_2 = 0$. From $a_r^2 x_1 = a_r(a_r x_1) = 0$ and $a_r^2 x_2 = a_r x_1 = 0$, we have a_r^2 is adjacent to both of x_1 and x_2 and hence $a_r^2 = 0$.

If $a_r x_1 = a_2(a_3)$ for $r \neq 2(3)$, then $a_2 x_1 = a_r x_1^2 = a_r x_1 = a_2(a_3 x_1 = a_3)$ and $a_2^2 = a_2 a_r x_1 = 0$ ($a_3^2 = 0$). In particular, if $a_3 x_1 = a_2$ ($a_2 x_1 = a_3$), then $a_3^2 x_1^2 = a_3^2 x_1 = a_2^2 = 0$ ($a_2^2 = 0$). Thus $a_3^2 \in \{0, a_2, a_3\} \cap \{0, a_1\}$ and hence $a_3^2 = 0$.

It is easy to verify that the binary operations satisfying the conditions (1)-(3) are associative. Thus they are the zero-divisor semigroups of Gn(2,1) for some $n \geq 5$.

Now we calculate the number of mutually non-isomorphic commutative zero-divisor semigroups determined by Gn(2,1) ($n \geq 5$). We divide it into two cases to discuss:

Case 1. Assume there exist a_p and a_q such that $a_p x_1 = a_2$, $a_q x_1 = a_3$ ($p, q \geq 4$; p; $p, q \neq t$ and $p \neq q$). let S_1 be a semigroup such that there are t_1 different a_i such that $a_i x_1 = a_2$ ($i \geq 4$ and $i \neq t$). Let S_2 be a semigroup such that there are t_2 different a_j such that $a_j x_1 = a_2$ ($j \geq 4$ and $j \neq t$). Then it is east to see S_1 is isomorphic to S_2 if and only if $t_1 = t_2$ or $t_1 = n - 4 - t_2$. Then the numbers of mutually non-isomorphic zero-divisor semigroups in this case is equal to the numbers of the solutions of the equation $i + j = n - 4$ (where $1 \leq i \leq j \leq n - 5$). If n is odd, it is (n -5)/2; If n is even, it is (n - 4)/2.

Case 2. Assume $a_p x_1 = a_2$ for any $p \geq 4$ and $p \neq t$. If $a_3 x_1 = a_2$, we only obtain one zero-divisor semigroup; if $a_3 x_1 = a_3$, we obtain three zero-divisor semigroups when a_3^2 take different number from $\{0, a_2, a_3\}$. Note that all the semigroups we obtained in this case are not isomorphic and are also not isomorphic to the semigroups we obtained in Case 1.

Hence if n is odd, then there are (n + 3)/2 mutually non-isomorphic zero-divisor semigroups determined by Gn(2,1); if n is even, then there are (n+4)/2 mutually non-isomorphic zero-divisor semigroups determined by Gn(2,1). We complete the whole proof.

Acknowledgement. The authors want to express their gratitude to the referees for their helpful suggestions and comments.

This research was partially supported by the National Natural Science Foundation of China (No.10971053), the Natural Science Foundation of Shandong province (No. 2010ZRE09006) and Youth Foundation of Henan University of Science and Technology (No 2010QN0036).

References

[1] DeMeyer, F., McKenzie, T., Schneider, K.: The zero-divisor graph of a commutative semigroup. Semigroup Forum 65, 206–214 (2002)
[2] DeMeyer, F., DeMeter, L.: Zero-divisor graphs of semigroups. Journal of Algebra 283, 190–198 (2005)
[3] Godsil, C., Royle, G.: Algebraic Graph Theory. Springer, New York (2000)
[4] Howie, J.M.: Fundamentals of Semigroup Theory. Clarendon Press, Uxford (1995)
[5] Wu, T., Lu, D.: Zero-divisor semigroups and some simple graphs. Comm. Algebra 34, 3043–3052 (2006)

The Design of Fusion Semantics Automatic Labeling and Speech Recognition Image Retrieval System

Lu Weiyan[1], Wang Wenyan[2], and Liu-Suqi[1]

[1] Guangxi Normal University for Nationalities, ChongZuo, China
[2] Guangxi Normal University, Guilin, China

Abstract. Existing Content-based image retrieval technology is research the underlying vision .There are still non-consistency with high-level semantics. This paper give one system of image retrieval that can relieve the question above. In the system, using semantic automatic tagging and Chinese speech recognition research. It is implemented by Matlab software, implementation of human-computer interaction interface. Do the experiments for small quantities of image set, the result show that the system is feasible, and more humane and intelligent.

Keywords: Content-based image retrieval, semantics automatic tagging, speech recognition

1 Introduction

With the multimedia technology and the rapid development of network technology, Digital Imaging and speech recognition in the society has been widely applied, classification, image annotation and retrieval of digital images are developed, and speech recognition technology is mature. In the traditional digital image retrieval in text keyword-based classification and retrieval technology, these technologies is extremely inefficient and the shortcomings of their non-standard words can no longer meet the needs of modern people, and later Content-based image retrieval technology will become the research focus of the researchers. Content-based image retrieval with traditional text-based image retrieval technology research is a great deal of non-uniform low-level visual features (color, shape, texture, etc.) and between high-level semantics, which is people call "semantic gap". Semantics-based image retrieval technology is the inevitable development of content-based image retrieval. Automatic image semantic annotation emerged, allowing people to alleviate some problems, has also been widespread concern.

Automatic image annotation in image retrieval is one very challenging work [2].The training image set has been marked, it can be described in a high-level semantics and the visual characteristics between the bottom of the corresponding model, and then use the model to complete the images that are not marked for automatic semantic annotation. The question of "semantic gap" description of between semantic visual features is effectively relieved.

As mobile broadband, mobile Internet and Internet of Things, people have higher requirements for the personality, intelligence, and simplistic. This paper presents a

Z. Du (Ed.): Proceedings of the 2012 International Conference of MCSA, AISC 191, pp. 215–221.
springerlink.com

fusion of automatic annotation of speech recognition and image semantic image retrieval system. The system is more humane and intelligent, more suitable for image retrieval system using the mobile and Internet of Things.

2 Existing Problems in Digital Image Retrieval System

1.Current the search of image engine focuses on semantic description of the content of the image, and the image is in focus on feature extraction of image content, focusing on the two are different, if combining both that the Web image retrieval technologies will make new progress, better meet the needs of people, more suitable for daily needs.

2. Different people on the same understanding of the content of the image and explain the differences.

3. About the existing image search engines, as for the automatic processing and tagging of the image information, we still have a long way to go. First of all, we need to develop a new algorithm and technique base on image tagging to improve the index function of the system, after that, we can expect a new software related to this algorithm, by the help of the software image search engines will be more automatic and intelligent

4. Currently, a number of content-based image retrieval systems and methods applied, such as fingerprint recognition, trademarks and medical image retrieval, a comparison of a wide range of applications. But for images on the Internet, which are different areas of image, and the quantity is large. In order to solve this problem about how researchers have put forward a lot of image retrieval techniques is provided, it was suggested that image contained excavated to Internet connected features, combining text classification, Adaptive text extraction technology, the pictures on the Internet more accurate automatic semantic annotation system[2].

3 Design of Image Retrieval System

For the above problem this article independently proposed integration of speech recognition technology and image semantic image retrieval system design of automatic dimensioning scheme.

The system can be achieved on image tagging, solves the traditional method of manually tagging the image, this will not only save time and effort, and more objective expression of unity and understanding of image content. Fusion speech recognition technology, solves the question that the user must master the key words or images related to the content of your image database, plan to comply with the characteristics of the original image .These question to inhabit traditional retrieval technology and Content-based image retrieval system .In other word, users cannot search the image(set) without sketch and original image and keywords. In the case of user discommodiously enter the draft for keywords and without anything about the image for search, the user just need to the voice signal about content of image what are objectively described by the user. This system is suitable for the future of smart into the mainstream and Internet of things era.

4 System Structure

This system is made up of three parts, speech recognition system, engine of image retrieval system and semantic tagging system, done by PC. System structure as shown in Fig 1.

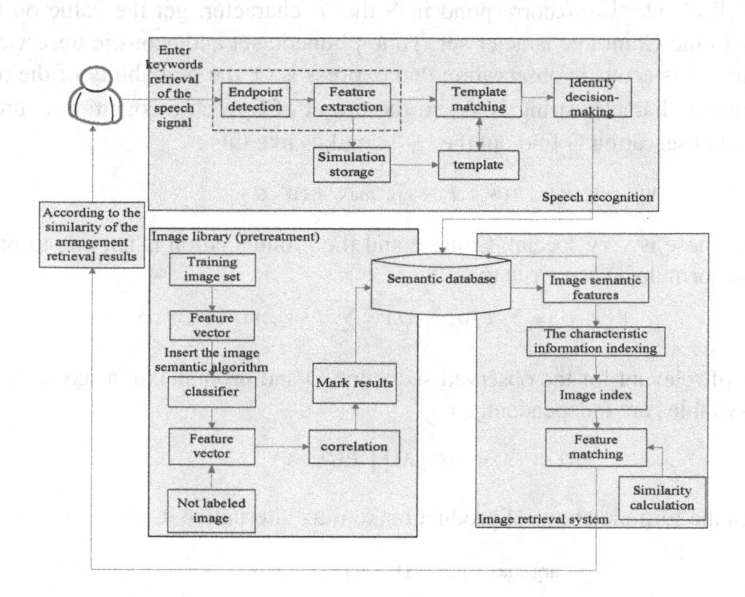

Fig. 1. System structure

4.1 Speech Recognition System

There are many characteristics and the merit for people's pronunciation, for example without the special training, not only make human's hand and the eye achieves liberation, but also more flexible and higher bit rate transmit information. Therefore in the context of space scope of the information is broad and complex, the speech interface that will be take a great of convenience to communication and management of information . In the speech recognition system, we can translation of the voice signal into writing sequence, then the result of recognition for the last output, or as the input of natural language processing system.

4.1.1 Chinese Speech Recognition System Hierarchical Strategy

This system has been used in literature [5], The Chinese have used speech recognition language model is based on the distribution of HMM Voice Recognition model. The identification system is divided into 3 layers: Carries on the transformation to the sound character、tune phonetic understanding、acoustic layer recognition. The input is enters the system from acoustics recognition level, then to has tune phonetic understanding, finally is carries on the transformation to the sound character, and then use the knowledge library and each level recognition algorithm, finally obtains the natural language Chinese character recognition result. The system hierarchical strategy for this:

Chinese sentences, Assume that the corresponding character string is $W=(W_1, W_2, \cdots, W_L)$ Corresponding to tone pinyin string is $A=(A_2, A_2, \cdots A_L)$, acoustic observations is $O=(O_1, O_2, \cdots O_L), W_i, A_i, O_i, i=1,2,\cdots,L$, correspond in S the i character, get the value on that respectively in the Chinese character set, Tune phonetic set and acoustic observation set .order $P(W/O)$ is acoustic observation that express is O, the probability of the mention of the Chinese character string is W, so the target of Speech recognition is under get the acoustic observation O find out the $\hat{W}$, makes like this :

$$\hat{W} = \arg\max_{W} P(W/O) = \arg\max_{W} P(W,O) \tag{4.1}$$

Because Chinese is very special, Chinese and the pronunciation is not one-to-one,so in front of the formula(1) can write this:

$$P(W,O) = \sum_{A} P(W,A,O) = \sum_{A} P(W,A)P(O/W,A) \tag{4.2}$$

Can think of relevant for the observed sequence O and pronunciation sequence A, and adjacent syllables are independent, so

$$P(O/W,A) = P(O/A)\prod_{i} P(O_i/A_i) \tag{4.3}$$

Usually, in the formula above of product make max alternative it, so

$$\hat{W} \cong \arg\max_{W} (\max_{A} P(W,A)P(O/A)) \tag{4.4}$$

In the formula $P(W,A)$ is the model of language in Chinese ,$P(O/A)$ is pronunciation model of Chinese . The task of speech recognition is find out the maximum of multiplication for value above (W,A), But the value range of (W,A) is too big, so it can't realize the full search. Therefore, using the hierarchical identification strategy to complete (4.2) to solve .

After slicing,(4.2)is written for this:

$$\hat{W} \cong \arg\max_{W} (\max_{A} P(W/A)P(A)P(O/A)) \tag{4.5}$$

$P(W/A)$ is sound character transformation model, P(A) is based on pronunciation understanding model, P(O/A) is pronunciation model.

After the introduction of the upper knowledge of the alphabet P(A) candidates alphabet selection criteria like this:

$$\hat{A} = \arg\max_{A} P(A)P(O/A) = \arg\max_{A} P(A,O) \tag{4.6}$$

4.2 Image Database Creation, Automatic Annotation of Image Semantics

Automatic image labeling is through computer analysis of images in the visual content, automatically generated by the corresponding image text information. It is the image retrieval research field is very challenging, it has been marked by learning set of images or other information can be obtained the semantic concept space and visual relation annotation, is also trying to image high-level semantic information and the underlying features to establish a mapping relationship. If it can realize the automatic semantic annotation of image, the question of image retrieval is from complex into simplify. It can provide the accuracy and objectivity of retrieval.

At present, the automatic image annotation method that based on multiple samples is according to the characteristic of the multi-demonstrations study the question. Extend existed the algorithms of machine learning, it can be applied to problems for multiple samples learning. Experiment for the design of this system, used the methods of automatic image annotation which based on packet-based layer multi-sample study in the literature [3]. The method is the image which describes with many demonstrations transforms to the demonstration level for the envelope single demonstration describes, then directly utilize by the algorithm of the labeling surveillance studies which realization automatic image labeling using. Using data sets of Corel 5k, as generic image database in the field of image annotation, the data set is divided into two parts: training and test set. Finally, practice image semantics automatic labeling and image database in this system.

The Algorithm which the methods of automatic image annotation which based on packet-based layer multi-sample study as follows:

Input : Training set to mark with a concept of the image collection L, set of test set (U) for the image to be marked;

Output: with tagging of the test set (U)

Initialization: $V = \Phi$ (null set)

Step1: Image segmentation and feature extraction: Segmentation of images L and U, then Extract low-level features of each region.

Step 2: Setting parameters n_0, d_0 .

Step3: Select the visual vocabulary

Step4:Repeat step3, until the election of the visual vocabulary of the semantic category L in all.

Step5: L and u in each image are converted to the c-dimensional feature vector.

Step6: L converted feature vectors training SVM classifier , Similarly, the feature vectors U converted to test the performance of SVM classification.

Step7: Clear V, change parameters n_0, d_0 , it can change the length of feature vector C , repeat step3~step5,get many SVM classifier.

Step 8: To determine the unlabeled set of images U label.

4.3 Image Retrieval System

At the first, image semantic feature extraction to achieve. Index the information which access to the feature from the image database that has been marked [8].Index image after get the feature information, then index the image that is according to the similarity computation .Judging from feature matching. Finally feedback back to the user what the results of retrieve which according to the similarity arrangement. We input one

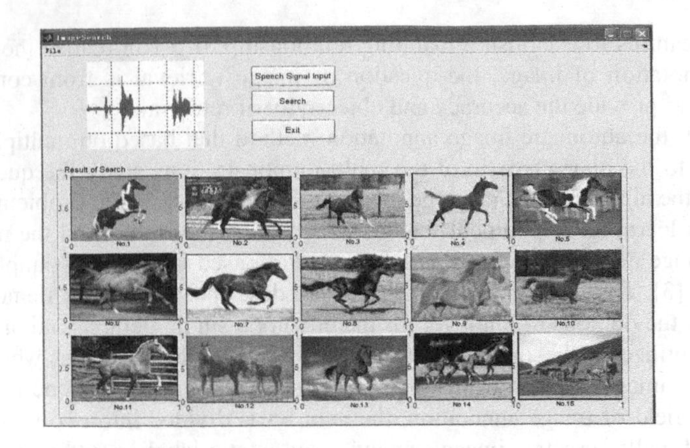

Fig. 2. System structure

voice signal like "horse, grass". Through the system of automatic recognition and retrieval, it got one image set which the content is about "horse, grass". The result as shown in Fig 2

5 Conclusions and Discussion

Design a system of image retrieval which fusion image semantic tagging and speech recognition technology in this article, used the methods of automatic image annotation which based on packet-based layer multi-sample study in the literature [3].And used method that the Chinese have used speech recognition language model is based on the distribution of HMM Voice Recognition model in literature [5], take them together then use in this system. We can practice the system by Matlab, it is just simple and complete system. Because this work is still initial stage , so had not study deeply for the technologies are contained in every module of the system. Just verify the feasibility of this system, and whether it is worth to study. It is exists many problems still, further research is needed, give a system which more complete and practical.

Acknowledgement. The work was supported by the Project of Guangxi Normal University for Nationalities of Natural Science Foundation (ybxm200911).

References

[1] Datta, R., Joshi, D., Li, J., Wang, J.Z.: Image retrieval:ideas, influences, and trends of the new age. ACM Computing Surveys 40(2), 1–60 (2008)

[2] Yu, F., Ip, Horace, H.S.: Semantic content analysis and annotation of histological images. Computers in Biology and Medicine 38, 635–649 (2008)

[3] Wang, K.-P.: Research on Key Techniques of Automatic Image Annotation. Beijing University of Post Telecommunication 04 (2011)

[4] Li, Z.-X., Shi, Z.-P., Li, Z.-Q., Shi, Z.-Z.: Automatic Image Annotation by Fusing Semantic Topics. Journal of Software 22(4), 801–812 (2011)
[5] Jin, W.: Parallel Optimization Method in Language Model for Mandarin Speech Recognition, vol. 05 (2010)
[6] Liu, Y.-H.: Studies on the Semantic Image Retrieval Based on Bayes Statistical Learning Theory. East China Normal University (October 2010)
[7] Zeng, H., Cheung, Y.: A new feature seleetion method for Gaussian mixture clustering. Pattern Recognition, 243–250 (2009)
[8] Zhang, D., Wang, F., Shi, Z.: Interactive localized content based image retrieval with multiple-instance active learning. Patten Recognition, 478–484 (2012)

A Method to Automatic Detecting Coronal Mass Ejections in Coronagraph Based on Frequency Spectrum Analysis

Zeng Zhao-xian[1,2,3], Wei Ya-li[3], and Liu Jin-sheng[3]

[1] Center for Space Science and Applied Research, Chinese Academy of Sciences
[2] Graduate University of Chinese Academy of Sciences
[3] Beijing 5111POB, China

Abstract. We present a new method of automatic detection of Coroanal Mass Ejections(CMEs) in Coronagraph. Different from brightness comparison that was widely used in currently methods of detecting CMEs, we detect CMEs by analyzing the sudden change of frequency spectrum in coronagraph. And this method shoud be more effective while detecting the faint CME whose brightness is so weak that maybe ignored by brightness comparison.

Keywords: Coronal mass ejection, Automatic detection, frequency spectrum analysis.

1 Introduction

Coronal mass ejection(CME) is a strongly energetic eruptive event that occur in the solar corona, and the earth-directed CME is the most important source of damaging space weather events which is extremely dangerous to modern technology system such as spacecraft. Early days, scientists detect CMEs by human observation, this method is very low efficiency and the detecting process is subject to human bias. The last century 70's, along with the continuous progress of space observations of the corona, a series of satellites with coronal imaging observation ability like OSO-7, P78-1, Skylab, SMM, SOHO were launched, artificial method has been unable to cope with so large amount of data which was got by this satellites. In this situation, automatic detecting methods were developed.

The first automated CME detection method introduced in the literature was the Computer Aided CME Tracking software(CACTus),which was introduced by Berghmans et al. in 2002[1][2], CACTus transforms the coronagraph to a 2D image and find CMEs in this 2D image. In 2005, Borsier et al. proposed a method which utilizes LASCO C2 images to detect CMEs[3], Liewer et al. proposed another scheme that uses STEREO and SECCHI images to look for CMEs[4]. In 2006, Qu et al. introduced a algorithm to detect, characterize, and classify CMEs used image segmentation, morphological methods[5][6][7]. In 2008, J. Zhang et al. presented the solar eruptive event detection system (SEEDS) which can detect a CME based on individual LASCO C2 coronagraph images by using advanced segmentation techniques[8].

Z. Du (Ed.): Proceedings of the 2012 International Conference of MCSA, AISC 191, pp. 223–227.
springerlink.com

The key of CME automatic detection is how to distinguish CME from other parts of the image. As shown in Figure 1, in the coronagraph image which is after noise filtering and image enhancement, CME usually appears as a bright, complex texture enhancement structure, with a lack of brightness of dark region; non-CME parts or parts of coronal streamers, are mostly simple radioactive linear cluster. The core of all these CME automatic detection system that mentioned above is brightness enhancement detection, this method usually has the following shortcomings:① the position of CME region is not accurate: a CME region includes both bright and dark parts, only based on the brightness detection will lose much of the dark region, and must cause CME location serious deviation;②multiple CME region cannot be detected: when the sun broke out two or more CME events succession, these CMEs may appear in the same coronagraph image. the traditional method only with a brightness analysis, will let the darker CME signal overwhelmed by the bright CME and cannot be identified; ③ some dark CME cannot detect: the brightness of some CMEs is relatively small, if separate the background and detect the CMEs only by the absolute value of luminance, these dark CMEs will not be identifid.

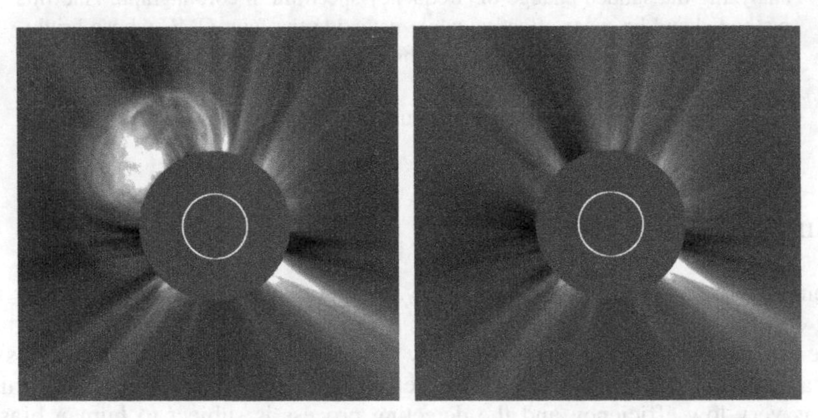

Fig. 1. Comparison between the CME image and non-CME image. (a) typical CME image. (b) non-CME image

In this paper, we present a new automatic CME separation and detection algorithm, It is mainly based on the frequency spectrum analysis to find one or multiple, different brightness of CMEs in one coronagraph image. The steps are described in Sections 2–3, and a conclusions is provided in Section 4.

2 Separating the CME Image

As mentioned above, in the coronagraph image, CME usually appears as a bright, complex texture enhancement structure, with a lack of brightness of dark region; non-CME parts or parts of coronal streamers, are mostly simple radioactive linear cluster. This means, The maximum visual differences of CME and coronal streamers is the spectrum distribution of coronal streamers is relatively uniform, but in the CME region often exist local sudden change of spectrum distribution. Therefore, we can separate CME based on spectrum analysisc, the process flow diagram is shown in Figure 2.

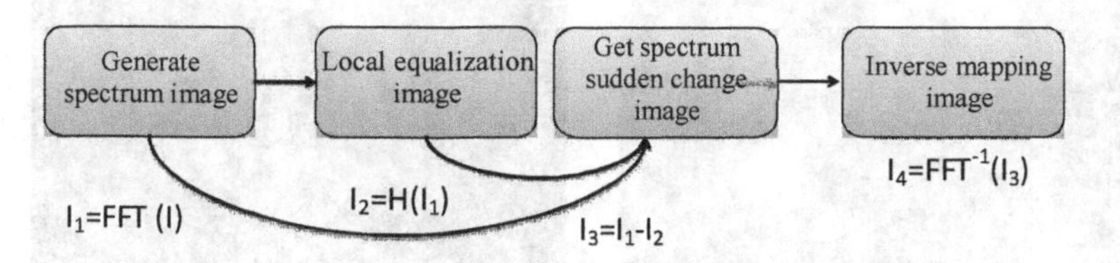

Fig. 2. Process flow of the automated CME separation based on the analysis of spectrum

Given I is the image to be analyzed, after Fourier transform we can obtain its corresponding spectrum, including amplitude and phase statistical spectral, and what we used is the amplitude spectrum I1. The low frequency portion contains the main information of the image, while the high frequency part reflects the detail information of the image. The main difference of CME and coronal streamers region is not low or high frequency, but how many sudden changes of the spectrum.

Therefore, we can analyze the spectrum distribution law of CME and background region (including coronal streamer and the actual background), and then use the "subtraction" way to acquire frequency spectrum sudden changing region in the spectrum-image. Its basic principles is: according information science and statistics research achievement, the image information is composed of two parts: redundancy information and significant information, and the redundancy information has statistical invariance, which can be forecasted through image analysis. Based on this principle, the image of the significant information can be obtained through "subtraction". In the frequency domain, the distribution of a non-mutation information's amplitude A(f) has such rules: the logarithmic representation of A(f) usually shows the similar curve, this curve is composed of several local lines, that is good local linear. So the significant information can be got by remove the non-mutation spectrum from the source image. The calculate formula was shown in Figure 2, in which the local equalization function H(I) is defined as follows:

$$H(I) = \frac{1}{n^2}\begin{pmatrix} 1 & 1 & 1 & \ldots & 1 \\ 1 & 1 & 1 & \ldots & 1 \\ \ldots & \ldots & & \ldots & \\ 1 & 1 & 1 & \ldots & 1 \end{pmatrix}$$

Finally, as the correlation between the spectrum and the actual image is not obvious, that's not suitable for further image analysis, so we use Fourier inverse transform to obtain the frequency mutation pixel's corresponding information I4 in the original image space. A sample of the processing result was shown in Figure 3.

Fig. 3. A process result of the automated CME separation and its source image. (a) source CME image. (b) result of automated separation

3 CME Outline Detecting

After get the preliminary CME separation, the next step is how to locate and distinguish the CME accurately. Although in the separation image, CME region often corresponds to the connectivity region whose most pixels have higher brightness, but the traditional method based on gray level threshold division has many limitations. For example, when the content of images and luminance changes it can often take a big deviation into the detection results, and it is also unable to correctly divide the bright-dark inclusions CME area. Therefore, in this paper, we propose a new method to correct positioning the CME area by detecting the local stable extremal region.

The Method of local stable extremal region detection is based on the watershed algorithm. Treat a grayscale image as a two-dimensional function according to a gray level mapping, and sort all the pixels according to their gray value. As the division threshold enlarge from 0 gradually, the divided connected region area will change Accordingly. By detecting the local minimum rate of the changing function of connectivity region, we can find the region that is more stable. Due to the adoption of the brightness relative ranking and the tree data structure, this connected region detection method can quickly find all the connectivity regions with different luminance in one image, and can effectively avoid the interference of noise. Figure 4 the diagram of connected region detection method based on the component tree, figure 5 is a detecting sample.

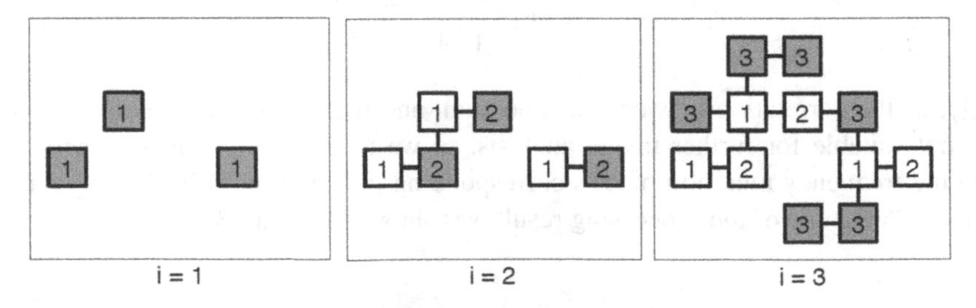

Fig. 4. The detection method of connected areas based on component tree

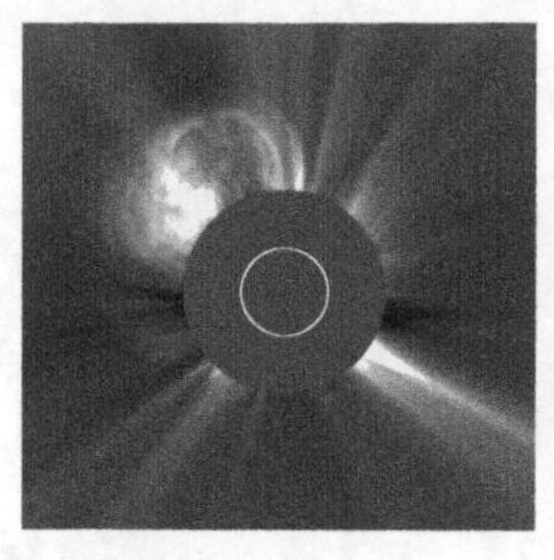
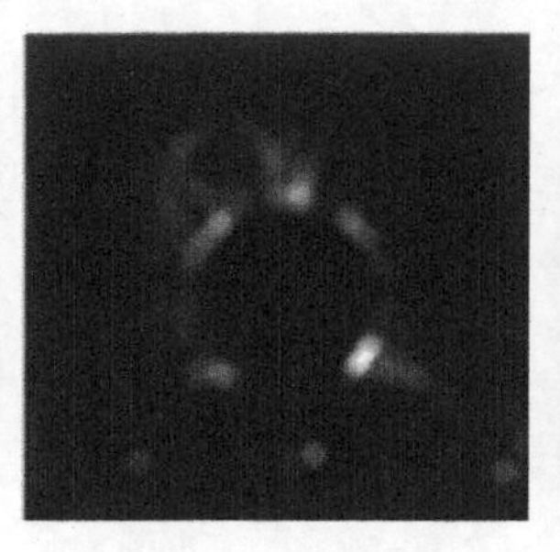
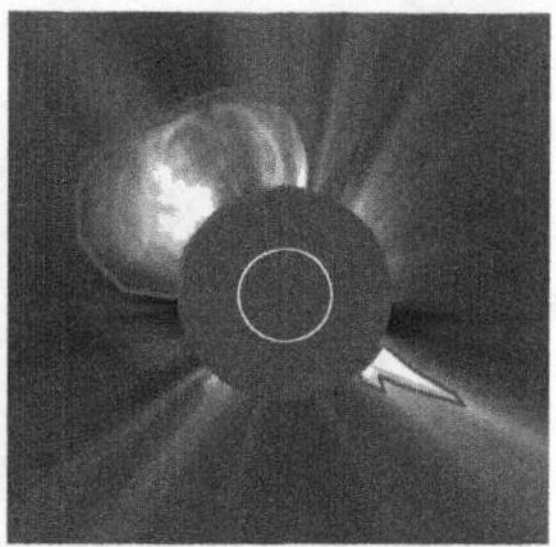

Fig. 5. The finally detection result of CME. (a) source image. (b) preliminary separation image. (c) finally detection image.

4 Conclusion

In this paper, an new automatic method to detect CMEs is presented. This method separates CMEs from coronagraph image by calculate the sudden changing of the image's spectrum. Compared to the previous CME detection methods which use brightness enhancement detection, we have provided an alternative approach for CME detection, and this method seemd to be more effective while handling the multiple CMEs or faint CMEs.

References

[1] Berghmans, D., Foing, B.H., Fleck, B.: In: SOHO 11, From Solar Min to Max: Half a Solar Cycle with SOHO, ESA, Noordwijk, vol. SP-508, p. 437 (2002)
[2] Robbrecht, E., Berghmans, D.: Astron. Astrophys, vol. 425, p. 1097 (2004)
[3] Borsier, Y., Llebariaf, A., Goudail, F., Lamy, P., Robelus, S.: SPIE 5901, p. 13 (2005)
[4] Liewer, P.C., DeJong, E.M., Hall, J.R., Lorre, J.J.: Eos Trans. AGU 86(52). Fall Meet. Suppl., Abstract SH14A-01 (2005)
[5] Qu, M., Shih, F.Y., Jing, J., Wang, H.: Solar Phys., vol. 217, p. 157 (2003)
[6] Qu, M., Shih, F.Y., Jing, J., Wang, H.: Solar Phys., vol. 222, p. 137 (2004)
[7] Qu, M., Shih, F.Y., Jing, J., Wang, H.: Solar Physics, vol. 237, pp. 419–431 (2003)
[8] Olmedo, O., Zhang, J., Wechsler, H., Poland, A., Borne, K.: Solar Phys., vol. 248, pp. 485–499 (2008)

The Study on Network Education Based on Java 2 Platform Enterprise Edition

Liu Chunli and Huang Linna

Department of Computer Engineering, Cangzhou Normal College, Cangzhou, Hebei, 061000 China
liuchunli@cssci.info

Abstract. Improve flexibility and interaction, and based on campus network education has become the most promising teaching methods; particularly in the information technology education (ITE) itself involved the network education development. This paper describes the characteristics of the information technology education and focus on the implementation of network education based on J2EE framework of the system, elaborated design and structure, the user classification and corresponding application mode. We discussed several key technology system developments and present some ideas of programming optimization.

Keywords: Network education, J2EE, STRUTS.

1 Introduction

Information literacy is the characteristics of the information age talents. The information quality education has become a common life education. In order to improve the information literacy of university students, they must be based on the information technology education. Information technology education often require an effective platform, the latest teaching resources are rich, keep up with the rapid expansion of more and more professional knowledge field. Network education, based on the identity authentication and authorization, provides such a mechanism, real-time interactive and dynamic added new facts and available courses can be achieved. Therefore, the development of the corresponding education network platform for information technology has become an active field of research.

2 The Structure Design

In this work, we use a based on J2EE distributed computing model, namely (Java 2 platform enterprise edition), not only has the advantage of high-Java platform-independence distributed processing power and safety, and has become a standardized structure network application development. The J2EE system using UML language data format and data structure, development and construction free from code to focus on the process logic, so as to improve the efficiency of development.

Z. Du (Ed.): Proceedings of the 2012 International Conference of MCSA, AISC 191, pp. 229–234.
springerlink.com

Given the application of the system and the requirements of the service, simplified as a J2EE tri-tier C/S/D (client/server/database) model with independent function [1] the specified services launched, as shown in figure 1, this simplified structure, function each layer to achieve the good maintainability, reveal and expansibility. The particular service with regard to each layer is described as follows:

1) Client layer. Web browser and the client application is running in the client provide service for layer, the end user, including administrators, teachers and students. Teachers and students can executive search operation and interaction, and managers in charge of system upgrade and maintenance.

2) Service layer. The network layer, business logic layer are included in service layer. The network layer, by running the network layer J2EE server and client unit layer, interaction and make corresponding reaction. Through the module, network and JSP Servile layer sends request terminal EJB components from the business logic layer, the results for processing and then back to the terminal network layer. The network layer provides two interfaces, namely system interface through which all users can access education resources and management interface, only approximates to the administrator.

3) Database Resource layer. The main function of this layer is to realize the database operation of the business layer by running the information system layer software on EIS (Enterprise information system) server. With client not able to access the background database directly, J2EE thus provides a module-based environment of application development by separating performance logic, business logic and background database [3].

3 The Key Techniques

Usually, the communication network system is based on the direct interactive server and client program. However, in the J2EE platform, the direct exchange client program limit and remote database, as security problems. In other words, the client and the server must be connected through the exchange between the middleware. In the work described in the platform, middleware is Serve and JSP, its operation process image shows. As a messenger and Server can indirectly interact with database returns to the program JavaBeans results and display their web site [5].

Teaching system, development and operation platform, is a dynamic scheduling, interactive and efficient web database server application, to achieve the server middleware and resources between database server to establish a interaction with background database. For client access, teaching platform using the JSP and XML web page and middleware makes the platform keep fast connection and various teaching resources database. Its key multimedia teaching platform is a multimedia database. Current object-oriented web database development tools, in addition with a powerful visual web database programming interface, tight, IIS and DB2 compatible.

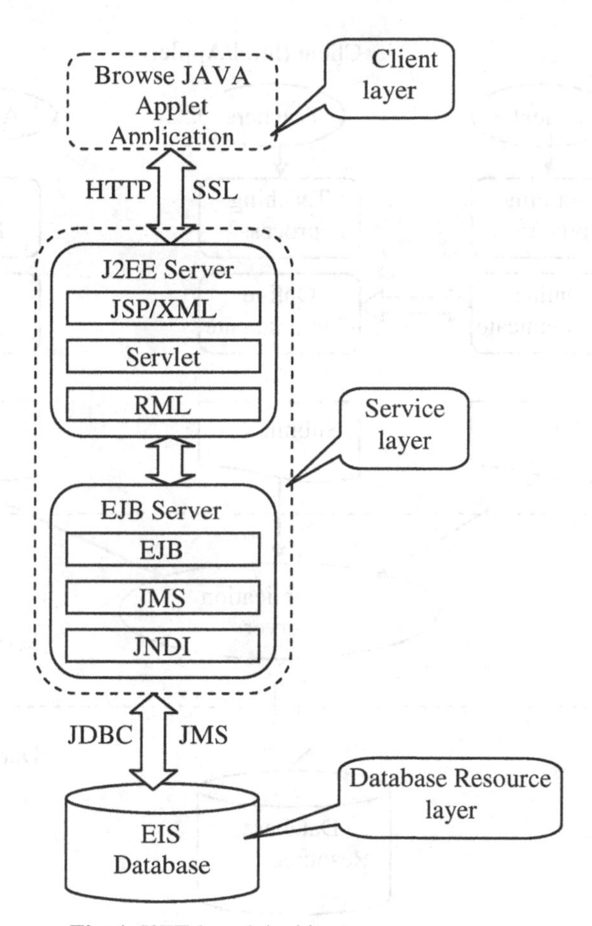

Fig. 1. J2EE-based Architecture

The guidance resource module assembly technology on the basis of [1], based on component model, description, and open system framework, through the assets for learning and get courseware exercise. It is not only an open source courseware and independently of teaching strategies and versions. The system can produce the focus of the form, facilitate web more rich and colorful and illuminating, because they can easily edit and deal with such as insert form and HTML code. Therefore, the editor platform and learning key can be seamlessly linked. The visual system dynamic link can be combined into assets module according to the requirement and create the automatic generation of teachers corresponding courseware. The system adopts the database and independent discipline method, that is, different link linkable library related specific counterparts, represent different teaching strategies. And these linkable library and courseware assets all managed whatever operating system. Commonly used tools provide comprehensive environment make data attribute is independent of the operating system and teaching assets can be distribution of individual network. The teacher and students can cooperation a courseware, make the integration of new functions with the courseware breeze.

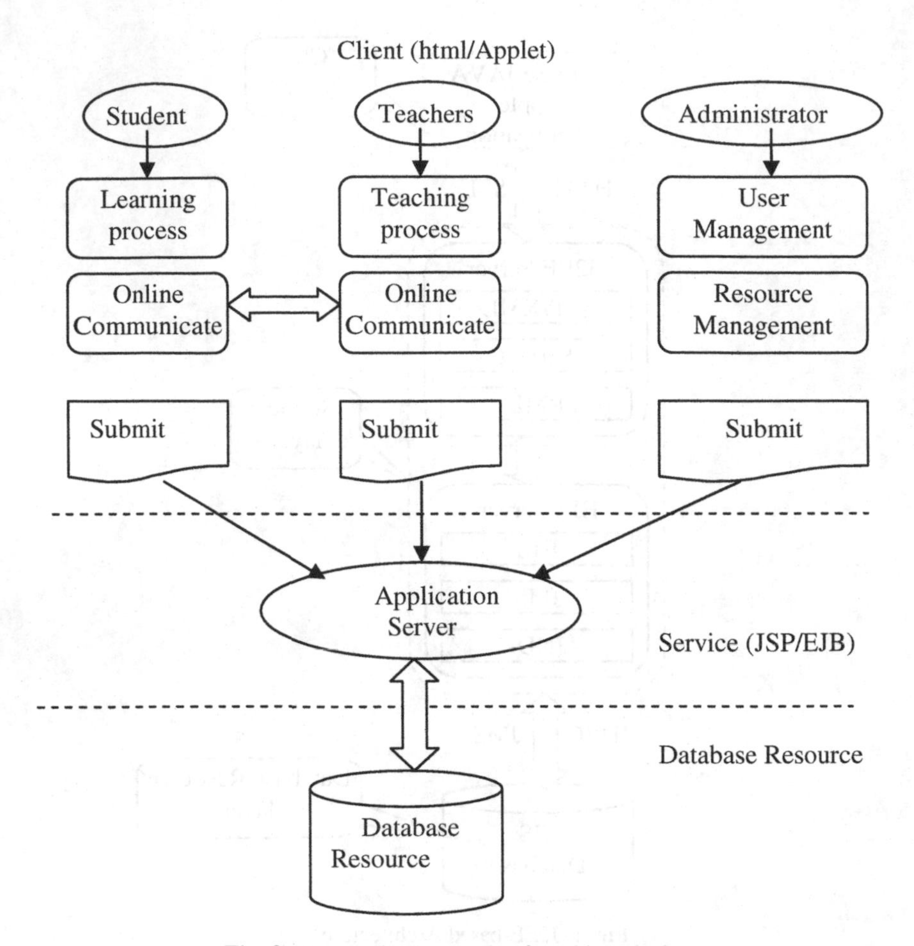

Fig. 2. The application mode of teaching platform

The courseware resources database is the vital resource of multimedia teaching platform[2]. The techniques of collecting abundant materials should be as handy as possible, such as downloading from the internet; other advanced methods of input such as scanning, voice record, digital imaging and recording should also be adopted. The materials should be categorized and stored in corresponding databases after being edited. The multimedia materials are inserted into web pages in the form of plug-ins, matched with corresponding activation hot zone and drivers. The overall framework, category navigation and dynamic Webpage design should be realized by visual programming tools. The web pages are stored in related folders according to the arrangements of the overall framework in the format of HTML. The file names are hyperlinked to courseware databases and switching among files and hotkeys and pages is realized by hypertext technology.

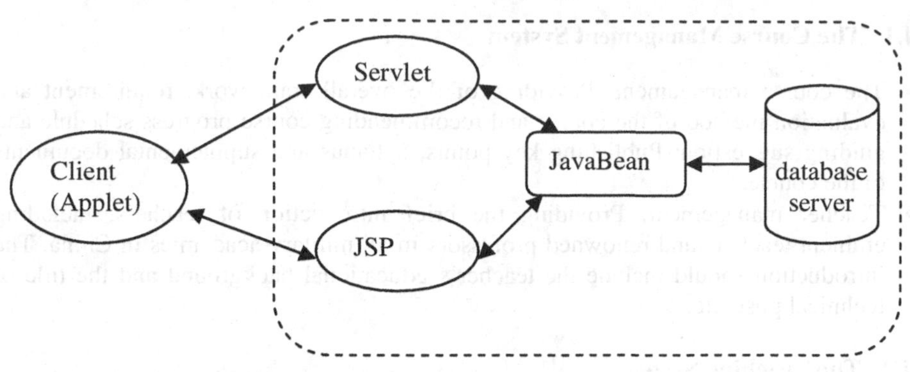

Fig. 3. The communication between clients and servers

4 The Implementation of Functions

As shown in Fig. 4, The Network platform of Information technology education is composed of teaching System, students system and management system [1]. The platform integrates many independent teaching service subsystems, which coordinate together to form a whole teaching system. The structure of one of the systems, the ITE application system which acts as the foreground of teaching platform, includes course management system, teaching system, students' independent learning system and courseware development system. The other system, the ITE resources system which acts as the background for the development and management of teaching platform, includes course resources, teaching preparation resources, courseware resources and exercise problems resources. The procedure in implementing the interactive teaching service is divided into two steps, first, teachers utilize the multimedia resources under the support of course management system and courseware development system to create related Network courseware according to course requirement. Then the interactive and customized teaching service is achieved with the support of teaching platform and independent learning system.

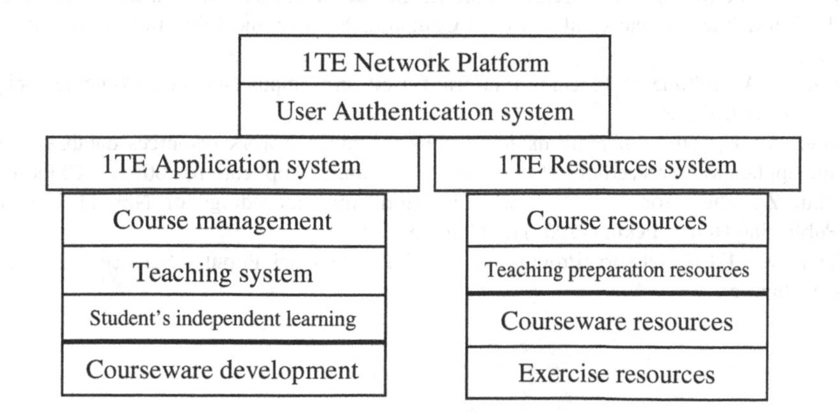

Fig. 4. The constitution of ITE Network platform

4.1 The Course Management System

- The course management: Providing of the overall framework, requirement and evaluation method of the course and recommending course progress schedule and guiding suggestion; Publishing key points, syllabus and supplemental documents of the course.
- Teacher management: Providing the brief introduction of teachers, including eminent teachers and renowned professors in all military academies in China. The introduction should include the teacher's educational background and the title of technical post, etc.

4.2 The Teaching System

- The courseware teaching platform: With navigation search set up, teachers are able to select distinctive learning path which is connected to corresponding courseware database to organize the teaching resources flexibly and exert teaching control efficiently.
- The homework management: Providing the interface for teachers to assign, check the homework and for students to obtain, submit and retrieve homework.

5 Conclusion

It has already proved, based on j2ee platform can reduce the network education application of the complexity of the design and shorten the development time. In addition, it also provides strong support system integration. This platform has been applied in the author's school and was found can greatly improve the quality of the products and efficiency of the information technology education.

References

[1] Liu, S.: The Multimedia Teaching Platform Based on Digital Campus. In: Proceedings of the Third International Conference on Computer Science and Education, China, pp. 1–6 (2010)

[2] Liu, S.: A Multimedia Teaching Platform Based on Campus Network. Computer Science (2009) (in Chinese)

[3] Guo, X.: The study of multi-media campus teaching network resources database design and application. Proceedings of the Study of Computer Application (2001) (in Chinese)

[4] Zhu, Z.: The tutorial of Network education. In: Proceedings of Normal University Publishing House, Peking (2001) (in Chinese)

[5] Tian, X.: J2EE Network Programming Guide. Shanghai Popular Science Press (2004) (in Chinese)

Study on a Learning Website of Self-directed English Study

Yan Pei

The Chinese People's Armed Police Force Academy
Langfang, Hebei, China, 065000
peiyan@cssci.info

Abstract. The new study requirement is put forward based on the rapid development and application of the new information technology and network technology. The theory of CALL is studied and a learning website is developed. Thus the individual and self-directed study is realized through this practice.

Keywords: self-directed, learning website, CALL theory.

1 Introduction

The basic requirements for college English teaching issued by the Department of Education of China (later referred to Basic Requirements) puts forward a new patter for English teaching, which is to advocate teachers to make full use of multimedia technology, especially websites to teach and guide students to do self-directed study. It requires that the technology of multi-media and websites be applied in English teaching. The new teaching method should take the place of the old traditional teacher-focused method. Teaching should be supported by modem information technology, especially network technology to make the English teaching fee of the limit of time and places. Teachers should focus more on developing students' ability in self study or individual study rather than on stuffing them with the language itself in the most traditional way. (Basic Requirements, 6)

2 Purpose and Significance of Network-Based Teaching

The patter of computer assisted language teaching and learning which was started in the late 1990s has been changed from the standalone courseware to varieties of multimedia network, such as using websites to assist language teaching. This new form of network-based language teaching has attracted more and more educators' attention. Different kinds of websites for English teaching have come into being in China and has been used more and more widely at home and abroad. Lots of web sites have been set up in colleges and universities to assist teaching in China. All this has greatly contributed to the improvement of language teaching in China.

Multimedia assisted teaching provides students with more chances and choices in their language study than any other media. Through careful design and development, the courseware used to assist teaching has the characteristics of abundant information,

Z. Du (Ed.): Proceedings of the 2012 International Conference of MCSA, AISC 191, pp. 235–239.
springerlink.com

highly integrated media, varieties of special effects and it is easily obtained and used in teaching and learning. It has many advantages over the traditional language teaching approaches. It provides an ideal atmosphere and environment for students to learn the language themselves. Learners can get the information in a very flexible way and individual way, such as choosing what to read, how long to study, how many times to read or practice, etc. By working on the network courseware, students can develop their own ways of study. It can also develop students' ability in self-study and self-management. Their responsibility and motivation can be enhanced. It is especially ft for teaching students at different levels. (Warschauer, 2006:96)

Since 2003 the authors began to study and develop a website to assist language teaching in Changchun Institute of Technology. The website was set up according to the characteristics of our own students. The research was supported by the Institute with finds and personnel in the field of technology. Both English teachers and IT technicians work together very hard for a year or two, with the English teachers design the columns of the website, make the software or courseware tailed to different students and different textbooks, while the IT teachers are in charge of the work related to IT technology. Shortly after it was completed, we tried our best to use the website to assist teaching. We use it in class a lot, we also encourage and guide the students to study by themselves on the website after class. Teachers in the research group devoted a lot of time and efforts supervising students' study, giving them advice and guidance, answering their questions, developing courseware and updated website regularly. We organize different kinds of activities, too, making the website a base for students to practice their English and improve their ability in using the language. We did all of this in order to improve students' English at different levels to a large extent.

3 Theories on Self-directed Learning

The self-directed study in learning a foreign language is not a new concept. It was put forward in the 1980s. Holec first applied this theory into foreign language teaching. Afterwards other linguistics, such as Dickenson and Little Wood all gave different definitions to self-directed study. The Chinese scholar Mr. Su Dingfang made the following definitions to it after he studied the foreign theories:

Learners adopt an active and eager-to-learn attitude to their study, which means they are responsible for their study and put an effort in their study.

Learners should develop their ability of self-study, self-management and their own study strategies so that they can finish their study separately to their own needs.

Learners should be given enough chances to be exposed to the real or stimulated environment to practice using the language and develop their abilities in real communications. (Shu Dingfang, 2004:34)

Network-based study is just to meet the need stipulated by the definitions. It can provide students with an environment for self-directed study. It takes the place of the traditional teacher-focused teaching method and makes it possible for students to study through active practice in learning a foreign language.

4 The Application of Network-Based Teaching

In order to use network to assist teaching, the priority is to set up a website which is proper for students to study.

Our website was designed and developed based on the requirements for the development of students' English ability and the purpose of using the website as a after-class base for students' English study, aiming at making it assist English teaching and learning in an efficient and effective way.

We want to make it ft and useful to different kinds of students at different English levels and to meet different needs in learning the language.

All the courseware is tailed to the students' need for study at different levels. Some of tem were developed by ourselves, while most of software were bought either from abroad or at home because developing them needs so much work and efforts, and our skills and abilities in developing this is very limited. For non-English major students, we provide them with "Experiencing English for College Students" (network courseware) published by Higher Education Press, "Multimedia Courseware for Practical English Study" also published by Higher Education Press. For English major students, we downloaded the software for Intensive Reading for College English published by Beijing Foreign Language Press. We also developed our own power points, courseware to assist daily teaching, among which those for translation course have been selected to the National Database of Excellent Courses. We bought some newly published British or American software for speaking or listening, such as "Small Talks", "Inter-reactive English", "Introduction to Britain and America", etc. These are for students to practice their language skills as well as introduce culture, history, geography, arts, etc. about the countries in which the language is being used. The students love them very much and the experiencing center(our 300-seat-computer room) is always full and students have to book the seats in advance in order to study there after class.

There is also a bank of exercises or tests for both English and non-English major students on the website. We developed a platform for the bank and entered lots of questions for different levels and different types. Both the teachers and students can come up exercises based on their own needs and purposes either for teaching or for practicing.

There is feedback to all the questions. Students may do the problems and get feedback immediately after they are done with the exercises. About the objective problems, students can get the results of the practice at once, knowing which part they do well, which part they need to improve. About the subjective exercises, such as translating or writing exercises, we attach the answers, the translated version or the model version for their reference. We also give comments on them so that students may know clearly about the feedback.

The students have to insert their users' names and passwords before they visit the website and study on it. By doing this, the computer automatically keep the record of their history of study on line so that the teachers may know whether they study or not before or after class and how often or how long they study on line. The teacher may adjust students' grades for regular study partly based on their on-line study performances. The teachers may also know how well their students study and they may give them individual help or guidance respectively.

The section for students to study on the website is made up of three parts, which are students' registering system, logging-in system and CAI software Study system. The registering system is designed to insert students' information and keep it in the data bank of ACCESS on the programmer's platform. The logging-in system is designed to verify the codes about the users, such as users' names, the language can be enhanced. Students may be immersed in the stimulated English environment and learn authentic English. Our website was attached to the main website of our college. All the network software was installed on the webpage of the Department of English of our institute.

The website is composed of the following sections: Software Study, Online Communication, English Forum, Bulletin Board, Related Connections and Entertainments, etc. All these sections were programmed through the software of passwords, real names, sex, classes, time span to study on line each time, etc. All the information is kept in each teacher's section. Teachers may check and supervise students' study on the website. The software, whether network courseware or power points are provided by teachers and sent to the services. Any teacher or student may visit them and use them at any time on the platform.

Besides the above self-study sections, there are on-line communication section and English forum section. The former is of the characteristic of a chatting room with a sound system. In this system, teachers may give on-line lessons, communicate with students, answer their questions or give guidance for their self study.

The section for English forum has more functions. Both teachers and students can send messages in this section, discussing the stuff they learn, the method they use, the problems they meet with and the solutions to them, etc. They can also exchange materials and opinions related to their studies. The number of the users who are working on line and the rosters of the registered users appear on the webpage at the same time, thus students' self study records are kept on the website automatically. Files or documents can also be sent in an attached form in this section. Pictures of users or complete short messages can be designed to one's taste and sent to the webpage. These can appear automatically with a voice prompt. The total number of people who visit the forum and the latest themes being discussed are shown on the webpage. The sticky post may be sent directly and they can be searched by inserting key words.

The hard equipment of the website is the server of IBM netfinit 5000, including a Gigabyte network card SCSI hard drive, and other equipment needed to get access to the website of the university. All these are provided by the network center of the university. Besides, streaming media equipment is installed so as to make it possible for videos to be played on the website, too.

5 Existing Problems and Strategies for Improvement

Network-based teaching has its own advantages over any other teaching approach. It has the features of teaching one to one, one to many and at different time and in different places. It can lengthen the classroom teaching unlimitedly and enrich the content of the classroom teaching. It has become more and more popular with college administrators and teachers and has been used more and more widely. However, in our application of website-assisted teaching, we have found some problems.

Though network-based teaching has become a major concern for many college teachers, a large proportion of teachers still stick to the most traditional teaching method---with one blackboard and a piece of chalk in class. They feed the students with a lot of stuff orally or in written form without any modem education technology for one reason or another. This makes students tired of language learning and their abilities in using the language are quite poor. Even though they can get a good mark in the English test for college students, they cannot communicate with foreign people in an effective way. So training these teachers in this field should be organized. Both theories and practice of multimedia assisted language teaching should be emphasized to them. Especially they should be exposed to network-based teaching theories, network courseware, on-line teaching equipment and application so that they may have a knowledge of modem technology and its application and know how to use it. (Chen lianlin, 2004, 30)

Both teachers and students should develop a good attitude towards on-line study. As a teacher, he or she should play the role of a main instructor and supervisor. Otherwise, the net-work based teaching and learning will be affected adversely and out of control, for modem education technology is not only education media, but also a tool for entertainment. If not properly used, the media will make students addicted to on-line games or chatting, etc. Thus, students' time is wasted and study is affected. Therefore, teachers should work hard to guide the students to study in a correct way and supervise them. As a student, he or she should not treat self-directed study as study without control or we say "fee study"---they may do whatever they like on line. When they are told to study by themselves, they are still required to do what they are required to and to work on their studies rather than on entertainment like chatting or games. "Teachers should let them aware that successful learners are usually those who are good at managing their own time and efforts in study.

References

1. Chen, J.: Theory of Network-based College English Teaching and Analysis of Its Application. Foreign Language Visual Education (6) (2004)
2. Gu, P.: Internet and Foreign Language Teaching. Foreign Language Visual Education (6) (2004)
3. Lei, P.: On Self-directed Teaching Mode in College. Joural of Harbin College (3) (2004)
4. Mo, J.: Study on the Development of Teaching Resources for Self-directed. Foreign Language Study at College, Foreign Language Visual Education (2) (2005)
5. Su, D.: Foreign Language Teaching Renovation: Problems and Strategies. Shanghai: Shanghai Foreign Language Press (2004)
6. Warschauer, M., Ker, R.: Network-based Language Teaching Concepts and Practice. Cambridge University Press, Cambridge (2006)

The Building of Network Virtual Laboratory for Physics Teaching

Baixin Zhang

Puyang Vocational and Technical College, Puyang, Henan, 457000 China
zhangbaixin@cssci.info

Abstract. Physics teaching is important to cultivate students' scientific inquiry ability and innovation ability. Based on the recognition of context for physics teaching and the discussion on the theoretical foundations for the design of virtual laboratories for teaching, this paper reported the construction, characteristics, and application scenarios of a network virtual laboratory for physics teaching in high school.

Keywords: Virtual Laboratory, Network, Physics Teaching.

1 Introduction

In the physical teaching, it has been widely recognized, this is one of the most important ways to improve the learning performance to provide students with more opportunity to observe, experimental and practical work.

Since the concept of "virtual laboratory (virtual laboratory)" first in 1989 by William professor from the university of Virginia, the U.S. has more and more researchers and equipment developers focus on [3] [4]. In recent years, with the rapid development of network technology, design and develop network virtual laboratory has become a hot topic. In contrast with the real laboratories, a virtual laboratory has its advantages to save test costs, admitted experimenting again and again, to provide a safe environment, etc.

2 The Context for Physics Teaching of High Schools

In a broad sense, on one hand, physical themselves change is the study of scientific body movement, is a highly laboratory-dependent theme. On the other hand, learn to learning ability, and found that the ability to analyze and solve problems of the survey is the basic quality of creative talent in modern society. This is of great significance to the traditional teaching methods reform and improves the students' learning ability and the innovation ability. In the basic education course reform under the background of China, especially since the emergence of the new curriculum standard based education for more efforts to change of teaching and learning methods to realize 3 d teaching goal, including improving knowledge and the skill, grasps the processing methods, training emotions, attitudes and values. In a narrow sense, physical experiments are often expected to stimulate students to find, propose, analyze and

Z. Du (Ed.): Proceedings of the 2012 International Conference of MCSA, AISC 191, pp. 241–246.
springerlink.com

solve problems, and then to cultivate the innovative spirit and practical ability of students. The importance of understanding how to support students' inquiry learning by means of various resources is a key factor in bringing about change in physics teaching. However, due to various experiment conditions, there are still some problems in current physics teaching of high schools, including:

For some experiments, it is difficult to complete all the steps because these experiments often need long time, or require high cost, or have high risk.

For some experiments, it is difficult to observe experimental phenomenon, and the experimental micro-phenomena can not be effectively displayed;

For some experiments, it is difficult for students to understand involved principles or phenomena;

For some experiments, it is difficult to combine involved principles with real life;

For some experiments, it is necessary to be repeatedly carried out or demonstrated at any time;

For some experiments, the experimental steps have to be determined based on repeated comparison of an experimental step, and it is difficult for students to master right experimental steps.

In some high schools, physics experiment teaching is mainly relying on conventional laboratory experiments. In some other schools, most physical experiments are just demonstration experiments used in class teaching.

As to the use of information technology for virtual experiments, most teachers prefer downloading experimental demonstration courseware from WWW for teaching. There are also some schools that have built digital labs, but most digital labs are just used for conventional experiments where sensors are utilized to collect data, and software is utilized to analyze and present data. There are no substantive changes in terms of the levels of students' understanding and application of experimental principles in such experimental teaching.

Computer-based network, especially the Internet, is promising to provide learners with an open, shared, personalized, multi-dimensional interactive learning space. In recent years, with the rapid development of computer networks and its wider application in the field of education, the term e-learning is being used with increased frequency in both academic and corporate circles.

The introduction of network virtual laboratories promises to transform the traditional experimental teaching model by breaking space and time separation to provide students with opportunities to carry out physical experiments when they need to do so. Through virtual experiments, students can analyze the process of experiment, sum up the feelings of their own experiments, and to conduct self-assessment on the level of the grasp of theory and analytical problem-solving capabilities. Students can also collaboratively promote virtual experiments. In this context, it is meaningful to adopt virtual laboratories to cultivate students' collaboration attitudes and abilities, and to reduce pressure on teachers.

More specially, network virtual laboratories promise to: 1) support individual or collective preparation of lessons for teachers; 2) provide teachers with a flexible teaching environment to help avoid repeated work; 3) support students to promote independent and collaborative inquiry by providing students an interesting, open, digital, networked and safe learning environment.

3 Theoretical Foundations for Virtual Laboratories Design

Virtual laboratory is a virtual environment design for experiments. Improve the function of the virtual laboratory and efficiency in education process; it is necessary to know its theoretical basis, the design process of participation. This study chooses three theories focused on the theoretical basis of the careful design virtual laboratory, including: social constructivism theory, distributed cognitive theory and cognitive context learning theory.

3.1 Social Constructivism Theory

Constructivism asserts that knowledge constructions have social origins and are promoted through interaction with other people, and that learning is a process of social negotiation or collaborative sense making, mentoring, and joint knowledge construction. Driscoll has summarized five major components of constructivism: (1) a complex and relevant learning environment; (2) social negotiation; (3) multiple perspectives and multiple modes of learning; (4) ownership in learning; and (5) self-awareness and knowledge construction.

In the light of social constructivism theory, several issues have to been considered in the design of a virtual laboratory, including: (1) it is of importance to build a constructive environment support for learners' active knowledge construction; (2) it is indispensable to mediating social interaction in learning processes; and (3) it is necessary to support learners' personalized learning experiences by taking learners personalized differences into account.

3.2 Distributed Cognition Theory

Distributed cognition refers to "the idea that cognition, knowledge, and expertise are not simply a property of individual minds or located and manifested in individual heads but are distributed among people and among people and cultural tools or artifacts". Karasavvidis suggested that cognition extended beyond individual, but was essentially distributed along two main dimensions: material and social. The material dimension of the distribution refers to the incorporation of all kinds of mental and physical artifacts in which cognition is encapsulated. The social dimension of the distribution of cognition refers to the involvement of social others who provide assistance in many ways during task execution by functioning as cognitive resources.

The introduction of distributed cognition theory helps the identification of the roles of virtual laboratories. On the one hand, a virtual laboratory itself is not a single aiding tool for learning and teaching, but can be seen as a cognition carrier, even a cognition subject with active functionality and reaction. On the other hand, a virtual laboratory should provide learners material and social cognition tools and according support.

3.3 Situated Cognition Theory

According to Lave & Wenger, learning as it normally occurs is a function of the activity, context and culture in which it occurs (i.e., it is situated). Accordingly, learning should not be simply regarded as the acquisition of certain forms of knowledge, but should be situated in social participation and interaction. Social interaction is a critical component of situated learning- learners become involved in a "community of practice" through legitimate peripheral participation. Cognitive apprenticeship turns to be an important part in social interaction based practices.

In the context of situated cognition theory, a virtual laboratory should support the real experience of learners and provide scaffolding for learners by differentiating learning situations, mediating social interaction, promoting personal and collaborative learning activities, and admitting learners' legitimate peripheral participation.

4 IWVL: A Network Virtual Laboratory for Physics Teaching in Junior High Schools

Experiments in IWVL are organized according to grade levels in junior high school, and are classified into two types: one is for virtual demonstration and group-based experiment, and the other is for inquiry-based experiments. The former type includes 72 experiments, and the latter comprises 34 experiments. IWVL consists of five components, including virtual demonstration and group-based experiment module, Inquiry-based virtual experiment module, experimental tool module, experimental platform module, virtual instrument module, and extension resource module.

- Virtual demonstration and group-based experiment module. This module focuses on providing virtual demonstrations for teachers' demonstration-based teaching and virtual guide for students' group-based experiment process, which encompasses several sub-modules as follows: virtual operation, operation demonstration, experimental instruments display, and experiment description.
- Inquiry-based virtual experiment module. This module includes four parts: support software for inquiry-based experiments, experimental instruments description, experimental programs and experimental guidance.
- Experimental platform module of IWVL. IWVL encompasses three main experimental platforms, including the physical electricity experiment platform, the physical optics experiment platform, and the physical pulley block experimental platform. More discussions in detail about this module will be provided later.
- Experimental instrument module of IWVL. The design and development of virtual experimental instruments is one of core parts in the construction of a virtual laboratory because all the experiments are designed and conducted on the applications and operations of experimental instruments. three strategies have been chosen for the design and development of experimental instruments in IWVL: 1) 3Dmax for 3D modelling of experimental instruments, which can be operated by users ; 2) Flash software for the 2-dimensional representation of experimental instruments; and 3) real photo presented to deepen students' perception of real experimental instruments. Users can choose different types of instruments for experiments.
- Extension resource module of IWVL. Various extension resources are designed in IWVL to expand students' vision in physics learning, to support students' to further explore knowledge and skills. For example, a typical extension resource in

IWVL is experimental instruments introduction that is committed to deepen students' understanding of the real forms and operation principles of instruments. It is helpful for combining virtual and real operations to support students easily to be faced with real experimental contexts.

5 Application Scenarios of IWVL In High Schools

IWVL can be used in various instructional contexts, such as application at real experimental room, application for collective preparation of lessons, application at multimedia classroom, application at networked classroom, application individually at home etc. This section presents two typical application scenarios of IWVL in high schools, including the usage of IWVL at real experiment room, and at multimedia classroom.

5.1 Scenario 1: Use for Collective Preparation of Lessons

IWVL can be used for collective preparation of lessons, which is an important approach to teachers' professional development and the improvement of teaching quality. A group of teachers can utilize demonstration experiments in IWVL to identify the goals, difficult points and focuses of teaching, to cooperatively carry out instructional design.

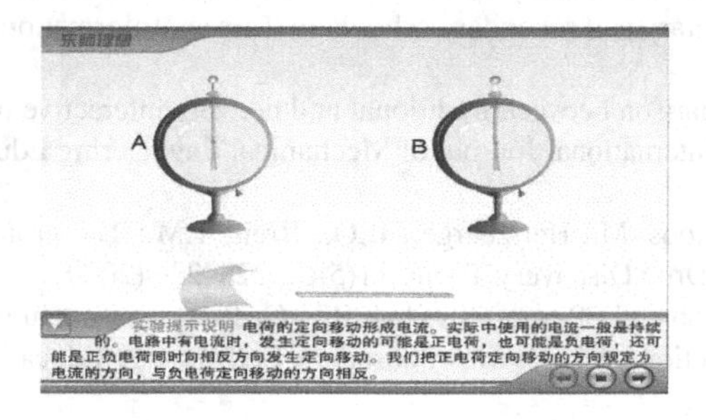

Fig. 1. Usage scenario of collective preparation of lessons

5.2 Scenario 2: Use for Teaching at Real Experimental Room

IWVL can be used in real experimental room for guiding students to carry out real experiments. Typically, teachers first demonstrate experiments on computers. Students can also operate experiment on computers and promote face-to-face discussions on experiments. Most importantly, students will practice experiments with real instruments. In the experimental process, students can reflect the principles, caution issues discussed in IWVL.

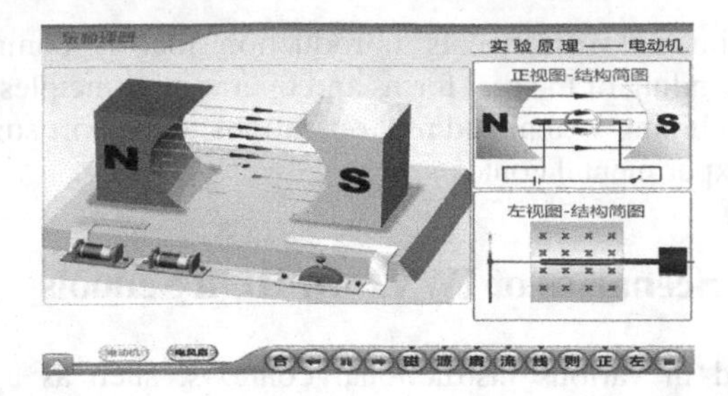

Fig. 2. Usage scenario of IWVL at real experimental room

6 Conclusion

This study focuses on the discussions of a network virtual lab for physics teaching in high school. We discuss the theoretical basis for the design of virtual laboratories for teaching. We present a framework report on the focuses network virtual lab. More efforts will be promoted for the actual application efficiency in future study.

References

[1] Ho, W.-H.: Research and analysis report on: taking action to elevate the Internet education quality at elementary and secondary schools in Taiwan. Information and Education, 14–25 (1998)

[2] Lee, H.P.: Comparison between traditional and network interactive manuals for laboratory-based subjects. International Journal of Mechanical Engineering Education 30(4), 307–314 (2010)

[3] Rauwerda, H., Roos, M., Hertzberger, B.O., Breit, T.M.: The promise of a virtual lab in drug discovery. Drug Discovery Today 11(5-6), 228–236 (2009)

[4] Gibbons, N., Evans, C., Payne, A., Shah, K., Griffin, D.: Computer simulations improve university instructional laboratories. Journal of Cell Biology Education 3, 263–269 (2004)

Intelligent Computer Aided Second Language Learning Based on Constitutive Principle

Wanzhe Zhao

Nanyang Institute of Technology
Henan, Nanyang China, 473004

Abstract. The article is in order to improve the learning model as a breakthrough point, comparing the difference of computer aided learning and intelligent computer aided learning, advocate establishing the teaching mode based on the constitutive principle and promoting the intelligent computer aided second language learning.

Keywords: computer aided.

1 Intrinsic Motivation and Extrinsic Motivation

Motivation is the process in learning second language whereby go-directed activity is encouraged and stimulated. One of the most general theories in the motivation theories is intrinsic and extrinsic motivation. Intrinsic motivation is to learn new things for the satisfaction of understanding in order to achieve pleasant experience ,challenges and stimulation . Extrinsic motivation is a kind of impetus in which the people are forced to learn L2 by rewards, threats, stress, pursuit or responsibility. It is necessary to set goals in order to achieve motivation, and as long as the goals are reasonable and specific , the motivation must be enhanced .At the same time, the intrinsic and the extrinsic interact with each other .

1.1 Previous Motivation and Goals

While there are certain universal aspects of second language motivation , other aspects of the process vary according to the individual and cultural environment .Most second language learners in non-English speaking country have strong motivation to learn English . Their intrinsic and extrinsic motivation are the followings:

Intrinsic motivation
_ to get high mark in the school
_ to obtain the better job with higher salary in the future
Extrinsic motivation
_ higher salary of being English teacher than others
_ more chances for English language experts
_ more awards given
_ English is a mandatory subject.

Z. Du (Ed.): Proceedings of the 2012 International Conference of MCSA, AISC 191, pp. 247–252.
springerlink.com

In the light of principle of achieving five language skills in listening , speaking , reading writing and translation, the goals are:

_ to learn and master 5,000 vocabulary
_ to grasp grammatical knowledge
_ to understand daily dialogue , simple news and academic lecture in the multi-media room.
_ to be able to communicate simple information with spoken English
_ to learn to use reading skills
_ to master writing skills
_ to translate accurately from L2 to L1 or from L1 to L2

Before analysing the motivation and goals above , it is useful to look at Bachman's view on components of language competence in Figure 1.

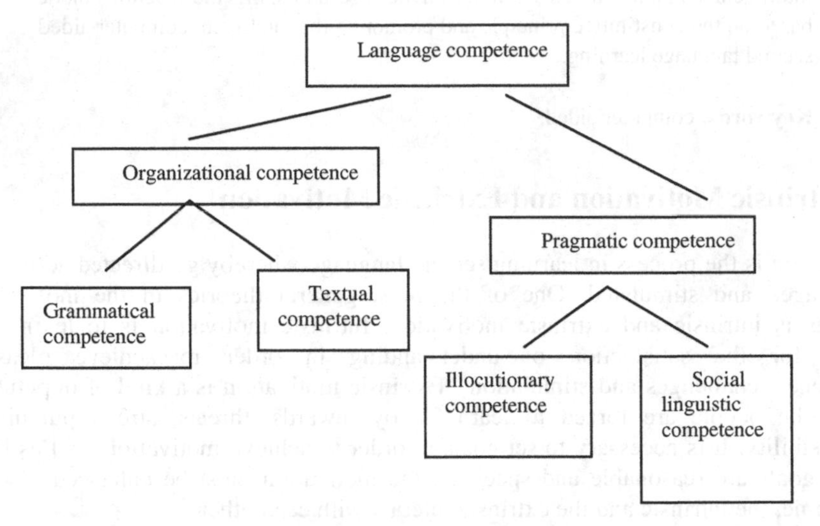

Fig. 1. Components of language competence Adapted from Bachman (1990:67)

1.2 Findings

From this Figure 1, it is obvious that setting goals only completes a partial process of the language learning , and the extrinsic motivation is stronger than the intrinsic .That's to say , it is easy to make the learners bored in the learning process because of lake of interest and inherent stimulation. To some extent, it must undermine the learning efficiency, and there is no way to achieve the translation between two languages.

1.3 Current Motivation and New Goals

In a sense, when the second language learners step into the English speaking country, they have been aware of deficiency in their language learning. In the domestic country, they learnt formal language . So when they hear some phrases in reality, such as " sweet as, cheers, kiwi buddy" they are confused. And due to diverse cultural background, sometimes what they say is not what they really want to convey. This is a so-called Chinese- English phenomenon, so pragmatic competence seems to be more important in the language learning.

Of course,their motivation is still strong. However, the intrinsic is stronger than the extrinsic .In this case, the learners personally want to discover the nature of the second language, experience the exotic culture and people, and to really know the methods to learn L2. Therefore they have the strong intrinsic drives to set new goals in the process of language learning:

_ to learn 25 new words a day.
_ to collect the expressions the native people often use every day
_ to listen to news stories from radio and television
_ to have a good accent
_ to develop the ability to speak fluently and accurately on the complex topic
_ to do intensive and extensive reading every day
_ to practise writing and seek feedback..

1.4 Findings

It's shown in Figure 2 that intrinsic motivation stimulates the learners to reflect on the process of language learning, and then consider how to improve language practice again, finally the learners establish new specific goals , which will satisfy the learners themselves.

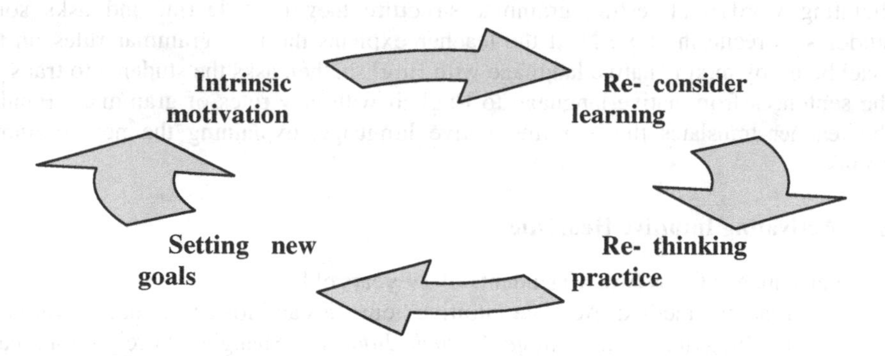

Fig. 2. Circular process of learning L2

Generally speaking, this is the specific short –term goal, but it can achieve a longer-term goal, that is to promote pragmatic competence, and complete the translation between two languages. Of course, the goal will potentially affect the motivation, in which the intrinsic and extrinsic interact each other. Hence, this is a productive circular process to provide an impetus that encourages the learners to improve language competence. However, the best method to learn English is also the key point.

2 Fostering Students' Language Awareness by Activating Intuitive Heuristics

In Learning English the teacher is main role, which is to foster students' language awareness.The teacher can use activating intuitive heuristics to design and create a rich linguistic environment for stimulating students to structure language context .Here to mention teaching English grammar as an example. Based on this principle, the two grammar classes are chosen as questionnaire participants for the discussion.

2.1 Traditional Grammar Teaching Method

- Participants: Class one, 40 students of 19 years old
- Teaching method: Traditional grammar teaching
- teaching material: *College English Book II* Shanghai Foreign Language Press (1999)
- Period: 12 weeks
- Purpose: Does traditional teaching method help students to learn grammar knowledge?
- Test taker: A female teacher of 35 years old with rich teaching experience
- Procedure:

Before starting the new lesson every time, the teacher checks the students mastery by dictating words and testing grammar structure they have learnt, and asks some students to recite the text. Next the teacher explains the new grammar rules on the blackboard by mixing native language with English, then asks the students to translate the sentences from native language to English with new rules of grammar. Finally, the teacher translates the text into native language, explaining the new grammar points.

2.2 Activating Intuitive Heuristics

- Participants: Class two, 40 students of 19 years old
- Teaching method: Activate intuitive heuristics and foster language awareness
- teaching material: *College English Book II* Shanghai Foreign Language Press (1999)
- Period: 12 weeks
- Purpose: Does activating intuitive heuristics and fostering language awareness help students to learn grammar knowledge?
- Test taker: A female teacher of 35 years old with rich teaching experience
- Procedure:

The teacher designs many activities (such as role- play, story telling, five minutes speech) to motivate the students to speak with grammar knowledge in the classroom. Before explaining new syntactic structures, the teacher firstly guides the students to listen to the text. Then the teacher encourages them to find and generate the grammar knowledge. Finally the teacher explains differences between English and native language in the grammar. Thus the students have strong acquisition to use grammar rules to convey their idea repeatedly.

In order to get explicit data, eventually the questionnaires (see following) were handed out in two classes to investigate what students have mastered in the grammar course. From the following table, it is clear that the difference can be easily understood by referring to the table blow which shows that the students in class one are lack of self-confidence during learning English grammar, they always make mistakes in conversations and writing. Whereas, the students in class two are confident to learn English grammar. They are able to use grammar properly in conversations, but they still have no ability to correct grammars in writing.

	Good ability to use grammar	Good control of syntactic structure in writing	Frequent and systematic errors in conversations	Structure conversations logically	Self-confidence
Class One	52%	43%	46%	22%	41%
Class Two	56%	31%	24%	58%	66%

3 Conclusion

By troditional teaching in class one, the teacher inputs the grammar rules into students regardless of the students' language ability. However, the students are passive to learn English without any concepts of the language environment. The students know grammar rules, but do not use in daily life (it is so- called dead language). Even though they can get high marks in their test by their retention and grammatical foundation, they have no ability to think and speak in English. The classroom ambience is boring. While, in class two,the teacher uses various forms of drill and practice exercises to facilitate the students to develop grammatical competence. The students are able to properly assimilate the generalized rules of grammar flexibly to a range of linguistic situation through lots of practice. Furthermore, the students learn to use English to think and speak. The classroom ambience is happy and relaxing.

In a word, once the second language learners set the goal driven by intrinsic and extrinsic motivation, the activating intuitive heuristics and fostering language awareness is more helpful than the traditional teaching method. These not only train learners listening and speaking competence, but help them enlarge English knowledge, which draws their attention to learn and know more grammatical rules, and foster the learners to have self- confidence and courage to use fluent English. Conversely, there is still a large gap in the learners' knowledge of correct pragmatics, how to improve deficiency in stimulating the learners' motivation and in teaching the second language needs further study.

References

1. Johnson, D.M.: Approaches to research in second language learning, Apply research as reseeing learning, pp. 6–7. Longman Publishing Group (1992)
2. Jordan, G.: Theory construction in secotablend language acquisition. Key terms and current problems in SLA, pp. 8–9. John Benjamins Publishing Company (2004)
3. Dornyei, Z.: Motivation in second and foreign language learning. Thames Valley University, London
4. Gunn, C.L.: Teaching, Learning and Researching in an ESL Context.Language awareness, pp. 42–43. University Press of America (2003)
5. Kunna, A.J.: Studies in Language Testing 9. An exploratory study, pp. 175–179. The Press Syndicate of University of Cambridge (2000)
6. Leeson, R.: Fluency and Language Teaching. Key factors in the FL Teaching Programme, pp. 151–153. Longman Group Limited, London (1975)

The Application of Computer - Based Training in Autism Treatment

Li Lin and Chunmei Li

Mudanjiang Medical University, Mudanjiang, Heilongjiang, 157011 China
linli@cssci.info

Abstract. In this paper, we reported the effective of the tangible user interfaces system to support training basic geometric shape for low-functioning children with autism. The results from the experiment clearly show that the children with autism who study from the TUI training system can learn more shape than conventional treatment. Furthermore, TUI was more attractive and offers a multimedia real time feedback that stimulates children's learning and also increases their attention span rather than the conventional treatment.

Keywords: Computer - Based Training, Autism Treatment, Evaluation.

1 Introduction

Autism is a barrier of brain function, appear very early in life before the age of three. Autism causes the damaged social intercourse, and communication difficulties, restricted or repeated activities and interests. People with autism often show the anomalies sensory stimulation (such as, touch, sound and light), there are usually moderate retarded, and have a higher likely to develop epilepsy. Some patients is with autism aggression and self-injurious ACTS of exhibition (for example, head, bite their). About a third of the autistic patients have the normal or close to normal intelligence business (IQ). There is no cure for autism, yet, with the appropriate treatment and education, children with autism can learn and develop a skills in order to live a normal life in society. Treatment depends on the patient's personal needs. In most cases, the treatment of combination method is required. In autism treatment methods, occupational therapy and physical therapy are two of the most need to autism training. Occupational therapy can improve the function of the independent and teach basic skills (such as buttoning a shirt, taking a bath, and basic shape, color). Physical therapy including the use of physical exercise and other physical variables (such as, massage, heat), help customers to control the body action [5].

And general computer use explosive growth, many researchers have developed using the technology to auxiliary treatment an autistic children and exchange. An important influence factors to the computer for the foundation of the efficiency of the training application is a choice of the user interface and interactive style, for example, Dautenhanh et al. [1] in their aurora project investigated is, how to encourage a child to autonomous mobile robot can be engaged in social interaction, one of the important aspects of the interaction between respectively. Bosseler et al. [2] the development and evaluation based on computer animation mentor to improve vocabulary learning and grammar ability children with autism. Some researchers investigated the

Z. Du (Ed.): Proceedings of the 2012 International Conference of MCSA, AISC 191, pp. 253–258.
springerlink.com

effectiveness of the virtual reality technology for the treatment of autism. For example, dorothy's history LanDing XiaoPing. [3] published two case studies examining whether will put up with autistic children dressed in virtual reality equipment and can respond generated world meaningful way. Leonard et al. [4] hired a virtual reality system. This system is used for education for autistic children's social skills to find a place to sit for a stranger. Mary Barry and Ian Pitt provide the basis of interactive software design based on the window icon menu and pointing device (incompetent user interface) for learners with autism. The research results show that children with autism may like to use computer interface interaction, because they can actively control it.

2 Autism Treatment

Autism is with surgery or drug therapy (although some people with autism treatment to improve some symptoms, such as aggressive behavior or attention... "). Instead, those who have autism education skills need to have a normal life. The best of the results are often see children start treatment, they were very young. Special education project children, for its tailored to the individual needs is usually the most effective form of processing children exchange teaching (sometimes by pointing to or using pictures or sign language) and interact with others. A treatment plan including basic life skills and the following ACTS: behavior modification, speech therapy, physical therapy, music therapy, changes in diet, drug, occupational therapy, listen to or vision therapy. The same experts to help diagnose condition usually come up with the best combination treatment common use in addition to education plan. [4] There is no single best treatment for autistic children's services. Little most professionals agree that early intervention is very important; the other is a most people to people with autism highly structured, professional program

Shape matching training is one method has been used in treatment of the basic skills. In the traditional the geometry of the training, and construct a occupational therapists display picture card includes a geometric shape or against, clearly represent geometric shape (figure 1), and then asked the patient found similar shape and video CARDS. In another way, the board of geometric shape is different hole introduced to a patient (figure 2). Then, the patient asked to choose the right shape piece for these loopholes to. In fact, these traditional treatments were found to be unattractive, autistic children and tired. At the same time, these methods need a sufficient number of occupational therapists therapy is effective.

Fig. 1. The snapshot of the child with autism during conventional training

Fig. 2. The snapshot of the shape block set that use for conventional training

3 Geometric Shape Training

Autistic children in the teaching of the basic geometrical figure known to display a occupational therapists photo CARDS contain geometric shape or against, clearly represent geometric shapes, and asked them to choose the right shape or use a block as described earlier.

According to the observation, autistic children have their own unique learning style they like playing computer system uses a specific user interface device, like a steering wheel used in a computer game. Then said such equipment may be more suitable for autistic children than the incompetence of standard equipment (a mouse and keyboard). On the other hand, computer games can rule out and help children around hate to focus on the task, they can learn from his mistakes without the power of occupational therapists. At the same time, price et al reported has the potential to provide innovation, the methods of physical let the kids play, learn, through the novel form of interaction and found.

A specific user interface is a interface, computer users use digital environment of the interaction of a specific input device. Molding matching training system development uses this type of user interface. It aims to provide a similar experience and other reality interactive tangible objects, such as toys. A system is by the hardware components as shown in figure 3. The operation is controlled by custom software running in computer CBT machine. In training operation, the case for the traditional drug therapy, a geometric shape is random selection and displayed in a translucent glass desktop, through the digital projector in the table. User requirements (through the narrative voice) choose a suitable shape with board and put it on the table over an area where a image geometric shape is projected (see chart 5 for the lens of program). Consumption level on the side of the digital camera is used to capture images with board of choice. To take photos, on the edge of color, and then the block is a image processing module (figure 4), realize as CBT software.

The developed training system real-time response and earnings reaction to provide children with reward children note if the chosen shape is correct. Given in a reward is a cartoon image applause voice and music (figure 6). If the wrong shape is selected, they were encouraged to "try again" by the computer narrative voice.

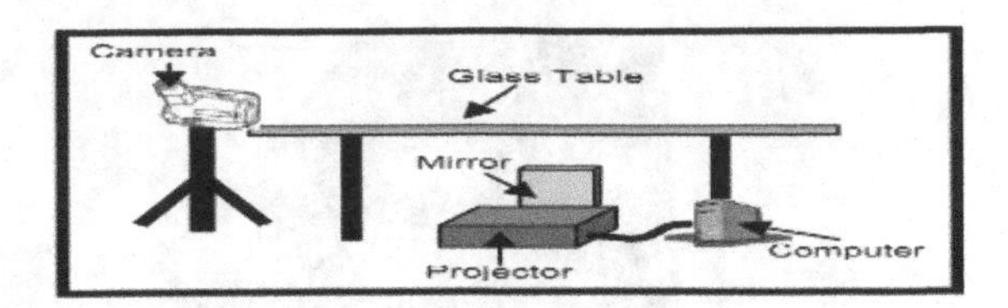

Fig. 3. Hardware components of the TUI-based training system

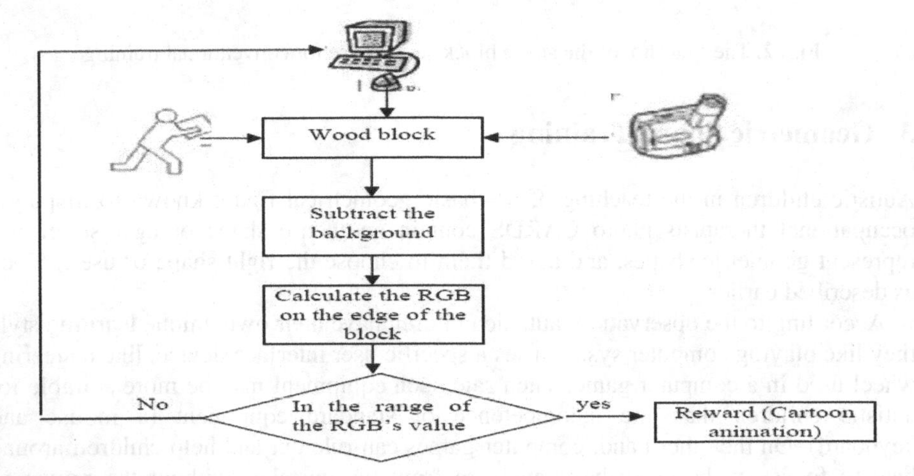

Fig. 4. Block diagram of the image processing module, implemented as part of the training software

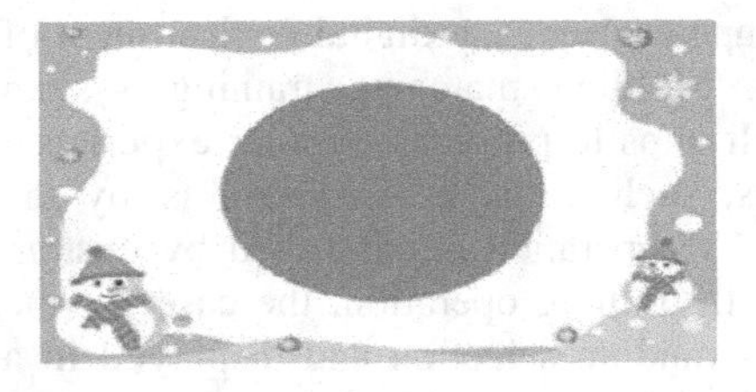

Fig. 5. The screen shot of the TUI-based training system

Fig. 6. The screen shot of the correct answer

4 Result and Evaluation

In our previous research we explore the ease of use comparison of WIMP and tangible Uis. The results from the experiment clearly show that tangible user interface is easier to use and suited for low-functioning autism than standard WIMP interface. Then to support these research we compare the ease of use of the two interfaces (WIMP and tangible Uis) as perceived by non autistic children by using shape matching skill for the case study. The results obviously show that tangible user interface is easier to use for the children with autism than the WIMP interface.

The difference between the two UIs is noted to be much more noticeable than the case of non autistic children.

Then in this paper the experiment was conducted to compare the two training methods in term of learning efficacy. The measurement was done by means of pre-test and post-test. Before the first session the participants were asked to do a pre-test. The test consists of 6 pictures of geometric shapes and the participants were asked to select the matching shape. After completing all training sessions, they were again asked to do a post-test. The same post-test was conducted again a week later. The tests were then analyzed. The difference between the pre-test and post test results was used to measure learning efficiency. The analysis results are shown in Table 1 and 2. From these tables, it was found that the participants from the group where the TUI-based training system was assigned could learn more shapes than those from another group using the conventional training method.

In supporting the hypothesis that the TUI-based system achieves higher learning efficiency (in terms of the number of learned shapes) as compared with the conventional method, the Two-sampling t- test is used (due to a small sample size with the assumption that the population is normally distributed). At the significant level of 0.05, for the pre-test, the calculated value of $t = 2.457$. Then the p-value for this upper- tailed test is 0.046. The p-value is less than 0.05. On the other hand, for the post-test the calculated value of $t = 6.807$ Then the p-value for this upper- tailed test is 0.000004. Therefore at the significant level of 0.05, the null hypothesis is rejected. It is thus conclude that the tangible user interface training system is more efficient for training basic shape skill than conventional treatment as measured by the number of learning object shapes.

Based on observation during the experiment, we also found that time to complete the task using the TUI-based training system was generally shorter than that of the conventional method. We noticed also that, while strongly expressing unwillingness to use the conventional method, all participants were joyful when training with the new system.

Table 1. The number of correct matching shapes of the pre-test and post-test for the grouped assigned with the TUI-based training system

Name	pre-test score/6	post-test score/6
T1	1	6
T2	0	6
T3	0	6
T4	0	4
T5	0	6
T6	0	5
T7	0	6
T8	0	6

Table 2. The number of correct matching shapes of the pre-test and post-test for the grouped assigned with the conventional method

Name	pre-test score/6	post-test score/6
C1	0	1
C2	0	2
C3	0	1
C4	0	2
C5	0	3
C6	1	2
C7	0	0
C8	0	1

5 Conclusion

Most of the research work, but the main focus on autism. No open literature study focused on using technology to train because the basic technology low, and in children with autism. According to the report, and the elements of the user interface is easier to use, because of the low than the standard of the incompetence of autism interface. However, how to effectively system, compared with the traditional methods of training has not discussed. This paper studies the role in the term training system learning efficiency compared with traditional methods, training basic geometry for children with autism.

Acknowledgement. The paper is supported by Department of Education of Heilongjiang Province (No.11551524).

References

1. Dautenhahn, K.: Design issues on interactive environments for children with autism. In: Proceeding 3rd Intl.Conf. Disability, Virtual reality & Assoc., Alghero Italy (2009)
2. Jerry, A.F., Allison, D., Gene, C., Sante, S., Wayne, C.: Child's Play: A Comparison of Desktop and Physical Interactive Environment. In: Proceedings of IDC 2005. Boulder CO, pp. 48–55 (June 2009)
3. What is Autism?, `http://www.nas.org.uk/nas/jsp/polopoly.jsp?d=211`
4. Autism., `http://www.kidshealth.org/teen/school_jobs/school/autism.html`
5. Neurology channel, `http://www.neurologychannel.com/autism/treatment.shtml`

Survey to CAT Mode Based on the Construction Theory

You Yangming[1], Wang Bingzhang[2], Chi ZhenFeng[3],
Zhao BaoBin[3], and Zhang Guoqing[4]

[1] Department of Physices and Electron Information, Cangzhou Teachers` College,
Hebei Cangzhou 061001

[2] Department of Machine and Electric Power Engineering Cangzhou Teachers` College,
Hebei Cangzhou 061001

[3] English Department, Cangzhou Teachers` College, Hebei Cangzhou 061001

[4] Department of Materials and Energy, Guangdong University Technology, Guangdong
Guangzhou 510006

Abstract. CAI is a new teaching mode combined construction theory with IT and modern education theory. Based on the construction theory, the new teaching modes under the principles of computerization, interaction, self-learning, cooperation and creation is surveyed..

Keywords: Constructivism Theory, Experimental Subject, Computer.

1 Introduction

In 21st century, great changes will take place in China's higher education. The symbol is that the traditional teaching method (a method majoring in passing on knowledge) are changing into a method of fostering students with the ability of bringing forth new ideas and overall quality. We should form the teaching principle---the students are the main body and the teachers are the guides. Writer thinks the theory of constructivism as a newest theory of the western educational psychology has become mature and it is our educational thought and principles of teaching of the day. Constructivism guides our study and exploration of Preceding Experimental Methods.

2 The Ideology of Constructivism

The Theory of Constructivism emphasizes that the students are the center of teaching. It requires that students should change from passive receivers of knowledge into the main body of information processing and active constructors of knowledge. And it requires teachers should transform from the ones passing on knowledge to the ones helping and motivating students. Thus, the roles of students and teachers have been changed. It means that teachers must adopt a new teaching model, new teaching methods and ideas. The teaching idea---learning as the center was put forward in the Constructivism learning situation, and constructivism learning situation lays the foundation for our new teaching idea. A famous Chinese educator, Professor He Ke

Z. Du (Ed.): Proceedings of the 2012 International Conference of MCSA, AISC 191, pp. 259–264.
springerlink.com

kang summarized the teaching idea as: knowledge outlook, learning outlook, student outlook, the position of teachers and students, learning situation and teaching principle etc. six [1-6].

The theoretical basis of constructivism is cognitive psychology. After updating David H. Jonassen and some others' extreme constructivism theory, constructivism theory is becoming ripe and perfect. Thanks to the strong support from the newest information technology to the learning environment required by constructivism, we can combine constructivism theory with teachers' teaching practice. So constructivism becomes the guiding ideology to deepen reform in education at home and abroad[7].

3 The Compatibility between the Guiding Ideology of Preceding Experimental Methods and the Teaching Ideology of Constructivism

The study of Preceding Experimental Method began in 1933. Its guiding ideology was not in accordance with the teaching ideology of Constructivism.[8-9] after rejecting extreme Constructivism, the theory of Constructivism got enriched and perfect. It has become the ideology of education and principle of education. Since 2000, Constructivism has been guiding the study and exploration of Preceding Experimental Method[10-14].

3.1 The Compatibility between the Guiding Ideology of Preceding Experimental Methods and the Teaching Ideology of Constructivism

Preceding Experimental Teaching Method is successful in the study and exploration into the Experimental Course. This is because the guiding ideology of Preceding Experimental Methods is in great agreement with the teaching ideology of Constructivism.

3.2 A Case Study in Preceding Experimental Method

For example, in "Modern Physics" teaching, when using spin-orbit interaction to explain the nuclear fission atomies energy level splitting of alkali metal and the fine structure of spectrum [10], we explain normal Zeeman Effect. When guiding students to do the normal Zeeman Effect experiment, we design a preceding one: in the experiment, we studied 5461Å spectral line normal Zeeman splitting in mercury electro-optical tube spectrum, then we further guide students to observe the complex unusual A normal Zeeman Effect) (as shown in the figure). And we remind the students that they can observe the phenomenon when the outside magnetic field is weak. When the magnetic field is strong and magnetic moment s=0, what they observe is three spectral lines with the same structure as normal Zeeman Effect. Why can not we observe the complex Zeeman Effect when the magnetic field is strong and magnetic moment S=0 while we can observe complex Zeeman Effect when the magnetic field is weak and $S \neq 0$? Is it caused by the difference between weak and strong magnetic field? As we know, complex Zeeman Effect and normal Zeeman

Effect have different selection rules. They both use coupling. But, owing to different g factors, their spectroscopic splitting is different and there are more than splitting spectral lines

When doing the experiment, instructors do not answer students' questions, but let students think about them. Thus, after observing the experiment phenomena and thinking about instructors' questions, students become greatly interested in the questions so as to mobilize students' initiative in previewing textbooks and teaching materials. It also provides excellent students with an opportunity to use their wisdom and foster their creative ability. In practical teaching, students really think that it is easier to have perception first and then to receive the explanation of theory. They do not fear abstract theory any longer.

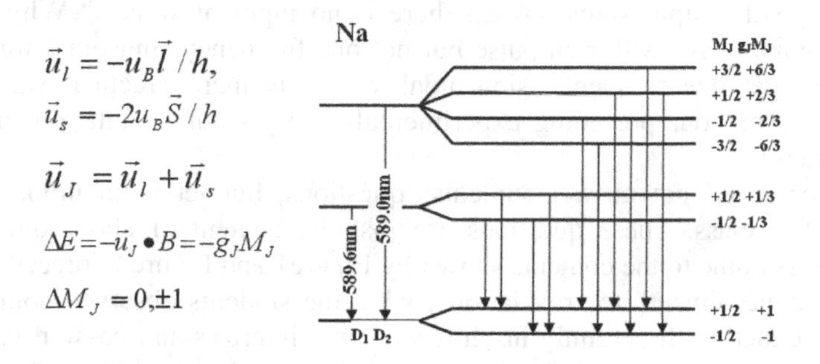

Fig. 1. Zeeman Effect Zeeman Effect Diagrammatic sketch

Another example: in teaching "Analog Circuits", when we come to the chapter "degenerative feedback amplifier", we design a preceding experiment. The tasks of the experiment is to study the effect of degenerative feedback on the performance of amplifier and to know about the ways of measuring degenerative feedback amplifier input and output impedance and the voltage amplification factor and pass band width. The experiment circuit conspectus (2) shows: different connections of the circuit produce four types of two-pole feedback.

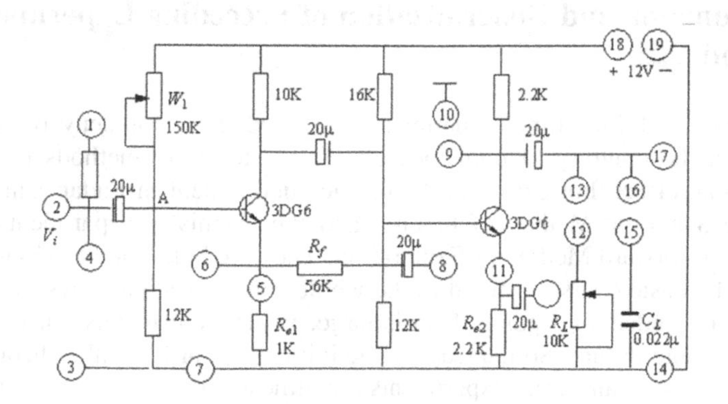

Fig. 2. Voltage shunt plus feedback circuit

After finishing the tasks of the experiment class, instructors ask students to keep the experiment circuit, for example, to keep the voltage series feedback circuit, that is (5)and (6), (8) and (9), (12) and (13) circuits. When we input signals through the two ends (2) and (3), the oscilloscope is connected with the two ends (14) and (17), then input sinusoidal signal of f =1KHz,Vi =5mV and adjust the circuit to get a non-distortion sinusoidal wave. After removing the signal, the sinusoidal wave on the screen of the oscilloscope will disappear. When adding the signal again, the sinusoidal wave will re-appear. If the instructor guides students to cut off (5) and (6), but connect network node (6) with node A, the circuit will become a voltage shunt plus positive feedback circuit. At this time, a cyclic pulse wave of tens of Hz appears on the screen of the oscilloscope. After removing the signal, the wave does not disappear from the screen. This shows the amplifier still outputs signals. Why does the amplifier still output signals when there is no input of signals? Why is output signals harmonic wave with rich pulse but not one frequency sinusoidal wave? Then how can we get one frequency sinusoidal wave? Is there practical value in this circuit? These are our preceding experimental theory-- auto excitation theory and vibrating circuit.

Instructors need not answer students' questions, but let them think about the questions after class. These questions are just the content of classroom teaching. When teachers come to the contents shown by Figure1 and Figure 2, preceding theory goes from "concealment" to "revelation", what the students observed coincide with theoretical deduction. It certainly inspires students' interests and passion in learning. And it provides an opportunity for them to use their wisdom and foster their creative ability.

So effective teaching should enable students acquire as much related knowledge and perception as possible before classroom teaching. The design of preceding experiment need not change our teaching plan for experiment class. And there is no need to design a preceding experiment purposely. According to a certain experiment class, teachers can insert a 10-15 minutes preceding experiment in an experiment class. Certainly, there are many occasions in which we can design a preceding experiment before a theory-teaching class[8-15].

4 The Function and Generalization of Preceding Experimental Method

The application of Preceding Experimental Method does not deny other teaching methods. On the contrary, it combines with other teaching methods to produce a better teaching effect. For example, the physics department in Fudan University put forward the Scientific Method of Inquiry, and the chemistry department in Beijing University put forward Method of Exploration. They are both good teaching methods. Especially the basic courses offered to the whole school, such as physics, chemistry; the number of students attending them is large, while experiments must be done by only a few people in turn. So the teaching activities are hard to link. Theory teaching can have a unitary plan, while experiments are almost carried out at the same time. It is hard to arrange an experiment before the theory teaching. So teachers are required to carry preceding demonstration experiment to achieve active effects. If we instruct on the basis of competency, combine classroom teaching and experiment teaching properly, flexibly use Preceding Experimental Method and strengthen the guide in every experimental section, we are sure to achieve better teaching effects.

If the school is set up for basic courses such as physics, chemistry, the large number of classes, while the experimental activities are generally carried out several groups of rotation, it is difficult in a time before class to arrange a unified theory of the experimental course, the experimental class convergence with theory is difficult. This requires teachers to lead classes on "demonstration experiments" can also be a positive effect.

During the first part of our research project, we carried out a survey among the students Shanghai University of Science and Engineering, Guangzhou Industrial University and Cangzhou Teachers` College. the effect is good. 91.7% of the subjects think the quality of education improved obviously. After using Preceding Experimental Method, the effect in classroom teaching is most outstanding. Its operability is suitable for experiment teaching in specialties of Science, Engineering, Medicine, and agriculture in university.

To sum up, the basic guiding ideology of Preceding Experimental Method is in accordance with the teaching ideology of Constructivism. They are the same in nature. The application and popularization of Preceding Experimental Method is a must in quality education, in China's new construction and in deepening teaching reform in higher education.

References

1. Jonassen, D., et al.: Constructivism and Computer-MediatedCommunication in Distance Education. The American Journal of Distance Education 9(2), 7–26 (1995)
2. Wilson, B.G.: Metaphors for Instruction: Why We Talk About LearningEnvironments. Educational Technology 35(5), 25–30 (1995)
3. Savery, J.R., Duffy, T.M.: Problem Based Learning: An InstructionModel and Its Constructivist Framework. Educational Technology 35(5), 31–38 (1995)
4. Dede, C.: The Evolution of Constructivist Learning Environments:Immersion in Distributed Virtual Worlds. Educational Technology 35(5), 46–52 (1995)
5. Toomey, R., Ketterer, K.: Using MultiMedia as a Cognitive Tool. Journal of Research on Computing in Education 27(4), 472–482 (1995)
6. He, K.K.: Constructivism: the theory foundation of reforming traditional teaching (Part 1). E-education Research (3), 3–9 (1997) (in Chinese); He, K.K.: Constructivism: the theory foundation of reforming traditional teaching (Part 2). E-education Research (4), 25–27 (1997) (in Chinese)
7. Pang, W.: Autonomic Learning the Principle and Policy of Study and Teaching, 1st edn., pp. 159–189. East China Normal University Publishing Corporation, Shanghai (2004)
8. Zhang, J., You, Y.: Application of Preceding Experimental Method in Electronic Technology. In: Liu, X., Pang, Z., Zhang, D. (eds.) Progress of Higher Education Teaching and Research New Proceedings, pp. 66–69. China Science and Technology Press, Beijing (1996)
9. Chai, S., You, Y.: Reform and exploration of Electrical electronics experiment. College Physics (Education Monograph) (2), 126–128 (1999)

10. Wei, L., You, Y., Ma, W.: Application of Preceding Experimental Methods in Modern Physics Experiments. International Journal of Engineering Education Theory and Practice 12(8), 37–39 (2003)
11. You, Y., Wei, L.: Application of Preceding Experimental Methods in Optics. Journal of Hebei Teacher's University 29(Supplement), 118–121 (2005)
12. You, Y., Wang, B., et al.: A Research Into Cultivating Creative Talents to Deepening Educational Teaching Reform The Preceding Experimental Teaching Method's Study and Exploring. Journal of Cangzhou Teachers, College 23(4), 43–46 (2007)
13. You, Y., Ma, W., et al.: The Preceding Experimental Teaching Method and Electromagnetism Experiments. Journal of Cangzhou Teachers, College 24(3), 33–35 (2008)
14. You, Y., Wang, B., You, S., et al.: Application and extension of Preceding Experimental Methods in practice instruction. Research on Higher Education(Journal of Nanjing University Philosophy and Social Sciences) (92), 47–49 (2007)
15. Zhao, K., Lo, W.: New Concept Physics Course,quantum physics, 1st edn., pp. 270–280. Higher Education Press, Beijing (2002)

Intelligent Distance Learning Model of Ethnic Minority

Jianru Zhang

Organization and Personnel Department Inner Mongolia Radio & TV University, Huhhot, Inner Mongolia, 010010 China
zhangjianru@cssci.info

Abstract. This paper takes the ethnic minority distance learners in Inner Mongolia as the research targets. Through the comparative research on the learning environments of the ethnic minority learners and the Han nationality learners, the following conclusions have been drawn: there are striking differences between the network learning environment and the unbeneficial learning environment of the ethnic minority distance learners and those of the Han nationality learners. An intelligent distance learning model is presented based on a combined courseware structure.

Keywords: Ethnic minority, distance learner, Intelligent distance learning model.

1 Research Method

This research bases on the learners, who are major in accounting, computer, and business management in Inner Mongolia Radio & TV University in 2009. The research carries out some questionnaires. The researchers take the David Kember distance learners dropping off school table as the major references. Questionnaires have advantages over some other types of surveys in that they are cheap, do not require as much effort from the questioner as verbal or telephone surveys, and often have standardized answers that make it simple to compile data. However, such standardized answers may frustrate users. Questionnaires are also sharply limited by the fact that respondents must be able to read the questions and respond to them. Thus, for some demographic groups conducting a survey by questionnaire may not be practical. The questionnaires are mainly divided into three parts: the first one is the basic information; the second one is the main body of the questionnaires and the third one is the open questions. There are altogether 1482 questionnaires and there are altogether 1179 questionnaires that have been back effectively. The returning rate is 79.6%.

The research makes use of the data from the questionnaires that have been returned effectively. A table has been summarized, which is "the data table of learning environment research on the distance learners". The data is to be analyzed using the SPSS17.0 software.

Z. Du (Ed.): Proceedings of the 2012 International Conference of MCSA, AISC 191, pp. 265–270.
springerlink.com

2 Data Analysis

2.1 The Analysis of Main Components of the Learning Environment of Distance Educational Learners

The "research data of the learning environment of the distance educational learners" is made use of. 27 project data are put together to make the Factor-Analysis-Based Dimension Reduction. Through the Rotation Method- Varimax with Kaiser Normalization, 3 major components have been draw. The testing result from the KMO and Bartlett prove that the Kaiser-Meyer-Okin is .933; the Bartlett's Test of Sphericity Approx. Chi-Square is 16066.92. The testing results are workable, explaining that the major components that have been abstracted are able to represent the corresponding potential factors. The five major components that have been abstracted making use of the major components stand for the following five factors: the educational institution environment, the network environment, the unbeneficial factor to learning, and the working environment as well as the social family environment.

Table 1. The testing of the major component KMO and Bartlett in the learning environment for distance learners

Enough Kaiser-Meyer-Olkin		.933
Bartlett's Test of Sphericity Approx	chi-squared approximation	16066.920
	Df	351
	Sig.	.000

2.2 The Difference Analysis of the Learning Environments for Different Nationalities

The five major components that have been abstracted making use of the major components stand for the following five factors: the educational institution environment, the network environment, the unbeneficial factor to learning, the working environment and the social family environment. The difference analysis of the learning environments for different nationalities has been made, which can be seen in Table 2.

Table 2. The difference analysis of the learning environments for different nationalities

		average	Standard differences	F	Sig.
Network environment	Han nationalities	.0378442	.94275615	7.983	.005**
	ethnic minorities	-.0629972	1.11875221		
Educational institutions	Han nationalities	-.0254766	.98355446	.074	.785
	ethnic minorities	.1114755	1.02824624		
Unbeneficial factors for learning	Han nationalities	-.0261379	1.03460034	7.026	.008**
	ethnic minorities	.1257428	.87915176		
Working environment	Han nationalities	-.0035240	.97675801	.001	.981
	ethnic minorities	-.0607993	1.02710490		
Support from families and friends	Han nationalities	.0225701	.99535740	.079	.779
	ethnic minorities	-.0054582	.97018686		
**. in .01 horizontal (both sides) related differences, *. in .05 horizontal (both sides) related differences					

Two independent samples are made use of to test and analyze the differences of the learning environment for distance educational learners from different nationalities. The results show that there are striking differences on the network environment and the unbeneficial factors between the distance educational learners from the Han nationalities and the distance educational learners from the ethnic minorities. The research assumptions have been testified. It has proved that there are great differences on the evaluations of the network environment and the unbeneficial factors to the distance learners of the ethnic minorities and the distance learners of the Han nationalities. The researches have shown that the evaluations on the network environment by the ethnic minority distance learners lay far behind the evaluations on the network environment by the Han nationalities. The researching results prove that the current networking environment construction still fail to meet the learning requirements of the ethnic minorities learners.

2.3 The Difference Analysis of the Network Environment of Different Nationalities

It can be seen from the T testing results that there are striking differences on the learning environment between the ethnic minorities and Han nationalities. The learning environment has included the following two aspects: the networking environment and the unbeneficial factors to learning. Proceeding to the next step, the research adopts the independent sample T to make analysis on the difference performance of different nationalities to the networking environment and the unbeneficial factors to learning.

Table 3. The difference analysis of the network environment of different nationalities

	Nationalities	Sample amount	Average values	Standard difference	F	Sig.
Convenience degree to access network	Han nationalities	669	4.04	.848	.732	.392
	ethnic minorities	172	4.06	.856		
Network condition home and at work	Han nationalities	814	4.03	.746	12.067	.001**
	ethnic minorities	176	3.99	.865		
Satisfactions for accessing network	Han nationalities	666	4.09	.749	1.135	.287
	ethnic minorities	171	4.02	.811		
Network interact skills	Han nationalities	660	3.98	.854	.888	.346
	ethnic minorities	173	3.92	.866		
Network language obstacles	Han nationalities	818	4.02	.734	8.997	.003**
	ethnic minorities	177	3.99	.839		
Network cultural background knowledge	Han nationalities	663	4.02	.839	.673	.412
	ethnic minorities	172	4.03	.830		

It can be seen that there are striking differences on the networking conditions at home and at work. In addition, there are great obstacles for network language usage for different nationalities. It proves that the ethnic minorities' learners do not locate in a good condition speaking of the networking conditions and the networking language usage.

2.4 The Difference Analysis of Unbeneficial Learning Environment for Different Nationalities

As for the unbeneficial condition of learning, it mainly refers to the following aspects: the negative effect of national language, the lack of communication between classmates and the relatively high tuition fees and the heavy burden of economy. Distance education is a field of education that focuses on teaching methods and technology with the aim of delivering teaching, often on an individual basis, to students who are not physically present in a traditional educational setting such as a classroom. It has been described as "a process to create and provide access to learning when the source of information and the learners are separated by time and distance, or both." Distance education courses that require a physical on-site presence for any reason (including taking examinations) have been referred to as hybrid or blended courses of study.

Table 4. The difference analysis of unbeneficial learning environment for different nationalities

	Nationalities	Sample amount	Average value	Standard difference	F	Sig.
The negative effect of national language	Han nationalities	662	3.53	1.275	19.584	.000**
	ethnic minorities	171	3.80	1.130		
Lack of communications between classmates	Han nationalities	822	3.56	.961	.080	.778
	ethnic minorities	177	3.76	1.006		
Relatively high tuition fees and heavy burden of economy	Han nationalities	667	3.51	1.158	9.263	.002**
	ethnic minorities	171	3.76	1.066		

**. in 0.05 horizontal related differences, *. In 0.01 horizontal related differences

3 Conclusions

The distance education has played an important role for promoting the equality between different nationalities, different sexes and different areas of education. Today, there are many private and public, non-profit and for-profit institutions worldwide offering distance education courses from the most basic instruction through to the highest levels of degree and doctoral programs. Researchers have found that there are striking differences on the aspects of networking environment and the unbenifical environment for learning between the ethnic minority distance learners and the Han nationalities learners.

In the ethnic minorities' area, in order to develop the distance learning and improve the hardware environment of network, the most important thing to do is to strengthen the network resources construction and construct much more fit learning resources for the ethnic minority's learners. This program is now known as the University of London International and includes Postgraduate, Undergraduate and Diploma degrees created by colleges such as the London School of Economics, Royal Holloway and Goldsmiths. Therefore, the concept of learning society should be strengthened and the recognition of distance learning for ethnic minorities.

The research has proved that the learners from the ethnic minorities face much more serious learning conditions that are unbeneficial. The project of "nationality language culture habits have negative effects on the distance learning." The existence and the teaching service of the distance education intuitions have satisfied the learning requirements of the ethnic minorities learners.

References

1. Tan, G.-D.: The educational research on the original citizens. Wunan book publishing corporation limited, Taibei (1998)
2. Michael, G.: Moore distance education system view. Shanghai Higher Education Electronic Video Publishing House, p. 11 (2008). May first Edition first time PVI
3. Chen, L.-P., Li, N.: Sixty Years for the Ethnic Minorities Education in our Country: Looking Back and Rethinking National Education Research, 1st edn., pp. 5–13 (2010)
4. Chen, X.-S.: The networking education learning adaptability research. Chinese Distance Education, 6–8 (2002)
5. Gong, C.-Q.: The brief discussion on the current situation, problem and reformation of the higher education for Inner Mongolia. Tsinghua University Education Research (3) (2002)
6. Huang, L.: The brief discussion on the development of the distance education for ethnic minorities in Sichuan area. Journal of Southwest University for Nationalities (June 2008)

Study on English Writing Based on Modern Information Technology

Wen Lai and Han Lai

School of Foreign Languages, Hubei University for Nationalities, Hubei Enshi 445000

Abstract. This paper concerns how the multimedia as a information technology (IT) aids the teaching of English writing. Based on the analysis of the challenges to and the aim of the writing for undergraduates, how the multimedia is employed during the teaching of writing is demonstrated.

Keywords: Multimedia, information technology, English writing.

1 Introduction

The teaching of English writing (for non-English majors) has long been viewed as a weaker link in the college English education due to its difficulties and the inefficiency. Many teachers even give up the training of writing because of the lack of teaching hours, especially when they have to make sure other teaching tasks to be completed first. As a result, the students tend to miss the chance of applying the new vocabularies and grammar structures they have learnt from intensive reading class into their writing in time. Without training, the students are not able to complete their writing task properly. For those who do attach importance to writing, what they do in training are just to give students some writing tasks and correct their errors. The students tend to be afraid of English writing. Furthermore, the training of writing has been regarded as a dull play which has nothing to do with the IT. Based on her practices in the writing class, the author illustrates in this paper how to carry out the teaching of English writing effectively with the aid of multimedia in order to improve the students' writing ability.

2 The Challenges for the Students in English Writing Based Information Technology

The challenges students are facing with in their writing are various. First is the use of words. Chinese being the first language, many students tend to compose their English by translating from Chinese directly rather than thinking of them in a real English way. Many other students, however, believe that it will be better if more complicated words are employed in writing. There are also many students who are poor in English especially in vocabulary. For Chinese students, the most popular errors in the use of words take place in the form of a word. The students may use the proper words but neglect changes of the form in a sentence, such as subject-predicate disagreement, wrong collocation, and the misuse of part of speech in word building, etc.

Z. Du (Ed.): Proceedings of the 2012 International Conference of MCSA, AISC 191, pp. 271–276.
springerlink.com

The second challenge for the students resides in their grammar. In the worst cases, the students have no sense of English grammar, even the basic elements and the structure of a sentence. As a result, they even cannot make a simple sentence correctly. But little else can be done. Their articles are full of individual simple sentences as the students are not able to improve them into effective ones by making any complex sentences.

A good structure of an article is another headache for most students. In many cases, their articles are not organized very well showing no clear structure or style. On the one hand, there is probably no clear theme in their articles. On the other hand, there appears no linkage between the sentences and paragraphs, as the students cannot make good use of linking words or transitional sentences.

3 Suggestions for the Teaching of English Writing Based Information Technology

"The result oriented teaching model for writing" (Lu, 2006) is an effective method in the teaching of English writing. Placing importance onto the grammar and discourse, this method develops manipulative training and adopts imitative practice. The aim of the training should be clarified first before it begins: narration,description,exposition or argumentation etc. During the training, it has been referred to the mode of operation of Audio-Lingual Method: training starts with the sentence making, which is followed with the training of paragraph structure, such as making topic sentence、main sentence and conclusion. This kind follow-to-leader practice focuses on the right article pattern from individual sentences, to paragraphs, and to whole articles (Lu: 123). This kind teaching method has been employed by existing author in her previous research (Lai, 2011). Based on previous research, and "the College English Curriculum Requirements(for non-English majors)" (2007), the author has explored further in her teaching practices, during which the multimedia, as an information technology, has been utilised together with the result oriented teaching model. It is found that, as an aid to the teaching of English writing, the multimedia is more positively related to the required teaching aims.

Words are the basic units of a sentence, while the sentences constitute a paragraph and further to an article. This implies a natural teaching sequence from words, to sentences, and to a whole article step by step. As imitation has been regarded as a crucial element in the training due to its importance for learning, three sub-steps are designed into each step, namely comprehending with explanation, patterning with paradigm, creative development. Based on the author's teaching practices, the following sections will introduce this effective method in details by taking the *"New Horizon College English (2nd Edition)"* (Zheng, 2008) for an example. This teaching material for non-English majors includes five books, in which 'Reading and Writing', 'Listening and speaking' are taken as the key books for writing during the author's teaching practices. Each unit in 'Reading and Writing' contains two texts and some exercises, such as vocabulary, collocation, word building, sentence structure, translation, cloze, text structure analysis and structured writing. In the book, 'Listening and Speaking', there are exercises like listening, speaking, movie clip and cultural talk.

3.1 Multimedia Presentations for the Words

It is very important as well as difficult to enlarge students' vocabulary in English teaching. Solely explaining the meaning of a word is not enough for students during teaching. The multimedia can assist teachers at ease in illustrating vocabulary item, different forms for word building, synonym, antonym, confused words and comparative method etc. For instance, when I was teaching the word 'assert', the multimedia was employed to present the related words: its noun, assertion; its adj., assertive; its synonyms, claim, declare; and confused words, insert, asset, etc. Besides, the teacher should consolidate how to use some key expressions with the multimedia. The teacher firstly offers examples, then asks the students to grasp the expressions and patterns with their own sentences, for example, the teacher should teach students some words that can collocate with the above words such as assert one's authority (independence, rights),assert oneself, and some sentence examples such as "She asserted that she was innocent". Meanwhile, the teacher asks students to drill learned words and expressions with part of vocabulary and cloze in the Reading and Writing, the multimedia screen can be displayed "fill in the blanks", which can strengthen students' comprehension for the words. Part of collocation companied, the teacher must remind students of paying attention to word form and collocation in writing. The multimedia can be utilized to present the collocation of words clearly, which is very straightforward and enables students to memorize. The utilizations of multimedia make a clearer presentation and therefore make an easier understanding for students. Furthermore, by means of multimedia, the teacher can easily display a sentence with an underlined word or expression, and then ask students to replace the underlined parts with other expressions but closely keep the original meaning. In this way, the teacher can train students how to use these words and expressions again. With the 'part of word bank and cloze' displayed by multimedia, the teacher can train students' comprehensive ability of words application and vocabulary expansion. These practices will get good preparation for sentence writing as students have solid words foundation.

3.2 Multimedia Examples for the Syntax

In the intensive readings, there are many polished sentences, key sentences, the sentences containing new words. They are actually good examples for English learning. By the use of multimedia, these examples can be presented clearly to students for their memorizing and imitation. A good command of sentence making lays a solid foundation for the composition of an article. Along with this, as the 'sentence structure' and 'translation' parts in the exercise can be utilized as training tasks for a clear purpose, sentence making. In this way, the useful sentences, sentence structures, and key expressions from intensive reading course can be commanded by the students. For example, the teacher asks students to combine each of the pair of sentences into one using 'while' and 'not only...but also',or to make sentences using 'communication with', 'have access to' ' despite', 'distract sb. from sth.' and 'sign up for',or complete sentences using' as' ' no matter'. These practices displayed on the screen can lay a good foundation for writing effective sentences.

The instructions and practices in grammar, and the training in sentence structure will improve the ability in sentence making, which assists students in making idiomatic English and correct grammars. Based on this, it is designed in the training that clause patterns are displayed to students for their imitation, during which the multimedia can demonstrate the examples clearly, such as the evolution from simple sentence to compound sentence or complex sentences, and the non-predicate verb. For example: firstly, the teacher asks the students to follow screen examples, they can write their own correct simple sentences, then the teacher asks the students to join simple statements together to make compound statements or complex sentences. While the teacher shows the students to go over some compound words, and pay close attention to the way they have been joined. Thirdly, the teacher shows the students original simple sentences on the screen, and ask them some methods to rewrite these sentences into expanded sentences, through which, the students can understand how to compose improved sentences from modifier of proper adjective, adverb, prepositional phrase, appositive phrase and "with" phrase to attributive and adverbial clauses.

3.3 Multimedia Samples for the Paragraph

After sentences are made, coherent paragraph can be formed by connecting them. In training this, the teachers will combine the grammar and syntax training by taking some examples and instructing students to compose some paragraphs via imitation. Coordinating with the writing examples in the 'text structure analysis' from the textbook, the students are required to compose their own paragraph which should be similar to the sample paragraph provided by teachers in their structures, for example, time sequence paragraph, the writer narrates the story according to the time sequence, that is the principle of "first thing first".

By imitating the sample paragraph, the students are trained to connect words to sentences, organize sentences to form a paragraph. It helps students make sense of what a well organized article is. According to 'the College English Curriculum Requirements" (2007), an undergraduate should be able to command the writing articles in different types, such as narration, description, patterns of exposition, summary, argumentation, report, and persuasion, etc. Based on this, different types of sample article will be provided to and analyzed by the students under the teacher's instruction via the multimedia, which includes general samples, how to write topic sentence, sentence opening, body and conclusion. The students will know what should be included in an article and how to explore it. After this special training, they will be able to deal with the different types of articles, and will not be afraid of writing any more.

3.4 Multimedia Stuffs for the Composition

Reading can also enhance writing ability; especially it can offer enough stuffs or materials for writing. As reading articles are full of the substantial contents, different ranges of subjects, and rich literature styles, students can collect writing materials and follow some key sentences, which is good for them to write effective sentences and proper compositions. In the reading activity, the teacher aims to enlarge the writing view, inspire their idea, and study writing approach, meanwhile, students' language foundation can be strengthened. Because reading activity is dull, the students do not know how to do with it for writing. However, the multimedia can be used for this teaching process, which will pave the way to meet the writing requirements. After reading, the teacher will utilize the multimedia to show charts, blanks according to reading contents, so that the students can practice some sentences and paragraphs. Also, the teacher can demonstrate some main pictures for students to make sentences, complete sentences, displace and combine sentences. At the same time, the teacher can use the multimedia to analyze the outline and structure of reading papers, and ask students to write summary, the same subject writing, paraphrase and abridgement. Besides, the teacher lists linking words or transitional words on the scream, and asks students to have writing practice with them. For example, there are two reading texts in teaching material, 'reading and writing'. The teacher can ask students to discuss and compare the writing goal and style between the Section A and Section B, and give them some key words to be organized for a whole guided writing.

3.5 Multimedia Flashes for the Free-Essay

Multimedia is used very often in listening and oral training. This can also be adopted for the writing. After the students listen to an article, they are asked to summarize what they have heard in a paragraph, during which they can use the original sentences from the article, or their own words. The summary does not have to be very long, as the aim is to train the students to imitate what they have heard from the article, to make sentences and organize a paragraph with a focus. For example, the listening comprehension part, 'movie clip' and 'cultural talk', are quite difficult for the students. After each talk, conversation, or monolog, there is a blank-filling. This is actually a good opportunity for the training of writing. When students complete the exercises, the teacher will discuss the related words and similar sentence structures with the students. After that, the students will be asked to make a similar and not-controlled paragraph by applying the vocabularies and syntax discussed previously. The 'cultural talk' can be served as a good sample for students as well. The theme in each lesson is the same, and different nations have different cultures, traditions and perspectives. The students learn the related expression as a start. Then they are asked to retell the main points under the teacher's instruction, and to talk about their own opinions consciously. This is actually a training of oral and free composition, during which the vocabulary and the syntactical structure get consolidated, and the students' ability to summarize and organize is improved. Furthermore, multimedia can feature some pictures or films, so that students will focus them for a writing paper with their own opinions and ideas.

4 Conclusion

The methods of the teaching of English writing for students are various. "The result oriented teaching model" places emphasis on the gradual development of students' writing skill, from words, sentences, paragraphs, to an article. From mechanical copy to controlled writing, and eventually to free composition, this method has helped students achieve good results. This paper explored the application of multimedia as an information technology in the teaching of writing for university students. The application of IT has dramatically improved the boring training and demonstrated many advantages: more straightforward teaching makes easier understanding; bigger amount of teaching content implies more intensive trainings; more systematic training makes better implementation of teaching; vivid sounds and animations increase students' interests. As a matter of fact, there is no fixed teaching method, nor can any methods be isolated from each other. It is a rule of thumb that different teaching methods should be considered according to different students and different context. Individualization is always the best.

References

1. Brown, H.D.: Principles of Language Learning and Teaching. Foreign Language Teaching and Research Press, Beijing (2001)
2. Wang, Q.: A Course in English Language Teaching, 2nd edn. Higher Education Press, Beijing (2008)
3. Lu, Z., Wang, D.: New English Teaching Theory. East China Normal University Press, Shanghai (2006)
4. Lai, W.: Practice on Result oriented teaching model for writing. Henan: Happy Reading 2(2), 69–70 (2011)
5. Department of Higher Education. College English Curriculum Requirements. Shanghai Foreign Language Education Press (2007)
6. Zheng, S.: New Horizon College English, 2nd edn. Foreign Language Teaching and Research Press, Beijing (2008)

The Study on College Sports Theory Teaching Based on Computer Intelligence Technology

Wenjie Zhu

The P.E Department of Xuchang University, Xuchang, Henan 461000 China
zhuwenjie@cssci.info

Abstract. The college sports theory teaching intelligent exam system has a simple interface which is easy to operate and even the students not familiar with computer can handle it easily. Now the senior school students begin to have computer class so they can handle the system only need to know the basic Windows operation and typewriting. Meanwhile, the operation system achieves good effects in practical application, such as convenient exam, rapid paper building, easy grading and high efficiency. The scores directly output can be easily handled or recorded by computer. This scientific design reduces teacher's work load of setting questions, grading and correcting exam papers and recording scores and then improves exam efficiency substantially, saves resources like paper, and enhances the fairness and justice of exams.

Keywords: sports, theory teaching, computer intelligence technology.

1 Introduction

At present, our country sports teaching model of college and university is divided into theories teaching and practice teaching. The research is part of the theory of traditional classroom teaching practice teaching key, physical exercise and learns the basic motor skills. The development of sports theory class teaching plays an important role in arousing students' interest in study, and masters the key points and improves physical exercise consciousness. The exam is education evaluation of important ways. Its most important role is to test students learning results, and is also a very important link of teaching quality evaluation. Therefore, if the exam of the implementation of the system can meet the requirements of China's education reform, the development trend of the available education evaluation.

2 Necessity of Exam System Reform

And the traditional exam mode, clever examination system has incomparable advantages. It shrinks in the link of traditional test process-building, the examination and approval, printing, records, issue, classification and filing papers, into one or two links. It almost eliminate the possibility of all direct intervention process examination handbook which saves a lot manual resource, material resource and financial resources, increase the test results the objectivity and impartiality of the many.

Z. Du (Ed.): Proceedings of the 2012 International Conference of MCSA, AISC 191, pp. 277–282.
springerlink.com

Especially sports teacher is very tired when class outdoors, but the traditional exam process and trifles and classification theory test paper noise is a work, workload is big, so the traditional exam mode become inappropriate for the current examination requirements.

3 Research Object and Method

3.1 Research Object

We chose two classes from Grade 2009 of Hunan University of Science and Technology, Hunan Province, China, and made random allocation. The computer exam group is Class One with 52 students (45 males, 7 females) and the written exam group is Class Two with 53 students (46 males and 7 females). The average scores of the two classes in the most recent final exam is basically no difference so they have comparability （$P>0.05$）.

3.2 Research Method

1)Document Literature ： Getting information of current research situation, existing problems and development tendency through relevant literature on research of computer aided system, to offer theory and practice basis for the design and implementation of this research.

2) Expert Interview ： In the process of subject selection and research, seeking for advice about development tools selection and research design from experts widely through interviews and informal discussions to ensure the smooth development of the research.

3)Mathematical Statistic : Making application study statistics of intelligent exam system of sports theory course by inputting all the data gained in the software of Excel and SPSS17.0.

4 Application of Intelligent Exam System of Sports Theory Course

4.1 Software Environment and Hardware Environment

Intelligent test system using a computer to the test, the test contents and use randomly chooses close-book exam type. The operation of the system hardware environment usually adopts microcomputer network room. As the three statistical software, software SPSS, SAS and SYSTAT, are based on the Windows operating system, use more convenient and quick than the network with a single computer transfer and collecting the test paper. Therefore, this system also choose interface to make them more direct set observation, friendlier and more humane.

Software of the sports course exam system can be divided into two parts, the examination system and grading system. The exam is a good student can avoid obtained through illegal means the answer. Another advantage is that it makes teachers reading and grade, realize the real subjective topic the defects and avoid the computer level.

4.2 Exam System

1) Design of question bank. The problem is with a bank to use Access database management powerful database for storage and management development platform. The basic design principles question bank is establishing a proper database, user requirements analysis on the test characteristics, building problems according to the bank in order to ensure that the overall knowledge test, and according to the question type standardization and standardization, make the problem set and exam easily, arrange boils inspection of the knowledge, to determine proper scoring distribution of each question. The difficulty is, the problem in the database design should be will avoid the students cheating in random, namely the problems in computer can bank, but their command is different set serial number, serial number, asked many questions separately.

2) Login system of examinee. Students can enter the client through server examination system. The examinee should enter his candidate number or student Numbers, the name and class and choose his own paper type choice of corresponding input box "choice" (for example, A, B or random) and click "login" with the mouse. The exam system will appear a page table basic information for his check and verify the students (such as name, gender, student Numbers, professional, etc.). The teacher can design a paper and try B so avoid cheating. Login system design of difficulty, if the database is not closed the transmission of paper, the database papers sent to the server will present 0 bytes, which means it cannot will test paper database of transmission failed. In order to solve this problem, we set the program that makes the force-close before all the database of transmission test paper.

3) Answer system of examinee. The exam questions mainly include seven parts, single selection, multi-selection, filling in the blanks, short answer, term interpretation, essay type and true or false. The database adopts ADO control to realize the real-time connection which can ensure that every revise of the student can be recorded in the database. In this situation, if the system halted unexpectedly, the student can still finish his exam in his second login. The difficulty in answer system design is that the system lock is realized through the function SystemParametersInfo() of Windows API. It can block the key combinations of Ctrl+Alt+Del and Alt+Tab to ensure that the students can't do any revise again.

4) Grading system. The grading system is composed by three parts, making standard answer, login system of grading and development & making of grading interface. The standard answer is also designed through Access database witch can go hand-in–hand with the design of question bank or made separately accordingly. We only need to bring the serial number of standard answer into correspondence with the serial number in question bank and then we can find the corresponding answer of each question. The login system of grading is build for the purpose of finding the designated exam paper. The file path on the left can be used to find the exam papers proposed to be graded. On the right there is some information which can be filled with grading person or question types needed grading and correcting. There are two approaches for selection. The first approach is grading according to given candidate number range. The second approach is automatic searching which needs a database of students for searching and the system will make grading exam papers one by one according to the candidate numbers provided by the student database. As to the

subjective questions, when the keywords of the answer input by the student are inconsistent with the standard answer, the software will “discuss with” the grading teacher besides “weighing” and “comparison” the answer with the one in answer bank. For instance, as to the question type of filling in the banks, it gives the answers in corresponding blanks, while for the question types of term interpretation and short-answer question, their answers are generally longer. But the database has a limitation to the text length that the longest text has 255 bytes, so this text selects and uses text files to keep answers. In which situation, students can input answers according to their abilities without the limitation of database. It has achieved good effect. Therefore, transferring text answer into grading system to execute grading is a characteristic of the grading system. As to the grading interfaces of short answer question and term interpretation, the answer keys are on the left for the teacher's reference and student' s answer is on the right for the teacher to grade. The teacher should grade according to answer of the student and fill the sore directly in the score column which can avoid the dogmatism of direct computer grading and realize of grading. The teacher should analyze the exam results when the exam is over so that he can take different measures to improve his teaching quality according to different situations. This function is mainly outputting the information like exam scores in bulk to assist teachers to analyze exam results. The difficulty in grading system design is using Jet database engine SQL of Microsoft. The structured query language (SQL) is the standard language when visiting the database. We could complete complex database operations through SQL without considering how to deal with the low-level details of the physical database. Meanwhile, SQL is an optimized language which can increase access speed to database through specialized database technology and mathematical algorithm.

5 Test and Analysis of TEACHING Experiment of Intelligent Exam System

In order to check the effect of this system, we randomly chose two classes from grade 2009 of Hunan University of Science and Technology, Hunan Province, China, and made random allocation. The computer exam group was Class One with 52 students (45 males, 7 females) and the written exam group was Class Two with 53 students (46 males and 7 females). The average scores of the two classes in the most recent final exam were basically the same so they have comparability ($P>0.05$). Computer exam group got familiar with computer operating system in one week before the exam and they could make practice by using the computers offered by the teacher at any time. The unified exam on computer was taken when the whole group was familiar with relevant operations. The written exam group used exam paper to complete the exam. The exam content and exam time of two groups are the same (120 minutes). Statistic method: Grading the scores into the excellent (over 85), the good (70-84), the pass (60-69) and the fail (under 60). We used the statistical software SPSS17.0 to analyze the data and the Wilcoxon rank sum test to compare the exam scores of two groups. Considering that it has a statistical significance with the difference of $P<0.05$. Table 1 and fig.1 presents the results.

Table 1. Results Distribution of Exams of Different Methods

Grouping	Score				Total
	Excellent	Good	Pass	Fail	
Computer exam	35	15	1	1	52
Written exam	25	24	3	1	53
Total	60	39	4	2	105

The result of computer exam group is better than that of written exam group and the comparative difference between two groups has statistical significance (P=0.038). The excellent percentage of computer exam group is 67.30% with four people more than that of the written exam group whose excellent percentage is only 53.00%. In the score rage of good (70-84), the number of people in written exam group is 22 which are a little more than 16 of computer exam group. In the score range of pass (60 -69), the number of people of computer exam group is 3 and the number of written exam group is 2. In the fail range (under 60), only one student in the computer group failed the exam.

From the comparison above and overall evaluation we can see that, first, the development tools of theory intelligent exam system of sports major are advanced with the characteristics of simple operation, good interface and comprehensive design, but in the combination with sports professional theory they can't change the answer ways used in traditional exam completely. For instance it can't realize the answer ways of choosing according to pictures, analyzing according to pictures, drawing and analyzing and analyzing according to video. This field needs further development and research. Second, the score proportions of each part of exam paper analysis in computer exam and written exam are basically the same. It indicates that it's feasible to used this research result directly in sports theory exam of regular institutions of higher learning though the five question types designed by this system can't meet the high level requirements of professional sports colleges and universities completely. Third, from the process between early training and later exam on computer, we can see that students can answer questions through computer smoothly and the man-machine communication pattern basically agrees well with the behavioral habits and communication pattern of the adolescent which can cause students' enthusiastic participations. It also indicates that the system with reasonable design and perfect operation is suitable for promotion and applications in colleges and universities. Fourth, from the unique characteristics of this system we can see that teachers don't need many training when use it. The one ever used office tools like Windows and Excel can use any function of serve-side (teacher management part) of this system through self-study. It also indicates that this system is suitable for promotion and application in colleges and universities.

It enables teachers to spend more time and energy on designs of classroom teaching and exam content. It improves teaching quality greatly, makes it possible to check students' ability of absorbing knowledge more objectively and promotes educational reform of sports major of our school. Therefore it has significant practical meanings and reference values.

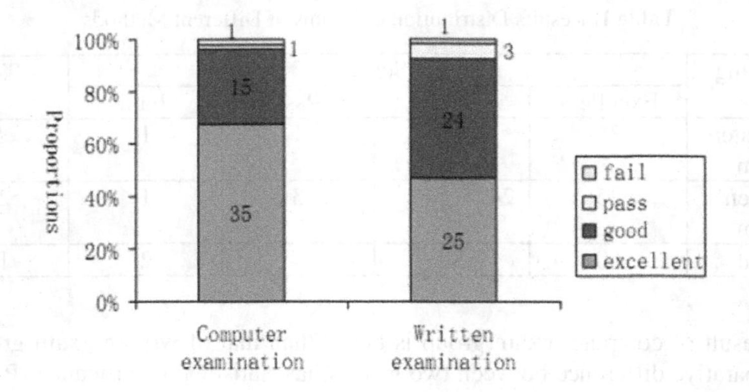

Fig. 1. Results Comparison of Exams of Different Methods

References

[1] Guo, C.: Research and Realization of Intelligent Question Bank System of Theory Course. Journal of Harbin Financial College 1, 12–14 (2008)

[2] Zhao, Y.: Reality Thinking of Question Bank Establishment of Sports Colleges and Universities. Harbin Physical Culture Institute 27, 75–76 (2009)

[3] Xiao, Z.: Research on Question Bank and On-line Exam System Based on Net. Computer and Digital Engineering 36, 35–37 (2008)

[4] Zeng, R., Lin, Y., Lv, Y.: Design and Realization of Management System of General Question Bank. Journal of Fujian Agriculture and Forestry University (Natural science edition) 37, 78–80 (2008)

Research on Multimedia English Teaching Model Based on Information Technology

Shuying Han and Xiuli Guo

Handan College, Hebei Handan, 056000 China

Abstract. Construction principle is the foundation of the multimedia English teaching mode based on information technology in theory. It reflects the interaction, maneuverability and realization in technology. It is an effective attempt to establish students' main position in the course of teaching.

Keywords: multimedia, information technology, teaching reform.

1 Introduction

With the profound development of Reform and Opening-up in China, the rapid development of social economy, China joining WTO, all trades and professions require more and more English talents urgently. Obviously, the quality of English education will directly influence the cultivation of English talents. From the end of the 20th century, some colleges at home were promoted to colleges with undergraduate students in order to meet the needs of enrollment expansion of colleges and universities. But in 600 common universities and colleges of undergraduate course in China, the running level of this kind of colleges is in the low level because of their short running history, the weakness of teachers, the lower grades of students when they are entered, the insufficiency of hardware facilities. Therefore, the newly upgraded colleges need further teaching reform, improve teaching quality so as to meet the need of teaching and evaluation. Handan College is one local college in Hebei province, upgraded in 2004, and English major is one major with undergraduate students. Then, how to make English major construction and teaching reform get rid of the constraint of previous construction, improve English teaching quality, make the undergraduate students catch up with nation-wide level and attain to the demand of syllabus? The teaching reform for English major is imperative.

On April 28-29 in 2005, a seminar on teaching reform for English major was hold in Shanghai Foreign Language Education Press. Many famous Chinese English educational experts, for example, professor Weidong Dai, professor Zhaoxiong he, professor Yuanshen Huang, etc. gave wonderful topic reports in the seminar, together with 300 leaders of English majors and teacher representatives from over 200 colleges and universities, communicated and discussed on the topics of standards of English talents' cultivation, English discipline orientation and construction, teaching materials, teaching model, curriculum design and teaching means. The ideas of the experts are generally accepted by the representatives, they had quite a lot in common in many problems, at the same time, they put forward some problems worth conferring further. This seminar made local colleges further realize that teaching

Z. Du (Ed.): Proceedings of the 2012 International Conference of MCSA, AISC 191, pp. 283–287.
springerlink.com

reform is suitable to social change, satisfy the market need and fits the demand of the time development. Reform is the only way for them to survive and develop in the market with hot competition. English major is certainly not an exception. Henceforth, the teaching for English majors in Handan College began its exploration and practice. Up to now, it has been over seven years, we devoted more, and we got more achievements. According to a survey to our teachers for English majors in Handan College, the following will talk about the exploration and practice of teaching reform since 2004.

2 Current Situation of Teaching Reform for English Majors

This part is the current situation of teaching reform for English majors in Handan College from goals and model of cultivation, teachers construction, teaching material construction, curriculum design and construction, teaching methods and means, practice teaching and evaluation system.

2.1 Goals and Model of Cultivation

The 21st century is an international economic time with high science and technology, information time. The more English talents they need, the higher the quality should be. This is challenge and test to foreign language education. Cultivating foreign language talents is the only way for English major to survive and develop. So, the teaching reform for English majors begins with English talents' goals and model of cultivation, then involves the comprehensive reform exploration and practice.

Because the need of society for foreign language talents varies from the past tendency to modern diversified tendency, the talents of the past single foreign major with basic skills can not meet the demand of market economy development. According to our college's teachers team, students resources, major condition, employment market, local needs, we work out composite talents' model of cultivation, the core is foreign language talents. Our present model of cultivation is "foreign language plus professional knowledge". When we make the cultivation goals, we study the teaching syllabus carefully and change the previous "following with experience or feeling", avoid the tendency of casualty. According to the demands of "Syllabus for English Majors" by State Board of Education, the current goals of cultivation for English majors in colleges and universities are "having strong English basis and wide culture knowledge, devoting themselves to be composite English talents of translation, teaching, management, research by using English skillfully in foreign affairs, education, economy and trade, culture, science and technology, military, etc", we work out the goals of cultivation in our college, that is to cultivate qualified middle school teachers and high-level talents for local economic construction with high political quality, meeting the need or our socialist construction, developing well in morality, intelligence, physical education and capability, having higher science quality and sturdy English basic skills and language essential technique, and having high application ability in English.

2.2 Teachers Construction

Teachers' quality problem is deciding the quality level of high education, and teachers construction is the key point of the development of high education. The weakness of teachers team is one problem of the current situation of local college. When we were upgraded, teachers with the degree of Bachelor of Arts occupied 80%, and most graduated from universities of Hebei Province, fewer from key universities in other provinces. Teachers with high professional titles are fewer. In a word, teachers for English majors are not ideal in age structure, educational background structure and professional titles. Teachers team construction is one important task of teaching reform for English teaching in our college. So, besides importing graduates with higher quality, we take many measures in order to strengthen the original teachers team, such as let teachers pursue further education in key universities, read for academic degree, etc. Thus, teachers can avoid the outmoded knowledge and improve teachers' professional level and teaching level. Meanwhile, invite some senior professors with high visibility from important universities to give lectures for our English majors. The practice proved that our measures are feasible.

The survey of our teachers shows that there are many differences in age structure, educational background structure, professional titles, scientific research and awards in 2004 and 2011 respectively.

For example, in age structure, in 2004, there were 6 teachers at 51-60 years old, 8 at 41-50, 5 at 31-40, 2 at 21-30; in 2011, there are 4 at 51-60, 7 at 41-50, 7 at 31-40, 4 at 21-30. In teachers' educational background structure, in 2004, one teacher had MA, 20 had BA; in 2011, 6 have MA, 16 have BA. In teachers' professional titles, there were professors, 4 vice professors, 8 lecturers, 7 assistants; in 2011, there are 4 professors, 7 vice professors, 5 lecturers, 4 assistants. In scientific research items and papers, they make great progress, too. In 2004, they had only one provincial scientific research item, 4 municipal research items, but in 2011, there are 16 provincial items, 9 municipal items, 2 colonel items. In 2004, there were 1 core journals, 18 provincial journals; in 2011, there are 85 core journals, 23 provincial journals. Meanwhile, teachers' awards increase a lot. In 2004, 1 teacher had municipal award, 2 colonel awards; in 2011, 2 teachers have national awards, 8 have provincial awards, 20 have municipal awards, 10 have colonel awards. In all, we can easily see that since 2004, we have made great progress in teachers construction, not only are our teachers' age structure, educational background structure, professional titles are ideal, but the achievements of our scientific research and awards of our teachers are great. Besides, we teachers have teaching and research activities regularly, organize teaching discussion, attend and judge instruction for each other, learn from others for improving teaching quality together. The teaching effect is better.

2.3 Teaching Material Construction

In order to guarantee English majors' teaching materials systematic, authority, complete and make students learn teaching materials efficiently, teachers edit or take in editing some relative referential documents, such as English News Listening, English Fast Reading 100, Selected Readings of English and American Novels, An Introduction to English Linguistics, An Introduction to English and American

Literature, Brief History of English and American Literature and so on. These teaching references' presentation to the public not only perfect teaching content, but add many valuable teaching documents for English majors.

2.4 Curriculum Design and Course Construction

Curriculum system and course construction are the important and difficult points for teaching reform for English majors. English major in Handan College plans and designs new curriculum system and course design according to foreign talents' goals of cultivation and composite talents' model of cultivation in New Syllabus for English Majors and our college' orientation for English major. According to New Syllabus, English majors must have strong basis of language skills, which is also the first importance of English teaching and carrying out in four years' teaching. Therefore, in the basic stage of Grade one and Grade two, we arrange Integrated Skills of English, Extensive English, English Listening, English Phonetics, English Grammar, Oral English, English Writing, English Lexicology etc, which are basic technical courses for English majors. In the high stage of Grade 3 and 4, we open Advanced English, Linguistics, English and American Literature, English Teaching Methods, French etc, which are obligatory courses for English majors. Selected Readings of English and American Poems, Pragmatics, English Stylistics, Cultural Communication are opened as selected courses. Theses courses not only improve students' correctness in using language, but cultivate their sensitivity towards culture differences and improve their cultural communication ability and pragmatic ability. In addition, students need select four public courses in our college, which are about humanistic discipline, scientific discipline etc., and can lay the foundation for English majors. Besides, we have one provincial excellent course, one colonel excellent course and two colonel key courses in our college, other courses are all qualified courses.

3 Problems and Countermeasures

In all, we have made great progress in teaching reform since 2004; many results can be used for references for other majors of similar colleges. But from the current teaching situation, we can see clearly that there are a lot for us to promote further. For example, in the aspect of teachers construction, we should introduce talents with Doctor of Arts, teachers need declare national scientific research items, issue more research papers with high level; in teaching materials construction, we should edit or take in the edition of national planning teaching materials; in curriculum construction, we have one provincial excellent course at present, we should work hard and strive for more courses like this; in evaluation system, although we put forward process evaluation, the ordinary score is not given scientifically, the form is single and lack of perfect evaluation system; in teaching equipment, we teachers can use multimedia classroom for teaching, but we can not surf the internet in the classroom and share the internet resources; in teaching content, besides language, culture teaching, we should add humanistic discipline teaching and improve students' humanistic quality, conform with the cultivation of world humanistic quality; in our library, the foreign documents are most common English novels, academic journals or teaching methods, lack of theoretical books, there are no materials for linguistics, literature, etc., which is not good for students to write their graduate thesis. In addition, in teaching management, our college should take some incentive measures for teachers, give more rewards and awards in teaching and scientific research and improve their work motivation. Consequently, the teaching reform for English majors need reform further, let us stand higher and see farther, base ourselves on the current situation, devote ourselves to the new teaching reform for English majors, make our college cultivate more talents with high qualities, make society know us, recognize us, let us serve local economy better.

Acknowledgments. This paper is one stage result of the Research Project of Hebei 2011 College English Teaching Reform——A Survey on Current Situation of Teaching Reform for English Majors and Countmeasure Research in Local College" (No. HB11Y0035).

References

1. English Group of the Teaching Guiding Committee for College English Majors under the Ministry of Education. Foreign Language Teaching and Research Press, Beijing. Shanghai Foreign Language Education Press, Shanghai (2000)
2. He, Q., Yin, T., Huang, Y., Liu, H.: Several Opinions about Foreign Language Education. Foreign Language World (4) (1998)
3. He, Z.: Several Considerations on Current Situation and Future of English Major. Shandong Foreign Language Teaching Journal (6) (2004)
4. Qu, J., Tang, Y.: Attempt to Graduate Teaching Reform for English Major in Colleges and Universities. Modern Education Science (5) (2004)
5. A Survey on National English Teaching Reform for English Majors. Foreign Language World (4) (2005)
6. Wu, L., Wang, Y.: The Current Situation and Improved Ways of College English Teaching. Gansu High Normal College Journal (1) (2010)
7. Zhang, C.: Considerations on Orientation of English Major. Foreign Language World (4) (2003)

The Study on University Library Information Service for Regional Characterized Economy Construction Based on Integrated Agent

Shuhua Han and Xuemei Su

Library of Hebei Normal University of Science and Technology, Qinhuangdao, Hebei, 066004 China
hanshuhua@cssci.info

Abstract. An Integrated Agent-Based University Information System (IABUIS) consisting of following four modules: Student Administration Management System (SAMS), Library Information System (LIS), Distance Learning System (DLS) and University Management Information System (UMIS), has been presented in this paper. An agent based testing subsystem as a part of the Distance Learning System (DLS) has been also described. It is better for regional characterized economy construction.

Keywords: integrated university information service, intelligent agents, regional characterized economy.

1 Introduction

Education world is changing rapidly. Communication and information technology is becoming a forever in the teaching and learning is an important part. The explosive growth of the web makes education institutions to become virtual and globalization.

Huge produced each year, the administration and the teaching materials, teaching material and the library collection of comprehensive university college level must be the knowledge and information system, fully standardized and national level compatibility. The process of integration, and introduce must improve online learning ideas and technology, the unity of content, digital library information management integrated standardization and to establish e-university environment and more close to the knowledge society and globalization.

2 The Model of IABUIS

In our model, each institute is represented by an Intranet, while all institutes together constitute the University Intranet. The University Intranet with a few external connections consist the University Extranet [5], which is connected to the Internet on several points. The model of IABUIS includes four processes: student administrative information management, library information management, distance learning information management and university administrative information management.

Z. Du (Ed.): Proceedings of the 2012 International Conference of MCSA, AISC 191, pp. 289–294.
springerlink.com

In Figure 1, the communication of these processes with the "external world" is given. The clients are HTTP oriented using web browsers. Special web-based programs serve the connection with the IABUIS elements: the free text, files and multimedia management system (FTFM); the data base management system (DBMS) and the independent software applications (ISA).

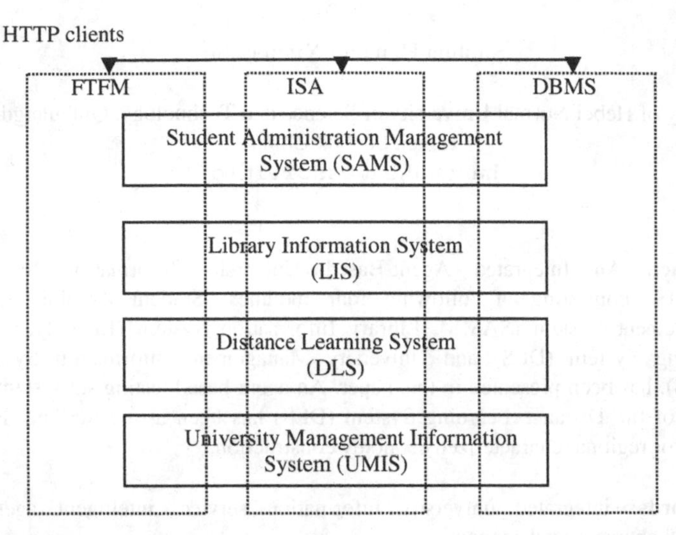

Fig. 1. Communication of the IABUIS processes with the "external world"

In Figure 2, cooperation among different servers and the connection with the web clients are given. Web servers represent the basic interface with the users. They are http based and communicate bi-directionally with the web clients. In a direct and/or indirect way they communicate bi-directionally with the FTFM servers and DBM (data base management) servers.

In our multi-agent system described in [1] we have proposed two types of agents: user-specific and system specific agents. The user-specific agents served the user's specific interests while the system-specific agents the system administrator and system management interests. We proposed three user-specific agents: user profile agent (UPA), personalized questionnaires and forms agent (PQFA) and personalized report agent (PRA); and three system specific agents: questionnaires and forms generation agent (QFGA), reports generation agent (RGA) and data base management agent (DBMA). These agents belong to two layers of abstraction: PFL – the personalized filtering layer and UISML – the university information system management layer. The agents in one layer cooperate by exchanging knowledge and learning rules. Between the two layers, the agents could also cooperate by exchanging knowledge. Learning mechanisms for two agents from different layers are different in spirit. The user-specific agents cooperate among themselves exchanging KQML messages. They are built in Java language using JATLite agent extension packages developed at Stanford University, and run as an applets in Java-enabled browsers. The system-specific agents are currently in implementation with JADE, they would be considered in a future.

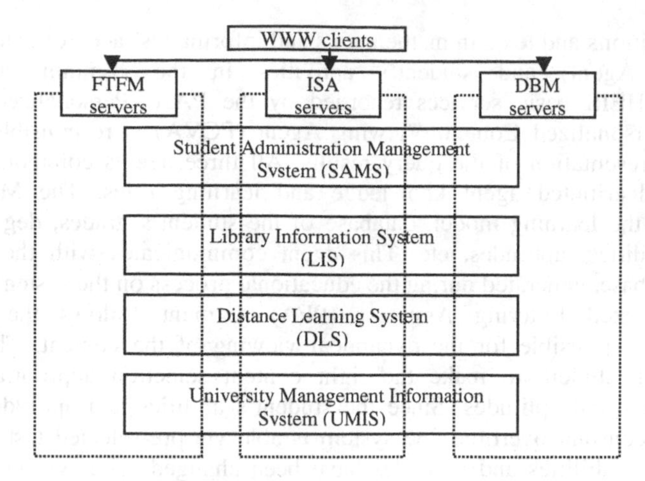

Fig. 2. The cooperation among servers

3 An Agent Based Testing Subsystem

In the following subsection a model of distance educational system with an embedded agent infrastructure will be presented. It is an agent based extension to the existing distance educational system MATHEIS (MATHematical Electronic Interactive System). This extension is an essential part of a Distance Learning System (DLS) and IABUIS.

MATHEIS is an educational system for learning mathematics and informatics for pupils and students [3], [4]. Recently, at the Faculty of Natural Sciences and Mathematics, this system has been successfully integrated into newly implemented ORACLE iLearning management system. The new intelligent iLearning/MATHEIS is agent-based consisting only of user-specific agents. It supports easy XML based import/export between applications, as well as SCORM and AICC based one-click import/export of the course metadata and content files. Students use various types of services: web based material treasury, e-mail communication, discussion forums, chat, testing services, etc. It supports different learning styles, from self-paced material, to scheduled or synchronous classes offline or online and collaborative and integrated learning. Similar integrations of intelligence in ORACLE iLearning management system were not detected in a literature.

An agent extension empowers the system with the following new services: monitoring the students behavior and interests at the system; determining the student's skill level; enabling cooperative task resolution among students; enabling different views of the services according to the student's skills; notifying the students when the newest tests for appropriate level are available; presenting the tests to the students and estimating the received results; automatically update of the student's levels depending on the estimated results, etc.

The model can be extended to any subject with an appropriate expert system for results estimation.

The agent-based structure of iLearning/MATHEIS is shown in Figure 3. The system is able automatically to assist into filtering of information (exercises, examples, definitions and texts in mathematics and informatics) according to the UPA (User Profile Agent) and student's activities in the communication with iLearning/MATHEIS basic services recorded by the PAA (Personalized Activity Agent). The Personalized Content Viewing Agent (PCVA) is responsible for the selection and presentation of the query results. All three agents collaborate among them sharing distributed agent knowledge and learning rules. The Mediator is responsible for the learning model, database of the student's grades, degree levels, preferences, abilities, aptitudes, etc. This agent communicates with the Mediator knowledge database, generated during the educational process on the system.

The Personalized Filtering Assistant (PFA) segment follows the student's activities. It is responsible for the dynamical viewing of the content. The PFA is trained for each student to make the right content selection appropriate to the student's abilities and aptitudes. Since the student's abilities and aptitudes are not assumed to be constant overtime, the system is able via pre-selected tests to notice that the student's abilities and aptitudes have been changed. The system adapts its behavior in response to these changes. Any new student on the system, after few pre-selected tests for evaluation of grade-level and estimation of student's abilities and aptitudes will be accompanied with appropriate personal set of agents adjusted to the student's knowledge.

The structure of MAS (iLearning/MATHEIS Agent Subsystem) is shown in Figure 4. Relations between the agents in MAS as well as messages they exchange will be presented through the following typical scenario: A user logs on the MAS which is accomplished through his User Agent (UA) who sends a connection message to the router consisting of username and password in order to get the acknowledgment from the router that the user is registered, so the access to MAS is granted. Afterwards User Agent sends a check_level message to the Level Agent (LA) asking for the latest skill level for the user. The Level Agent checks the User levels database, and replies the User Agent with the reply message consisting of the level for the corresponding user.

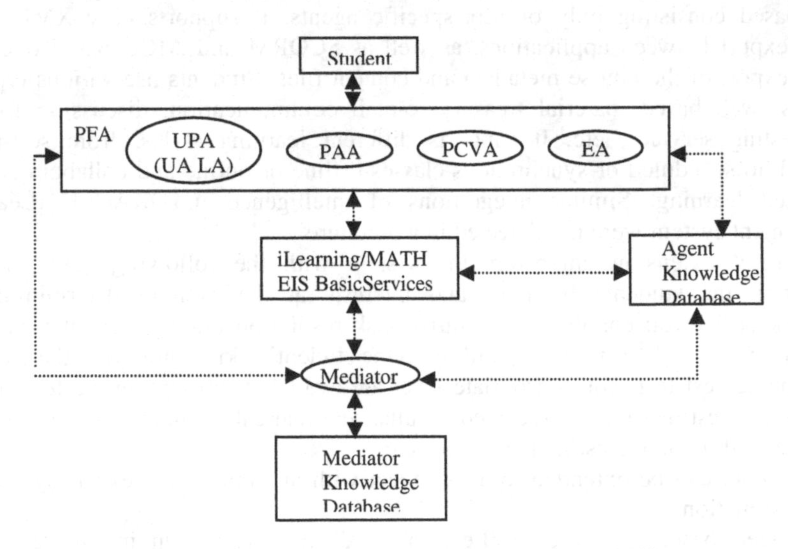

Fig. 3. Agent-based structure of iLearning/MATHEIS

In the first case PFA presents the user a view of MATHEIS services appropriate to the returned level. In the later case the User Agent sends to the Supervisor Agent (SA) an ask_for_test message, asking tests available for the school year . The test is required in order to estimate the user's starting level. The user is supposed to do the test and send the results for assessment to the Expert Agent (EA).

The User Agent, Level Agent and Expert Agent are part of the PFA segment, while the Supervisor Agent represents the Mediator (see Figure 3).

Four major subsystems are identified at MAS: The User Agents Community; the Level Maintenance Subsystem; the Supervisory Subsystem and the Fuzzy Expert Subsystem. We will describe shortly the role of each of these subsystems bellow.

The User Agent Community subsystem is consisting of the agents representing every particular user of the system. It serves as an interface between the user and MAS. It sends messages to the rest of the agents in MAS on behalf of the user, and receives the messages addressed to specified user.

The Level Maintenance Subsystem is consisting of a single agent called Level Agent responsible for maintaining user levels and User levels database.

The Supervisory Subsystem consists of a Supervisor Agent monitoring and coordinating other agents, making suggestions about user levels, notifying for new test arrival, maintaining the test database, and analyzing the results database.

The Supervisor Agent occasionally observes the Results database analyzing the user's behavior and interest at the system. It might suggest the Level Agent correction of the level for a user.

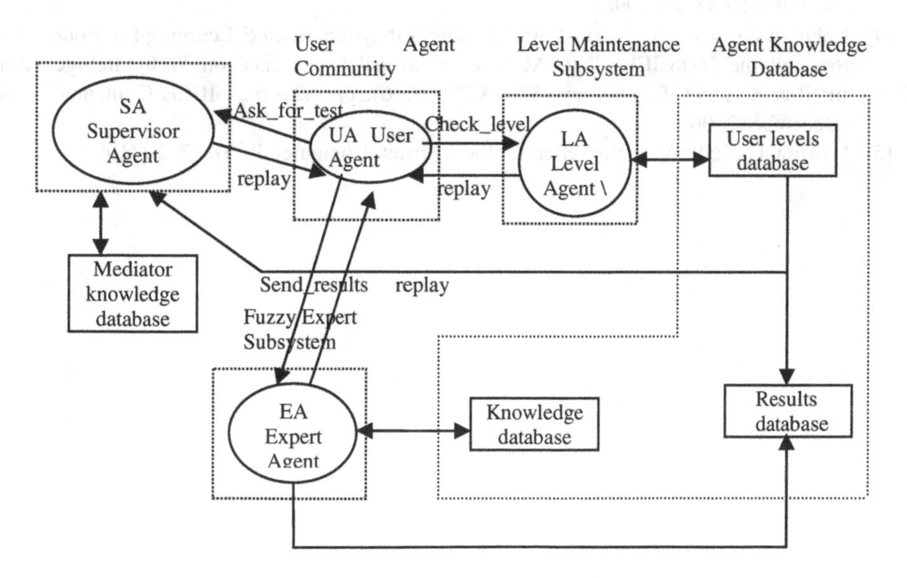

Fig. 4. The structure of MAS subsystem

4 Conclusion

An IABUIS is available at the Faculty of Natural Sciences and Mathematics in Skopje, but there are no limits to be used by other universities in Macedonia or in some neighboring countries.

Acknowledgement. Science and Technology VS Development and Planning Scientific Research Issue in 2011 in Qinhuangdao Municipal Science and Technology Bureau.Name of Scientific Research: Existing Problems and Countermeasures Resear ch on the Colleges and Universities Library and Archives in Service of the Special Di stricts' Economic Construction.The Scientific Research No. : 201101A446.

References

[1] Davcev, D., Cabukovski, V.: Agent-based University Intranet and Information System as a Basis for Distance Education and Open Learning. In: Proc. of the 1st UICEE Conference of Engeneering Education – Globalization of Engineering Education, pp. 253–257 (1998)

[2] Dahanayake, A., Gerhardt, W.: Web-Enabled Systems Integration: Practices and Challenges. Idea Group Publishing (2003)

[3] Cabukovski, V., Davcev, D.: MATHEIS (MATHematical Electronic Interactive System): An Agent-Based Distance Educational System for Learning Mathematics. In: Proc. International Conference on the Teaching of Mathematics, pp. 59–61. John Wiley & Sons Inc. Publishers (July 1998)

[4] Cabukovski, V.E.: An Agent-Based Testing Subsystem in an E-Learning Environment. In: Proc. of the 2006 IEEE/WIC/ACM International Conference on Web Intelligence and Intelligent Agent Technology (WI-IATW 2006), pp. 622–625. IEEE Computer Society (December 2006)

[5] Lawton, G.: Extranets: Next Step for the Internet. Computer 30(5), 17–20 (1997)

The Research on Information Commons Based on Open Access in College Library

Xiaohui Li, Hongxia Wang, and Hongyan Guo

Library of Hebei Normal University of Science and Technology, Qinhuangdao, Hebei, 066004 China
lixiaohui@cssci.info

Abstract. Information commons, the new service model, brings a new development chance for the service innovation of the university library in China. The practice and study abroad have been in a more mature development stage, then, how to introduce and learn from foreign experience, and build the IC suitable for the actual situation of China's library has become an important task for service transition of university library at present. Beginning with introducing the general overview of the information commons, in terms of the general procedure of IC constructing, the physical layer, virtual layer and support layer are deigned which can provide some guidance to the IC plan of university library in China.

Keywords: information commons, university library, open access.

1 Introduction

Information sharing space (IC) is a new type of service mode of university library in the United States, with the development of computer technology, multimedia technology, network technology and modern communication technology, the school teaching reform and change people's study method. Integrated circuits provide continuous service users from making a study plan finally production; it has a very good reference sustainable development and expand their service, to the library.

Integrated circuit and open access practice of a library of information resources sharing ideas, the extension of library service and innovation of inevitable direction service [1]. Information processing in China is developing very quickly from the library of quantity and size is becoming a very important question. Establish IC can not only for users to provide more abundant resources and more convenient and effective learning and research environment, still can effectively improve the depth and breadth of the library information service. Then, integrated circuit has very important practical significance and good application prospect, improve the library management service function, promote the share of information resources, improve the service performance.

Z. Du (Ed.): Proceedings of the 2012 International Conference of MCSA, AISC 191, pp. 295–300.
springerlink.com

2 Basic Knowledge of IC

2.1 Definition and Characteristics

In the late 1990s, the user to library rate and the paper-based circulation rate of many libraries in America decline generally, IC was brought out in this background which is a new service model designed to attract users[2]. IC can provide users a one-stop service by optimizing and integrating the space, resource and service of the library.

Although many experts and scholars have different interpretations about IC, but the basic point of view is consistent, that is, IC is an especially designed, assured open access one-stop service facility and collaborative learning environment. IC can integrate the Internet, computer hardware and software, knowledge base resource including print type, digital, multimedia, and so on), and it can cultivate the reader information literacy, promote the readers to learn, exchange, collaborate and study under the support of the skilled library consultants, computer experts, multi-media workers, and teachers.

IC has several important characteristics.

a) ubiquity: Each computer has the same interface and uses the same software and electronic resources;
b) utility: to meet the need of various users; c) flexibility: to adapt the environmental change and technical development;
d) community: to provide a comfortable space to work together.

Because of these characteristics, IC is very popular and highly welcome in a number of university libraries.

2.2 Service Model

With the application and development of IC practice, the service model has changed to mature gradually, which is based on the functions integration of various departments. On the one hand, the cooperation is improved based on the level of organization and management; on the other hand, the function integration of technical level is deepened. IC model is shown as Fig. 1.

As shown in Fig. 1, the functions of various departments are integrated effectively. The librarians should have strong flexibility, adaptability, sufficient ability and knowledge to support the new service models. At the same time, they can cooperate closely with other departments to provide users with high-quality reference and consulting services as a whole.

The information service desk is the center in the IC service model, which is still the first link a user access to the library and its primary task is to provide the basic information of library and school and correspond with other service desks and various regions at the same time[3]. When the user's requirements are so complex that the workers can not answer, they will recommend the user's need to the relevant desk or other professional librarians or departments.

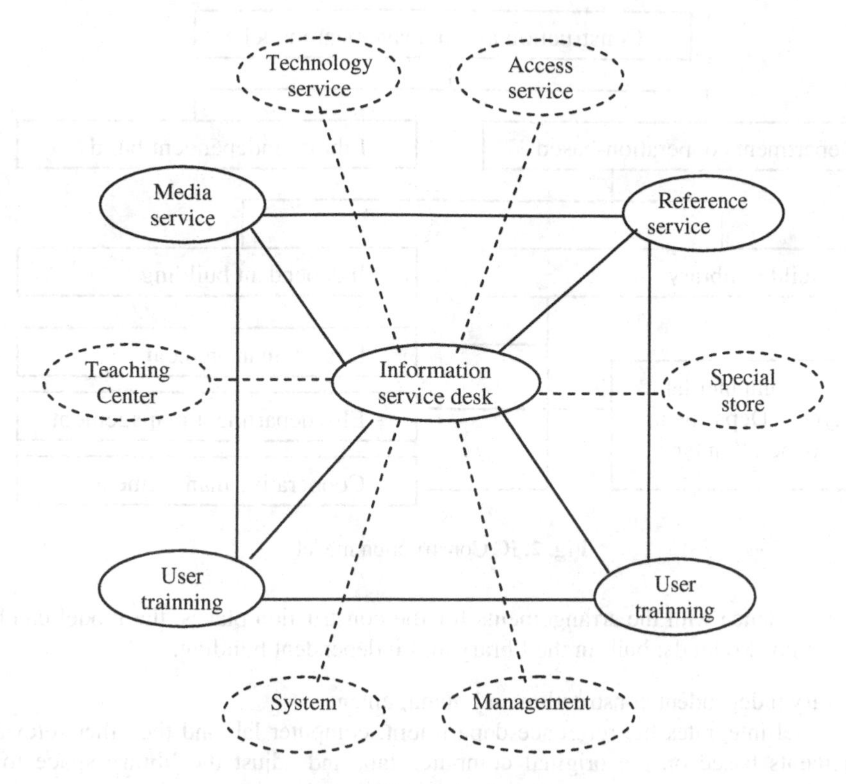

Fig. 1. IC service model

3 IC Design of University Library

By analyzing the data of mature IC pattern and combining with the practice situation in China, the plan and construction of the physical layer, virtual layer, operation guarantee model are discussed.

The physical layer is also called entity layer, including the construction model, arrangement of physical location and design and distribution of service region[4].

According to the difference of the constructional foundation of IC and the differences of construction site arrangement and management format, the types of IC construction model can be shown as Fig. 2.

From the construction base, IC can be divided into two categories:

a) Based on cooperation of different departments
Cooperative construction of library and the other relevant department is the main construction mode in the foreign universities. It is usually built by library, computer center, information technology service department, and so on, which has better computer hardware and software and strong technologies.

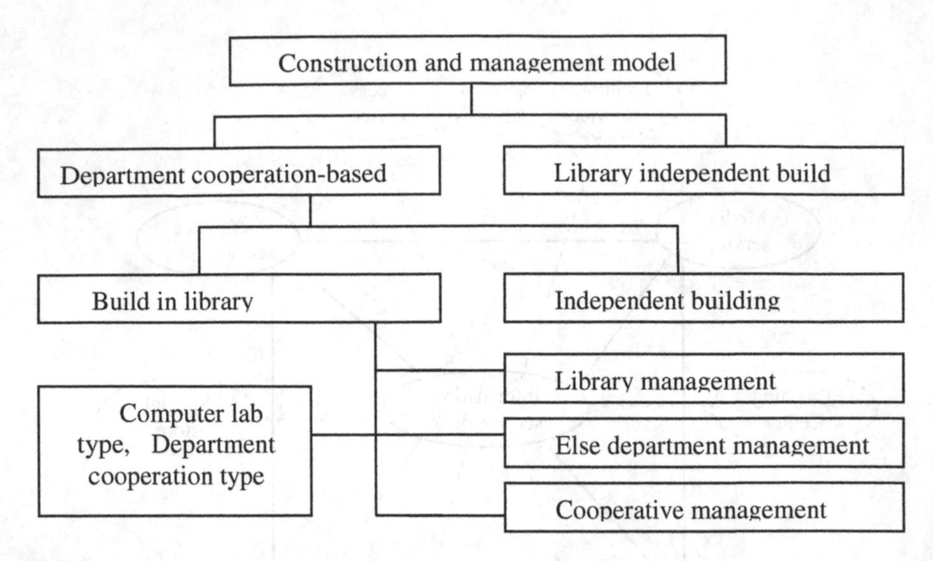

Fig. 2. IC Construction model

In accordance with the arrangements for the construction places, the model can be divided into two kinds: built in the library and independent building.

b) Library independent construction and management
This model integrates the reference department, computer lab and the other relevant departments based on the original computer lab, and adjust the library space to a certain degree.

For the construction of IC, the most important is to provide the physical space of study and communication to meet the user's need according to the difference of the user's learning and studying methods and the need of exchanges. The division method is shown as Fig. 3.

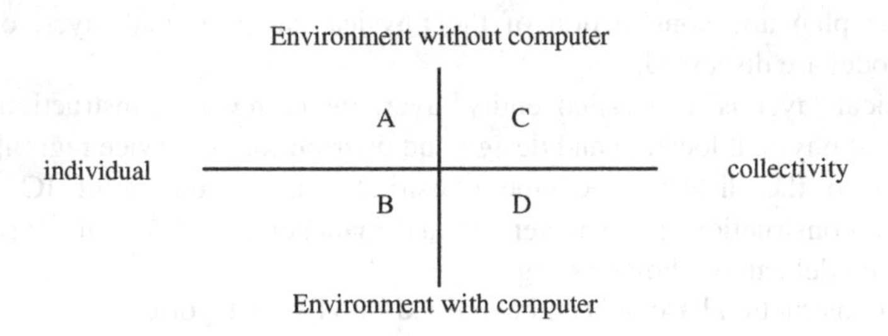

Fig. 3. Division of service region

a) Space A: Individual space without computer environment.

This personal silent study area without computer function is adapted for individual to read, think and learn without any interference.

b) Space B: Individual space with computer environment.

It is equipped with computer and wireless network which is an individual learning space to provide computer support.

c) Space C: Collective space without computer environment.

It is adapted for studying team and student discussion team to discuss and exchange face-to-face.

d) Space D: Collective space with computer environment.

The space can provide computer equipment for collective members, and support members communicate through the network computer.

In the four types of space, space D is the most important space in IC, for it can provide users with a collaborative environment to learn or exchange.

The design of IC virtual layer mainly includes the construction of the information resource system and the virtual environment, and the software support that the information services need.

The information resource system in the library construction is still the basis of IC services. We can make full use of the existing fruit in the construction of digital library, consult the correlative study theories, develop the electronic resources and network resources to make rational allocation of traditional print-based resources and digital resources based on the information resource system.

IC is an integrated digital environment. It is needed to install a variety of application software and professional software with the exception of the software accessing the electronic resources.

The operation and protection layer of IC, also called support layer, includes three aspects, management of IC, personnel allocation, evaluation of service quality.

IC has various organizational models, which is mainly divided into two types: inter-departmental cooperation and independent construction[6]. The two types have different constructional sites and management bodies. Effective management is the base of the success of IC. The essence of IC is regarded as a new form of integrated services. The major management problems that the library faces are the intellectual property issues and the diffusing right of information networks in the construction of resources, which are the study base of IC management.

It is the key point for IC success to equip the well-trained and with reasonable knowledge structure attendants. IC requires a service team with high-quality and integrated services capacity. When selecting the workers, we should think over the cooperation of resource services and technical personnel. At the same time, we should establish relatively immovable cooperative relation with the book suppliers, software suppliers and maintenance providers, equipment suppliers and maintenance providers, so as to ensure coordination and stability of the corresponding information resources, tool resources, equipment resources, consulting jobs and training job. In the personnel system, we can also fully consider and exploit the users' advantages and rationally make use of student assistants to promote the mechanism of mutual aid between the users.

How do we evaluate the service quality is an important step in effective operation after the successful construction of IC. In addition to the use of existing formal or informal evaluation methods, we can also develop the new evaluation method.

4 Application Example of IC

The west campus library is one part of Linyi normal university library, which primarily provide services for engineering students. The library follows the idea of reader first, service first, and it makes contributions for the teaching and research of the engineering course. The west campus library induces IC to further enhance service quality.

5 Conclusion

The practice of information commons is carried out which can provide a certain amount of experience and reference to resolve the planning and constructing of information commons.

References

[1] Creth Sheila, D., Lowry, C.B.: The information arcade: playground for the mind. Journal of Academic Librarianship 20, 22–24 (1994)
[2] Bailey, R., Tierney, B.: Information Commons Redux: Concept, Evolution, and Transcending the tragedy of the Commons. The Journal of Academic Librarianship 28, 277–286 (2002)
[3] Jennifer, C., Vaughan, J., Wendy: The Information Commons at Lied Library. Library Hi Tech. 23, 75–81 (2005)
[4] Beatty, S., White, P.: Information Commons - Models for E-literacy and the Integration of Learning. Journal of E-Literacy 2, 2–14 (2005)
[5] Andrew, R.A.: Campus library 2.0. Library Journal 129, 30–34 (2004)
[6] Beagle, D., Bailey, D., Tierney, B.: The Information Commons Handbook. Neal-Schuman, New York (2006)

The Development Trend of the 21st Century Computer Art

Weiming Deng

Department of Art, Sanming University, Sanming, Fujian 365000 China
dengweiming@cssci.info

Abstract. The twenty-first century is an era, which integrates technology and art together. Also, in this era, the computer art plays an incredibly important role in the people's life. In this paper, the authors discusses the relationship between the computer technology and art and also analyzes the orientation of the computer culture attribute and the connotation of the computer art culture, and then explores the cultural trend of the computer art development.

Keywords: Technology, Culture, Art, Computer.

1 Introduction

Computer and the peripherals including the Internet attaining a great development around computers in the information age have a very close relationship with the daily life of people at the present time. In the mean time the human beings have adapted to and even have indulged in such a state that it is hard to imagine what they world will be if there were no computers or networks in daily life. The computers in the 21st century are exerting an influence on the life and work of human beings with the integration of technology and art.

2 Technology and Art

Technology is the general name of all kinds of activities, means and methods which are mastered by the human beings in the process of using the nature, transforming the nature, and promoting the social development. Art is a kind of cultural phenomenon, which is created by people in order to meet the needs of comforting the subjective regrets as well as the behavioral needs in emotional organs. Also, in a sense, the life process of the human beings is a process for technology to be artistic. Technology progress and aesthetic improvement are the catalysts for technology transformed to art. From this point of view, the coexistence between technology and art can be seen.

3 Orientation of the Computer Culture

Computer culture is a new cultural form, which sources from the fundamental changes of the ways of life of the human beings due to the wide application of

Z. Du (Ed.): Proceedings of the 2012 International Conference of MCSA, AISC 191, pp. 301–305.
springerlink.com

computers. In1981, the "computer literacy" became the most important theme at the Third World Conference on Computer in Education (WCCE). If the "computer literacy" was still in the primary stage at that time, the art of a true sense should be the "computer culture".

From the broad sense, the computer culture is the general name of the computer resources and the corresponding human concepts, which emerged from the development of the computer technology [1]. Culture is a process for the human beings in the world to gain knowledge, skills, experience and ideas, beliefs, as well as values in the communication between each other. A culture age is named with the most representative technical achievement in general.

Besides, one of the most important characteristics of culture is the reform of the ways of communication in the human society, namely, the information processing revolution.

In addition, an important property of culture is the existence of language. Therefore, if the language of a culture disappeared, the existence of this culture will lose the basis for supporting and extending the spirit.

Hence, it is necessary for people in the modern times to make a continuous upgrade to the computer language to a higher platform. In other words, it is necessary for people in the modern times to recognize that the computer language is virtual and universal in the realities, and also is different from the human languages of the traditional sense [2].

4 Culture Connotation of Computer Art

Computer art is using the computers for carrying out the artistic creation. The computer art can be traced back to the 1950s. In January 1965, German mathematicians Frieder Nake and George Nees and American mathematician Michael Noll and others held the first digital electronic computer drawing exhibition. The exhibition was held in the art Gallery of Stuttgart University of Germany [3]. Thus, for the computer art which takes up a very important role in the computer reference technology, the cultural property of it can be defined as the art-influenced image technology.

Generally speaking, the art-influenced technology image and the artistic technology visualization are two completely different concepts. A great number of people in the modern time will often draw a conclusion on the artistic technology visualization or digitalization when they are talking about the computer art with other people at most cases. Admittedly, in the computer art of the present time, the modern people can indeed often see such a phenomenon—the computerization of the works of art.

During the period from the end of 1980s to the early of 1990s, the computer personalization age came to the world. And one of the most important characteristics of this age was the application and development of the computer graphics operation interfaces. Subsequently, the computer art or the digital media art came into being in the world.

4.1 Features of Computer Art

4.1.1 Virtual Feature

Generally speaking, it is known to all that the computer art is a virtual art. And this can be feature can be conclude from two aspects.

On the one hand, the computer art can be converted by the people in the modern times into the digital results eventually. On the other hand, the artistic effect showed in the computer art can rely on only a small number of natural materials to imitate the natural materials.

4.1.2 Transcendence in Time and Space

Using the computer technology as well as the computer art can realize the breakthrough in time.

On the one hand, this feature can help solve the limitations of the music as a time art.

On the other hand, as for the transcendence in space, the space effect which is displayed by the computer art breaks through the local space of painting or sculpturing.

4.1.3 Rapid Communication

With the help of the current network, the computer art can transfer and show the artistic works in the shortest time. And this relies on the computer art's virtual feature and transcendence in time and space.

4.1.4 Replication

Different from other art forms, the vast majority of contents of the computer art can be replicated. However, such a replication is still on the technical level, because concept and thinking can never be copied. Therefore, it is an ideal way for the computer technology at the present time to transform to an artistic form by combining with art. This is also one of the rulers to measure the concept of the computer culture.

4.2 Culture Connotation of Computer Art

4.2.1 Popularization of the Art

The current world can be compared to a world for the computers. This can't be achieved by other kinds of l cultural forms. The popularization is the first element of the cultural connotation of the computer art. However, such a popularization exactly launches challenges to the elite models of the traditional art.

Therefore, increasingly more people can create the artistic language belonging to them in the process of operating these software applications. Maybe the writers can't be called as artists at the present time. However, the influences from the art at least can be felt by the artists in the process of creating the artistic works.

Compared with other art forms, the popularization of the computer art may be a revolution in art.

4.2.2 Art-Oriented Technology

In the modern time, along with the rapid development of computer technology, the integration of both art and science has changed into a general trend of the development of culture. As a matter of fact, it is necessary for the innovation of technology to rely on the thinking and imagination of art. However, for the artistic creation, it is necessary to have new media as well as new vision to give expression to the in-depth influence of the computer art on the languages of the human beings and the materials which are used in technologies.

When the computer becomes an essential tool and technology in life, the artists will change into the users of such a tool and also are very possible to be endowed with the Art-oriented techniques.

5 The Cultural Trend of the Humanities Spirit Is the Development Direction of the Computer Art

Generally speaking, it has been known to all that the integration of art and computer technology can produce a comprehensive beauty, which goes beyond both art and computer technology.

For this reason, the modern aesthetics can be transformed into a kind of the visual aesthetics so outstandingly [4]. This can be analyzed from the aspects in the following.

On the one hand, the integration of art and computer technology gives expression to the reliance of the art forms on technology. This can not only enrich the existing original forms of art, and also making the art structure change with the new ways and new forms.

On the other hand, the integration of art and computer technology is a frequently-happening process in the interaction between the two sides. It not only gives the possibility and new driving force to art to attain a development in the new age, and also reveals the computer art is an existing, variable but continuously developing art in the new century. Computer art is not the simple integration of art and technology. It is an integration which is aiming at reflecting the humanistic spirit at the premise of the computer technology. The highest aim of the computer art is promoting the balanced development of the material and spiritual civilizations in the mutual coordination and complement of science and humanity and achieving the liberation of the people themselves [5].

The integration of technology and art has changed into an orientation of the human beings from the perspectives of mentality and psychology in the modern times. In this aspect, it is necessary for the cultural feature of the computer art to be especially outstanding and important.

Subsequently, the outstanding works with the cultural connotation, the artistic style, and the features of the times emerged all over the world, and simultaneously have changed into the most important trend of the development of the computer art in the modern times.

It has been an existing fact that technology tends to be artistic and will exist in the life of the people all over the world. A lot of people are worrying and thinking that the computer technology will eventually replace the "hands" of artists by influencing art gradually.

Brecht used to say: "the common point between art and science lies in that the meanings for the existence of the two sides are reducing the burdens of human life, and one manages the livelihood of human beings, while the other manages the entertainment human beings" [6].

6 Conclusion

From the above analysis, it can be known that the most fundamental quality of the computer art and all other forms of art is acting on people with the beautiful things. The beautiful function of art is the embodiment of the humanistic spirit. This is also the most fundamental starting point for the people to make computer develop toward cultural quality in the modern times.

Therefore, it is necessary for the computer art to use the humanist spirit as their own inner spirit and develop towards the cultural characteristics. Only insisting on the people-oriented idea, the computer art can really own the enduring strength and eternal value.

Maybe it is too early to make a prediction on the tomorrow of the computer art at the present time.

However, in this paper, the author is sure that one point is beyond all doubts that the humanities spirit cultural trend will be certainly the development direction of the computer art.

References

1. Bai, Y.: Illustrated English-to-Chinese Dictionary of computer science, p. 369. Shanghai Jiaotong University Press, Shanghai (2001)
2. Wang, Z.: Postmodernism Dictionary, pp. 370–371. Central Compilation and Translation Press, Beijing (2005)
3. Zhenyu, X.: Computer Science and Engineering Encyclopedia, p. 277. Tianjin Science and Technology Publishing House, Tianjin (1999)
4. Bell, D.: Cultural Contradictions of Capitalism. Translated by Yifan ZHAO, p. 155. Joint Publishing, Beijing (1989)
5. Hu, Z., Liu, C.: Challenges and responses—Education objective views of the 20th century, p. 423. Shandong Education Publishing House, Jinan (1995)
6. Wu, L.: Selection of Modern Western Literature works, p. 155. Shanghai Translation Publishing House, Shanghai (1983)

On Foundation Database System of Education MAN Based on Three-Layer Structure

Xiaohong Xiong

School of Social Development, Huzhou Teachers College, Huzhou, Zhejiang 313000
xiongxiaohong@cssci.info

Abstract. By analyzing the present condition of the application software's data resources within education MAN, this paper puts forward the assumption that the establishment of uniform database platform can fully realize resource sharing and sustainable application, and the design project of the foundation database system of education MAN based on three-layer structure. It also puts forward some principle-based suggestions on the construction of foundation database.

Keywords: Education Metropolitan Area Network, Foundation Database, Data sharing.

1 Moral Values

Morality is the differentiation among intentions, decisions, and actions between those that are good and bad. A moral code is a system of morality and a moral is any one practice or teaching within a moral code. The adjective moral is synonymous with "good" or "right." Immorality is the active opposition to morality, while amorality is variously defined as an unawareness of, indifference toward, or disbelief in any set of moral standards or principles. Ethics is that branch of philosophy which addresses questions about morality. The word ethics is commonly used interchangeably with morality, and sometimes it is used more narrowly to mean the moral principles of a particular tradition, group, or individual.

Personal moral concept is closely related to the soul structure of a person. As for Eduard Spranger, he uses the word "attitude" to describe the personality of the soul structure of human beings. He has divided the attitude into six kinds. Eduard Spranger's contribution to personality theory, in his book Types of Men were his value attitudes. The Theoretical, whose dominant interest is the discovery of truth; The Economic, who is interested in what is useful; The Aesthetic, whose highest value is form and harmony; The Social, whose highest value is love of people; The Political, whose interest is primarily in power; The Religious, whose highest value is unity; Those six in more detail are shown as the followings: in the theoretical aspect, it refers to a passion to discover, systemize and analyze and a search for knowledge. In the aspect of utilitarian, it is a passion to gain a return on all investments involving time, money and resources. In the aspect of aesthetic, it is a passion to experience impressions of the world and achieve form and harmony in life as well as the self-actualization. In the social aspect, it is a passion to invest myself, my time, and my

Z. Du (Ed.): Proceedings of the 2012 International Conference of MCSA, AISC 191, pp. 307–311.
springerlink.com

resources into helping others achieve their potential. In the individualistic aspect, it is a passion to achieve position and to use that position to affect and influence others. In the traditional aspect, it is a passion to seek out and pursue the highest meaning in life, in the divine or the ideal, and achieve a system for living. This paper makes explorations on morality, education and the relevant problems of moral education of Eduard Spranger.

Collective morality is a kind of social view of ethics. It is a kind of function of the subjective social culture. The moral question is personal, individual, of every person. Anytime, in the case of reality, the essential is freedom. Man is responsible, and makes choice of your life, maintain it to the measure where circumstances allow it, but the project itself is, based on social circumstances and something that cannot repair chances that often. There are accidents, its survive and things that have nothing to do directly with project on individual, but wondering, however the chance does not want to be suppression of project coherence, as each person that takes it, turns oneself, or digests, convert them to another ingredient, but assimilated in life. In the meanwhile, man lives in a society, has a personal life combined with collectiveness, and therefore of moral life is apparently submitted to the socio-cultural situation in which you live.

Collective morality concept is the wisdom that has been passed away generations after generations. Freedom is always crucial and decisive. But society has a great diffused pressure, in some way: within life, social customs, creation of one's life, a part of their autonomy. It regalements its automaker, what makes easier, but bears the modes of actual life, in particular nowadays, in reason of the promotion of strong communication. Man receives different interpretations of reality that often has a moral character: ways of life, human relationships, family, morality that are presented under the form of attitudes often considered being normal. That identification is very dangerous as it tends to consider what is common and normal; they are illicit or legally authorized, but not morally. After a long period of discussion, it cannot get rid of the conflicts between the personal value orientations and the collective principles as well.

2 Moral Education Belief

From the perspective of Eduard Spranger, human beings are the completely open animals that keep on studying. The openness of human beings shows in the freedom that he is pursuing. Eduard Spranger conceptualized a psychological performance as part of a meaningful life totality that requires knowledge and understanding of the psychological-mental whole.

Eduard Spranger thinks that whether the current schools are able to cultivate the citizens with senses of responsibility depends on the views of ethics of the school education. Under these circumstances, the current schools should examine the education ethics that it bears.

The educational believes require people to understand that teenagers are completely open. Holistic education is a philosophy of education based on the premise that each person finds identity, meaning, and purpose in life through connections to the community, to the natural world, and to humanitarian values such as compassion and peace. Holistic education aims to call forth from people an intrinsic reverence for life and a passionate love of learning. As for education, if it is lack of believes, the students will be got rid of the confidence, imagination, and creativity. The hopes of education have pointed to the future. Creating prospects lighten the future life of the students.

3 Moral Senses of Responsibilities

The cultivation of senses of responsibilities is an indispensible link in the school education.

From the perspectives of Eduard Spranger, the moral senses of responsibilities of human beings are closely related to the following two aspects: "Id" and "ego". The id is the set of uncoordinated instinctual trends; the ego is the organized, realistic part; and the super-ego plays the critical and moralizing role. Even though the model is structural and makes reference to an apparatus, the id, ego and super-ego are functions of the mind rather than parts of the brain and do not correspond one-to-one with actual somatic structures of the kind dealt with by neuroscience. The id comprises the unorganized part of the personality structure that contains the basic drives. The id acts according to the "pleasure principle", seeking to avoid pain or displeasure aroused by increases in instinctual tension. The ego acts according to the reality principle; i.e. it seeks to please the id's drive in realistic ways that will benefit in the long term rather than bringing grief.

Under this circumstance, how to cultivate the senses of responsibilities of human being? As for Eduard Spranger, he thinks that every human being has senses of honor and senses of recognition from psychology. Ego has many meanings. It could mean one's self-esteem, an inflated sense of self-worth, or in philosophical terms, one's self. Ego development is known as the development of multiple processes, cognitive function, defenses, and interpersonal skills or to early adolescence when ego processes are emerged. Honor and respect are interactive. As for a child, when the adults show respectively that they need him and depend on him, the child will improve the id awareness from the psychology. With the age of the children and the maturity of the mental intelligence, they will feel a change in their inner heart. They will get the order of "pay attention to you" and thus change to "pay attention to respect yourself". As for teenagers, their moral obligations of inner heart will as well experience two stages, changing from passive responsibilities to active responsibilities. At the second stage, the teenagers will undertake the responsibilities out of honored freedom and without the monitor of others. From the perspective of moral concepts, those who can reach the second stage can be called as "the complete stronger".

1) In the first place, Eduard Spranger considers that the school education in Germany at the period of time has turned the students to "learn together" instead of the situation that "live together". This kind of situation caused the students to be separated from each other. They can hard cooperate with each other.
2) In the second place, the adolescence is a key link in the development of a human being. It has played a key role in the comprehensive development of a

human being. As for Eduard Spranger, he considers that the adolescence is of great importance. The key to the adolescence lies in the situation that the formation of id carries out in this particular period of time. As for this period, the id development may have troubles. Therefore, in the adolescence, human beings should pay attention to the cultivation of morality. As for teenagers, they are located in the middle period that produces naturally. That is to say, for teenagers who are from sixteen to twenty, it is easy for them to form a teenager community. As for this kind of communities such as the "Teenager Red Cross Society", the "boy scout" and so on, they have their own moral believes. Human beings are able to learn the moral connotations in this kind of communities and take the responsibilities so as to improve the reputation in the communities of the same kind. In this case, they can not only stay firm their principles and objectives, but also do something beneficial to the nationwide which is of values.

3) In the last place, as for Eduard Spranger, he considers that when dealing with human relationships, people should do as the followings: What is meant by "making the thoughts sincere." is the allowing no self-deception, as when we hate a bad smell, and as when we love what is beautiful. This is called self-enjoyment. Therefore, the superior man must be watchful over himself when he is alone. There is no evil to which the mean man, dwelling retired, will not proceed, but when he sees a superior man, he instantly tries to disguise himself, concealing his evil, and displaying what is good. The other beholds him, as if he saw his heart and reins;-of what use is his disguise? This is an instance of the saying -"What truly is within will be manifested without." Therefore, the superior man must be watchful over himself when he is alone. The disciple Zeng said, "What ten eyes behold, what ten hands point to, is to be regarded with reverence!" Riches adorn a house, and virtue adorns the person. The mind is expanded, and the body is at ease. Therefore, the superior man must make his thoughts sincere. Considering the current circumstances, it has pointed out that there are three conditions that are beneficial to realize the human communications. In the first place, it should have the time. It is the precondition of all other moral behaviors. In the second place, it should be "good at listening to others" and in the third place, it should be "take care of others".

4 Several Inspirations

Considering the above statements, we have made above analysis on the moral senses of values, the views on education and his views on moralities and responsibilities of Eduard Spranger. It can be found that there are certain logical relationships between the three aspects. The moral values are the theoretical premise to the cultivation of moral senses of responsibilities while the cultivation of moral senses of responsibilities is the manifestation of the moral education belief. The moral education belief shows the social moral values on the other hand as well. As for these three parts of views by Eduard Spranger, we should make analysis from the perspective of historical materialism.

On the one hand, the moral values of Eduard Spranger build on the basis of idealist philosophy. It is full of mysticism. Eduard Spranger has upgraded the morality to a spiritual product that surpasses every individual. He considers that morality exists before human beings. This is absolutely against the points of views of Marxism, who considers that morality comes from human social practice. His points of view are without doubt a kind of religious idealism. However, he thinks that there are different morality values for different nationalities and different cultural systems. New cultural structures will definitely produce new morality values, which is consistent with the concepts of historical materialism.

On the other hand, there is something for us to learn from in the concepts of education and moral senses of responsibilities by Eduard Spranger. In the first place, the educational views of ethics have provided references to our school education. Our country is now in the stage of constructing socialism market economy. As an organically integrated system, Deng Xiaoping's educational thoughts are the most remarkable and characteristic achievements in modern China's educational theories. With its rich connotation, "Three Directions to Face (the three directions towards which education should be geared: modernization, the world, and the future)" takes modern construction as its centre and has the goal of promoting People's comprehensive development and bringing forth a new generation of well-educated and self-disciplined people with lofty ideals and moral integrity. "Three Directions to Face" has established a completely new value orientation for China's reform and development and has become the most original and most important part in Deng Xiaoping's educational thoughts. Last but not least, the cultivation method of moral senses of responsibilities, which is as well brought out by Eduard Spranger, is worth for us to learn. The m oral senses of responsibilities come from the communications between human beings and nature and society. Therefore, the students are encouraged to walk out of the classrooms and be closed to the nature and participate in the social activities. Only in this way can the students get more touch with the nature and others. They can realize that all living creatures in the world including human beings have their unique existence values. Respect nature and respect others are to respect and improve ourselves.

References

1. Tan, G.-D.: The educational research on the original citizens. Wunan book publishing corporation limited, Taibei (1998)
2. Michael, G.: Moore distance education system view, May 2165, p. 11. Higher Education Electronic Video Publishing House, Shanghai (2008), May first Edition first time P VI
3. Chen, L.-P., Li, N.: sixty years for the ethnic minorities education in our country: looking back and rethinking. 1st edn., pp. 5–13. National Education Research (2010)
4. Chen, X.-S.: The networking education learning adaptability research. Chinese Distance Education (3), 6–8 (2002)
5. Gong, C.-Q.: The brief discussion on the current situation, problem and reformation of the higher education for Inner Mongolia. Tsinghua University Education Research (3) (2002)
6. Huang, L.: The brief discussion on the development of the distance education for ethnic minorities in Sichuan area. Journal of Southwest University for Nationalities (June 2008)

The Practice of Physical Training Based on Virtual Trainer Concept

Bin Tian

The P.E Department of Xuchang University, Xuchang, Henan 461000 China
tianbin@cssci.info

Abstract. Physical training is considered the best non-pharmacological approach to reduce the incidence of heart diseases and maintain an independent life-style even in old age. The most helpful approach to improve cardiovascular function is aerobic training, independently of exercise mode. Oxidative stress is a metabolic status in which equilibrium between pro-oxidants and anti-oxidants is broken. Virtual Trainer (VT) Concept is a technological platform for the controlled practice of physical training to achieve better cardiovascular fitness results that includes a motivational environment crated to support the long term compliance with the training program.

Keywords: physical training, virtual Trainer, cardiovascular fitness.

1 Introduction

Adjust the risk factors including smoking, high blood pressure, blood fat level, obesity, diabetes and physical inactivity, understanding the body the lack of activity as a level of activity less than those who need to maintain good health. The development of sports activities, prevent cardiovascular disease (CVD) and other cardiovascular disease also successfully modified risk factors, including high blood pressure, blood lipid levels, insulin resistance, and obesity. Physical activity is very important treatment or those with the risk of cardioascular disease increase cardiovascular disease development, including patients have high blood pressure, chronic stable angina or peripheral vascular disease, or previous myocardial infarction, or heart failure [1, 2, 3, 4, and 5). Although the physical activity of the famous benefits, a huge population of some of the leading a relatively sedentary lifestyle and not active enough to realize these health benefits. The report's main problem is lack of the people support in practice and difficult to adapt to the sports activities in their daily life.

Virtual training (VT) realize the concept is in my heart is-2002-507816 engineering [6] by the European commission for cardiovascular disease prevention combat and early diagnosis using advanced information and communication technology (ICT) system. Virtual training is an interactive information environment concept oriented cardiovascular disease primary prevention provide the user a tool to achieve the better cardiovascular health results through the interesting and stimulating experience in the exercise, according to their preferences and professional training goal.

Z. Du (Ed.): Proceedings of the 2012 International Conference of MCSA, AISC 191, pp. 313–319.
springerlink.com

Three important issues have been addressed in the project:

A. The technological platform: the tool that the user will use anywhere and anytime to perform controlled exercise.
B. The fitness condition assessment: the methodology used to provide feedback to user regarding the progress over his/her cardiovascular fitness condition.
C. The motivational ambient: the multimedia environment created to motivate the user to achieve the exercises goals and to support the long term compliance with the training program.

This paper describes the results achieved during the first year of the project, on the technological platform and the fitness condition assessment.

2 Materials and Methods

It is composed by four main components: the training static bicycle, the biomedical sensor, the data processing and user interaction module, and the VT service centre. See Fig.1

1) Training static bicycle. We used a PRECOR [7] model for the construction of the VT prototype. This is a semi-professional training bike used in fitness centers.

2) Biomedical sensor. Wearable biomedical sensors built on electronic textile fabrics with sensor function and interconnections woven into them. The development of suitable biomedical sensors is part of the R&D activity within My Heart project. VT Concept uses a two-layer band with 4 textile electrodes on the internal layer, in contact with the skin, and one temperature sensor. The electronics is removable and inserted in a pocket on the external layer.

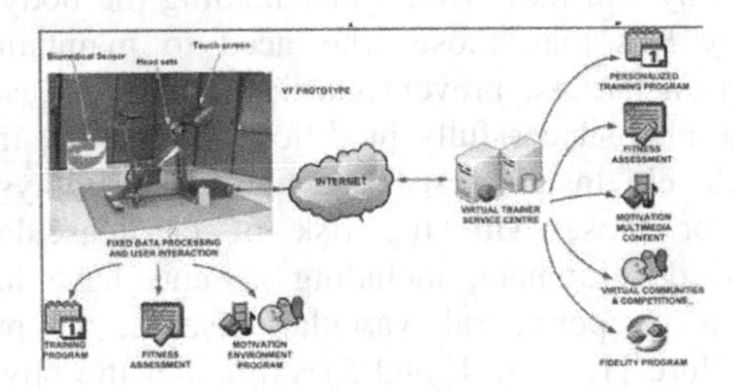

Fig. 1. Architectural view of the components and communication links

Bluetooth technology is used for the data link. The monitored signals are 1 ECG (Heart Rate), 1 respiration (impedance) and 1 skin temperature [8].

3) Data processing and user interaction (DPUI). These are hardware and software components providing computing, communication, local data storage, and user interface functionality. We used two DPUI platforms: a) a mobile DPUI implemented over a personal digital assistant HP iPaq Pocket PC mod hx2415 w/ Windows Mobile 2003 edition, 64MB SDRAM, Bluetooth wireless technology, 802.11b WLAN [9]; b) a fixed DPUI on a PC w/ Pentium 4 processor @ 1GHz,

512MB RAM, 40GB HD, 10/100 MHz Ethernet network adapter and USB Bluetooth interface. A touch screen monitor is screwed on top of the bike handle.

The software was developed within Visual Studio .NET 2003 [10] using C# as programming language. Among the various reasons for this selection, we emphasize the following: i) previous knowledge of language and environment, ii) reuse of code between programming PC version and Pocket PC version, iii) libraries and components available, iv) code and data capacity of migration to any language or data structure. The local database was implemented on Microsoft SQL Server, using XML syntaxes to send data via web services.

4) Service Centre. It is a web server providing VT services to clients. It is under development at the time of wiring this paper.

The aim of the fitness condition evaluation is to provide intelligent, evaluative feedback in a way that the user has an objective measurement of his progress and performance. Fitness status is based on the study of oxidative stress that is produced during exercise and its formation appears to be related to the intensity of exercise [11, 12]. BRT is based on measured heart rate during a constant load-incremental pedaling frequency with the aim to assess antioxidant capacity and blood redox status in order to evaluate the improvement of user physical status.

Methdology of the test: it is an incremental exercise test (IET), based on power increases and Heart Rate (HR) continuous measurement, following a simple protocol:

a) First time: determination of MaxHR = 208-0.7*age;
HR1 = 0.65*MaxHR; HR2 = 0.8*MaxHR
b) During exercise:

- pedaling during 3' increasing power to get HR1.
- pedaling 3' at HR1 and then stop and rest during 3'
- pedaling during 3' increasing power to get HR2.
- Pedaling during 3' at HR2 and then stop and rest

The BTR is based on the quantification of the recovery of HR after a sub-maximal effort. The HR recovery curve kinetic can be characterized by the following parameters:

- τ = time of recovery
- A_{max} = Maximum amplitude of recovery
- A_0 = Minimum value of HR

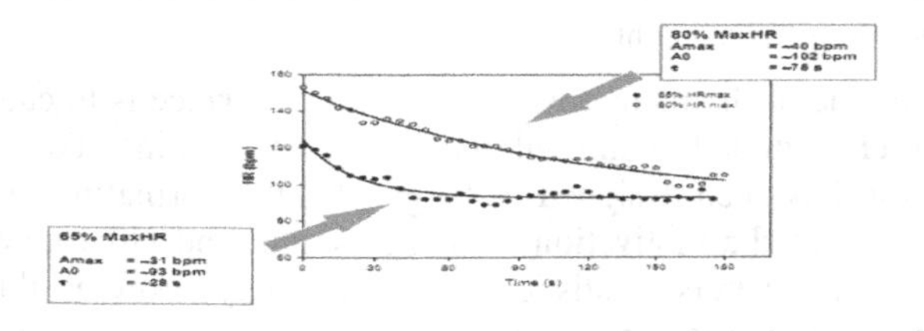

Fig. 2. Example of HR recovery kinetics

3 Results

According with the software realization plan of the project, the following modules were implemented and tested:

a) Virtual Trainer Repository and Data Manager: the repository is composed of the different databases: User Profile DB, Multimedia Content DB, Virtual Trainer Support DB, Sport Knowledge DB and Health Knowledge DB. The data structure designed is based on the characterization of session-exercises and all their relations with the rest of components like the user profile, motivation and alarm definitions, media files, etc.. This structure stores all measurements acquired and data generated during the exercise in order to be processed afterwards, and recorded as historical backup for futures analysis. Thanks to the database design, any exercise can be reproduced at any moment in the same way that it was made by the user.

b) Communication Module: for the first prototype, while the actual biomedical sensor was available, an On-body Electronics Simulator has been developed in order to generate the needed biomedical signals for the VT application. Its main features are:

HR simulation producing HR values within the range 60 - 250 bpm following increase-recovery patterns

Parameter adjustments through GUI interface
Communication over RS232 and Bluetooth

c) RPM sensor: It is a speed sensor for the bike that consists of a Bluetooth enabled optical detector that counts the revolutions of the wheel and sends the data to the application. It's installed in the chassis of the bike near the edge of the wheel.

d) Training Session Module: software application that controls the execution of the exercise session. The main components are:

- Graph component
- N-polygon interpolation/regression for summary graph estimation.
- Handling of error and alert states.
- Status indicators during session (calories, velocity, respiration animation.)
- Summary report of completed exercise
- Questionnaire for feedback evaluation
- VTControls: Auxiliary components to manage the application, such as VTPlayer to reproduce simultaneously different music files during the session; VTSpitch to show the rhythm of the user's heart rate; VTDisplay can reproduce videos or show photos randomly or under control of the motivation & alarm module; VTGraph to draw in real time the measurements of the sensors.

e) User Interface: One of the main objectives of this interface is to catch the attention of the user, in order to guide him through the process of training. To achieve this goal, an attractive layout has been designed to fit graphical information on: real time HR and respiration, multimedia motivation contents (sound and video), system messages (voice and text), and exercise statistics. The user may browses through different screens during his training sessions.

f) Motivation & Alarm Algorithm: It consists of two main components:

- Evaluation component: real time feedback of user performance according to given objective. This is based on the BRT algorithm

- Candidate selection algorithm: Rule based selection algorithm for determining which motivation events are triggered during the different moments of the exercise.

The objective of the alert algorithm is to inform the user on both, the system's status (i.e. connection loss, transaction error...) and the user's real-time measurements (i.e. when the user leaves recommended training zones).

The motivation algorithm objectives are:

- Advise and inform users about things he/she ought to be aware of or might find useful during the excise session.
- Inspire the user to keep up even when he/she gets exhausted.
- Entertain the user and give him/her a unique training experience.
- Motivate user to continuous use of the VT so that his/her physical condition improves.

Evaluation

Previous of the software implementation of the algorithm for fitness condition assessment and its inclusion within the VT system, a test was conducted to verify that the HR recovery kinetics after sub-maximum exercise may represent a parameter for the control of the fitness level and this parameter is sensitive to variations in the cardiovascular fitness. 30 people (female n=15; male n=15), average age 31.3 SD 5.9 (range 23-46); mass: 69.2 kg SD 8.1 (range 50-98); height: 170.9 cm SD 1.2 (range 154-182); Vo2max 2581.2 ml/min SD 184.1; followed a protocol of training during 8 weeks and the BRT and the Test Incremental Maximal (TIM) based on the direct measurement of the VO2 and concentration of lactate in blood during the exercise were applied at the end of week 4 and week 8.

The results of the BRT during the control condition and the end of week 4 (Fig. 3) are similar to the results of the TIM. The HR have decreased 7% and the concentration of lactate in 13% for a similar effort demand equivalent to 4 MET which represents the aerobic capacity needed to step a stair or ride a bicycle. This represents an increment in the cardiovascular fitness capacity produced by the training program designed for the VT concept. In the same period of time, the A_{max} value have increased a 69% respect to the control condition for HR1, and have increased 30% for HR2 after 3 minutes of recovery. This preliminary results shows that the BRT indicators may be suitable for implementation within the VT concept.

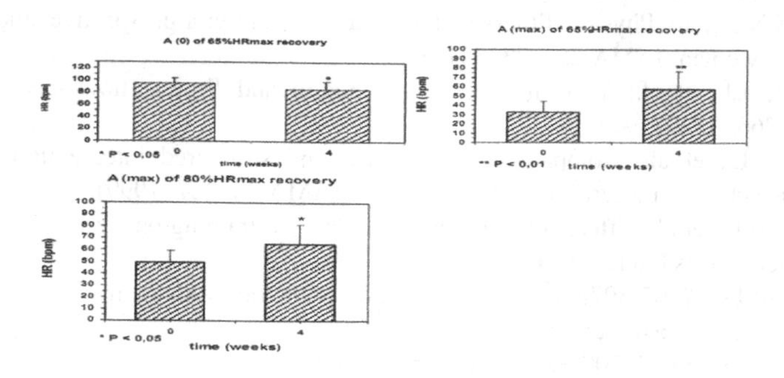

Fig. 3. Results of BRT at the end of week 4

4 Conclusion

Scientific evidence supports the ability of the organism to increase the antioxidant defences with a well conduct exercise training. VT concept within project My Heart is an environment for the methodological practice of aerobic exercise addressed to healthy people to fight against sedentary and to reduce CVD risk factors. VT strategy is oriented in two objective lines: for one part, to create a tool that motivate the user to perform regular practice of physical activity throughout a long term training program, to achieve better fitness results within a fun and motivational experience. The sense of achievement is a strong value for the VT users. They need to feel that they are progressing; they need to see real results in the medium and long term. VT provides an intelligent feedback of objective data that helps them to understand their performance and a sense of working with reliable information. Motivation was also high lined as a strong value that helps users to perform better the routine exercise and commit them in a continuous active life. VT aims at creating a friendly and motivating exercise environment tailored to the user special preferences and fitness goals.

The second strategic objective line is the coaching. This is based on the design of personalized training programs adapted to user's characteristics and preferences, and of a continuous assessment of the actual fitness status and the intelligent feedback through the motivational environment created for this purpose.

In the first year of the project we have built the first prototype of the concept and designed a method for the assessment of the cardiovascular fitness status of the user. The evaluation of this method is under way but the first results show that it may be suitable the VT concept. This period is now ending with the first evaluation of the stakeholders feelings: users, professional trainers and fitness club owners. The second phase in years 2005-2006 will be devoted to the consolidation and stabilisation of the technical platform, the validation of the algorithms and the evaluation in real use conditions.

References

[1] Anderson, R.E., et al.: Effects of lifestyle activity vs. structured aerobic exercise in obese women. JAMA 281, 335 (1999)
[2] Blair, S.N., et al.: Physical fitness and all-cause mortality: a prospective study of healthy men and women. JAMA 262, 2395 (1989)
[3] Duncan, J.J., et al.: Women walking for health and fitness: how much in enough? JAMA 266, 3295 (1992)
[4] Dunn, A.L., et al.: Comparison of lifestyle and structured interventions to increase physical activity and cardiorespiratory fitness. JAMA 28, 327 (1999)
[5] Hurley, B.F., et al.: Effects of high intensity strength training on cardiovascular function. Med. Sci. Sports Exerc. 16, 483 (1984)
[6] My Heart IST-2002-507816 project (2004), Information available in, http://www.cordis.lu/ist
[7] Precor Incorporated (2005), http://www.precor.com
[8] Lauter, J.: My Heart project Deliverable D2: Concepts Realisation Plan & State-of-the-Art (2004); Available on request to My Heart project coordinator

[9] Hewlett Packard Development Company (2005), http://www.hp.com
[10] Microsoft Visual Studio Developer Centre (2005), http://www.msdm.microsoft.com/vstudio/default.aspx
[11] Bailey, D.M., Young, I.S., McEneny, J., Lawrenson, L., Kim, J., Barden, J., Richardson, R.S.: Regulation of free radical outflow from an isolated muscle bed in exercising humans. Am. J. Physiol. Heart Circ. Physiol. 287(4), H1689–1699 (2004)
[12] Sastre, J., Asensi, M., Gasco, E., Pallardo, F.V., Ferrero, J.A., Furukawa, T., Vina, J.: Exhaustive physical exercise causes oxidation of glutathione status in blood: prevention by antioxidant administration. Am. J. Physiol. 263(5 Pt 2), 992–995 (1992)

Development of Graduate Simulation Experiment Teaching System Based on Matlab Language

Chunhua Tu

Postgraduate Department,
Jiangxi Science and Technology Normal University,
Nanchang, Jiangxi, 330013 China
tuchunhua@cssci.info

Abstract. Facing the existent problems in graduate teaching especially experimental teaching for general university, a new idea of developing a simulation experimental teaching system for self-study is proposed. The simulation system is introduced in detail.

Keywords: Matlab language, simulation, graduate teaching, measuring and testing technique

1 Introduction

At the present time, China has become a big power in graduate education. Making an improvement on the education quality is one of the most important tasks and goals for the reform and development of China's graduate education currently, while making an optimization on the graduate education structure is a premise to increase the quality of China's graduate education.

The adjustment of China's graduate education structure is meeting the need of the social economic development. Therefore, it is imperative to make a reform on the graduate training model of China at the present time. Also, it is necessary to make a new review on the orientation of the graduate student training model of China.

2 Current Situation and Reform Reasons of Graduate Student Training Model of China

2.1 Problems in Graduate Student Training Model

There are some problems still existing in the graduate education of China. These problems are mainly reflected from the aspects in the following.

First of all, there is a deviation in the training idea. For a very long time, as the types of the trained talents are relatively simple, the graduate-student training idea is oriented at the "academy first" obviously. In the mean time, the training of the application-oriented graduate student is thought lightly by people to some extent.

Second, the examination-oriented education model which has been lasted for a very long time gives rise to the low ability of the graduate student in innovation.

Z. Du (Ed.): Proceedings of the 2012 International Conference of MCSA, AISC 191, pp. 321–326.
springerlink.com

Third, some course systems for the graduate students are still not sound, so that there are no strict differences among specialties, and the talent training modes are similar in many graduate programs in China.

Fourth, the instructors who undertake the researches of some important subjects usually have not too much energy to give guidance to the graduate students, while the instructors who have no important subjects to study are unlikely to give guidance to the graduate students at the forefront of science.

Fifth, the education for graduate students attaches too much importance to the speculative theory training but thinks lightly of the training of the practical ability training. In the mean time, the result of the teaching model which lays a stress on the academic studies but looks down on the practical application has given rise to the divorcement of the talent training from the needs of the society with a gradual step.

Sixth, generally speaking, it is necessary for the graduate students to be confronted with the three major pressures (i.e. survival, development and employment) after they begin to study a graduate program. For this reason, the graduate students do not inject the sufficient dedications and investments to learning and scientific studies.

Seventh, the management department for the graduate students makes no standards and limits for the instructors to train the graduate students, so that the "batch training" model for the graduate students appears very frequently.

Eighth, the educational conditions and levels of some schools are quite low. As a result, the conditions for admitting the new graduate students and the training quality are reduced imperceptibly.

Ninth, the non-specialization of the employment of the graduate students gets more and more obvious with each passing day. And this exerts an impact on the enthusiasm of the graduate students in the learning of their specialties.

Tenth, the structure of training the graduate students at higher learning schools losses the balance.

2.2 Necessities of Training the Application-Oriented Graduate Students

The employment of the graduate students has transformed from the "seller market" to the "buyer market". Such a huge change force to have to think of and recognize the orientation of the training objectives of the education for the graduate students. Therefore, it changes into a mission from the times to train the application-oriented special talents at high levels vigorously.

First of all, it is necessary for the graduate student training to change from the academy-orientation to the application-orientation. As the graduate education attains a very rapid development in China and especially the scales of the programs for the doctors of philosophy in art and design are expanded greatly also higher learning schools and scientific research units put forward a higher requirement on the quality and educational backgrounds of the faculty and research personnel, the PhD graduates have changed into the main force of the teaching and scientific research at higher learning schools day by day, and also the historical mission of the graduate education taking the dull model of training the high-level specialized technological talents goes to the extreme.

From the employment situation of the recent years, the proportion of the graduate students achieving a master's degree to work at the higher learning schools and the scientific research units is less than 20%. Therefore, it is necessary to make an orientation on the training objective of the graduation education again, changing from the dull academy-oriented training model to the training model giving consideration to both application-oriented graduate student training and academy-oriented graduate student training.

Second, it is one of the most important guarantees to train the application-oriented graduate students and meet the development of the social economy and the need changes on the high-level talents. At the present time, obviously, the objectives of the employment of the graduate students have changed from the professional technical positions to the o the party and government management positions and the enterprise management positions. This can fully suggest that it is necessary for China to make an adjustment to the training objective of the graduate education in time according to the new changes of the requirement of the society, make a change to the model of training the graduate students, practically make an enhancement to the high-level application-oriented training, and also promote the training of the high-level talents to develop harmoniously with the needs of the economic and social development.

Third, the training of the application-oriented graduate students is one of the mature experiences in the development of the graduate education in the world. In the developed countries, the specialized degree education which is cored at the training of the application-oriented graduate students has taken up a very important position in graduate education and especially at the master's degree level.

Such a graduate student education structure can not only effectively give consideration to the development of academy and the satisfaction of the social needs, and also can reflect the ideas which are supposed in the graduate education of the modern times in depth. In other words, the purpose of the development of graduate education is not merely for the sake of the academy and to meet the needs of all kinds of industries in society on the high-level talents at the same time.

3 Reform Path of the Application-Oriented Graduate Student Training

3.1 Optimizing the Structure of Curriculum

It is necessary for the curriculum offering for the application-oriented graduate students to be oriented by concentrating on the objectives of the application-oriented talents and the needs of the employment market. In the mean time, it is necessary to take the actual application as the guidance, the professional demands as the goal and the improvement of the comprehensive quality as the core, and also attach importance to the dynamic combination of the theoretical courses and the application-oriented courses. The application-oriented graduate students put forward a higher requirement on the curriculum offering.

First of all, it is necessary to take great energy to make the close courses integrated, make a promotion to the optimized combination of the specialized courses, reduce the repeating between the contents of the specialized courses, and also attach higher importance to the forefront and application of the contents in the courses.

Second, it is necessary to lay a stress and make an expansion to the foundation of the specialties of the graduate students, and then strengthen the intensity of offering the courses between the close disciplines.

Finally, it is necessary to increase the credits of the elective courses at the premise of ensuring the simplification and optimization of the required courses, and then set some courses at the forefront of disciplines, the disciplinary courses, the method-oriented courses, and the technology-oriented courses and so on.

3.2 Changing the Teaching and Guiding Methods

It is necessary for the application-oriented courses to use the course bidding method and promote the "made-dish" teaching model in which the multiple specialized teachers are responsible for a lecture under the joint efforts. In the mean time, it is necessary for all these teachers to convey the most important contents and experiences of the personal learning and studies to the students, for the purpose of practically making an enhancement to the quality of the course teaching for the graduate students.

In the teaching method, it is necessary for the speaking teachers to make use of the heuristic research-based teaching and flexibly apply the expository method, the research and discussion method, the presentation method and experiment methods and so on, for the purpose of stimulating the students to thinking way of the students and also driving them to have interest and enthusiasm in learning.

The guiding way which is used by the instructors in China graduate education is obviously directed at the "students are under the leadership of the teachers". The personal tutor system is actually implemented in the graduate education.

On the one hand, such a guiding way is not beneficial for the graduate students to generate a reasonable knowledge, ability and quality structure.

On the other hand, such a guiding way is easy to give rise to the divorcement of the graduate-student training model with the needs of society on the talents.

3.3 Strengthening the Practice Training

Practice training is one of the most important methods of training the graduate students to get innovation in spirit, improve the practice skills, and also enhance the social practice experiences.

It is necessary to orient the application-oriented graduate students at the high-level applied talents for all kinds of the industries. It is highly necessary to make an expansion to the forms and channels of the practice education for the graduate students further, establish a wider and smoother practice platform, and also make an enhancement to the training of the practices.

4 Establishing a Diversified Graduate Education Quality Evaluation System

4.1 Reforming the Student Admission Way

It is necessary to make a reform on the student admission way.

On the one hand, it is necessary to make an enhancement to the intensity of evaluating the knowledge in the practical application and also improve the evaluation on the practical operation.

On the other hand, it is necessary to set the different exam ways and time to hold an exam for the academy-oriented graduate students and the application-oriented graduate students, allow the students to make a choice according to their own interests, and bring the learning enthusiasm of the students into full play, for the purpose of selecting and training the talents who can meet the needs of all kinds of industries.

4.2 Adjusting Graduation Standard

With the purpose of training the application-oriented graduate students, the academic standard can't be used to make an evaluation on the quality of the graduate students all the time.

However, it is necessary to make use of the diversified quality ideas and a diversified evaluation system to replace the academic standard. Therefore, it is beyond all doubts that it is necessary to make an evaluation on the academic graduate students who receive a continuous education and are engaged in scientific research by adhering to using the scientific research ability, academic innovation, theoretical innovation as the main criteria.

However, as for the majority of the application-oriented graduate students, it is necessary to make use of the practical ability, the operation skills, and the technology development ability, and the strain capacity, the pioneering spirit and the comprehensive quality as the main evaluation standards.

5 Conclusion

Therefore, in order to meet the needs of training the high-level application-oriented graduate students, it is necessary for Chinese higher learning schools to make an adjustment to the requirements of the students on graduation, make use of the diversified evaluation standard which is concentrated on the training of the practical ability to replace the dull academic standard, and attach higher importance to the evaluation on the innovation ability, and the operation skills, the technical development ability, the market development ability, the comprehensive quality, etc.

In the mean time, it is necessary for degree thesis to be diversified. Except the academic papers, the research overview, the professional design, the project planning book, the investigation report, the inventions and creations, the technical development, the literature and art works, the book reading experience and others can be replaced with something else.

Acknowledgement. Fund Project: This Paper is Supported by Jiangxi Social Science Planning Project—Research on Higher Education Popularization and Graduate Training Orientation.

Reference

1. Wang, H.: Postgraduate Education at Crossroad by Taking the Postgraduate Education of Higher Education Discipline for Example. Research on Education Tsinghua University (October 2008)

Study on the Application of Fuzzy KNN to Chinese and English Recognition

Cuiping Zhang

Shenyang Ligong University, Shenyang, Liaoning, 110015 China
zhangcuiping@cssci.info

Abstract. Automated Chinese and English recognition is defined as the process of assigning category labels to new documents based on the likelihood suggested by a training set of labelled documents. A study on the performance of fuzzy KNN for Chinese and English recognition is presented. Four famous approaches to text feature selection are adopted. The improved fuzzy KNN method is compared to the conventional KNN method and the KNN method based on similarity-weighting. and the experimental results show that the proposed method can weaken the impact of the disparity of training samples in distribution on categorization performance with different feature selection methods selected. Also, the categorization accuracy is improved, and the sensitivity to Kvalue is reduced to some extent.

Keywords: Chinese and English recognition, Fuzzy KNN, Feature selection.

1 Introduction

The different cultural characteristics give rise to the different language characteristics. There is a very close connection between names and cultures. It is known to all that names play a part in the daily life of people with the help of cultures. Name is not only one of the symbols that are used by people to make identification, and will also receive the restrictions from its own culture and will give reflection to the characteristics of the corresponding culture. As English names and Chinese names are subject to two different cultures (i.e. Chinese culture and western culture), there is a great difference between English language and Chinese language in characteristics, and hence the cultural meanings which are reflected from the two kinds of names have the great difference completely.

As is known to all, there are the obvious differences between English names and Chinese names in the multiple aspects such as phonetics, structure, morphology, etymology and semantics. The huge differences just give reflection to the different cultural characteristics of the eastern world and the western world. Also, the differences have a natural connection with their own cultures, and concentrate the different thinking ways and behavioral characteristics of the diffident kind kinds of people from the different t cultures.

In this paper, from the perspective of these differences, the author carries out a discussion on the great differences between the people of the two kinds of cultures in the thinking behaviors.

Z. Du (Ed.): Proceedings of the 2012 International Conference of MCSA, AISC 191, pp. 327–332.
springerlink.com

2 Similarity and Diversity

After a comparison between English names and Chinese names, it can be found that there are the obvious differences between English names and Chinese names just like English language and Chinese language.

Generally speaking, from the perspective of structure, Chinese names can be divided into two types, which are the compound names with a single surname and the single name with a single surname.

However, English names are completely different from Chinese names. The structure of English names shows a very flexibility and is not t qualitative. English people like to make a choice on the length of their own names according to the actual interests. Therefore, the names of some people include a middle name, while the names of some people do not. More strangely, the names of some people include the first name, the second name, and the third name and so on. The last name in English name is called as the surname.

However, from the forms of English names, it can be clearly seen that there are no unity and regular rules in English names. And this characteristic of English name can reflect that English-speaking people often make a choice on the different ways according to their own interests.

3 Sense of Family and Personal Ideas

A significant difference between English names and Chinese names in the structure can be described below: in a full Chinese name, the surname (the last name in English) is placed at first, which is followed by the second name; in a full English name, the surname is placed at last. Such a difference between English names and Chinese names is fundamentally caused by the two different kinds of cultures.

There is a common tradition among Chinese people that the higher importance is attached to the surname. This is because the surnames stand for not only the symbols and also for the whole family as well as the blood relation.

In Britain, the surname of the native people is placed after the personal name. That is to say, within a very long time in history, the names of British only included the first names actually, with their surnames absent. Only to the modern times, the British finally began to take advantage of the surnames when a great number of the people apply the same first names.

Therefore, it can be seen that the personal values are placed at the front of the families in English name system. However, the surnames are always placed at the front of the personal values in Chinese name system, and this characteristic has its own historical reasons, which are reflected in the sequence of the first name and surname.

4 Religious Beliefs and Confucianism

English nations and Chinese nation have their own beliefs and cultural awareness. Christianity has an extremely great influence on the social life of the western world. For this reason, when English people name their children, they often make a choice on the names of gods, such as Abraham, David, Adam and John. These names usually source from bible.

However, the Confucianism has been rooted in the minds of Chinese people. For this reason, Confucianism has profound education significance in the behaviors of Chinese people for a very long time. In the mean time, the Confucianism exerts an extremely great influence for people on how to handle the human relationship with others in families and society.

The connotations of benevolence, righteousness, propriety, wisdom and honesty which are advocated in Confucianism receive a very good expression in Chinese names, such as LI Zhongren, SUN Jieyi, HOU Xinli, CAO Ruizhi and WANG Shouxin.

5 Comparison between Chinese and English Family Names in Word Meaning, Origin and Source

5.1 Word Meanings of Family Names in Chinese Language and English Language

In the "Analytical Dictionary of Characters" of Xu Shen, it is recorded that the family name stood for the blood relation of a child with his family members, and was continued from his mother, and this meant that the family name sourced from the women who gave birth to the child. Therefore, it can be seen that the origin of the surname had a very close relationship with the action of a woman to give a birth. The children who were born by the same mother owned the same family name.

The family name in English language means "surname", which is on top of the name". In the western languages such as French, Germany, Roman and Latin, the family name stands for the name of a family.

Therefore, from the word meanings, it can be known that the family name is defined by people from the perspective of the relationship with a family, or is used to describe the relationship with the same names but not to attach importance to the blood relation with a family.

5.2 Origins of Family Names in Chinese Language and English Language

In China, the family names were originated from the matriarchal society of the ancient times, and owned a history of more than five thousand years. At that time, the family names were used to make a distinction among the different families as well as the different marriages in China. Therefore, it can be seen that family name was a symbol of the blood relationship all the time.

In the ancient times of Britain, the ancestors (Anglo-Saxons) of the British only had their names but did not have the family names. Generally speaking, the names of Anglo-Saxons were very simple. And most Anglo-Saxon names were composed of a generally-used name. In 1066, the Normans made an incursion in Britain, and brought the family name system of Frances to Britain at the same time. During the initial period, the family names were only used in those feudal aristocracies. In the following 500 years, however, the British formed their own complete surname system with a gradual step.

5.3 Sources of Family Names in Chinese Language and English Language

5.3.1 Types of the Sources of the Family Names of Chinese People

The family name system of Chinese people mainly takes the family names of the Han-nationality as the basis. According to the analysis, the types of the sources of the family names of Chinese people can be concluded as follows.

(1) The surnames of the mothers or the totems of the ancestors (these were the oldest family names such as JIANG, YAO, JI, YUN, LONG, NIU, MA and YANG)
(2) The titles of the ancestors in that society, such as XUANYUAN and GAOYANG
(3) The flag number, designated names, characters or names of the ancestors, such as TANG and WEN
(4) The living places, such as DONGMEN, GONGGE, XIMEN and NANGONG
(5) The kingdoms and surname fiefdoms, such as QI, LU, QIN and JI
(6) The ranks of nobility, such as WANG, GONG and HOU
(7) The official names, such as SIMA, SITU, SHUAI and WEI
(8) The occupations, such as TAO, WU, SHANG and Yue
(9) The family names of the emperors, such as the "LI" which was given by the Tang dynasty emperors to their meritorious statesmen (the descendants of these meritorious statesmen continued to use LI as the family name)
(10) The transliteration of the family names of minorities people, such HUYAN, YUWEN, BAI and BAO

5.3.2 Types of the Sources of the Family Names of English People

According to the analysis, the types of the sources of the family names of English people can be concluded as follows.

(1) The occupations, such as Carpenter, Thatcher and Tailor, which source from the original occupations (e.g. mill-man, the persons covering the roof and the persons making clothes)
(2) The topography and landforms near the living places, such as Moor, Hill and Lake, which mean wild place, small mountains and lakes respectively
(3) The names of the living places, such as York, Kent and London, which are originally the British places but changed to the surnames in English later
(4) The bodies of the individuals and the individuality characteristics, such as Small and Long which mean small men and tall men respectively
(5) Some surnames in English are composed of the first name and the "son", such as Johnson, Wilson and Jackson, which are used to give expression to a relationship between parent and child
(6) Surnames are composed of two words in English, such as Loyd-Jones and Bartle-Smith
(7) Surnames emerged due to the population migration (e.g. the outsiders migrated to England are usually called as Scot, Wallace, Scott, which become the surnames of these new immigrants; some new immigrants take Newcome and Travelers as surname)
(8) Some people establish unconventional or unorthodox names with the purpose of expressing the worship of the great men

5.4 Forms of Family Names in Chinese Language and English Language

The "family names" of Chinese people give priority to the single syllables. Therefore, most family names are composed of only one word. However, there are double-syllable words in Chinese family names, and also there are two-word, three-word, and four-word compound surnames.

In the form sense, the surname in an English full name contains only one word no matter how many syllables it contains. From the morphology and word property, the surnames in English include nouns (such as Hill) and adjectives (such as Long). In English countries, the surnames of children source from the last names of their fathers; the surnames of women will change to the surnames of their husbands after getting married. For example, May Smith gets married with John Robinson, but needs to change to Mary Robinson later.

6 Cultural Differences

Name is a special cultural language. Through the comparison between English names and Chinese names, it can be known that there are some similarities between English names and Chinese names, but the differences between the two sides are very obvious. Considering the differences between English names and Chinese names in depth, the following enlightenments can be obtained.

First of all, Chinese family name system has a longer history than English family name system. Chinese family names are quite more complex than English names in forms, and also take up an extremely important position in society or the minds of people. China's "family name" was originated very early. And priority was always given to the names. Thus, Chinese family names were placed before the second names, indicating Chinese people placed the family names in an extremely important position and also family name is a symbol of a family. However, the second name is only the symbol of individuals. In other words, family has an irreplaceable position in the minds of Chinese people. And the second name of individuals is inferior to family name. Therefore, for thousands of years, the patriarchal clan idea of Chinese people was very profound. Chinese people worship the large family, and have a strong sense of carrying on the family line. For this reason, there are some backward ideas in Chinese people.

Second, names can reflect the different reactions of Chinese people and English people in face of authorities. In China, from the ancient times to now, people think that the names of deities, emperors, sages and meritorious statesmen are sacrosanct, so people do not call their names directly when speaking with them.

Third, name is also a tool which is used to record the cultures, and can reflect the national history and cultural development to some extent. Therefore, it is with the spirit of the times. Britain is not a country which underwent a large national revolution, but was evolved from the continuous foreign invasion. It is not a country like China, the United States and France to emerge by going through years of revolutionary struggles.

Also, Chinese names are with the characteristics of the different times and also shine with history, such as the names (YOUWEI, KEQIANG and JUEMIN) in the late Qing dynasty; the names (WEIDONG, XUEWEN and XUEWU) in the Cultural Revolution. All these names are the most concise words which are utilized to suggest the history.

References

1. Jin, L.: Names and Social Life. Shaanxi People's Publishing House, Xi'an (1989)
2. Zhang, H.: Language and Name Culture. China Social Sciences Press, Beijing (2002)
3. Zhang, L.: Foregin Names. China Youth Publishing House, Beijing (1996)

Study on English Converting to Chinese of Database Fields

Cuiping Zhang

Shenyang Ligong University, Shenyang, Liaoning, 110015 China
zhangcuiping@cssci.info

Abstract. As we develop the software of database program, it refers to various of condition searches, in order to get lots of information data. Although modern relationship databases support table s fields defined with Chinese character code, but fields usually define with English character code. A big database application always makes up several hundred tables and several thousand fields. However fields English have to show with Chinese language, if we convert English to Chinese using 'as' with SOL command, it will increase our work, furthermore if the application changes in the future, we must modify code in the program. Thus it will decrease database system's flexiblity. In this paper, to solving these questions, we create auxiliary table that storages fields English to Chinese comparing records and create general data getting function.

Keywords: Dataset, Data table, Converting.

1 Introduction

Name is a kind of language phenomenon. It has a natural relationship with the literature works. Generally speaking, the literature works can't exist without the people's names. Also, the people's names often include the literature styles. The words of people become more wonderful due to the wide popularity of some names in the literature works.

In this paper, by starting from English and Chinese names, this author carries out a discussion on the association of English and Chinese names with the literature works from three perspectives.

2 Names Source from Literature Works

One of the effective methods which are used by people to give names to others is establishing a certain connection with something in the literature works. In general, people are used to directly make use of the names in the literature works or draw the essence of the literature materials as part of the names.

For example, people take utilization of the names in the ancient Greek and Roman mythologies in the processing of naming themselves or their children, such as Hercules, Apollo, Prometheus and Helen.

Z. Du (Ed.): Proceedings of the 2012 International Conference of MCSA, AISC 191, pp. 333–338.
springerlink.com

Chinese nation is with five thousand years of history. Chinese people often take advantage of the essence of Chinese culture when they name their young children. This can be embodied specifically in the names, which source from the excellent poetries, prose and idioms and are endowed with the profound cultural connotations. This is a distinctive way which is frequently used by Chinese people. In the mean time, the sources for people to give names to others are different as well. For example, Jiazhen derived from a Chinese idiom and means a subject with great familiarity.

3 Names Are Utilized by Writers as a Special Writing Technique in Literature Works

In the literature works, the writers often create the personality characteristics of the different characters and give implications to the end of the stories through multiple methods, such as describing the historical and cultural backgrounds, depicting the mental state of the figures and making an ingenious design for all kinds of names.

In English, "the Pilgrim's Progress" can be a well-known example. The readers can make a guess on the personalities of all characters through the meanings of the names in this work.

In Chinese literature works, "A Dream in Red Mansions" is a typical example as well. The names in this classic novel contain the personality characteristics of the different characters, such as JIA Baoyu (Chinese homophonic tone of this name is implied it is the artificial treasure jade), JIA Hua (Chinese homophonic tone of this name is implied this man likes telling lies), ZHEN Shiyin (this name means the real thing are hidden, and is used to imply this man conceals the truths), BU Shiren (Chinese homophonic tone of this name is implied this man is not human and has some improper behaviors).

4 Names in Literature Works Enrich Daily Expression

A great number of the names in the literature works have entered the daily life of people all over the world, and change into part of the words that are frequently used by people in the daily life.

In the famous work "Hamlet" of Shakespeare, the protagonist Hamlet is a typical weak-minded character. At the present time, "Hamlet" has been converted into one of the nouns that are widely known among people and hence have entered the languages of the daily life of people all over the world. It is used to express that a man is always in two minds about his decisions.

Like English literature works, Chinese literature works also contain some names which have been very well-known among Chinese people. Hence, the language of Chinese people is enriched by these names as well.

For example, ZHU Geliang, a central figure of the classical work (Romance of the Three Kingdoms), is a well-known military strategist with great wisdom to all Chinese people. For this reason, Chinese people often make use of ZHU Geliang to give expression to the great wisdom, rich experience and discretion that a man has.

5 Implied Meanings in Literature Works

5.1 Homophonic Implied Meanings

It is extremely common for writers to often take advantage of the homophonic implied meanings to give names to the characters in both English and Chinese literature works.

In English and American literatures, Geoffrey Chaucer, a British writer, named the mother superior of the convent as Eglantyne in "the Canterbury Tales". Eglantyne is similar to the homophonic tone of eglantine and has a close form with eglantine, implying the romantic and seductive charming of this mother superior.

Also, the homophonic implied meanings were used by the writer for the names in "A Dream in Red Mansions" (a Chinese classic literature work), for the purpose of giving expression to the different attitudes of the writer towards the different characters and destinies.

For example, JIA Hua, characterized as flying time and also named as Yuchun, was born in Huzhou. JIA Hua can be understood as "telling lies", "irrespective of goodness and badness", "speaking false statement" and "telling fabricated wild tales" from the perspective of the homophonic implied meanings of Chinese language; JIA She can be understood as "eroticism" from the perspective of the homophonic implied meanings of Chinese language; WANG Ren is read like "forgetting the benevolence" from the perspective of the homophonic implied meanings of Chinese language; FU Shi is read like "pleasing and flattering the wealthy and influential persons" from the perspective of the homophonic implied meanings of Chinese language and is implied that this character is a base person.

5.2 Sourcing from Classical Allusions

Another way that is used by the writers for implying the meanings of the names in the literature works is citing the classics from the famous works of the previous times. Namely, the writers take advantage of the ancient codes and records of the figures that have been highly well known among people to give names to the characters in the literature works in accordance with the actual needs, for the purpose of arousing all kinds of the associative meanings from the minds of the readers. The most important point for the writers of both Chinese and western literature works to use the classical allusions is relying on the obvious things to give expression to the subtle implied meanings, so as to reach a result that a thought can be entertained deeply and the influence can be always in existence at the same time.

In both English and Chinese literature works, a large number of the writers draw support from the associative meanings of the classical allusions in the minds of people and give names to the characters according to the different personality characteristics of the characters or the different themes of the works.

"Moby Dick", a masterpiece of American writer Herman Merville contains a character, namely, the narrator (Ishmael) of the story. In the "Bible · Genesis", it is recorded that Sarah, the wife of the earliest ancestor (Abraham) of Jewish people, ill-treated Hagar (a concubine of Abraham) and her son Ishmael and evicted the mother and son, and then they led a vagrant life in desert but were rescued by god later, and therefore Ishmael became a symbol for people living a wandering life but pursuing a good life. Also, the reason why the narrator of the story of "Moby Dick" went to the "Bertrand" whaleboat was that he hoped to escape from the situation that made him miserable.

5.3 Giving Meanings as the Names of Characters Imply

In the literature works, the most direct way that is used by the writer for implying the names of the characters is giving expression to the personality characteristics, the thought states as well as the destinies of the characters according the literal meanings of the names of the characters. The "Tender Is the Night", a literature work written by F. Scott. Fitzgerald of the United States contains a character (Diver) that is desperate and disheartened and descends to a doctor in a rural place. The name of Diver gives implication to the decadency of the character and the end to the Jazz age.

6 Translation for the Names in Literature Works

Generally speaking, there are three ways used by people to translate the names in English literature works. They are transliteration, free translation and the combination of transliteration and free translation.

6.1 Transliteration

The transliteration method is seen the most commonly in general, and is one of the most important translation means in the translation world. The names in a great many of British and American literature works are translated through the transliteration method. Under the premise that the translators get a correct understanding of the original purposes of the writers and with the purpose of giving expression to the original cultural characteristics and images, the names in British and American literature works can be translated through the transliteration method, and also the original character images in the literature works can be used. As a result, the sense of foreign areas as well as the sense of the newness can be enhanced in the minds of the readers.

For example, JIA Baoyu and LIN Daiyu in "A Dream in Red Mansions" can be translated with phonetic letters.

6.2 Free Translation

Free translation for the names is selecting an appropriate translation according to the personality characteristics of the characters in the literature works, or the ends of the destinies, or the themes that are expressed by the writers. Some of the names in English literature works are the real names in history or source from the figures in the legends and mythologies of Bible, ancient Greek and Roman. The implied meanings of these names have exceeded the original meanings of these names. They are used by the writers to extensively refer to or imply a sort of human beings or a sort of behaviors. Moreover, some of these names change from the proper nouns to the common nouns, and even have changes in property.

For example, Hamlet, a character in the drama of Shakespeare, was always in making decisions; Shylock, a character in the drama of Shakespeare, depended on lending out money at an exorbitant rate of interest to support his life; Epicurus, a Greek philosopher, believed that the purpose of life was to pursue happiness, and therefore his name changed into a noun for hedonism, etc.

Therefore, you can make the following processing if meeting the following conditions.

a) I am no Hamlet.
   This means you will never be irresolute about making decisions.
b) His income derived from illicit activities—bookmaking, gambling, shylocking, and questionable union activities.
   This means the economic income sources of a person are the horse racing registration bet, gambling, lending out money at an exorbitant rate of interest as well as some questionable group activities.
c) The old man said that he was an Epicurean for all his life.
   This means the old man said he was a hedonist in this whole life.

6.3 Combination of Transliteration and Free Translation

In English and American literature works, some names are composed of the words which have a certain meaning. The meanings of these words can give reflection to the personality characteristics of the characters in English and American literature works. For this reason, the Chinese translations for these English names sound neither rhyming nor reasonable if these names are translated only through free translation method. Therefore, for translating such a kind of the character names, the combination of transliteration and free translation can be the quite ideal way beyond doubts. This combination of transliteration and free translation can not only make up the shortcoming that the double meaning expressed by equivoque is not sufficient in the transliteration process, and also can promote the translated names through transliteration to keep the voice meanings as well as the implied meanings of the original works.

For example, Mr. YANG Bi translated Rebecca Sharp in "Vanity Fair" (one of the representative works of William Thackeray) to XIA BO, which can be an optimal and valuable translation. In English, Sharp suggests quickness of perception and a certain clever and perhaps tricky resourcefulness. However, in Chinese version, the Chinese word "BO" (transliteration) can endow a woman with the rude, unreasonable and shrewd qualities, and can give expression to the female figures that can make use of any kind of means for the sake of fame and gain. Thus, it is the best word to express Sharp in Chinese. At the same time, XIA BO is quite similar to Sharp in pronunciation, and therefore this word is used by YANG Bi very rightly in translation.

However, "sharp" as an electrical trademark is translated as Xiapu". Pu means popularization and promotion in Chinese, so it gives expression to the true meanings of the trademark "sharp".

7 Conclusion

In Both English and Chinese, the names have a very close association with the literature works. Name is one of the most important tools which are necessarily utilized by human beings in communication, while literature is the clever application of the languages. Names exist not only in communication with the people in the real life and also in the plots of the literature works.

The application of the names in the literature works is one of the means which are used by the writers to give expression to the themes. Name is seen very commonly. The literature works make the names rich in the elegant styles. Literature makes the spiritual world of the people rich.

The well-known names in the literature works make an enhancement to the language expression effects of people. Some names in the literature works change into part of the language of the actual life after entering the daily life of people, and hence make an enhancement to the influencing force of the languages.

References

1. Jin, L.: Names and Social Life. Shaanxi People's Publishing House, Xi'an (1989)
2. Liu, Z.: Different Views on Names. Qilu Press, Jinan (2000)
3. Zhang, H.: Language and Name Culture. China Social Sciences Press, Beijing (2002)
4. Zhang, L.: Foregin Names. China Youth Publishing House, Beijing (1996)

Cognitive Function Analysis on Computer English Metaphor

Cuiping Zhang

Shenyang Ligong University, Shenyang, Liaoning, 110015 China
zhangcuiping@cssci.info

Abstract. The daily life of computer language,exploration and research computer metaphor and its basic structure of cognitive function is analyzed. Through the corpus of analysis, we found that metaphor in the field of computer are widespread and reflected the people to the understanding of the computer, this kind of understanding not only expressed in computer language is also present in the daily man-machine communication..

Keywords: Computer metaphor, Missile, Cognitive law.

1 Introduction

On the basis of the two different languages, English names and Chinese names show highly obvisous diffireces between each other. The diffireces are emboded not only in the structures of the two different languages and also in the cultural meanings, which are deeply relfected from the two sides.

As a mattter of fact, name is not merely one of the symbols which are applied by the people to tell the difference among the diffirent people, and also has its own in-depth foundation in the cultures.

At the same time, it has been known to all that there are signifiance diffirences between English names and Chinese names in pronunciation, strcture, morphology, the origin of a word, semantics and other aspects. These great diffirences between English names and Chinese names properly relfct the diffirent cultural characterics of China and the western countries.

Thererfore, the diffirent cultural meanings which are relfected from English names and Chinese anmes will be discussed by the author in this paper from the following aspects.

2 Relationship betwween Individuals and Families

Name is not solely a communication tool among people and also is one of the symbols of culture in human society. The family names in English language are called as the last name by English-speaking people all the time.

However, the family name in Chinese language is generally the combination of the word "NV" (meaning women in Chinese) with the word "SHENG" (meaning giving birth to a child in Chinese). Therefore, it is not difficult to see that the family name in

Z. Du (Ed.): Proceedings of the 2012 International Conference of MCSA, AISC 191, pp. 339–343.
springerlink.com

Chinese language stands for a profound blood relationship between individuals and families. From the perspective of the structure of names, a highly apparent difference between English names and Chinese names is that the family name is placed at first in a full Chinese name, but is put at last in a full English name.

Importantly, the difference in the sequence of the family names between English language and Chinese language makes the cultural difference between the two sides known well.

In the long run, Chinese people lay a high stress on the tradition that it is necessary for them to value their own family names. The reason why Chinese people keep the family names so importantly is that family name stands for not merely a symbol and also for the in-depth blood relation between families and individuals.

On the contrary, unlike Chinese nations, English-speaking nations do not have such a sense of family. In English-speaking countries, a name is on behalf of only a person. Therefore, English names (given names) representing the values of persons are placed at first in a full English name in general, while the family names are put at last although standing for the blood relation between individuals and families.

Therefore, family names are called as surnames which mean additional names in English language, or the last names which mean the names placed at last. From this perspective, it can be known that the individuals receive higher importance than families in English names.

3 Different Faiths and Ideas

For example, Abraham, David, Adam and John are all selected from the bible actually.

In general, unlike the western people, Chinese nations do not receive a profound influence from the religions. However, there is also a very complete ideological system among Chinese people for a very long time—Confucianism. The connotations of the benevolence, righteousness, propriety, wisdom and honesty qualities which are advocated in Confucianism are well embodied in the names of Chinese people.

For example, Chinese names LI Zhongren, SUN Jieyi, ZHOU Xinli, CAO Ruizhi and WANG Shouxin actually give vivid expression to the connotations of some ideas of Confucianism.

4 Different Ways of Thinking

Just like English culture and Chinese culture, English names and Chinese names have a great number of the clear differences as well. On the whole, Chinese names can be classified into two types: one is composed of a one-word family name and a one-word second name, and the other is composed of a one-word family name and a two-word second name.

On the contrary, there are highly apparent flexibility and uncertainty in the structure of English names.

The comparison between English names and Chinese names in the similarities and diversity of structure makes known the great differences between English-speaking people and Chinese in the ways of thinking under the influences of English and Chinese cultures.

5 Other Cultural Differences

The family names in Chinese language mainly refer to the surnames of the Han-nationality. The Book of Family Names of China mainly includes the family names of the people from the Han-nationality in general. At the same time, the family names in Chinese language are very complex, but can give expression to the cultural characteristics of the people from the Han-nationality in depth, as well as the different expectations of the parents on the futures of their own children.

However, many Chinese words repeatedly appear in the names during a certain period although there is a very wide selection for the parents to name their children. Therefore, the names which are given by parents to their own children change along with the changes of the different times.

5.1 Expecting the Sons Could Grow Healthily and Had a Very Long Life

(1) Some Chinese people expected their sons could grow healthily and had a very long life and hence named their sons as YONG NIAN, HE NIAN and QIAN QIU (meaning a person lives a very long life in Chinese), while other Chinese people named their sons as TIE DAN ER, GOU ER, DA HAN and SHI TOU (meaning the solid substantial things in Chinese) to hope their sons can grow healthily and have a very long life.

(2) Chinese people had a low expectation on their own daughters in general, only hoping the girls can be as pretty as flower and jade, and hence made use of the names of beautiful birds and flowers to call their daughters, such as JU HUA (chrysanthemum), QIU YUE (white moon in autumn), XIANG LIAN (lotus flower), YUE E (Chinese goddess in the moon) as well as YU DIE (beautiful butterfly).

5.2 Using the Words with Characteristics of the Times to Name the Children

(1) Chinese people often used the words "ZHONG", "HUA", "YING", "MING" and "GUO" as names or part of names during the early period of the People's Republic of China, for the purpose of expressing that new China was founded and the people became the masters of the country.

(2) Chinese people often chose "HONG", "WEN", "KE", "JUN" and "WEI" as names during the Cultural Revolution period (1966-1976), which had a very close connection with the political movement at that time.

(3) Chinese people can name their children more colorfully since China's reform and opening up policy, and especially use the transliterations of the foreign names to name the boys as DA WEI (David), YUE HAN (John) and TANG MU (Tom) and the girls as MA LI (Mary), ZHI ZI (Japanese name) and SHA LI (Lisa), which give expression to the distinctive characteristics of the different times.

5.3 Using the Names of Symbolized Things to Name Girls

Chinese people name the girls with the names of flowers, birds, and weak animals, which are rarely used to call the boys, because they think that it will damage the male images if the weak animals are used to name them.

5.4 Using the God Names to Name Boys

Gods in the minds of people are generally the strong men, so the names related to religions and gods are the male names most, and few of them are female names.

5.5 Using the Meaningful Words to Name Boys

People are also used to make use of the fight, weapons or the related words as the names of boys.

People are also used to make use of the words which express the braveness, scholarly attainment, authorities and reputation as the names of boys.

6 Associative Meanings of English and Chinese Men and Women Names

In English language and Chinese language, a great number of names have owned a certain inherent associative meaning due to the social and history environment as well as the accumulation of the cultural tradition. In other words, many names have marked with obvious bad and good meanings.

For example, CHANG'E, Chinese goddess in the moon, is good at elegantly dancing with long-sleeves in heaven, making the vicissitudes of life not to exist in the world of human beings. Therefore, Chinese parents are willing to name their daughters with the combination of CHANG or E.

At the same time, Chinese parents are willing to use the words in the names of the Four Great Beauties (DIAO Chan, SI Shi, YANG Yuhuai and WANG Zhaojun). However, few of people are willing to use LIN Dayu (a beauty in A Dream in Red Mansions) as the names of daughters, because this name is easy to make people think of the sentimentality, unfinished life and misfortunes of the character.

Also, the male names such as WU Dalang (short and fat body and ugly appearance), XI Menqing (gambling and drinking playboy), and CHEN Shimei (ungratefulness and loving the new and loathing the old) are with a bad meaning and sound very disgusting. Therefore, it is nonsense to use these names.

However, RAO, SHUN, KANG Youwei, LIANG Qichao and ZENG Guofan are the names with a good meaning. For this reason, a great number of Chinese parents like to use these names to call their own sons.

English names endow people with a series of associations at the same time. For example, Methuselah can make people think of the oldest person (969 years old) in bible; Alexander, protector of human beings, can make people think of the Alexander the Great (356-323B.C). Macedonia kings and people liked using these kinds of names to name their sons.

However, Judas (who sells Jesus) and Adolf and Mussolini (who launched a crazy massacre on Jew and Jewess) still make people hair stands now. Thus, these names have been the taboos in social life, and nobody uses these names to name their own sons. There are also a great number of the female names, which can make people think of the beautiful things, such as Helen and Helena. These kinds of names can make people think of the ancient Greek goodness Angela in Trojan War as well as the beautiful angels. Thus, parents are highly willing to use these pleasing names to call their own daughters.

However, Padora (who was beautiful but hypocritical in ancient Greek mythologies and also brought the disasters to human beings) and Methuda (who was a very ugly and extremely cruel demon in ancient Greek mythologies) will never be used by people.

7 Conclusion

Name is not only a linguistic phenomenon, and also is a cultural phenomenon. In the final analysis, name is cultural phenomenon, and can give reflection to the profound cultural connotation and also has a very close relationship with human behaviors, social psychology and cultural traditions.

The diversity of English name forms gives expression to the flexibility of the ways of thinking of human beings.

However, in the mean time, the normalization of Chinese name forms properly reflects the same psychology of human beings in the behavior conducting process.

In both origin and structure of names, family name is always put at last in a full English personal name and suggests the cultural psychology that the individual value is in the leading position, while family name which represents a profound blood relation in a full Chinese name is always placed at first and shows the strong sense of family in Chinese people.

The different beliefs and cultural consciousness (Christianity and Confucianism) in two different cultures have the profound influences on English names and Chinese names respectively. English names give expression to the worship of god, while Chinese names are a kind of self-improvement way under the influence of Confucianism.

From the above comparative study on English names and Chinese names at the cultural level, it can be seen that these two different name systems can be in sharp contrast in many aspects and have the great differences between each other.

References

1. Jin, L.: Names and Social Life. Shaanxi People's Publishing House, Xi'an (1989)
2. Zhang, H.: Language and Name Culture. China Social Sciences Press, Beijing (2002)
3. Zhang, L.: Foregin Names. China Youth Publishing House, Beijing (1996)

ASP.net-Based Boutique Web Design and Implementation of Curriculum

Wang Lifen

Jilin Agricultural Science and Technology College Jilin 132101 China

Abstract. This website is developed with ASP.net programming language SQL database, based on B / S structure. The test and analysis shows that the system works stable and reliable with a certain practical value.

Keywords: Quality courses, website, database, resources.

1 The Current Situation of Classic Course

The school can push forward the innovation of education and deepen its educational reform by through the classic course and its website construction[1], widen the use of modern information technique in teaching, share teaching resources, comprehensively improve the teaching quality and level of education. But through the investigation of the network and contrast, we discover that almost all classic courses websites are without optimization, and don't have good user experiences, nor have optimization of the website rankings. The classic course website like this can not keep a customer, and the flux is less.

2 Implementation Methods

In order to solve the problem of most of websites without optimization, this design will introduces two thoughts of optimization, one thought is SEO, the other is UEO.

The development of website adopted a more popular technique and tool, the webpage design adopt the ASP.net(the adoption vb.net language) design[2], as well as the tool adopted Dream weaver, the database adopted the system of Microsoft SQL Server of .The interaction between teachers and students of the website adopt the database technique and have the ability to manage and protect the system perfectly. The main part of the system adopt B/S structure, and the client can use it through theIE browser and the IIS for server.

3 Systemic Function

The aim of this system design is to make teaching more convenient; students can study at home or dormitory independently, which in turn can improve students' ability of self-study, and save time of both. This system is an on-line teaching website, which

Z. Du (Ed.): Proceedings of the 2012 International Conference of MCSA, AISC 191, pp. 345–349.
springerlink.com

can update in time, easy to manage and maintain. Therefore, the system's function has several modules.

4 The Design Concept of Database Model

This system adopt "Entity Relationship Model"(ER model) to describe the structure and semantics of database to carry on the first abstraction of the real world. ER model directly abstract the entity type and entity relationship out of real world, then mean data model with the ER diagram.

4.1 User Information Entity E-R Chart

Fig. 1. User information entity E-R chart

4.2 Curriculum Information Entity E-R Chart

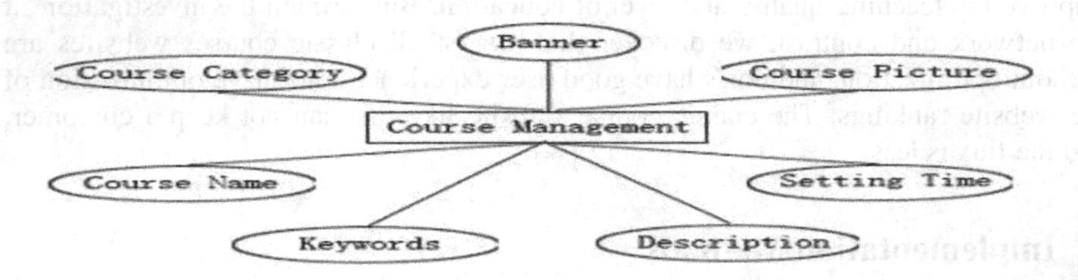

Fig. 2. Curriculum information entity E-R chart

4.3 Material Management Entities E-R Chart

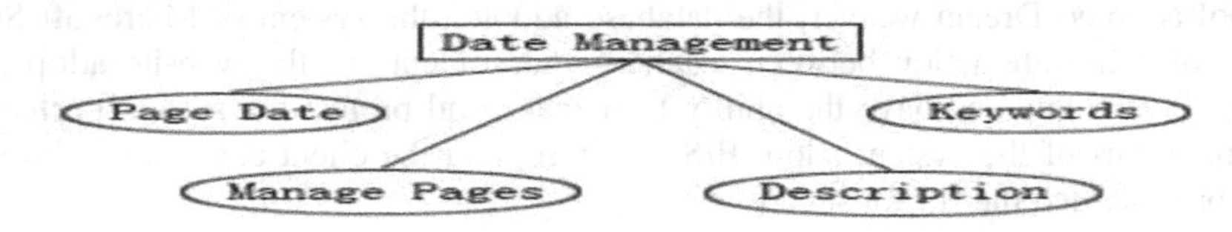

Fig. 3. Material management entities E-R chart

5 System Design

5.1 Module Chart

The system module chart, mainly divided the outline of the module of the classic website, including curriculum management module, material management module and the others. The module chart is showed in figure 4.

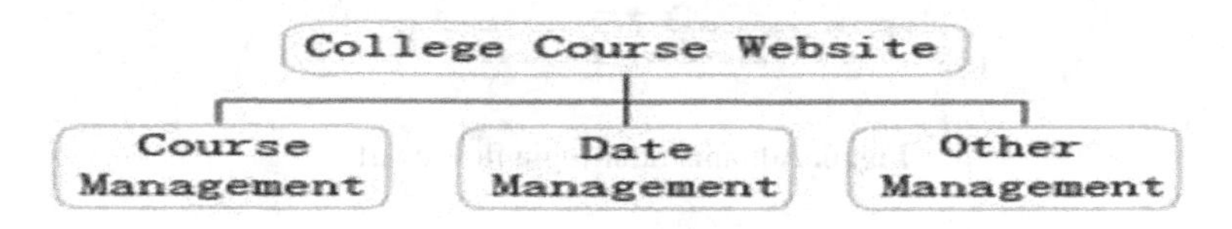

Fig. 4. Module chart

5.2 Basic Flowcharts

As the classic course website is a study platform that students can study independently, a communication platform that school can show the achievement of curriculum reformation. The classic course website ought to contain display type column, such as, declaration of material, achievements and awards, as well as teaching videos, test papers and so on. The system flow chart is showed in figure 5:

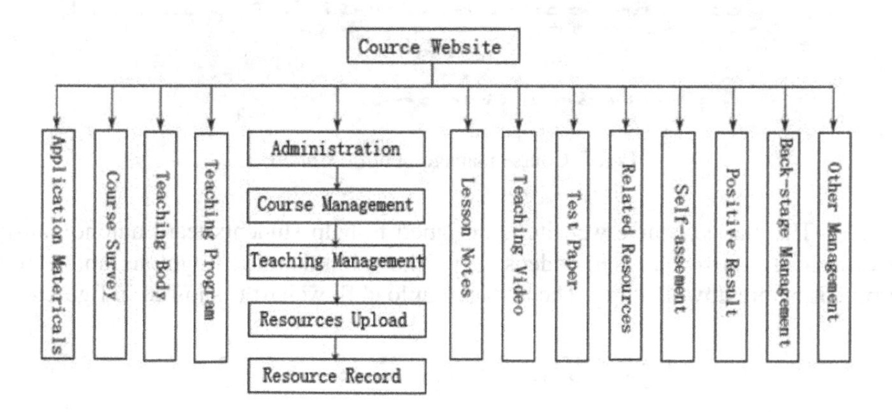

Fig. 5. System flow chart

(1) The backstage management is the main part of classic course website, and need to login first, then carry on a management through a backstage. The administrator login flow chart is showed in figure 6.

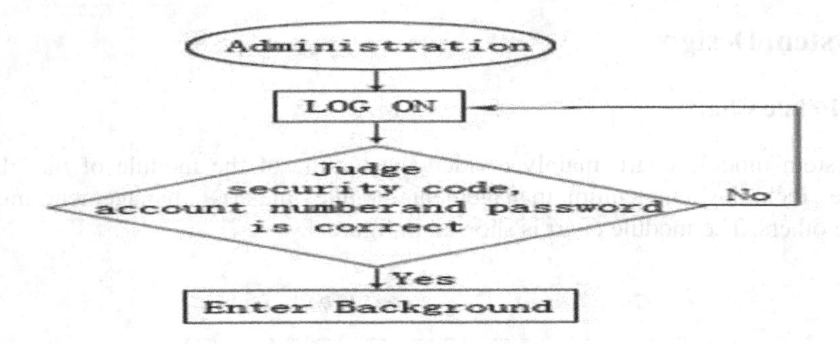

Fig. 6. Administrator login flow chart

(2) After entering into the backstage, users can work on a course management, including add, delete, edit , the course management flow chart is showed in figure 7.

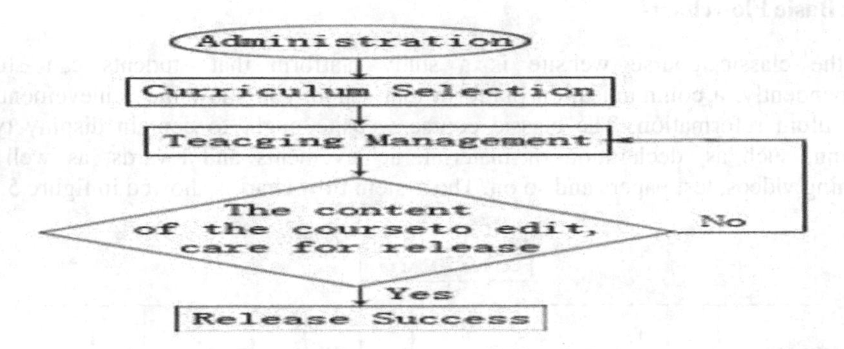

Fig. 7. Course management flow chart

(3) The classic course website is designed to help students learn autonomously to download courseware and videos, etc. The administrator ought to upload the resources before downloading. The resource upload flowchart is showed in figure 8.

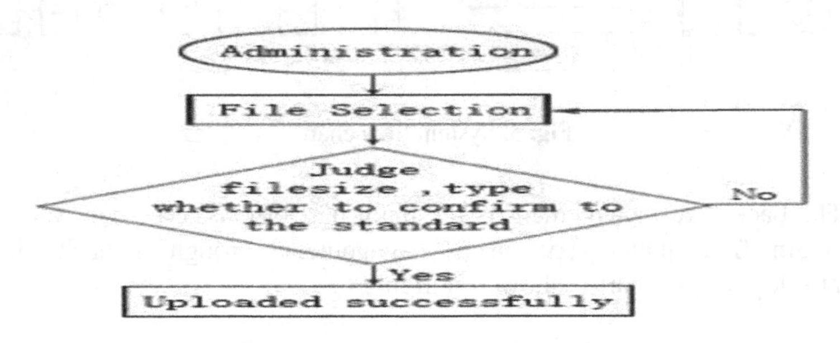

Fig. 8. Resource upload flowchart

6.1 System Homepage

The navigation adopts JavaScript with div + CSS, which is more vivid and artistically decorated.

6.2 Teaching Video Page

To provide students with the teaching videos help students to study more valid for they can study their weak parts on the basis of themselves repeatedly. Teaching videos provide students with real models, which can help them study independently .

6.3 Login Function Realization

Login pages are used to enter the management interface, where users can manage the website. The checking page is to deal with the submitted data and find out whether a corresponding user is checking the password.

6.4 Course Interface

The administrator can directly get into to manage interface after registering, the interface can display all information, including the management information. The interface is designed by applying Dream waver 8.0, the main course approaches with

6.5 Add Editing Curriculum Page

In this page, users can click "edit" or "add" to enter the page to modify, including name, grade and so on.

6.6 Detailed Material Edit Page

This page provides the editor's issued page, user can choose the edited course and edit, In this page[4], users can choose the title, orders, key words, description, section, content, etc for editing. This page applies strong backstage for edit and process system, which is more convenient.

This system is designed with Dreamweaver8.0 use, in order that represents a more magnificent interface, the backstage adopt ASP.NET, which is in harmony with Dreamweaver8.0, making the design with high proficiency.

References

[1] Yan, Y.: The design of teaching website based on ASP.net. In: CNKI, vol. (05) (2007)
[2] Sheng, W., Wang, H.: An analysis of applying ASP.net technology to establish network teaching system. Techno Plaza 12, 237–238 (2008)
[3] Xu, Y.: An analysis of the combination of Flash and ASP,net of website design. Computer Knowledge and Technology (03) (2009)
[4] Ferraiolo, D., Cugini, J., Kuhn, D.R.: Role Based Acces Control:Features and Motivations. In: Annual Computer Security APPlieations Conference, IEEE CSP (1995)

The Research on Objectives Design of Vocational CNC Machining Process

Shen Weihua

Sichuan Engineering Technical College, Sichuan, China

Abstract. According to the teaching objectives of vocational CNC machining process design background, performance analysis to determine the problem and determine the design criteria of effective teaching objectives by clarifying teaching objectives, the goal of teaching is also constant feedback to refine and improve.

Keywords: Vocational CNC machining process teaching objectives performance analysis to clarify the teaching objectives.

1 Introduction

Vocational digital processing technology course is vocational NC students a very important backbone of professional courses, students would be able to master this course, to their future can be better engaged in the profession, such as operator of CNC machine tools, machining workshop, craft workers and related occupations.

Teaching goal setting is a teaching design the most critical link. If the teaching objectives determined unreasonable, no matter how good teaching may be unable to meet the real needs of students and businesses. There is no accurate teaching objectives will lead to non-existent demand to design teaching.

2 Vocational Digital Processing Technology Courses Teaching the Target Design Background

Through the collection to the number of areas, dozens of companies thousands of CNC professionals and job demand information, statistical analysis, is mainly engaged in the post of CNC graduates: Operating Status craft positions, management positions and other positions four broad categories such as jobs, the first post positioned on the front line of production operations shown as Table 1.

With the school curriculum content divided according to the above questionnaire, the ability of the NC process completion of the course students should meet the basic requirements should be as follows: In order to achieve the goal of production responsibilities, in accordance with the requirements of the pattern and process technology, the use of CNC machine tools and process equipment, the finished product parts of the processing task. This requires that the process of teaching employment should focus on practical application in the teaching process, not to overemphasize the theoretical argument derived as well as a system of rigorous and systematic teaching content cannot be subject.

Z. Du (Ed.): Proceedings of the 2012 International Conference of MCSA, AISC 191, pp. 351–356.
springerlink.com

Table 1. Job setting of the graduates, competency requirements and distribution of research

Survey Survey Project	Post	Basic capacity requirements	Job distribution (%)
Expert survey	Operating positions	In order to achieve the goal of production responsibilities, in accordance with the requirements of the pattern and process technology, the use of CNC machine tools and process equipment, to complete the processing tasks of the product components; assume the operation of CNC machine tools, the confirmation process, the preparation of the processing procedures and responsibilities; to complete some of the more simpleCNC Machining: turning, drilling, three-axis milling and simple processing of two linkage programming; can call in advance has been entered into the program within the machine tool control system, and use of detection equipment to ensure the quality of machined parts	73.2%
	Craft positions	According to the processing requirements of the parts by hand or use computer-aided manufacturing software for CNC machining technology and programming staff. Analysis of customer demand collaborative product development engineer, responsible for new product development process in the formulation of the process; in accordance with process requirements and processes responsible for the more complex parts CNC machining program to write, debug, and sample test cut; responsible for new product tooling design selection and tool life management; responsible for the optimization of processing procedures to improve product quality and production efficiency; responsible for the operation of the CNC operator, theoretical training and guidance; responsible for organizing the implementation of technological adaptation and innovation	7%
	Management positions	In order to achieve the goal of production responsibilities, in accordance with the drawings and process documentation requirements, responsible for production planning and arrangements, organization, management and work instructions; responsible for the daily management of the production team.	10.1%
	Other jobs	Primarily for the sale of machinery-related products and technical services. Knowledge of CNC products, process characteristics and product costing, common faults and maintenance of CNC machine tools. With a good marketing ability	9.7%

3 Teaching Goal of Performance Analysis CNC Machining Process

The goal of teaching is the most critical part of the instructional design process, it is to determine students' completion of teaching, and I hope they can do? This does what must be linked and career orientation. If the teaching objectives vague and unreasonable, and will be teaching to deviate from the designer's intent. To determine the method of teaching objectives are four basic ways: subject matter experts France, content outline law, executive order, and law and performance method.

By subject matter experts (SMEs) to determine the teaching methods, and often contain about "understanding" and "master" on the teaching content requirements, which is a kind of built on assumptions based on the "Students need to learn the knowledge of subject matter experts above, which emphasizes the information passed on to students by teachers in the teaching process will, this approach to determine the teaching objectives may be subject matter experts based on their own knowledge and assessment, consider the secondary school in this subject area to run counter to the actual demand.

The second common method to determine the teaching objectives for the content outline method. This method is that: there is convincing evidence that the existence of performance problems, and often the circumstances leading to performance problems because the learners have not mastered correctly, a sufficient amount of knowledge. The danger faced by this approach is determined by the teaching objectives are many and complex, key and vocational needs little or nothing to do is not sufficient to resolve the issues.

Table 2. Main point of Robinson (1995) Performance diagram

Performance analysis of the problem	Performance analysis of the answer
What is the problem of the initially proposed?	Mainly located in the NC graduates jobs, teaching objectives, design must meet the basic ability to CNC production line operators have, taking into account the needs of the second post.
Two problems with the goals of the organization?	It is relates to NC school training professionals to meet the needs of the community.
To achieve organizational goals set up the operation target?	Is. 1.The teaching objectives of the course: one familiar with the concept of machining accuracy and improve the ways and means of processing quality 2. to grasp the meaning and concept of the basic terms of the machining process; 3.Able to correctly analyze and judgment moderately complex parts drawings of the structure of technology, processing technology; 4.Familiar parts of a typical surface processing methods, a typical part of the process and process characteristics; 5. development of moderately complex parts machining process planning; 6. Have basic machining process implementation capacity; 7. With the initial capacity of the production technology.
4 Operational needs?	Need to re-design of digital processing technology course teaching objectives, the operator positions the basic competency requirements match
5 Job performances achieved?	Not fully achieved. Students after the subsequent Practice and enterprise feedback information back to that ideal and the real state of the gap.
6 Have a solution to meet performance issues?	Professional positions to re-determine the teaching objectives, and then determine the goal of teaching the child, and then choose to match the course content, to enable the appropriate teaching strategies.

The third method is the executive order law. The danger is that managers with the help of his authority to start teaching this method may be due to the managers of the various cognitive limitations of the discipline and often lost their original direction.

The fourth method is Performance Technology Act. The teaching goal of using this method to determine the main issues and opportunities in response to tissue. The purpose of performance analysis is to seek to identify problems and seek solutions to the problem.

An ideal way to determine the teaching objectives with the performance analysis process to determine the teaching out where the problem statement, and then teaching objectives.

4 To Clarify Teaching Objectives of the Course of the CNC Machining Process

Only a teaching goal of the teaching modules in the original teaching of the CNC machining process target is divided into two shown in Figure 1 (sub-goals).

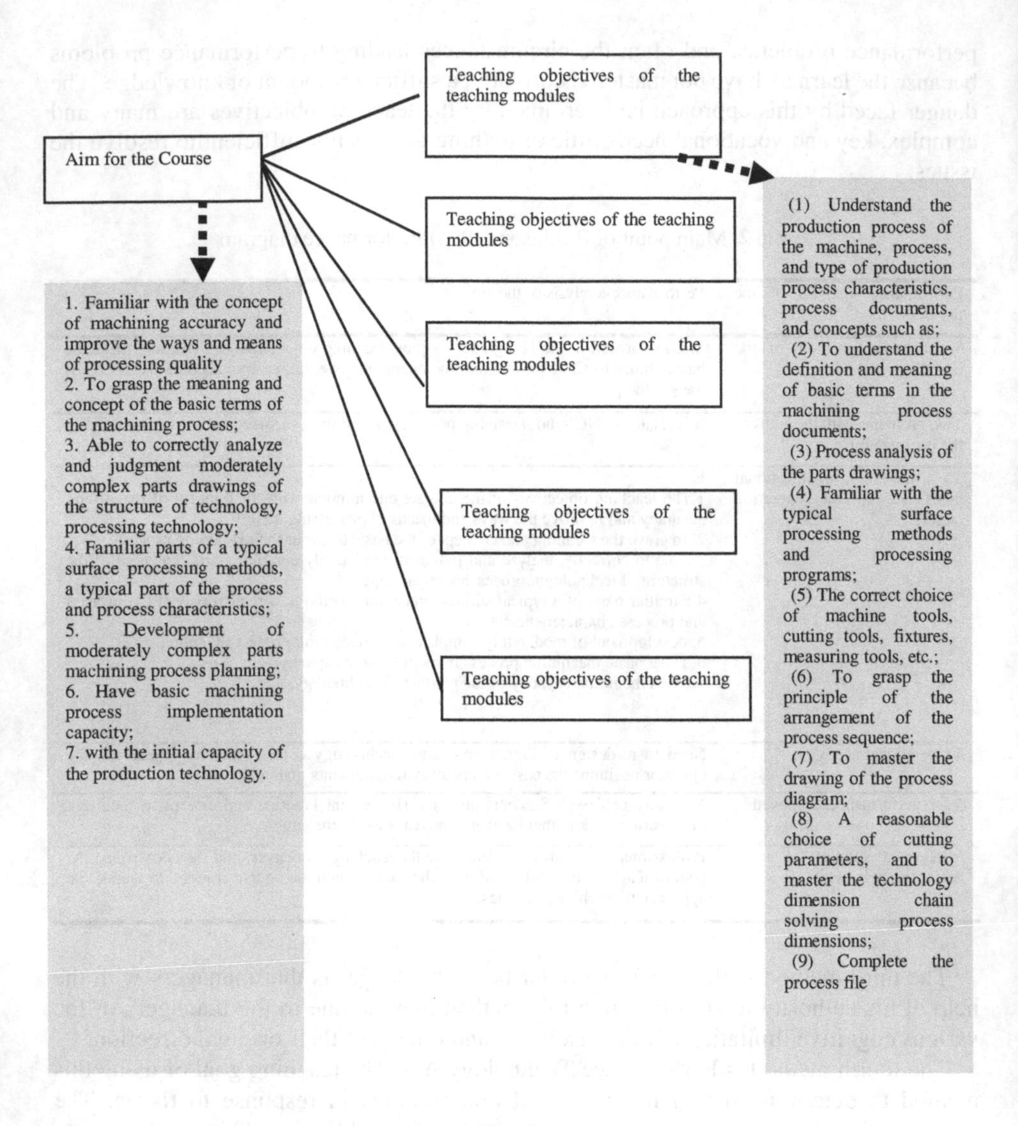

Fig. 1.

On the one hand, can be seen from Figure 1 in the formulation of teaching objectives, whether it is the overall goal or sub-goals (teaching unit a description of the target) "Understanding," "familiar," "master" terms such as description, these expressions are a form of language, the abstract description of the requirements for student behaviour, teaching those who do not really know what they mean, because there is no indicator description if the students reach this goal, he will be able to what to do. Instructional designers assume that, as long as the successful completion of this course, students will be able to show they have reached the target, but if the goal is clear, the success of behaviour makes it difficult to judge.

Table 3.

Teaching objectives of the first teaching unit	A teaching goal of the teaching unit in the modified
(1) understand the production process of the machine, process, type of production process characteristics, process documents, and concepts	(1) different parts of a given production batch machining process card, students from a given process card information as defined in the concept of the type of production, comparative judgments of various types of production process characteristics. Understanding of process from the process issued cards, processes card file.
(2) Understand the definition and meaning of basic terms in the machining process documents;	(2) The student should be able to accurately from the process in the processing of parts to determine process composition; process has to write the process route. Write accurately install the contents of a given process, according to a given process content to determine the composition of the working steps; accuracy rate of 85%
(3) The parts drawings process analysis;	(3) Students from the machined surface characteristics of the machined surface of the relevant technical requirements, other the three parts diagram.
(4) Be familiar with the typical surface processing methods and processing programs;	(4) Students according to the given parts diagram, with the typical surface processing method selection table according to the choice of economic machining accuracy to parts machined surface processing methods.
(5) The correct choice of machine tools, cutting tools, fixtures, measuring tools, etc.;	(5) Students according to the structure and shape of the work piece machining accuracy to select the corresponding size of the machining, the machining accuracy.
(6) To grasp the principle of the arrangement of the process sequence;	(6) students according to the given process card, summary of the chronological arrangement of machining processes should follow the principle; annealing, normalizing, quenching and tempering, hardening, carburizing, nit riding, the aging treatment stage in the processing the location of points; crossed check, go to the general arrangement of the burrs and other auxiliary processes.
(7) To master the drawing of the process diagram;	(7) Students, according to a process in the specified process card, as required by the draw process diagrams of the process diagram drawn in the process card process diagram area.
(8) Can be a reasonable choice of cutting parameters, and to master the technology dimension chain solving process dimensions;	(8) Students to deal with the solving of the linear dimension chain and planar dimension chain processing often met. (Cutting the amount of a reasonable choice of arrangement here is inappropriate, should be classified to the back of CNC lathes, CNC boring and milling requirements of the teaching modules.)
(9) Will complete the process file	(9) Students according to a given part of the process to fill out the card process and process card.
Lack of locating datum to select teaching objectives and requirements	(10) students to process card based on a given part or process card determine the crude, refined benchmark; and then use simple parts of the principle of locating datum is a reasonable choice of the positioning surface of the parts used in each installation.

If the overall goal can be expressed, use generous including words, then the sub-goals should be appropriate, clear and specific objectives to elaborate.

5 Implementation of Conclusions

Through the teaching of classes in recent years, the teaching is effective; the student process design capability has been greatly improved. The goal of teaching is also constant feedback has been refined and improved.

References

[1] School-related teaching materials
[2] Dick waiting, W., Weiguo, P.: Systematic instructional design, 6th edn. East China Normal University Press (January 10, 2007)
[3] Tian, W.C., Cao, Y.R.: HFSS Simulation of Reconfigurable Multi-band Antenna Bands Based on RF Switch. IEIT Journal of Adaptive & Dynamic Computing (1), 1–4 (2012), DOI=10.5813/www.ieit-web.org/IJADC/2012.1.1
[4] Zhao, Z.L., Liu, B., Li, W.: Image Classification Based on Extreme Learning Machine. IEIT Journal of Adaptive & Dynamic Computing (1), 5–11 (2012), DOI=10.5813/www.ieit-web.org/IJADC/2012.1.2

The Research on Online Examination System of PE Theory Courses

Xing-dong Yang

North China Electric Power University, China
xingdongyang1@yeah.net

Abstract. With the continuous development of network technology, teaching and examinations have changed Significantly. For PE courses, online exam is a hot topic of the reform of current education mode and development trends. This article is in view of the research on online examination system of PE theory course, using the NET and SQL for research and development, firstly, describe the demand analysis of the system, secondly design the details of the exam system, finally, how to implement the system. The research and development of online examina-tion will improve teaching quality and perfect the education system.

Keywords: PE theory, exam, network.

1 Section Heading

For a long time, people's cognition in PE courses is how to exercise physical quality and how to master the skills of action, and neglect the PE theory. With the constant deepening of China's education reform, in the teaching link, the traditional education mode of teacher repeated demonstration to deepen the impression of students was broken, and making full use of network, video and multimedia to guide students, but there is a blank in the exam of PE theory courses. The application of PE theory courses online examination system can improve the test paper production process, and at the same time reduce the use of paper, make the re-sources sharing among the schools convenient, and evaluate the student more scientificly, fairly and objectively.

2 Demand Analysis of the System

2.1 The User Operation Flow of the Exam System

In the development process of computer software system, we must consider the is-sues in the actual user's position, so we must have full understanding and research to the actual operation of the user's.

On online examination system of PE theory courses have to accord with the teaching objectives, take full account of the keys and difficulties of teaching out-line, arouse enthusiasm of the students, and strengthen the communication and interaction between students and teachers. Throughout the system, mainly related to three

Z. Du (Ed.): Proceedings of the 2012 International Conference of MCSA, AISC 191, pp. 357–362.
springerlink.com

categories, are: students, teachers and administrators. Students take the exam in the network front desk, they use user names and passwords the administrators given to them to login, the system will distribute questions to the student automatically, and start the countdown, make the students to complete the test within the stipulated time, and submit the answers to the server. According to the form of the exam, the questions are divided into objective questions and subjective questions, papers are corrected by the system and teachers. Final results are aggregated by the system, students can check the scores on the system, so as to achieve the paperless office, and break the time and space constraints of traditional test. Specific students process is shown in Figure 1:

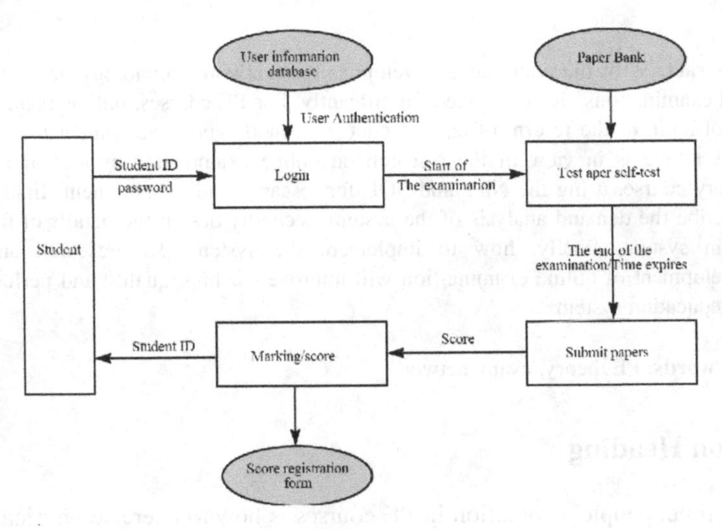

Fig. 1. Operation flowchart of student examination system

For teachers, they refine and revise the test questions, and evaluate the subjective questions of the papers; the administrators ensure that the user accounts of teachers and students are safe and reliable.

2.2 The Functional Requirements

Currently, functional requirements of all schools on online examination system of PE theory classes mainly reflected in the following areas. On administrators, are subject management, scores queries, user management, system management, pa-per management, course management, class management, audit papers; On students, the main functions are the test pages and information modification; On teachers, main function are question bank management, papers production, questionnaires management, information modification and so on. For some schools, the boundaries between teachers and administrators are not very obvious, teachers paly the role of system administrator in a certain period of time, this requires schools to carry out the specific arrangements according to their actual situation. For the design of some modules, such as results inquiry, after landing, the students can view their own scores. Specific functional requirements are shown in Figure 2:

2.3 Performance and Operating Environment Demands

After the system putting into use, its performance must meet the principle: ease of maintenance, practical, safe, reliable and beneficial to expansion. On the operating environment, I use Visual Studio 2005 as a development tool, the database is Microsoft SQL Server 2005, server operating system is Windows 2003.

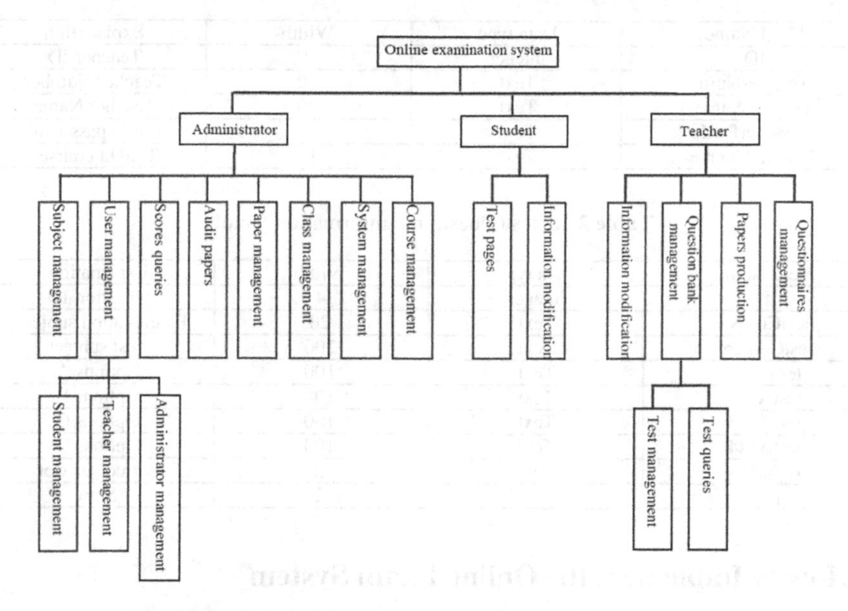

Fig. 2. The functional requirements diagram of system

3 The Design of Online Examination System

3.1 Design of System Architecture

In the current design of system architecture, the most popular is to use a layered structure, it can decompose complex problems into several smaller problems to be resolved, can maximize the utilization of the codes. For each module, as long as its interface definition is clear, for the specific content of its upper layer or lower layer module ,is not the concern of the module.

3.2 Design of Database

Database is the foundation of the entire system, all operations are around the data-base to expand, the design of the database is the top priority of the entire system. According to the demand analysis, the design of the database include the following:

(1) Management information: role number, role name, permission number, authority name, the authority value, menu number and menu name.

(2) Basic categories of information: teacher information (teacher ID, name, type, and taught courses, etc.), student information (student ID, name and test subjects, etc.),

information of exam papers (examinations nature, type of questions and exam time) and information of test scores.

Here, we give the descriptions of the actual field of the table, which are specifically shown in Table 1 and Table 2 below:

Table 1. tb_Teacher (Teacher imformation table)

Field Name	Data type	Width	Explanation
ID	Integer	4	Teacher ID
TeacherNum	Text	20	Teacher Number
TeacherName	Text	20	Teacher Name
TeacherPwd	Text	10	Login password
TeacherCourse	Text	10	Taught courses

Table 2. tb_test(Questions imformation table)

Field Name	Data type	Width	Explanation
ID	Integer	4	Question ID
testCourse	Text	20	Examination subjects
testContent	Text	200	Test subject
testAns1	Text	100	Options A
testAns2	Text	100	Options B
testAns3	Text	100	Options C
testAns4	Text	100	Options D
rightAns	Text	2	correct answer
pub	Text	2	Sign

4 How to Implement the Online Exam System

4.1 How to Implement Test Management

After teachers Landing system through their own accounts, they can look over and modify the questions they input previously at any time, and on the basis of the original questions, they can add and delete questions to further improve the questions. The entire design and implementation are around the database, through the retrieval of tables in the database, we can search in according to the classification of subjects, the results will be bound to the GridView control, and be displayed. It's necessary to affirm that: Although the questions of the subjects are mostly given by classroom teachers, in the query, teachers can not remember all of the topics, they must use fuzzy inquiry.

The core codes is as follows:

```
protected void TestAdd_Click(object sender, ImageClickEventArgs e)
    {
        ZWL.BLL.TestPaper Model = new ZWL.BLL. TestPaper ();
        Model. testCourse = this.TextBox1.Text;
        Model. testContent = this.TxtContent.Text;
        Model. testAns1= this.TextBox2.Text;
Model. testAns2= this.TextBox3.Text;
Model. testAns3= this.TextBox4.Text;
Model. testAns4= this.TextBox5.Text;
```

```
    Model. rightAns = this.TextBox6.Text;
    Model. rightAns = this.TxtPub.Text;
        Model.Add();
        ZWL.Common.MessageBox.ShowAndRedirect(this, " add questions
Successfully ! ", "MyTest.aspx");
    }
```

4.2 How to Implement the Management of Exam Results

In the design management of this module, after classroom teachers submitting results, they can always login the system to view the scores of the class they taught, but they can not modify or delete the scores, If you must modify the results have been submitted, you have to send application to higher authorities or network managers ,In the case of permission, network managers will modify and delete the scores, the main purpose is to ensure the safety and reliability of student achievement. So after the teachers inputting scores of student, it's necessary to check the results very carefully, after confirmation saving and submitting, so that they can avoid to make errors, increase their amount of labor and cause unnecessary trouble to their work.

The core code is as follows:

```
protected void ImageButton2_Click(object sender, ImageClickEventArgs e)
    {
        string IDList = "0";
        for (int i = 0; i < GVData.Rows.Count; i++)
        {
            Label LabVis = (Label)GVData.Rows[i].FindControl("LabVisible");
            IDList = IDList + "," + LabVis.Text.ToString();
        }
        Hashtable MyTable = new Hashtable();
        MyTable.Add("TitleStr", " Course Name");
        MyTable.Add("UserName", " classroom teachers");
        MyTable.Add("CanLookUser", " Instructor class");
        MyTable.Add("TimeStr","UpdateTime");
ZWL.Common.DataToExcel.GridViewToExcel(ZWL.DBUtility.DbHelperSQL.GetD
ataSet("select TitleStr,UserName,CanLookUser,TimeStr from ERPWorkPlan where
ID in (" + IDList + ") order by ID desc"), MyTable, "Excel Statements");
    }
```

5 Summary

This paper on the research of online examination system of PE theory courses, is due to space limitations, can not show all the description of the database table and the specific implementation of the system module completely, I hope readers with interest can improve it.

References

[1] Xu Zeng chun, Hu Ping.The resarch on network test system based on B / S three-tier structure, The computer and modern. 23 (12): 50- 51,2003

[2] Liu, X.: The research on online examination system. Science and Technology Information of HeiLongjiang Province 22(10), 176–178 (2008)

[3] Wu, Y.: Design and research on test system in open and dynamic network environment. University of Science and Technology, Beijing (2002)

[4] Chen, Z., Zhang, J.: Design and realization of online examination system. Journal of Henan University (Natural Science) (3), 69–71 (2003)

[5] Li, X.: Design and Application of Sport and Health Management System in Colleges. Journal of Ningxia University(Natural Science Edition) 24(4), 371–374 (2003)

[6] He, L.: System Analysis and Design of Sport Performance Management. Modern Business Terade Industry 4, 231–232 (2012)

[7] Shen, M.: Design and realization of p.E. Teaching management system in Colleges. Journal of Wuhan Institute of Physical Education 38(3), 167–169 (2004)

Regression Method Based on SVM Classification and Its Application in Influence Prediction of a Liberalism Case in International Law

Gang Wu[1] and Jinyan Chenggeng[2]

[1] Tianjin Radio and TV University, Lecturer, Yingshui Road. 1, Nankai District 300191 Tianjin, China
[2] Tianjin Foreign Studies University, Lecturer, Xuefu Road. 60, Dagang District 300270 Tianjin, China
LNCS@Springer.com

Abstract. For non-linear problem, the forecasting technique of pre—classification and later regression was proposed, based on the classification approach of Support Vector Machine(SVM). Compared with other forecasting techniques and their forecasting results, this algorithm outperforms others in influence prediction of a liberalism case in international law.

Keywords: Support Vector Machine (SVM), Classification, Regression, Liberalism, influence, international law.

1 Introduction

Modern liberal worldview is influenced by David Hume and Jean-Jacques Rousseau. A more important architect of liberalism Immanuel Kant advocated a federation of democracies and global citizenship to safeguard international peace.[1] Contrary to realism, liberalists believe that man is essentially "good" although subset views may vary in many theoretical aspects. What makes a man do sinful things is bad institutions rather than evil human nature. As to their perspectives in international politics liberals maintain that war is not inevitable. States are more likely to collaborate to prevent conflict and war. Unlike realists who insist the anarchic nature of the international community, liberal international relations scholars are looking for institutional arrangement on the international plane which is expected to eradicate the international anarchy eventually. Likewise, international law and institutions, liberals and neoliberals believe, play a key role in the peaceful settlement of international disputes and prevention of war.

2 Liberalism and Neoliberalism in International Relations

Liberalists after World War One are sometimes referred as idealists. Although liberal scholars share similar attitude towards realists, they differ from each other for subset views of international relations. Debating on how to reform the international political

Z. Du (Ed.): Proceedings of the 2012 International Conference of MCSA, AISC 191, pp. 363–368.
springerlink.com

system and how institutions help states prevent war and foster stability, they fell into one of three groups. The first group of scholars directly confronts the issue of how to prevent war by advocating the creation of international institutions to end the anarchic and war-prone balance of power system. The collective security adopts the alliance arrangement to protect weak states from being attacked by potential antagonists. Unlike realists, collective scholars believe states should reject the use of force, and they join together to suppress the temptation of aggressors by deterrence of overwhelming force[2].

The second group emphasizes the use of international legal process to settle disputes between states peacefully. They prefer the role of mediation and arbitration by a third party[1]. Their effort can be found in the creation of the Permanent Court of International Justice in 1921 and passing of Kellogg-Briand Pact outlawing any war.

The third group concentrates on overall disarmament by states. Supporters of this group rely on the notion that disarmament is the ultimate way to prevent war. They were working hard on the naval disarmament conferences in the 1920s.

Although Wilsonian idealism was criticized for being impracticable and failing to come into reality it has provided much fuel to the liberal international political literature. And it has been revived since the end of the Cold War in the 1990s. Liberal idealists now find Wilsonian idealism applicable and appropriate in the post-Cold War world. Wilson's ideas and ideals now appear less unrealistic and more compelling. Although many liberal IR scholars may vary in terming the newly revived Wilsonian idealism (Fukuyama's "neo-Wilsonian" idealism, Kober's "idealpolitik", Nye's "neoliberalsim", and Grieco's "neoliberal institutionalism"[3]), they agree that Wilsonian idealism is key to the creation of a "transformed international system"[3]. They are thus categorized as neoliberal scholars. Their central theme is the prospects for progress, peace and prosperity. The core components of neoliberal literature include:

First of all, democracy promotes peace. Neoliberals justify this cause by stating that democratic states never engaged war on each other and it is "as close as anything we have to an empirical law in international relations"[4]. And neoliberals are certain that Wilsonian idealism is rooted in Kantian liberalism by suggesting domestic politics is much important to the shaping of national security policy[5]. The democratization of domestic political system would critically reduce the possibility of military conflict with other democratic states.

Secondly, the rise of non-state actors in international politics such as multinational corporations (MNCs), international organizations (IOs), and other quasi-political entities has challenged the traditional view of realism that nation-states are the principal actors of international politics. Self-determination, the principle purported by Wilson more than eight decades ago is once again on the highlight. The so-called supranational and subnational units are confronting the authority and legitimacy of nation-states[6]. However, this challenge is regarded by liberal political theorists and jurisconsults as an unprecedented opportunity to the expansion of the jurisdictional power of international law. The ever increasing role of non-state actors on the world plane has enlarged the possibility of those supranational and subnational entities to acquire full international personality and, therefore, become subjects of international law.

Thirdly, idealists are criticizing realists for their failure to explain "the changing nature of modern states and the kinds of aspirations they entertain"[7]. Former imperialist states since the end of the Second World War have been giving up their title to the colonies more or less voluntarily, and former colonies and semi-colonies claimed their independence from their colonial powers under the doctrine of self-determination. The rest of colonies which have yet to claim their independence have been brought under the supervision of the U.N. Trusteeship Council. Such development is contradictory to the keystone of classic realism that stronger powers followed the instinct of maximizing their power and control weak ones.

The last but not the least, although realists argue that international law and organizations are merely tools of strong powers exploiting weak states[8] IL and IOs have been gaining influence on international relations. States for much of the time more or less voluntarily bring themselves under the jurisdiction of international law to solve bilateral or multilateral disputes rather than simply launching war against each other over the past several decades. International law and international institutions are more and more likely to step in the spheres which used to be taken care of by nation-states. International law is eroding the jurisdiction of national legal system and developing supranational jurisdictional power which national government is not in a position to enforce. Examples can be found in the universal protection of human rights, international efforts in the protection and exploration of *res communis* (territory which is not capable of being controlled by any single state, Antarctica, for instance.) and outer space, and prosecution of international war criminals and criminals against humanity.

3 Theoretical Development in International Law

Debate in academic realm relating international law lasts for hundreds of years. What is international law? Hedley Bull tries to define that "international law may be regarded as a body of rules which binds states and other agents in world politics in their relations with one another and is considered to have the status of law"[9]. Although many international relations theorists and jurisconsults vary in defining international law, and some of them even doubt the existence of such a legal system, the majority of scholars agree that there are some rules applied in international politics to govern the behaviors and relations of sovereign states.

Modern international system emerges under the Westphalian system of nation-states. In such a system, a basic principle is that of independence, autonomy, and equality of the "sovereign" states. The system, or the "society of nation-states" is characterized by what Robert Keohane and Joseph Nye described the vulnerability and sensitivity of interdependence between states.[10] Vulnerability and sensitivity has brought unpredictability and challenge which a state is to prevent when interacting with other states for the self-interest. Predictability can only be obtained through the adherence of certain behavioral rules in the community. Consequently, this necessity encourages some scholars such as Hugo Grotius and Samuel Pufendorf to create a theoretical framework of the law of nations to regulate and stabilize the anarchic society of nation-states.

Where does international law come from? There are two schools arguing the origin of international law: the naturalists and the positivists. The naturalist scholars hold the view that there are "transcendent, universalistic principles (natural law principles)" which apply to all people at all times under all circumstances[11]. They originate from divine sources or general principles of the nature, and independent from human consciousness.

The other camp of scholars takes a positivist approach to explain the origin of international law. The positivist school dismisses the natural law principles by arguing that law is what people say is law. Positivists emphasize viewing events as they occurred and discussing actual problems that had arisen. The states of the international community shall abide by international behavioral rules for the sake of self-interest. The keystone for the effectiveness of international law is the general principle of *pacta sunt servanda* (the idea that international agreement must be binding upon the contracting parties). The positivist approach is now dominant in modern international law although some of international law scholars are taking effort to merge the two approaches to deal with such issues as universal protection of human rights.

4 Liberalism in International Law

Liberals are very optimistic of the role of international law in preserving the order of the society of nation-states. They not only recognize the existence of international law, but also realize its different nature from that of the domestic legal system. Moreover, they emphasize the utmost importance of international law to nation-states and other political entities on the global arena, and it is particularly so in the post-Cold War world.

The subjects (states, the principal actors of international politics in realist view) of international law are sovereign, and there is no real coercive authority to enforce the law to sovereign states. States create law to keep themselves under self-constraint. However, liberal scholars believe that through the creation of such norms and laws, the society of nation-states is bettered, and the order of the system improved. States can do what they ever wanted to do more efficiently and "to do so in such ways as to enhance the status of states"[11].

The collapse of the Soviet Union and its Eastern Bloc has encouraged international relations scholars to think about the future of international law. Liberals are encouraged by the increasing role of international law. The 1990s and the first years of the 21st century saw ever growing interdependence between states. The supreme authority of national government, according to liberals, has been undermined by the challenge of non-state actors on the global arena. One of general principles of international law is reciprocity. States are willing to bring themselves on self-constraint not only because they want to create a good order of society which provides members stability and predictability but also because they fear retaliating response from others if they fail to act by rule. Thus, respecting the "rules of games" means to states the protection of self-interest. By abiding by the rules of games states can also set up a good reputation of behaving in good faith and in adherence to law and order, and thus obtain positive reaction from others. The principle of reciprocity does make sense particularly in the Post-Cold War years. International politics becomes far more complex since the early 1990s. The future of the world is so much unpredictable in a vulnerable "non-polar" or "multi-polar" international system. International legal system at least will provide some element of certainty and predictability to states. This argument is closely connected to the nature and functions of international law.

Starr believes that the basic functions of international law is conflict-related functions ranging from instrument of direct control to limiting the conditions under which a justified conflict can originate, to regulating the legal means of conflict, and to serving a central role in the range of processes involved in conflict prevention, management, and resolution[11]. Such functions provide a means of bargaining between states so that the ultimate means of war can be prevented and international order is preserved. A series of international mechanisms are offered to states to deal with their disputes, including the processes of diplomacy, arbitration and adjudication. Since the last several decades, a majority of states of the international community have more or less accepted the peaceful means as their foremost conflict-solving choice instead of threat or use of force.

Communication is regarded as another basic function of international law. William Coplin asserts that international law serves as an instrument of communication and, "as such, perhaps the primary device for socializing policymakers as to the nature of the prevailing consensus in the international system and its changing expectations regarding the rights and duties of international actors" [12].

For liberals, international law is much like civil law in the domestic legal system with facilitative function. It does not provide any element consisting of command or coercion but some sort of guidance of procedures or means which a citizen or a member of society can obtain when he or she wants to do something. This civil law model coincides with the need of states when their cooperation and coordination between each other is of utmost importance. Uneven distribution of resources necessitates the exchange of goods, services and capital. International law, therefore, functions as the "rules of roads".

Besides the aforementioned functions, liberals/idealists also highly regard the role of international law in the enhancement of the general welfare of mankind. They passionately advocate the adoption of specialized international conventions to take care of the universal protection of human rights. Such effort can be found in the advocacy of adopting a binding International Bill of Rights by the U.N. General Assembly and creation of an international court of human rights. Part of the dream has been realized by a series of progresses in the codification of international conventions on human rights protection, including the 1948 Universal Declaration of Human Rights, the International Covenant on Civil and Political Rights of 1966, and a number of specialized multilateral agreements against slavery, apartheid, and various degrading treatment on people. The international community also speeds up its work on the creation of international institutions on human rights protection in the first years of the 21st century. Such institutions include International Criminal Court in 2002, and U.N. Human Rights Council in 2006.

5 Conclusion

The arguments of liberalists in international relations and international law represent their worldviews. For liberals/idealists, people are good in essence, and the future of mankind is always bright although there may be some negative episodes. As their views applied to international politics, the actors of the world community, large or small, strong or weak, will tend to collaborate with each other to ameliorate mutual relations and prevent conflict or war. Liberal/idealist view does not consider nation-states as the sole actors in modern global politics. Instead, they are calling for active participation of non-state actors in world politics in the Post-Cold War era. They are also taking their effort to enhance the general welfare of mankind. The striving of liberal publicists for the creation of universal legal mechanism to protect human rights is the implementation of this divine mission.

References

1. Kegley Jr., C.W., Wittkopf, E.R.: World Politics: Trend and Transformation. Beijing University Press, Beijing (2004)
2. Mearsheimer, J.J.: Realism, the Real World, and the Academy. In: Brecher, M., Harvey, F. (eds.) Rea-Inst. LNCS, pp. 23–33. The University of Michigan Press, Ann Arbor (2002)
3. Kegley Jr., C.W.: The Neoliberal Challenge to Realist Theories of World Politics: An Introduction. In: Contro-Inter-Rel-Theo. LNCS, St. Martin's Press, New York (1995)
4. Levy, J.S.: The Causes of War: A Review of Theories and Evidence. In: Tetlock, P.E., Husbands, J.L., Jervis, R., Stern, P.C., Tilly, C. (eds.) Beha-Soci-Nucl-War. LNCS, vol. I, Oxford University Press, New York (2001)
5. Bruce, B., Siverson, R.M., Woller, B.: War and the Fate of Regimes: A Comparative Analysis. J. Ame. Poli. Sci. Rev. 86, 638–646 (1992)
6. Writon, W.B.: The Twilight of Sovereignty. Charles Scribner's Sons, New York (1992)
7. Fukuyama, F.: The Beginning of Foreign Policy. J. New Rep. 207, 24–32 (1992)
8. Morgenthau, H.J.: Politics Among Nations: The Struggle for Power and Peace, 6th edn. Alfred A. Knopf, New York (1985)
9. Bull, H.: The Anarchical Society. Columbia University Press, New York (1977)
10. Nye Jr., J.S.: Understanding International Conflicts: An Introduction to Theory and History. Beijing University Press, Beijing (2004)
11. Starr, H.: International Law and International Order. In: Kegley Jr., C.W. (ed.) Contro-Inter-Rel-Theo. LNCS. St. Martin's Press, New York (1995)
12. Coplin, W.D.: The Functions of International Law. Rand McNally, Chicago (1966)

The Establishment of the Database of Mathematics History

Xie Qiang

Pingdingshan University, School of Normal Education, Henan Pingdingshan 467000

Abstract. The establishment of the database of mathematics history has become possible when computers , communications , multimedia facilities , superchips and the Internet are available. The digitized documents are the symbol of the level of digital technology of a given country or region. A database system of mathematics history is designed and implemented that promote and raise the educational approaches and methods to improve the mathematics history study.

Keywords: mathematics history, investigate and analyze.

1 Introduction

The main form of Current mathematics education is planed and organized class teaching, the teacher is the activity planner and the student is the activity executor. Therefore, the teacher has main responsibility to pass on mathematics knowledge designedly and selectively in their activities during the classes. Whether the college students need to study mathematics, they think what kinds of mathematics they need for their improvement and further development, how to conduct the effective and fruitful mathematics education in college school and universities, through what kinds of teaching method and approaches can make college and university students gain and reap more and more mathematics knowledge, what is current situation and today's condition in college history of mathematics education. Based on these questions and problems, the author in this paper launches a investigation and research on the started history of mathematics courses in PINGDINGSHAN University in the first half of 2011.

2 Questionnaire Design

The contents and details of the questionnaire mainly include four parts: (1) the common knowledge of history of mathematics, coming from the teaching material college and university students used in daily classes. (2) Energetically get to know college and university students how to face the history of mathematics study and how to study efficiently and productively. (3) Comprehensively understand and get to know the mathematics study condition and current situation of college and university students (4) Realize and understand the recognition of college and university students in the aspect of starting history of mathematics course in the college and universities.

Z. Du (Ed.): Proceedings of the 2012 International Conference of MCSA, AISC 191, pp. 369–374.
springerlink.com

3 Implement the Survey

The research adopts and uses the self-designed questionnaire to investigate and research the college and university students who take part in the examination and contests. After the receiving the related questionnaire, we select painstakingly to get the valid questionnaires. In the process of implementing the survey, we energetically attempt to allow college and university students reflect their true thoughts and true feelings in the questionnaire through the correct and rational leading. In order to ensure and guarantee the truth of questionnaire, we has told the college and university students who take part in the research activity the questionnaires will not be treated as any valuate evidence and estimate basis in or out of college and university in the future, this makes and allows the college and university students answer and fill related questions without any anxiety and worry.

4 Investigation Results

We issued 107 pieces questionnaires, and receive 100 pieces which account for 97% of total issued questionnaires. Among them, finally we got 100 pieces valid questionnaires.

4.1 First Part Analysis on Related Questions

Questions one to ten are given a score, each is given ten points, total scores are one hundred. And the situation is below:

Question number	1	2	3	4	5	6	7	8	9	10
The number of students who do correct answer	74	83	80	54	92	89	95	69	85	76
rate	74%	83%	80%	54%	92%	89%	95%	69%	85%	76%

Fig. 1.

The Figure one shows that the whole condition is good.

Figure-2 The score distribution of mathematics

The scores	Excellence (90-100)	good (80-89)	pass (60-79)	fail
The number of students	48	28	15	9
rate	48%	285	15%	9%

Fig. 2.

The figure two indicates that:
(1) The number of students whose performance is above good is seventy-six, accounting for 76, and the scores is intensive.
(2) The number of students whose performance is fail is nine, fail rate is 9%.
Figure-3 The comprehensive situation of student's mathematics history level

Total number	Average score	The median	mode	The standard deviation	fail number
100	82	80	80	12.4	9

Fig. 3.

Through the statistic analysis and systemic analysis of investigation results and research conclusions, we can conclude that the comprehensive level of college and university students' history of mathematics study. It mainly reflects in the standard deviation of 12.4, and indicates the degree of dispersion is not so obvious and apparent. The median is 80, while the seven-six students whose score are all over 80 have proved that the score has showed and displayed the intensive trends and concentrated tendency. However, the twenty-four students whose score are all below the average scores, it mainly indicates and proves that the history of mathematics level and standards of college and university students still needs to further improve and deep development in the future.

4.2 Collect and Analyze the Second Part of Questionnaire

Figure-4 The approaches and modes college students get mathematics history knowledge

The approaches and modes	Introduction in class	library and internet	promotion	others
rate	96%	50%	16%	18%

Fig. 4.

The results and statements in figure-4 displays and shows that the college and university students' initiative of history of mathematics study and learning has been enhanced and strengthened gradually, they vigorously acquire the history of mathematics knowledge and information not only through the teaching classes, college and university libraries and the internet and other media. From the view of investigation results and the perspectives of research results, ninety-six percent students vigorously consider the history of mathematics knowledge is mainly coming from the daily teaching classes in college and universities, and intrigue energetically their interest and potential to study and learn, and lead they go to college and university libraries and surf and check the internet to collect and seek the information and related knowledge of history to mathematics. Therefore, the teachers have played an important, vital and crucial role in history of mathematics study and learning among today's college students and current university students. The teaching classes still are important, vital and crucial parts in passing on and imparting the history of mathematics knowledge. It is urgent and serious to change and convert the class teaching materials of history of mathematics into the form and pattern the college and university students easily accept, study and adopt. The investigation and research also displays and indicates that there are still sixteen percent to eighteen percent college and university students acquire and reap history of mathematics knowledge through the promotion, bulletin boards and other approaches and ways. It also can vigorously intrigue the college and university students' interest to make and allow the history of mathematics knowledge and information integrate into campus and school culture through actively and unremittingly publishing and issuing some history of mathematics knowledge on bulletin boards.

Figure-5 The college students' recognition of the educational function of mathematics history knowledge

The option	Big	A little	None	Other
Rate	32%	65%	2%	1%

Fig. 5.

The figure-5 displays the students' recognition degree of mathematics history educational function is high.

From the research results and the investigation conclusions, the college and university students in our country are so interested in the set up of history of mathematics courses and curriculums College and university students mostly think and consider that study history of mathematics knowledge can vigorously intrigue and greatly strengthen their interest in learning and studying mathematics, improve self quality actively, energetically promote self characteristics, increase and enhance the force and motivation for studying and learning mathematics, vigorously enrich and greatly develop their mathematical structure, comprehensively master the method and approaches of mathematical thoughts and mathematical ideas, most importantly reap and acquire some life philosophy through learning and studying the mathematician spirit and mathematician great thoughts. And at the same time, students in colleges and universities in our country also consider and think the class teaching materials are out of date, too old and not good for today's research and current study, moreover, the teaching methods and approaches are backward and can not fit the current condition and today's situation, and the teaching is not systemic and so on.

5 Suggestions for College Mathematics History Education

As we all know, mathematics is the marginal and middle discipline which is between the discipline of liberal arts and science. Moreover, mathematics education will play its medium and important roles if we want to destroy and clear up the separation and gap between the nature science and social science, attempting to build and construct a harmonious bridge and cooperative relationship between them. In recent years, the history of mathematics has been quickly developing and improving fleetly with the education reform and teaching innovation, and has already vigorously produced some influence and effects on the mathematics education process and mathematics courses through many-year efforts and unremitting persistence in our country. And moreover, we have already put the history of mathematics study and history of mathematics learning into current teaching program and our today's syllabus. And in nowadays, many universities and colleges in our country have already started and set up the elective courses of history of mathematics and selective curriculums of history of mathematics, but the goals and targets are still obscure and misty of history of mathematics course and curriculums in many college schools and universities in our countries. And the class teaching contents in history of mathematics study and history of mathematics learning have been limited in historical material introduction before eighteen century and merely involve little about modern mathematics development.

Base on the college and universities cultivation goals and training targets, we should ensure and guarantee the mission and the basic task of history of mathematics courses and curriculums, making it better fit and more suitable to cultivate qualified college and university students in our modern society. The teaching period of history of mathematics is just one semester in many colleges and universities in our country. It means that we have to make a difficult and hard choice of history of mathematics materials itself. The investigation and research we have conducted also shows and indicates that college and university students in our country common realize and think that the history of mathematics materials we used now are too simple and too old and out of time. For the arrangement and lay-out of history of mathematics teaching materials, we do not need to purse the knowledge systematic and organize the contents according to logical system of history of mathematics development. Therefore, in aspect of class teaching materials and activity materials, we should energetically promote and vigorously raise the contents that are related to history of mathematics study and learning, actively issue more books and magazines that are suitable as history of mathematics teaching materials and publish some popular science readings related history of mathematics. Based on the teaching materials of history of mathematics, sometimes we introduce and bring some foreign knowledge about history of mathematics, it can not only enrich and expand the students' study contents in college and universities, but also increase and promote the college students' interest of learning and studying mathematics, in addition it also can help them explore and probe initiatively, and get rid of bad emotion and lower sentiments of mathematics study and learning. Through the guidance of teachers in colleges and universities, enrich and expand their study and learning contents, increase and raise the mathematics knowledge and related information and transform the college and university students' study interest from mathematics history to mathematics itself. At the same time, we should launch some research and study activities based on history of mathematics knowledge, allowing college and university students to feel and experience the perfect and harmonious relationship between mathematics and our daily life.

With the further promotion and unremitting reform of modern education, news media and computer technology have insensibly entered into our teaching classrooms in their unique ways in colleges and universities around our country, going against the traditional mathematics teaching approaches and conventional mathematics education modes. When teachers in colleges and universities in our country are passing on and imparting mathematics knowledge and mathematics information, they will make best use of modern media technology to assist their teaching. It can greatly illustrate the development of history of mathematics, and provides direct materials dynamically, leading to better teaching results.

6 Conclusions

In the twenty-first century, reform of basic education demand a higher requirement for teachers' quality, making the teacher cultivation become a profound and lasting work. Different math perspectives can have different effect on college students through teaching. Therefore, it can let college students better and further understand the teaching value of math, math thoughts and finding process of math through teaching the mathematics history. This is vital and more important for them. Both explore and discuss the importance of mathematics history course and promote the teaching quality of math specialty. The mathematics history as an important part of mathematical culture, we must pay more attention to education of mathematics history in order to reap integrant and conducive math knowledge, develop and get culture promotion in the math study.

References

1. Sun, Y.-Q.: Chinese mathematics education reform should face twenty-first century. Course Materials and Teaching Method 10, 33–35 (2004)
2. Zou, B., Guo, X.-Y., Jia, G.: Talk about mathematics history education. Journal of Mathematics for Technology (3) (2000)
3. Yuan, X.-M.: Introduction of mathematics history thoughts, vol. 232. Guangxi Educational Pubishing, Nanning (1991)
4. Deng, M.-L., Chen, X.-M.: Focus on the function of mathematics history in mathematics education. Shuxue Tongbao 12 (2002)

Research on Real-Time Monitoring and Test of Download Threads of the Literary Fiction Website —Take "The Woman Warrior" as an Example

Yanhua Xia

School of Foreign Languages, China West Normal University, Nanchong, Sichuan 637009
China
xiayanhua@cssci.info

Abstract. Digital resources play increasingly important role in today's life and research. With the increased use and increase the value of literature, malicious download digital resources are growing as well. By analyzing the use of netflow, to construct a specific data source based on controlling single ip flow and the number of traffic monitoring system, in order to control malicious download and avoid other risks.

Keywords: Malicious download, Literary fiction, Traffic monitoring.

1 Brief Introduction to Orientalism

Orientalism refers to the researches of the people from the western world on the social cultures, languages and humanities of the Near East and the Far East. It can be translated into the imitation and description of the writers, designers and artists from the western world on the oriental world. Also, it can refer to the sympathy appreciations of these western writers, designers and artists on the oriental cultures.

However, a depreciative meaning was covered on the Orientalism since Edward W. Said (1978) pointed out in his highly controversial work "Orientalism" that the scholars of the western world held the attitudes of European imperialism of the eighteenth and nineteenth centuries to get an understanding of the oriental world or the understanding of the foreigners on the oriental cultures was out-of-date and prejudicial in general [1] and the "Orientalism" should belong to one of the products of the western constructivism and was aiming at establishing an obvious dividing line between the western world and the eastern world and hence attached higher importance to the superiority of the western and eastern cultures [2].

Therefore, the Orientalism began to contain a negative meaning, which referred to a concept which had been accumulated in the "western" knowledge, institutions and political and economic policies for a very long time and assumed and constructed "the oriental world" as a heterogeneous, divisive and "alienated" world.

Under the influence of the Orientalism thinking, the understanding of the people of the western world on the oriental world was embodied in the xenophobic attitude.

Actually, Orientalism at that time was nothing else than a cultural tool which was used to serve for the colonial rulers during the imperialism period. Edward W. Said

Z. Du (Ed.): Proceedings of the 2012 International Conference of MCSA, AISC 191, pp. 375–379.
springerlink.com

also pointed out that it was just the encouragement of the literary works which were written by the Orientalism scholars of French to drive Napoleon to occupy Egypt and hence start up the process of European colonial rulers to seize the Middle East for the purpose of establishing an honor in history.

New York Times used to declare that the "Orientalism" of Edward W. Said would not be eliminated and the immutability of this book laid in that it was not only an academic work and also a political declaration no matter how many mistakes were in existence.

2 Orientalism Narration and Motivation of the "The Woman Warrior"

The novel "the woman warrior" was a maiden work of Chinese American Tingting Tang (English name Maxine Hong Kingston) and was published in 1976. The "the woman warrior", by relying on the delicate and periphrastic writing style, integrates the real life of Chinese Americans who lived in China towns of the United States, and therefore can be equivalent to the female hero stories about the miraculous ghosts, immortals and free battles which are very popular in Chinese Mainland. The real life of these Chinese Americans gives a vivid expression to the oppression, poverty as well as instability. However, in the "the woman warrior", the story background is transferred to the exotic China from the normal society of the United States. Therefore, the "the woman warrior" becomes one of the typical novels in which the writer made use of the eyes of the people of the first world to examine and observe the destiny of the women of the third world.

Generally, there are five chapters in "the woman warrior" in total. A majority of the contents which are narrated in this novel were the stories happening in China. Also, the stories happening in the United States were between native Chinese people and Chinese Americans. The beginning sentence in the stories that you can't tell what I told you to anyone [3] has changed into one of the commonplace sayings among the students of the higher learning schools of the United States.

Naturally, the bright spots in the novel "the woman warrior" can be the "drama narrations" about the traditional Chinese stories such as the "Mulan Joins the Army". In "the woman warrior", Tingting Tang combined the stories such as "Mulan Joins the Army", "Legend of Yuefei", "Peasant Uprising" and martial arts chivalries together to make the narrations. Such a kind of writing method is totally an unrestrained and vigorous style that brims with talent, and makes the stories very interesting at the same time.

In the novel "the woman warrior", the objects which mainly receive the criticism are those habits and prejudices in the traditional Chinese culture. For example, it's an ill bird that fouls its own nest, which is a very popular traditional Chinese saying [4]. Owing to the influence from the foreign thinking way of the Orientalism, Chinese people and Chinese American suffer from a great number of the stereotyped negative images in the society of the United States. However, Maxine Hong Kingston, the writer of the "the woman warrior", transferred the story background to the exotic China from the normal society of the United States in this novel, and therefore such a writing style increases the negative influence of such an image. Through the related analysis, the reason for the emergence of the stereotyped negative images laid in that a great many of Chinese workers made use of the identity certifications which were from others to enter the United States during the period in which Americans had the sense of excluding Chinese. During that period, Chinese workers in the United States were expelled from the country as they were unwilling to reveal the real basic information, and they often kept silence at the public places.

However, in the novel "the woman warrior", Kingston made such a projection on the reasons why Chinese Americans kept silence and looked so mysterious and unfathomable in the following: with the purpose of confusing the deities and transferring their curses, Chinese immigrants took advantage of the ambages in the traditional stories and false names to mislead the deities [5].

Kingston was supposed to make full use of the gorgeous history or her own family stories to make a correction on the fallacious understanding of the image that Chinese Americans did not like speak too much and looked very mysterious and unfathomable. However, she showed her ignorance of the subsequent influence of the American policy of discrimination against the Chinese on Chinese Americans latter. She made a choice on avoiding the facts in history and was in conformity with the prejudices of the mainstream society of the United States on Chinese Americans. And this made an enhancement to the conceptualization image on Chinese Americans in the society of the United States further.

Under the context of Chinese culture, the well-known literary quotation which is called by Chinese people as the "mother of Yue Fei tattooed characters". The action of the mother is sublime. Therefore, Chinese people will not get an understanding that the woman (the mother of Yue Fei) placed a persecution on the man (Yue Fei). However, when the writer transplanted the story "mother of Yue Fei tattooed characters" to Mu Lan, it can be found that the "the father of Mu Lan tattooed characters on Mu Lan" was from being sublime and would change into the evidence of the persecution which was placed by the man on the woman if this story was read under the context of the feminism.

In the mean time, it was also believed by Chinese American feminism critic YingMin Lin that the detailed story which is utilized to narrate Yue Fei can be more appropriate for the narration on the story of Mu Lan, because it can be powerful to symbolize the mental suffering that Chinese women endure for a very long time [6].

In addition, under the context of the feminism, the bound feet wrapped by traditional Chinese women is the most stable evidence to prove that traditional Chinese culture put oppression on Chinese woman in both physical and mental states [7]. In this aspect, Tingting Tang carried out a pungent criticism on such a bad custom in the novel "the woman warrior" [8].

In 1977, the "New York Times Magazine" used to make use of the "Duck Boy" as the title and release part of the early manuscript of the novel "the Woman Warrior". Seen from this released manuscript, it could be known that the story in the "the Woman Warrior" was supposed to make use of the society of the United States as the background. However, what was the motivation to drive Tingting Tang to transfer the background of the stores to the exotic China during the last period in which the writing of "the Woman Warrior" came to end and take advantage of the Orientalism narrative strategy? It was thought by Chinese scholar David Li that the reason why Tingting Tang to transfer the background of the stores of "the Woman Warrior" to the normal society of the United States to the exotic China was that she carried out an artistic modification on the manuscript with high consciousness for the purpose of publishing the "the Woman Warrior" [9]. In other words, the reason why the novel "the Woman Warrior" received a so high welcome among the mainstream readers and landed on the ranking list of the bestsellers in the United States has a very close relationship with such a background transfer.

3 Ethical Literary Theory and the Responsibility of the Writer

Like there is a discrimination against the gender in the western culture for a very long time, there is a discrimination against the gender in the traditional Chinese culture indeed. There is nothing to be said against the criticism of Kingston on the traditional culture of her motherland though she followed the example of the May 4th Movement that modern Chinese intellectuals pretended to be an ill bird that fouls its own nest. However, the problem is that it may seem empty and pointless that Kingston made a comparison between China of the feudal times and the United States of the twentieth century.

For example, the book "my country and my people" which was written in English language by Yutang Lin aimed at making a criticism on the traditional Chinese feudal code of ethics so there were a great number of the bad customs exposed to the whole world. The ultimate purpose of this book was for the benefits of the modern China. However, the feminism which was expressed in the novel "The Woman Warrior" seemed to reach a common view with the superiority of the third-world Orientalism which was intensive to make the traditional customs of the third world exposed to the people all over the world, but did not care about the objective situation of Chinese women of the modern times [10].

In 1977, ethnic Chinese scholar Kathy Feng used to write an open letter to Tingting Tang, and made a criticism on the distortion which was used by Tingting Tang on the history of China and Chinese Americans and strengthening the conceptualization image of Chinese Americans in the society of the United States [11].

Chinese ethical literary criticism expert Zhenzhao Nie used to put forward that the ethical responsibilities and duties of the writers in the writing area or the ethical responsibilities and duties of the critics to make a criticism on the literatures all belong to the category of the ethical literary criticism [12].

Therefore, the judgment standards of the literature values and the purposes of the literary creation should take the continuous development and expansion of the ethics and morals as the ultimate purposes. Thus, the value of the literature works lies in whether the works contain the consciousness of undertaking the due responsibilities and duties t, but is not only concentrated on the aesthetic entertainment functions and meeting the needs of the general readers; the ultimate value of the literature values should rest with the care on the humanistic spirits [13].

4 Conclusion

From the above analysis, it can be known that Chinese American writers have the responsibilities to always keep in mind that they are also part of the weak minority group which is oppressed by the racialism in the society of the United States in which the white race takes the leading position. Also, it is necessary for them to get rid of the infatuation with the Orientalism narration, strive for the internal solidarity and friendship in the minority group, and make every effort to breaking away from the control, temptation and constraint together.

Therefore, it is necessary to give consideration to the needs of the minority group on the discursive power struggles when the dross of the culture of the ancestors is being criticized, firmly stick to the principles of ethical literary theory, deem the construction of the equal discursive power for the minority group as an indispensable social responsibility. Only based on these, Chinese American literature can develop into one of the literary forces with true maturity in the society of the United States.

References

1. Said, E.W.: Orientalism, p. 3. Pantheon Books, New York (1978)
2. Said, E.W.: Orientalism, p. 12. Pantheon Books, New York (1978)
3. Kingston, Hong, M.: The Woman Warrior: Memoirs of a Girlhood among Ghosts, p. 1. Vintage Books, New York (1977)
4. Kingston, Hong, M.: The Woman Warrior: Memoirs of a Girlhood among Ghosts, p. 1. Vintage Books, New York (1977)
5. Kingston, Hong, M.: The Woman Warrior: Memoirs of a Girlhood among Ghosts, p. 6. Vintage Books, New York (1977)
6. Ling, Amy: Between Worlds, p. 160. Pergamon Press, New York (1990)
7. Shen, D.: An Unofficial History of Wan Li Period, vol. 23
8. Zhao, W.: Chinese American Literature and Feminism Orientalism. Contemporary Foreign Literature (03), 50 (2003)
9. Li, D.L.: Representing the Woman Warrior. In: Skandera-Trombley, L.E. (ed.) Critical Essays on Maxine Hong Kingston, p. 200. G. K. Hall & Co, New York (1998)
10. Zhao, W.: Chinese American Literature and Feminism Orientalism. Contemporary Foreign Literature (03), 53 (2003)
11. Fong, K.M.: An Open Letter/Review. Bulletin of Concerned Asian Scholars 9, 69 (1977)
12. Nie, Z.: Ethical Approach to Literary Studies: A New Perspective. Foreign Literature Studies (05), 20 (2004)
13. Li, C.: Contemporary Northern Ireland Culture Integration and Reflection from the Perspective of the Ethical Literary Criticism. In: Nie, Z., Zou, J. (eds.) Ethical Literary Criticism: Discussion on New Methods of Literature Researches, pages 125, 126. Central China Normal University Press, Wuhan

Design of Forest Fire Monitoring System in Guangxi Zhuang Autonomous Region Based on 3S Technology

Yuhong Li[1,2], Li He[1,2], and Xin Yang[1,2]

[1] GuangXi Institute of Meteorology, Nanning, China 530022
[2] Remote Sensing Application and Test Base of National Satellite Meteorology Centre, Nanning, China, 530022

Abstract. With the development of 3S technology, 3S integration technology combining closely remote sensing (RS), geographic information systems (GIS) and global satellite positioning system (GPS) has shown great future. In this paper, we design the forest fire monitoring system based on 3S technology for forest fire forecasting in real-time, forest fire monitoring and decision supporting. The system will be helpful for Guangxi meteorological department to improve service on forest fire monitoring.

Keywords: forest fire monitoring system, design.

1 Introduction

Forests are an important component of the Earth ecosystem. In recent years, due to high temperature, high wind, lightning, man-made outdoor fire violations and other reasons, forest fire happen in high frequency. Forest fires not only caused huge economic losses, but also take a negligible impact. On the ecological environment and climate. Therefore, strengthening the monitoring capacity of forest fire and monitoring accurately the fire point are the prerequisite to fight forest fire fastly and efficiently. It's also the important part to protect the forest resources. Establish an effective and reliable forest fire monitoring system is the important work for forest fire prevention that has important theoretical and practical value.

Guangxi cross tropical and subtropical area where contain high temperature, enough heat and abundant rainfall. The region is rich in forest resources. The frequent forest fire happening each year in Guangxi result in enormous economic and environmental damage. The forest fires usually occurred in the region of high mountains with dense forests, steep terrain, harsh natural environment, thus it's Random and unexpected features. Furthermore, since the forest fire monitoring mode drop behind, all those bring about many negative factors. To implement measures for forest fireproofing. With the rapid development of technology, 3S technology (including satellite Remote Sensing (RS), geographic Information System (GIS) and Global Positioning System (GPS)) has become an important supplementary means for forest fire monitoring.

Satellite remote sensing (RS) has many advantages, such us the coverage is wide, information cycle is punctual, channel information are rich, the information source is stable and processing technology is mature etc. With the satellite remote sensing

Z. Du (Ed.): Proceedings of the 2012 International Conference of MCSA, AISC 191, pp. 381–386.
springerlink.com

developing to high space resolution, high spectral resolution and high time resolution, it'll undoubtedly increase the level of forest fire monitoring greatly. Geographic Information System (GIS) has the function to manage, collect, analysis, and roll out a variety of spatial information, which have been widely used in many industries. As the main information source from GIS, satellite remote sensing data combined with GIS and GPS can exploit their advantages and cover their shortcomings.

2 System Goal

The system will take satellite remote sensing data as the main source of information. Utilizing GIS efficient data management and analysis capabilities, taking advantage of RS quickly accessing to data and getting help from GPS, the system can achieve functions as real-time forest fire forecasting, monitoring and decision support. Thus Guangxi meteorological departments can improve their service capacity of forest fire monitoring.

3 System Operating Environment

Hardware: 512M or more memory, 800 and above CPU frequency, 10G and above hard disk capacity.

Software: Widows 2000 or later operating system; SQL Server2000 or above database platform, SuperMapObject5.2 GIS development platform, application development languages: C #, Application Development Platform: Visual studio 2008.

4 Overall System Design and the Main Module Function

Guangxi forest fire monitoring service system consists of four parts including hardware, software, spatial database and management staffs. The system mainly fulfill the functions such as data preprocessing, forest fire automatic identification, service products production, fire information processing etc.

4.1 Overall System Functional Requirements

The system should automatically identify whether there is a new forest fires remote sensing products and automatically find the latitude and longitude of the fire point, fire point area, the fire point happening time, the fire point administrative region.

4.2 System Process Flow

- Receive satellite data and preprocessing.
- Automatic identify and read the fire point information while displaying the data on the system screen and store the satellite data.

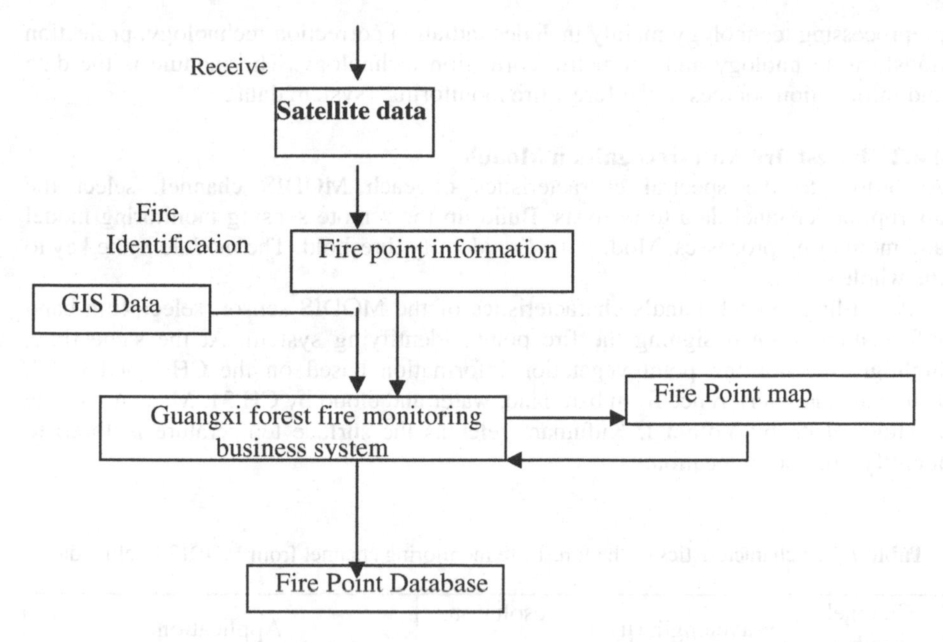

Fig. 1. Guangxi forest fire monitoring business system flow chart

4.3 Data Organization and Management

4.3.1 System Database Platform

The database of Forest fire monitoring business system is the base for the GIS systenm running which take a very important role. The databse is a comprehensive database while be divided into several sub-databses according to different function,including geographic information, forest area flammability classification information, fire remote sensing information and product generation. The database platform is SQL Server2000.

4.3.2 System Data Organization

System data include three categories: static data, Guangxi remote sensing map (Modis or TM), Guangxi, administrative map, Guangxi traffic map etc; dynamic data, remote sensing products from forest fire monitoring system; database Data, Save the fire point infromaton etc.

4.4 System Features Modular Design

4.4.1 Data Preprocessing Module

The module receive the MODIS satellite data throuth DVB-S system. Currently, the received MODIS satellite data are L1A origianl data as pds files. These data must be unpacked and generate L1B data as hdf file. After the L1B data be preprocessed and tranfer into LD2 files such as radiation correction,projection transform, geometric correction they can be used for the fire mornitoring service products. MODIS satellite preprocessing technology mainly includes radiation correction technology, projection transform technology and geometric correction technology. The module is the data and information sources to the forest fire monitoring system data .

4.4.2 Forest fire Auto-recognition Module

According to the spectral characteristics of each MODIS channel, select the appropriate channel data to analysis. Build up the remote sensing monitoring model and monitoring processes. Modify the fire identify threshold. The module is the key to the whole system.

According to each band's characteristics of the MODIS sensor, select 4um band information when designing the fire point identifying system. At the same time, highlight the burning point vegetation information based on the CH.1 and CH.2. Eliminate the interference from bare land, water and cloud by CH.31. Made use of the module set up by Yofam J. Kufuman (refer as the surface temperature method) to identify fire point location.

Table 1. The characteristics of the forest fire monitoring channel from MODIS satellite data

Channel number	Wavelength (μ)	Resolution	Application
CH.1	0.62~0.67	250m	Fire area, smoke
CH.2	0.84~0.87	250m	Fire area, smoke
CH.6	1.62~1.65	500m	Fire point detection, fire area estimation
CH.7	2.10~2.13	500m	Fire point detection, fire area estimation
CH.20	3.66~3.84	1000m	Fire point detection, fire area estimation
CH.21	3.92~3.98	1000m	Fire point detection, fire area estimation
iCH.22	3.92~3.98	1000m	Fire point detection, fire area estimation
CH.23	4.02~4.08	1000m	Fire point detection, fire area estimation
CH.24	4.43~4.49	1000m	Fire point detection, fire area estimation
CH.25	4.48~4.54	500m	Fire point detection, fire area estimation
CH.31	10.7~11.2	1000m	Fire area and burned area estimation
CH.32	11.7~12.2	1000m	Fire area and burned area estimation

4.4.3 Product Producing Module

GIS will provided the analysis result to the manager in the form of visual image or sheet. Producing process:Firstly, superimposed the identified fire point information and geographic information together; Secondly , locating the fire point and estimate of burned area. Thirdly, provide the relevant forest fire monitoring product to the decision-making department which including fire point map and information file in TXT format.

4.4.4 Fire Information Processing Module

a) Monitor the specified folder automatically or select interactively the output file from the remote sensing forest fire identification system. Store the the fire point information into the database automaticly. Check the data in the datebase to avoid duplicate storaging.

b) Find the fire point information file in the specified folder automatically when the system restarting . Stotage the new files in the folder automatically. This function can also be interactively realized when the software running .

Open and display the latest forest fire product automatically when the system switch on.Display; automatically the fire point information on the map; The fire point information can be showed in sheet or be drawn on the map.; The query and statistical results can be output to the EXCEL.

5 Conclusion

Guangxi forest fire monitoring business system based on 3S which combining GIS, RS and GPS can monitor the high temperature fire point efficiency and automation through remote sensing. At present, the system has been basically completed, try run is ongoing. It's most important characteristic are:

Monitor the remote sensing forest fire and analysis the output product automatically. Storage the information into the database and mark up the fire point on the map. Displaying clearly the fire point's location and the geographical environment. the system facilitates really the meteorological department work on forest fire monitoring.

References

1. Li, Lifu, S., Xiaorui, T.: A Summary of the forest fire research(IV)-The current situation and the development trend of forest fire management based on GIS. J. World Forestry Research 17, 20–24 (2004)
2. Yin, H.-W., Kong, F.-H., Li, X.-Z.: Rs and Gis-Based Forest Fire Risk Zone Mapping. J. Chnese Geographical Science 14, 251–257 (2004)
3. Xiao, S., Zhi, B.B.C., Xiaoke, W.: GIS application research in Forest Fire Management. In: ArcGIS cum ERDAS users in China Conference Proceedings. Earthquake Press, Beijing (2002)

4. Tian, X., Li, C.: GPS and GIS technology in forest fire prevention. J. Forest Fire Prevention 4, 31 (2006)
5. Jiang, X., Jin, A.-X., Tianhe: The design of forest fire prevention management system based on GIS. J. Forestry Science and Technology 21, 72–74 (2007)
6. Weihua, L., Guizhou, Z., Zhen, H.: Design and Implementation of Forest Fire Information Management System based on GIS. J. Computer Development and Applications 8, 34–36 (2004)

Analyze the Film Communication under Internet Circumstances

Yang Qi

Chongqing Normal University, Institute of Media, Chongqing 400047

Abstract. The internet plays a more and more important role in film and movie communication. This mainly embodies the internet has gradually broken the technology limits and it has surpassed the tradition film media in the aspects of speed, quality and quantity of digital image transmission, such as the network movies, mobile movies, city movies and other electric image publishes which arise in recently years, have been greatly increasing. It indicates that the internet has an important influence on film communication. The paper mainly discusses the film communication under the internet circumstances, and the study and its conclusions have realistic and instructional meanings.

Keywords: the internet, film communication, strategy, image.

1 Introduction

In the great development times, economy globalization and culture diversification have becoming the inevitable trends and tendency. With the globalization and international environment, mass media has played more and more important roles, especially film arts. As one of the most influential art creation and culture dissemination approaches, films have a large number of mass, wide and broad covering areas and most influential of culture during the cross culture dissemination. The researches in domestic and abroad have proved that movies with high quality is not only good for expanding wider economy markets, but also have irreplaceable effect on promoting national culture and thoughts influence and raising the nation image in the world. With the development of our society, the film communication influence has become more obvious, adding the increasing popular and universal of the internet, the film arts have entered into thousands of families. The internet continuously breaks the technology limits, and has excelled the traditional film media in speed, quality and quantity of digital image dissemination. This has been reflected into the network film, mobile film and city film arising in recently years. The paper mainly discusses the film communication and dissemination with the internet, and it has realistic and instructional meanings.

Z. Du (Ed.): Proceedings of the 2012 International Conference of MCSA, AISC 191, pp. 387–391.
springerlink.com

2 The Statement about Film Communication

2.1 The Basic Concept of the Film

Film is a comprehensive art mixing with various kinds of art elements such as literature, drama, acting, fine art, music and architecture and so on. Therefore, film has been called the seventh art. With image as basic element and mixed with sound and color together as basic language, film creates visual and direct art images and artistic conception on the screens, and brings the audio visual feeling and experience to the audience. Thus the film becomes a kind of mass art with big influence.

In addition, the film is also an intermediary of mass dissemination. The film producer states a story or the belief to the audience through screens based on the media itself, finishing its dissemination procedures and producing effect. The figure is below:

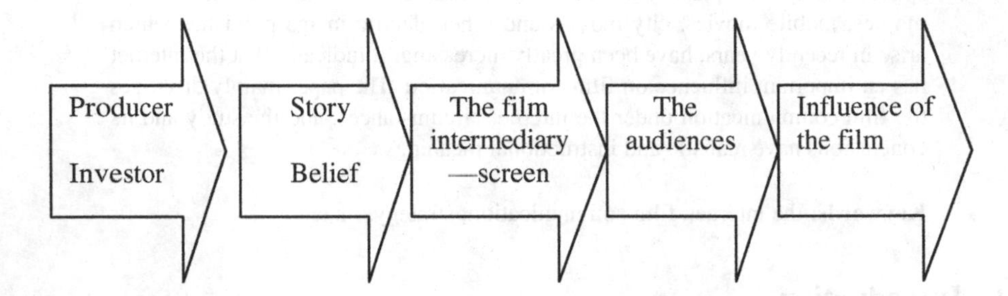

Fig. 1. The process of film communication

From this point, firstly films exit as one kind of communication approach and mode. Secondly is one kind of art. Briefly, films are the propagating art and the art communication approaches. Therefore, we can define the film concept as following: films are the mass communication intermediary based on continuous visual images and dissemination signs of sound and images, and modern acousto-optic technology.

2.2 The Communication Features of the Film

As one kind of comprehensive art, films have features and characteristics with gathering creation and expression approaches, mixing the sound and images in art languages and freedom in space change. In addition, when professional film producers show movies to audience and the public in cinemas, such certain places, as the mass communication intermediary, films' disseminate attributes reflects the common features of mass communication. As the results of technology and arts, films has unique materials and communication activities, making this electronic intermediary embodies self features and characteristics in the activities, especially compared to televisions that also belonging to auto visual media, films has advantages of self communication, and they are reflected in followings:

(1) Sound gathers with images

Film is adopted to make movies. It can allow the pictures in the movies more clear and color more saturation. Moreover, latest recording equipments can make the sound effects in the movies meet to high standards, especially the equipments of multi-track and stereophonic system, it can create an lifelike acoustics for the audience as if they are in the movies. Lifelike sound and images and strong audio-visual shock allow the audience as if in the fantasy movies.

(2)The particular seeing environment of films makes the audiences are more likely to concentrate on movies.

The movies are always showed in professional cinema, it reduces more disturbs and other interference when audiences watch movies. Under this kind of watching environment, audiences can focus on the movies effectively and add more emotion and feel more deeply, achieving to better communication results.

(3)Easy to watch

All the films have a long running period. Movies copies can be reserved and duplicated and it also can be showed repeatedly at any time.

Because of these benefits and advantages above, films have already become a priority for entertainment and recreation, and the popularization of the internet leads the films communication to a wider and broader areas.

3 Film and Movie Communication under the Internet Circumstances

Great development of science and technology and popularization of the internet are bringing new approaches and modes for film communication and dissemination. It makes the movies issue and communication no more depend on cinemas and video tapes. With the internet, modern films and movies embody a series of new features:

3.1 Fast Communication and Mass Storage Provide Security and Guarantee for Film and Movie Communication

The internet has special and important features with mass storage, and this is the reason why network films and movies can be communicated and disseminated so quickly. Through the fast communication and dissemination and mass storage, various kinds of films and movies on the internet, they can meet different appreciate demands for different audiences. Among these movies and films, we can find some non-mainstream and alternative films and movies, thus new social thoughts and ideas can also be spread and disseminated at the same time. With the conventional and traditional communication modes, films and movies with special form and non-mainstream contents can not be accepted immediately by audiences, thus they can be showed in the cinemas and theatres. But since the appearance of internet, those films and movies with non-mainstream and alternative contents will find the audiences who are fond of them, and their value will be spread and reflected.

Furthermore, in aspects of transmission speed, films and movies with 6G capacity can compressed into 500M according to new computer technology. The compression offers possibility for film and movie communication and dissemination. As long as the network can reach 200kb or so, the transmitted images can meet the requirements which audiences ask for on the screen.

3.2 The Film and Movie Audiences Has Changed Greatly

After adopting internet technology for film and movies communication, audiences' initiative has been liberated and their negative status is converted. When people and audiences watch movies in the cinemas, they sit politely and obey the cinemas' rules and regulations. While watching movies and films through the internet, people become initiatives, and they can choose films and movies casually just based on their own like and time. With these advantages and environment, people can totally steep into the movies and enjoy audio-visual feeling. From this point, the internet brings new communication approaches and modes for films and movies, and further leads to audience initiatives which never exist before.

In addition, the audience characteristics get freedom and liberation. Using the internet to watch films and movies, people can play and pause just according to their own likes. The computers and movies are all under their control. Because of the freedom of their characteristics, they will love films more and understand movies more, therefore, the films and movies can be communicated and disseminated wider and boarder.

3.3 Related Information and Comments on Films and Movies Are Greatly Increasing

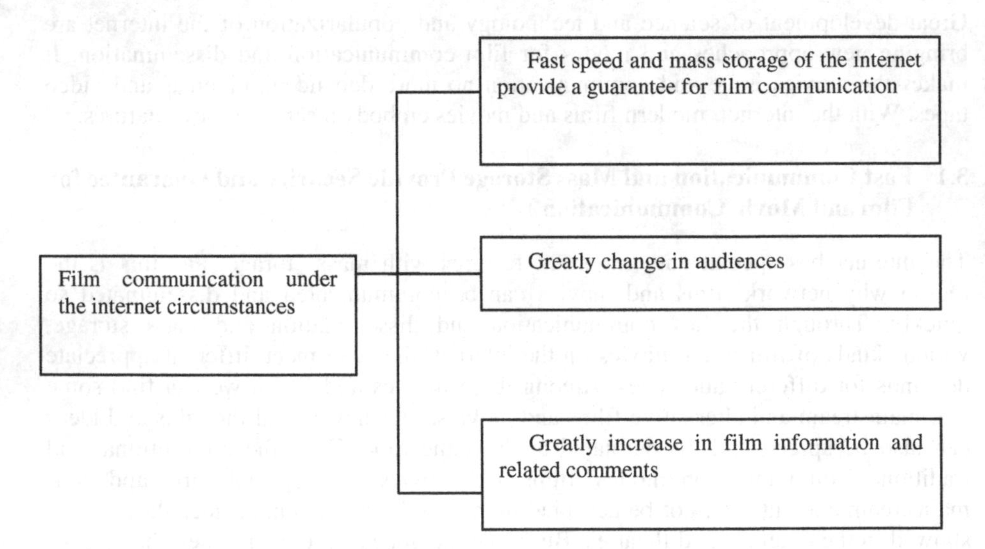

Fig. 2. The film communication under the internet circumstances

Because of the popularization of the internet, the approaches and modes of information distribution such as film introductions before showing, have changed and converted greatly. In the traditional communication ways, the film cinemas will take a control of movie resources, and the media such as newspapers and televisions always dominates the film and movie information. Thus we can make conclusions that audiences always keep negative and stay the last positions to know the information of films and movies. But with the popularization of the internet, the dominant status of cinemas and mass media have been damaged, and cinemas are not the only places for audiences to watch movies and films, mass media is no more the only lead to comment the films and movies. The audiences have become freedom and liberal to choose movies and films. In a word, the internet has broken the limits of life and time, and strengthened the audience initiatives and brought the new joy and pleasure to our life.

4 Conclusions

The rapid development of computer technology and network technology not only lead us to information explosion time, but also bring us the new approaches and modes of film and movies communication. The internet has given people more private space and provided more choices. The audience initiative and their characteristics have been greatly liberated, self-conscious is waked up. Under the internet circumstances, film and movie communication truly achieve the goal that audiences and films respect each other.

References

[1] Yang, X.-R.: Research on film communication in the internet age-elements, change and features of film communication. Film Literature (07) (2011)
[2] Li, L.: Films meet communication-analysis on film communication status. Journal of Chongqing Institute of Technology (09) (2005)
[3] Yan, C.-J.: Some special forms of film communication since the foundation of China. Modern Communication (Journal of Communication University of China)
[4] Yuan, Y.: Understand Gone with the wind through the information dissemination processes. Journalism Lover (10) (2011)
[5] Cao, Y.-M.: On the characteristics of film art in the turn of the century. Press Circles (04) (2008)
[6] Zhang, X.-D.: Digitization: the deconstruction and reconstruction of China's televisions. Modern Communication (Journal of Communication University of China) (01) (2010)

Research on Data Mining in Remote Learning System of Party History Information

Ruifang Liu[1], Changcun Li[1], and Qiwen Jin[2]

[1] Organization Department of the Communist Party, Hebei Union University, Tangshan, Hebei, 063009 China
[2] School of Humanities and Laws, Hebei Union University, Tangshan, Hebei, 063009 China
liuruifang@cssci.info

Abstract. After analyzing the insufficiency which existed in current most of long-distance education Websites, this paper introduces Data Mining technology, which is widely used in the field of the electronic commerce and so on, into the remote learning system of party history information. A kind of system model based on agent is proposed, and how to construct a multi-dimensional data cube to evaluate the studies of students is introduced, which can be realized by OLAP Analysis Service tools provided by Microsoft SQL Server.

Keywords: Data mining, Agent, Remote learning system, Party history.

1 Introduction

In China, the student Party members are one of the advanced groups as well as one of the dominant forces in all students at the higher learning schools. For this reason, how to make the special group get an enhancement to the awareness of the Party members and also exert the exemplary vanguard role is one of the most important tasks for Chinese higher learning schools to make an improvement on the construction of the Communist Party and to maintain the advancement of the members of the Communist Party at the present time and in the future.

2 Connotation of the Awareness of the Party Members

With the purpose of carrying out the awareness education of the student Party members at Chinese higher learning schools, it is first necessary to get a full understanding of the connotation of the awareness of the Party members. As a result, such a purpose can be realized in realities and also gets increasingly clearer and more definite.

2.1 Visual Theory of the Awareness of the Party Members

First of all, the awareness of the Party members refers to the consciousness of the advancement of the Communist Party members. Specifically speaking, the awareness

Z. Du (Ed.): Proceedings of the 2012 International Conference of MCSA, AISC 191, pp. 393–398.
springerlink.com

of the Party members means that the spirit of a Communist Party member can be clearly reflected from the daily behaviors, and can be utilized immediately at the key moment and also is ready to risk everything at the danger and disaster time [1].

Secondly, the awareness of the Party members is the reflection of the unique social identity, status and role of the Party members in the concept, and comes into being on the basis that the Party members have gotten common recognition on the guiding principle and task of the Communist Party, the responsibility and obligation they are necessary to undertake the organization and disciplines of the Communist Party [2].

Third, the awareness of the Party members is the ideological self-consciousness of the Communist Party members, and also is a kind of inner and core Party spirit consciousness which is used by the Communist Party members to remember their own social status all the time and ask themselves to do everything in accordance with the strict standards of the Communist Party.

2.2 Levels of the Awareness of the Party Members

The awareness of the Party members can be mainly reflected from the four aspects in the following.

First, it is necessary to own the initiative standard awareness. Only possessing the clear awareness of the Party members, these student Communist Party members at Chinese higher learning schools can take initiative to ask them to do things in accordance with the strict standards in the aspect of conducting themselves. On the contrary, it is necessary to reduce the standards and also loosen the transformation on the world outlook.

Second, it is necessary to own the initiative organizational awareness. Only possessing the clear awareness of the Party members, these student Communist Party members at Chinese higher learning schools can get a clear understanding that all of their words and actions give expression to the image of the Communist Party of China. Subsequently, these student Communist Party members can own a strong organizational discipline concept as well as a collective sense of honor, and hence can take initiative to abide by the all kinds of the rules and regulations which are formulated by the Communist Party of China.

Third, it is necessary to own the initiative responsibility awareness. Only possessing the clear awareness of the Party members, these student Communist Party members at Chinese higher learning schools can firmly keep the historical mission which is undertaken by the Communist Party in mind all the time. Then, these student Communist Party members can study diligently and make the learning practices with painstaking, and also become very brave to open up new things. As a result, these student Communist Party members can make an improvement on the ability to know and transform the world.

Fourth, it is necessary to own the initiative example awareness. Only possessing the clear awareness of the Party members, these student Communist Party members at Chinese higher learning schools can take initiative to set themselves an example to others and strictly exercise self-disciplines.

3 Awareness Education of the Party Members Is an Internal Requirement to Strengthen the Construction of the Student Party Members Group

First, some student Party members get a fuzzy understanding of the Party member standards. The standards of the Communist party member are the objective foundation for the Communist party members to form the ideological conception as well as the psychological quality which is required by the Communist party.

Second, some student Party members are in shortage of the sufficient understanding of the Communist party. Some student Party members at Chinese higher learning schools have a very great difference after and before joining the Communist Party. Generally speaking, part of the student Party members is in shortage of the understanding of the Communist Party when they are applying for joining in the Party. For this reason, there is a great difference in these higher learning schools students after and before joining the Communist Party. And this exerts an impact on the lofty image of the Communist Party members in the minds of the general students at Chinese higher learning schools.

Third, some student Party members are not solid in the ideal, faith and awareness of the Party members. In the information age in which the economic globalization has reached a high level, the socialism education which takes a leading position in Chinese higher learning schools all the time as well as the communism ideal and faith education has received an intense impact. As a result, this gives rise to the confusion, loss and disorientation of some student Party members at Chinese higher learning schools in the ideal and faith.

Fourth, some student Party members obtain a diversified value orientation. part of the student Party members at Chinese higher learning schools is diversified in the value orientation. On the one hand, the sense of competition of these student Party members is promoted to attain an enhancement further. On the other hand, part of the student Party members at Chinese higher learning schools tends to be practical in the pursuit of the goals and becomes diversified in value orientation at the same time [3].

4 Strengthening the Entry Point of the Awareness Education of the Student Party Members at Chinese Higher Learning Schools

With the purpose of making an enhancement to the awareness education of the student Party members, it is highly necessary to start with the ideological, political and moral foundations of these students. Then, the awareness of the Party members can be enhanced effectively and actually without a stop.

4.1 Making Every Effort to Strengthening the Scientific Theory Quality Cored at the Guiding Principle of the Communist Party

First, it is necessary to transform the ideology firmly by the scientific theory. Through an in-depth understanding of the essence of the scientific theories, the negative influences on the ideology, such as the "the hope for the realization of the communism is distant and indistinct" and the "the ideal of the socialism is fuzzy and not clear", can be overcome. As a result, these student Party members at Chinese higher learning schools can make full use of the scientific theories, so as to observe, analyze and solve these problems.

Second, it is necessary to build up the image firmly by combining the practices. The scientific theories of the Communist Party have endowed the students at Chinese higher learning schools with the powers of both truth and personality in the modern times.

Third, it is necessary to persist in seeking truth from facts and keeping pace with the times. It is highly necessary for the student Party members at Chinese higher learning schools to get a clear understanding that the guiding principle of the Communist Party of China is scientific and keeps pace with the times at the same time. More importantly, it is very necessary for these student Party members to always seek the truths from the facts when they are faced up with the diversification of the social economic components, the organizational forms, the employment ways, the benefit-based relationships as well as the allocation ways along with China's reform and opening to the outside world and the development of the socialist market economy.

4.2 Making Every Effort to Strengthening the Political Quality Cored at the Firm Ideal and Faith

First, it is necessary to correctly recognize the communism is not the castle in the air. First of all, it is highly necessary to get a real understanding that the socialist society of China is still being in and will be in the initial development stage for a long term. Second, it is highly necessary to get a real understanding that the communism is a senior state that the socialism will reach after undergoing a development for a long time. Third, it is highly necessary to correctly view all kinds of the twists and turns which are encountered by the socialism all over the world.

Second, it is necessary to correctly recognize the right motivation for joining the Communist Party. First of all, it is highly necessary for the student Party members at Chinese higher learning schools to get a real understanding that the only and right motivation for joining the Communist Party is making dedication to the construction of the undertakings of the communism and serving for the people better. Second, it is highly necessary for the student Party members at Chinese higher learning schools to deepen the understanding of the great undertakings of the Communist Party by use of the personal experience continuously and hence make the motivation for joining the Party upright constantly.

Third, it is necessary to continuously making the motivation for joining the Party upright in the practices. A student Party member at higher learning school can possess an upright motivation for joining the Communist Party only when he gets a clear and profound understanding of the undertakings of communism as well as the routes, guiding principles and policies of communism [4].

4.3 Making Every Effort to Strengthening the Moral Cultivation Cored at Serving for All People Wholeheartedly

The substantive characteristics of the Communist Party members are serving all people heart and soul, building the party serving the interests of the people and also exercising the state power in the interest of the people. For this reason, it is highly necessary for the student Party members at Chinese higher learning schools to persist in making an enhancement to the moral cultivation which is oriented at serving for all people wholeheartedly, for the purpose of strengthening of the awareness of the Party members.

First, it is necessary to firmly build up the correct world outlook, life philosophy and values. The student Party members at Chinese higher learning schools can firmly keep the purpose of the Communist Party in minds and make the selfless dedication to the country only if they can build up the correct three outlooks, which are the world outlook, life philosophy and values. Subsequently, these student Party members can be confronted with the complex social realities by easy stages and take initiative to resist the impacts from the negative common practices in society, and also can make the knowledge and actions integrated dynamically to a higher degree.

Second, it is necessary to firmly build up the excellent traditions and practices of the Communist Party. The practices of the Communist Party give expression to the purpose of the Communist Party and also have a very close connection with the image of the Communist Party, whether the Communist Party wins or loses the loyalty of the people with whom popular sympathy lies, as well as the survival of the Communist Party and the country.

Third, it is necessary to firmly build up the awareness of using the standards of the Party members to check the practices. It is pointed out by Hu Jintao that whether a Communist Party member can persist in serving for all people with heart and soul all the time is a most fundamental standard to measure whether the Communist Party member is qualified or not.

4.4 Making Every Effort to Strengthening the Ability and Quality Cored at the Cultures and Business Knowledge

First, it is necessary to look at the profound ideals and starting from the actual practices. Just as Wen Jiabao hopes that, it is highly necessary for the student Party members at Chinese higher learning schools to not only look at the profound ideals and also attach importance to starting from the actual practices, and also learn how to be a good person, how to do things and how to make use of the knowledge and skills, for the purpose of becoming a person who is concerned about the fate of both the world and the state.

Second, it is necessary to have both ancient and modern learning and taking history as a mirror. The student Party members at Chinese higher learning schools can get a real understanding of the history and get a control on the realities only if they can seriously learn the knowledge in both society and history.

Third, it is necessary to respect science and setting example for others. It is highly necessary for the student Party members at Chinese higher learning schools to not only show the respect to science and also set themselves an example to others first. Also, it is necessary for these student Party members to make a very close connection between the enhancement to the awareness of the Party members and the construction of the learning spirit, make full use of the excellent spirit civilization achievements to enrich themselves so as to make them become a typical example which can be do things like the versatile experts, hold a sincere desire all the time and also take the first to excellence and then bide by the rules and principles which are formulated by Chinese higher learning schools. As a result, these student Party members at Chinese higher learning schools can exert a modeling role in all aspects.

5 Conclusion

Possessing the awareness of the Party members is the most fundamental requirement of the Communist Party on the student Party members at Chinese higher learning schools. Therefore, it is highly necessary for all functional departments including the organization department of the Communist Party to attach higher importance to the awareness education of the student Party members at Chinese higher learning schools and establish the sound and related party member development and education training system. First of all, it is necessary to standardize the working flow which is oriented at developing the qualified Party members. Thus, the development of the Party members can be strictly implemented, and the quality of the development of the Party members can attain an improvement at the same time. Second, it is necessary to establish a long-term effective mechanism for the education and training of the student Party members at Chinese higher learning schools. Thus, the education and training of the student Party members can be standardized, scientific and regular with a gradual step, and also the intensity of training the awareness of the student Party members at higher learning schools can be strengthened actually.

References

1. Zhong, Z.: The Awareness of the Party Members is the Foundation of the Advancement of the Communist Party. People's Daily, ed. 1 (July 26, 2008)
2. Li, J.: The Communist Party Members should have the Awareness of the Party Members. Studies on the Construction of the Communist Party (Theory) (02) (2005)
3. Zhuang, S.: Cultivation and Intensification of the Awareness of the Student Party members at Chinese Higher Learning Schools—By Taking the Current Situation of the Student Party members at Art Schools for Example. Xiangchao (the Second Half Month) (Theory) (11) (2008)
4. Zhang, X.: Training Materials for Joining the Party. Xinhua Publishing House (March 2010)

The Study on Oil Prices' Effect on International Gas Prices Based on Using Wavelet Based Boltzmann Cooperative Neural Network

Xiazi Yi and Zhen Wang

China University of Petroleum (Beijing), Beijing, 102249 China
yixiazi@cssci.info

Abstract. In this paper, we build up WBNNK model based on wavelet-based cooperative Boltzmann neural network and kernel density estimation. The international oil prices time series is decomposed into approximate components and random components. The approximate components, which represented the trend of oil price, are predicted with Boltzmann neural network; the random components are predicted with Gaussian kernel density estimation model. In this paper, we analyzed the time-frequency structure of dubieties wavelet transform coefficient modulus for international natural gas and crude oil price time series, and predicted the oil price with cooperative Boltzmann neural network and Gaussian kernel density estimation model.

Keywords: boltzman neural network, oil price, international gas price.

1 Introduction

Most of these methods hypothesis international crude oil prices sequence of linear relationship between the means, but in fact, crude oil prices are very complex nonlinear time series, it is not only by the mercy of the objective economic laws, but also be political and pricing system, appear very complicated characteristics. Therefore, it is difficult to establish an effective prediction model based on the general time series analysis. The international natural gas prices could affect the entire world pricing system. In this paper, we establish WBNNK (wavelet boltzman collaborative neural network and kernel density estimation model. The international natural gas advertising crude oil prices time series down into approximate composition and random component. Approximate composition, representing the oil price forecast the trend of neural network, and boltzmann cooperation ability and the international natural gas prices; forecast composition and gaussian random kernel density estimation model. In this paper, we analyzed the time-frequency structure of dubieties wavelet transform coefficient modulus for crude oil price time series, and predicted the oil price with Boltzmann neural network and Gaussian kernel density estimation model. The results show that the model has higher prediction accuracy. International crude oil prices International crude oil prices rose to 78.4 U.S. dollars per barrel in July of 2006.But at he end of 2006, the crude oil prices fell to around 60 U.S. dollars per barrel. Organization of Petroleum Exporting Countries (OPEC)

Z. Du (Ed.): Proceedings of the 2012 International Conference of MCSA, AISC 191, pp. 399–404.
springerlink.com

announced that OPEC's crude oil output will decreased from 27.50 million barrels per day down to 26.30 million barrels per day from November 2006. Taken as a whole, the international market curde oil prices remained high from 2006 to 2007. Due to the American financial crisis, the global economic situation has not improved from October 2008. And the international oil prices have been in a recession, because a serious shortage of market confidence. Table.1 and Fig.1 show the international crude oil prices from 2007 to 2009.

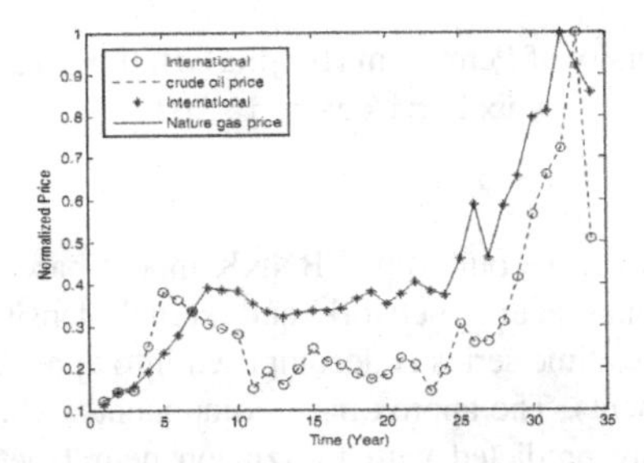

Fig. 1. International crude oil prices and international nature gas prices

Table 1. International crude oil spot prices (US dollars per barrel) and International nature gas price US Dollar /MMBTU

Year	Time	Crude Oil Spot Prices	Nature Gas Price	Year	Time	Crude Oil Spot Prices	Nature Gas Price	Year	Time	Crude Oil Spot Prices	Nature Gas Price
1976	1	12. 23	1.64	1988	13	15.97	4.63	2000	25	30.37	6.59
1977	2	14.22	2.04	1989	14	19.68	4.74	2001	26	25.93	8.43
1978	3	14.55	2.23	1990	15	24.5	4.83	2002	27	26.16	6.63
1979	4	25.08	2.73	1991	16	21.54	4.81	2003	28	31.07	8.4
1980	5	37.96	3.39	1992	17	20.57	4.88	2004	29	41.49	9.41
1981	6	36.08	4	1993	18	18.45	5.22	2005	30	56.59	11.42
1982	7	33.65	4.82	1994	19	17.21	5.44	2006	31	66.02	11.65
1983	8	30.3	5.59	1995	20	18.42	5.05	2007	32	72.2	14.35
1984	9	29.39	5.55	1996	21	22.16	5.4	2008	33	100.06	13.2
1985	10	27.98	5.5	1997	22	20.61	5.8	2009	34	50.6	12.3
1986	11	15.1	5.08	1998	23	14.39	5.48	2010	35		
1987	12	19.18	4.77	1999	24	19.31	5.33	2011	36		

2 Wavelet Transform of Oil Prices Series

The Continuous Wavelet Transform (CWT) is used to decompose a signal into wavelets, small oscillations that are highly localized in time, which is an excellent tool for mapping the changing properties of non-stationary signals.

The CWT of oil prices series $p(t)$ is defined as[2],

$$P_f(a,b) = |a|^{-1/2} \int p(t)\overline{\psi\left(\frac{t-b}{a}\right)}dt = < p(t), \psi_{a,b}(t) > \quad (1)$$

The CWT is a convolution of the data sequence with a scaled and translated version of the mother wavelet, the psi function:

$$\psi_{a,b}(t) = |a|^{-1/2} \overline{\psi\left(\frac{t-b}{a}\right)}, a,b \in R, a \neq 0 \tag{2}$$

In these equations, a is the scaling factor; and b is the translating factor; $P_f(a,b)$ is the wavelet coefficients.

The commonly used mother wavelet functions are Mexican hat wavelet, Morlet wavelet and so on. In this paper, based on analysis of the characteristics of oil prices, we selected daubechies wavelet. Daubechies wavelet satisfied the admissible condition as follow,

$$C_\psi = \int \frac{|\hat{\psi}(\omega)|}{|\omega|} d\omega < \infty \tag{3}$$

In (3), $\hat{\psi}(\omega)$ is the fourier transform spectrum of $\psi(t)$. And he corresponding inverse transform of eqation (1) is ,

$$p(t) = C_\psi^{-1} \int\int P_f(a,b)\psi_{a,b}(t)\frac{da}{a^2}db \tag{4}$$

Because the negative frequency of $\hat{\psi}(\omega)$ is no meaning, so (4) can be turned into,

$$p(t) = 2C_\psi^{-1} \int_0^\infty \int_0^\infty P_f(a,b)\psi_{a,b}(t)\frac{da}{a^2}db \tag{5}$$

Figure.2 is the CWTof oil prices time series. From the diagram, we can see that CWT increased the data redunancy of crude oil prices series, and brought more analytical content.

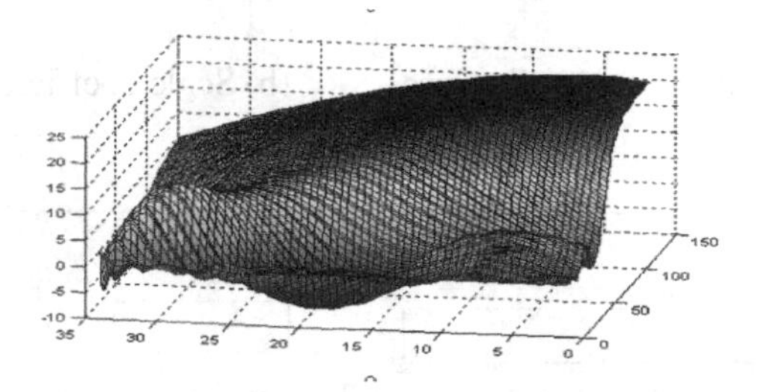

Fig. 2. The CWT of international crude oil prices series

CWT increased the data redundancy of crude oil prices series, and reduced the speed of operation. In order to accelerate the computing speed, while reducing the data redundancy, we can scale up the scale and displacement parameters of CWT, and define discrete mother wavelet functions as follow,

$$\psi_{j,k}(t) = a_0^{-j/k} \psi\left(a_0^{-j} t - k b_0\right), j, k \in Z \tag{6}$$

And the discrete wavelet transform (DWT) of oil price series is defined as,

$$P_f(j,k) = a_0^{-j/k} \int p(t) \overline{\psi\left(a_0^{-j} t - k b_0\right)} dt \tag{7}$$

Correspond to different frequency components a_0^{-j}, the sampling step is $b_0 a_0^j$, which is adjustable.

Define $a = 2^{-}\langle j \in Z \rangle$, the equation could be turned into binary wavelet transform,

$$P_f(j,b) = 2^{-j/2} \int p(t) \overline{\psi\left(\frac{b-t}{2^j}\right)} dt =< p(t), \psi_{j,b}(t) > \tag{8}$$

Figure 3 are the details and approximate discrete binary wavelet transforms coefficients. From the approximate coefficients of international crude oil prices, we can see: the international oil price began to show declining trend from the 20th month, until the 27th month, turned into an upward trend. But the approximate coefficients of international natural gas prices had begun to show declining trend since the 14th month, which is six months earlier than international crude oil prices, and the international natural gas prices turned into an upward trend form the 22nd month, which is almost six months than earlier than international crude oil prices showing upward trend. the General details of coefficients show the decrement is not obvious. But from the details coefficients, the decrement of crude oil prices is obviously come out.

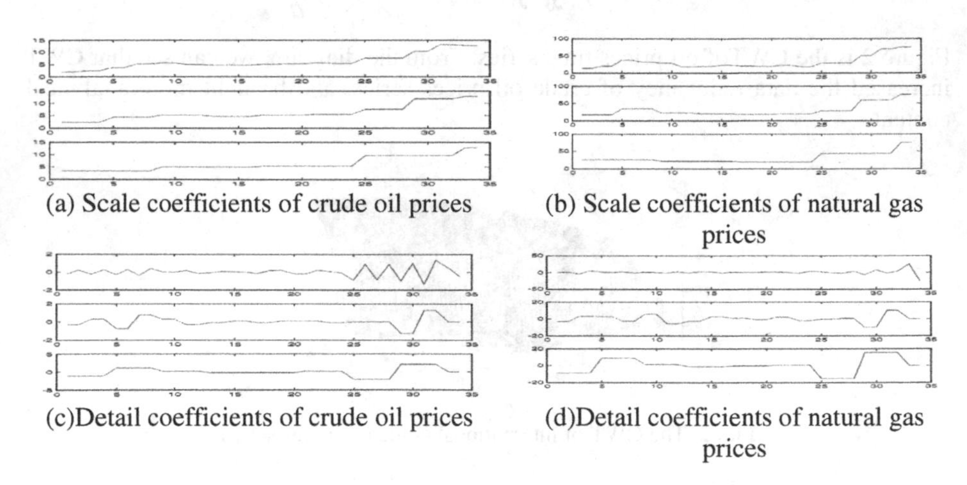

(a) Scale coefficients of crude oil prices (b) Scale coefficients of natural gas prices

(c)Detail coefficients of crude oil prices (d)Detail coefficients of natural gas prices

Fig. 3. DWT coefficients of international crude oil prices and international natural gas prices

3 Time-Frequency Analysis of Oil Prices Serices

As can be seen from the above data, the oil prices change are not only global but also with local features, and show many time-scale structures. The time-frequency analysis of the crude oil prices will provide more background information to study the different time-scale changes of oil prices. Changing $P_f(a,b)$ with a and b, we could get time-frequency structure of Daubechies wavelet transform coefficient modulus for international crude oil prices time series. Figure.4, at the same scale, the wavelet transform coefficients changes reflects the changes of the characteristics of international crude oil prices, from the 1st month to the 17th moth, the positive wavelet coefficients correspond to the high oil prices period; from the 17th month to the 27th month, the negative wavelet coefficients correspond to the low oil price period; from the 27th month to the 30th month the positive wavelet coefficients show another high oil prices period . The wavelet coefficients around zero is correspond to a mutation point of oil prices; while the 15th and the 21th month the significant absolute value of wavelet coefficients indicates that the crude oil prices characteristics change significantly.

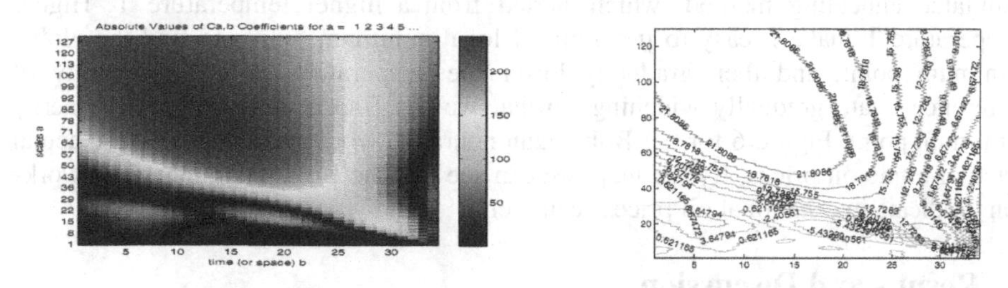

Fig. 4. Time-frequency coefficients structures of crude oil prices

In order to observe the distribution of the fluctuation energy of crude oil prices with the scale. Define international crude oil prices wavelet variance:

$$F(a) = \int (P_f(a,b))^2 db \tag{9}$$

From Figure 5, the wavelet variance distribution, we can find two main period at least : 20a and 80a

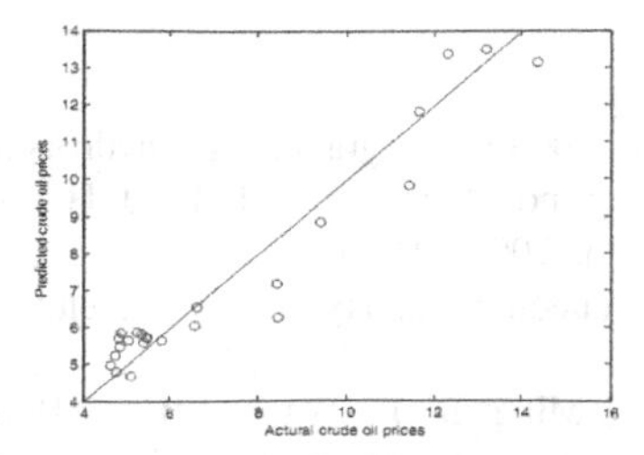

Fig. 5. The prediction of Cooperative Boltzmann neural network

4 Boltzman Neural Networks

Boltzmann neural networks is a random neural networks, is also a feedback neural network, which similar to Hopfield networks in many respects. Crude oil Boltzmann neural network works as follows:

$$\begin{cases} p_i^k = \sum_{j=i-m-1}^{i-1} w_{ij} O_{k,j} + \sum_{j=i-m-1}^{i-1} w_{ij} G_{k,j} - \theta_i \\ s_i^k = \dfrac{1}{1+e^{-p_i^k/T}} \\ \overline{p}_i = \dfrac{1}{K}\sum_{k=1}^{K} p_i^k \end{cases} \tag{10}$$

In (10), i is the discrete time; j is the discrete translating factor; and k is the discrete scaling factor; $O_{k,j}$ is the discrete wavelet coefficients of international crude oil prices; is the average crude oil price of all scales at i ; $G_{k,j}$ is the discrete wavelet coefficients of international natural gas prices; s_i^k is the Boltzmann probability of the corresponding value of scale factor k and translation factor j . In order to find global mina of the network energy function effectively, we used simulated annealing method, which started from a higher temperature T. Higher temperature T make it easy to jump out of local minimum points to find the global minimum point, and then gradually lower the temperature T, the probability of differences state gradually widening, so that we can find a more accurately energy minimum point. Figure 6 for the Boltzmann neural network prediction and the actual price of crude oil prices. By the map, we can see that the Boltzmann neural networks can predicate international oil prices accurately.

5 Results and Discussion

The international oil price forecast needs comprehensive analysis of the overall trend components and local oscillation random component. Wavelet analysis can be decomposed international crude oil prices time sequence into approximate coefficient waveform features and detail coefficients are different mechanism, and the trend of the future is not the same. So the approximate coefficient waveform features and detail coefficients have different future value of the contribution of oil prices. Use boltzmann neural network estimation model, we can predict the contribution of approximate coefficient of different waveform characteristics and detail coefficients, and finally to consider the two factors prediction accuracy get a oil prices.

References

[1] Ball, M., Wietschel, M., Rentz, O.: Integration of a hydrogen economy into the German energy system: an optimising modelling approach. Int. J. Hydrogen Energy 32, 1355–1368 (2007), doi:10.1016/j.ijhydene.2006.10.016

[2] Wensheng, W., Jing, D., Yueqing, L.: Hydrology wavelet analysis. Chemical Industry Press, Bejing (2005)

[3] Jinliang, Z., Mingming, T., Mingxin, T.: Wavelet-Based Boltzman Neural Network and kernel density estimation model portfolio in international crude oil prices prediction. In: International Conference on Future Computer and Communication (2009) (To be published)

Forecasting of Distribution of Tectonic Fracture by Principal Curvature Method Using Petrel Software

Li Zhijun[1], Wang Haiyin[2], and Wang Hongfeng[2]

[1] Chongqing University of Science and Technology, Chongqing 401331
[2] The Research Institute of Petroleum Exploration and Development (RIPEE) of Tarim Oilfield Company, Xinjiang Korla 841000

Abstract. Principal curvature method is preferably reliable in forecasting of tension fracture of uniform rock structure. In this paper, the structural plane curvature is calculated based on forecasting of fracture distribution and development level using Petrel software widely used at present. The principal curvature based structure calculation in Tarim Basin indicated that this method can be used to calculate curvature value of structural plane in an accurate manner and features simple, efficient and convenient use and favorable application value.

Keywords: Petrel software, curvature method, principal curvature, fracture forecast, fracture distribution.

1 Introduction

Curvature method was first adopted for tectonic fracture forecast in Sanish oil field in Dakota, USA, by Murray (1968) and introduced into China later, and has been progressively developed and improved. Curvature value can be calculated by various methods, among which extremum value principal curvature method [1~5] has been widely accepted, and used for fracture distribution forecast successfully.

Principal curvature method is presently one of the three major methods for fracture forecast (curvature method, finite-element method and seismic coherence cube), and the calculation of principal curvature is considered the precondition and key to the mastery of this method. By using this method together with the result of FMI logging interpretation, the feature of inter-well fracture can be forecasted easily [6]. In recent years, this method has been applied by lots of scholars in fracture distribution forecast[3].

Since the forecasting of horizontal distribution of reservoir fracture is critical during the study of reservoir properties and development program for the field with fracture-pore reservoir, it is extremely necessary to make study on techniques relating to fracture distribution forecast. In this paper, base on the structural setting information of a gas field, Petrel software is adopted to provide technical support for fracture distribution forecasting.

Z. Du (Ed.): Proceedings of the 2012 International Conference of MCSA, AISC 191, pp. 405–411.
springerlink.com

2 Computing Formula

Previous researchers have established the computing formula of principal curvature method[1~9]. The formulas mentioned in consulted references are summarized as follows:

There are infinite orthogonal curvatures at some certain point in the plane; the maximum curvature is called K_{max} (i.e. the principal curvature), and the curvature that is perpendicular to K_{max} is called minimum curvature K_{min}.

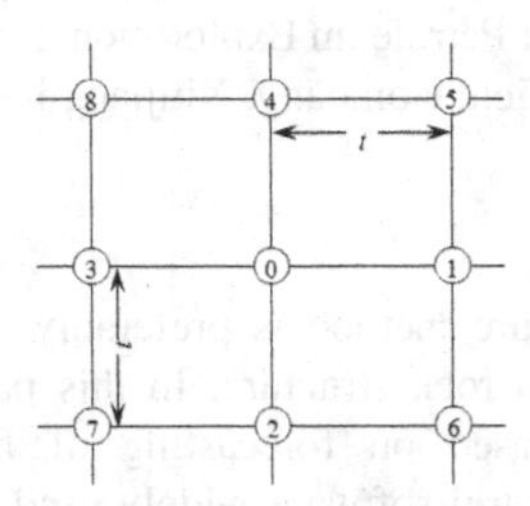

Fig. 1. Diagram of determination of curvature by difference method

As is shown in Fig. 1, in case the elevations of points 0, 1, 2, 3, 4, 5, 6, 7 and 8 are given, the principal curvature Kmax of point 0 is expressed as:

$$K_{max} = \frac{1}{2}\left(\frac{1}{R_x} + \frac{1}{R_y}\right) + \sqrt{\frac{1}{4}\left(\frac{1}{R_x} - \frac{1}{R_y}\right)^2 + \left(\frac{1}{R_{xy}}\right)^2} \tag{1}$$

Parameters in the equation can be obtained through equations (2)-(4) by difference method[9]:

$$\left(\frac{1}{R_x}\right)_0 = \left(\frac{\partial^2 Z}{\partial x^2}\right)_0 = \frac{Z_1 + Z_3 - 2Z_0}{t^2} \tag{2}$$

$$\left(\frac{1}{R_y}\right)_0 = \left(\frac{\partial^2 Z}{\partial y^2}\right)_0 = \frac{Z_2 + Z_4 - 2Z_0}{t^2} \tag{3}$$

$$\left(\frac{1}{R_{xy}}\right)_0 = \left(\frac{\partial^2 Z}{\partial x \partial y}\right)_0 = \frac{(Z_6 + Z_8) - (Z_5 + Z_7)}{4t^2} \tag{4}$$

Where, Z0, Z1, Z2, Z3, Z4, Z5, Z6, Z7 and Z8 are heights above sea level of points 0, 1, 2, 3, 4, 5, 6, 7 and 8 respectively, m; t represents grid step length, m; $\frac{\partial^2 Z}{\partial x^2}, \frac{\partial^2 Z}{\partial y^2}, \frac{\partial^2 Z}{\partial x \partial y}$ represent formation curvatures (1/m) in X direction, Y direction and the direction of XY bisector respectively.

After the principal curvature of structural plane is obtained, the distribution of fractures could be forecasted by calculating fracture porosity (Φ_f) and fracture permeability (K_f) through equations [10] (5)-(6), i.e.

$$\phi_f = \frac{1}{2} H \cdot K_{max} \tag{5}$$

$$K_f = 2\times10^5 \cdot \frac{(H \cdot K_{max})^3}{D_{lf}^{\ 2}} \tag{6}$$

Where, Φ_f: fracture porosity (%); K_f: fracture permeability (mD); H: rock stratum thickness (m); D_{lf} : the fracture density (number / m).

3 Implementation Method

In previous studies, principal curvature was usually calculated through programming [1~9]. Programming design has high requirements on interpolation algorithm, graphic display and other techniques, and data exchange with other software also involves in practice. As the currently popular commercial software, Petrel incorporates many advanced technologies and thus has the ability to calculate structural curvature.

It can be observed from the calculation method of principal curvature that, to obtain the curvature value at point 0, it is necessary to calculate the height values of eight grid points around point 0 (Figure 1). Petrel software is applicable only to arithmetical calculation of points with the same grid coordinates. In other words, to obtain the curvature of point 0, the coordinates of the eight points around point 0 must be converted to point 0.

To calculate principal curvature value, it would be necessary to use Petrel's grid translation function which is the key to the calculation of principal curvature value. As is shown in Fig. 2, translate point 1 in (I-) direction by one grid step to obtain the coordinate value of point 1 at point 0; translate point 2 in (J+) direction by one grid step to obtain the coordinate value of point 2 at point 0; the coordinate values of points 3 and 4 are calculated by similar method. Point 5 shall be translated by one grid in (I-) and (J-) directions, and point 6 is translated by one grid in (I-) and (J+) directions; the coordinate values of points 7 and 8 are calculated by similar method.

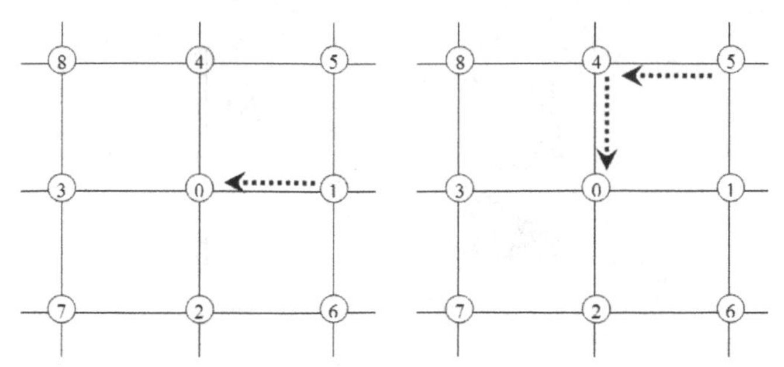

a. Peak value translation of adjacent grids b. Peak value translation of diagonal grids

Fig. 2. Schematic diagram of grid translation

To calculate the principal curvature value of a certain structural plane (plane 0) in practice, it is necessary to create 8 planes (designated as plane 1, plane 2…and plane 8) that are the same as plane 0. The principal curvature value of plane 0 can be calculated by the abovementioned translation method using the grid coordinate translation function of Petrel software.

4 Application Effect

Comparative study was made on calculated results of curvature by taking into account grid direction and step length based on structural data of a block in Tarim Oilfield.

(1) Analysis of Grid Direction Effect
Calculate the principal curvature with a step length of 50m in the directions parallel to and perpendicular to structure axis direction. The comparison of calculation results in the two grid directions (Figures 4, 5, 6 and 7) showed that the difference between the results was not significant and the curvature difference was 0~0.0003/m. It can clearly be seen that grid direction has a relatively small impact on calculated result of curvature, and that only grids in structural plane were influenced, which brought about partial loss. Comparatively speaking, grids that are parallel to axis experienced relatively small loss (Figure 4-a).

(2) Analysis of Effect of Step Length
Calculate the principal curvature value in the direction that is parallel to axial grid of the structure at step lengths of 50m, 100m, 150m and 200m respectively; comparison (Figures 4, 5, 6 and 7) shown that the calculation results at step lengths of 50m and 100m showed relatively poor regularity and mussy curvature trend (Figures 4 and 5); the calculation at step lengths of 150m and 200m reflected favorable structural curvature results and showed the clear trend (Figures 6 and 7). However, it is observed from finished chart that less interpolation loss was detected at the edge of grid plane at the step lengths of 50m and 100m, and that more interpolation loss was detected at the edge of grid plane at the step lengths of 150m and 200m.

In sum, the curvature result calculated in the grid direction that is parallel to tectonic axis at the step length of 150m in the block under study is satisfactory.

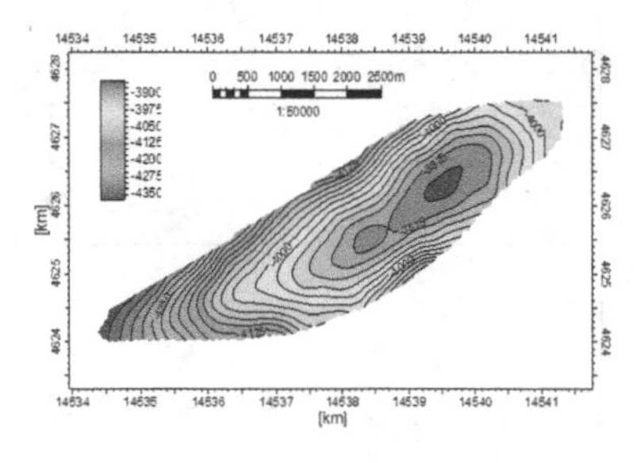

Fig. 3. Top tectonic surface of area under study

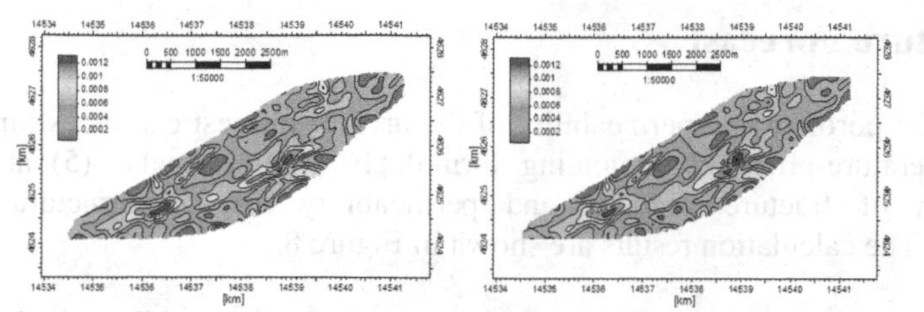

a. Grid direction is parallel to structural axis   b. Grid direction is perpendicular to structural axis

Fig. 4. Distribution of principal curvature at the step length of 50m

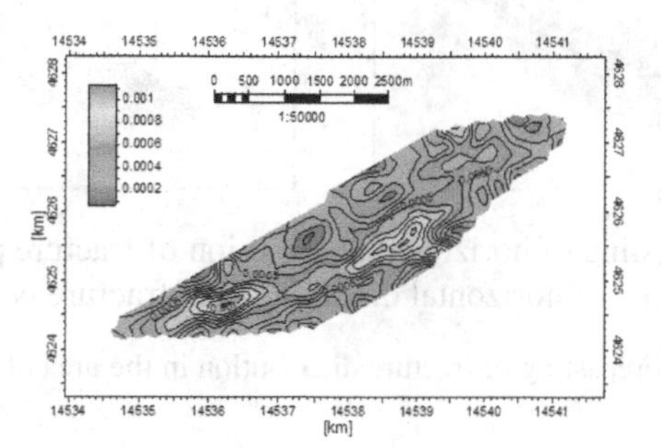

Fig. 5. Distribution of principal curvature at the step length of 100m

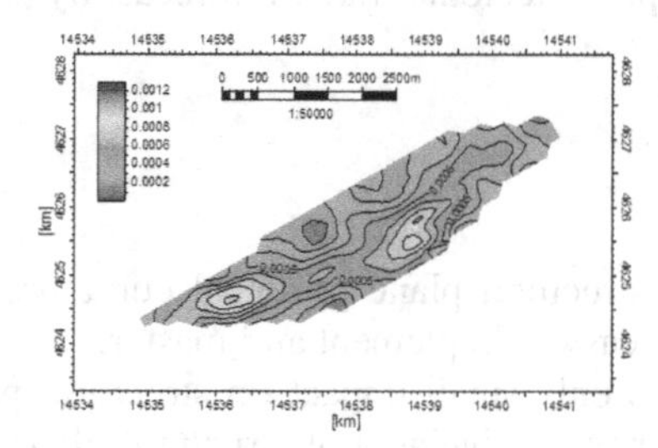

Fig. 6. Distribution of principal curvature at the step length of 150m

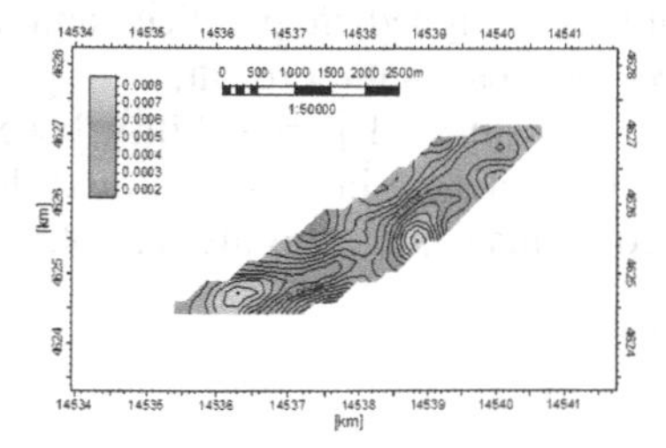

Fig. 7. Distribution of principal curvature at the step length of 200m

5 Fracture Forecast

The fracture porosity and permeability of the area of interest can be estimated based on the literature-provided computing formula[10] (i.e. formulas (5) and (6)) for calculation of fracture porosity and permeability through structural principal curvature. The calculation results are shown in Figure 8.

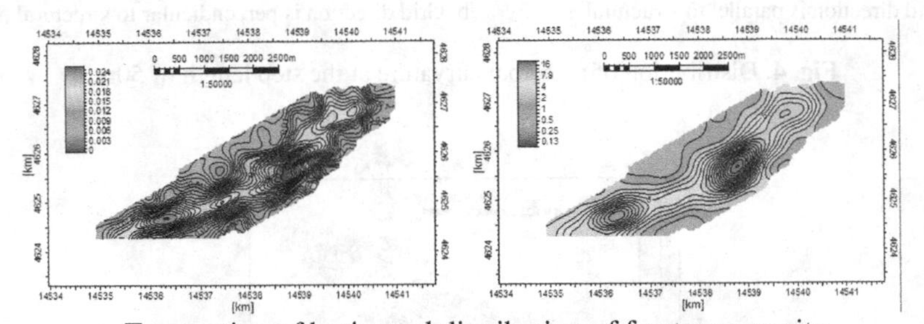

a. Forecasting of horizontal distribution of fracture porosity
b. Forecasting of horizontal distribution of fracture permeability

Fig. 8. Forecasting of fracture distribution in the area of interest

It is observed from Figure 8 that, the fracture porosity and permeability estimated based on principal curvature values are favorably consistent with structural feature, complies with the principle of tectonic fracture forecast by principal curvature method and is favorably reliable.

6 Conclusions

(1) The calculation of structural plane principal curvature using Petrel software is intuitive, convenient and easy to implement and master;

(2) Grid direction exerts only small impact on structural plane curvature value and influences only the integrity of edge area of structural plane; the grid direction that is parallel to axis would bring about favorable results in practice;

(3) Grid step length exerts significant impact on structural plane curvature value; different step lengths could be designed in practice, and the step length shall be determined based on relative ideal calculation result;

(4) The estimated fracture porosity and permeability based on structure principal curvature are consistent with the change in structure, and the forecasting of fracture distribution by method stated in this paper is highly reliable.

References

1. Sun, S.: Comparison of the two curvature methods for forecasting of reservoir fracture. Geological Science and Technology Information 222(4), 71–74 (2003)
2. Li, Z., Zeng, Z., Luo, W.: New exploration of principal curvature method for fracture forecast. Petroleum Exploration and Development 30(6), 83–85 (2003)

3. Peng, H., Xiong, Y., Sun, L., et al.: A study on application of principal curvature method in forecasting of tectonic fracture at gas reservoir of carbonate rock. Natural Gas Geoscience 16(3), 343–346 (2005)
4. Guo, K., Xu, Z., Ni, G.: Fractured reservoir study by principal curvature method. Geophysical and Geochemical Exploration Technology 20(4), 335–337 (1998)
5. Kang, Y., Ren, W., Zhu, Y., et al.: Method for and practice of forecasting of natural fracture development zone using extremum value principal curvature method. Oil and Gas Well Testing 11(4), 22–26 (2002)
6. Chang, B., Zhang, C., Qu, T., et al.: Predictive study on dolomite fracture in deep recessed area of Miyang. Oil & Gas Journal (of Yangtze University) 31(3), 202-206 (2009)
7. Hu, Y., Lu, G., Yang, Z.: A study on quantitative prediction method of reservoir fracture. Southern China Oil & Gas 19 (2-3), 43–45 (2006)
8. Tao, H., Gong, J., Zhang, X., et al.: Application of stratum principal curvature in research of reservoir fracture development. Inner Mongolia Petrochemical Industry 3, 138–139 (2006)
9. Luo, M.: Quantitative reservoir geology. Geological Publishing House, Beijing (1998)
10. Van, T.D., Golf, R., Chen, R., Jin, L., Qin, T.: Fracture reservoir engineering fundamentals. Petroleum Industry Press, Beijing (1989)

Study on Bulk Terminal Berth Allocation Based on Heuristic Algorithm

Hu Xiaona, Yan Wei, He Junliang, and Bian Zhicheng

Container Supply Chain Engineering Research Center, Shanghai Maritime University, China

Abstract. Operational efficiency is the key factor of terminal's profit .This paper is intended to improve the efficiency of berth allocation and establish the mathematical model of berth allocation strategies including berthing location and berthing time according to the principle of quayside's limitation and the loading capacity of ships. On the condition that maximizes terminal's throughput, all the ships has the shortest time delayed when they depart from the port . A case study of a Terminal in Tianjin Port, heuristic algorithm is used to obtain the solution for this model, the result has the feasibility and practicability.

Keywords: Continuous quayside, bulk cargo terminals, berth allocation, heuristic algorithm.

1 Introduction

With the increasing competition among various ports, one of the key factors is how to maximize berth utilization, accelerate the speed of loading and unloading of ships, upgrade the port's operational efficiency and the service level. As an important resource in port transportation, berth is also one of the key factors that influences the port's development. Therefore, berth allocation problem becomes one of the important issues in relative field. On the premise of ensuring the service level, improving the throughput of the terminal on a certain day act as a key factor.

The domestic and foreign scholars have already had a mature research on the berth allocation. Cai Yun .etc[1] established simulated optimization model to minimize the overall time of in-port ships for the container terminal berth allocation and quay crane scheduling problem and used genetic algorithms to generate and evaluate the solution to berth allocation; Zhang Haibin .etc[2] changed the berth scheduling problem into special 2d packing problem with constraints, established nonlinear programming model for continuous berth scheduling problem, and then used heuristic algorithm to solve berth allocation problem; Wang Hongxiang .etc[3] set up the mathematical model for dynamic berth allocation strategy on the basis of heuristic algorithm, they simulated and optimized this model with a container terminal in Shanghai as an example ; IMAI .etc[4] constructed linear programming model and a genetic algorithm was adopted, the model was proposed with the target of shortest loading and unloading time as to mixed berthing problem for big ships and barges; CHEN .etc[5] put forward the Network Pattern that could be selected and established a network model for berth allocation with the consideration of time and space; GUAN .etc[6] solved this kind of

Z. Du (Ed.): Proceedings of the 2012 International Conference of MCSA, AISC 191, pp. 413–419.
springerlink.com

large-scale problems by combining binary tree and heuristic algorithm, minimizing the total time. This paper established the maximum throughput and the navigation delay time the shortest mathematical model to Tianjin coal terminal as an example, the use of heuristic algorithm to find the solution to the model.

2 Problem Description

With the increasingly competitive business environment, both the bulk terminals and container terminals are trying to reduce costs through the effective utilization of resources, including human resources, berths and all kinds of terminal equipments. The operational procedure for bulk terminal is that: the arrival of ships, preparation for berthing, judgment on whether the quay length is in accordance with the berthing conditions, if it accords with the judgment, the ship berths at the certain location and processes loading or unloading works, otherwise the ship must wait.

Generally, he following factors are considered during berth scheduling: the size, arrival time and loading capacity of the ships, both with the type and location of the cargo. This paper is to maximize the daily throughput of the terminal, and at the same time to minimize the delay of ships' departure.

3 Model Established

3.1 Model Assumption

This model is based on the following assumption: (1) the machines that serve each ship are enough; (2) each ship berths at the preferred section; (3) ships arrive at the port on time according to the schedule;(4) the distance between two in-port ships is 1.2 times of the ship's length; (5) conflict between ships will not happen.

3.2 Parameter Definitions

s- number of arriving ships in 24 hours;
L – quay length;
$i,\ j$ – the ship i,j=1,2,…,s;
l_i– length of ship i ;
C_i– deadweight of ship i;
At_i–time of arrival of ship i;
Et_i –time of departure of ship i;
T_i– handling time of ship i;
E_i–ending berthing position of ship i;
Dt_i –estimated time of departure of ship i;
S_{ij}–if ship i and j overlap in time schedule, $S_{ij} = 1$, otherwise $S_{ij} = 0$;
P_{ij}–if ship i and j overlap in berthing location, $P_{ij} = 1$, otherwise $P_{ij} = 0$;
R_i–if $Et_i \le Dt_i$, $R_i = 0$, otherwise $R_i = 1$;
S_i- decision variables, beginning berthing position of ship i;
St_i–decision variables, specific time that ship i berths;
X_i- decision variables, if ship i berths, $X_i = 1$, otherwise $X_i = 0$;

3.3 Objective and Constraints

$$f_1 = \max \sum_{i=1}^{s} X_i C_i \tag{3.1}$$

Formula (3.1) is the optimization objective, requiring that the total deadweight of ships within one day is maximized.

The objective function is normalized as below.

$$f_1 = \sum_{i=1}^{s} C_i / \sum_{i=1}^{s} X_i C_i \tag{3.2}$$

$$f_2 = \min \sum_{i=1}^{s} R_i \left(Et_i - Dt_i \right) \tag{3.3}$$

Formula (3.3) is another optimization objective, requiring that the summation of delaying time of ship's departure is minimized.

The objective function is normalized as below.

$$f_{22} = \left[24s - \sum_{i=1}^{s} R_i \left(Et_i - Dt_i \right) \right] / 24s \tag{3.4}$$

$$f = \min\{\omega_1 f_{11} + \omega_2 f_{22}\} = \min\left\{ \omega_1 \sum_{i=1}^{s} C_i / \sum_{i=1}^{s} X_i C_i + \omega_2 \left[24s - \sum_{i=1}^{s} R_i \left(Et_i - Dt_i \right) \right] / 24s \right\} \tag{3.5}$$

Formula (3.5) is normalized objective function. Assume that the largest throughput within one day means all ships have been berthed, while the smallest means they aren't berthed; Each ship's maximum delayed departure time is 24 hours, while the minimum is no delay. Weight ω_1 is greater than ω_2.

$$S_i \geq 0 \tag{3.6}$$

Formula (3.6) is the constraint, ensuring that the beginning position of all ships is within quay line;

$$E_i \leq L \tag{3.7}$$

Formula (3.7) is the constraint, ensuring that the end position of all ships is within quay line.

$$\left(E_i - S_j \right) \cdot \left(S_i - E_j \right) \cdot S_{ij} \geq 0 \tag{3.8}$$

Formula (3.8) is the constraint, ensuring that berthing ships don't conflict in space horizon under non-overlapping time schedule.

$$\left(Et_i - St_j \right) \cdot \left(St_i - Et_j \right) \cdot P_{ij} \geq 0 \tag{3.9}$$

Formula (3.9) is the constraint, ensuring that berthing ships don't conflict in time horizon under non-overlapping space schedule.

$$\left(E_i - S_i \right) \geq 1.2 L_i \tag{3.10}$$

Formula (3.10) is the constraint, ensuring the distance that quay line left is more than 1.2 times of ship's length.

$$St_i - At_i \geq 0 \tag{3.11}$$

Formula (3.11) is the constraint, ensuring that in-port ships can be served.

$$\left(Et_i - Dt_i\right) \leq T_i \tag{3.12}$$

Formula (3.12) is the constraint, delayed departure time is no more than the handling time of ship *i*.

4 Heuristic Algorithm

4.1 Model Transformation

Before the ship arrives at the port, scheduling staff usually arrange ship's berthing position and berthing time according to such corresponding information as ships' arrival time and planned departure time. This arrangement can be shown through figure1.

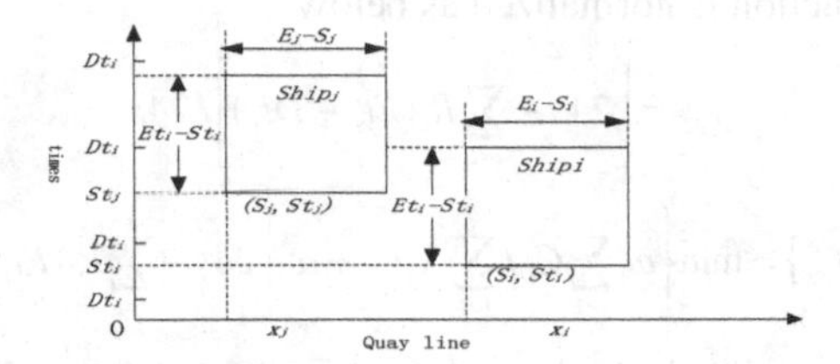

Fig. 1. Transformation model of berthing ships

Each ship's berthing operation can be expressed by a rectangular [7], rectangular's horizontal edge E_j-S_j is the occupied quay length(including clearance distance), its vertical edge Et_i-St_i represents for loading and unloading time, (S_j, St_j) respectively means the beginning berthing position and the berthing time, At_i is the arrival time of ships, Dt_i is for the expected departure time. And so, they can be changed into rectangular arrangement problem with specific constraints to get the solution. In practice, experienced scheduling staff may select several berthing order as an initial strategy in most cases, effectively arranging ships' loading and unloading, this is very favorable to get a better satisfactory solution.

4.2 Processing Steps

1 Assume that *s* ships arrive at the port in 24 hours, among which the maximum length is *M*, the minimum one is *N*. From the total quay length and the smallest ship, we can know the largest number of available berths.

2 According to initialized parameters, and the waiting for the arrival of the ship will be expected to rectangular, check the berthing ships and ships waiting at the anchorage, and give the order number according to the ships' estimated time of arrival ship.

3 Arrange the order according to ships' deadweight, namely rank them from big rectangular to small one.

4 Select the item with the greatest height (the largest deadweight is presumed as *i*) into box's lower left corner.

5 Transversely cutting the line that is above i , and separate the rest of space into two parts, the space below the tangent is the first layer for packing(berthing ships of the first round).

6 Find the item of minimum height difference compared with i in the rest of s (loads don't largely differ from each other); If there are many small article of minimum height difference, then choose the item with smallest order number, presuming j on the first layer, rank them close to i to right direction, continue to scan s, and look for items with minimum height difference and arrange them to the right direction, until the item can't be place into the space.

7 Transversely cutting the line that is above j, rescan s, look for minimum height difference of items compared with i and j, set the sort on upper right part of j , repeat the above process until no access to the first layer .

8 In the arrangement of the next layer, check whether ships can be put into this layer according to ship's estimated time of arrival and then update s, repeat step 3--7, choose appropriate rectangular items into two dimensional space.

In the above process, in order to guarantee the maximum daily throughput, after ensuring the first ship of each layer, it calculates arrival time of all ships in the next planning horizon, if it appears that ships in the next layer have greater loading capacity than that in the first layer, the certain ship will be arranged with higher priority. And meanwhile, in the planning process, we check the expected time of arrival to see whether they could set ships into one layer and then update the results.

4.3 Example Analysis

Take coal terminal of Tianjin port as an example, quay length is 1147 meters, Because the water depth and berthing capacity are same, the coal terminal has a continuous berth. Assume that continuous berth water depth is greater than maximum water depth of all arriving ships. Choose ships' data on from October 10, 2011 to October 11, 2011, in order to meet required time format and convenient calculation, the berthing time starts from 0:00. Ships' data is shown in table 1.

Table 1. Initial data of berthing ships

Numbers of ship	Occupy berth length	Carrying quantity(t)	arriving anchorage time	Numbers of ship	Occupy berth length	Carrying quantity(t)	arriving anchorage time
1	174	16000	2011-10-10 12:00	16	123	5000	2011-10-11 02:00
2	140	8000	2011-10-10 13:00	17	142	10000	2011-10-11 04:00
3	224	27000	2011-10-10 14:00	18	168	12000	2011-10-11 05:00
4	174	16000	2011-10-10 15:00	19	174	16000	2011-10-11 06:00
5	224	27000	2011-10-10 16:00	20	140	8000	2011-10-11 08:00
6	174	4000	2011-10-10 16:00	21	168	12000	2011-10-11 08:00
7	149	8500	2011-10-10 16:00	22	226	27000	2011-10-11 10:00
8	140	8000	2011-10-10 18:00	23	120	5000	2011-10-11 11:00
9	198	20000	2011-10-10 18:00	24	168	12000	2011-10-11 12:00
10	150	3000	2011-10-10 19:00	25	179	17000	2011-10-11 14:00
11	142	10000	2011-10-10 20:00	26	161	11000	2011-10-11 16:00
12	130	9000	2011-10-10 20:00	27	167	16500	2011-10-11 16:00
13	123	5000	2011-10-10 22:00	28	161	10400	2011-10-11 17:00
14	176	16000	2011-10-10 00:00	29	120	5000	2011-10-11 17:00
15	150	13000	2011-10-11 01:00	30	120	5000	2011-10-11 17:00

Heuristic algorithm is applied to solve this example, and finally the solution with several constraints is figured out, as shown in figure 2, figure 3 is the berthing allocation as to the first come first service rule.

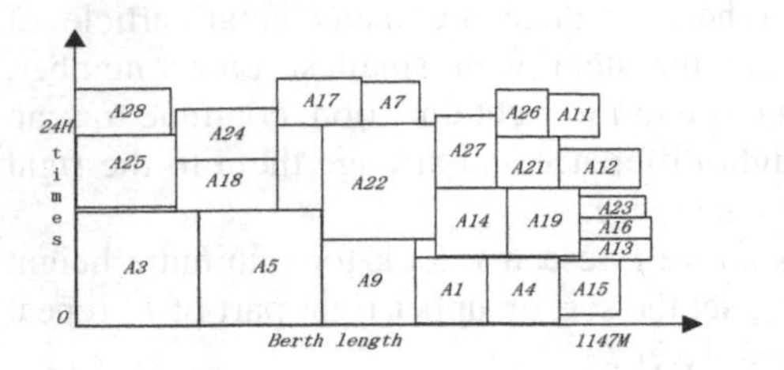

Fig. 2. Berth allocation with heuristic algorithm

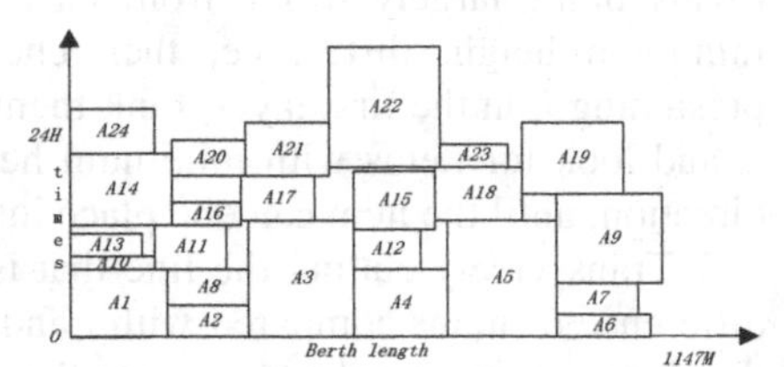

Fig. 3. Berth allocation with FCFS rule

From the above mentioned two figures, the terminal's daily throughput can be calculated. And it is shown in table 2.

Table 2. Two different scheme comparison

methods	Daily throughput(t)
heuristic algorithm	331400
first come first service	297500

Many tests show that, with the extension of time and berth length, the heuristic algorithm can more effectively enhance berth's daily throughput.

5 Conclusion

This paper first offers dynamic berth allocation model based on continuous quay line, and then in the case of coal terminal of Tianjin port, heuristic algorithm is used to solve the problem, and the results show that this algorithm has an obvious effect on the increase of daily throughput and the reduction of total time that ships spend in the port. The model and algorithm is proved to be feasible and practical.

Acknowledgement. This work sponsored by National 863 plans projects (2009AA043001), supported by Shanghai Education Committee Projects (J50604), supported by Ministry of Communications Research Project (2009-329-810-020 & 2009-353-312-190), supported by Shanghai Maritime University Research Project (20100130 & 20110019). supported by Shanghai Science & Technology Committee Research Project (09DZ2250400).

References

[1] Cai, Y., Zhang, Y.: Simulation optimization for berth allocation in container terminals. Chinese Journal Of Construction Mechinery 2(4), 228–232 (2006)

[2] Zhang, H., Zhang, J., Xuan, J.: A Study On heuristic algorithms with Container berths of scheduling. Journal of Qingdao University 25(4), 57–60 (2010)
[3] Wang, H., Yan, W.: Quaywall-length-based berth allocation strategy on heuristics algorithm and simulation optimization. Journal of Shanghai Maritime University 29(2), 19–22 (2008)
[4] Imai, A., Nishimura, E., Papadimitrious, S.: Berth allocation at indented berths for mega-container ships. Eur. J. Operational Res. 179(2), 579–593 (2007)
[5] Chen, C.Y., Hsieh, T.W.: A time space network model for the berth allocation problem. In: 19th IFIPC7 Conf. on Syst. Modelling & Optimization, Cambridge,UK (1999)
[6] Guan, Y., Cheung, R.K.: The Berth allocation problem:models and solution methods. OR Spectrum 26(1), 75–92 (2004)
[7] Andrew, L.: The berth planning problem. Operations Research Letters 22, 105–110 (1998)

The Research of Intelligent Storage Space Allocation for Exported Containers Based on Rule Base

Yan Wei, Bao Xue, Zhao Ning, and Bian Zhicheng

Logistics Engineering School, Shanghai Maritime University, China

Abstract. This paper mainly solved the problem of the storage space allocation for export containers based on the rule base. It putted forward an intelligent method which can easily lay the outbound container on the object container. The analysis based on the rule base which made the plan of container yard accomplished more easily. It also met the demand of the real operation to the maximal limit. Meanwhile, the improved utilization ratio of container yard and the increased mechanical efficiency were beneficial to arrange the operation of shipping. At last, the efficiency and practicality of this intelligent system was proved by practical case study. So the settlement based on rule base is more reasonable for intelligent storage space allocation.

Keywords: Container Terminal, Export-Container, Storage Space Allocation, Rule Base.

1 Introduction

With the development of economic globalization and trade internationalization, the throughput and loaded and unloaded rate of container terminal, as a hub in the shipping field, directly impact the economic benefit. The process of choosing a space to allocate objective container is before the operation of concentrating containers in the yard. According to advanced marshaling plan, it is a decision making process of allocating containers on the appointed space. Therefore, it is an important process of explored business. The results of this system is better whether or not determined machines operation and the rate of shipment.

Many researchers have made headway about Problem on the resource allocation of container terminal. KozanE and PrestonP [1] established the comprehensive model of container transportation and container allocation in the yard to solve the relevant problems. They determined the optimal cooperative dispatch by using iterative search algorithm. Tao Jing-Hui and Wang Min [2] established a hybrid storage model and solved it by using heuristic algorithm. Bazzazi Mohammad,Safaei Nima and Javadian Nikbakhshp [3] researched on the allocation problem of extent storage space in container terminal. And they solved it by using an effective genetic algorithm. Zhou Pengfei had a research on the container allocation in container terminal under the indeterminate conditions. He established a dynamic model to solve problems and proved the feasibility of relevant algorithm by simulation experiment.

In this paper, the main research is to solve the problem of choosing a proper allocating space for explored containers after entering the yard. Based on analyzing

Z. Du (Ed.): Proceedings of the 2012 International Conference of MCSA, AISC 191, pp. 421–425.
springerlink.com

and concluding the mature and faults of the system about intelligent storage space allocation for exported containers, for the purpose to propose better rules for this process, the study about process of operation of transporting explored container is made further. These rules can improve the theory and method of choosing space to allocate explored containers. Moreover, the case analysis takes further steps to prove the flexibility operation and advantage of this method.

2 Problem Description

2.1 Prerequisite Hypothesis

The expression of rule base needs some hypothesis. One is the completive container yards operation schedule. The other is containers of other ships will not grab the location for planed containers.

2.2 Expression for Common Rule Base

The process of selecting location for export container includes three models are : "put the container on others top or side", "grab the location of other containers", "hybrid stack model". Every model has its own rules about location selection. However, different models have some common rules. These common rules as follows.

The heavy container lay on the light container. If some containers have the same characters on voyage number, size, port of discharge and tons empty, they need follow this rule. Meanwhile, the weight of objective container is close to the pressed container will have the priority to locate.

The rule of concentration. When the operation needs a new slot for locating objective containers, these containers should possibly guarantee that they have the same characters.

Classified location. If the schedule places in the yard are enough to locate, different types of containers need different location places.

Distributed location. If the number of some characteristic container is large, these containers are better to locate in distributed yard region.

2.3 The Expression of Concrete Rules

The flow chart of process put container on another above or laterally is shown in Fig.1. The concrete rules are as follows.

Rule1: After container trucks entering the container terminal crossing, related personnel will get the characters of objective by RFID.

Rule2: Searching for the containers' information mainly includes the voyage number, the port of discharge and the size of container, which is similar to the objective containers'. If results are irrelevant , the rule6 will be carried out.

Rule3: Making a determination whether the found container is appropriate one. The container which can be pressed must have not been put on the top story and it is not in the immutable region. If the result meets the demand of rule3, the system will go on, or else, the system will carry out Rule5 directly.

Rule4: The weight of pressed container must be less than the weight which is equal to the objective containers' and the weight limit value between light container and pressed heavy container. The target of this rule is to guarantee the feasibility of the container stack in the processing of shipment. If the result meets above conditions, this system will carry out the Rule7.

If there are several feasible pressed containers, the objective container will have the priority to lay on when the weight difference between objective container value and pressed container value is the least.

Rule5: Within a certain range in the container yard, there are empty row whether or not. If there is a new row , then Rule7 will be carried out, or else, the system will go on. Use of this rule turns the process of finding a container can be pressed to finding one which can put a objective container on laterally.

Rule6: Opening up a new bay. This rule which refers to the container yard operation schedule will be carried out, when other rules are not reasonable. To get the schedule of objective container by reading information from container yard operation schedule. Next step is to find an empty bay within this schedule to put the objective container.

Rule7: Finding a reasonable location to put an objective container on it based on the system.

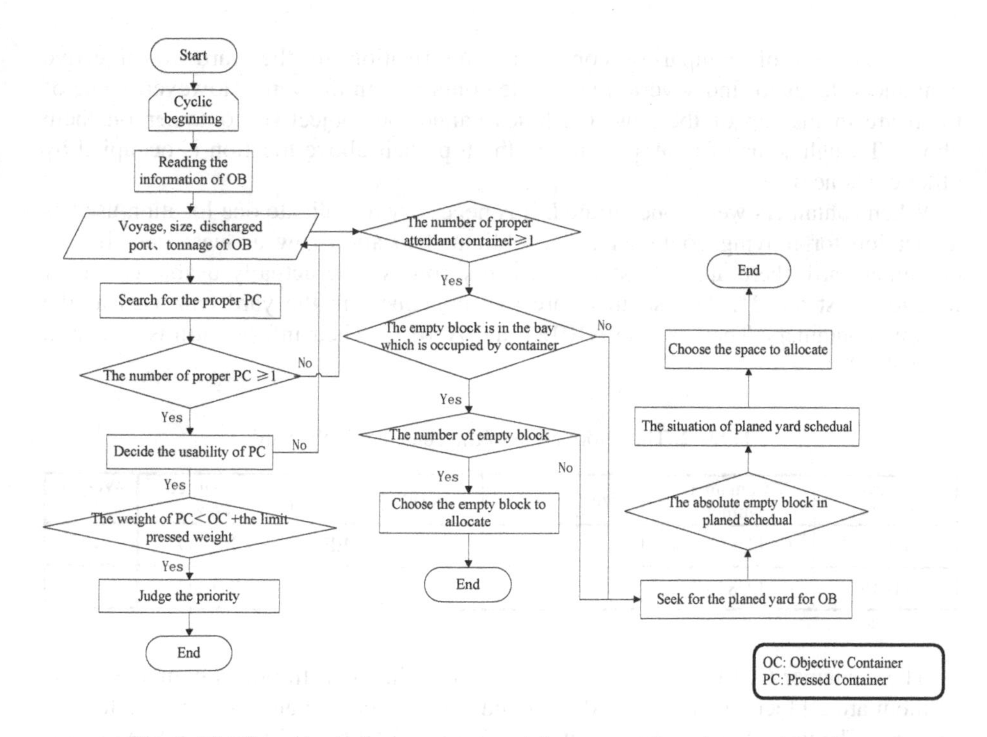

Fig. 1. The flow chart of process put container on another above or laterally

3 Case Analysis

According to the information got from the operation in container terminal, the analysis of the process of concentrating containers in container yard to prove the practicability and validity of the rule base. Based on the export ships manifest, containers have their own operation schedule for different port of discharge and tons. In the container yard operation schedule of chose ship, there are four container block: A1, A2, A3and A4. In A1, the first, third and fifth bay are used to put 20 feet containers. In A2, the first and third bays are used to put 20 feet containers which will be discharged in Busan. In A3, the seventh and ninth bays are used to put 40 feet containers. In A4, the forty-seventh bay is used to put .refrigerated containers.

According to rules the system will simulate the process of choosing location for object containers. The information of first container is shown in Table1.

Table 1. The information of first export container

Serial Number	Container Number	Size	Style	Earmark	Mode	Discharged port	Weight (T)
1	CCLU9875934	20	GP		EF	CNTAO	15

The process of comparing containers' information in the yard to objective containers' leads to find several reasonable containers in the yard. However, some of them are in the top of the row, machines cannot put objective container on them above. Though some of others are not in the top, their above location is occupied by other containers.

When containers were concentrated, it is necessary to obligate one location used as a location for moving containers. The obligated location row at least one. In this container yard, there are four stories in every row, so the actually usable container locations just are 21. At last, there are two containers in the yard can used as the pressed containers. They are the A10121 and A10152. Their information is shown in the Table2.

Table 2. The information of the container in the yard

Yard Number	Container Number	Size	Style	Earmark	Mode	Discharged Port	Weight (T)
A10121	CCLU9878396	20	GP		EF	CNTAO	14
A10152	CCLU8509396	20	GP		EF	CNTAO	12

The container CCLU9878396 in the A10121 is similar with object container. Both of them are 20 feet container, the discharged port is Qingdao and the tonnage level is the first. The weight of presence container CCLU9878396 is 14 tons which is one ton lighter than the weight of objective container. It meets the rules of laying heavy container on the lighter one.

The next step is to analyze the container CCLU850939 in the A10152. Compared to the objective container, they also have same size, discharged port and tonnage. But the weight of container CCLU850939 is 12 tons, which is 3 tons less than the objective container. So it also match the rules.

According to concentrate judgment, both container A10121 and container A10152 are favorable for being as pressed containers. However, the different between container A10121 and the container in CCLU9878396 is much less than the different between container A10121 and the container in CCLU8509396.

With analysis and cooperation, it is clear that the container CCLU9878396 has priority to be pressed by the objective container. So the block A10122 will to be the terminal chooses.

4 Conclusion

According to a series theory research, this paper establish system based on a rule about putting an objective container to some container which in the yard. Moreover, this system proved the flexibility and practicability by analyzing the case. This system can make the rate of moving container lower. However, the large randomness of the allocation mode in the yard leads to different occupancy rate. These rules have limitations under the high occupancy rate. Therefore, the target of further research is to study relevant rules under the high occupancy rate. Then, the system can be further improved.

References

[1] Kozan, E., Preston, P.: Mathematical modelling of container transfers and storage locations at seaport terminals. Journal of the OR Spectrum Part 2, 87–105 (2006), doi:10.1007/978-3-540-49550-5_5

[2] Tao, J.H., Wang, M.: Assign problem of container yard section based on mixed storage model. Journal of Transportation Systems Engineering and Information Technology 29, 185–192 (2009)

[3] Mohammad, B., Nima, S., Nikbakhsh, J.: A genetic algorithm to solve the storage space allocation problem in a container terminal. Journal of Computers and Industrial Engineering 56, 44–52 (2009)

[4] Zhou, P.F.: Research on Slot Optimal Allocation of Export Container in Container Terminal under Uncertain Conditions (The Master's excellent thesis, Institute of Technology of Dalian (2010)

Study on Evaluation System of The Quayside Container Crane's Driver

Liu Xi, Mi Weijian, and Lu Houjun

Shanghai Maritime University, Shanghai 201306

Abstract. Due to the operation of quayside container crane's driver includes many uncertain factors, this article adopt to FAHP to create the evaluation system of the quayside container crane's driver. Firstly, the indicator of system should be formulated. Secondly, the weight of each indicator should be made by AHP. Finally, the membership function model of system should be created and combine the fuzzy judgment matrix to get the result of the whole system.

Keywords: Quayside container crane, Evaluation system, FAHP, Judgment matrix.

1 Introduction

Recently, with the development of social economy and economic globalization, the import and export trade of our country is increased quickly, which raises a higher requirement for maritime transport. Cargo-handing is the start and end of maritime transport and the key point of freight. Because the transportation of general cargo exists lots of drawbacks, such as the low efficiency of loading and transport, the sever break of cargo etc, which bring a big loss to the companies of the cargo and shipping, as well as the port enterprise. The best way to solve this problem is to adopt to the container transport, which cause a fact that containerized cargo is in the account for a large proportion of the total seaborne cargo. How to achieve faster speed of cargo-handing and reduce the loss of cargo will be related to the development and benefit of the enterprise. According to the survey, the main equipment of loading the containers is quayside container crane, so the operation level of drivers is the important factors that will affect the speed of cargo-handing and the loss of cargo. So it is necessary to study the evaluation system of quayside container crane's driver , which will be the reference to train drivers to improve the benefits of port enterprise. AHP, a multi-criteria decision-making methods, which is widely used in the analysis of complex systems and decision-making. this article uses the AHP to make a qualitative and quantitative analysis about the operation of drivers .This article put forward the evaluation model based on the FAHP by formulating the fuzzy judgment matrix ,ensuring the weight of each indicator and creating the membership function model.

Z. Du (Ed.): Proceedings of the 2012 International Conference of MCSA, AISC 191, pp. 427–432.
springerlink.com

2 Establishment of the Indicator and Scoring for Quayside Container Crane's Driver

Establishing the indicator is the basis of evaluation of the driver's operation is good or bad. In the production process, the port enterprise usually pursues two indicators, which are safety and efficiency. According to the requirement of the port enterprise, the evaluation of driver should be included by safety indicator and efficiency indicator. But, this two indicators can't fully reflect the real situation of driver. The value of safety indicator and efficiency indicator will vary greatly. In order to evaluate the driver perfectly, this article sets some other indicator that can reflect the quality of driver.

2.1 Determination of the Evaluation Set

Setting the target indicators $W=\{W_i\}, i=1,2,3,4$;the factor indicators $W_i=\{W_{ij}\}$, j represents the indicator that will be included by target indicator i .

Firstly, Safety indicator(W_1)includes : The number of collisions(W_{11})、 The fastest speed of collisions(W_{12}) and The operating steps(W_{13}).(1)The number of collisions: A good driver should try to reduce the rate of collisions, because collisions will break the cargo ,destroy the agency of equipments and even cause personal injury or death.(2)The fastest speed of collisions: if it is inevitable to collide, the driver should reduce the speed of collision. The fastest speed, the more lost will be caused.(3)Operating steps: Due to quayside container crane is a large and complex organization, there is no action in some organization, no the next job, or some device is not functioning at the same time. Although the protection of electrical system has been done quite well and can prevent the accident, however, the driver develops such a poor operating practice, which will also affect the useful life of equipment and also will cause danger.

Secondly, efficiency indicator(W_2) includes: Handing efficiency(W_{21}). With the increasingly fierce competition between the word container port, improving the efficiency of container terminal production has gradually become the main way of ports to enhance market competitiveness. while the main measure to improve the production efficiency is to train skilled drivers of quayside container crane.

Thirdly, Other indicator(W_3) includes: Proficiency degree(W_{31})、 Response capability(W_{32}) and The level of tension(W_{33}), which can reflect the driving skill and mental endurance of drivers.

2.2 Scoring Standard of Each Indicator

(1)Scoring standard of safety indicator : The expression $W_1=\{W_{11},W_{12},W_{13}\}$ is the set of factor factors of safety indicator. The two factor indicators W_{11} and W_{12} can be merged into collision situation.

a. The factor set of collision situation is $\{W_{11}, W_{12}\}$ and the rank set is U={Excellent, Well, Medium, Eligible, Failure}.According to research data, as well as the experience of the old master drivers of quayside container crane, we can get the following scoring standard(table 1 and table 2).

Table 1. The score sheet of the number of collisions(40 TEU)

The number of collision	0—2	3—5	6—8	9—10	11 and above
rating	Excellent	Well	Medium	Eligible	Failure

Table 2. The score sheet of the fastest speed of collisions

The fastest speed of collisions(m/s)	0—5	5—10	10—14	14—16	16 and above
rating	Excellent	Excellent	Medium	Eligible	Failure

b. The factor indicator of operating steps is W_{13}. According to the quayside container cranes safe operation standards, we have developed six criteria in the evaluation system. If the drivers in the operation failed to comply with the six standard, the evaluation system should be to make the appropriate deduction, happening once, on 10 points. Finally, the score of 90 or more is excellent, more than 80 is well, more than 70 is medium, more than 60 is eligible, and the rest failed. The specific content of the six standard as follows:

①Drivers must fasten seat belts before the job ; ②no-load test must be carried out.
③Only the spreader top pin light , the drivers can turn on or turn off the spin lock switch.
④Cart is anchored to the state, to prohibit the operation cart run handle.
⑤Only selected fixed guides, the guides up or down operation.
⑥Over alarming to press the emergency stop button.

(2)Scoring standard of efficiency indicator: The expression $W_2 = \{W_{21}\}$ is the set of influencing factor of efficiency indicator. Rank set is U={Excellent, Well, Medium, Eligible, Failure}. According to research data, as well as the experience of the old master drivers of quayside container crane, we can get the following scoring standard(table 3).

Table 3. The score sheet of handing-efficiency

Handing-efficiency(TEU/h)	36 and above	31-35	21-30	11-20	10 and below
rating	Excellent	Well	Medium	Eligible	Failure

(3)Scoring standard of other indicator: The expression $W_3 = \{W_{31}, W_{32}, W_{33}\}$ is the set of influencing factor of other indicator. The rank set is U={Excellent, Well, Medium, Eligible, Failure}. The specific score of each factor is given by instructor.

3 Determination of Weight and Establishment of Membership Function Model

This article uses AHP to determine the weight of each indicator and gets a fuzzy judgment matrix by judging each indicator, then conducts related calculation by using the membership function model and get the result of whole system.

3.1 Determination of Weight

Solving weights is the key measure of the comprehensive evaluation. AHP is an effective method to determine the weight and applies to complex problem. Specific steps includes:(1)Define the objectives and evaluation factors: with n-evaluation, $P=(P_1,P_2,...,P_n)$.(2)Structure judgment matrix: the value of factor of the judgment matrix reflect the relative importance by adopting to the scaling method varying from 1 to 9. P_{ij} expresses the importance degree of P_i relative to P_j. Get the judgment matrix $S=(P_{ij})_{n\times n}$. (3)Calculate the judgment matrix: The software of Mathematic will calculate the maximum characteristic root $\lambda_{\max}$ and the relative eigenvectors $A=(\omega_1,\omega_2,...,\omega_n)$,which will be the ranking of importance of each indicators. That is weight coefficient.(4)Consistency test: if C.I=$\frac{\lambda_{\max}-n}{n-1}\le 0.1$,the matrix is satisfactory.

3.2 Establishment of Membership Function Model

Firstly, the set of factors W should be judged. Secondly.we can get the membership R_{ij} that is the rank U_j of the single factor W_i. Lastly, the fuzzy subset of evaluation R_{ij} will be ensured.

$$R_{ij}=(R_{i1},R_{i2,...,}R_{in}) \tag{1}$$

By combining m single-factor evaluation of the fuzzy subset, we can get a fuzzy evaluation matrix R.

$$R=\begin{pmatrix} R_{11} & R_{12} & ... & R_{1n} \\ R_{21} & R_{22} & ... & R_{2n} \\ ... & ... & ... & ... \\ R_{m1} & R_{m2} & ... & R_{mn} \end{pmatrix} \tag{2}$$

If the weight vector of evaluation factors is $A=(\alpha_1,\alpha_2,...\alpha_m)$, we can get the set of fuzzy comprehensive evaluation

$$B.\ B = A \cdot R = (\alpha_1, \alpha_2, \ldots, \alpha_m) \cdot \begin{pmatrix} R_{11} & R_{12} & \ldots & R_{1n} \\ R_{21} & R_{22} & \ldots & R_{2n} \\ \ldots & \ldots & \ldots & \ldots \\ R_{m1} & R_{m2} & \ldots & R_{mn} \end{pmatrix} = (b_1, b_2, \ldots, b_m) \quad (3)$$

b_i express the membership degree of rank fuzzy subset for whole system. This article take the Maximum membership degree principle.

4 Case Study

Through counting the training record data of a driver during one mouth, we get the fuzzy matrix R_1, R_2, R_3, respectively on behalf of the safety indicator, efficiency indicator and other indicator.

$$R_1 = \begin{pmatrix} 0.10 & 0.20 & 0.50 & 0.10 & 0.10 \\ 0.05 & 0.10 & 0.50 & 0.25 & 0.10 \\ 0.20 & 0.20 & 0.30 & 0.20 & 0.10 \end{pmatrix} \quad R_2 = (0.10, 0.30, 0.40, 0.10, 0.10) \quad R_3 = \begin{pmatrix} 0.30 & 0.30 & 0.20 & 0.10 & 0.10 \\ 0.20 & 0.20 & 0.20 & 0.30 & 0.10 \\ 0.20 & 0.30 & 0.25 & 0.15 & 0.10 \end{pmatrix}$$

According Delphi method, we can get the judgment matrix S,S1,S3 respectively on behalf of the target layer indicator and the factor layer indicator of the safety indicator and the other indicator. We can get the maximum vector of each matrix. That is $A = (0.40, 0.40, 0.40)$, $A_1 = (0.167, 0.5, 0.333)$, $A_3 = (0.333, 0.333, 0.333)$.

The set of fuzzy comprehensive evaluation of the factor layer indicator will be got by formula $B_i = A_i \cdot R_i$.

$$B_1 = A_1 \cdot R_1 = (0.1083, 0.1500, 0.4334, 0.2083, 0.1000)$$
$$B_2 = R_2 = (0.1000, 0.3000, 0.4000, 0.1000, 0.1000)$$
$$B_3 = A_3 \cdot R_3 = (0.2331, 0.2664, 0.2165, 0.1832, 0.1000)$$

The set of fuzzy comprehensive evaluation of the target layer is shown as B.

$$B = A \begin{pmatrix} B_1 \\ B_2 \\ B_3 \end{pmatrix} = (b_1, b_2, b_3, b_4, b_5) = (0.1300, 0.2333, 0.3767, 0.1600, 0.1000)$$

After the above calculation processing•the general practice is determined in the maximum membership degree principle. We should get the most value of B, so the rank of the driver is medium.

5 Conclusion

The goal of safety and efficiency is introduced in the evaluation system of the quayside container crane's drivers and we put forward the evaluation model of the quaysi-de container crane's drivers based on FAHP. The model will quantify and compare the indicators, then get the weight of each indicator, which has strong operability. The pr-oblem of the evaluation involves lots of facets. In fact, we advice the port enterprise to select available factors to build database, which will work well in evaluation. In the actual evaluation project, introducing some appropriate coefficient will make the evaluation system scientific and reasonable.

Acknowledgement. This work sponsored by The National Natural Science Foundation of China (71101090), supported by Shanghai Education Committee Projects (J50604), supported by Shanghai Municiple Education Commission Project (12ZZ148), supported by Ministry of Communications Research Project (2009-329-810-020) supported by Science & Technology of Shanghai Maritime University.

References

[1] Wang, Z., Liang, G., Liang, C.: Container crane training simulator. Journal of System Simulation, 904–906 (2002)
[2] Yang, P.: The establishment and realization of scoring models of the shore container crane simulation training system. Shanghai Maritime University, Shanghai (2005)
[3] Cheng, L.: Design and Implementation of the driver operating expert system of the quayside container crane. Shanghai maritime university, Shanghai (2005)
[4] Zhang, G.: Fuzzy Comprehensive Evaluation in Simulation and Training System. Journal of Computer Simulation (10), 31–32 (1999)
[5] Li, Y., Shi, G.: The fuzzy comprehensive evaluation model of the small and medium sized coal mine safety inventory

Using a Hybrid Neural Network to Predict the NTD/USD Exchange Rate

Han-Chen Huang

Department of Leisure Management,
Yu Da University, Miaoli County, 36143 Taiwan

Abstract. In the financial market, although foreign exchange options or foreign exchange forward contracts are available for corporations to hedge risks, reports of profit losses due to foreign exchange losses remain common. This study employs a multilayer perceptions (MLP) neural network with genetic algorithm (GA) to predict the New Taiwan dollar (NTD)/U.S. dollar (USD) exchange rate. The GA is used to determine the optimum number of input and hidden nodes for a feedforward neural network, the optimum slope of the activation function, and the optimum learning rates and momentum coefficients. The empirical results show that the ability of the proposed model to predict the NTD/USD exchange rate is excellent. The absolute relative error between the predicted value and the actual value was 0.338%, and the correlation coefficient was 0.995885.

Keywords: Exchange Rate, Neural Network, Genetic Algorithm.

1 Introduction

During the first half of 2011, the NTD depreciated by 1.01% against the USD; however, it appreciated by 6% in the second half of 2011. Business owners struggle to maintain business operations when encountering rapid fluctuations in the exchange rate. For trading priced using the USD, depreciation of the NTD increases the cost of payments in NTD, and appreciation of the NTD decreases the profits receivable in USD; thus, both reduce corporate profit and even lead to operational losses. Foreign exchange forward contracts or foreign exchange options can enable corporations to hedge exchange rate risks. However, the crucial factor is not the risk hedging tool selected; instead, it is the ability of corporations to predict fluctuations in the exchange rate. By employing a tool that can predict exchange rates, corporations can retain their existing business profits and create exchange gains. Therefore, this study employs multilayer perceptions neural network with genetic algorithm to predict the NTD/USD exchange rate.

The remainder of this paper is organized as follows: In the next section, we review the application of GA in optimization of artificial neural network's architecture and learning parameters. The hybrid methodology is presented in Section 3. In Section 4, we report the empirical results. Finally, our concluding remarks are provided in Section 5.

Z. Du (Ed.): Proceedings of the 2012 International Conference of MCSA, AISC 191, pp. 433–439.
springerlink.com

2 Neural Network Modeling Approach

In time series forecasting, past observations of a variable are collected and analyzed to develop a model for extrapolating the time series into the future[1]. Over the past decades, many researchers focus on the development of time series forecasting models. One of the popular time series models with widely usage in engineering, economic and social applications is auto-regressive integrated moving average (ARIMA). Although ARIMA models perform well over a short period of time and their implementations are easy, they have two kinds of limitations: "linear limitation" and "data limitation"[2].

In recent years, artificial neural networks have been employed in many fields. The main advantage of neural networks is their flexible capability for nonlinear modeling[3-6]. Numerous studies have incorporated ANNs with fuzzy theory or GAs to improve the analysis; these ANNs are called “Hybrid Neural Networks”[7]. To improve the predicative performance of ANNs in NTD/USD exchange rate forecasting, this study employs GA to determine the optimal network structure and parameters. Thus, the number of hidden nodes of a multilayer perceptrons neural network, the slope of the activation function, the values of the learning rates and momentum coefficients in the hidden and output layers, and the number of features (MLP inputs) are determined through optimization.

GA improves the performance of ANNs by selecting the optimum input features, optimizing the network parameters, and modifying the slope of the activation function and determination of weights. The optimization process[8] is described below:

- Randomize the population.
- Evaluate the fitness function of each individual in the population.
- Select the first two individuals with the highest fitness values and copy them directly to the next generation without genetic operations.
- Select the remaining individuals in the current generation and apply crossover and mutation genetic operations to produce the next generation of individuals.
- Repeat the process from the second step until all individuals in the population meet the convergence criteria.
- Decode the converged individuals in the final generation to obtain the optimized parameters.

In this study, we use different operators[9] for selection and crossover operations (Table 1).

3 Neural Network Model Implementation

A multilayer perceptrons with one hidden layer and the sigmoid activation function (Eq.1) was selected as the base neural structure for this study:

$$f(x) = \frac{1}{1+exp(-\beta X_i)} \tag{1}$$

β in Eq.(1) is the function slope.

Table 1. Description of different operators for selection and crossover operations in GA

Operation	Operator	Description
Selection	Roulette	The chance of a chromosome getting selected is proportional to its fitness.
	Top percent (x)	Randomly selects a chromosome from the top x percent of the population.
	Best	Selects the best chromosome.
	Random	Randomly selects a chromosome from the population.
	Tournament	The winner of each tournament is selected for crossover.
Crossover	One point	Randomly selects a crossover point within a chromosome, interchanges the two parent chromosomes at this point to produce two new offspring.
	Two point	Randomly selects two crossover points within a chromosome, interchanges the two parent chromosomes between these points to produce two new offspring.
	Uniform	Decides (with some probability-know as the mixing ratio) which parent will contribute each of the gene values in the offspring chromosomes.
	Arithmetic	Linearly combines two parent chromosome vectors to produce two new offspring.
	Heuristic	Use the fitness values of the two parent chromosomes to determine the direction of the search.

The Newton algorithm with a momentum term was used as the learning function (Eq.2)

$$\Delta W_i(n+1) = -\eta \nabla W_i + \rho \Delta W_i(n) \tag{2}$$

in which η was the learning rate, and ρ was the momentum coefficient.

Mean square error (MSE) during the training of network is calculated using equation as follow：

$$MSE = \frac{\sum_{j=0}^{P}\sum_{i=0}^{N}\left(d_{ij}-y_{ij}\right)^2}{N \times P} \tag{3}$$

in which N is the number of training patterns and P is the number of output nodes. y_{ij} and d_{ij} are actual output and desired value, respectively. The conceptual architecture of the model is shown in Figure 1.

The exchange rate data used in this study was obtained from the Central Bank of Taiwan (http://www.cbc.gov.tw). This data comprised 1,255 working day observations of the NTD/USD exchange rate from January 2007 to December 2011. Figure 2 shows the NTD/USD exchange rate data.

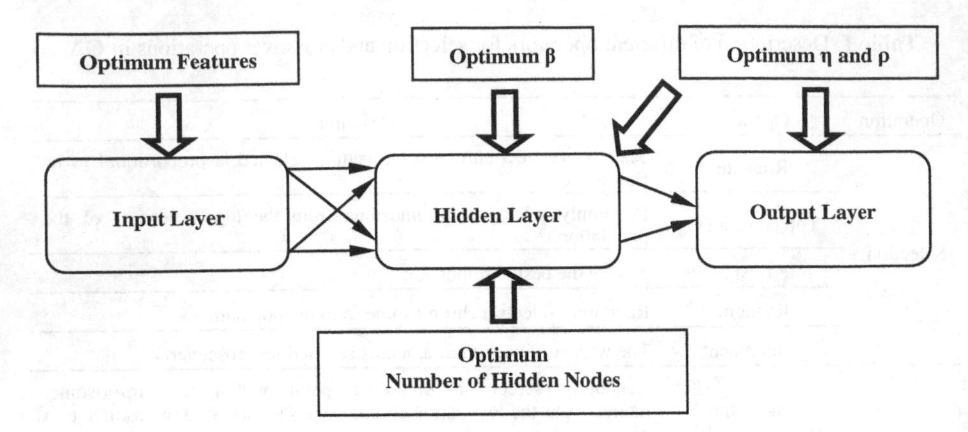

Fig. 1. Conceptual architecture of the hybrid model

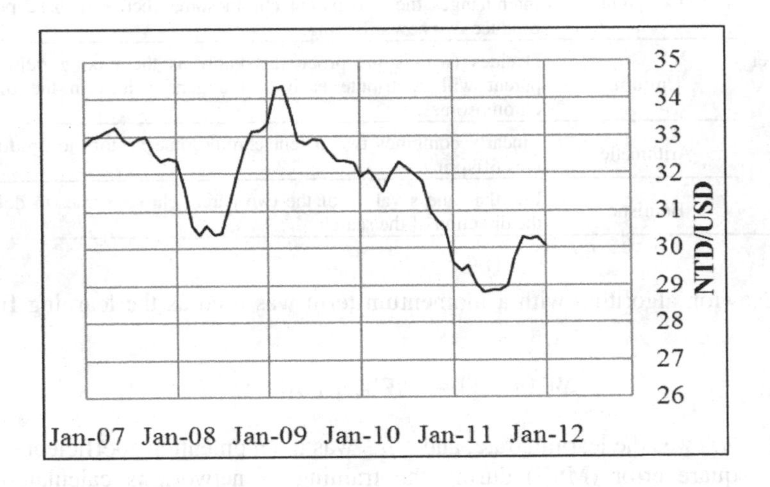

Fig. 2. NTD/USD exchange rate from January 2007 to December 2011

The exchange rate data was divided into three groups; 65% was used for training; 15% for validation; and 20% for testing. The input and target values were normalized and training data was applied in random sequence to eliminate the effects of current trends on the output. NeuroSolutions[9] Toolbox is used for simulations in this paper. Table 2 lists the neural network input and output index.

4 Empirical Results

The empirical results based on the optimum values of the MLP neural network parameters, which were determined using the GA, are presented in this section. To evaluate the ability of MLP neural network to predict the NTD/USD exchange rate, the values of MSE after training and for various optimum conditions are presented in Table 3.

Table 2. Input and Output Index for Training GA-MLP Neural Network

Type	Index
Input	The close price at 4 PM (Taiwan time) for the following 13 currencies: JPY/USD; USD/GBP; HKD/USD; KRW/USD; CAD/USD; SGD/USD; CNY/USD; USD/AUD; IDR/USD; THB/USD; MYR/USD; PHP/USD and USD/EUR.
Output	The NTD/USD close price of next working day.

Table 3. Performance of the proposed model in NTD/USD exchange rate prediction

Selection operator	Crossover operator	Number of inputs	Number of hidden nodes	β	MSE
Best	One point	6	14	0.51	0.0059
	Two point	4	24	0.27	0.0046
	Uniform	5	14	0.37	0.0027
	Arithmetic	9	11	0.74	0.0047
	Heuristic	7	8	0.53	0.0036
Top percent (20)	One point	4	30	0.73	0.0037
	Two point	8	29	0.21	0.0041
	Uniform	5	7	0.48	0.0051
	Arithmetic	9	22	0.13	0.0055
	Heuristic	7	26	0.37	0.0030
Roulette	One point	11	20	0.56	0.0049
	Two point	9	16	0.77	0.0058
	Uniform	7	18	0.24	0.0026
	Arithmetic	9	6	0.77	0.0025
	Heuristic	6	21	0.53	0.0039
Random	One point	4	16	0.10	0.0040
	Two point	6	19	0.73	0.0050
	Uniform	6	5	0.64	0.0056
	Arithmetic	8	12	0.72	0.0029
	Heuristic	9	24	0.11	0.0049
Tournament	One point	8	9	0.17	0.0046
	Two point	4	20	0.22	0.0048
	Uniform	8	17	0.64	0.0023
	Arithmetic	7	5	0.59	0.0044
	Heuristic	9	16	0.43	0.0031

The results show that different operators were used for selection and crossover operations in the GA. The optimum values for the number of input and hidden nodes and the slope of the activation function for each combination of operators are also shown in

Table 3. When a combination of "Tournament" and " Uniform " operators is employed, the mean square error (MSE) of the predicted NTD/USD exchange rate is minimized.

A sensitivity analysis was conducted to distinguish the impact of each input index on NTD/USD exchange rate. The importance of each factor to exchange rate, from highest to lowest, is SGD/USD (43.83%), THB/USDD (8.96%), CAD/USD (7%), MYR/USD (6.12%), KRE/USD (5.16%); USD/EUR (5.12%); HKD/USD(4.70%), USD/GBP(4.55%), CNY/USD(4.49%); PHP/USD (4.31%), USD/AUD (3.21%); IDR/USD(1.66%), and JPY/USD(0.89%). The SGD/USD is the most important feature. The JPY/USD is the least important feature in the prediction.

Table 4 shows the training, verification, and predicted results of the NTD/USD exchange rate using the proposed model. The absolute relative error (ARE) of the predicted result was 0.338%, and the correlation coefficient was 0.995885, which indicates that the prediction model developed in this study can precisely predict changes in the NTD/USD exchange rate. Figure 4 shows a scatter plot of the actual exchange rate and the exchange rate predicted using the model.

Table 4. The training, verification, and testing results of the NTD/USD exchange rate using the GA-MLP model

	Training data			Verifying data			Testing data		
	Target	Output	ARE	Target	Output	ARE	Target	Output	ARE
Mean	31.754	31.754	0.262%	31.599	31.632	0.337%	31.638	31.636	0.338%
Std. Dev	1.424	1.419	0.221%	1.465	1.477	0.321%	1.529	1.488	0.297%
Correlation	0 .9970717			0.995105			0.995885		

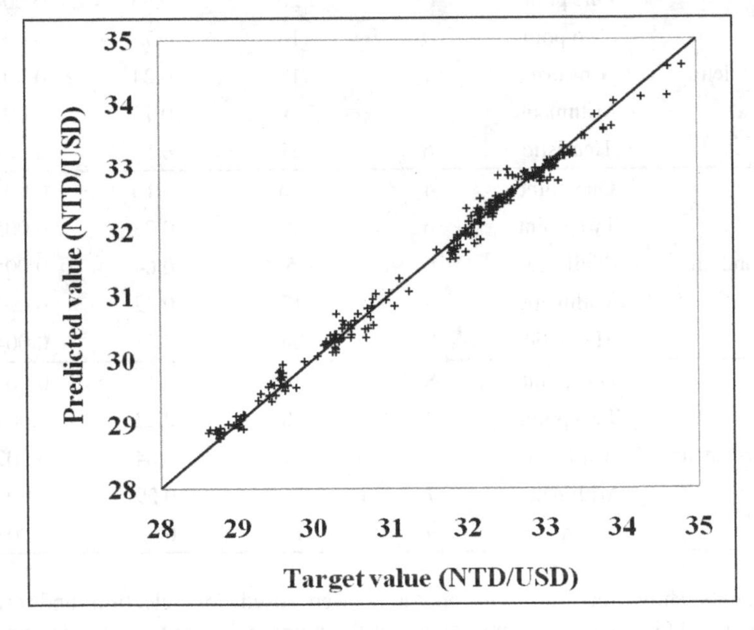

Fig. 4. Scatter plot of predicted values from the NTD/USD prediction model and actual values

5 Conclusions

Dramatic changes in exchange rates present a significant challenge to Taiwanese enterprises that engage in import and export trade. Incorrect predictions of exchange rate fluctuations can reduce the expected profits and even result in operational losses. However, financial markets provide numerous methods for corporations to hedge the risks of exchange rate fluctuations. Nevertheless, a model for predicting exchange rate fluctuations can enable business owners to make more appropriate judgments.

In this study, we used MLP neural network and GA to construct a NTD/USD exchange rate prediction model. The empirical results show that the ability of the model developed in this study to predict the NTD/USD exchange rate is excellent. The absolute relative error between the predicted value and the actual value was 0.338%, and the correlation coefficient was as high as 0.995885.

References

[1] Box, P., Jenkins, G.M.: Time series analysis: forecasting and control, Holden-Day (1976)
[2] Chen, K.Y., Wang, C.H.: A hybrid ARIMA and support vector machines in forecasting the production values of the machinery industry in Taiwan. Expert Systems with Applications 32, 254–264 (2007)
[3] Katijani, Y., Hipel, W.K., McLeod, A.I.: Forecasting nonlinear time series with feedforward neural networks: a case study of canadian lynx data. Journal of Forecasting 24, 105–117 (2005)
[4] Harpham, V., Dawson, C.W.: The effect of different basis function on radial basis function network for time series prediction: a comparative study. Journal of Neurocomputing 69, 2161–2170 (2006)
[5] Jain, A., Kumar, A.M.: Hybrid neural network models for hydrologic time series forecasting. Applied Soft Computing 7, 585–592 (2007)
[6] Giordano, F., Rocca, M.L., Perna, C.: Forecasting nonlinear time series with neural network sieve bootstrap. Computational Statistics and Data Analysis 51, 3871–3884 (2007)
[7] Hornik, K.M., Stinchcombe, M., White, H.: Multilayer feedforward networks are universal approximators. Neural Networks 2, 359–366 (1989)
[8] Sheikhan, M., Movaghar, B.: Exchange rate prediction using an evolutionary connectionist model. World Applied Sciences Journal 7, 8–16 (2009)
[9] Principe, J., Lefebvre, C., Lynn, G., Fancourt, C.: Neuro Solutions Documentation. Neuro Dimension Incorporation (2009), http://www.neurosolutions.com

Research on the Influential Factors of Customer Satisfaction for Hotels: The Artificial Neural Network Approach and Logistic Regression Analysis

Han-Chen Huang

Department of Leisure Management,
Yu Da University, Miaoli County, 36143 Taiwan

Abstract. This study conducts a customer survey and performs an analysis using factor analysis, artificial neural networks, and logistic regression analysis to gain an in-depth understanding of the impact of hotel service attributes on customer satisfaction. The results show that among the four hotel service attributes (personnel services, room quality, dining quality, and business and travel services), "personnel services" has the greatest impact on customer satisfaction, whereas "business and travel services" has the lowest impact. Artificial neural network is more accurate than logistic regression analysis for predicting customer satisfaction. Artificial neural networks achieved an accuracy rate of 93%. Although logistic regression analysis has an accuracy rate of 87% for predicting customer satisfaction, it only scores 23.77% for predicting "unsatisfied customers." Artificial neural network is more suitable than logistic regression analysis for predicting customer satisfaction.

Keywords: Service Attributes, Customer Satisfaction, Neural Network.

1 Introduction

Studies on customer satisfaction are beneficial. Numerous studies have shown that customer satisfaction is a key factor that directly impacts customer intention to repurchase or recommend to others[1-5].Customer repurchase or word-or-mouth behavior increases corporate profit and market share[6]. Therefore, measuring customer satisfaction is a crucial task for businesses. The assessment results can motivate the service industry to improve their behavior and enhance their service quality. Customer satisfaction assessment is an indispensable measure for obtaining competitive advantage in an industry.

This study investigates the service attributes that affect customer satisfaction and the degree of impact each service attribute has on customer satisfaction. Customer opinions are obtained through questionnaires and the data is analyzed using artificial neural networks (ANN) analysis. The analysis result is compared with logistic regression analysis (LRA). The results enable hotel management to understand future marketing directions and the service attributes that customers prefer and that may require improvement.

Z. Du (Ed.): Proceedings of the 2012 International Conference of MCSA, AISC 191, pp. 441–448.
springerlink.com

2 Artificial Neural Networks

ANN employ mathematical simulation of biological nervous systems in order to process acquired information and derive predictive outputs after the network has been properly trained for pattern recognition. A neural network consists of numerous layers of parallel processing elements or neurons. One or more than one hidden layers may exist between an input and an output layer. The neurons in the hidden layer(s) are connected to the neurons of a neighboring layer by weighting factors that can be adjusted during the model training process. The networks are organized according to training methods for specific applications. Figure 1 illustrates a three layers neural network consisting of four neurons in the input layer, five neurons in the hidden layer, and two neurons in the output layer, with interconnecting weighting factors, W_{ij}, between layers of neurons.

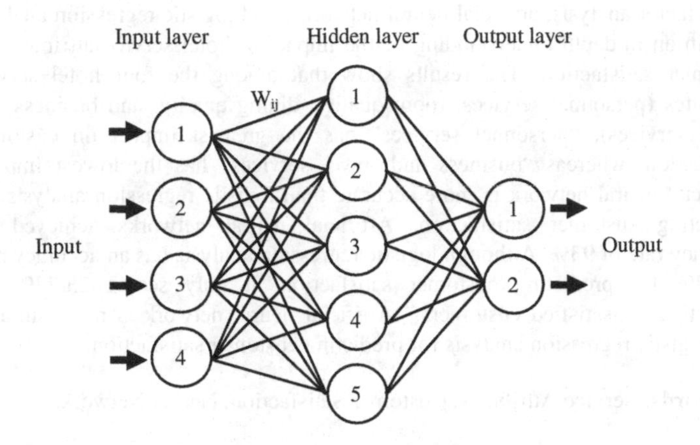

Fig. 1. Three layers feedforward artificial neural network configuration

"Training" of an ANN model is a procedure by which ANN repeatedly processes a set of test data (input-output data pairs), changing the values of its weights according to a predetermined algorithm in order to improve its performance. Back-propagation is the most popular algorithm for training ANN[7]. It s a supervised learning method in which an output error is fed backward through the network, altering connection weights so as to minimize the error between the network output and the targeted output.

3 Data Collection and Analysis

This study uses a questionnaire that is composed of two sections. The first section obtains customer evaluations for each service provided by hotels. The questionnaire was designed based on previous studies[2,4-6]. The 25 questions in the first section use the 5-point Likert scale for assessment (l is very unsatisfied, 5 is very satisfied). One question in the second section is used to determine if customers are satisfied (yes or no).

We distributed 500 copies of the questionnaires to customers in six hotels in northern Taiwan (two five-star hotels and four four-star hotels) between July and

September in 2011. Surveyed customers had already completed their service interaction with the hotel. Four-hundred and fifty-six copies were returned (91.2%), although only 425 copies were valid (85%). This study used the SPSS software package to perform factor analysis, and LRA. The Alyuda NeuroIntelligence software was used to establish ANN for predicting customer satisfaction.

3.1 Factor Analysis

Factor analysis (FA) is employed to reduce the number of observation variables to a smaller number of factors. Researchers can use these factors to replace original variables and conduct further analysis. We performed FA for the first set of 25 questions in the questionnaire (Bartlett's test of sphericity=6837.41, $p=0.000$; KMO measure of sampling adequacy=0.911). The results show that the questionnaire is suitable for factor analysis. Table 1 shows that the 25 variables can be categorized into four factors. The cumulative total variance explained is 70.48%. The factors are titled F1: Personnel services, F2: Dining quality, F3: Room quality, F4: Business and travel services.

3.2 MLPs Neural Network

This work uses a multilayer perceptrons (MLPs) neural network with supervised learning. This approach allows the network to compare its actual results with the desired output and then computes the error during the training process. The error is presented to the input through the back-propagation algorithm. This process continues until the actual outputs approach the desired output. Alyuda Neuro Intelligence software is employed to construct the required MLPs neural network. The constructed network is composed of an input layer, a nonlinear hidden layer, and an output layer. The hidden layer and the output layer apply the tanh transfer function.

The MLPs neural network has been trained based on 250 pieces of questionnaire data. The cross validation process of the network uses a data set of 75 pieces of questionnaire data. The testing process is defined as data used to evaluate performance after training is complete. The trained network was then tested using 100 pieces of questionnaire data that were not used in the training set and cross validation set. The number of nodes of the network is the number of exemplars of the training set equal to 250 and 1,000 epochs. The number of nodes is configured automatically by the software.

To validate the proposed model, positive predicted value (PPV) was computed as:

$$PPV=(\text{Correct results} / \text{All results}) \times 100\% \quad (1)$$

Figure 2 shows the MLPs neural network learning curve. The active cost curves approaches zero which means that classification of the dataset was carried out correctly. The results are shown in Table 2. Overall PPV for the test is 93%, indicating effective predicting ability. A sensitivity analysis was conducted to distinguish the impact of each input factor (F1 to F4) on customer satisfaction (Figure 3). The importance of each hotel service attribute to customer satisfaction, from highest to lowest, is personnel services (34.89%), room quality (29.95%), dining quality (28.65%), and business and travel services (6.51%).

Table 1. Factor analysis results

Code	Question	F1	F2	F3	F4
Q2	Service speed of check-in/check-out procedure	0.895			
Q1	Service accuracy of check-in/check-out procedure	0.853			
Q3	Providing relevant information	0.794			
Q5	Front desk service	0.766			
Q4	Luggage service	0.760			
Q22	Friendliness of service personnel	0.760			
Q24	Service vocabulary of service personnel	0.740			
Q23	Behavior of service personnel	0.684			
Q25	Appearance and expressions of service personnel	0.610			
Q16	Food variety in restaurant		0.805		
Q18	Comfort of restaurant environment		0.748		
Q17	Taste of food in restaurant		0.724		
Q19	Cleanliness of restaurant environment		0.715		
Q21	Thoroughness of restaurant services		0.644		
Q20	Prompt service in restaurant		0.597		
Q10	Room quietness			0.729	
Q7	Convenience of room amenities			0.727	
Q8	Completeness of room amenities			0.701	
Q9	Room safety			0.679	
Q11	Room cleanliness			0.677	
Q6	Bed comfort			0.579	
Q12	Meeting room arrangement service				0.812
Q13	Travel arrangement service				0.774
Q15	Transportation and shuttle service				0.732
Q14	Business facilities				0.548
	Variance	0.438	0.116	0.082	0.067
	Cumulative	0.438	0.554	0.636	0.703

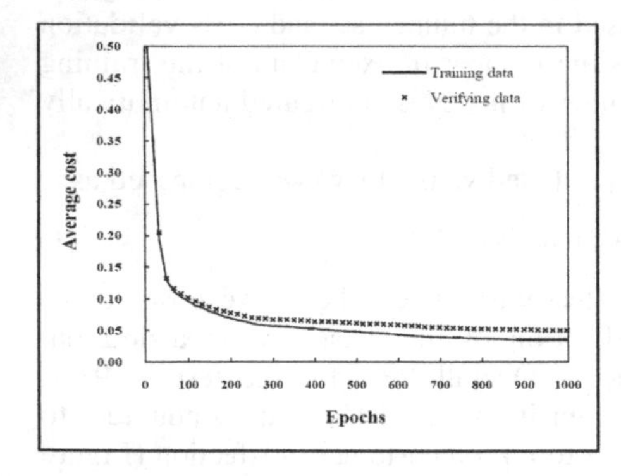

Fig. 2. The MLPs model learning curve

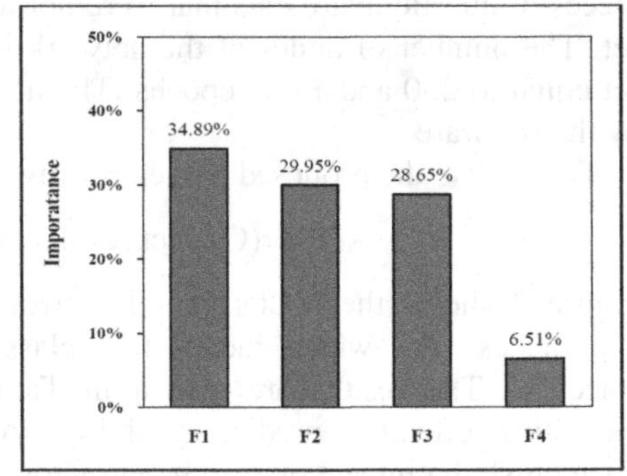

Fig. 3. The importance of input variables to output variables

Table 2. The results of MLPs model

	Training data set	Verifying data set	Testing data set
PPV	97.2%	98.67%	93%

3.3 Logistic Regression Analysis

To compare the results derived using MLPs neural network, we used binary logistic regression analysis to establish the relationship between service attributes and customer satisfaction. The results are shown in Table 3.

Table 3. Binary logistic regression analysis results

	β	S.E.	Wald	*P*
Personnel Services	0.389	0.092	17.873	0.000
Room Quality	0.601	0.113	28.288	0.000
Dining Quality	0.142	0.068	4.403	0.036
Business and Travel Services	0.116	0.105	1.222	0.269
Constant	-26.212	4.974	27.770	0.000

Table 3 shows that the *P* value of the factor “Business and Travel Services” is 0.269. It is greater than 0.05, indicating that “Business and Travel Services” does not have a significant impact on customer satisfaction. Table 4 shows that the Overall PPV for the test is 87%.

Table 4. The results of LRA

	Training data set	Verifying data set	Testing data set
PPV	86%	85.3%	87%

4 Discussion

4.1 Service Attributes That Influence Customer Satisfaction and Their Impact Levels

The importance of each hotel service attribute to customer satisfaction, from highest to lowest, is personnel services (34.89%), room quality (29.95%), dining quality (28.65%), and business and travel services (6.51%).

The personnel services attribute has the most significant impact on customer satisfaction. The face-to-face contact between personnel and customers is a crucial factor that affects customer satisfaction. This result is in accordance with the view of

Cadotte and Turgeon[8] that when customers evaluate overall service quality, personnel service is the most significant factor. In the hotel industry, which includes intense competition, the amenities and services provided by each hotel are similar. The enhancement of personnel services should be a critical management subject for hotels attempting to differentiate themselves from their competitors. Management should address the vocabulary, behavior, manner, and appearance of service personnel.

The impact of room quality on customer satisfaction has the second-highest significance for customer satisfaction. Although there is debate on whether room quality affects customer satisfaction[9], this study shows that room quality still had a reference value for hotels participating in the study. Management should consider customers perspectives and use hotel resources to enhance room quality factors such as room cleanliness, full amenities, and room safety. When creating a promotional plan, management should place emphasis on tangible evidence of hotel quality such as hotel room amenities to retain existing customers and attract new customers.

The importance of dining quality to customer satisfaction is 28.65%. McCleary[10] argued that the dining service quality in hotels is not significant to customers. Most customers do not think that dining in a hotel is necessary. This viewpoint differs from the results of this study. When staying at a hotel, customers have the opportunity to dine in the hotel. Even one dining experience affects customer evaluation of the hotel. Therefore, management should make efforts to improve dining quality, increase the variety of the food, enhance the taste of the food, maintain a comfortable and clean dining environment, and provide prompt and thorough services.

This attribute has the least impact on customer satisfaction, with an importance of 6.51%. This study presumes that the reason for this phenomenon is that customers have arranged their business or travel activities before checking into the hotel; therefore, hotels only assist to fulfill or complete these arrangements. When there is a demand from customers, prompt and thorough assistance can satisfy customers' expectations.

4.2 Comparison of MLPs Neural Network and LRA

Table 2 and 4 shows that MLPs neural network have greater PPV than LRA for training, verification, and testing. MLPs have greater overall accuracy than LRA. Table 5 shows the detailed prediction results of MLPs neural network and LRA for the test data. For customer satisfaction, out of the 100 test data, 87 customers rated their hotel as "satisfactory," and 13 customers rated theirs as "unsatisfactory." MLPs neural network and LRA have the same accuracy for predicting customers who gave a "satisfactory" rating, which was 96.55%. However, when predicting customers who gave a rating of "unsatisfactory," the accuracy of MLPs neural network was 69.23% and the accuracy of LRA was 23.08%, which is was unacceptable.

Table 5. Testing data set classification

Model	Class	Satisfactory	Unsatisfactory	Percentage
MLPs	True	84	9	93%
	False	3	3	7%
LRA	True	84	3	87%
	False	3	10	13%

5 Conclusions

The study results show that of the four categories of hotel service attributes (personnel services, room quality, dining quality, and business and travel services), personnel services has the greatest impact on hotel customer satisfaction, followed by room quality, dining quality, and business and travel services. To increase customer satisfaction, management should improve personnel services first by implementing measures such as personnel training, authorization, performance appraisal, reward and punishment systems, improving work environments, and respecting employees. Management should also allocate resources to improve room quality.

Room factors valued by guests include room comfort, full amenities, safety, quietness, and room cleanliness. Management should focus on these factors to maintain high standards of service quality. For marketing strategies, hotels should advertise based more on their competitive advantages of better personnel services and room quality to differentiate themselves from their competitors.

For analyzing the impact of service attributes on customer satisfaction, this study used MLPs neural network and LRA. For predicting overall customer satisfaction, the accuracy of MLPs neural network is superior to that of LRA, the overall accuracy of MLPs neural network is 93%. Although the overall accuracy of LRA is 87% for predicting customer satisfaction, the accuracy is only 23.77% for predicting unsatisfied customers, whereas the accuracy of MLPs neural network is 69.23%. Therefore, MLPs neural network is more suitable than LRA for predicting customer satisfaction.

References

[1] Engel, J.F., Blackwell, R.D., Miniard, P.W.: Consumer behavior. Doyden Press, Hinsdale (1990)

[2] Jaksa, K., Inbakaran, R., Reece, J.: Consumer research in the restaurant environment, Part 1: a conceptual model of dining satisfaction and return patronage. International Journal of Contemporary Hospitality Management 11(5), 205–222 (1999)

[3] Anton, J.: Customer relationship management: making hard decisions with soft numbers. Prentice-Hall, Upper Saddle River (1996)

[4] Oh, H., Parks, C.S.: Customer satisfaction and service quality: a critical review of the literature and research implications for the hospitality industry. Hospitality Research Journal 20(3), 36–64 (1997)

[5] Zeithaml, V.A., Berry, L.L., Parasuraman, A.: The nature and determinants of customer expectations of service. Journal of the Academy of the Marketing Science 21(1), 1–12 (1993)
[6] Lewis, B.R.: Getting the most from marketing research. The Cornell Hotel and Restaurant Administration Quarterly 25(4), 82–96 (1985)
[7] Lippman, R.: An introduction to computing with neural nets. IEEE ASSP Mag. 4, 4–22 (1987)
[8] Cadotte, E.R., Turgeon, N.: Key factors in guest satisfaction. The Cornell Hotel and Restaurant Administration Quarterly 34, 45–51 (1988)
[9] Gilbert, D.C., Morris, L.: The relative importance of hotels and airlines to the business traveler. International Journal of Contemporary Hospitality Management 7(6), 19–23 (1994)
[10] McCleary, K.W., Weaver, P.A., Hutchinson, J.C.: Hotel selection factors as they relate to business travel situations. Journal of Travel Research 32(2), 42–48 (1993)

Research on Analysis and Monitoring of Internet Public Opinion

Li Juan, Zhou Xueguang, and Chen Bin

College of Electronic Engineering, Naval Univ. of Engineering, China

Abstract. The purpose of the research is to collect and analyze the internet public opinion, understand the public voice and feed back in time. Basing on the understanding of public opinion and its related concepts, the paper explained the developing process of public opinion and internet public opinion in their life cycle. Next we the monitoring model of internet public opinion has been built according with the change rule of public opinion. The original information is webpage in internet. Hot topic would be obtained with passive monitoring and active tracking. Furthermore, the situation of internet opinion is discovered. Leading and blocking, these two ways can be used to guide the internet public opinion positively. The research of this paper would be beneficial to quickly grasp and understand internet public opinion and make effective control and guide. It has important practical significance to construct the socialist harmonious society.

Keywords: Internet Public Opinion, Monitoring, Public Opinion Information, Hot Topic.

1 Introduction

Along with the rapid development of internet all over of the world, internet media has already recognized as fresh media. Internet communication has the characteristic of real time and interaction. It provides a convenient platform for public to express their opinion and attach themselves to economy and polity society. Convenient Expression, multi-information and interacting mode gives internet a big advantage over those tradition media. But, some problems has also bring. Internet opinion is non-rationality and emotionality. It is difficult to distinguish verity of the internet information [1]. So, it is necessary to build perfect social public opinion information feedback network to find various crisis factor in unaccomplished-state. It is favorable for us to judge accurately and control the situation roundly. Especially, the straightway channel for government to seize the social dynamics and opinion is ensured.

2 Public Opinion and Internet Public Opinion

Public Opinion, in brief, means peoples attitude to the social conditions. It is public subjective reflection on certain social reality and phenomenon. Public opinion is general representation of population consciousness, idea, attitude and request. Manifold of different emotions, intentions and ideas are staggered combined. All

Z. Du (Ed.): Proceedings of the 2012 International Conference of MCSA, AISC 191, pp. 449–453.
springerlink.com

these factors interlace and impact each others, so that lead to public opinion show dynamic and anfractuous [2].

The lifecycle of public opinion is shown in figure 1.The specific object of public opinion is public affair which peoples give their attention to or closely relate to their interests. They include social event, social issue, social conflict, social activity, words and actions of public figures and so on. The public affair is stimulation source which inspire the occurrence of public opinion. The subject of public opinion is public. The attention publics paying to public affair leads to form of public opinion. Furthermore, communicating, discussing and disputing between the different viewpoints and mutual influence of these attitudes promote the development of public opinion. Another important factor driving the development and movement of public opinion is in the event itself, such as related events arising. Notably, when the development of event influence public opinion, public opinion will feeds back into the development of event as well. Finally, with the end of the event, the interest of the public will transfer and the public opinion falls into the declining period.

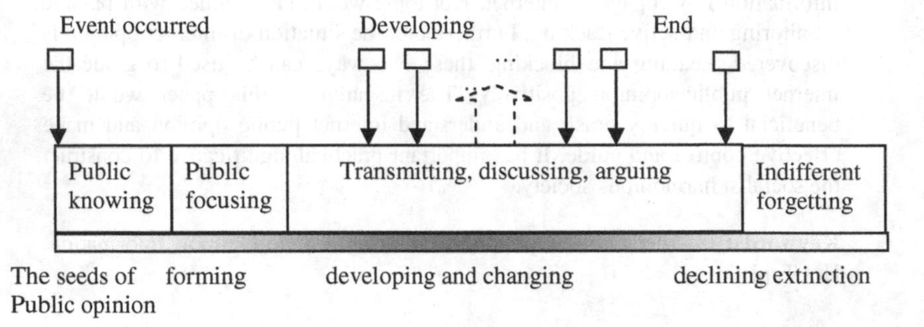

Fig. 1. Life Cycle of Public Opinion

Public opinion has strong timeliness, because the social affairs and public reflects are changing with time.

Because the Internet has the characteristic of virtual, anonymous, emanative, permeability and random, more and more Internet user would like to express views and transmit ideas through this channel. Then, the internet public opinion comes into being. The Internet public opinion just is people's different emotions, ideas and views expressed and spread through the Internet [4]. Internet public opinion also stems from the reality of social affairs, but place or channels for it expressed and communicated is extended to the internet.

As a kind of emotion, attitude and view, public opinion cannot be measured directly, but describe and reflect by dint of public information. The information of internet public opinion is involved in many web pages. With its content always sensitive, the information of internet public opinion would be paid attention to by most of the readers. It mainly includes the information which can reflect on emotions, attitudes and views of some public affairs. It can be expressed in the forms of text, images, audio, and video and so on. In addition, the information of internet public opinion also includes the information that reflects the Internet user behavior, such as click, excerpt, comment, and collections etc.

As an expression form of public opinion, internet public opinion has its specific representation in its life cycle. When web pages which have sensitive contents, relate to interests of the people, or reveal some social problems appeared, network public opinion is in its infancy. With the network users reading, reprinting and reviewing, more and more web pages relating to the affair appear. Many Internet users are concern to relevant events via the Internet. Then the Internet public opinion is formed. Internet users understand the affair from different angle and different channels, and published their own expression on the internet to express their attitudes. All these different emotions and views are mutually colliding and influencing, which promote the further development and dynamic change of internet public opinion. The information of related topics continues to increase, and by clicking, forwarding, replying and so on, more and more people involved. Finally, with the end of the affair, people gradually no longer update and attend the related internet information, and the network public opinion is to decline.

3 Monitoring Model of Internet Public Opinion

Using modern information technology to collect and analyze the internet public opinion, can mine and draw an inference about the content, direction and intensity of public opinion, understand the public voice timely then. Further, facing on the arisen public opinion on internet, it should be feed back in time, and make effective control and guide. The basic structure of the internet public opinion monitoring is shown in figure 2.

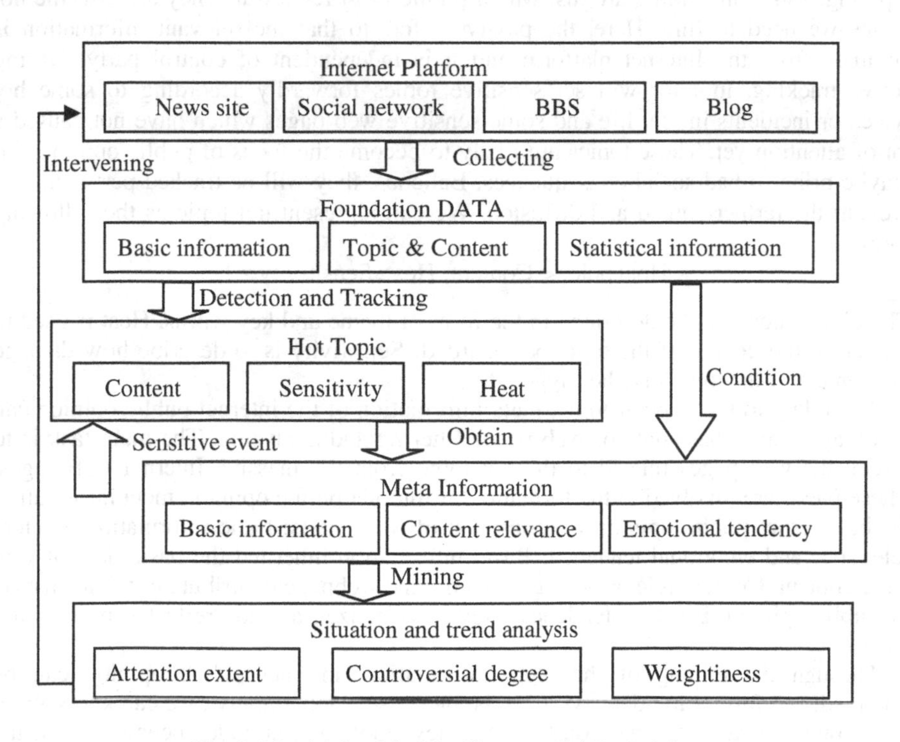

Fig. 2. Monitoring Model of Internet Public Opinion

The birthplace of internet public opinion is Internet. The main source of public opinion information including news sites, social networks, current affairs forum (BBS) and network diary (Blog). News website is the internet image of the news media in real life. Social network allows users to freely publish, reproduce, reviews various information. Current affairs forum will usually have the reflection to the hot news. Network diary allows users on the Internet to publish information in their own spaces, and to reproduce and browse information from others.

Same with other network information, the most basic form of information of internet public opinion is web pages which contain a wealth of information. Web pages are the basic elements to obtain public opinion. Then, we must first extract the basic data from the webpage relating to public opinion. It will be convenient for internet public opinion analyzing and mining. These data mainly include the elements such as the basic information the webpage, theme and content, as well as the relevant statistical information and so on. The basic information of the webpage contains URL, published time, and reproduced sources, etc. The theme and the contents have the most abundant basic information of public opinion. Statistical information such as read quantity and reproduced quantity will reflect attention degree of Internet users.

To find what does the people concern about and discuss is the chiefly task of internet public opinion monitoring, i.e. find hot topic. There are two ways to discover the hot topic: passive monitoring and active tracking. According to the basic data of web page, passive monitoring is focus on the readers' attention. Themes which majority of users reading and topics many netizen participating in by clicking, replying, and commenting are just which public most regardful. They are also the hot topics we need to find. Here, the passive refers to that the relevant information is obtained from the Internet platform and it is independent of control party. During active tracking, monitor will set sensitive topics forwardly according to some hot issues or incidents in real life and some sensitive web pages which have not caused a lot of attention yet. These topics are likely to become the focus of public opinion and maybe bring to bad social consequences. Enhence, they will be tracked pertinently to prevent the farther spread and diffusion. We can represent hot topic as the following form.

HotTopic: < Content, Heat, Sensitivity>.

Therein, content can be described in the form of theme and key words. Heat is used to reflect on the degree of the topic be regarded. Sensitivity is to describe how damage the content of the topic may be triggered.

According to the popular topic, meta information of the internet public opinion can be obtained after base data of web pages filtering and analyzing. The main task is to screen the web pages related to the hot topic from the massive Internet web pages. These pages are involved in the formation of internet public opinion. meta information of the internet public opinion is consisted of three parts: basic information, content relevance and emotional tendency. Basic information inherited the basic data of web page content. Content relevance suggests how the webpage contribute to the formation of public opinion. Emotion tendency shows the netizen attitude reflected by the web page.

Through the mining of the meta information, internet public opinion can be obtained, and further trend analysis can be processed. accordingly , we can know What and how is crowds pay attention to, what viewpoint and attitude people hold, if the dispute intense and whether the opinion may be cause adverse consequences. The situation of the internet public opinion can be described as attention extent, controversial degree and weightiness. Attention extent is to reflect how the netizens concern to. Controversial degree reflects how users' attitude is different and controversial. Weightiness is used to reflect the influence on the society.

The monitoring of internet public opinion is designed to understand and grasp the voice of the people in time and be controlled and guided further. Especially for those high-profile, sensitive, strong controversial topic, it can be intervened appropriately and guided positively to ensure a harmonious network environment. From the point of view of technology, the platform of internet public opinion intervening is same to the platform of it producing, namely internet. Intervention can be run in two ways: leading and blocking. On the one hand, we can release the positive and correct information of public opinion via internet timely and make response quickly to the development of public opinion to guide the users to understand the truth, to quell the netizen emotions, and lead public opinion toward harmonious development. On the other hand, Bad information such as Internet rumors, seditious speech and unhealthy webpage should be filtered by artificial or technology modes. We should try to block all the avenues it appeared in the internet.

4 Conclusion

With the progress of science and technology and the development of network, Internet has become the main carrier of the social public opinion reflecting. Public opinion is collected much easier and more extensive than before. Explosion of the internet participators led to public opinion forms rapidly and has huge impact on [5] society. So, it should be attached great importance to.

If network monitoring is well employed, it will be beneficial to quickly grasp and understand internet public opinion. Accordingly, the appropriate solutions will be presented to meet the requirement s of various departments. Related research has important practical significance to construct the socialist harmonious society and the socialist democratic politics.

References

[1] Laihua, W.: Public Opinion Research Concept-Theory, Method and Hotspot of Reality. Tianjin Academy of Social Sciences Press (2003)

[2] Gao, H.: Analysis of Netizen's Affective Tendency on Public Opinion. JCIT: Journal of Convergence Information Technology 5, 120–124 (2010)

[3] Zeng, J.: Predictive model for internet public opinion. In: Proceedings - Fourth International Conference on Fuzzy Systems and Knowledge Discovery, FSKD, vol. 3, pp. 7–11 (2003)

[4] Yin, C.: Research on analysis technology of Internet public opinion based on topic cluster. In: 2nd International Conference on Information Science and Engineering. IEEE Computer Society, 445 Hoes Lane - P.O.Box 1331, Piscataway, NJ 08855-1331, United States.2010:6002-6005 (2010)

[5] Guan, Q.: Research and design of internet public opinion analysis system. In: Proceedings IITA International Conference on Services Science, Management and Engineering. IEEE Computer Society, 445 Hoes Lane - P.O.Box 1331, Piscataway, NJ 08855-1331, United States.2009:173 - 177 (2009)

Nonlinear Water Price Model of Multi-Source for Urban Water User

Wang Li[1], Ligui Jie[2], and Xiong Yan[3]

[1] School of Electronic and Information Engineering,
University of Science and Technology Liaoning, China
[2] Anshan City Plans Water-Saving Office, China
[3] School of Science, University of Science and Technology Liaoning, China

Abstract. This paper considers the water supply, sewage treatment and the affordability of urban water users, the nonlinear water price model for urban water user is established, in which the objective function is maximum the total social surplus. The Lagrange multiplier method is used to solve the model. The relationship between the ratio of the local water supply and the water price of urban water users is discussed. At last, the numerical example of a city is given.

Keywords: Nonlinear Price Model, Multi-Source, Lagrange Multiplier Method.

1 Introduction

In this paper, the system contains multi-source of water, urban water supply enterprises and urban water users, so water price is necessary to consider the costs, but also to analyze the mutual constraints of the water, enterprises and users. Factors affecting the water price in the following areas: demand and the affordability of urban water users, the degree of abundance or lack in the local water resources, the relationship between diversion and local water, et al. As a general commodity, water also follows certain laws of the market.

Many scholars have done lots of researches for water pricing problem. Ioslovich considered that the water price is the water allocation tools, a water extension model for urban, agricultural and other water users is proposed, and the allocation scheme and shadow price of water sources are obtained [1]. Qdais studied the impact of water pricing policies on water-saving effect in Abu Dhabi [2]. Schneider revealed the relationship between water price elasticity and water-saving, based on user water demand elasticity [3]. Murdock studied the role of socio-economic and demographic characteristics in the water forecast [4]. Using marginal cost model and the average cost model, Michael studied the water price in the southern and western regions of the United States [5]. Elnaboulsi studied the determination of the optimal nonlinear pricing rules for water supply services, a standard water supply distribution system model and a separate wastewater recycling system model are proposed [6]. Literature 7 represented the demand for different water users, from the meaning of the probability density, and a nonlinear water price model of the multi-source is constructed.

Z. Du (Ed.): Proceedings of the 2012 International Conference of MCSA, AISC 191, pp. 455–460.
springerlink.com

2 Problem Descriptions

Water, based on the nature of the urban use, can be used in living, administration, enterprise, hotel, restaurant, and other places. In this paper, the city water prices are unified classified as urban water price. The water prices transferred to the water source, are affected by urban water price and demands. Assume that city has only a water division project, and the local water supply price is known.

2.1 Some Symbol

Following are some symbols used in the model:

$1-\alpha$, α: proportion of the division and the local water source; $Q_L, Q_{L\max}$: the supply amount and the largest supply amount of local water; $Q_T, Q_{T\max}$: the supply amount and the largest supply amount of division water; Q_S: water demand; ΔQ_S: changes in water demand; Q_1: per capita water consumption; Q_2: 10 thousand yuan industrial output water consumption; C: division water supply costs; C_L: local water supply costs; p_t: division water price; p_l: local water price; p: urban user water price; ΔP: changes in water price; h: unit additional cost; S_{P_0}: minimum producer surplus; C_A: water price affordability of water user; A_1: disposable income; A_2: GDP of the industrial enterprises; ω_1, ω_2: the proportion of residential and industrial water; K_1, K_2: water charges account for a proportion of disposable income and gross industrial production; M: sewage treatment fee; M_1, M_2: standards and excessive pollution charges; R_1, R_2: standards and excessive pollution emission equivalent of the various pollution factors; R: charges price of pollution equivalent.

In the multi-source city water supply system, urban water demand is meeting by division and local water supply quantity:

$$Q_T = (1-\alpha)Q_S, \; Q_L = \alpha Q_S, \; Q_S = Q_T + Q_L$$

2.2 Demand Price Elasticity

Demand price elasticity measures the level of demand response to price changes. Let the demand price elasticity coefficient be E_s, then

$$E_s = \frac{\Delta Q_s / Q_s}{\Delta P / P} = \frac{dQ_s}{dP} \cdot \frac{P}{Q_s}, \text{ or } \frac{dQ_S}{Q_S} = E_s \cdot \frac{dP}{P}, \text{ so}$$

$$Q_S = K \cdot P^{E_S} \tag{1}$$

Where K is constant. Formula (1) is the mathematical model between water demand and water price.

2.3 Sewage Treatment Fee

Sewage treatment fee are calculated by actual emissions pollution equivalent.

$$M = M_1 + M_2 = \sum R_1 W + \sum n R_2 W \tag{2}$$

Where $n = R_2 / R_1$ is a penalty factor, which denotes excessive level.

3 Water Price Model

In this paper, we consider the interests of both sides of the city water users and water enterprises, some other factors such as development needs, as well as water supply, sewage treatment industry and social capacity. The objective function maximizes the total social surplus of the urban water users and urban water supply enterprise.

(i) Consumer surplus of urban water users expressed as:

$$CS(p) = C_A \cdot Q_S - p \cdot Q_S - M$$

$C_{A_i} = (\sum \omega_i \cdot A_i \cdot K_i)/Q_i$ ($i = 1,2$) represent living water and industrial water, respectively.

(ii) Producer surplus of water supply enterprises expressed as:

$$\begin{aligned} PS(p) &= p \cdot Q_S - (p_t + h) \cdot Q_T - (p_l + h) \cdot Q_L \\ &= p \cdot Q_S - (p_t + h)(1-\alpha) \cdot Q_S - (p_l + h)\alpha \cdot Q_S \end{aligned}$$

(iii) The objective function of the water price model is:

$$\max CS(p) + PS(p) \tag{4}$$

In the model constraints, the water price affordability of water users and marginal profit of water supply enterprise are considered, so we have:

$$PS(p) \geq S_{p_0} \text{ , and } p \leq C_A$$

4 Water Price Mathematical Model Solving

The Lagrangian function L is:

$$L = CS(p) + PS(p) + \lambda(PS(p) - S_{p_0}) = CS(p) + (1+\lambda)PS(p) - \lambda S_{p_0}$$

$$\frac{\partial L}{\partial p} = \frac{\partial CS}{\partial p} + (1+\lambda)\frac{\partial PS}{\partial p}$$

$$\frac{\partial CS}{\partial p} = C_A \cdot \frac{\partial Q_S}{\partial p} - \frac{\partial p \cdot Q_S}{\partial p} = C_A K E_S \cdot p^{E_S - 1} - K(E_S + 1) \cdot p^{E_S}$$

$$\begin{aligned} \frac{\partial PS}{\partial p} &= \frac{\partial p \cdot Q_S}{\partial p} - (p_t + h)(1-\alpha)\frac{\partial Q_S}{\partial p} - (p_l + h)\alpha \frac{\partial Q_S}{\partial p} \\ &= K(E_S + 1) \cdot p^{E_S} - (p_t + h)(1-\alpha) K E_S \cdot p^{E_S - 1} - (p_l + h)\alpha K E_S \cdot p^{E_S - 1} \end{aligned}$$

If $\frac{\partial L}{\partial p}=0$, then

$$C_A K E_S \cdot p^{E_S-1} - K(E_S+1)\cdot p^{E_S} + (1+\lambda)K(E_S+1)\cdot p^{E_S} = (1+\lambda)KE_S \cdot p^{E_S-1}\left[(p_t+h)(1-\alpha)+(p_l+h)\alpha\right]$$

So we have

$$\lambda K(E_S+1)\cdot p^{E_S} = KE_S\cdot p^{E_S-1}\left\{(1+\lambda)\left[(p_t+h)(1-\alpha)+(p_l+h)\alpha\right]-C_A\right\}$$

Hence the water price p is:

$$p=\frac{E_S}{\lambda(E_S+1)}\left\{(1+\lambda)\left[(p_t+h)(1-\alpha)+(p_l+h)\alpha\right]-C_A\right\} \quad (5)$$

Let $\varphi=(p_t+h)(1-\alpha)+(p_l+h)\alpha$, then the above result may be simplified to:

$$p=\frac{E_S\left[(1+\lambda)\varphi-C_A\right]}{\lambda(1+E_S)}$$

Let $\frac{\partial L}{\partial \alpha}=0$, that is

$$\frac{\partial L}{\partial \alpha}=\frac{\partial CS}{\partial \alpha}+(1+\lambda)\frac{\partial PS}{\partial \alpha}=(1+\lambda)\left[(p_t+h)Q_S-(p_l+h)Q_S\right]=0$$

then $\lambda=-1$, so the urban water users water price is:

$$p=\frac{E_S\cdot C_A}{(1+E_S)}$$

From (5) we have

$$p=\frac{E_S}{\lambda(E_S+1)}\left\{(1+\lambda)\left[(p_t+h)(1-\alpha)+(p_l+h)\alpha\right]-C_A\right\}$$
$$=\frac{E_S}{\lambda(E_S+1)}\left\{(1+\lambda)\left[(p_t+h)+(p_l-p_t)\alpha\right]-C_A\right\}$$

Because $p_l \le p_t$, urban water price p shows a decreasing, with α increases. It can be said for local water supply, under the case of determinate demands, the greater the proportion of urban water users, the lower the water price.

5 Practical Examples

In this paper, the proposed model calculated on the water price on a city's water users in China. In 2008, urban sewage discharges are shown in table 1.

Table. 1. A city into the river of waste water and discharge of pollutants scale 2

serial	sewage into the river	Emissions of pollutants					
		COD	BOD5	Ammonia	SS	other	total
1	5456	2641	1080	660	23187	0.71	27569
2	8735	5032	1136	594	75562	4.19	82328
3	1646	1241	591	86	9202	0.36	11120
total	15837	8914	2807	1340	10795	5.26	121017

(1) Measurement of the sewage treatment fee

According to Table 1, the first three pollutants are suspended solids, chemical oxygen demand things, and biochemical oxygen demand material, which belong to the second class of water pollutants.

(2) The second class of water pollutants equivalent

Suspended solid (SS) is 4; biochemical oxygen demand (BOD5) is 0.5; and chemical oxygen demand (COD) is 1. By calculating, we have:

SS=26987750, COD=8914000, BOD5=5614000.

Sewage treatment fee is $M = 0.7 \times (26987750 + 8914000 + 5614000) = 29061025$.

A city in 2003 and 2007, two sets of data are shown in Table 2 and Table 3.

Table 2. Urban water supply in 2003 (unit: cubic meters, no sewage reuse)

classification	groundwater		surface water	total
	within borders	outside borders	outside borders	
municipal WSS	2785	3130	4694	10609
corporate WSS	0	15660	2055	17715
owned water source	584	0	0	584
total	3369	18790	6749	28908

Logarithmic both sides of formula (5), we have

$$\ln Q_S = \ln K + E_s \ln P$$

The linear regression analysis results are: $E_S = 0.6$, and $K = 0.25$.

Table 3. Urban water supply in 2007 (unit: cubic meters, sewage reuse 110380)

classification	ground water		surface water	total
	within borders	outside borders	outside borders	
municipal WSS	2844	4121	5212	12177
corporate WSS	0	15439	2641	29118
owned water source	263	0	0	263
total	3107	19560	7853	41558

Based on above dada, water price affordability is $C_A = (\sum \omega_i \times A_i \times K_i)/Q_i = 6.6$, so water price may be calculated by formula (11) as $p = 2.48$.

6 Conclusions

The principle of the urban water user water price model developed taking into account the interests of both supply and demand, maximizes total surplus of social goals, and ultimately the maximization of social welfare.

References

[1] Ioslovich, I., Gutman, P.O.: A model for the global optimization of water price and usage for the case. Mathematics and Computers in Simulation 56, 347–356 (2001)
[2] Qdais, H.A., Nassay, H.I.: Effect of pricing on water conservation: a case study. Water Policy 3, 207–214 (2001)
[3] Sehneider, M.L.: User-specific water demand elasticity. Journal of Water Resource Planning and Management 1, 45–52 (1991)
[4] Murdock, S.H.: Role of sociodemographic characteristics in projections of water use. Journal of Water Resource Planning and Management 2, 117 (1991)
[5] Michael, L.: Estimating urban residential water demand: effects of price structure, conservation and education. Water Resource Res. 3, 12–15 (1992)
[6] Elnaboulsi, J.: Peak-load pricing for water and wastewater public services. Meeting of the Canadian Economics Association, Canada, pp. 5–8 (1997)
[7] Lingling, Z.: Research on nonlinear water price modeling of multi-source in water market. Dissertation, Hehai University (2007)

A Collaborative Filtering Based Personalized TOP-K Recommender System for Housing

Lei Wang[1,2], Xiaowei Hu[3,1], Jingjing Wei[1], and Xingyu Cui[1]

[1] Department of Management Science and Engineering,
Nanjing Forestry University,
Nanjing 210037, China
[2] School of Computer Science and Engineering,
Southeast University,
Nanjing 211189, China
[3] School of Informant Management,
Nanjing University,
Nanjing 210093, China

Abstract. Electronic information resource has become the main way for users obtaining information. Facing the huge amount of information in the real estate market, traditional methods are difficult to meet the users' effective information needs. How to dig out from the mass of information to the appropriate information is a difficult and time-consuming problem for anyone. How can personalized recommender system solve the problem? In this paper, the authors proposed an algorithm named Collaborative Filtering Based Personalized TOP-K Recommender system for Housing (CFP-TR4H), and a personalized recommender system based on CFP-TR4H is also designed in this manuscript. A case study on Nanjing (a city in China) real estate market is also conducted to discuss and validate the effectiveness of our method.

Keywords: personalized recommender system, preference, collaborative filtering, space vector, TOP-K.

1 Introduction

Plentiful information not only brings convenience to humans but also some problems: the first problem is it is difficult to obtained total information for its huge amount, let alone extracts the effective information from it; The second one is to separate true information from false; The third one is the variance of information form leads to it is difficult to handle unitively. Therefore, people began to put up a new slogan: "learn to abandon information" (Cheung, et al, 2003). People began to consider such a problem, how to find useful information timely and improve our efficiency of using information but not submerged by superfluous information. Personalized recommender system arises at the historic moment.

Personalized recommender system is an intelligent service system which is facing to users, which could overcome many problems and insufficiency of traditional information system. It reflects the connotation and requirements of knowledge

Z. Du (Ed.): Proceedings of the 2012 International Conference of MCSA, AISC 191, pp. 461–466.
springerlink.com

service, it is also an important aspect and application of information acquiring (Cho, Kim, & Kim, 2002). But when the personalized recommender system applied on a specific aspects it always exist several problems that is hard to overcome (Thai-Nghe, et al, 2010). The personalized recommendation system for housing in Nanjing area (a city in China) can be a good example.

This research has important theoretical significance and practical value: first, meeting the trend of information service personalized and representing the application and development direction of information technology under the knowledge service circumstance; Secondly, it also promote effective technical support for the development of personalized recommender technology, and offer reference for related research.

2 Our Proposed Method

The algorithm put forward in this paper named Collaborative Filtering Based Personalized TOP-K Recommender for Housing (CFP-TR4H) is based on the prerequisite that "people who are similar to each other always have similar preference" (Yang & Li, 2009; Lee, Cho, & Kim; Barragáns-Martínez, et al, 2010).

Pearson similarity (Coelho, Braga, & Verleysen, 2010) is popular used as equation (1).

$$\mathrm{Sim}(i,j)=\frac{\sum_{i=1}^{n}(R_{k,i}-R_i)(R_{k,j}-R_j)}{\sqrt{\sum_{i=1}^{n}(R_{k,i}-R_i)^2}\sqrt{\sum_{i=1}^{n}(R_{k,j}-R_j)^2}} \tag{1}$$

In equation (1), $R_{k,i}$ is the mark that the user numbered k gives to the item i, R_i and R_j represent the average mark of item i and j that users give.

Definition 1. We define R_i as the average mark of item i that users give.

Set, $\{R_i(1), R_i(2), \cdots, R_i(n)\}$ is the set of the marks that users give to the item i, so

$$R_i=\frac{1}{n}\sum_{k=1}^{n}R_i(k) \tag{2}$$

Definition 2. We define R_i^+ as the value of upper bound of R_i, and R_i^- as the lower bound of R_i, so we can get:

$$(1)\ \ R_i^+=\max(R_i(k)),\ (1\le k\le n,\ n\ \textit{is the history user number}) \tag{3}$$

$$(2)\ \ R_i^-=\min(R_i(k)),\ (1\le k\le n,\ n\ \textit{is the history user number}) \tag{4}$$

Definition 3. We define a thresthold value T_i ($0\le T_i\le 1$) for item i. T_i is used to judge the similar preference users' mark for item i.

Set $R_{k,i}$ denotes the mark that the user numbered k gives to the item i. By thresthold value T_i and definition 2, we can get:

$$(1)\ \ R_{k,i}^1=R_k^i+(R_i^+-R_i^-)\times T_i \tag{5}$$

$$(2)\ \ R_{k,i}^2=R_k^i-(R_i^+-R_i^-)\times T_i \tag{6}$$

Here $R^1_{k,i}$ and $R^2_{k,i}$ is the value of of R^i_k explores up and down. But $R^1_{k,i}$ or $R^2_{k,i}$ may exceed the upper bound or lower bound. So, with these dedinitions and equations, we can define the value of $\widetilde{R_i}$.

Definition 4. We define $\widetilde{R_i}$ as the similar preference mark by thresthold for item i (we set, the number of the values in $[R^1_{k,i}, R^2_{k,i}]$ from the set of $\{R_i(1), R_i(2), \cdots, R_i(n)\}$ is n), and $\widetilde{R_i}$ can be calculate as equation (7):

$$\widetilde{R_i} = \begin{cases} R_i, (if\ (R^1_{k,i} \geq R^+_i)\ or\ (R^2_{k,i} \leq R^-_i)) \\ \frac{1}{n}\sum[R^1_{k,i}, R^2_{k,i}], otherwise \end{cases} \tag{7}$$

Hence, the calculation similarity from user i and j can be calculated as equation (8),

$$\mathrm{Sim}(i,j) = \frac{\sum_{i=1}^{n}\left(R_{k,i} - \widetilde{R_i}\right)\left(R_{k,j} - \widetilde{R_j}\right)}{\sqrt{\sum_{i=1}^{n}(R_{k,i} - \widetilde{R_i})^2}\sqrt{\sum_{i=1}^{n}(R_{k,j} - \widetilde{R_j})^2}} \tag{8}$$

In the equation (8), $R_{k,i}$ is the mark that the user numbered k gives to the item i, $\widetilde{R_i}$ represent the average mark of item i the similar preference users give.

The core thinking of our proposed algorithm CFP-TR4H is described as follows:

Input data:

(1) Item-mark matrix R(m,n);
(2) The number of nearest neighbors SNN;
(3) The threshold of mark H;

Output data:

(1) TOP-K recommend results
(2) The similarity of system users compared with customers

The process of CFP-TR4H:

- STEP 1:

Input user's basic information, and convert the basic information into space vector R_k according to the discretizing and generalizing standard;

- STEP2:

Traversing all users' information in system, and calculating the similarity between R_k and record in system one by one, then write consumers' number and similarity into array sim [m] [2], m represents for the consumers' amount that saved in system. The first column in array records user Number, second column records similarity. We use the equation (8) to calculate the similarity.

- STEP3:

Arranging the array a [m] [2] according to the value of second column by descending order, and deleting array [m] [2] except for the first SNN row;

- STEP4:

Searching for the ID of housing according to the consumer's ID in first column of array [m] [2], then inquiring housing information in the database and output the housing information as TOP-K recommendation results to users.

3 Case Study: A Framework for Personalized Recommender System in Nanjing Based on CFP-TR4H

The whole structure of system design is shown in figure 1.

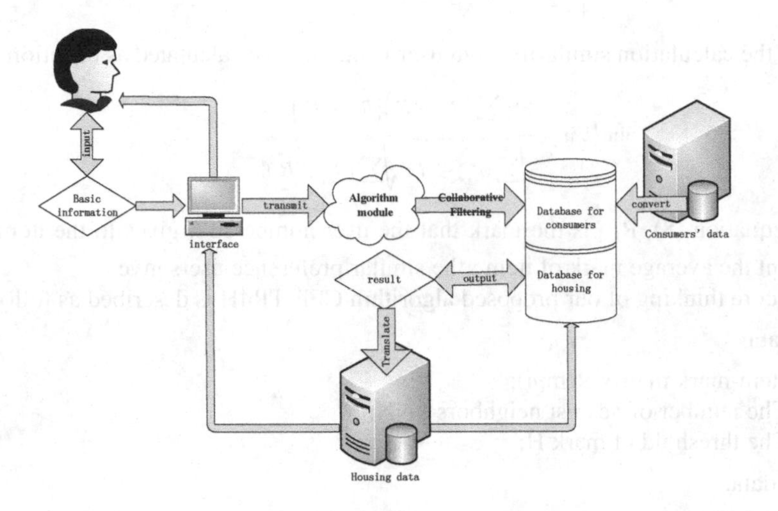

Fig. 1. The framework of CFP-TR4H

The data for this experiment is collected by questionnaire, we extract 3 groups of consumers' data from 5 000 groups of consumers' preference data randomly. Through correlation test between the consumers' preference and the result recommended by the system, we can analysis the recommender precision of system. The experimental data extracted from each group includes two parts, the first part is customers' basic data, which would be transmitted into system, the second part is the customers' housing data, which would be compared with results to test the accuracy of recommendation. Because the experiment data have already existed in system database, detecting the data that the similarity is 100% is one of the standard that measure whether or not the system can run normally. The space vectors of housing which is recommend and the similarity is shown in table 1.

Table 1. Vector Recommended by System and the Precision

Input data	Vector Recommended by System	Precision
female, 32 years old, two numbers in family, doctor, self-employed, household disposable income 7800 Yuan per month	(9,1,3,3,3,2,1,2)	1.000
	(9,2,1,2,1,4,4,3)	0.902
	(8,3,2,2,1,3,2,3)	0.812
female, 32 years old, two numbers in family, doctor, self-employed, household disposable income 7800 Yuan per month	(8,3,2,2,2,1,3,2)	1.000
	(7,2,2,2,1,1,3,3)	0.825
	(5,2,2,2,1,1,2,3)	0.816
female, 52 years old, two numbers in family, undergraduate, state-owned enterprises, household disposable income 48000 Yuan per month	(10,2,4,3,1,4,1,3)	1.000
	(11,3,3,2,2,2,1,2)	0.891
	(11,4,3,2,1,1,2,1,2)	0.887

From the table 2, it can be seen that, the system are able to detect the data in database and make TOP-3 recommendation, it can prove the normally operation of system.

At the same time, traditional collaborative filtering algorithm fills item-mark matrix (by equation 1) with the average mark of items , this method is always inaccuracy and seems unuseable, and we are hard to get accurate nearest neighbors. The improved collaborative filtering algorithm of CFP-TR4H could avoid the scarcity of data in item-mark matrix problems through reasonable and effective data collection.

4 Conclusion

To solve the defect in user-based collaborative filtering algorithms and item-based collaborative filtering algorithms, this paper aims at constructing Collaborative Filtering Based Personalized TOP-K Recommender for Housing (CFP-TR4H) and proposes "space vector similarity-based" collaborative filtering algorithm, this algorithm tries to disassemble complex things according to space dimensions, and gives a reasonable mark for each dimension of each item, and then fills the item-mark matrix, which could avoid the problem of data scarcity in traditional collaborative filtering algorithm effectively. However, how to divide the space vector more reasonable to improve the precision of recommendation is still a hard problem needed to keep on researching.

Acknowledgement. The work was supported by the Education Department of Jiangsu province college students' practice and innovation project, China.

References

Cheung, K.-W., Kwok, J.T., Law, M.H., Tsui, K.-C.: Mining customer product ratings for personalized marketing. Decision Support Systems 35(2), 231–243 (2003)

Cho, Y.H., Kim, J.K., Kim, S.H.: A personalized recommender system based on web usage mining and decision tree induction. Expert Systems with Applications 23(3), 329–342 (2002)

Coelho, F., Braga, A.P., Verleysen, M.: Multi-Objective Semi-Supervised Feature Selection and Model Selection Based on Pearson's Correlation Coefficient. In: Bloch, I., Cesar Jr., R.M. (eds.) CIARP 2010. LNCS, vol. 6419, pp. 509–516. Springer, Heidelberg (2010)

Lee, S.K., Cho, Y.H., Kim, S.H.: Collaborative filtering with ordinal scale-based implicit ratings for mobile music recommendations. Information Sciences 180(11), 2142–2155 (2010)

Thai-Nghe, N., Drumond, L., Krohn-Grimberghe, A., Schmidt-Thieme, L.: Recommender system for predicting student performance. Procedia Computer Science 1(2), 2811–2819 (2010)

Yang, J.-M., Li, K.F.: Recommendation based on rational inferences in collaborative filtering. Knowledge-Based Systems 22(1), 105–114 (2009)

An Express Transportation Model of Hub-and-Spoke Network with Distribution Center Group

Xiong Yan, Wang Jinghui, and Zheng Liqun

School of Science, University of Science and Technology Liaoning, China

Abstract. Hub-and-spoke network is the main form of express transportation at present. The logistics optimization of such network is the core to get better benefits for express company. According to the transport characteristic of hub-and-spoke network, combined with hub-and-spoke theory, the express transportation model is proposed based on the lowest total transportation cost in this paper. Then the genetic algorithm for the model is given, by which we can obtain the division scheme of distribution centers and the vehicle scheduling plan. Finally, an example is solved.

Keywords: Express Transportation, Mathematical Model, Genetic Algorithm, Hub-and-Spoke Network.

1 Introduction

In recent years, with the number of online shopping orders quickly rising the express company business scale is continually expanding. Weather, rising oil prices, the policies of road transportation and other factors increase the cost of logistics and transport. So the optimization method of express logistics transportation network has been paid more and more attention. The hub-and-spoke network is a novel logistics transportation system, through which express packages can be adjust in the axis point temporarily, thus disordered and backlog of packages distribution are improved, and the express company benefits from economies of scale. Early hub-and-spoke networks were mainly used for long-distance and large scale cargo transport services, such as air transport, shipping and road transport [1-4]. Now the networks are widely adopted by express transportation, and play a very important role [5-7].

In this paper, hub-and-spoke network with distribution center group is applied to the express transportation problem and for obtaining the division scheme of distribution centers and the vehicle scheduling plan, a mathematical model based on the lowest total transportation cost is established. Finally, the genetic algorithm and an example are given for the model.

2 System Description

2.1 Structure and Characters

Express transportation system based on hub-and-spoke network with distribution center group is shown in Figure 1. A transshipment center is denoted by H , called

Z. Du (Ed.): Proceedings of the 2012 International Conference of MCSA, AISC 191, pp. 467–472.
springerlink.com

hub. Let point $N_i(i=1,\cdots P)$ represents freight station, called spoke. Distribution center is expressed as point $M_j(j=1,\cdots,M)$. According to different transportation methods, the network is divided into two regions, namely center region and distribution region. Since the large freight and long transport distance in the center region, efficient transport tools are used, such as train, plane, ship, and so on. We called trunk line transport. While the opposite is the case in distribution regions, small vehicle can be used well, such as small truck or package car. We called branch line transport. The distribution center is only connected to one freight station. All the express packages collected from the service area are sent to the freight station. If both the start distribution center and the objection distribution center belong to the same freight station, the express packages will send directly without passing the hub.

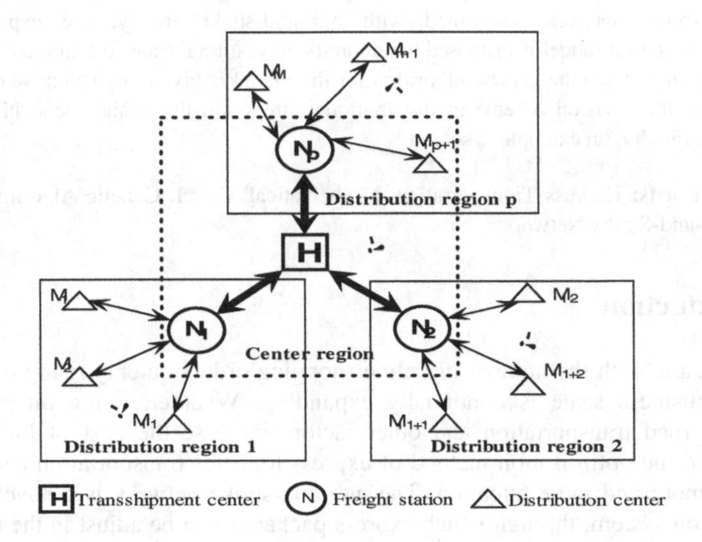

Fig. 1. Structure of hub-and-spoke network with distribution center group

2.2 Cost Analysis

The total cost of the express freight is composed of four parts, namely the transport cost of branch lines, the handling costs of freight stations, the transport cost of trunk lines, and the handling cost of transshipment center. In the center region, the transport cost of trunk lines is related to the transport methods and volume of packages. While in the distribution region, the transport cost of branch lines is associated with the cost of vehicle maintenance and depreciation, fuel cost, parking fee and toll fee. The handling cost either in the freight station or in the transshipment center is about storage, load and unload, and sort.

3 Mathematical Model

3.1 Assumptions

The following assumptions are needed to formulate the mathematical model.

1. There is only one transshipment center in the transportation system.
2. Transport between distribution centers belonging to the same freight station doesn't pass the transshipment center.
3. Each distribution center is serviced by just only one freight station.
4. Distribution center doesn't transport directly to other distribution centers.
5. Each freight station owns the fixed types and numbers of vehicles. The vehicles must go back to the freight station after service.

In order to describe the model some notations are given below.

N_i^k -Number of the k-type vehicle belonged to freight station i;

r_k - Unit transport capacity of the k-type vehicle;

c_0 - Handling cost of unit weight of package in the transshipment center;

c_i - Handling cost of unit weight of package in the freight station;

c_{i0} -Average transport cost of unit weight of package between freight station i and the transshipment center;

c_{ji}^k - Average transport cost of the k-type vehicle between distribution center j and freight station i;

y_{jq} - Transport volume from distribution center j to distribution center i;

z_{i0} - Transport volume from the freight station i to the transshipment center;

$\overline{z}_{0i}$ -Transport volume from the transshipment center to freight station i;

x_{ji} - Transport volume from distribution center j to freight station i;

$\overline{x}_{ij}$ - Transport volume from freight station i to distribution center j.

Then we defined the decision variables as:

n_{ji}^k - Number of the k-type vehicle from freight station i to distribution center j;

$$\delta_{ji} = \begin{cases} 1, & \text{distribution } i \text{ connected with freight station } j \\ 0, & \text{otherwise.} \end{cases}$$

So the transport volume can be expressed as:

$$z_{i0} = \sum_{j=1}^{M} \delta_{ji} [\sum_{\substack{q=1 \\ q \neq j}}^{M} y_{jq} \left(1 - \delta_{qi}\right)]; \ \overline{z}_{0i} = \sum_{j=1}^{M} \delta_{ji} [\sum_{\substack{q=1 \\ q \neq j}}^{M} y_{qj} \left(1 - \delta_{qi}\right)]$$

$$x_{ji} = \delta_{ji} \sum_{\substack{q=1 \\ q \neq j}}^{M} y_{jq}; \ \overline{x}_{ij} = \delta_{ji} \sum_{\substack{q=1 \\ q \neq j}}^{M} y_{qj} \qquad i = 1, \cdots\cdots, P$$

3.2 Model Establishment

With the assumptions, a formulation of this problem is stated as:

$$\min C = c_0 \sum_{i=1}^{P} z_{i0} + \sum_{i=1}^{P} c_i [\sum_{j=1}^{M} (x_{ji} + \bar{x}_{ij}) \delta_{ji}] + \sum_{i=1}^{P} c_{i0} (z_{i0} + \bar{z}_{0i}) + 2 \sum_{i=1}^{P} \sum_{j=1}^{M} \sum_{k=1}^{K} (n_{ji}^k c_{ji}^k \delta_{ji}) \quad (1)$$

$$s.t.\ \max\{\sum_{j=1}^{M} x_{ji}, \sum_{j=1}^{M} \bar{x}_{ij}\} \le \sum_{k=1}^{K} n_{ji}^k r_k \quad i = 1, \cdots, P \quad (2)$$

$$\sum_{k=1}^{K} \sum_{j=1}^{M} n_{ji}^k \delta_{ji} \le \sum_{k=1}^{K} N_i^k \quad i = 1, \cdots, P \quad (3)$$

$$\sum_{i=1}^{P} \delta_{ji} = 1 \quad j = 1, \cdots, M \quad (4)$$

$$x_{ji} \ge 0, \bar{x}_{ij} \ge 0,\ n_{ji}^k > 0 \quad i = 1, \cdots, P;\ j = 1, \cdots, M;\ k = 1, \cdots, K \quad (5)$$

The objective function (1) minimizes the total cost of the express freight of hub-and-spoke network. Transport capability can meet the transport volume between freight station i and distribution center j by constraint (2). Constraint (3) denotes the total number of each type vehicle used by serviced distribution centers can't exceed the vehicle number owned by the freight station. Constraint (4) limits each distribution center is serviced by just only one freight station.

4 Solution Algorithm

Genetic algorithm (GA) is a very popular algorithm for searching a near-optimal solution in complex spaces. Therefore a hybrid genetic algorithm is proposed to solve the model in [8]. The algorithm adopts hybrid-coding, and the vehicle scheduling algorithm in the freight station is embedded to calculate the transport cost of branch lines. In order to test the validity and practicality of the method for our model, an example is shown below.

There are 3 freight stations and 6 distribution centers in the hub-and-spoke network. Each freight station has two type vehicles, namely Ⅰ and Ⅱ, and each type has 5 vehicles. Detailed simulation date is shown in table 1 and table 2.

Table 1. Daily average transport volume of packages form the start distribution center to the purpose distribution center (unit: ton)

Objective Distribution Center	Start Distribution Center					
	1	2	3	4	5	6
1	--	0.3	0.5	0.6	0.1	0.2
2	0	--	1	0.1	0.4	1.5
3	0.4	0	--	0.6	0.8	1
4	0.3	0.2	0.5	--	1	0.1
5	0.1	0	0.6	0.7	--	0.3
6	1.5	0	0	0.1	0.3	--

Table 2. Average transport cost from freight station to distribution center by different vehicle (unit: Yuan)

Freight Station	Distribution Center											
	M_1		M_2		M_3		M_4		M_5		M_6	
	I	II	I	II	I	II	I	II	I	II	I	II
N_1	50	25	70	35	130	65	150	76	100	50	180	100
N_2	90	70	80	50	30	15	60	35	60	35	80	40
N_3	150	100	200	130	170	80	75	35	75	55	20	15

The division scheme of distribution centers is given in future 2, and the vehicle scheduling plan is shown in table 3. The total cost is 10050 Yuan.

Table 3. Vehicle scheduling plan

Freight Station	N_1	N_2		N_3		
Distribution Center	2	1	6	3	4	5
I	2	1	1	1	0	0
II	0	1	2	1	3	3

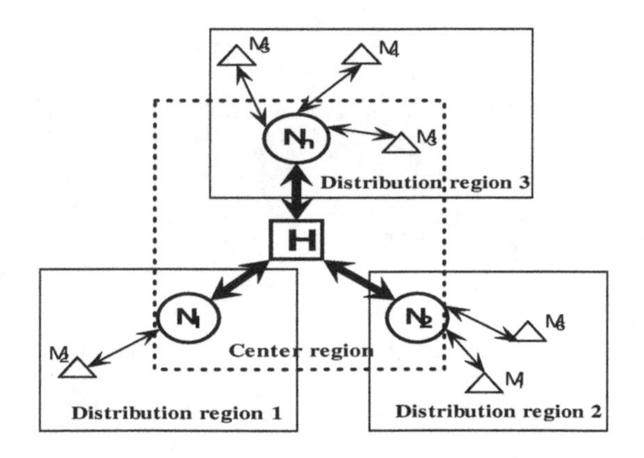

Fig. 2. Division scheme of distribution centers

5 Conclusion

In this study we proposed an express transportation model of the hub-and-spoke network, through which both the division scheme of distribution centers and the vehicle scheduling plan can be obtained. There also many questions involving complex hub-and-spoke network structure in practice. Thus, those networks will be the next research focus.

Acknowledgement. The work was supported by Natural Science Foundation of Liaoning Province under Grant 20102097.

References

[1] Hanyi, Y.: The application of hub-and-spoke network in Chinese aeronautic transportation. Journal of Beijing Institute of Technology 12(2), 27–30 (2010)
[2] Chuanxu, W.: Optimization of hub-and-spoke based regional port cluster two stage logistics system network. Systems Engineering Theory and Practice (9), 152–158 (2008)
[3] Zhihong, J., Yuzhen, X., Yang, L., Jun, H.: Scheduling optimization problems of feeder line container ships. Navigation of China 31(4), 415–419 (2008)
[4] Deborah, L.B., O'kelly, M.E.: Hub-and-spoke networks in air transportation: an analytical review. Journal of Regional Science 39(2), 275–295 (1999)
[5] Shixiang, Z., Jiazhen, H.: Research of city group logistics distribution system planning in Changjiang river delta based on hub-and-spoke network model. Chinese Journal of Management (4), 194–199 (2005)
[6] O'kelly, M.E.: Routing traffic at hub facilities. Netw. Spat. Econ. (10), 173–191 (2010)
[7] Jian, Z., Yaohua, W., Pei, L., Yanyan, W.: Hybrid hub-and-spoke network planning of road express freight. Journal of Shandong University 38(5), 6–10 (2008)
[8] Yan, X., Jinghui, W.: Hybrid genetic algorithm for the hub-and-spoke network of the express transportation problem. Journal of University of Science and Technology Liaoning (2012)

Developing a Small Enterprise's Human Resource Management System -Based on the Theory of Software Engineering

Chen Yu

Wuhan Polytechnic, 430074, China
chenyupaper@163.com

Abstract. First, this paper analysis the existing problems in Chinese software development. Second, the software engineering theory(especially the waterfall model) is discussed in this paper. Third, a small enterprise human resource management system is taken as an researching object. Finally, through using of waterfall model in the software development, the small enterprise's human resource management system is developed. Through implementing the system, some suggestion is provided for us at last.

Keywords: Software engineering, human resource, water fall, design.

1 Introduction

China have carry out the policy of Informatization construction since the end of the last century. Therefore the development of Chinese information industry has made a gratifying progress. Software industry is the pillar industry of Information industry. It had been realized the purpose of narrowing the technology gap with the world's advanced software industry gradually. With these gratifying achievement ,we also found that there are some deep problems in chinese software industry. These problems are showed as follow: While software development enterprises develop software, part of enterprises do not follow the inherent laws of software, they pure pursuit economic benefits. So the whole development process is disordered. Therefore, the products which were produced by these enterprises is not stable. If these problems cannot be solved well, it is likely to lead to Chinese information industry (especially software industry) into the state of crisis. We take a software development as an example. In this example, we will focus on a small enterprise's human resource management. Through using software engineering theory[1], we construct a standardized software development process.

2 Basic Theory

In the theory of software engineering, the software life cycle is described with the waterfall modeled. The model comprises seven steps concretely. They are showed as follow: Feasibility Analysis, Demand Analysis, Overall design, Detail design, Coding,

Z. Du (Ed.): Proceedings of the 2012 International Conference of MCSA, AISC 191, pp. 473–477.
springerlink.com

Test, Maintain. The above seven steps organically constituted the entire software life cycle[2][3].

Feasibility analysis is determining whether the problem can be solved with the the shortest possible time and the minimum cost. Feasibility analysis includes technical feasibility analysis, economical feasibility analysis, operational feasibility analysis. The basic task of Demand Analysis is to answer the problem what the system must do accurately. At the same time, it should put forward complete, accurate, clear, specific requirements for the target system. Overall design is also called the outline design or initial design, its basic goal is to answer the question how to implement the system? The fundamental goal of Detailed design is determining how to achieve specific required systems. That is to say, after this stage of designing work, the target system should be draw out. So, during the encoding phase ,the idea of detailed design can be translated directly into a written procedures which was wrote by some programming languages. Coding[4][5] is to translate the software design results into computer-understandable form. The purpose of the testing is to find faults as much as possible. The software testing is still the main step of ensuring software quality Software maintenance is the final phase of the lifecycle. After the software had been delivered for using, some mistakes were found in the software. We must correct these mistakes to meet the new needs.

The object of this researching is a small enterprise's human resource. We will use informatization method, we will using the theory of the waterfall model in software engineering. Finally we will develop a small enterprise's human resource management system. In the next chapter, we will carry out these comprehensive process.

3 Total Process

In this chapter, we will use the waterfall model in the software engineering to develop small enterprise's human resource management system. We will begin our researching work which was based on the waterfall model. The first step is to begin feasibility analysis.

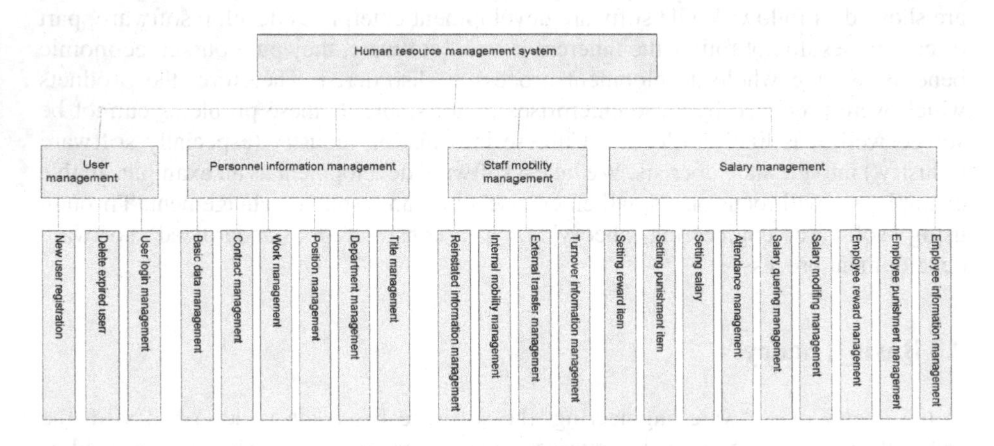

Fig. 1. System structure

3.1 Feasibility Analysis

In this section, the feasibility analysis will include technical feasibility analysis, economical feasibility analysis, operational feasibility analysis. We only focus on the technical feasibility for example.

3.2 Demand Analysis

Through surveying the small business , we found that, the small and medium-sized enterprise in the human resources management has the following four aspects of the basic needs. First, the system must be able to support the informatization of enterprise's internal human resources information. Second, the system must be able to support the querying of enterprise's human information and support the recording of specific operation record. Third, the system must be able to support the informatization of enterprise's staff adjustment. Fourth, the system must be able to support the auditing of all personnel in the system.

3.3 Overall Design

At the stage of overall design, we focus on the designing of system structure, the designing of system process, the designing of database. At the stage of system structure design , an overall system structure of human resources management was obtained. It was shown as below:

3.4 Detail Design

At the stage of the detail design, this paper is focused on the database detailed design. We will take the designing of staff basical information table as an example to discuss the detail Design.

Table 1. Staff basical information

Field name	Field description	Data Type	Length	Is NUll
ser_id	Serial number	Varchar	8	Not null
cust_name	Name	Varchar	10	Not null
cust_sex	Sex	Varchar	2	Not null
cust_id	Identity card number	Varchar	30	Not null
marry_statue	Marital status	Varchar	4	Not null
register_person	Registered person	Varchar	10	Not null
register_date	Registered date	Datetime	8	Not null
education	Education	Varchar	20	Not null
other	Other	Text	100	

3.5 Coding

At the stage of coding, the staff basical information table will be taken as an example in this paper. Below, the staff basical information table is created.

```
Create table Employer_basic_infor
( ser_id          varchar(8)      is not null,
cust_name         varchar(10)     is not null,
cust_sex          varchar(2)      is not null,
cust_id   Identity   varchar(30)     is not null,
marry_statue      varchar(4)      is not null,
register_person   varchar(10)     is not null,
register_date     datetime        is not null,
education         varchar(20)     is not null,
other             text(100)
)
```

3.6 Test

Through the above five steps, we have completed the system design and development work of a small enterprise human resource management. Then we will begin the internal testing and external test. When the product meets the test requirements, the product will deliver to the user directly.

4 Conclusion

Through researching of human resource management system, we can found the software development is a system engineering. If we want to develop high quality software products, the entire development process must be accorded to the theory of software engineering.

References

[1] Braude, E.J., Bernstein, M.E.: Software Engineering: Modern Approaches, pp. 345–547. Wiley (April 5, 2010)

[2] Lu, L., Quan, X.: Optimization of software engineering resource using improved genetic algorithm. International Journal of the Physical Sciences 6(7), 1814–1821 (2011)

[3] Xie, J.Y.: Introduction to human resource development, vol. 5. Tsinghua University Press, Beijing (2005)

[4] Stewart, R.B., Wanhua, Q.: Foundation of Value Engineering Method. China Machine Press, Beijing (2007)

[5] Zhou, R.: Study on Human Resource Management Based on Value. Journal of Zhongnan University of Economics and Law (4), 129–134 (2006)
[6] Zhang, D.: Development and Management of Human Resource, 2nd edn. Tsinghua University Press, Beijing (2001)

Research on Applying Folk Arts' Color in Computer Based Art Design

Hu Xiao-ying

Xinxiang University, Henan Xinxiang, 453000, China
huxiaoying555@163.com

Abstract. The boundaries of folk art are subjective, but the impetus for art is often associated with creativity, regarded with wonder and admiration along human history. The goal of our approach is to explore novel creative approaches of applying color into folk art with the help of computer-aided art design. A creative system is designed to aid the creative processes, and the system is able to generate novel effect on folk art color. The sequences are combined and selected, based on their different characteristics, in forms of reflecting the color styles. The system is of great significance for folk art color in computer-aided art design process.

Keywords: folk art, color application, computer, art design.

1 Introduction

Folk art encompasses art produced from an indigenous culture or by peasants or other laboring tradespeople. In contrast to fine art, folk art is primarily utilitarian and decorative rather than purely aesthetic. Folk art is characterized by a naive style, in which traditional rules of proportion and perspective are not employed. Closely related terms are Outsider Art, Self-Taught Art and Naive Art[1]. As a phenomenon that can chronicle a move towards civilization yet rapidly diminish with modernity, industrialization, or outside influence, the nature of folk art is specific to its particular culture. The varied geographical and temporal prevalence and diversity of folk art make it difficult to describe as a whole, though some patterns.

Computer-aided art design is developing and becoming more common, changing our perception of what traditional art is and what it will become. Not only have traditional forms of art been transformed by digital techniques and media, but entirely new forms have emerged as recognized practices. Computers may enhance visual art through ease of rendering, capturing, editing and exploring multiple compositions, supporting the creative process. Artists express their creativity in ways intended to engage the audience's aesthetic sensibilities or to stimulate mind and spirit, sometimes in unconventional ways[2]. There is a trend that more and more people use mobile devices on the way home or during waiting bus. The traditional wired network can't meet high demands on mobility, it's easy damaged, hard to locate and expand its network. It's easy to bring and access to internet by wireless network, easy to be installed and low cost. There are more and more corporations realize the importance

Z. Du (Ed.): Proceedings of the 2012 International Conference of MCSA, AISC 191, pp. 479–483.
springerlink.com

of art design and add them to folk art. Now days, enterprise and personal applications will take more and more important role in folk art, and with the high development of color design and devices, it will bring a deep revolution for people's life. Contract to the rapid development of art design applications, the security is the bottleneck of color management. The attacker can pretend to be a valid identity and then access to folk art to attack and theft information. Art design is more dangerous than conventional art for its open policy. Therefore, it's important and necessary to research on the folk art color with computer-aided art design.

2 The Significance of the Folk Art Color

With the diversification of the society, the traditional art of color is not fixed. The aesthetic consciousness, impacted gradually by people, becomes more and more important in daily life. But the overall trend of folk art color was followed for thousands of years by culture precipitation. In the traditional Chinese mind, red color is the main one. It represents the harvest, peace and best wishes, able to meet people's hope and expectation. Therefore, in the present generation of folk art color, in order to auspicious happy means, red base can be used; to express romantic mood, light purple is optimal. If spring comes, green color can be choosen[3]. The application of color will undoubtedly reflects the aesthetic psychology of people and customs, cultural design requirements. Folk art color affects people not only in daily life, but also ethnic affinity. It is just because of this ancestral culture and identity feeling, making our folk art colors in bidding design more capacious.

3 Experimental

The initial example is generated by composing elements with randomly chosen segments, or according to specific rules. One such rule is that the segments in each group keep the order they had in the original state. Another rule is to satisfy specific criteria for each segment, defining a structure and color system for the fork art. In addition, one can specify if segment repetition is allowed in the same group. Other rules can be defined. The characteristics of the color has to be pre-defined, either a fixed style or not. In this case, the system will work with variable color effect. Evaluation of each individual is done by a fitness function which only takes into account annotations associated to each segment in the folk art. This allows a faster processing.

There is a balance to be made regarding the color variations. To apply the folk art color to the computer involves a significant processing time and consequently a slow process. An alternative approach is to have a process of novel development to apply them to produce the actual color remediation. On the other hand, this alternative requires that we change the color whenever producing a major structural modification. It makes do with a simpler and processing times may be acceptable at the folk art color application.

4 Results and Discussions

4.1 Representation and New Features

At the syntactic level, fork art has a structure defined by the order of their nature, reflecting the applying effect in the individual. Different individuals can further be combined in a higher level structure to compose color and convey further art design.

It should be noted that since the color pieces correspond to the art design segments, we maintain a design system of all the segments, which constitutes the domain of the alleles. The normal operation of the handling process does not need more than that. Only in the end, for the art design process, we need to generate the color characteristics, indicated in Figure 1.

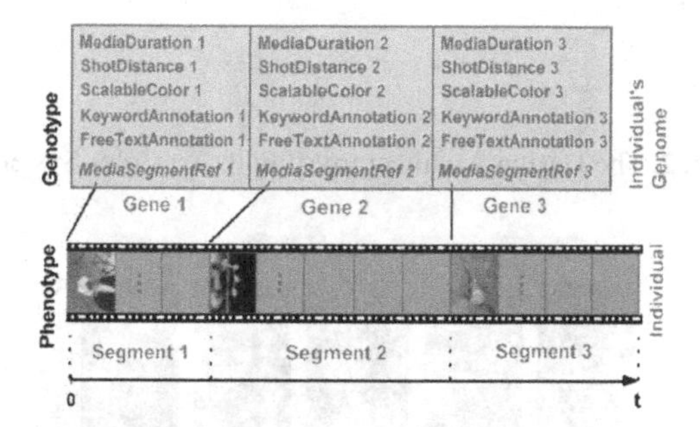

Fig. 1. Example of the Folk Art Fragement

4.2 The Handling Process

First of all, select elements to generate the effect from the pool of computer system. This is the phase where the structural characteristics of the folk art is defined. Then, three selection plans are used: random selection, not restricted; ordered selection, where the segments must respect certain order in relation to the original color elements selected, where more data can be defined for each segment. As shown in Figure 2.

When the treatment process begins, the evolving factor is presented after each process. In Figure 3, we see a group of five individuals, with three segments each. The calculated fitness value is presented on top of each individual. The user may choose to play/pause each individual, providing a preview for better examination and choice. The user can eliminate one or more individuals in each group, influencing the next step and thus the whole process. In the current version, the user can also select one individual as the optimal and presenting the best individual as the final generation[4]. When the step is finished, the treatment process is performed without intervention, and the final solution obtained with the pre-defined criteria and parameters is presented, just like in Figure 3, but with the final solution-a single individual. At any time, the process can also be restarted. In the example of Figure 4, the goal includes turtle as the semantic keyword, and the dominant color blue at the lower level, in a pool having several segments in the modified art design process.

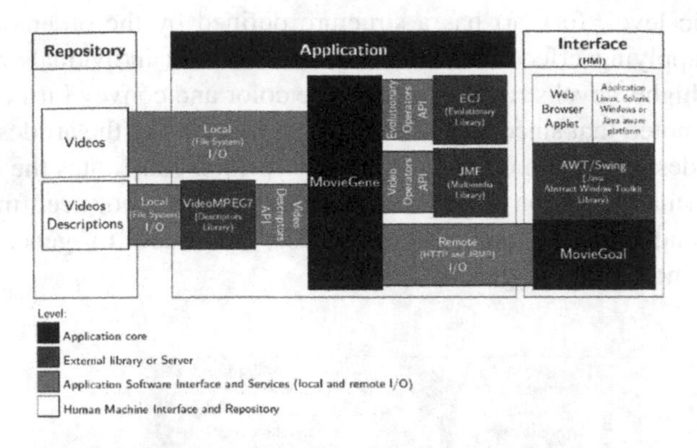

Fig. 2. The Design Architecture of Folk Art Color System

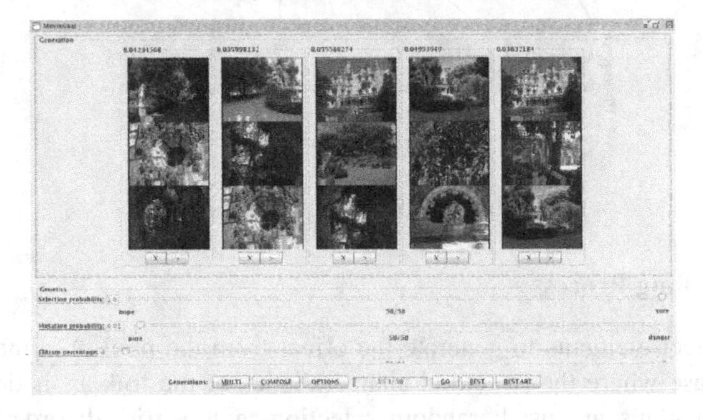

Fig. 3. The Applying Effect of Folk Art Color with Computer

As in the previous examples of Figure 3 and Figure 4, fitness is defined as a similarity measure, to this goal. The initial group of three individuals was created by random selection. No repetition was allowed for the example, mutation was chosen to benefit precision over the difference. Consecutive generations of individuals explore different combinations of segments with elements in several colors and styles, in a search for examples with dominant color blue and featuring turtles. In this example, the best individual after 30 steps has the highest fitness concerning these parameters. But again, while approaching the intended goal, the search process provides us with combinations of different individuals with different properties, often of aesthetical and narrative value[5].

Fig. 4. The Final Applying Result of the Color Modification

5 Conclusions

A novel approach of folk arts' color into computer system was proposed and developed. The prototype were extended to include more flexible and structural options, increasing the available solutions for creation of new art design styles, by proposing innovative combinations which might be subjectively selected from the process of arriving at more satisfactory or artistic solutions. Results obtained so far showed the technical viability as a creative approach, although experimented in simple scenarios. Interaction with this new prototype reflects a higher flexibility and richness of solutions.

By exploring color approach based on the intentions and options, the color application can be used in the production of a final piece of folk art, through the interactive definition of criteria and selections that influence the color process. The process may also be regarded as the piece of folk art on itself, stimulating the viewer senses and mind as the possibilities are explored. Future directions include experiments with richer color sources in the selection and composition stages, which may also help the user to think in more structured ways. This will enrich the creative exploration of folk art color system with computer-aided design.

2012 soft science research project of Henan science and technology bureau(122400450124)(Heritage and development of Henan folk art under the background of cultural industry).

References

[1] Yoo, H.-W., Cho, S.-B.: Video scene retrieval with interactive genetic algorithm. Multimedia Tools and Applications 32(3), 353–359 (2007)

[2] Adams, A., Blandford, A., Lunt, P.: Social empowerment and exclusion: A case study on digital libraries. ACM Transactions on Computer Human Interaction 12(2), 174–200 (2005)

[3] Ellis, D.: A behavioural approach to information retrieval system design. Journal of Documentation 45(3), 336–345 (1989)

[4] Wilson, T.: Models in information behaviour research. Journal of Documentation 55(3), 249–270 (1999)

Detecting Companies' Financial Distress by Multi-attribute Decision-Making Model

Li Na

Anhui Audit Professional College, Hefei 230601, China
lina23423@163.com

Abstract. In recent years, financial crisis events related to the public companies have obtained more and more attentions. In this paper, we proposed a novel multi-attribute decision-making model to forecast the possibility of companies' financial distress. The proposed model contains two main phases, which are attribute weight computing and financial distress detecting. Experiments conducting on 20 companies' from shanghai or shenzhen stock markets show that the proposed model performs better than other methods in most cases, and our model can effectively detect the financial distress in early stage.

Keywords: Financial distress, Multi-attribute decision-making, Decision matrix, Attribute weight.

1 Introduction

Financial distress detecting is important for business bankruptcy avoiding, and several methods using financial ratios have been proposed. As is stated in Wiki, Financial distress is a term in corporate finance used to indicate a condition when promises to creditors of a company are broken or honored with difficulty. Sometimes financial distress can lead to bankruptcy. Financial distress is usually associated with some costs to the company; these are known as costs of financial distress.

However, if a corporate can not detect financial distress and take effective measures at an early stage will run into bankruptcy. Recently, some corporates encounter financial crises in the international marketing, such as Enron, Kmart, and Lehman Brothers. Hence, in order to improve the performance of the financial distress prediction, we propose a novel model to detect companies' financial distress in advance.

Recently, one of the most attractive business news is a series of financial crisis events related to the public companies. Some of these companies are famous and also at high stock prices, originally (e.g. Enron Corp., Kmart Corp., WorldCom Corp., Lehman Brothers Bank, etc.). In consequence of the financial crisis, it is always too late for many creditors to withdraw their loans, as well as for investors to sell their own stocks, futures, or options. Therefore, corporate bankruptcy is a very important economic phenomenon and also affects the economy of every country.

The application of financial distress detecting is firstly started with univariate models that relied on the predictive value of single financial ratio[1]. Afterwards, multivariate models were developed from multiple discriminant analysis (MDA) [2], which drove financial distress early warning research into the period of prediction with multiple variables.

Z. Du (Ed.): Proceedings of the 2012 International Conference of MCSA, AISC 191, pp. 485–489.
springerlink.com

Recently, researchers on financial distress prediction have gathered great importance in recent years. Kumar et al. [3] have endeavored to comprehensive reviews of recent researches on bankruptcy prediction, Xu et al. [4] and Hua et al. [5] undertook the state-of-the-art researches on financial distress prediction. Therefore, corporate financial distress forecasting has become an important and widely studied topic since it has a significant impact on lending decisions and the profitability of financial institutions.

The rest of the paper is organized as follows. Section 2 introduces the proposed financial distress detecting model. Section 3 presents the model to detect financial distress by multi-attribute decision-making. In section 4, we illustrate a case to show the effectiveness of the proposed model. In Section 5, we conclude the whole paper.

2 Overview of the Proposed Financial Distress Detecting Model

The propose framework of multi-attribute decision-making for financial distress early forecasting is illustrated in Fig.1. The multi-attributes set is input to our model, and financial distress possibility are output as the final results. The proposed model contains two main phases, which are attribute weight computing and financial distress detecting. The whole process is judged by all the experts.

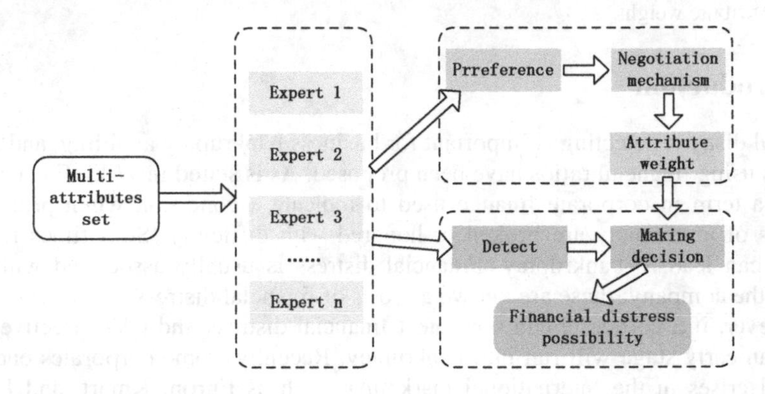

Fig. 1. Framework of multi-attribute decision-making for financial distress

Based on the above decision information shown in Fig.1, we apply the dynamic intuitionistic fuzzy weighted averaging (DIFWA) operator, the dynamic weighted averaging (DWA) operator, intuitionistic fuzzy TOPSIS method and the hybrid weighted averaging (HWA) operator to give an interactive method to rank and then compute financial distress possibility.

3 Detecting Financial Distress by Multi-attribute Decision-Making Model

Multi-attribute decision-making is one of the most common problems in intelligent computing, which means to choose the optimal result from a finite of alternatives with respect to a collection of the predefined attributes. The attribute values given by a decision maker over the alternatives under each attribute may take one of the various forms. In this section, we will propose a multi-attribute decision-making model which is suitable to detect companies' financial distress.

Firstly, we use the DIFWA operator to aggregate all individual decision matrices $R(t_k^l)$ at time $t_k(t \in [1,p])$ into the decision matrix $R^{(l)} = (r_{ij}^{(l)})_{m\times n}$ for $e_l(l \in [1,q])$. Then, we apply DWA operator to aggregate the weights of the attribute $w_j(t_k)$ of p different time interval $t_k(t \in [1,p])$ into the weight of attribute $w_j(j \in [1,n])$. Afterwards, Calculate the individual relative closeness coefficient of each alternative by intuitionistic fuzzy TOPSIS method and obtain the individual ranking of alternatives for $e_l(l \in [1,q])$

Secondly, we define $x^{(l+)}$ and $x^{(l-)}$ as the intuitionistic fuzzy positive ideal solution and the intutionistic fuzzy negative ideal solution. Then, let j_1 and j_2 be the benefit attribute and cost attribute. Therefore, $x^{(l+)}$ and $x^{(l-)}$ can be calculated as follows.

$$x^{(l+)} = (r_1^{(l+)}, r_2^{(l+)}, \ldots\ldots, r_n^{(l+)}) \tag{1}$$

where $r_j^{(l+)} = (u_{r_j^{(l+)}}, v_{r_j^{(l+)}}, \pi_{r_j^{(l+)}}), (j \in [1,n])$

$$x^{(l-)} = (r_1^{(l-)}, r_2^{(l-)}, \ldots\ldots, r_n^{(l-)}) \tag{2}$$

where $r_j^{(l-)} = (u_{r_j^{(l-)}}, v_{r_j^{(l-)}}, \pi_{r_j^{(l-)}})(j \in [1,n])$

The individual relative closeness coefficient of an alternative x_i with respect to the IFPIS x^+ is computed by the following equation:

$$c^{(l)}(x_i) = \frac{d(x_i, x^{(l-)})}{d(x_i, x^{(l+)}) + d(x_i, x^{(l-)})} \tag{3}$$

where $c^{(l)}(x_i) \in [0,1]$ and $l \in [1,q]$, $i \in [1,m]$.

Afterwards, The decision that the possibility of the financial distress of the given company could be calculated as follows.

$$p = 1 - \frac{n\sum_{i=1}^{m}(r_i^A - r_i^{(l)})^2}{m(m^2-1)}, \forall l \tag{4}$$

4 Experiment

We choose 20 companies from Shanghai and Shenzhen Stock markets as the testing dataset from 2005 to 2010 as time interval. Six quantitative financial distress prediction methods including MDA, logistic regression (logit), BP neural networks (BPN), DT [6], similarity weighted voting CBR [7], SVM [8] are applied. The sample data were obtained from China Stock Market & Accounting Research Database. The average accuracy on 20 data sets and their corresponding mean, variance and coefficient of variation are listed in Table 1.

Table 1. The accuracy of different methods on 20 datasets

Dataset	MDA	Logit	BPN	DT	SVM	CBR	Our approach
D1	83.62	86.41	77.41	77.41	78.77	78.77	86.39
D2	78.14	83.58	80.12	84.20	74.69	81.48	86.35
D3	75.40	82.17	78.77	73.33	78.77	81.48	84.15
D4	82.25	80.75	81.48	77.41	81.48	77.41	86.10
D5	78.14	83.58	84.20	81.48	78.77	80.12	87.08
D6	78.14	85.00	73.33	70.61	77.41	82.84	83.69
D7	78.14	83.58	80.12	78.77	84.20	84.20	87.57
D8	83.62	92.09	82.84	81.48	80.12	84.20	90.32
D9	82.25	93.50	84.20	73.33	86.91	84.20	90.33
D10	71.28	89.25	80.12	82.84	80.12	71.97	85.17
D11	87.73	86.41	82.84	76.05	82.84	84.20	89.55
D12	80.87	92.09	85.55	71.97	80.12	82.84	88.37
D13	80.87	87.83	84.20	86.91	81.48	86.91	91.01
D14	78.14	89.25	82.84	84.20	80.12	81.48	88.83
D15	82.25	90.67	80.12	78.77	82.84	85.55	89.58
D16	79.51	87.83	74.69	81.48	84.20	85.55	88.34
D17	82.25	86.41	80.12	85.55	80.12	81.48	88.81
D18	80.87	83.58	81.48	80.12	78.77	80.12	86.84
D19	75.40	86.41	80.12	82.84	82.84	84.20	88.07
D20	74.03	77.92	78.77	73.33	81.48	81.48	83.63
Mean	79.64	86.42	80.66	79.10	80.80	82.02	87.51
Standard Variance	3.82	4.03	3.13	4.80	2.73	3.37	2.22
Coefficient of variation	0.048	0.047	0.039	0.061	0.034	0.041	0.025

As is shown in Table.1, we can see that our approach performs better than other methods in most cases. The reason lies in that our approach can utilize each company's special circumstances, experts' experiential knowledge and all kinds of financial and nonfinancial information to detect financial distress diagnosis. Using our multi-attribute decision-making based method, the company's financial distress possibility as well as various errors, loopholes or hidden dangers which may induce financial distress can be detected in advance. Moreover, the experiment results demonstrate that financial distress detecting by our approach can remedy the limitation of pure quantitative financial distress prediction methods based on financial ratio data, even when there is little distress symptom at the beginning.

5 Conclusion

To avoid the limitation of quantitative financial distress prediction methods, it is very important to utilize experts' experiential knowledge and all kinds of financial or non-financial information to detect financial distress. In this paper, we propose a novel multi-attribute decision-making method to detect companies' financial distress. Experiments conducting on 20 companies' from shanghai or shenzhen stock markets show that the proposed approach performs better than other methods in most cases.

References

[1] Beaver, W.: Financial ratios as predictors of failure. Journal of Accounting Research 4, 71–111 (1966)

[2] Altman, E.I.: Financial ratios discriminant analysis and the prediction of corporate bankruptcy. Journal of Finance 23, 589–609 (1968)

[3] Kumar, P.R., Ravi, V.: Bankruptcy prediction in banks and firms via statistical and intelligent techniques: a review. Eur. J.Oper. Res. 180, 1–28 (2007)

[4] Xu, X., Sun, Y., Hua, Z.: Reducing the probability of bankruptcy through supply chain coordination. IEEE Trans. Syst. Man Cybern. Part C Appl. Rev. 40(2), 201–215 (2010)

[5] Hua, Z., Sun, Y., Xu, X.: Operational causes of bankruptcy propagation in supply chain. Decis. Support Syst. 51(3), 671–681 (2011)

[6] Sun, J., Li, H.: Data mining method for listed companies' financial distress prediction. Knowledge-Based System 21(1), 1–5 (2008)

[7] Bordogna, G., Fedrizzi, M., Passi, G.: A linguistic modeling of consensus in group decision making based on OWA operator. IEEE Transactions on Systems, Man, and Cybernetics Part A, Systems and Humans 27, 126–132 (1997)

[8] Hui, X.-F., Sun, J.: An Application of Support Vector Machine to Companies' Financial Distress Prediction. In: Torra, V., Narukawa, Y., Valls, A., Domingo-Ferrer, J. (eds.) MDAI 2006. LNCS (LNAI), vol. 3885, pp. 274–282. Springer, Heidelberg (2006)

Study on Advertisement Design Method Utilizing CAD Technology

Ru Cun-guang

Xinxiang University,
Henan Xinxiang, 453000, China
rucunguang552@163.com

Abstract. In this paper, we propose a novel advertisement design method based on CAD technology, which could enhance the quality and effectiveness of advertisement design. After analyzing superiority of CAD technology in detail, we illustrate our two-layer advertisements design model, which is made up of source layer and design layer. Source layer consists of well organized materials database including trademark logos, advertisement database, images, artistic designs and so on. In design layer, five function modules are utilized to conduct advertisement design based on the materials collected from the source layer.

Keywords: CAD, Advertisement design, Computer graphics, Image processing.

1 Introduction

Advertisement is a form of communication used to encourage or persuade a viewer or reader to buy the product or support a viewpoint. In addition, Advertisement are almost paid for by companies and viewed through various traditional media including mass media such as newspaper, magazines, television commercial, radio advertisement, outdoor advertising. Advertisement design mainly should meet to people's psychological requirement, and more and more advertisements infuse the human emotional demands.

However, traditional advertisement design is usually made by hand drawing. The drawbacks of the traditional methods mainly lie in that the quality and effectiveness is not satisfied. Hence, in this paper, we propose a new approach to conduct advertisement design by computer-aided design(CAD) Technology. As is shown in Wikipedia, CAD is a kind of industrial art extensively adopted in many applications, such as automotive, shipbuilding, and aerospace industries, industrial and architectural design, prosthetics, and so on. On the other hand, CAD is also widely utilized to produce computer animation for special effects in movies, advertising and technical manuals.

The rest of the paper is organized as follows. Section 2 introduces superiority of CAD technology. Section 3 presents our advertisement design method based on CAD Technology. In section 4, we conclude the whole paper.

Z. Du (Ed.): Proceedings of the 2012 International Conference of MCSA, AISC 191, pp. 491–494.
springerlink.com

2 Superiority of CAD Technology

Computer-aided design (CAD), also known as computer-aided design and drafting (CADD) aims to use computer to assist in the creation, modification, analysis, or optimization of a design. CAD software is designed to increase the productivity of the designer, improve the quality of design. The products of CAD are often organized in the form of electronic files for print or machining operations. Particularly, CAD software usually utilizes either vector based graphics to depict the objects of traditional drafting, or may also produce raster graphics which could show the overall appearance of designed objects.

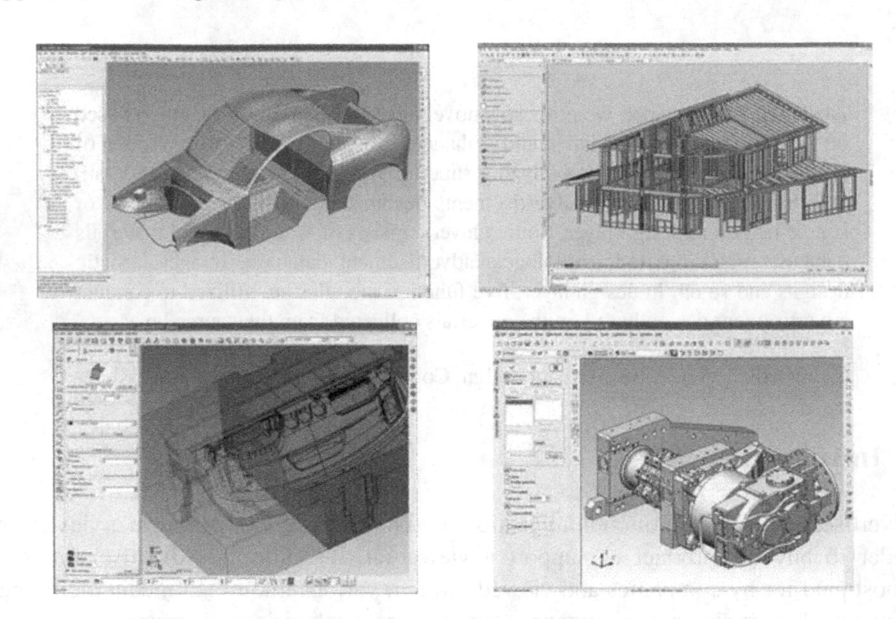

Fig. 1. Examples of CAD products

As in the manual drafting of technical and engineering drawings, the product of CAD must convey useful information which includes materials, processes, dimensions, and tolerances. CAD may be used to design curves and figures in two-dimensional space or in three-dimensional (3D) space.

As is well known that design is a complex activity. The most important goal of CAD system is to relieve designers from the burden of drawing. Design model is the key for any CAD system, that is, when discussing CAD, we should make at least a crude model of the design process. As is shown in Fig.1, we illustrate four CAD products including several application fields. It can be seen from Fig.1 that integrating CAD technology in design process, high quality design can be easily achieved and the operations of design is fairly easy.

3 Advertisement Design Method Based on CAD Technology

The development of computer graphics and image processing technology have made it possible to replace the traditional advertisement design approach CAD technology. Currently, many commercial CAD software systems have been developed, such as CorelDraw and PhotoShop. Particularly, Photoshop is mainly for image editing and image processing, while CorelDraw is for graphics editing and generation. However, these CAD software are developed from the idea of graphics system and image system. For example, although Photoshop can easily edit or process an image, users still feel inconvenient to edit or combine two images. The main reasons lie in that advertisements design process can not be easily conducted on Photoshop-like software.

Therefore, for the advertisements designer, a more suitable CAD technology based advertisement design method should be researched at the point of view of advertisements design field. Based the above analysis, we develop a novel CAD technology based advertisement design approach. The main characteristics of our method lie in the following aspects. For the first aspect, the application fields of our approach are fairly wide, and it can enhance the advantages of computer vision. For the second aspect, we propose a design model for advertisements designing by dividing the whole design work into two layers.

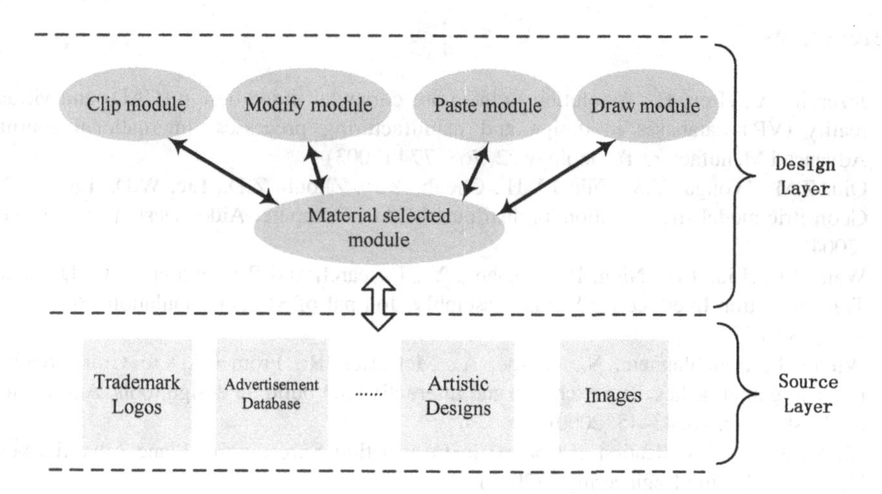

Fig. 2. Framework of two-layer advertisements design model.

As is shown in Fig.2, a framework of two-layer advertisements design model is illustrated which includes design layer and source layer. In the source layer, materials database used in advertisements design are collected and organized in order. The materials database we used includes trademark logos, advertisement database, images, artistic designs et al, and these materials can effectively support advertisements design. The upper layer is denoted as design layer, which works based on source layer. Some function modules are allocated in design layer, such as "Material selected module", "Clip module", "Modify module", "Paste module" and "Draw module".

"Material selected module" could automatically choose suitable materials from the materials database in source layer. This process is the basic step of the whole advertisement design. Afterwards, other functions can be executed as follows. "Clip module" can select object in order to get a proper shape. Functions of "Modify module" are made up of scaling, rotating, warping and any special effects, and this module is superior to the traditional methods of advertisement design. "Draw module" can easily draw an object interactively, and this module is easy for users to conduct objects in advertisements.

4 Conclusion

This paper present a new advertisement design method based on CAD technology, and our method is superior to traditional advertisement design methods. We give a two-layer advertisements design model, which includes source layer and design layer. Source layer is made up of trademark logos, advertisement database, images, artistic designs and so on. Five function modules are design in design layer to make high quality advertisement design.

References

[1] Jezernik, A., Hren, G.: A solution to integrate computer aided design (CAD) and virtual reality (VR) databases in design and manufacturing processes. International Journal Advanced Manufacture Technology 22, 768–774 (2003)

[2] Qiu, Z.M., Wonga, Y.S., Fuh, J.Y.H., Chenb, Y.P., Zhoub, Z.D., Lic, W.D., Luc, Y.Q.: Geometric model simplification for distributed CAD. Computer-Aided Design 36, 809–819 (2004)

[3] Wan, B.L., Liu, J.H., Ning, R.X., Zhang, Y.: Research and Realization on CAD Model Transformation Interface for Virtual Assembly. Journal of System Simulation 18(2), 391–394 (2006)

[4] Whyte, J., Bouchlaghem, N., Thorpe, A., McCaffer, R.: From CAD to virtual reality: modelling approaches, data exchange and interactive 3D building design tools. Automation in Construction 10, 43–45 (2000)

[5] Li, Y., Xu, R.: Application of Context in Information Spreading of Plane Advertisement Design. Packaging Engineering 3 (2007)

[6] Huang, Z., Chen, R., Wang, X.: Discussion on the Application of Symbol Image in Commercial Advertisement Design. Packaging Engineering 6 (2007)

[7] Li, M.A.: On Aesthetic Construction of Commercial Outdoor Advertisement Design. Packaging Engineering 12 (2009)

[8] Wang, T., Xu, M., Zhang, W.: Computational multibody systems dynamics virtual prototype oriented CAD design model. Journal of System Simulation 20, 2068–2073 (2008)

[9] Studnicka, C.: Results of introducing and applying computer-aided design (CAD) in rail vehicle engineering. Zeitschrift fur Eisenbahnwesen und Verkehrstechnik 114, 42–46 (1990)

A Solution Research on Remote Monitor for Power System of WEB Technology Based on B/S Mode

Guo Jun, Yu Jin-tao, Yang Yang, and Wu Hao

Xinyang Power Supply Company in Henan Province,
Xinyang Henan 464000, China
guojun3322@163.com

Abstract. The internet supervises and controls to the embedded system is one of the most important disrection in the embedded system.We have studied the basic requirements for embedded monitoring system and analysed the implementation mechanism of monitoring system. A new idea of componentization design is proposed in this paper to make remote monitoring of the electric power system base on B/S mode.A TCP/IP protocol stack ,named LWIP is ported on the embedded RTOS C/OS and a new type of embedded file system is designed.Then the workflow of CGI and HTTP servers is also improved to enhance the stability of system.

Keywords: embedded system, remote monitor, component, Web Server.

1 Introduction

The development of telecontrol technology has taken powerful measures to electric power systems and it is ensured by the communication and computer technology. When telecontrol technology is combined with internet that will greatly enhance the ability of power system automation. The service of Web is widely used in electric power system now and people need real-time monitoring for system by remote technology and simple devices. Among the existing methods, the embedded system has become an important technology to monitor and manage the power system effectively. It is hardly to apply the mature technology of PC directly to make real-time data collecting because of the limited resources in embeded Web servers. So we need improved the existing scheme when studying the work flow of the Web service. Considering the actual need in power monitoring system based on B/S, we give an approach of both hardware and software technology. The experiment shows, compared with other measures, our method has better performance in real-time, stability and functionality.

2 The Architecture of Embedded Monitoring and Control System

The remote monitoring system mainly consists of three parts: the device for field data collecting, embedded Web server and remote monitoring host. The especial device

Z. Du (Ed.): Proceedings of the 2012 International Conference of MCSA, AISC 191, pp. 495–500.
springerlink.com

collect the field data and transfer data by some of the field bus protocol; embedded Web server collect and release the field data and it is also the gateway to connect the control network and data network, which can exchange the field bus protocol with the Ethernet protocol; The monitoring hosts get real-time field data by the Ethernet to make a remote supervise. We disassemble the functions of monitoring system to describe the problem clearly and the modules are defined as components of system. The components include: WebServer, CGI, serial communication, data storage and etc. Figure 1 describes the logical relationship of these components.

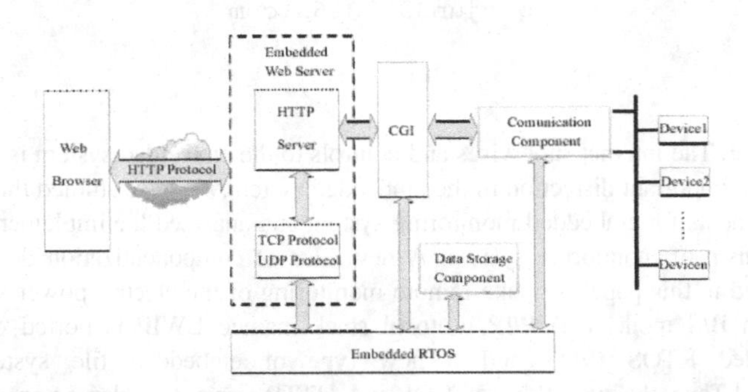

Fig. 1. The main function components of embedded monitoring system

Web server is the module which can react and deal with the user's request, and make independent communication based on the TCP/IP protocol. Web server interacts with the browsers and return results to users.CGI component is the independent program modules designed according to CGI standard. It supplies an interface for the interaction between embedded Web server and external expanded application programs. The field data is transferred to embedded Web server real-timely by CGI, which establishes a dynamic relation between the remote users and the field devices. Serial communication components are the independent modules to transfer data from field devices to CGI components. They can cooperate with CGI components to renew the data of embedded Web servers. Data storage component saves the collected field data and it is a kind of function module which offers the method to read and write for other components. The implementation form of data storage component can be either file mode or embedded database mode(such as SQLite, Solid).

3 Key Technologies of System Design

3.1 Design of Hardware Modules

The core processor of system is P89V51RD2 chip. It is a type of enhanced microcontroller and can work under full duplex mode. Each chip has two flash storage modules, whose module 0 has 64k bytes to store the user's codes; the other modeule 1 has 8k bytes, which is used to store ISP/IAP programmes.NIC takes RTL8019s made in Realtek company as the Ethernet controller because of its stable performance. DS1820 of DALLAS company is used as an temperature sensor for perceiving. It has the advantage of small size and measuring range of -55°C~ +125°C, which can be changer to 9-12 digits A/D conversion accuracy.Its temperature measurement resolution can reach 0.0625°C so it satisfies the actual need of system.

We use synchronous AC sampling technology in system and the operation follows the sampling theorem. The AC voltage signals are sampled non-uniformly in a period by fixed sampling frequency. After A/D exchange the discrete sampling data are used to calculate the effective value of power frequency voltage and sub-harmonic content by Fast Fourier Transform algorithm. Then we can get the other parameters such as MVA power allocation, MVA loss allocation and power factor. Sound and light alarm will work when the voltage is over or below the standard value. As the frequency of power frequency signal is changeable, the frequency of sampling signal must change synchronously in case spectrum leakage occurs, which will cause measurement error. The circuit is constructed by Shaped Circuit and phase-locked frequency circuit. First, the voltage signals are changed into square wave signals which have the same frequency. Then we use phase-locked loop to control 6-channel sampling/retainer after the signal of the other end is frequency-multiplicated.

3.2 Software Design of System

3.2.1 Network Protocol Stack Module

The network protocol stack for embedded application need to be modified to save the resource, based on keeping general TCP/IP protocol stack. So we transplant a lightweight TCP/IP protocol stack LWIP on our system. The key point to use LWIP is to minimize the memory consumption while keeping the basic protocol of TCP/IP. Therefore, when transplanted to μC/OS, the first to design is the interface between LWIP and μC/OS, and the driver programme of NIC. Since C/OS has offered some synchronization mechanism and communication mechanism, we just re-encapsulation them and add management functions of the information queue to meet the need of LWIP. In C/OS there is only the conception of task instead of thread. So the task function OS-TaskCreate() of C/OS is needed to be re-encapsulated and the priority of each task in LWIP need distribution in advance. Then we get the set- up thread function Sys_Thread_New().The task for the Ethernet monitoring, protocol stack arrange and application layer is established in accordance with the level of priority.

3.2.2 Modules of File System

We design a small file system to save the historical data and users' message. The file system is defined by storage media, the format data kept in media and API for access. Referred to FAT32 file system format, we keep part function of the file system and make modification for special application. The file system standard is implemented with a specific embedded chip based on Nand-Flash.FAT32 storage volumes consist of 5 areas: MBR, DBR, FAT area, root directory area and data area. The MBR can be omitted because of no reference to Flash chip partition.AS Figure 2 shows, four-layer structure is taken and they are, from top to bottom in order, driver layer, file system buffer layer, file table management and API layer.

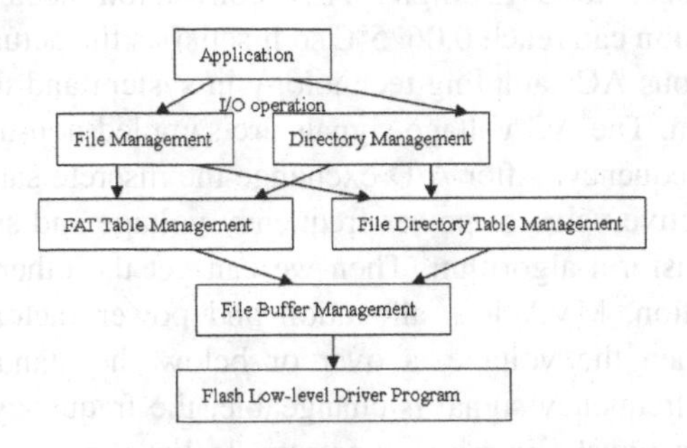

Fig. 2. The structure of embedded file system

4 Design and Implementation of Web Server

4.1 The Improvement on CGI

Traditional CGI specification includes three parts: environment variable, standard input and output. When clients request for CGI script, the server will establishment extra processed to support CGI codes, which greatly increases the burden on server. In our system, the standard CGI works as single executable file and is handled by the form of functions inside the server, keeping original interface with performance function. The grammar to submit information to server with HTML form is usually defined as: <FORM METHOD="GET/POST" ACTION=URL>. Beacuse POST method has better performance than GET method in security and in the amount of submitted data, we set the mode of all submission is POST. Then the grammar above is changed to "<FORM METHOD="POST"ACTION="key word"> ".The dynamic loading function AddDynamicHandler() is used to relate the keyword and corresponding operation function. When POST request occurs, the analysis engine analyzes the parameter value in packet and excutes related operations. Thus the CGI which is implemented by single executable file is excuted as the mode of function dynamic loading.

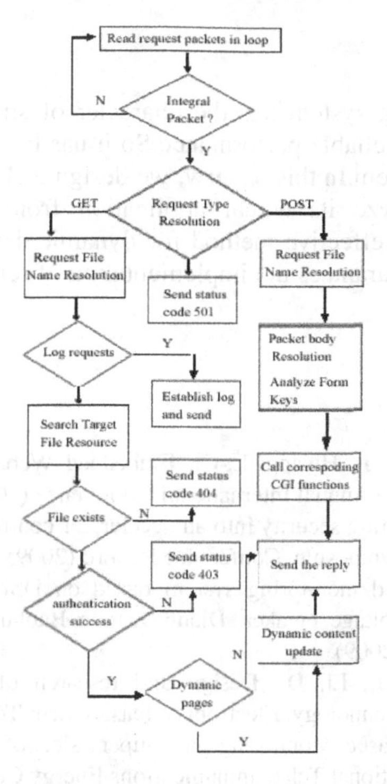

Fig. 3. The handling process of HTTP request

4.2 The Workflow of HTTP Server

Web server is based on the socket programming and class BSD socket provided by LWIP protocol is used in this system. The server establish an socket based on TCP connection and get in monitoring status after binding the IP address and 80 port of the host. When the client sends HTTP connection request the server accept request and begin handling HTTP programmes. Figure 3 shows the flow of this process. The servers accept request packets in loop and determine the integrity by ending flag. If the packet entirety is correct then servers decide the request method token according to the packet request lines. If it is a log request, the server establishes log for client and send it; or according to the file name of object, it begins searching corresponding filesin Flash by file system API. If the files exist, then privilege authentication starts referred to the privilege attribute of files. Once the clients pass, the server must decide whether to update dynamically and response, by the query of if there is dynamic tag in files. If the request from client is POST, the request file name is analyzed first. Then the servers accept the form parameters by analyzing the packet body and trigger the appropriate functions to be handling.

5 Conclusion

The embedded monitoring system has the character of small size, long continuous working time, stable and reliable performance.So it has become the hot spot of study on remote monitoring system.In this paperw, we design a electric power system based on B/S mode and analyze its actual application from hardware and software technology.It provides an effective method for dynamic data exchange in embedded monitoring system and guarantees the implementation of remote real-time monitoring system of network.

References

[1] Chen, Y.-H., Cheng, M.-J., Hwang, L.-C.: Embedded Web server based power quality recorder. IEEE Region 10 Annual International Conference (2005)

[2] Robson, C.C.W.: Integrating security into an accelerator control systems web interface. In: IEEE Nuclear Science Symposium Conference Record (2009)

[3] Xiong, X.-F.: Distributed monitoring system based on DSP and ARM for mechanical characteristic of high voltage breaker. Dianli Xitong Baohu yu Kongzhi/Power System Protection and Control (2009)

[4] Niu, D., Qu, G., Qu, J., Li, D.: Design and research of power remote monitoring systembased on GPRS technology. Electronic Measurement Technology (2006)

[5] Costa, N.F.: Ethernet Based Monitoring and Supervision of Telecommunications Power Supply Units. In: International Telecommunications Energy Conference (2003)

The Research of Supply Chain Management Information Systems Based on Collaborative E-Commerce

Ruihui Mu

College of Computer and Information Engineering,
Xinxiang University,
453003, Xinxiang, China
mrh112@163.com

Abstract. Lasting competitive market environment, enterprises have been the fo-cus of attention. Collaborative e-commerce is the present and future of virtual op-erating mode, the line with the notion of its ideological and supply chain manage-ment. Article first to analyze the problems of the existing supply chain management information systems, management model on the basis of the percep-tion and response, based on the concept of collaborative e-commerce supply chain management information systems, and on this basis, a complete introduction to building programs based on collaborative e-commerce supply chain management information systems. The paper presents the research of supply chain management information systems based on collaborative e-commerce.

Keywords: collaborative E-commerce, collaborative information centre, sense and response.

1 Analysis of Supply Chain Management Information Systems

As global economic integration, market competition form from the competition between enterprises of the competition [1], multinational corporations and multinational corporations, the evolution of competition between the supply chain and supply chain. In the case of such a drastic change, the existing supply chain management information system the following problems.

The dispersion of information systems, the heterogeneity of large enterprises of supply chain nodes can not be smooth interaction, the current supply chain nodes in a decentralized, autonomous status, the system heterogeneity is very prominent. All units use the system with different hardware platforms and implementation technologies, even in the basic state of the "fragmented", the obstacles are more different enterprises interact.

The alliance of the traditional supply chain flexibility, companies can not reduce the operational risk. Supply chain system is an open complex system, with the level of diversity, and numerous functions inside and outside the association, and functional

Z. Du (Ed.): Proceedings of the 2012 International Conference of MCSA, AISC 191, pp. 501–506.
springerlink.com

requirements of the large differences in the degree of information technology requirements of characteristics. In this system, the traditional strategic partnership established on the basis of the companies directly connected to reduce the number of node enterprises, reduce costs, but the ever-changing market, the lack of such a relationship dynamic, and the establishment of this relationship investigated the accumulation of a certain time is required.

2 Based on Collaborative E-Commerce Supply Chain Management Information System

2.1 Based on Collaborative E-Commerce Supply Chain Management Information System Concept

Collaborative e-commerce (the Collaborative E-Commerce) is a new business model, coincides with the idea of integration and collaboration with the core concept of supply chain management [2]. Collaborative e-commerce is in real-time data transfer process to help the production department and sales department, purchasing department coordinated, integrated mode to another business with a business-driven enterprise with suppliers, businesses with consumers, enterprises with cooperation business model of coordination between the partners, businesses with employees, businesses with third-party logistics enterprises, enterprises with the fourth party logistics enterprises.

2.2 Based on the Significance of Collaborative E-Commerce Supply Chain Management Information System

Based on collaborative e-commerce supply chain, is a collaborative e-commerce theory under the guidance of, in order to adapt to changes in information technology, supply chain management integration, information technology means.

From the collaborative e-business point of view, based on collaborative e-commerce supply chain management information systems, enterprise collaborative development of e-commerce systems and external collaboration in the supply chain on the basis of unified outside the enterprise collaboration and internal coor-dination. At the same time to build a unified information platform, and partners into an in-tegrated, unified information management system, the implementation of efficient sharing of information and business links.

$$x = \sum_{i=1}^{n} x_i + \prod_{j=1}^{m} x_j ' \tag{1}$$

From the point of view of supply chain management information system, based on collaborative e-commerce supply chain management information system is a means to achieve the integrated management of the supply chain. Collaborative thinking supply chain management across multiple links in the supply chain to co-ordinate the planned mechanism, it has changed the customer management, pro-curement, pricing, and measure the internal operation mode, the entire business process reengineering the supply chain logistics, capital flow and information flow in the integration of traditional business activities in an unprecedented way, while helping companies focus on their core business, they do not have a competitive advantage to outsource.

3 Collaborative E-Commerce Supply Chain Management Information System Program

3.1 System Construction of Ideas and Requirements

The current market environment and business trends require supply chain management information system must be reconfigurable, heterogeneous, and agility of the system by adjusting the structure of the system, function of the supply chain to follow the total standard hardware resources on the basis of all members to achieve the information exchange, sharing, and thus has the ability to quickly adapt to changes in demand [3]. The existing supply chain IT applications and collaboration problems faced by, based on collaborative e-commerce supply chain management information systems can be either the existing information system after the transformation, but scratch the new system.

1) Application-aware and respond to the management model.The establishment of integrated supply chain management, it is strategic thinking, and each node enterprise and consumer integration as a whole, application-aware, and response management model to achieve flexible operation. The basis to develop appropriate security mechanisms and trust systems, cross-organization and coordination of external relations, all aspects of trust in a strategic partnership through the establishment of collaborative information center different flexible-level coordination to adapt the supply chain network, on the one hand, the value chain node enterprises to achieve self-synchronization to adapt, through integrated business, information processing work together to create joint value of the optimal; the other hand, no longer pursuing the best optimization of the supply chain, but highly flexible and flexible, and can at any time respond to possible changes in the supply chain [4].
2) Analysis, extraction, and respond to demand from the different dimensions. No longer primarily dependent on the forecast, and rely on the perception and ability to respond to changes in demand, so the methods and means of perception and response is particularly important. From the collaborative level, it can be divided into demand collaboration and supply synergies. Increase the transparency of demand information between supply chain and a member of the upstream and downstream, and can improve demand visibility in the supply chain and plans to eliminate the "bullwhip effect". Its core is an integrated forecasting, reduce inventory and eliminate the additional labor of unworthiness. Explicitly supply information to enhance the transparency of the purchased product in transit, to facilitate the supply and demand sides to grasp the product of dynamic information.

Table 1. The collaborative supply chain table

Analysis	extraction	the different dimensions	transparency
Collaborative scope	75	154	supply chain function nodes
supply chain	125	362	e-commerce supply chain management
dynamic information	production, sales, time and attendance, human resources management sector collaboration	1120	horizontal and vertical synergies

Collaborative scope can be divided into horizontal and vertical synergies. Horizontal synergies in two stages, one is the internal collaboration, including internal management decision-making, planning and scheduling layer and generate the control layer of the synergies as well as budget, procurement, production, sales, time and attendance, human resources management sector collaboration, effectively reduce the internal costs, improve efficiency and effectiveness, and prepare the ground for the enterprise and external collaborative [5]; Second, external collaboration, enterprises and other nodes in the supply chain and supply chain synergies of the external social sector, internal intranet and external supply chain linked to other nodes in the enterprise on the Internet, online demand and inventory information sharing, synchronization planning, collaborative business processes, contracting, trading and settlement activities, in order to reduce the external costs, Depends on Demands, real-time The purpose of control, etc.

$$V_j^2 = V_j \otimes V_j, W_j^2 = (V_j \otimes W_j) \oplus (W_j \otimes W_j) \oplus (W_j \otimes V_j)) \quad (2)$$

3.2 System Framework

Collaborative mode of information management and information platform, it has established a collaborative information center, it can be independent of the outside of each node of the supply chain function nodes, operations centers, can also be integrated into the management of large enterprises information systems, their fundamental purpose is to let the data distinguish between privileges within the authorized range, unimpeded flow and summary at the top fusion, and to promote fair competition, to further reduce the cost of doing business. Based on the collaborative e-commerce supply chain management is information system architecture [6].

3.3 System Operation and Control Mechanisms

Based on collaborative e-commerce under the concept of collaborative information center, collaborative ideas in the Internet under the conditions of standing on the height of the virtual supply chain partner companies, enterprise supply chain integration of ideas and business models which solidified in the software, the value chain the overall management of the core enterprise within the sector, the various partners and other relevant departments closely linked to help companies achieve international social management. It integrates all available resources, supply chain members to arrange for the production forecast supplement joint design and implementation plan, share information for the specific production and operational measures, and interactive approach to the management of enterprises and partners and collaborative Commerce [7].

All members of the supply chain can be found in the Cooperative Information Center provides information suited to their suppliers, third party logistics and fourth party logistics companies. The exchange of data between them the standard (CML11.0,) is without going through the Collaborative Information Center. Completion of the mission of the supply chain, companies can suspend, interrupt, ter-minate this relationship, re-construct a new supply chain, which increases the sup-ply chain agility and openness.

The Common Object Request Broker Architecture (CORBA) application across the differences between the same systems, it is barrier-free through the firewall of independent supply chain members. It solves the compatibility prob-lems of system platforms, operating systems, programming languages, databases, network protocol and application version, and ultimately to seamlessly connect a variety of data sources and database applications. Structure of this agency, infor-mation, technology and product data in the supply chain integration, it is "user" needs throughout all stages of the product life cycle. Collaborative information center, in addition to data sharing needs to meet members of the supply chain business, but also to a certain extent to meet their decision-making information needs. Collaborative Information Center to develop dynamic indicators such as customer waiting time, satisfaction for customers and supply chain members of the respective standards, product availability, arrival rate, the time required to complete the redeployment of inventory time, returns processing situation, the customer satisfaction rate, and so on. Suppliers, third party logistics and fourth party logistics companies use different indicators to evaluate, and publishing the evaluation results, the establishment of an open assessment system for enterprises looking for partners to provide a fair and objective basis.

Suppliers in the supply chain is responsible for the integration of client companies demand information, and coordinate distribution; the supplier's customer is responsible for real-time demand information sent to the collaborative information center, and the Collaborative Information Center of the feedback information to adjust demand and procurement program. In addition, different companies will send feedback to the services and products of collaborative information center to provide a basis for evaluating suppliers.

4 Conclusion

Based collaborative e-commerce supply chain management information system is a strong source of power lead to improved mode of operation of the modern enterprise, but also to protect the supply chain, various products and services to a steady stream of high-speed operation information platform. Based on collaborative e-commerce supply chain will be members of the connection between the means of connection between the value chain aggregates with the target end-user success in the future management model.

References

[1] Zhou, X., Ma, F., Chen, X.: An Information Sharing Model in Dual-channel Supply Chain in E-commerce Environment. AISS 3(7), 232–243 (2011)

[2] Zhang, D., Zeng, X., Chen, H., He, W.: Evaluation of Customer Value in E-Commerce with 2-tuple Linguistic Information. JDCTA 5(11), 95–100 (2011)

[3] Tseng, K.-C., Hwang, C.-S., Su, Y.-C.: Using Cloud Model for Default Voting in Collaborative Filtering. JCIT 6(12), 68–74 (2011)

[4] Li, Y., Shu, C., Xiong, L.: Information Services Platform of International Trade Based on E-commerce. AISS 3(1), 78–86 (2011)

[5] Liu, D.: E-commerce System Security Assessment Based on Grey Relational Analysis Comprehensive Evaluation. JDCTA 5(10), 279–284 (2011)

[6] Ashwin, B.K., Kumaran, K., Madhu Vishwanatham, V., Sumaithri, M.: A Secured Web Services Based E-Commerce Model for SMME Using Digital Identity. IJACT 2(2), 79–87 (2010)

[7] Riad, A.M., Hassan, Q.F.: Service-Oriented Architecture??A New Alternative to Traditional Integration Methods in B2B Applications. JCIT 3(1), 41 (2008)

Analysis on Automation of Electric Power Systems Based on GIS

Jiang Chunmin and Yang Li

Sichuan Electric Vocational and Technical College, China 610072
jcmmoon@163.com, yanglitz@126.com

Abstract. The application of the Geographic Information System (GIS) in power system is the only way of digital and informatization of power system. Through the analysis on the domestic and foreign GIS application status, relevant technology and typical power distribution automation system project implementation example, this paper expounds the integrated design idea of the development and GIS application. Then the application framework of GIS in power system based on web services is proposed, and the further development direction is indicated.

Keywords: automation of power system, Geographic Information System, component technology, space database, web.

1 Introduction

The production management of electric power enterprise has the characteristics, such as space distribution, control object diversity and complex . The resources of electric power enterprise, such as power generation, transmission, distribution and electricity substation, scatter in the wide space area. The spatial data is the key factor for electric power enterprise application, and other new applications, seamless integration can be easily built. At the same time, as the production, supply and consumption of electric energy should finish at the same time, it has requirement for high to power dispatch, monitoring, and management in the reliability, efficiency, safety and economy. The introduction of GIS technology can not only provide a platform for the the electric power enterprise based on geographic information of maintenance and management , but also can make the SCA-DA, DMS, MIS, CIS, ERP (or SAP) systems in electric power enterprise integrated unified.

2 Application Example

2.1 Power Distribution Automatic System

Currently, most power supply enterprises have built management information system (MIS). To realize GIS and distribution management system (DMS) function mutual confluence, the body design principles that AM/FM/GIS/SCADA/DMS underlying

Z. Du (Ed.): Proceedings of the 2012 International Conference of MCSA, AISC 191, pp. 507–510.
springerlink.com

data integration, and function integration design distribution and graphical interface integration must be implemented. There are two kinds of concrete ideas. One is to develop power GIS operating system (GIS-OS), all the real-time control are in space object for the target. This kind of design patterns can minimize the complexity of the system development. However, at present, the space is not the object technology to provide perfect data definition and function, which is restricted the GIS-OS development and it is also one of the future research direction. For second, it is based on the current situation, independently used in power distribution system development leopard SCADA system and AM/FM/GIS, and make two sets of system in power sharing data model, under the premise of the integrated design. This can make full use of the present DMS, as well as the GIS retractility, extensibility.

2.2 Space Resources Management System

The electric power enterprise always reflects characteristics, such as space and complexity, from power generation, transmission and distribution of electricity to any link. These features are not only reflected in the electric power enterprise's space equipment management, but also in the final customers and power system production and operation management. Therefore, how to effectively manage these complex spatial resources in the electric power enterprise is the most big production management and operation management challenges.

The electric power enterprise management always exists contradiction between regions and concentration, so be accurate, comprehensive and timely mastering the electric power enterprise of various resources information is necessary. With the deepening of the modern management and technology, spatial resources planning (SRP) will be applied extensively. SRP is a set of system for dynamic industry, and it can also combine and extend traditional GIS and enterprise resource planning (ERP) system organically. As an advanced technology, the electric power enterprise SRP is a foundation of AM/FM/GIS, and its implementation can adapt to the development of the electric power enterprise.

3 The Application of the Advanced Technology

3.1 System Modeling and Data Share

In power GIS technology development process, the system modeling mostly concentrates in the geographical spatial attribute, and the control object application of electric power system has a complex power physical structure. Therefore, it is necessary to establish the analysis model for power system uniquely. In data sharing, it should have a power system as a basic model between different sectors of the data sharing foundation.

3.2 Improving System Integration

3.2.1 Object-Oriented Technology

Because the customer attributes always include some kinds of data, such as power poles, transformers, breaker line or other equipments. In order to facilitate the user operation and use, it is necessary to enhance the data interoperability. Also because implementing power GIS ultimate aim is to realize enterprise informatization, so the mixture of various information and technology in one of the huge enterprise information system is essential.

3.2.2 Web Services

The most typical characteristics of the electric power enterprise is the spatial distribution requires GIS platform provides distribution application services. As the distributed database has different ways, each prefecture (city) bureau can maintain and manage the jurisdiction of data. Also through the scattered data storage and management, data can be ensured real-time, and securely.

3.2.3 Space database Technology

Using the database to store and manage spatial data needs security mechanism and data backup mechanism. But the concurrent operation mechanism is the file management mode and incomparable. The present new development system in relational database management system (RDBMS) is given priority to using its powerful management advantages, asynchronous buffer mechanism and set up the multi-level index retrieval methods can effectively reduce the network load and rapid positioning to target inquires. The multidimensional space object space order is beyond description.

3.3 System Stability and Security

Electric power enterprises in the development of national economy plays a backbone, and it is a real-time operation system. So the safety and reliability are the most important. Power system stability and the safety of electric power equipment and operation should focus on security, network security and data safety operation, etc.

In order to ensure the system not damage intentionally or unintentionally, also the system should have complete access control mechanism. For electric power GIS , a big feature is the traditional centralized control system has been gradually be replaced decentralized control. Therefore, network security technology will become an key point for great power GIS application.

4 Intelligent Development Direction

The future information system in intelligent direction development will appear cyber space concept, and it will integrate computer technology, modern communication technology, network technology and virtual reality technology of comprehensive application as the foundation. Scientists predict that, people in the future will be in the information in the cyber space in the life. From one node to another node, from one source to another, exchanging information to realize the mutual communication, trade and education activities.

Computer software technology will further develop to the intelligent software. Software Agent is a further software design result, which is adapted to the widely distributed computing environment and development up software technology direction.

As a software agent, the spatial intelligence body will realize the intelligence of spatial data acquisition, processing, storage, search, performance and decision support. This kind of space agents have two kinds of important ability. One is the use of space knowledge reasoning, and another kind is surviving evolution.

In the cyber space, cyber GIS system is the constituting module. It automatically receive users with high-level language description of instructions, and use of it can sense and used in place of cybernetic space "ability" through the agent's interactive for users to find in cyber space with the information they need. We can foresee that power GIS will also face such direction.

5 Conclusions

From the perspective of power system, in the next few decades, GIS data will develop toward standardization, system integration, system diversification intelligent network and application platform, and the direction of socialization development.

The existing power GIS application is not stable, and the database design is not standard. Therefore, making our own power GIS technology standard has become very necessary. From the bottom of the network communication protocol to the database management and data model and customized means have meet certain standards, so the power GIS will obtain more economic and social benefits.

References

[1] Lixi, Y., Jiayao, W., et al.: Application of GIS and fuzzy pattern recognition theory in location of substation. Automation of Electric Power Systems 27(18), 87–89 (2003)

[2] Chen, T.H., Cheng, J.T.: Design of a TLM application program based on AM/FM/GIS system. IEEE Trans. on Power Systems 13(3), 904–909 (1998)

[3] Goodchild, M.F.: Geographical information science. International Journal of Geographical Information System 6(1), 31–45 (1992)

[4] Chyscmthou, Y., Slater, M.: Computing Dynamic Changes to BSP Trees. In: Proceedings of Computer Graphics Forum, Cambridge, vol. (2), pp. 321–332 (1992)

[5] Thibault, W., Naylor, B.: Set Operations on Polyhedra Using BSP Trees. In: Proceedings of ACM SIGGRAPH, vol. (1), pp. 153–162 (1987)

[6] Paterson, M., Yao, F.F.: Binary Partitions with Applications to Hidden Surface Removal and Solid Modeling. In: Proceedings of the 5th Annual Symposium on Computational Geometry, vol. (1), pp. 23–32 (1989)

The Empirical Study on the Relationship of Enterprise Knowledge Source and Innovation Performance

Dan Zhu, Ailian Ren, and Fenglei Wang

Glorious Sun School of Business and Management, University of Dong Hua, Shanghai, China
School of Management Henan University of Finance and Economic Zhengzhou, China
Tourism business Department, Zhengzhou tourism College, Zhengzhou, China

Abstract. Knowledge is an important expenditure of the enterprise innovation. The knowledge of enterprise innovation originates from the interior production and the exterior knowledge. These two kinds of knowledge sources have different influence on the innovation performance. In this paper based on the data form the questionnaire survey and using DEA, it is measured by the production function. The result indicates that the production efficiency of innovation knowledge is not high for all the sample enterprises. The exterior knowledge, especially coming from the industrial cluster and university knowledge, is seriously wasted and the patent output is very low and so on. In the end, advices are put forward for improving the knowledge innovation performance.

Keywords: Knowledge source, Innovation performance, DEA.

1 Introduction

Enterprise innovation knowledge comes from two sources: one is derived from the internal R & D; another is from the enterprise external knowledge spillover. Many research conclusions and innovation practice make clear that knowledge exerts positive effect on enterprise innovation performance. But these researches rarely consider the influence of innovation knowledge sources on the enterprise innovation performance, and mainly through the establishment of C-D function or extended knowledge production function to make regression analysis. Furthermore, these functions itself has defects that is difficult to overcome. Whether knowledge production can be expressed by dominant function form is questioned by lots of scholars. In this paper, based on research data and using Data Envelopment Analysis (DEA) the influence of different knowledge sources on innovation performance is measured.

2 Literatures Review

Literatures of study enterprise or regional innovation from the angle of knowledge mainly focus on the discussion of knowledge production function. Pakes and Griliches(1980),Griliches(1990) carry out empirical analysis. Both results show that

Z. Du (Ed.): Proceedings of the 2012 International Conference of MCSA, AISC 191, pp. 511–516.
springerlink.com

R&D expenditure has significant positive effect on the number of patents [1] [2]. Acs and Audretsch(1988), Koelle(1995) show that the expenditure of R&D has significant positive effect on the amount of innovation [3] [4]. Andrea Conte & Marco Vivareli(2005) use knowledge production function to measure the relationship between innovation input and output of innovation [5]. But these studies basically are limited to the impact of knowledge stock on performance such as the R & D investment including R&D input and R&D personnel input. Kevin Zhengzhou(2010) examine the role of technological capability in product innovation. Haibo Zhou (2011) find weak evidence that larger and older firms have higher new product sales than do younger and smaller firms[6][7]. Andrew M. Hess(2011) find resource combinations that focus on the same parts of the value chain are substitutes due to knowledge redundancies[8].

These findings indicate that the contribution of knowledge to innovation performance is positive. Is there a significant difference of sources of knowledge on enterprise innovation performance? Currently, quantitative studies on this question are few. But whether the expression of knowledge production is able to use dominant function form was questioned by many scholars. Based on the concept of relative efficiency, Data Envelopment Analysis values the relative effectiveness of multi index input and multiple index output economic system. In this paper this method is used to analysis the impact of different sources of knowledge on the enterprise innovation performance.

3 Multi-target DEA Model Construction

3.1 Choice of Variables

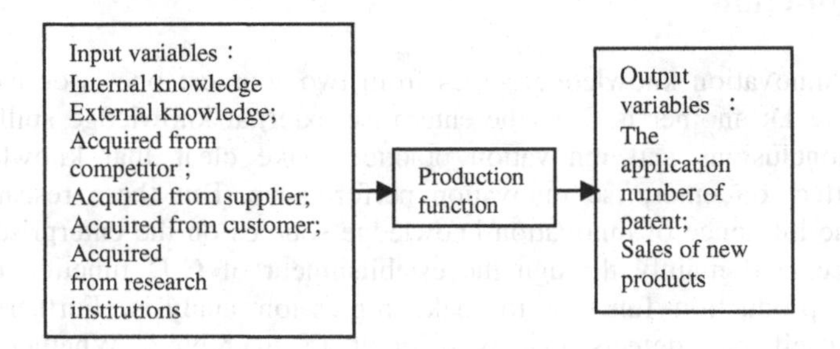

Fig. 1. Input and output variables

Innovation performance is mainly affected by innovation input. This model mainly studies the influence of different sources of knowledge on innovation performance. Therefore, the knowledge sources makes up the model input variables, and innovation performance is output variables. Knowledge has two main sources: one is the internal exchange of knowledge; another is from external knowledge. According to the different objects from which knowledge is acquired, it can be divided into knowledge from industrial clusters, knowledge from the supplier, knowledge from customer, and knowledge from the university and other research institutions, and other external knowledge acquisition.

So far, there is no a method recognized universally to assess enterprise innovation performance. Follow the valuation index construction principle of science, comparability, availability, comprehensiveness and representation, this paper uses the number of patent applications and sales of new products as the enterprise innovation performance index. All the name of input, output variable, code and relationship are in figure 1.

3.2 Construction of Multi-target DEA Model

Assume that a decision making unit (DMU) is efficient and is located in the efficient frontier. Any relative degree of deviation from the front surface is the DMU efficiency score. Based on this principle, we construct the following model:

Suppose that the number of a group of unit examined is m, namely the number of DMU is m. Each DMU input vector and output vector distribution is:

$$\sum_{j=1}^{m} x_j \lambda_j \leq X$$

$$\sum_{j=1}^{m} y_j \lambda_j \geq Y$$

$$\sum_{j=1}^{m} \lambda_j = 1, \lambda_j \geq 0$$

$$j = 1,2,3 \ldots m$$

Introduce slack variable $s^{+T} = (s_1^+, \ldots, s_k^+)$ 、 $s^{-T} = (s_1^-, \ldots, s_n^-)$ and Non Archimedes infinitesimal ε , get DEA model:

$$\min[\theta - \varepsilon(\sum_{i=1}^{n} S_i^- + \sum_{r=1}^{k} S_r^+)]$$

$$\sum_{j=1}^{m} x_{ij} \lambda_j + S_i^- \leq \theta x_{ij}$$

$$\sum_{j=1}^{m} x_{ij} \lambda_j + S_i^- \leq \theta x_{ij}$$

Of which x_{ij0} and y_{rj0} are the i the input and the r the output vector of the $j0$ th DUM, respectively.

4 Empirical Study

In this paper, based on related literatures to design the preliminary questionnaire, according to the data from preliminary questionnaire and on the basis of the reliability and validity being tested as well as many modifications being made the formal questionnaire was designed. The objects of investigation are innovative pilot enterprises, high-tech enterprises and small and medium sized enterprises of science and technology in Shanghai involving chemical industry, manufacturing industry and software industry, total 263 enterprises. 263 questionnaires were sent out, the recovery rate is 63.12%. After screening, the reliability coefficient of the questionnaire is 0.6 above, which indicates that it can be used to analyze.

Put the data into the equation and get solution by running EMS1.30 software. The average of each index was selected. (see Table 1 .)

Table 1. Result of knowledge productivity

	Efficiency score	$\theta=1$	$\sum_{i=1}^{m}\lambda_j$	S+ ls	S+ pa	S ex	S- cl	S- su	S- cu	S- un
Mean	0.8966	64	0.9927	0.0367	0.8462	0.5268	0.6974	0.4366	0.3778	0.9871

The calculation results indicate:

There are 64 enterprises located on the efficient frontier, namely 64 enterprises innovation performance from knowledge input is significant, which shows knowledge innovation production efficiency of sample enterprises is not high.

$\sum_{i=1}^{m}\lambda_j$ is equal to 0.9927, which means that a weak decreasing returns to scale exists for the innovation knowledge production of the sample enterprise. This finding is consistent with the findings of Yan Chengliang (2008), Tao Changqi, Qi Yawei (2008) studies which show that the knowledge production function of R&D enterprise in China exhibits a decreasing returns to scale.

The knowledge from external is wasted to a certain degree. The utilization rate of knowledge from different sources varies. The utilization rate of knowledge obtained from suppliers and customers reaches more than 50%. Nearly 70% of the input is redundant, which suggests that knowledge sharing and cooperation of sample enterprises need to be deepened further.

Internal knowledge redundancy rate is 52.68%, and utilization rate is 47.32%, which indicates that the communication inside enterprise is inadequate. The reasons may be that the communication among departments is not smooth, R & D and market are out of line; communication channel among R & D, production and sales departments is not free; the degree of integration of the relevant functional departments is not high.

In terms of the output of the enterprise knowledge innovation production, the new product sales deficiency rate is only 3.67%, while the deficiency rate of patent application quantity reaches 84.62%, which shows that new products could meet the demand of market and recognized by customers.

5 Conclusion Advices on How to Promote the Knowledge Productivity

According to the above conclusions, the following measures should be taken to enhance the innovation of knowledge productivity by enterprises in China:

Strengthen cooperation between enterprises, and construct the collaboration system. Enterprises should improve the cooperation innovation consciousness, strengthen cooperative innovation among relevant enterprises, and make full use of the cooperative enterprises and even competitors to enhance innovation performance. Improve the production and research system. The government should strengthen the scientific research institutions and enterprises jointly, build reasonable integration system, build cooperation platform. Improve the cooperation between Scientific research institutions and enterprises, create opportunities for cooperation, and promote scientific research achievements conversion rate. Strengthen internal communication. R & D, production, sales and other functional departments should strengthen communication, to obtain more knowledge through communication, make full use of the knowledge, in the innovation process to promote the knowledge utilization. Promote the enterprise consciousness of intellectual property protection. The empirical study shows that the proportion of innovation patent application quantity is very low. Government shall establish a sound system of intellectual property protection. Enterprises should enhance the consciousness of intellectual property protection.

6 Conclusion

Enterprise innovation knowledge originates from the enterprise in-house production and external knowledge input, while different sources of knowledge have different effects on innovation performance of enterprises. China's enterprise production efficiency of knowledge innovation is not high. All knowledge has some redundancy. External knowledge, especially from the industrial cluster and university research institutions, is seriously wasted. More than half of the internal knowledge cannot get full use. There is a serious shortage of patent output. Therefore the enterprise should adopt certain positive measures to enhance the utilization rate of knowledge in innovation.

Acknowledgement. The project was supported by the Soft Research Project of Shanghai Science and Technology Development Funding (No. 11692100800).

References

[1] Pakes, A., Griliches, Z.: Patents and R&D at the Firm Level: A First Report. Economics Letters 5(4), 377–381 (1980)
[2] Griliches, Z.: Patent Statistics as Economic Indicators: A Survey. Journal of Economic Literature 28(4), 1661–1707 (1990)
[3] Acs, Z.J., Audretsch, D.B.: Innovation in Large and Small Firms: An Empirical Analysis. American Economic Review 78(4), 678–690 (1988)
[4] Koeller, C.T.: Innovation, Market Structure and Firm Size: A Simultaneous Equations Model. Managerial and Decision Economics 16(3), 259–269 (1995)
[5] Conte, A., Vivarelli, M.: One or Many Knowledge Production Functions? Mapping Innovation Activity Using Micro-data. IZA DP Working Paper (2005)

[6] Zhengzhou, K.: Technological capability, strategic flexibility, and product innovation. Strategic Management Journal 31, 547–561 (2010)
[7] Zhou, H.: Flexible labor and innovation performance: evidence from longitudinal firm-level data. Industrial and Corporate Change 20(3), 941–968 (2011)
[8] Hess, A.M.: When are Assets Complementary? Star Scientists, Strategic Alliances, and Innovation in the Pharmaceutical Industry 32, 895–909 (2011)

E-Business Diffusion: A Measure of E-Communication Effectiveness on the E-Enabled Interorganizational Collaboration

Wu Lu and Min He

School of Business, Zhejiang University City College, Hangzhou 310015, China

Abstract. Globalized business forces many organizations apply the e-business technologies to facilitate collaboration relationships and the integration all business partners' activities and their systems. The role of e-business management is not only to facilitate information sharing, but also to enhance communication and collaboration. Mismeasured and ineffective collaboration occurs when e-enabled e-communication and e-business innovation are not included in the measurement package. The suitability of e-business diffusion was developed from Diffusion of Innovation (DOI) theory and assessed in a case study of Hangzhou Alibaba Electronic Organization. Alibaba runs one of the largest online e-commerce platforms in China – Taobao. Based on the case study and survey on e-collaboration technologies in outward collaboration via Alibaba's e-communication channel, related analyses (such as validity tests, reliability and factor analysis) were performed to understand the role of e-business diffusion in measuring interorganizational collaboration. The study also discusses the appropriateness of e-business diffusion theory as a measure of collaboration success and reorganizes the expanded role of e-communication in interorganizational collaboration. The most notable attributes are relative advantage, complexity, trialability and observability.

Keywords: e-business diffusion, e-communication, interorganizational collaboration.

1 Introduction

The globalized development of collaboration requires support from communication–centered intraorganizational interactive systems to smooth organizational change. Without the help of computers, the Internet, e-mail, telecommunications, share-database and other web-based Internet technologies and applications, there will be no multinational project [1]. Hence, the role of e-business management is multi-faceted. It is used not only to facilitate information sharing, but more importantly, to enhance the communication and collaboration.

The rapid application of e-business innovation has a great impact on the traditional communication channels and collaboration manners in business [2]. Amidst the recent global economic meltdown, the complexity over how to optimize worldwide business resources and collaborate at an efficient cost has continued to be an issue [3].

Z. Du (Ed.): Proceedings of the 2012 International Conference of MCSA, AISC 191, pp. 517–521.
springerlink.com

Furthermore, it becomes an inevitable trend that many organizations apply the e-business technologies, mainly e-communication technologies (including software and hardware), to facilitate collaboration within the organization as well as business partners outside the organization [1, 3]. For example, one of the most sophisticated real-time e-collaboration office solutions uses electronic whiteboards and e-communication applications, where each participant can create or edit a joint document in real-time [2].

The focus of this paper is to understand the role of e-business diffusion in measuring interorganizational collaboration. Basing on the case study of Alibaba's e-communication channel, this paper seeks to provide constructive advice on how to improve collaboration performance. With the use of semi-structured interviews and questionnaires in Alibaba and its business partners (online buyers and sellers), the study also discusses the appropriateness of e-business diffusion theory as a measure of collaboration success and reorganizes the expanded role of e-communication in interorganizational collaboration. The instrument was originally developed by innovation academics to assess general diffusion of innovation (e-business diffusion).

2 Research Methodology

The rate of diffusion for e-communication and e-business technologies innovations is affected by their perceived attributes. This concept was investigated in an extensive series of focus group interviews. The pilot study was a questionnaire used to assess the diffusion of e-business technologies perception of CS staff members (customer service department). Twenty items regarding the e-business diffusion (relative advantage (A1-A4), compatibility (B1-B4), complexity (C1-C4), trialability (D1-D4) and observability (E1-E4)) and twelve items regarding collaboration elements (F1-F12) were edited and condensed by focusing on the clarity of questions, dimensionality of scale and the reliability of the components. 300 staff members and business partners from Alibaba in the customer service department and partner relationship department were asked to rate each item on a seven-point scale, ranging from strongly disagree (1) to strongly agree (7). 260 usable questionnaires were collected via internal mail, and the response rate was 85% in Alibaba. From the results, factor analysis and stepwise were conducted in SPSS 16.0.

3 Case Analysis of Alibaba

3.1 Background

Alibaba Group is the first e-commerce organization in China and unanimously regarded as the e-commerce network leader. Since 1999, Alibaba Group has grown quickly, and established seven subsidiaries, including Alibaba (B2B website), Taobao Mall (C2C/B2C website), Alipay (e-payment solution) and others. Over 98 million members were registered in 2008 and the annual turnover was 99.96 billion Yuan, 80% of China's e-commerce market share. As an e-commerce organization, Alibaba pays more attention to the collaboration, especially online collaboration, to serve the high number of small and medium-sized enterprises and consumers. Alibaba will guide partnered organizations using internet services, such as QQ, mail system, blogging and online trade training. For example, Taobao's website integrates the role of online shopping environment provider and after-sales service (third-party quality guarantee) to forge a safe and effective e-business platform. Effective communication promotes the adoption of interorganizational collaboration. However, in order to promote a transparent, fair and open platform, the adoption of such e-business technologies creates conflict at a cultural level, operational level and trust level. It is of vital importance for Alibaba to investigate the factors that affect the diffusion of e-business technologies. One of the crucial e-business issues faced by the company is the adoption of e-enabled collaboration with customers and partners.

3.2 Assessment of E-Business Diffusion's Validity in Collaboration

The first part assessed content validity and refers to the instrument covering the range of meanings included in the concept. The thoroughness of the focus group suggests that e-business diffusion does in fact measure the adoption of e-communication technologies. This article could not discern any unique features of collaboration that make the dimensions underlie e-business diffusion appropriate for measuring collaboration effectiveness.

The second part tested the reliability of each dimension by Cronbach's alpha. For e-business diffusion, the alpha showed that deleting either component will significantly influence the total value. The result was reliable (all the covariance values were greater than 0.05, Cronbach's alpha was 0.837 (>0.8)) and deemed suitable for Factor Analysis. The data was analyzed by the principal components with varimax rotation (Loading ≥ .35). Analysis showed that the five-factor model can cumulatively explain 70.049% of variables (see table 1). Each of the four factors in e-business diffusion reached their expectation with appropriate nomological validity (when items to be loaded together in a factor analysis) and discriminant validity (when items underlying each dimension loads as different factors).

Table 1. Factor analysis for e-business diffusion (A1-E4) and collaboration (F1-F12)

Rotated Component Matrix[a]

	Component				
	1	2	3	4	5
A1	.827				
A2	.793				
A3	.828				
A4	.783				
B1			.738		
B2			.719		
B3			.842		
B4			.895		
C1		.761			
C2		.734			
C3		.828			
C4		.818			
D1					.618
D2					.665
D3					.792
D4					.663
E1				.786	
E2				.647	
E3				.685	
E4				.677	

Rotated Component Matrix[a]

	Component			
	1	2	3	4
F3			.863	
F4			.790	
F5	.825			
F6	.762			
F7	.878			
F8	.916			
F9		.808		
F10		.748		
F11		.843		
F12		.845		
F2				.690
F1				.852

3.3 Discussion of the Relationship between E-Business Diffusion and Collaboration

For collaboration elements, the Cronbach's alpha showed that deletion of either component will significantly affect the total value. Assessment of the results (all covariance values >0.05, Cronbach's alpha is 0.858 >0.8, the KMO-Meyer-Olkin from KMO and Baetlett test was 0.855 (>0.8), Sig.=0.000) means these components are suitable for Factor Analysis and reliability was obtained. Using principal components with varimax rotation (Loading ≥ .35) showed that the four-factor model could cumulatively explain 65.456% of the variables (see Table 1). The four components contributing to the collaboration were E-Technologies, Knowledge and Resources Sharing, Risk Management, and Information Sharing.

From the results of the stepwise regression test in SPSS 16.0 (as shown in Table 2), the Adjusted R Square and F value show a significant relationship among the attributes of e-business diffusion and elements of collaboration. Using e-communication channels has a positive influence on the collaboration performance, including information, knowledge and resources sharing. Compatibility appears to be less effective on the collaboration performance. This may be due to a Chinese organization that is less sensitive to incompatibility by comparing the relative advantages of innovation or change.

Table 2. Stepwise Regression for the e-business diffusion and Collaboration

Independent Variables	Dependent variables			
	E-Technologies	Knowledge and Resources Sharing	Risk Management	Information Sharing
Relative Advantage	.221**	.233**	.318**	.340**
Compatibility		.641**		.324**
Complexity	.201**	.176**	.215**	.221**
Trialability	.186**	.246**	.257**	.264**
Observability	.99**	.151**	.142**	.160**
Adjusted R Square	0.451	0.306	0.235	0.146
F	34.311**	25.256**	29.821*	11.101*

** Correlation is significant at the 0.01 level (2-tailed)
* Correlation is significant at the 0.05 level (2-tailed)

4 Summary and Limitations

In the era of globalization and e-business innovation, the adoption and diffusion of e-communication technologies provides valuable channels for the participants to facilitate collaboration. An e-business-based collaboration is one alternative for satisfying the need for success. The values in Table 2 indicate that e-business diffusion is a precursor of collaboration or e-collaboration effectiveness.

One major contribution of this study was that it demonstrated that e-business diffusion, an extensively applied innovation instrument, is applicable in the e-communication of interorganizational collaboration. This study highlights e-communication as a fundamental e-business innovation and a valuable component of diffusion of innovation that should be involved in interorganizational collaboration. It is believed that the satisfaction or effectiveness of a collaboration unit can be partially assessed by its capability to diffuse e-business to its business partners.

In this report, the relationship between the attributes of e-business communication and collaboration has been tested. The four attributes of relative advantage, complexity, trialability and observability all have significant relationships with collaboration. There is no significant correlation with compatibility. In China, innovation is regarded as incompatibility and the relative advantage from the innovation is the determining attribute.

References

[1] Azadegan, A., Teich, J.: Effective benchmarking of innovation adoptions: A theoretical framework for e-procurement technologies. Benchmarking: An International Journal 17(4), 472–490 (2010)

[2] Zhang, H.: The influence from informal organization on formal organization. Journal of Zhengjiang Sociology 2, 59–60 (2010)

[3] Fan, X.: The informal organization and management of communication. Journal of Management science 22, 427–428 (2008)

[4] Banade, C.: Perfecting communication channels for enterprise development and create harmonious atmosphere. Journal of Management science 5, 15–18 (2009)

[5] Wang, S., Archer, N.: Supporting collaboration in business-to-business electronic marketplaces. Information Systems and e-Business Management 2, 269–286 (2004)

[6] Busi, M., Bititci, U.: Collaborative performance management: present gaps and future research. International Journal of Productivity and Performance Management 55(1), 7–25 (2006)

[7] Riemer, K., Steinfield, C., Vogel, D.: e-Collaboration: On the nature and emergence of communication and collaboration technologies. Electron Markets 19, 181–188 (2009)

[8] Vaidya, K., Sajeev, A.S.M., Callendar, G.: Critical factors that influence e-procurement implementation success in the public sector. Journal of Public Procurement 6, 70–99 (2006)

The Application Research of Small Home Appliance Product Based on Computer Aided Ergonomics

Dong Junhua

Fushan Polytechnic, China

Abstract. With the continuous improvement of people's living standard and lifestyle, the product with only basic features cannot meet people's needs, therefore the designers have pay more attention to humanize design and the spiritual and cultural need. This trend prompted the companies of the design of product innovation, appearance, ergonomics, etc. refer to a new height when proceed with new product development for small home appliances, adhere to the concept of people-oriented, take full account of the psychological and physiological aspects, realize humane design through the product's shape, color and function, and give users a better use experience. This article described the application of computer-aided ergonomics in the design of small household electrical appliances in detail and guide the develop direction of small household electrical appliances in the future.

Keywords: small home appliance, ergonomics, humane design.

1 Introduction

In recent years, China has introduced the appliances to the countryside, trade for new, energy-saving projects and other policies to encourage the development of China's home appliance industry [5]. Demand for household electrical appliances continue to increase in the context of rapid economic development, small household electrical appliance has already become an essential articles for daily use, and also has a very big impact on people's quality of life and lifestyle [2]. With People's continuous improvement of consumption concept, and the ever-increasing demands on product quality, performance, appearance, ease of use, etc., which present unprecedented challenges for product development. China's small household electrical appliances manufacturers have been gradually mature in the design and the technical production of small household electrical appliances[1], have already produced a variety of products to meet the general requirements, modern appliances has not only stay in the functional requirements, but gradually closer to the cultural tastes and personal preferences. The vast number of small household electrical appliances manufacturers is necessary to put time and effort in product design to meet people's urgent consumption needs[2], this would require combine ergonomics with product design, with the help of developed science and technology to improve the product's features and production technology, develop and design the product to meet the people's aesthetic demands to enhance the competitiveness of enterprises, promote the development of enterprises[5]. Future products must be designed to the primary

Z. Du (Ed.): Proceedings of the 2012 International Conference of MCSA, AISC 191, pp. 523–528.
springerlink.com

condition of creativity and innovational, the only truly useful and practical goods will stand out in the market, so that what makes consumers feel intimate and affordable is the perfect weapon to win the business [4]. Ergonomic, human-based design is the most real and frontier trends, is the manifestation of the humanistic spirit, people and products perfectly harmonious combination. Human-based design truly reflects the respect and concern for human beings.

2 The Concepts and Application of Computer Aided Ergonomics

Ergonomics is disciplines that study the interaction among people, machine, and environment; it can make technology humanize. It involves psychology, physiology, medicine, anthropometry, aesthetics and engineering technical fields. The goal is to improve the efficiency, safety, health, comfort and other aspects when operating. Ergonomics is composed of technical science, human science and environmental science; Technical science includes: industrial design, engineering design, safety engineering and systems engineering etc.; human science include: physiology, psychology, labor hygiene, anthropometry and body mechanics etc., environmental science include: environmental protection science, environmental medicine, environmental hygiene, environmental psychology and environmental detection science etc. Thus, in all areas of people's daily lives are closely related to ergonomics, it is necessary to cause sufficient attention. " Human-based design " , whether it is an industrial product or a piece of art will it eventually serve for people, adhere to the guiding philosophy of " people " as the core is to bring the convenience and comfort life quality and efficiency in the design, this is the existence purpose of ergonomics and all goods. Human beings is the master of all things in the physical world, all the things we have created will establish relationship with people through all kinds of channel what we call human-machine interface. The interact whole composed of human and materials we called man-machine system. Narrow ergonomics contains human and machine; generalized ergonomics the environment in which people survive into consideration, constitutes a human-machine-environment system. The ergonomics' significance lies in the coordination of the relationship among the human-machine-environment system, on one hand make the product design to achieve optimal, on the other hand make a higher efficiency by adopt certain training aspects. Ergonomics has a wide and complex range, also contains a rich knowledge. It contains the body science, technical science, environmental science from the disciplinary system, each system includes its own branch discipline, these knowledge systems constitute a strong knowledge system of ergonomics by influence and restraint each other.

(1) The research of microenvironment: The microenvironment constitute of temperature, humidity, airflow, light, etc. has long been concerned. The good light environment can improve visual conditions and visual environment to raise productivity and reduce error rate.

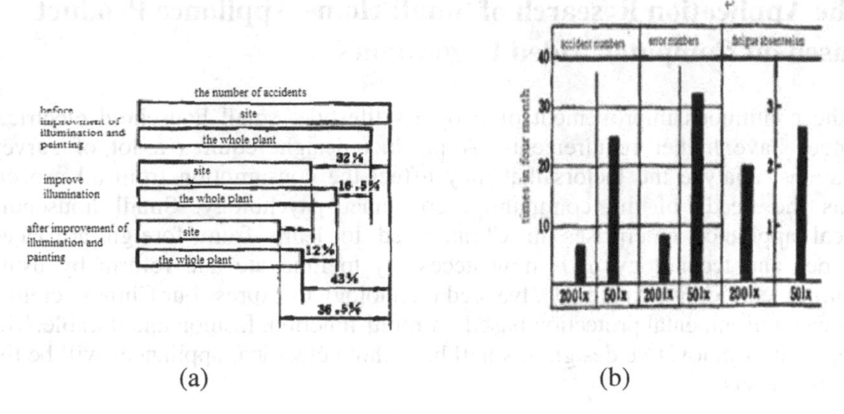

Fig. 1. The relationship between light and the accident rate

Fig1.(a) shows accidents decreased by 32%, the whole plant accidents decreased by 16.5% after improved lighting, accident reduction is more significant if improving both lighting and painting. Fig1.(b) shows good lighting makes significant reduction of accident numbers, error numbers and absent numbers.

(2) Switch setting: In a large space is often configured with a lot of lighting equipment, but Sometimes don't need to open all the lighting equipment, but only require a limited open, for example, a very large classroom with few people. But often the lamp position and the switch is not a regular column, Cannot find the habit of corresponding relations, try several times in order to open the lights is a waste of time and a waste of energy, set the switch according to the dislocation of ergonomics can be categorized.

(3) Operation platform design: People often need work in a sitting position including observations and control, ergonomic sitting position operation platform shown in figure 2. If don't meet the design principles, operation and observation will be uncomfortable, affecting the accuracy of the operation and observation.

(4) The design of engineering logo: On many occasions to express some kind of information briefly while not using too much text or do not have time to read, only to use graphics or symbols to express specific information.

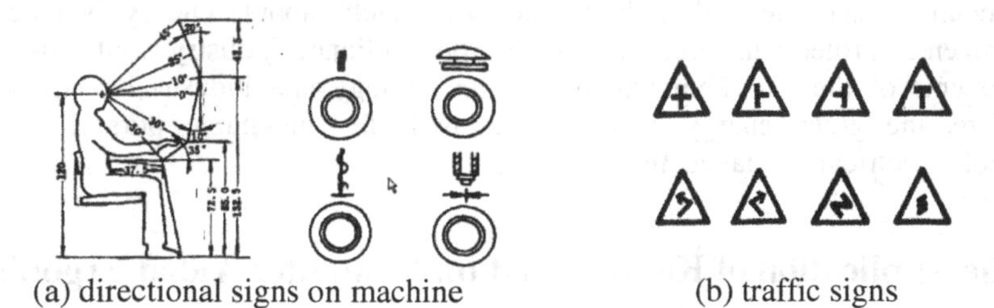

(a) directional signs on machine (b) traffic signs

Fig. 2. Operation platform **Fig. 3.** Directional signs

3 The Application Research of Small Home Appliance Product Based on Computer Aided Ergonomics

With the continuous improvement of people's lifestyle, small household electrical appliances have higher requirements. A product design requires a lot of survey preparations; analyze the factors that may affect the consumption from all aspects such as the needs of the community, consumer psychology. Small household electrical appliance enterprises in China used to learn from foreign advanced experience and technology, it is now necessary to innovate and reform by using ergonomics, employ international advanced technology to express our Chinese culture and green environmental protection based on multi-function, fashion and durable. The development of innovative design of small household electrical appliances will be the following aspects.

3.1 The Health and Harmless of Product

With the improvement of people's living standard, health problem is gradually becoming more and more concerned, healthy function of the product has already been the main direction of the innovative design of small household electrical appliances. Small household electrical appliance enterprises strive to design products that can be helpful to human health and reduce or avoid pollution caused to indoor.

3.2 The Fashion and Individualization of Product

Personality characteristics of the modern home appliance become more and more clear, small household electrical appliance's fashion and individualization become one of the most important criterion for consumers to buy. When considering add or replace small household electrical appliance, instead of considering the quality and price, now more concern about whether the style suit the room. The majority of manufacturers need to rack their brains to improve the fashion and individualization to meet the different tastes of consumers.

3.3 The Green and Energy Saving of Product

Nowadays, the energy and resources gradually become short of supplies worldwide, each country and general public concern much about energy saving and environmental protection. Green design of the appliance industry conforms to the requirements of the environmental protection strategy and industrial restructuring. Therefore, the green energy saving products is the inevitable pursuit of small household electrical appliance manufactures.

4 The Application of Kettle Based on Computer Aided Ergonimics

The kettle is a household appliance for boiling water which rated capacity of not more than 10 liters. Generally, the kettle is fitted with a temperature limiter what can cut off the circuit automatically when the water is boiled and protect the kettle in abnormal states such as dry burning. It widely used in home and hotel rooms with the features of fast, efficient and convenient. Consumers tend to note the following when buying a kettle: (1) whether the input power meet the standard requirements; (2)whether resistant to moisture, because the substandard products are easy to cause a risk of electric shock; (3)whether the creep age distance and electrical clearance are eligible. Because when the kettle is away from the base, if not a sufficient distance guaranteed, it is easy to cause a risk of electric shock when the users are cleaning, moving or accidental touching the base. (4) whether the grounding measures are qualified. (5) whether the sign and instruction are eligible, whether marked rated voltage, rated input power and model specifications.

4.1 The Choice of Material

With the promoting of consumers' health awareness, most Chinese consumers would like to believe metal is much more safe to contact with water. As a result, the proportion of kettle that with stainless steel container is rising, the proportion of kettle that with plastic container is declining. Most kettles use 304 stainless steel container as container or body. A lot of European brands tend to use metal container in the choice of materials. For example, Philips almost completely abandoned the plastic liner and switch to stainless steel in recent developments.

4.2 Convenient to Use

4.2.1 Analysis of Hand

The general size and length of the hand are as follows:

Table 1. Size of the hand

Size number	The palm measurement/mm	The length of palm/mm
6	190	160
7	196	170
8	201	180
9	205	190
10	210	200
11	213	210

4.2.2 The Strength of Grasp

People are familiar with dynamometer and determination of the strength of grasp, noted that the result measured by common used dynamometer is isometric contraction of the muscle rather than isotonic contraction. Generally to show normal grip strength with grip index: $I=\frac{S}{W}\times 100$ I means grip index, S means grip strength, W means weight. Normal grip index should be greater than 50, according to the investigation of Swanson, handedness grip is 5%-10% greater than non handedness. Women's grip strength is usually only 1/2-1/3 of the males. Man after the age of 50, woman after 40 often decrease by 10%-20% than the youth.

4.2.3 The Analysis of Pour Water

When pouring water, first you must to hold the kettle handle, this need the size of the hand and activities of the wrist, keep your hand and the kettle into 45 degrees is the most comfortable posture.

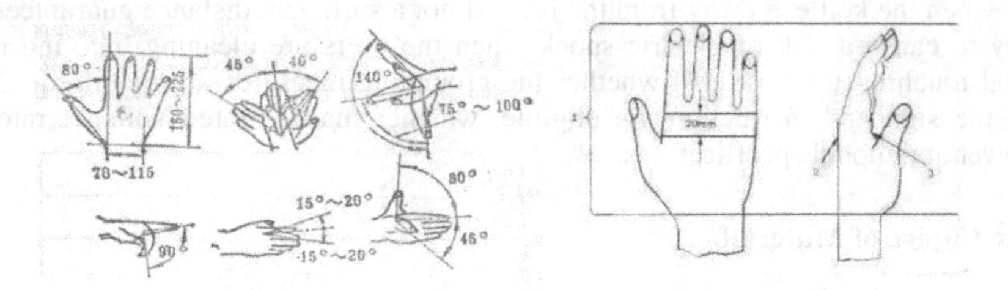

Fig. 4. The size of the hand and the activities range of wrist

This little has a certain slope that is very comfort for pouring water, the handle has a medium size convenient for both children and old man, the above is small and the bottom is large is easy for water to flow out.

5 Conclusion

This paper studied and analyzed the kettle based on ergonomics, not only to consider the structural characteristics of the hand, but also the other parts of the body's physiological characteristics, understanding the links among them and design the most appropriate and humanized product. So it is of great significance to carry out the application research of ergonomics vigorously.

References

[1] Nichol, D.: Environmental changes within kettle holes at Borras Bog triggered by construction of the A5156 Llanypwll link road, North Wales. Engineering Geology 59(1-2), 73–82 (2001)

[2] Cook, T.H.: Burning issues impact kettle purchases Metal Finishing 103(4), 46–50 (2005)

[3] Hadi, H.A.: Predictions of in-tube circulation models compared to performance of an experimental kettle reboiler bundle. Institution of Chemical Engineers Symposium Series 1(129), 197–207 (1992)

[4] Määttä, T.J.: Virtual environments in machinery safety analysis and participatory ergonomics. Human Factors and Ergonomics In Manufacturing 17(5), 435–443 (2007)

[5] Van der Sar, A.: Applying ergonomics. Ergonomics 3.(1-3), 313–315 (1990)

[6] Jones, G.: Ergonomics - An update. Gatfworld 16(1), 92–95 (2004)

Pro/E-Based Computer-Aided Design and Research of Small Household Electrical Appliances Mold

Li Baiqing

Fushan Polytechnic, China

Abstract. Traditional design of shaped small household product, mold design and its processing was accomplished according to the drawing data, therefore it always takes more time to modify the mould and costs a lot. The software of Pro/Engineer was adopted for 3D design of the household electrical appliances' shell, which can shorten the period of product development, mould design and manufacture, and lower the product cost. The characteristics of Pro/E were described; the design method of mobile phone shell, injection mould design and making with CAD/CAM system was introduced.

Keywords: household electrical appliance, Pro/E, injection mould design.

1 Introduction

In recent years, Chinese government has introduced the electrical appliances to the countryside, trade for new, energy-saving projects and other policies to encourage the development of China's household appliance industry [1]. Demand for household electrical appliances continue to increase in the context of rapid economic development, people's consumption concept continue to improve [3], and the ever-increasing demands on product quality, performance, appearance, use convenient, etc., this presented unprecedented challenges on product development. The traditional mold design usually with three view graphics, this representation is fast and convenient to express the designer's ideas, but for some complex products, it is difficult to express the shape and some of its characteristics clearly, not only does it increase the labor of the designer, but also add a lot of difficulty to the readers[2]. In this paper, three-dimensional mold design ideas are adopt instead of previous graphic design, in order to express the product structure and other requirements much more clearly, making the samples more intuitive, more texture while the graphics do not have these characteristics [5]. A mature enterprise must have a mature design methods, then get a better product quality and development speed, achieving the optimization design. So small household electrical appliances mold design cannot separate from the advanced technology such as CAD/CAM [5]. Pro/ENGINEER is a 3D solid model design system developed by PTC-company in 1989. It is a set of mechanical automation software working from design to produce where have a modification cause automatic changes of other relevant local, It is a parameterized solid modeling system based on characteristics which only with a single database system[6]. In the mold design process, using the solid modeling technology of Pro/E can quickly get an imaging understanding on the parts which are going to be

Z. Du (Ed.): Proceedings of the 2012 International Conference of MCSA, AISC 191, pp. 529–534.
springerlink.com

processed. With the Mold module of Pro/E, we can directly begin the mold design based on the Three-Dimensional model after analyzing the model. The Mold module of Pro/E provides a set of powerful Three-Dimensional mold design plug-in EMX (expert mold base extension), it offers a very rich mold database for the Three-Dimensional total assembly design.

2 The Research of Small Electrical Appliances' Shell Injection

Injection molding process is the most common molding process for making small electrical appliances' plastic parts such as shells. Generally, plastic injection molding design includes plastic product design, mold design, and injection molding process design, all of which contribute to the quality of the molded product as well as production efficiency. This is process involving many design parameters that need to be considered in a concurrent manner. Mold design for plastic injection molding aided by computers has been focused by a number of authors world-wide for a long period. Various authors have developed program systems which help engineers to design part, mold, and selection parameters of injection molding. During the last decade, many authors have developed computer-aided design/computer-aided engineering (CAD/CAE) mold design systems for plastic injection molding. Developed a collaborative integrated design system for con-current mold design within the CAD mold base on the web, using Pro/E. Low et al. developed an application for standardization of initial design of plastic injection molds. The system enables choice and management of mold base of standard mold plates, but does not provide mold and injection molding calculations. The authors proposed a methodology of standardizing the cavity layout design system for plastic injection mold such that only standard cavity layouts are used. When standard layouts are used, their layout con-figurations can be easily stored in a database. Lin at al. describe a structural design system for 3D drawing mold based on functional features using a minimum set of initial information. In addition, it is also applicable to assign the functional features flexibly before accomplishing the design of a solid model for the main parts of a drawing mold. This design system includes modules for selection and calculation of mold components. It uses Pro/E modules Pro/Program and Pro/Toolkit, and consists of modules for mold selection, modification and design. Develop a CAE system for mold design and injection molding parameters calculations. The system is based on morphology matrix and decision diagrams. The system is used for thermal, rheological and mechanical calculation, and material base management, but no integration with commercial CAX software is provided. Huang et al. developed a mold-base design system for injection molding. The database they used was parametric and feature-based oriented. The system used Pro/E for modeling database components. The previously mentioned analysis of various systems shows that authors used different ways to solve the problems of mold design by reducing it to mold selector. They used CAD/CAE integration for creating precision rules for mold-base selection. Many authors used CAE system for numerical simulation of injection molding to define parameters of injection molding. Several also developed original CAE modules for mold and injection molding process calculation. However, common to all previously mentioned systems is the lack of module for calculation of mold and injection molding parameters which would allow integration with the results of numerical simulation.

This leads to conclusion that there is a need to create a software system which integrates parameters of injection molding with the result obtained by numerical simulation of injection molding, mold calculation, and selection. All this would be integrated into CAD/CAE-integrated injection mold design system for plastic products.

3 Pro/E-Base Computer-Aided Designs and Research

The cell phone shell is an important component of mobile phone, it is composed of complex curved surfaces, its design will directly affect the quality of cell phone. The next let us have a detail description about how to use Pro/E for mold design and the general process and specific steps for Three-Dimensional assembly design with the application of EMX.

Table 1. Shell structural features and their corresponding modeling tool

Phone shell structural feature	***Modeling tool***
Thin-walled	Shelling
hole	Stretch
Circular bead and chamfer	Rounding
Repeat	Array

3.1 The Establishment of Mobile Phone Shell Entity Model

Click the stretch button, activate stretch characteristic, click place to activate sketch menu, select the front plane as sketch plane, click rectangle drawing tool to draw the rectangle shell. After finish the drawing use rounded tool to complete rounding the rectangle shell and modify the sizes according to stipulated requirements. Then click the finish button, enter the stretching distance so we can get the mobile phone shell solid model shown in figure 1.

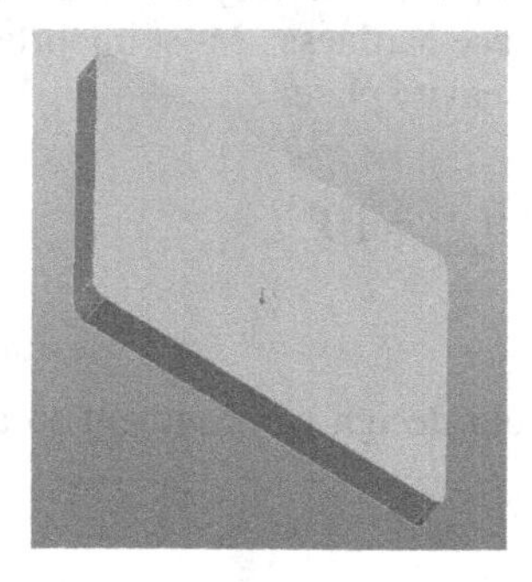

Fig. 1. Shell entity model

Fig. 2. Shell rounding

3.2 Shell Rounding

Click the rounding button, select the edge to rounding and enter rounding radius. Rounding feature is shown in figure 2.Click the shelling button, choose the unrounded plane of the model, enter the wall thickness 2mm.Shelling feature is shown in figure 3.

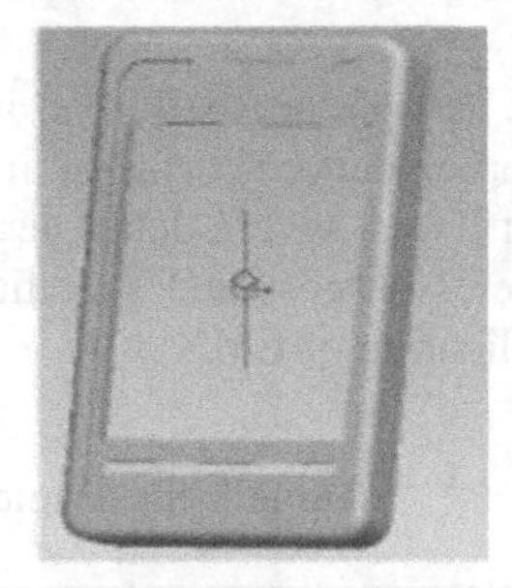

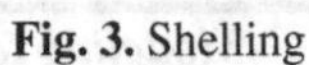

Fig. 3. Shelling **Fig. 4.** Screen window **Fig. 5.** Dial hole

3.3 Establish the Phone Screen Window Hole

Click the sketch button, choose the unshelled plane as sketch plane, draw a rectangle screen window and rounding, modify. Then click the stretch button, pick to remove the material, choose to stretch into the internal of the model, then we get the phone screen window hole shown in figure 4, and then establish the phone dial hole, steps are similar to 3.4,we can get figure 5.Then save.

4 Mold Design of Mobile Phone Shell

4.1 The Choice of Material and New Mold Model File

This cell phone shell select thermoplastic ABS, it has a very good Comprehensive performance, a higher impact strength and surface hardness. Most important, it has a good dimensional stability. Maximal flexure of cavity plate:

$$f_{\max} \leq \frac{p_k \cdot d_{kt}^2}{s_k}\left(\frac{d_{kt}^2}{32 \cdot E_k \cdot S_k^2} + \frac{0.15}{G}\right) \leq f_{dop} = 10^{-3},$$

Solution is $f_{\max} = 0.158 \cdot 10^{-3}$.Enter the mold design environment to activate the mold menu shown in figure 6.

4.2 Transfer the Phone Shell Model and Create the Shell Mold Work-Piece

Click mold model in the mold menu, select "positioning reference parts" in the drop-down menu, pick the established shell model, choose "the same model" in the dialog box of create a reference model, then opt "single layout" in the layout dialog box, so we can transfer the phone shell model.

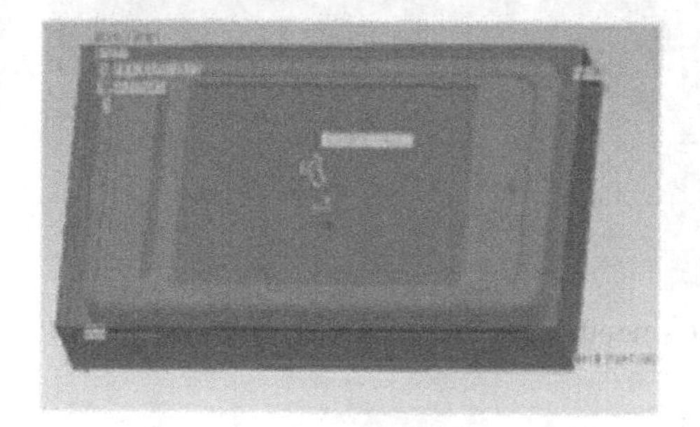

Fig. 6. Shell mold work-piece

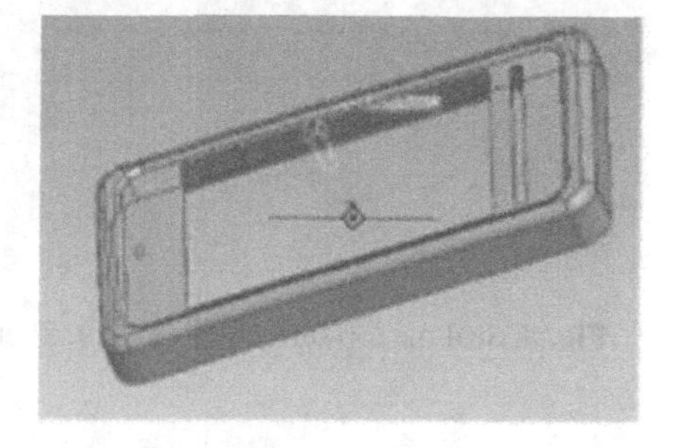

Fig. 7. Whole external

Click create in the drop-down menu of mold model to activate the drop-down menu of mold model type, pick "work-piece" then choose "manual create" to activate the dialog box of components to create shown in Figure 6.Choose part in Type and entity in Sub type as Fig 8,name then close. Next, pick "create feature" then "ok" back to the drop-down menu of mold model, opt "plus materials" and accept other default options, now we begin to create the entity model of phone shell mold. Choose the model' side as sketch plane, draw a rectangle that can accommodate the shell, pick to create entity of the shell mold. Choose "stretch to both sides", enter the stretch distance then we get the phone shell mold model shown in figure 6.

4.3 Define the Parting Surface

Parting surface is a complete plane that can split the mold along the part's contour. There are two steps:(1) copy and close the outer surface of the phone shell mold; (2) Continuation of the copied outer surface along the underside. The concrete procedure of the first step: click "parting surface" in the drop-down menu of "mold model", pick "create" and enter the name of the parting surface. Activate the drop-down menu of "curved surface define", pick "increase" and then "copy" to activate the dialog box of "curved surface replication", choosing the whole external surface of the shell as figure 7. Then double-click the "filling ring" to get the drop-down menu of "polymerization filling", pick the upper surface contains the screen hole, it will display lightly, so we seal the shell surface as figure 8. The second step is to create a plane that can split the work-piece along the bottom plane and combine with the sealed shell surface to form the final parting surface. The concrete procedure: click "increase" in drop-down menu of "curved surface define", pick "smooth" to activate the dialog box of "curved surface: smooth", choose the bottom plane as the sketch plane, use to draw a rectangle can include the work-piece completely as figure 9. Next click "ok" to close the drop –down menu of "curved surface: smooth", pick "merger" in drop-down menu of "curved surface define", choose the two planes established above to generate the final parting surface as figure 10.

Fig. 8. Seal the top surface

Fig. 9. Sketch smooth plane

Fig. 10. Shell parting surface

5 Conclusions

Pro/E not only has strong modeling capabilities, but also excellent mold design and processing capabilities. Using the mold design module of Pro/E, mold components can be easily extracted, greatly improved the small electrical appliance mold design efficiency and reduce design costs.CAD/CAM has changed the method of mold design, integrate the mold design process through digital renderings and models. But regardless how powerful CAD/CAM is, can it never replace the person's dominant position, only combine the dominant factor of human beings and CAD/CAM can it play a better role in mold design.

References

[1] Belolsky, H.: Plastics Product Design Process Engineering. Hanser Publishers, Germany (1995)

[2] Hodolic, J., Matin, I., Stevic, M., Vukelic, D.J.: Development of integrated CAD/CAE system of mold design for plastic injection molding. Mater Plastice 46(3), 236–242 (2004)

[3] Lin, B.-T.: Application of an integrated RE/RP/CAD/CAE/CAM system for magnesium alloy shell of mobile phone. Journal of Materials Processing Technology 209(6), 2818–2830 (2009)

[4] Ozcelik, B.: Warpage and structural analysis of thin shell plastic in the plastic injection molding. Materials and Design 30(2), 367–375 (2009)

[5] Matin, I.: A CAD/CAE-integrated injection mold design system for plastic products. International Journal of Advanced Manufacturing Technology, 1–13 (2012)

[6] Nee, A.Y.C.: Determination of optimal parting directions in plastic injection mold design. CIRP Annals - Manufacturing Technology 46(1), 429–432 (1997)

The Key Technology Research of Sport Technology Learning in the Platform Establishment That Based on ASP.NET

Tang Lvhua

Jiangxi Science&Technology Normal University, Nanchang, Jiangxi, 330038
jxnctanglvhua@126.com

Abstract. Based on the deepen research on present sport learning platform and the further developing requirement, this article designs the structure model that can be in common use for the platform of sport technology learning. It uses the ASP. Net development tool, adopts component technology during the platform establishment. In order to increase the component expandability and the reusability, here we use the object-oriented (OOP) method to package the component. We use the custom entity to realize the physical model. The platform frame uses B/S mode and the learning platform establishes the multiple data security based on four stages. Through the improvement of these key technologies, we can develop the software extensibility, flexibility and the security.

Keywords: ASP. Net, OOP, B/S mode, sport technology.

1 Introduction

The learning platform education of sport technology is the new teaching mode. It is based on the multimedia, computer network, and the international network. Expand the education from the classroom to the campus network or the internet. The teaching resources can share in the school, the country, even the entire globe. At the same time, the teaching platform of sport technology is the expression of the combination between network technology and the reality teaching. It breaks the limitation of time and space. Students can learn the sport technology any time anywhere. Through the learning platform of sports technology, students can look at the essential action through many times and multiple views. After the practice, they will rapidly grasp the relative skills.

It is very unlucky that the learning platform is limited in the theory teaching. It is lack of application in the technical practice. The developed sport network courses are most the form of video and text, or make the PPT of course contents in order to spread it on the network. Moreover, record the class and send it on the network. Therefore, you can imagine the course resource quality in the sport-learning platform. This cannot express the course characteristics. The platform is boring and students have low learning initiative. The other one is the method cannot combine or compare

Z. Du (Ed.): Proceedings of the 2012 International Conference of MCSA, AISC 191, pp. 535–540.
springerlink.com

the teaching content and practical competition. At the same time, there has no interactive between teacher and students. Aiming at the above questions, this article designs the platform that suitable for the sports technology learning. Take full advantage of the platform generality to expand the sport teaching will increase the learning interest of the students.

2 The Entire Structure Model of the Sport Technology-Learning Platform

The learning platform of sport technology is the generic terms of learning software, teaching resources and the teaching activities of the network technology. The implement of design implement needs to realize by the skill-learning platform. Therefore, the important circle of course development is the entire design of the learning platform.

2.1 The Foreground Function Design of Learning Platform

The design of the writer is combining with foreground and background. The combination is figure 1.

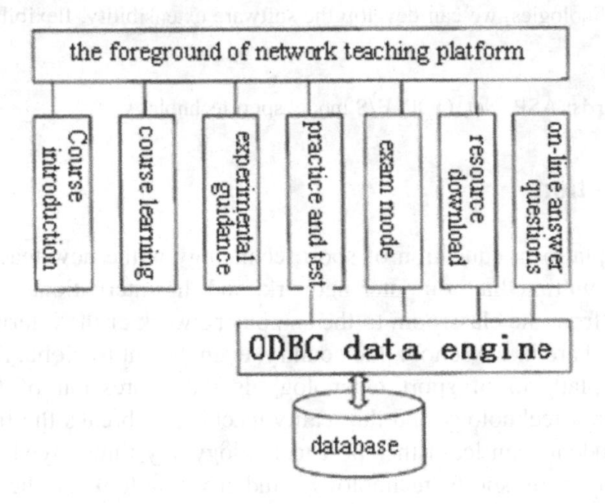

Fig. 1. Foreground structure of learning platform in sports technology

In there, the module content of course brief introduction is introducing concise short and clear that aiming at the basic course of college computer. The introduction contents include the teaching program of basic course, experimental program, and the exam topics. At the same time, the teaching progress chart and experiment progress chart can be inquired in this module. The course-learning module stores the teaching contents and other similar sport videos. Through this module, students can review the teaching content. At the same time, they can select the learning content that aims at their interest and learning process. This can increase the learning initiative. The experiment guide module includes the case demo video and provides the detailed experiment operation guide that can convenient for the students to grasp the skill of the experiment. Through the module application, students can learn that combine with the case. They can rapidly know well about the experimental environment, then to grasp the operation skill. The practice test module can solidify students' knowledge points. Through the ACTIVEDATAOBJECT application of the ASP technology, the module realizes the dynamic interaction of the database and the paging display. The database stores relative exercises of each chapter and display five questions in one page. Students can randomly select the page to do the practice. Each page has one deliver button. When the students finish the questions and click the button, the system will judge the answers and provide the correct answer by the compressed files. The imitate test provides the fixed, random course comprehensive test and the MS office imitate papers. The module uses the ADO technology of ASP and achieves the dynamic interaction with the database. Moreover, it realizes the random paper extract of the fixed test paper in the different database background and the random test paper. At the same time, their uses the SESSION object of ASP technology, store the different special test information into the background database. When the user finishes the question and deliver the answer, the system will log in the user performance of background basement and create the analysis report with the correct answer to convenient for the user research. The resource download module provides learning material's download. It includes relative learning materials and the grade examination resources. The on-line answer questions module accepts the combination of WEB pages and ASP technology. The user message can add to the database and dynamically display on the page for the reply and require. The application of this module promotes the communication between teacher and students. Moreover, it also promotes the communication among students.

2.2 The Background Function Design of the System

The background of this sport technology-learning platform design manager, teacher, students, and anonymity the four entry modes. Based on the different entry modes, the user permissions are various. The background structure of learning platform in sport technology is figure 2.

In there, the manager has the highest permission. He can add or delete the users, publish, modify or delete the announcement. He can upload and delete all the courseware and the applied teaching materials, he can add, modify and delete all the test questions. Meanwhile, he can delete all the invitation and reply. The teacher user can publish, modify and delete the independent announcement, upload courseware and teaching materials. He can modify or delete the courseware and teaching materials that uploaded by him. It can add test question, modify, and delete the added test questions. Moreover, it can post and reply, even modify and delete the reply.

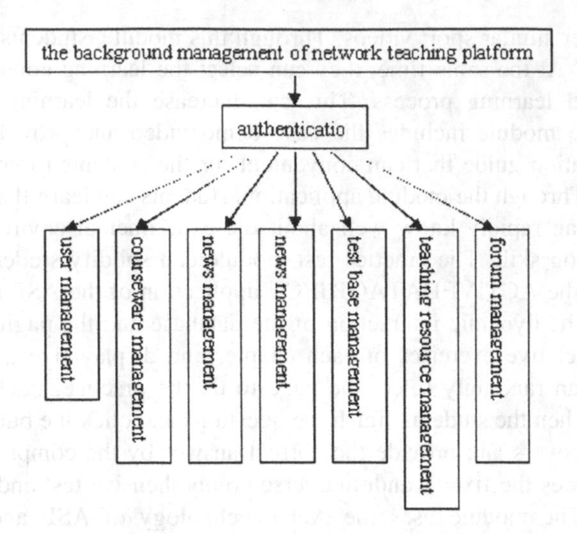

Fig. 2. Background structure of learning platform in sport technology

3 The Key Technology of Platform Development in Sport Technology Learning Platform

3.1 The Technology of Relative Component

In the learning platform of sport technology, the user control includes class list, major list, course list, and other combined list. The user controller of this system has the extension of .ascx that only can operate the small WEB page in the .aspx page. It means the operation cannot operate on the independent web. The service controller has logged in module, custom validator and the service controller is the programmed class basement. Add the required class basement into .NET toolbox. The operation only needs to drag the service controller onto the WEB page. Without compiling, this is same with the other standard controller and service for the service end.

3.2 The Data Access Technology of ADO.NET

ADO.NET is the data access technology that based on .NET frame and environment. The ADO.NET is forming for the distributed application. It can use in the data storage and provide relative data, XML, and applied programs data access to support the various development requirement which includes establish application program, tool, language, or the costumer end and middle layer object of internet browser. The data access technology of ADO.NET has the following characteristics. The first, ADO.NET is designed for the web application program that base of information. Use ADO.NET can avoid the user contest for a database resource to realize the access maximize of the shared data. The second is ADO.NET provides data order and data reader that can convenient for the direct communication of data source. Such as to require and storage process, establish database object, directly update and delete. The third is ADO.NET can provide special functions as well as support all the datasets transform to XML file. The fourth is the ADO.NET can support multiple databases. ADO.NET can realize this character through the management program. The management process research data from data source, and work on the bridge between application program and data source. For example, .NET frame provides multiple management support such as SQLserver、oracle、ODBc. Moreover, it achieves the data connection between the database and the data upload instep in the database.

3.3 The Page Base Class Design and Object Oriented Technology

The evaluation of object oriented and the design method application is the theory process that independent of the programming design language. The advantage is helping the group users to establish the abstract principle and provide one set of method to build the files in order to transform the complicated system into the principle mode that the computer can accept. The object oriented evaluation and design method is based on the objective fact. Through the class abstract and inheritance, we can build the connection between solid to solid. Use ASP.NET to develop the system, it is necessary to need the thinking of object oriented that can naturally degrade the system combination, reach the natural thinking of establishing the question domain mode. Moreover, makes the question domain mode has the encapsulation, inheritance and polymorphism. In there, the encapsulation expresses the model character as well as the data and operation package is an object, only the operation interface to accept the information from the object during the process. The user does not need to care about the internal model structure and programming language. Inheritance is as well as the reusability. Through inheriting, the basic class data and method definition of the object can satisfy the application requirements. Make the most inheritance to be the father and son. The common objects put into the super class, the subclass object can inheritance the super class object. Then, add the special object on this basement. The polymorphism is the flexibility of freight polymorphism and dynamic binding. The different object receives the same information can finish the various operations. Therefore, in the super class and subclass objects, the same operation has the different meaning. The operator can use the dynamic select operation.

3.4 Security and Identification Technology

The learning platform of sport technology has four levels of multi-media source security strategy. It includes global user GU, global role GR, local role LR, and the local user LU. The GU uses for the platform user research. When establishing one user, the GU will give one GR. If the GR want to access every data source, it needs to reflect on the special LR (LR1, LR2,..., LRn) from the detailed data source. Moreover, transform it into the LU set of (LU1, LU2, ..., LUn). Then it can achieve the data access. The LU has two methods to control the operation permission of the detailed data. The first is the large-scale database system supports the data. On the surface layer, the management tool of database system sets the LU access permission. In the second, the data of other file formal can establish the access comparison table on the data source end, control the access permission of LU. The security strategy application makes the learning platform system of sport technology has the perfect flexibility and maintenance. The GU and GR will integrate with the platform, the LR and LU will be managed by the local data source. At the same time, the security strategy applies two-dimensional table to store the data. That includes GU comparison table, GR-LR comparison table, and the LR user comparison table. The manager is the global user, the teacher, students, and the anonymity is the local user.

4 Summary

This article starts from the technical view, describes some key technologies on the learning platform development. However, it is just one side. Therefore, while developing this system or another system, the developers need to grasp the above key technologies. Moreover, they need to increase the familiar of system business process and the evaluation of system requirements in order to determine the developmental system quality. Through the platform system analysis of sports learning, the writer designs the structure, database, and the UML model. Use the above techniques, the writer develops the sport-learning platform that based on ASP.NET and obtains the perfect effect during the practical application.

References

[1] Mao, S., Zhu, G.: The discussion of using the network platform to apply the edicatopm reformation of computer application basis. IT Education (8), 45–46 (2009)

[2] Gong, Z., Zong, Y.: The research of course development mode in the college sport that based on the distance education environment. Journal of Capital Institute of Physical Education 20(2), 87–90 (2008)

[3] Wang, G.: Through network course to assist the research and design of education platform. Computer Knowledge and Technology (Academic Exchange) (12), 3193–9194 (2010)

[4] Lei, L.: The developmental course of ASPNET network application. Posts and Telecom Press, Beijing (2009)

[5] Zhang, Y., Zhou, K.: The middleware technology application and research. Computer and Communication (1), 31–37 (2001)

[6] Li, X., Li, H.: The research and application of office automation system that based on ASP.NET. Computer & Information Technology (3) (2009)

Computer Application in the Statistical Work

Yang Liping

Jiangxi Science&Technology Normal University, Nanchang, Jiangxi, 330038
262360334@qq.com

Abstract. With rapid development of the computer network technology, how to utilize modern computer technique to realize the statistic electronization and networking has been the important work to improve the office automatic and statistic information service. This is satisfying the modern social development. Moreover, this is the significant subject for the statistic industry. Different level statistician must good at various data statistics. They need to prepare well about the marketing policy collection and arrangement that can change the passivity statistic to the active service. Therefore, the computer application is imperative during the statistical work.

Keywords: computer, statistical work, application.

1 Introduction

The statistic work needs to provide the effective data support for the social development. This requires the mass data collection. However, this is difficult to depend on the manual work and cannot satisfy the statistic requirement. We need the computer to support the great function of algorithms, programming, and storage, which is heavy to the statistician. Therefore, how to take full advantage of the computer technology to achieve the electronization requirement for the statistical work, we need to realize the information networking, improve the service of office automation and statistical information in order to adapt to the modern social development. This is the important content of standardization during the statistical work.

2 Computer Development during the Statistical Work

Computer technology applies very earlier in the statistical work than in other fields. Our country has applied the technology of computer network information in the statistical work since 1988. This can save the human power to a limited tent and effectively improve the work efficiency. However, the computer technology was low at that time. It cannot process the information in the fast speed with the long exploration term. It is commonly realized the computer processing in some special professional parts and the electronization is not high.

With the ceaseless attention from various levels of the managers about the computer statistical technology, we realize the information management and

Z. Du (Ed.): Proceedings of the 2012 International Conference of MCSA, AISC 191, pp. 541–545.
springerlink.com

information management modernization and informatization. This can provide better service for the group company management and staff. The common statistical system completes the engineering from top to bottom. Many departments connect to the secondary unit LAN (local area network) in a primary. Moreover, they update the computer software and hardware that can process the teletransmission and data crisis for the second unit. Many statistical statements will report by U disk or E-mail. This can make correct and timely statistical data to be the reference of the decision. Moreover, we can avoid the not timely or high error score of the statistical report.

3 Questions about Computer Application during the Statistical Work

3.1 Office Automation and Information Statistic Backward

The statistical standing book setting, original record, and the statistical report management has not reached the defined standard. Moreover, there has no standard basement for the statistical information that we cannot take a full application of the management about the modern computer technology. The statistical office only realizes the office software to process the characters that cannot reach the information statistic networking and automation.

3.2 Data Processing Has Not Reached the Standard

Effective data processing is very important during the statistical work. The present computer electronization application needs higher improvement. Except the data processing of the professional report, other works are limited in the computer teletransmission and storage that besides the computer programming. Many works still need the manual operation without new technology and leads the low efficiency of part statistical work. At the same time, the computer hardware technology obtains the rapid development with the fast processing. However, the application software falls behind. It cannot form the coordinates development and cannot take full advantage of the information sources and computer technology.

3.3 Defect of Computer Network Management and Technology Level

The rapid development of the network and computer technology will increase more requirements during the statistical work. Compare with the requirement, the knowledge level update of computer technician still under the slow condition. They are lack of professional knowledge training. If do not develop the business and learning, they cannot satisfy the work requirement. Only the statistical technology level improvement can bring the development of other computer workers.

3.4 A Defect in the Network Security Management

With the development and popularity of the computer technical application, material statistic by computer, computer hardware resource, and the network security of the statistic department is the urgent question. Once ignored, many Trojan and malice virus will have a loophole to exploit and break some lost.

4 Question Strategy of Computer Application during the Statistical Work

4.1 Strengthen Computer Technology Training of the Workers

Effectively promote the technical application level of the statistician can determine the informatization establishment. With the network establishment and computer technology perfection, expand the knowledge of statistician and improve the operation ability. The worker can familiar with the computer operation and further understand the self-position in order to achieve learning with a lively mind.

4.2 Improve Network Manager and Computer Technology Training

We need to train the network manager termly for improving the network management and computer application. The present computer technology is under the rapid development, it brings the ceaseless requirement during the statistical work. This needs learning and develops the technology application and managed to satisfy the increasing job specification.

4.3 Strengthen Office Automation and Statistical Informatization

The information automation is very difficult. The system engineering needs to start from quantity. On the one hand, effective data informatization, data processing electronization and the system design networking. On the other hand, expand office automation gradually. No paper office, standardizes the file processing, and save office cost can develop the work efficiency.

4.4 Strengthen Establishment of Network Database and Program Development of Computer Application

Software is the key point of the statistical computer technology. Hardware is the basic. Application is the final target. Take full use of the present information resources and strengthen the software design. At the same time, develop and perfect the processing software. Moreover, strengthen database of the comprehensive statistic, directory basement, and other professional basement, develop the resources, promote the database usage, and increase the computer technology development and utilization.

4.5 Develop Network Security Management

Cultivate the habit of backup data during the practical work. The disc basement can protect data loss. We need to backup the carrier filing management, searching and killing virus, set firewall and avoid illegality hacker and Trajan.

4.6 Ceaseless Promotes Management System of Statistical Work

Ceaseless promotes the management system of statistical work, form standardization, institutionalization, and routinization. Statistician perfects the work processing, positive provide various requirement during the statistical work. At the same time, cooperate with the computer technician and optimize the programming. They need to be patient to the computer technician, positively discuss with the statistician and make the electronization and routinization development.

5 Computer Application during the Statistical Work

5.1 Application in Excel

The present manufacture in our country decides the heavy statistical work base of the productive property. Many years ago, most statistical work finished by manual operation with many errors, low efficiency and difficult checking. However, after applying Excel, the great statistic function expresses the convenience. It can satisfy the enterprise requirement of data summarization, collection, and evaluation. Moreover, it can provide effective materials for the leading decision. The details include three steps. The first, build the basic database module of each unit and input production name, quantity, prickle, unit price, stock, production value and other indexes in each table. For every report, it is the small database. We can storage the table in the computer module under the requirement for the further statistical work. If it is needed, we can directly insert it into the required table and screen the production and specification. Moreover, set formula in each database. After input data, the statistic table can automatically generate. In the second, establish the summary databases in the same workbook, effectively complete the collection and through the data reference of each department, we can collect data based on the classification. At last, we can under the requirement of upper department to build data chart for the outside report. Then, apply the copy table , paste the relative data, calculate based on a formula, and generate statistical report automatically.

5.2 Application of Electronic Chart Formwork

The common statistical work is cyclic. How to decrease the repeat work and improve the work efficiency is the important question to the statistician. The formwork of the electronic chart provides great help about this problem. It can separately divide the statistical work into the unique mode. We can insert the required formwork into the present table under the requirement. After the tiny modification, the work will complete. The formwork application can effectively promote the management and develop the informatization management.

5.3 Statistical Chart Application by Word Automatic Generation

First, open the Word and use Excel to insert the spreadsheet. At the same time, distribute the science editor. Use some advantages such as convenient of the software and solve the questions that cannot settle in Excel. For example, when settling the question of numerator and denominator, we can fill in the numerator and the system can automatically obtain the collected data and denominator. The Excel can use the functional relationships automatically. It can accumulate the complicated formula, change it into simple, and realize the data collection with high quality.

Take fully advantage of the LAN to the computer, we can standardize the database format. Input relative information and data on the website of each department then automatically generate the required report in the statistical comprehensive department. At the same time, we can use paste function in the Excel, edit required information, and process information in the second time. Then there will automatically generate needed statistic report.

6 Summary

apply the computer technology can realize the data automatic self-checking and control in order to avoid the repeated errors. It can save time and effective reach data sharing that brings convenient for the statistics and provide valid scientific evidence for the leading decision. Moreover, it can classify storage the report content and realize the internal data sharing. Effectively promote the work productivity, improve the worker quality, and determine the data quality can speed up data transmission and improve the further company management.

References

[1] Gaoyan, G.: Some experiences about the basic statistical work based on network. Medical Information (1) (2011)

[2] Shi, P., Zhou, R., Chen, L.: Reasonable application discussion of hospital network data during the statistical work. Medical Information (5) (2009)

[3] Xiu, J.: Computer application of traffic transmission management during the statistical work. Modern Economic Information (4) (2010)

[4] Liu, L.: Statistical informationization of transportation management. New Finance Economics (10) (2011)

[5] Zhao, L.: Information and statistic-brief experience discussion of computer application during the statistical work. Sci-Tech Information Development & Economy (12) (2008)

[6] Li, H., Feng, J.: Our statistical work thinking base on HIS. Chinese Society of the 15th National Conference on Medical Information (2009)

[7] State Statistics Bureau. people bring the establishment, grasp the application promote development-informatization survey of the national statistics. 25th Anniversary of the China Computer Users Association General Assembly (2008)

Study on Algorithm Design of Virtual Experiment System

Liu Jianjun and Liu Yanxia

School of Information Science and Engineering, HeBei North University, China

Abstract. The virtual experiment study is becoming a hot issue of network education study and attracts widespread attentions. This paper describes the important position of network virtual experiment system in the education field. It can conform to the requirements of remote education system and get rid of the restraints of traditional laboratory mode, so as to achieve the full sharing of teaching resources. Combined with the actual development work, this paper introduces the system architecture and drawing engine layer of virtual experiment system, the realization of the simulation frame layer and simulation realization layer. With the development of virtual reality technology and artificial intelligence, the virtual experiment system uses more and more applications of these advanced technologies to improve the user experience. And this is the major development trend of virtual experiment system.

Keywords: Virtual experiment system, algorithm design, MVC design.

1 Introduction

In traditional teaching, the experiment is an important part of teaching. Through experiments, we can train students the practical ability, and the ability to resolve the problem in real life, and these abilities have very important significance for these students handling with the society, and these abilities can be achieved simply through the classroom. With the development of computer technology, the simulation technology and virtual technology have made great progress as well, and how to apply these techniques to modern network education in order to solve the virtual experiment problems preventing its development has become a hot study issue. Making experiments in the laboratory, due to the equipment cost and personal safety and other factors, dangerous and destructive experiments are usually not allowed, while the virtual experiments can give full play to the imagination through the software, students can do the experiment which is impossible in the laboratory. With this background and advantage, virtual experiment study becomes a hot issue of network education study, which attracts widespread attention.

2 Overall Architecture of Experiment Platform

The experiment system consists of three functional layers: drawing engine layer, simulation frame layer, simulation realization layer. The drawing engine layer is responsible for the underlying drawing, common interface drawing, underlying information collection, operation of information processing pump, call of information

Z. Du (Ed.): Proceedings of the 2012 International Conference of MCSA, AISC 191, pp. 547–552.
springerlink.com

processing function and the overall initialization and destruction. Simulation frame layer provides a general frame for a variety of the simulation system, the frame is realized based on the engine layer, and this simulation frame layer is responsible for the simulation process operation, its provides the specific realization of upper layer for the engine layer function calls.

2.1 Design of Drawing Engine Layer

This layer provides interface for the upper drawing, which encapsulates the drawing work related to the system and the underlying information collection.

Engine module: This module is mainly used to provide a globally unique several resources, updates the trigger redrawing in every second, modification of various mouse, keyboard, attribute data, display update.

2.2 Design of Simulation Frame Layer

Frame unit module (Engniunit): the frame unit module has two main units, one is the frame unit in experiment area, the other is the frame unit in attribute bar, they almost have the similar functions, so they can be abstracted for a base class, the main function is used to set and record a variety of controllers, graphics modules, mouse and keyboard events, as well as various operating commands.

Controller: This module mainly handles the incidents from the system mouse and keyboard, and calls the interfaces provided by the mouse and keyboard to modify the specific data and drawing information. Each controller can perform one function only, so there are many controllers in the system, such as mobile equipment controller, equipment adding controller, connection controller and so on.

2.3 Design of Simulation Realization Layer

Experiment module: experiment module is the core of the system, it is responsible for the initialization of equipment buffer, initialization of attribute bar and equipment bar, initialization of equipment management class and initialization of algorithm class, and is responsible for recording the most of the current experiment, including various operational information of the experiment, save and read of the experiment script, listening the experiment topology structure, whether the equipment attribute information is changed, and then update the experiment. Update process includes: firstly update the modified equipment attribute, and then update the whole algorithm, and finally update the dial display data. Because the experiment is the only instance of the system, so when designing the module, we use the singleton mode.

Algorithm module: This module is mainly responsible for the implementation of experiment circuit algorithm, topology structure analysis, and the appropriate circuit algorithm for different circuit structure.

3 Design of Virtual Equipment Algorithm Library

Equipment is the core object of a virtual experiment system. The virtual equipment in the virtual experiment system is the direct mapping of real equipment in the objective reality, and it is an important component of the virtual experiment system and also the key of the virtual experiment system design. The equipment library in the digital circuit virtual experiment system is mainly consisted of various required component models in the digital logic experiment, such as various types of door devices, triggers, registers, counters, display equipments. These equipments are the objects composing the virtual equipment algorithm library. During the model design, we use the object-oriented design method, which abstracts the common characteristics of the virtual equipment and becomes Sub Component class. Each virtual equipment class inherits the class and derives their own virtual equipment class. They achieve their algorithms in this class.

We can take the 74LS00 with most simple logical as an example and observe the implementation structure of specific equipment algorithm. The program code is as follows:

```
Class CLS00: public SubComponent
{
public:
CLS00 ();
virtual ~ CLS00 ();

virtual void update (); / / overloaded method
/ / ~ Partial codes has been ignored hereby ~
;} / / ... ...

/ / overloaded method
void CLS00:: update ()
{
        / / Implements the non-logic of 74LS00
```

In the realization process of the equipment library, we pay attention to the complete expression of the actual equipment characteristics. Such as the simulation of digital LED fault code, which makes the students get similar feelings and exactly the same experiment results of the real experiment environment even in the virtual experiment system.

4 Design of Circuit Refresh Algorithm

In the simulation realization layer design of digital circuit experiment system, the design focuses on the logic algorithm of the circuit, and it is related to the result of digital circuit simulation.

In the design, we firstly conducted an abstraction for digital circuit. In the program implementation, we regard each entity set up by the user in the digital circuit as a collection of the node (node). And the equipments in the digital circuit can be seen as different collections of the node; in particular, the circuit in the wire can be seen as two nodes always in homogeneous composition. And then they consist of a larger node collection by the equipment (the node collection). In this way, we simplify the data structure of the circuit by the combination relationship of our design, and also facilitate the processing of circuit algorithm.

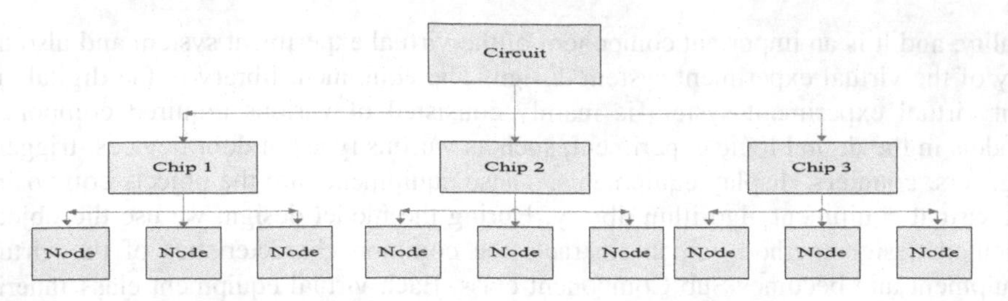

Fig. 1. Data structure hierarchy relationship of the storage circuit

Let us study the common attribute of the nodes. The attribute is summarized in the experiment entity, and it records the dynamic and static state of a node in the circuit. It constitutes the most basic rules of circuit refresh algorithm.

We define three nodes for each attribute:

Node type
Value of node state
Value of node voltage

The node type is the static attribute of node. It is decided by the inherent attribute of the equipment where the node is. The node type can be divided into output type, input type and other type.

The node of output type is usually the output pin of equipment, or the signal pin of power, signal generator, etc.; the node of input type is usually the input pin of equipment; other type is used to define the node with node type can not be determined.

The definitions of the program implementation examples are as follows:

```
enum
{
IN_NODE, / / input node
OUT_NODE, / / output node, such as the power pin, signal generator pin, ground pin
OTHER_NODE / / non input / output node type, such as power or ground terminal;
;} / / ... ...
```

The node type is the first attribute in circuit refresh algorithm, it has a highest priority, that is, when the node types are different, the mutual influence relation of node is only decided by this attribute.

There are two dynamic attributes for the node: value of node state and the value of node voltage.

The value of node state means the node is one of the following types:

Power type: means that this node is directly impacted by the power pin through non-linear equipment. The definition of "never through non-linear equipment" means a direct connection or never passing through a cable and connectivity board.

Pulse type: means that this node is directly impacted by the pulse pin through non-linear equipment.

Electrical type: this is the most widely used node in the circuit. It is the common type without occurrence of other types.

Floating type: the equipment where the node located has been powered, this node is not connected to any equipment with a floating state.

Non-input type: the equipment where the node located has not been powered.

These five types are in the following order (power type> pulse type > electrical type> floating type> non-input type) and they are mutually affected, that is in the same node type, the previous value of node state will change the following value type, but not the reverse.

The definitions of the program implementation examples are as follows:

enum
{
POWER, / / power, the node value can be 0 or 1
PULSE, / / trigger pulse, the node value can be 0, 1
NORMAL, / / with the electrical sense, the node value can be 0, 1, with a lower priority than power, the value of node state when the equipment works
;} / / ... ...
SPARE, / / floating input when the chip normally works, the node value of 1 must
NOINPUT / / no electrical sense value, the node value must be 0, the value of node state when the equipment does not work

There are two values of node voltage: high value of node voltage and low value of node voltage. It is the lowest priority attribute to determine the status of the node.

The definitions of the program implementation examples are as follows:

enum
{
LOW, / / low voltage

High, / / high voltage
;} / / ... ...

5 Summaries

Now, the vast majority of virtual experiment systems are used as a secondary means of teaching the traditional experiment, which can not play a leading role or completely replacement of the real experiment environment. But with the development of virtual reality technology and artificial intelligence, the virtual experiment system uses more and more applications of these advanced technologies to improve the user experience. And this is the major development trend of virtual experiment system.

References

[1] Duan, M.-D., Gao, Z.-B., Ma, W., Li, J.-S.: Application of RE in the Development of Tractor Covering Parts. Tractor & Farm Transporter, 86–88 (2007)

[2] Guo, L.-G., Chen, Y., Liu, X.-W.: Study of the Application of the Reverse Engineering Technology in the Design of the Complex Cavity-Surface Dies & Moulds. Die & Mould Industry 1, 12–17 (2008)

[3] Gamma, E., et al.: Design Mode: the Basis of Reusable Object-oriented Software. Machinery Industry Press (2009); translated by Li, Y., Ma, X.-X., Cai, M., Liu, J.-Z.
[4] Martin, R.C.: Principles, Mode and Practice of Agile Software Development. Tsinghua University Press (2008)
[5] Jacobson, I.: Software Reuse Structure, Process and Organization. Machinery Industry Press (2009); translated by HAN Ke
[6] Virtual Circuit Analysis Experiment. Documentation and Code, Virtual Experiment Program Team, Modern Education Technology Institute, Network Education College, Beijing University of Posts and Telecommunications (February 1, 2009)
[7] Freeman, E., Freeman, E., Bates, B., Sierra, K.: Head First Design Patterns (U.S.). O 'Reilly (November 2005)

The Application of GIS in the Real Estate Management System

Ru Qian

China University of Geosciences, Beijing, China
108300@qq.com

Abstract. Geographic information system (GIS) is the collection, processing, storage, management and analysis of geographic information software system, divided into the GIS platform software, application platform software, GIS and engineering applications of GIS industry development services. Housing space for geographical attributes based on the core functions of the GIS Real Estate Management is the natural attribute information, links to information of social and human attributes. The paper proposes the application of GIS in the real estate management system. GIS-based property management system is to provide information services for real estate administrative department.

Keywords: Geographic information system (GIS), management system, Real Estate Management.

1 Introduction

Towards regionalization and globalization at the same time, the application of GIS has been penetration into all walks of life, involving thousands of families to become production, living, learning and small as an indispensable tool and assistants. At the same time, the GIS from stand-alone, 3 dimensional, closed to open networks (including Web GIS), multi-dimensional direction.

In recent years, GIS technology has made leaps and bounds [1]. The introduction of GIS technology is to improve the traditional property management of the basic needs of the real estate management level basically rely on manual registration, record them, query and archiving way, there is the error rate, data query cumbersome and time-consuming and difficult to preserve and lack of unity norms, the application of modern information technology to real estate management, in order to get rid of the traditional management situation, is to achieve specialization, one of the indispensable measures of modern property management.

Geographic information system (GIS) is the collection, processing, storage, management and analysis of geographic information software system, divided into the GIS platform software, application platform software, GIS and engineering applications of GIS industry development services. Traditional infrastructure software and operating systems, database systems, middleware and office software as the core plays the role of the basic platform of the GIS base platform software in the geographic information industry, the industrial competitiveness of the strategic high ground. As supporting platform software, GIS software of the underlying platform technology

Z. Du (Ed.): Proceedings of the 2012 International Conference of MCSA, AISC 191, pp. 553–558.
springerlink.com

requirements higher, also has a pivotal role in connecting and strong industry pull, can drive down the operating system, database and middleware, and other types of infrastructure software, application development, up e-government can support the development of city normalization construction of public services, mobile applications, the core part of the whole industry, but also the independent intellectual property rights with core competitiveness.

2 The Research of Geographic Information System (GIS)

GIS can not only output the total factor map, according to user needs, hierarchical output a variety of topics map, all kinds of charts, diagrams and data. In addition to the above five features, user interface module, for receiving user instructions, program or data, the user interacting with the system tools, including user interface, programming interface and data interface [2]. Complex function of geographic information systems, and the user is often non-computer professionals, the user interface is an important part of the GIS applications, geographic information systems as open systems of the human-computer interaction.

Dedicated Geographic Information System is a stand-alone GIS operating system, GIS system development toolset and workflow for a particular area of expertise and business development aimed at using GIS tools targeted to address specific problems. It is consistent with areas of expertise or business sector _T workflow, targeted, and GIS products to the professional development of the product t plays an important role in expanding the influence of GIS products, as is shown by equation1.

$$L^k = G^k - 4\omega * [G^{k+1}]_{\uparrow 2} \quad k = 0 \ldots N-1 \tag{1}$$

GIS is an interdisciplinary role of computer and spatial data analysis methods developed in many related disciplines. These areas include surveying, photogrammetry learn, cadastre and land management, topographic mapping and thematic mapping, municipal engineering, geography, soil science, environmental science, urban system planning, utilities, network, remote sensing and image analysis. GIS is developed on the basis of geosciences disciplines, with a database management system (DBMS), computer graphics, computer-aided design (CAD), computer-aided mapping (CAM) and computer technology related disciplines binding. Both connections and differences between the GIS and all disciplines and systems, they should not be confused. Here only in respect of people prone to confusion in understanding the system is presented by one figure as equation 2.

$$\begin{aligned} P^j f(x,y) \in V_j^2 &\Leftrightarrow P^j f(x-2^j k, y) \in V_j^2 \\ P^j f(x,y) \in V_j^2 &\Leftrightarrow P^j f(x, y-2^j k) \in V_j^2 \\ P^j f(x,y) \in V_j^2 &\Leftrightarrow P^j f(x-2^j k, y-2^j k) \in V_j^2 \end{aligned} \tag{2}$$

Geographic Information System has the following three characteristics: (1) the ability of the acquisition, management, analysis, and output a variety of geospatial information; (2) the powerful spatial analysis and multi-factor analysis, and dynamic predictive capability, and can produce high levels of geographic information; (3) the support of computer systems is an important feature of the geographic Information

System, which makes geographic information systems to complex geographical systems for fast, accurate and comprehensive spatial orientation and dynamic analysis, completion of the human itself could not do the work [3]. However, due to the need to establish a link between the professional information and GIS platform, the secondary development, as is shown by equation3.

$$\begin{aligned} &\{2^{-j}\phi_j(x-2^j k_1\,,\, y-2^j k_2) \mid (k_1\,,k_2)\in Z^2\} \\ &=\{2^j \phi(x-2^j k_1)\, \phi(y-2^j k_2) \mid (k_1\,,k_2)\in Z^2\} \end{aligned} \tag{3}$$

GIS and CAD (computer-aided design), computer technology, computer-aided design (CAD) is used for mechanical, architectural, engineering, product design and industrial manufacture of products, it is mainly used to draw a wide range of applications of technical graphics, large to small aircraft to micro-micro arrays, CAD is mainly used to replace or auxiliary engineers have conducted a variety of design work can also be used together with computer-aided mapping (CAM) system for real-time control products plus T, as is shown by figure1.

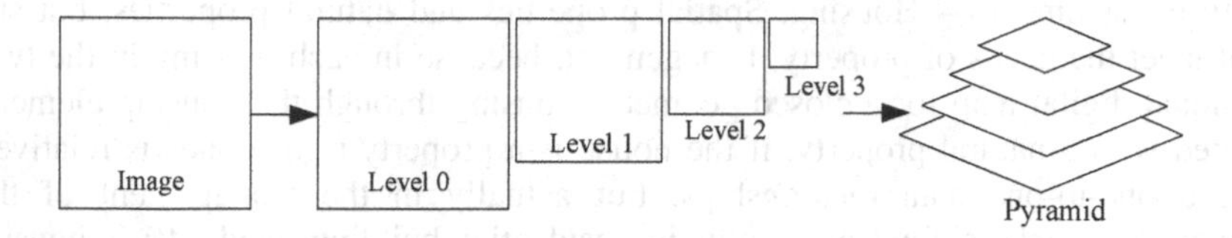

Fig. 1. Development of GIS and CAD (computer-aided design)

The common feature of GIS and CAD systems, both the reference system, can describe the topological relationship of the graphical data, but also can deal with non-graphic attribute data, their main difference is that the CAD processing rules of geometry and its binding strong graphics capabilities, property database feature is relatively weak [4].

3 Application of GIS in the Real Estate Management System

GIS is in Estate Management Information System. Application of GIS technology to meet the needs of real estate management mapping, such as real estate subdivision plan, sub-mound plan, and individual household division floor plan of the production and output of maps. In particular, is to achieve the warrants automatic generation of drawings and graphics Taoda. The application of state of GIS technology to achieve effective management of real estate, historical information, the establishment of the historical evolution of the real estate space and attribute information, and provide historical data back through history, to resolve property disputes, as is shown by equation4.

$$p(s(k)) = \frac{1}{\left(2\pi\sigma^2_{s(k)}\right)^{1/2}} \exp\left[-\frac{\left(s(k)-S_0(k)\right)^2}{2\sigma^2_{s(k)}}\right] \tag{4}$$

GIS of the real estate space information database [5]. Property mapping is complete, the data standard for the CAD drawing software for surveying and mapping results of data quality checks meet GIS requirements, data editing, and editing of CAD format files, spatial data provided through the GIS platform engine function to convert the storage into the GIS database, and to determine the accuracy of GIS set value (precision), the size of the coordinate translation value (, Yshifts) and the spatial grid index (Grid index) to form a real estate geographical information system (GIS) spatial information database.

Spatial data: spatial data is also known as graphical data refers to the target location information data, topological relationships. It is an important part of the Geographic Information System, is the object of systems analysis processing, GIS expression, real-world model abstraction of substantive content. Spatial data sources of various scale topographic maps, thematic maps, data documentation, statistical reporting, remote sensing image, the measurement data, digital photogrammetry data, database.

Based on GIS property management, the core is to create housing space geographic attribute information, the natural properties of information, and links to information of social and human attributes. Digital property maps can vividly reproduce the objects of property management --- Housing, Spatial properties and natural properties, but still can not meet the needs of property management, because in each housing in the two-dimensional digital map for a closed geometry housing through the housing elements associated with a natural property, if the house is a property right issues is relatively simple, a one-to-one data relationships, but actually in the management of this phenomenon is only a few, especially in residential buildings and office housing dozens or even hundreds of property rights simply can not be represented in the digital real estate map, even if the housing elements associated Uk natural attributes and then a large amount of information, but also unable to meet the needs of property management, as is shown by equation5.

$$\begin{aligned} \mathrm{M} = \sigma_{\mathrm{s|a}}^2 &= \operatorname{cov}(s(k) \mid a(k)) \\ &= E\{[s(k) - \mu_{s|a}]^2 \mid a(k)\} \\ &= \beta(k)^{\mathrm{T}} \Sigma_{\varepsilon(k)}^{-1} \beta(k) + \sigma_{s(k)}^{-2} \end{aligned} \tag{5}$$

Space geographic attribute information, the natural properties of a set of property management property management database (ie, housing, social and human attribute information), making housing, social and human attribute information associated with the map control room, in order to pipe cardin order to pipe files, in order to pipe the core of the industry. According to conventional stratified household diagram digital property map on the housing and the housing to create a one-to-one association to establish one-to-one association, and then by stratified household figure of household and property management database [6]. During a large area of real estate mapping and real estate information system for the first time, as long as the human, material, financial and time and other conditions allow, or to carry out the work in the small and medium-sized cities, surveying and mapping a relatively small area, it is entirely possible to achieve these objectives.

However, due to the large amount of stratified household survey and mapping work is on the property previously in mind the hierarchical splitting diagram mapping would involve a strong policy is difficult to synchronize in a large area of real estate mapping.

The case of hierarchical splitting map surveying can not be completed in a timely manner; resulting in lengthy construction period, in accordance with the above method can not even housing geospatial attribute information, the natural properties of information, social and human attribute information correspond to the association. To this end, the association must be based on GIS technology and large databases create a unique encoding, encoding increase the input of the housing, each housing property owners in the property management database, and the formation of much association this can be fully cross-access from the diagram to the file from the file to the map. Thus completing the housing space attribute information, the natural properties of information, links to information of social and human attributes, Figure tube housing and tube card in order to, in order to pipe the file, in order to pipe industry. The system is developed using Java on Java SDK 3.4.5 platform, as is shown by figure2.

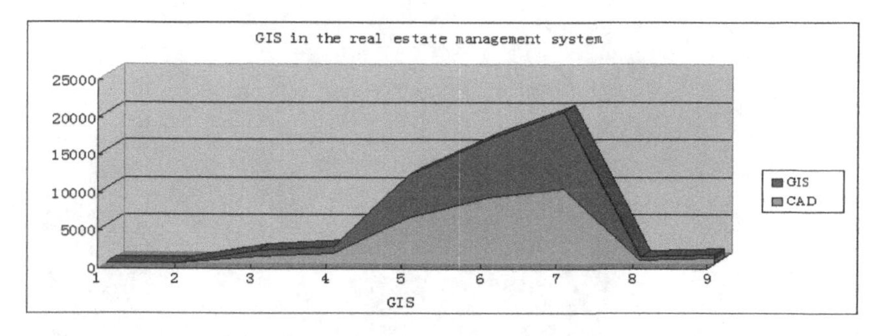

Fig. 2. Application of GIS in the real estate management system

4 Summary

GIS introduction of Real Estate Management is a breakthrough in the traditional real estate management methods, it has changed the traditional real estate management information system software to focus on the attribute information, and to ignore the shortcomings of graphical information marks the real estate management to a more mature, and more standardized way. As long as do the latest graphics and attribute data entry building a database, modify, update the original data, will enable the system to become a convenient query and output forms the new system, in order to achieve the dynamic management of the system, better to provide information services for real estate administrative department.

References

[1] Li, X., Li, S.H.: Research and Development of Virtual 3D Geographic Information Systems Based on Skyline. AISS 4(2), 118–125 (2012)

[2] Wu, Y., Huang, J., Yuan, Y., Cui, W., Zhan, Y.: Simulation of Land Use/Cover Change for Wuhan City Based on GIS and CA. JCIT 7(1), 253–260 (2012)

[3] Jiang, G., Zheng, X., Hu, Y.: Forest health assessment in the Badaling Forest Farm of Beijing based on GIS and RS. JDCTA 6(10), 87–93 (2012)

[4] Li, J.: Study on the Key Techniques of Emergency Logistics Visualization System Based on GIS. JDCTA 5(11), 87–94 (2011)
[5] Zhang, J., Zhu, Y.Q., Wang, J., Sun, J., Xu, Y.: Flash based WebGIS System and its Application in Monitoring and Evaluating China's Regional Development. JDCTA 5(5), 285–295 (2011)
[6] Chu, T.-H., Lin, M.-L., Chang, C.-H., Chen, C.-W.: Developing a Tour Guiding Information System for Tourism Service using Mobile GIS and GPS Techniques. AISS 3(6), 49–58 (2011)
[7] Haldar, R., Mukhopadhyay, D.: An Attempt to Verify Claims of University Departments by Text Mining its Web Pages. JCIT 5(5), 192–202 (2010)

Analysis and Implementation of Computer System of College Students' PE Scores Management

Jianfeng Ma

Sports Department, The South Campus, Xi'an International University, Xi'an City, Shanxi Province, China
majianfeng1@yeah.net

Abstract. Computer system of college students' PE scores management is intended to take the place of teachers in calculating and counting scores of PE courses, and implement information management of scores of PE courses. The system implements PE scores management by B / S mode, this mode solved the drawback that it was essential to enter the specialized laboratories to access the system by C / S mode, it facilitates the storage and query of students' scores. The computer system has completed tedious calculations and statistics of sports scores automatically, achieved information management of sports scoers, alleviated working pressure of the PE teachers, improved the working efficiency of PE teachers.

Keywords: C / S mode, PE scores, computer management.

1 Introduction

At present, the scale and level of sports development are important symbol to measure the progress of a country and a society. College students represent the motion of development of a country's future, so the physical education of university students is taken seriously by the society. Generally, in the university curriculums, PE courses are public and also compulsory. Students of Physical education teachers are from different classes even different grades, the number of them is relatively numerous. Therefore, the workload of calculations and statistics of the physical education curriculum is relatively large, and the management is relatively complex.

In oder to lessen the work intensity of physical education teachers, reduce the workload of them, we take the advantages of the computer, program computer management system to calculate and count the PE scores, complete the tedious calculations by computer, and complete the summary calculation of student scores and other operations automatically. Computer management systems also need to implement dynamic queries, screening, statistics and other operations for sports scores, physical education teachers can have a unified and Intuitive view of sports scores of each grade, each specialty and each sports event.

Z. Du (Ed.): Proceedings of the 2012 International Conference of MCSA, AISC 191, pp. 559–563.
springerlink.com

2 Requirements Analysis of Sports Scores Management System

Sports scores is the core content of sports scores management systems. Sports score is an important part of a student's sporting life, also is an important symbol which measures the physical quality and sports skills of a student. Sports scores management system must be convenient to store and inquire the scores of students, with the informationization of the sports scores management, sports scores management computer system will be an integral part of school network system. To complete the system, first we must make the requirements analysis well, which is the key in the success of a project development.

Questionnaire survey, research, conversations,etc are the general processes of requirements analysis to find out the requirements more clearly , in oder to make the system achieve the desired purpose. The main object-oriented of requirements analysis are students, teachers and administrators and other people who will use the system. After researching deeply, sorting out requirements, analyzing and summarizing, the system producted can be more convenient for teachers and students.

After analysis and research, summing up the four main functional modules, there are sports course management, sports information management, physical education teachers management and student scores management. Different schools have different physical education curriculum requirements and different students situation, according to these factors to expand and select the modules, the system will be constantly improved by analyzing specific issues, the following are the main functional modules:

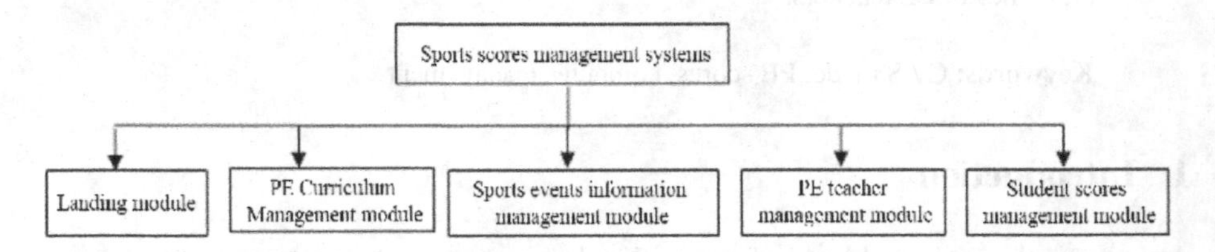

Fig. 1. Modules composition of sports scores management system

The landing module. Not all the staff can access to the system. The main function of the landing module is to keep the security of the system. Depending on the login permissions, landing module personnel are divided into three parts: students, teachers and administrators. Lander students: Query the scores of sports course. Lander physical education teachers: Complete the entry, modify, display and print of students information.

Lander administrators: Input student information, teacher information, add, delete, update and maintain the information of the system, manage the landing personnel, they have the highest administrative privileges.

The Physical Education Curriculum Management module. In this module you can add, modify, delete, print, query the information of students; can add, delete, print the physical education curriculum; can specify change and generate the student roster of some sports class of the semester.

The sports events information management module. This module is divided into the sports events management module and sports scoring table management module. In the sports events management module, you can add, modify, delete, print the list of sports. In sports scoring table management module, you can score on the PE program, generate the scores sheet, and also edit, query, print, the score sheets, etc.

The physical education teachers management module. The module manages the qualification, grade, length of service, good deeds, honor of physical education teachers, you can also list what kind of sports, which courses the teachers have taught and the specific time in that semester and so on.

The student scores management module. This module is main to input, query, modify, calculate the Individual subject score of student, and analyze student scores of the whole school year, and compare it to previous academic year, and print them on a comparison chart.

3 Software Technology Used to Implement the System

In order to make students and teachers access to sports scores management system at anytime, anywhere, the system adopts B / S mode, users can access the system just on internet, that solved the drawback that it's essential to enter the specialized laboratories to access the system by C / S mode. The front-end is developed by jsp, the development tools is Jbuilder, the backstage database is mysql database with the most cost savings. The following are explanation and brief introduction to the software terminology mentioned above.

B / S mode: B / S mode is the abbreviation for the Browser / Server. It is actually an extension of the client / server(C / S) mode. It was born with the rise of the Internet. In this structure, the user interface is achieved by the IE browser. B / S model has the advantage that as long as the network can be accessed you can Login the system, the system maintenance is relatively simple. The drawback of it is slower than the C / S mode, depending on the network environment strongly, if for some reason the network outages, the system by B / S mode can not be accessed.

Jsp: JSP stands for Java Server Pages. JSP is written in JAVA programming language, is similar to the XML tags and script applet scriptlets. JSP packages it and produces processing logic of dynamic web pages. JAVA language program segment can be inserted into the JSP script, to operate the database, redirect,ect. When the web server accessing to the JSP page, the first implementation is the JAVA program segment, and then insert the returned results into the JSP file, return them to the browser for the user to view. JSP technology separates logic and design of web pages, the components are reusable, it can shorten the development cycle to achieve the rapid development of dynamic Web pages.

Tomacat: Tomcat is an open source and free jsp server. The superiorities of it are small system occupation and good scalability, so it is so loved by programming enthusiasts. Tomcat supports load balancing and the development of e-mail services. Its codes are completely open, so programming enthusiasts can add new functions on basis of tomcat, the flexibility is high. Tomcat also has the ability to deal with html page, and is a big container of servlet and jsp.

Mysql database: MySQL was developed by MySQL AB in Sweden, and bought by the Sun later. It is a small relational database management system, is open source and free of charge. MySQL database can save development costs, is the first choice database for many small and medium-sized enterprises. At the same time, MySQL database is small system occupation, fast,ect. The MySQL database is a relational database management systems, it can classify and save the information to the table of different meanings, it improved data accessing speed, increased the flexibility of data access. MySQL supports the SQL language, SQL is the most common standardized language used to access the database.

The framework design diagram is shown below:

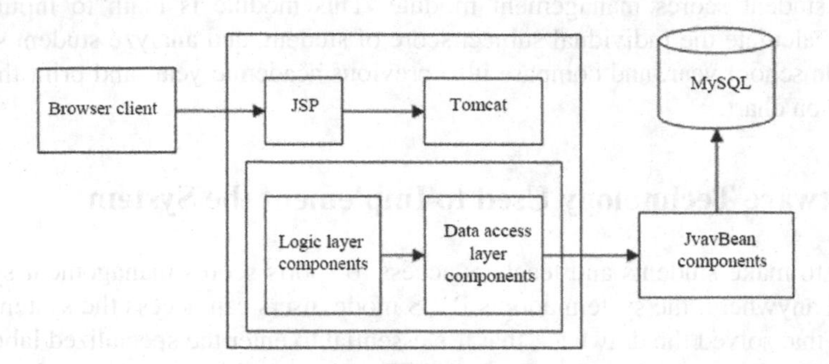

Fig. 2. Framework design

4 Detailed Design of Database

4.1 Table Design

According to requirement analysis, there are several main tables in the mysql database: the table of user login management is user_info, the table of student information management is student_info, the table of PE teacher information management is teacher_info, the table of student score management is score_info, the table of sports course information imanagement is course_info .

User_info table: user name is user_name, password is password. "User name user_name " is primary key.

Table student_info: student ID is sno, name is sname, sex is ssex, age is the sage university Department is sdept, class number is sclass. "student number sno" is primary key.

Tables teacher_info: Teacher ID is tno, Name is tname, gender is tsex, birth date is ttime,years of working is tage, job title is tprofessional_title, courses taught is tallcourse, this semester teaching courses is tnowcourse, honors is tEvent. "Teacher tno" is primary key.

The table score_info: course number is cno, student number is sno, teacher ID is tno ,scores is sgrade. "Course No. cno, student number sno, teachers No. tno" are primary keys.

The table course_info: Course No. is cno, course name is cname, class hour is chour, credits is cNum, teacher is cteacher. "Course No. cno " is primary key.

The table relations: table student_info and table score_info set relationship as outer key by student number sno. The table course_info and table score_info set relationship as outer key by course number cno. Table teacher_info and table score_info set relationship as outer key by teachers No. tno.

4.2 Key SQL Statement

/ / Query all sports teachers Number of female teachers, name, date of birth.
select tno as teacher number, tname as name, ttime as date of birth from teacher_info where tsex = 'women';
/ / Query results ranked the top three names of the students
Select top 3 sname as Name From score_info, student_info
where score_info. sno = student_info. sno
Order by sgrade desc
/ / Insert a record into the student_info table
Insert into student_info (of sno, sname, ssex, the sage, sdept sclass) values (101, 'Zhang Li', 'female', 19, 'freshman', ' class47').
/ /Update cno teacher of 121 courses, changed it to Li Jianguo in course_info table.
the update course_info set cteacher] = 'Li Jianguo,' the where cno = 121
/ / Remove a record of course number 1 , student number 2, teacher number 20 in table score_info
Delete from score_info where cno = 1, sno = 2, tno = 20

5 Summary

In summary, the implemention of management system of the university students' sports scores can relieve the work pressure of the physical education teachers, improve the working efficiency of the physical education teachers, and also can strengthen contacts and cooperation among students, teachers and all levels of educational administraters. At the same time, computer management system of sports scores makes management of the sports results standardized and network gradually. Before the system actually running, it needs to run for a bug test, it can be put into use when the system is stablely running for some time. We can also improve and expand the system based on specific needs.

References

[1] Wang, Q.: Design and implementation of College Sports performance management system. Office Automation (08) (2010)

[2] Cheng, Y., Zheng, D., Zhang, Y., Ken, C., Yongsheng: Development and Implementation of College Physical Education Management System. Higher Education Research (Chengdu) (04) (2007)

[3] Wang, L., Xie, X., Sun, J.: Design of integrated management system of the university PE results. Wuhan University of Technology (07) (2006)

[4] Jiang, C.: MySQL technology insider-InnoDB storage engine. Machinery Industry Press (2011)

[5] Kofler, M.: The Definitive Guide to MySQL5. People's Posts and Telecommunications Press (2006); Translated by Yang, X.-Y.

[6] Sun, W., Li, H.: Development technology Detailed of Tomcat and JSP Web, pp. 1–205. Electronics Industry Press (June, 2003)

The Study Based on IEC61850 Substation Information System Modeling

Yu Xian

School of Electrical and Electronic Engineering North China Electric Power University, Beijing, 102206, China

Abstract. The series of international standards IEC 61850 substation communication networks and systems to provide a basis for interoperability, the establishment of a seamless communications network of the power system. This study is the substation automation technology has entered the network development process, the study of IEC61850 based on the interoperability of substation and the substation communication network performance.

Keywords: IEC 61850, Substation, Information Systems, Modeling.

1 Backgrounds

One of the key technologies of intelligent substation construction of intelligent substation standard information systems and communications platform. A wide variety of communication protocols in the power system, the most widely used in conventional substations are oriented point communication protocol. The so-called oriented, ie, the measured value of the division and state of a switch, a line voltage, current, power alone as an information value. Rely on a variety of data address "point" value to distinguish between different information. These addresses are allocated in advance by remote engineers, haphazard, there is a change is not convenient, the communication rate is low, a waste of channel resources and other shortcomings[1].

With the rapid development of network technology, the development of broadband network, the Ethernet technology and the Internet, and object-oriented technology, widely used, this situation is difficult to adapt to the needs of the development of the situation, it is necessary to adopt a unified communication protocol from the source of information to the dispatch center to establish a unified communication model. The International Electrotechnical Commission TC57 group prepared a series of global unified communications substation standards of IEC 61850.

2 IEC 61850 Communication Protocol

The IEC 61850 standard is divided into 10 parts, consists of 14 standard. Including device model, the substation configuration language Level (SCL Substation Configuration Language), an abstract communication interface (ASCI,

Z. Du (Ed.): Proceedings of the 2012 International Conference of MCSA, AISC 191, pp. 565–570.
springerlink.com

AbstractComunication Service Interface), Specific communication service mapping (SCSM, the Specific CommunicationService Mapping), conformance testing and target requirements for content. IEC 61850 first substation automation system points 3 way.

1) process layer: the interface of a device, such as to open into the out.
2) interval layer: Using data to interact with a device.
3) substation level: the use of a spacer layer control equipment, and communication with the remote control, HMI.

IEC 61850 the IED into the servers and applications. The server contains the logical device, logical device contains a logic node, the logical node contains a data object, data object contains data attributes. Among them, the logical node is the smallest part of the exchange of data capabilities on behalf of the server of a function or a group of device information. Logical nodes can exchange information and perform a specific operation, which consists of data objects, data attributes. IEC 61850 defines 13 categories A total of 88 logical nodes[2].

Agreement compared to the IEC 61850 object-oriented thinking and point-oriented CDT ,60870-101 ,60870-104, etc.

Has the following advantages:

1) object-oriented modeling techniques;
2) Using the hierarchical structure of the distribution system;
3) Achieve interoperability between intelligent electronic devices;
4) Abstract communication service interface ACSI and special communications services to mapping SCSM technology;
5) to provide self-describing data objects and their services;
6) has a future-oriented, open architecture. These advantages make the IEC 61850 in the smart substations standard Information System has the following advantages:
1) unified the substation automation Association

Proposal to solve the compatibility issues of various types of equipment within the substation, suitable for the application of intelligent substation;

2) substation widely adopted

IEC 61850 protocol can reduce the investment costs, shorten the construction period of the substation;

3) secondary devices using the IEC 61850 protocol to further improve the automation and management level;

4) between protection and protection, protection, and monitoring and control, monitoring and control, and monitoring and control, protection and monitoring and control and monitoring system via optical fiber and cable to replace the complex control and interconnected cables, improve the accuracy of construction ;

5) using optical fiber and cable to enhance the protection and the measurement accuracy of the measurement and control devices. Improve the timeliness, accuracy and reliability of signal transmission: 6) on-site commissioning, greatly reducing the workload of system integrators and product suppliers in the actual work site. Shorten the construction period, concentrated in the factory with adequate human and technical resources for the FBI, to reduce the risk of the project. Ensure consistency

and standardization of the various subsystems interconnected, to facilitate later expansion of substation new features and extended scale.

Analysis can be seen from the above advantages, the application of IEC 61850 protocol on the operational features and mechanisms of the substation has a full range of increase, the use of intelligent substation information system modeling. Substation with a single Statute, high-speed efficient, safe and economic operation. The following IEC 61850 object-oriented idea of intelligent substation information system modeling and analysis[3].

3 Based on the IEC61850 Substation Smart Device IED

With the development of microprocessor technology, an increasing number of microprocessor-based intelligent electronic devices (IED) used in the substation automation system. IED is characterized by high-performance computing power and processing speed, the use of a single IED can be finished the functionality of multiple devices in the past, such as protection, measurement, control, fault recorder, etc., thereby reducing equipment space, simplifies wiring, reduce the initial investment and maintenance costs. More important is that the IED communication capabilities, design a network interface, to become a network node in the substation automation system communications network. At this time the IED from different manufacturers or the same manufacturer in different periods of production to interoperability is very important[4]. The object model and object modeling mechanism provided in the IEC 61850 standard, for each IED in the substation automation system to provide a basis for interoperability. This chapter first introduces the definitions and types of logical nodes in IEC 61850; interval division and then a typical 220kV substation, each interval of internal equipment and substation level equipment; through the decomposition of the device capabilities logical nodes reassembled to complete the process layer, the modeling of various types of IED in the bay level and substation level.

3.1 Division of Logic Node

IEC 61850 international standard and the function of the substation standardization, the function of each IED system is not fixed, the distribution of functionality and usability requirements, performance requirements, price factors, the level of technology and communication and information needs and other factors closely related to, the standard is to support and encourage functional free distribution. In order to achieve the purpose of the function of free allocation and to achieve interoperability between the IED, of IEC 61850 substation automation system features the smallest part of the decomposition of the functions of data exchange - Logical node. Each logical node in the standard object model by a number representing the application-specific meaning of data and data exchange. Common data classes, logical node in the data has been defined by the standard data objects and data attributes are standardized; the interaction between the logical nodes, data exchange through standardized communication services to complete, thus to achieve interoperability between the logical nodes, so that interoperability between features and equipment by the free distribution of logical nodes[5].

IEC 61850-7-4 section defines more than 90 kinds of logical nodes, divided into 13 logical groups of nodes. The standard provides for the logical node name should be the standard defined in the logical node, basically covering all of the features of the substation automation system provides the basis for the modeling of IED.

3.2 Substation Logical Point of Modeling

Free distribution of the logical node instance. IED modeling process is divided into two steps, the first step will be split into various functions of the IEC 61850 standard logic node, the second step a combination of logical nodes into a logical device, the new IED. Instance of the newly formed IED logic nodes are standard logical node object model, so that the IED with interoperability. Each logical device must have a logical node 0 (LLN0) and physical devices, logical nodes (LPHD), LLN0 used to describe the logical device itself, has nothing to do with the logical node contains the data it contains, does not involve the other parts.

3.3 The Object of Study Analysis

Construction of substation automation system IED modeling and communication networks and substation type. Substations of different voltage levels, different sizes of the amount of equipment and substation automation system to complete the functions are very different, which led to the communications network, the number of network nodes, network load size difference between the IED modeling choice of network topology, network performance requirements. Therefore, select a simulation analysis of the typical 220kV transmission substation to substation automation system targeted the establishment of communication networks and network performance[6].

3.4 The Role of Process Level

Layer device in the line interval in the process is mainly responsible for the implementation of command-line real-time operating data acquisition and control. Current, voltage, the amount of the acquisition and transmission by the merged cells to complete the implementation of the switching and protection tripping action of the line through the intelligent circuit breaker with a communication interface to. The transformer interval process layer device in addition to completion of the transformer is high, the low pressure on both sides of the data acquisition and switch command execution, but also the implementation of voltage and reactive power control commands, such as switched capacitor, reactor, adjust the transformer tap switch[7].

The bay level equipment is to use the data within this interval to operate a device within this interval. Line spacing and transformer interval interval layer device should protect the equipment in this interval, to achieve automatic reclosing the same period, check, fault range finder and protection-related control functions, lines, transformers, switching, load transfer run operation, complete real-time data of the first systematic measurement and the measurement of the electrical energy. Whether the power voltage regulator transformer interval set, so the transformer within the interval of the bay level equipment but also on the capacitor, reactor protection and control to adjust the transformer tap[8].

4.5 The Impact of the Substation Equipment

Substation level equipment to collect real-time operating data of the station and switch status information, the operator can use these data to man-machine interface to monitor the operating status of the station and make the necessary control, when failure occurs the system will generate automatic alarm signal. Substation level equipment to the station of historical data to archive records, in order to access. Substation level equipment also provides a communication interface for data exchange with the remote control center, either real-time data transmission substation to the control center running, you can also receive the remote control center, remote adjustment commands to operate the equipment.

4 Substation Unified Communications Network Simulation Study

Communication for substation level communication network based on IEC 61850 has been a lot of research, and has been put into operation project examples, this paper is no longer a separate substation level communication network simulation, but the process level within the interval of each line extension process layer communications network within the interval of the network and the transformer is connected with the station level IED. Due to the process layer communications network within the substation level communication network and the interval of a unified network infrastructure technologies - Ethernet, so that the whole substation automation system the IED unified Ethernet data exchange is very convenient, and because the IED of the whole station logical node in the standard unified modeling, each IED interoperability, and achieve seamless communication.

Simulation object substation studied a total of ten intervals, including two line line interval, interval 6 outlet line interval and two transformers; equipment at the substation level monitoring and recording IED and a distant communication: two stations IED. Station level IED station level switch and the process level within the network switch connected to form a unified substation communication network. The network links continue to choose the specification supports 100Base-FX fiber-optic link, to maintain consistency with the process layer of network technology. OPNET provides a multi-layer subnet nested network topology. The process layer network within the interval of each line and transformer interval can be in the subnet, and then each interval subnets station level switch and the substation level IED linked to form a unified communications network of the whole station. This network topology makes the network structure is simple, clear, but the process of network simulation and simulation results are not affected.

5 Conclusion

First defined in IEC 61850 logical nodes are introduced, and then in a typical substation on the 220kV interval divided based on the analysis of the substation automation system features, and finally the use of object-oriented modeling, by the function decomposition, has been the object of the IEC 61850 logical node and re-combined to form the logical node instance the process layer, the bay level and substation level in various types of IED, IED has the interoperability of these standardized.

References

[1] Brand, K.P., Ostertag, M., Wimmer, W.: Safety Related, Distributed Functions in Substations and the Standard IEC 61850. In: IEEE Bologna PowerTech Conference, Bologna, Italy, pp. 315–319 (June 2003)
[2] Moldovansky, A.: Utilization of Modern Switching Technology in Ethernet/IP Networks. In: Proceedings of 1st Workshop on Real-Time LANs in the Internet Age, pp. 25–27 (June 2002)
[3] Montgonery, S., Neagle, B.: Connectivity:The Last Frontier:The Industrial Ethernet Book (Summer 2001)
[4] Quinn, L. B., Richard, G.R.: Fast Ethernet. John Wiley&Sons (1998)
[5] IEEE802.1p, Standard for Local and Metropolitan Area Networks Supplement to Media Access Control(MAC)Bridges:Traffic Class Expediting and Dynamic Multicast Filtering (1998)
[6] IEEE802.1Q, Standards for Local and Metropolitan Area Networks: Virtual Bridged Local Area Networks (1998)
[7] Thomesse, J.P.: The Fieldbus. In: Proceedings of the International Symposium on The Intelligent Components and Instruments for Control Application, pp. 13–23 (1997)
[8] Glanzer, D.A.: Interoperable Fieldbus Devices:A Technical Overview. ISA Transaction 34(2), 147–151 (1996)

Axis Transformation and Exponential Map Based on Human Motion Tracking

Yu Xue

College of Physical Education, Yan'an University, China

Abstract. Human motion tracking is a special important research domain in the field of computer vision. It is widely used in security monitoring, video compression, video editing, sports training and so on. Therefore, this paper utilizes the parameters of exponential map to express all parts of the body posture. It mainly consists of the following parts: establish the objective function based on the video image, seek to reach the minimum parameters by iteration method and utilize the minimum parameters to obtain the place of all parts of the body posture in the current frames. The experimental results indicate the presented algorithm can track human motion effectively and accurately.

Keywords: human motion tracking, Axis Transformation, Exponential Map.

1 Introduction

The detection, tracking and analysis of motion objects are important fields of computer vision. They are widely applied in security monitoring, video compression, video edit, sport training and many other fields [1]. This article conducted a research in depth on two important points of human motion analysis which are motion tracking.

2 Exponential Map in 3D Space

The posture of rigid body in three dimensional can be showed by rotation and translation in the camera coordinate [2]. We can use the homogeneous coordinates to show the posture of rigid body in three dimensional and can further be represented as:

$$q_c = G \cdot q_0 \text{ with } G = \begin{bmatrix} r_{1,1} & r_{1,2} & r_{1,3} & d_x \\ r_{2,1} & r_{2,2} & r_{2,3} & d_y \\ r_{3,1} & r_{3,2} & r_{3,3} & d_z \\ 0 & 0 & 0 & 1 \end{bmatrix} \quad (1)$$

where $q_0 = [x_0, y_0, z_0, 1]^T$ and $q_c = [x_c, y_c, z_c, 1]^T$ means respectively that the point in the object coordinate system and camera coordinate system. The coordinator of the point in the image coordinate is $[x_{im}, y_{im}]^T = s \cdot [x_c, y_c]^T$ when the scaling factor s in

Z. Du (Ed.): Proceedings of the 2012 International Conference of MCSA, AISC 191, pp. 571–575.
springerlink.com

the orthogonal projection. We can see that the rotation matrix in the G only has three degrees of freedom and all of G only has six degrees of freedom [3].

If we utilize the axis transformation as $\xi=\left[v_1,v_2,v_3,w_x,w_y,w_z\right]^T$ to express the above transformation, we can see the matrix of G can be represented as:

$$G=e^{\xi}=1+\xi+\frac{\xi^2}{2!}+\frac{\xi^3}{3!}+... \tag{2}$$

So, the attitude transform of the objects can be expressed as $[s,\xi]$, the transformation that from point in the object space to point in the image space can be represented as:

$$\begin{pmatrix} x_{im} \\ y_{im} \end{pmatrix}=\begin{pmatrix} 1 & 0 & 0 & 0 \\ 0 & 1 & 0 & 0 \end{pmatrix}\cdot s\cdot e^{\xi}\cdot q_0 \tag{3}$$

The motion vectors of the point (x_{im},y_{im}) from t frame to t+1 frame in the image sequence can be expressed as:

$$\begin{pmatrix} u_x \\ u_y \end{pmatrix}=\begin{pmatrix} x_{im}(t+1)-x_{im}(t) \\ y_{im}(t+1)-y_{im}(t) \end{pmatrix}$$

$$=\begin{pmatrix} 1 & 0 & 0 & 0 \\ 0 & 1 & 0 & 0 \end{pmatrix}\cdot(s(t+1)\cdot e^{\xi(t+1)}\cdot q_0-s(t)\cdot e^{\xi(t)}\cdot q_0)$$

$$=\begin{pmatrix} 1 & 0 & 0 & 0 \\ 0 & 1 & 0 & 0 \end{pmatrix}\cdot((1+s')\cdot e^{\xi'}-1)\cdot s(t)\cdot q_c \tag{4}$$

Where, the last step might utilize the formula as follow:

$$\xi(t+1)=\xi(t)+\xi'\ ,\ s(t+1)=s(t)\cdot(1+s') \tag{5}$$

Utilizing the first order Talor expansions, we can obtain a new formula as follows:

$$(1+s')\cdot e^{\xi}\approx(1+s')\cdot 1+(1+s')\cdot\xi \tag{6}$$

Commanding:

$$w(t+1)=w(t)+\frac{1}{1+s'}\cdot w' \tag{7}$$

$$v(t+1)=v(t)+\frac{1}{1+s'}\cdot v' \tag{8}$$

Then the formula (4) can be written as

$$\begin{pmatrix} u_x \\ u_y \end{pmatrix}=\begin{pmatrix} s' & -w_x' & w_y' & v_1' \\ w_x' & s' & -w_x' & v_2' \end{pmatrix}\cdot q_c \tag{9}$$

Commanding $\phi = \left[s', v_1', v_2', w_x', w_y', w_z' \right]$, and putting it into the optical flow equation

$$I_t(x, y) + \left[I_x(x, y), I_y(x, y) \right] \cdot u(x, y, \phi) = 0 \tag{10}$$

We can see

$$I_t + I_x \cdot \left[s', -w_x', w_y', v_1' \right] \cdot q_c + I_y \cdot \left[w_x', s', -w_x', v_2' \right] \cdot q_c = 0 \Leftrightarrow I_t(i) + H_i \cdot \phi = 0 \tag{11}$$

where $I_t(i) = I_t(x_i, y_i)$, $I_x = I_x(x_i, y_i)$, $I_y = I_y(x_i, y_i)$, when there is N image points, we can obtain above N equations, then they may be unified as:

$$H \cdot \phi + I_t = 0 \tag{12}$$

Obtaining the least squares solution of above equation, we can determine the relative motion o the image block.

3 Kinetic Chain and Human Motion Tracking

Above for an object, we derive from the shaft transformation to the formula of motion state. If the end of the object connects another sub-object, we can consider the motion of the sub-object as an axis of rotation that rounds the space of the parent object [4]. We commend the direction of axis as w_1, there is a point q_1 in this axis and it corresponds an principal axis transformation as $\xi_1 = \left[(-w_1 \cdot q_1)^T, w_1^T \right]^T$, hence, the axis rotation angle θ can be represented as: $g_1(\theta_1) = e^{\xi_1 \cdot \theta_1}$ .So, the mapping that by the sub-object space to camera space can be written as: $g(\theta_1) = G \cdot g_1$, $q_c = g(\theta_1) \cdot q_0$.

Hence, when the length of kinematic chain is k, the transformation that from the object of K to the camera space can be represented as [5]:

$$g_k(\theta_1, \theta_2, ..., \theta_k) = G \cdot e^{\xi_1 \theta_1} \cdot e^{\xi_2 \theta_2} \cdot ... \cdot e^{\xi_k \theta_k} \tag{13}$$

Kinetic chain consisting of M objects, above equation can be written uniformly as:

$$\left[H, J \right] \cdot \Phi + I_t = 0 \tag{14}$$

Where,

$$J_{ik} = \begin{cases} \left[I_x, I_y \right] \cdot \begin{pmatrix} 1 & 0 & 0 & 0 \\ 0 & 1 & 0 & 0 \end{pmatrix} \cdot \xi_k \cdot q_c \\ 0, \textit{pixel i is not affected by } \xi_k \end{cases} \tag{15}$$

The equation and least squares solution is:

$$\Phi = -\left([H,J]^T \cdot [H,J]\right)^{-1} \cdot [H,J]^T \cdot I_t \tag{16}$$

This way, we can obtain all parts of the body relative motion parameters between before and after the frame.

4 Implementation

Based on the above method and under different illumination and complex background, we have made a lot of tests on various positions of person. To prove that the system has preferable universal, the system uses the same tracking parameters during the experiment. Figure 1 is a set of tracking results; the import is a resolution of 640*480 image sequence. From the experimental results can be seen that the algorithm can effectively track human motion.

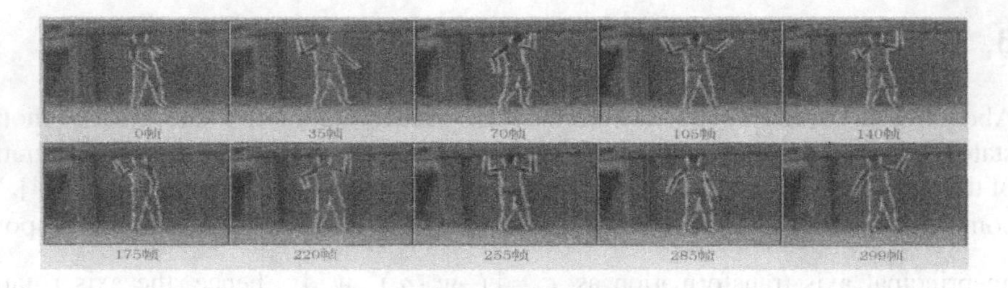

Fig. 1. A set of tracking results

5 Conclusion

Through the axis transformation and exponential map, the paper puts them into the tracking human motion. The experiment result indicates the presented algorithm can track the human motion effectively and automatable. Besides these, the algorithm what we have presented can make the initial contour as close as possible to the body contour and enhance the robustness of the entire system. There are no special requirements on the light environment, the background environment and the dress of the tracked objects etc. And the algorithm can be widely applied to other fields of human-computer interaction, virtual reality, animation and modeling and so on.

References

[1] Hogg, D.: Model-based vision: a program to see a walking person. Image and Vision Computer 1(1), 5–20 (1983)

[2] Aggarwal, J.K., Cai, Q.: Human Motion Analysis: A Review. Computer Vision and Image Understanding 73(3), 428–440 (1999)

[3] Kakadiaris, I.A., Metaxas, D.: 3D Human Body Model Acquisition from Multiple Views. In: Proceeding of the Fifth International Conference on Computer Vision (1995)
[4] Wren, C., Dynaman, P.A.: A Recursive model of human motion. MIT Media Lab, Cambridge (1998)
[5] Bbrno, M.G.: Bayesian S. Methods for multi aspect target tracking in image sequences. IEEE Transactions on Signal Processing 52(7), 1848–1861 (2004)

Research on Characteristics of Workload and User Behaviors in P2P File-Sharing System

Baogang Chen[1] and Jinlong Hu[2]

[1] College of Information and Management Science, Henan Agriculture University, Zhengzhou, China, 450002
bgchen_henau@163.com
[2] Communication and Computer Network Laboratory of Guangdong Province, South China University of Technology, Guangzhou, China, 510641
jlhu@scut.edu.cn

Abstract. Nowadays, Characteristics and user behavior of P2P file sharing system attract more attention. In this paper, we investigated system characteristics and user behavior in P2P file sharing system by using user log data of Maze system. We firstly confirmed that the users' request and the workload in P2P file sharing system have strong time characteristic. And then we analyzed composition of workload in P2P system. Finally we pointed out that different users have different roles to contribute the workload and summarized the characteristics of these users online time in P2P system.

Keywords: P2P file sharing system, MAZE, Workload, User behavior.

1 Introduction

Recently, with user scale and network traffic continually growth, P2P file sharing system has become one of mainstream applications on the Internet. P2P file sharing system is a self-organizing, autonomous system, users' involvement and behavioral have a great impact on system performance. With user number of P2P file sharing system increasing and its influence expanding, characteristics and user behavior of P2P file sharing system has been attracting more and more attention, and there have been great growth for research on P2P file sharing system.

In this paper, we analyzed Maze system load and users' behaviors through Maze system user log. We first examined characteristics of Maze file-sharing system for changes and composition of traffic load, and then analyzed the traffic characteristics for different types of user behavior as well as characteristics of active time, and finally summarized the main contribution of the paper.

2 P2P Measurements

Because P2P file sharing system is large-scale distributed collaboration network based on Internet, so the problem become very difficult that measurement and analysis of P2P

Z. Du (Ed.): Proceedings of the 2012 International Conference of MCSA, AISC 191, pp. 577–582.
springerlink.com

file sharing systems are involved in data acquisition and information storage. Many of the current research [1-3] modify P2P file-sharing client software to increase measurement functions that detect nodes operation and obtain some of the system situation. This approach helps to understand partial information and local operation situation, but it can not capture fully operational information, and at the same time bring additional burden to the system. In addition, some studies obtain network traffic by passively measuring, and decode the traffic by application layer network traffic analysis [4]. However, there are still problems that the approach can not obtain comprehensive information and it's not accurate.

We analyzed the operating situation and user behavior of P2P file-sharing system based on recently obtained seven consecutive days users log of Maze system. When Eytan Adar[5] assessed the Gnutella system, he found "free rider" users abound, where a small number of users share most of files. The result of data analysis shows that 70% of the Maze system users never upload files, we name this part of users "free rider" user;10% of users upload but not download files, called as "server" user, 20% of users both upload and download files, called as "servent" user.

3 Characteristics of Load Variation

Figure 1 shows the time characteristics changes in traffic loads, the load will reach a minimum in the morning (4:00-8:00), and usually achieve at the highest value about 12:00 or 0:00. This shows that the traffic load changes with a clear "day mode", which is related with the users' habit to use network. In order to accurately understand the system traffic load changes periodicity, the measurement data is handled by using FFT to study time series in the time domain characteristics, and analyzes its main cycle components and impact. Using fourier-transform can realize the time sequence in time domain and frequency domain transform. If Xt is the system load observed in P2P file sharing system within an hour, the system load can be used time series (X0, ... Xt, ... , XN) to describe. Supposed total observation time N = 168 hour is set as the cycle for system traffic load, that is, Xt = Xt + N, then there are corresponding N values (X0, ... Xk, ..., XN) of fourier transform, in which Xk is defined as:

$$X_k = \frac{1}{N}\sum_{t=0}^{N-1} x_t e^{-ik2\pi/N} \qquad k=0,\cdots,N-1 \tag{1}$$

Where Xt can be expressed as

$$X_t = \sum_{k=0}^{N-1} X(k) e^{ik2\pi/N} \qquad t=0,\cdots,N-1 \tag{2}$$

Xk correspond K harmonic that cycle is Tk = N / K. Amplitude of K harmonic At= | Xk | , and power spectral density (PSD) Pk= | Xk | ^{2}.

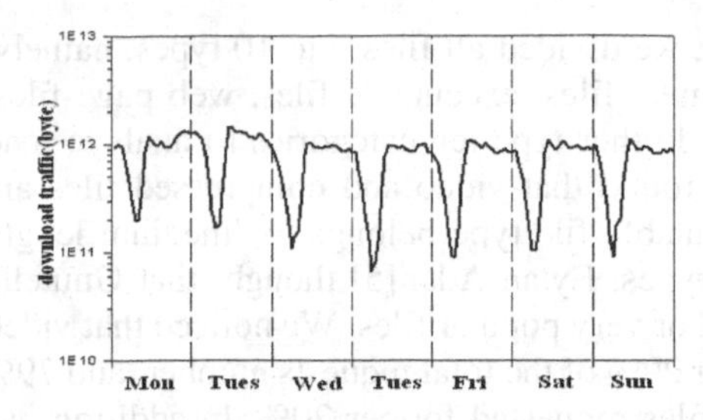

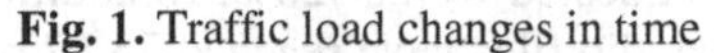

Fig. 1. Traffic load changes in time

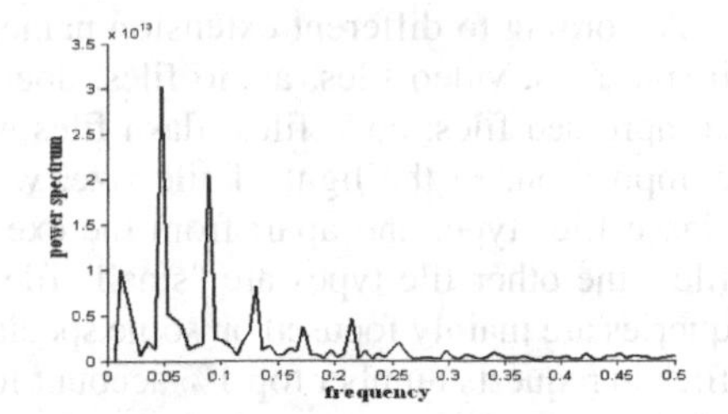

Fig. 2. Power spectral density of traffic

The results shown in Figure 2, the system traffic load have a significant peak at a frequency f = 0.0417, i.e. show a strong periodicity at the time T = 24 hour. In addition, the system traffic load also have the distinct peaks at frequency f = 0.0833, i.e. show a certain periodicity at time T = 12 hour. Graphics and calculations for changes of download requests number and the number of users putting forward download request also show the result similar to the situation in system traffic load.

Krishna P. Gummadi et al. [4] thought load types are hybrid of various loads in P2P file sharing system. He classified the files in Kazaa system as "small file" (size less than 10MB), "medium-length file" (size from 10MB to 100MB), "large files" (size longer than 100MB) to examine the load source. In his analysis, he found that users requests of "small file" occupy 91% of all files requests, but the traffic generated by "small files" downloading is less than 20%, while the number of requests from "large files" less than 10%, but its download traffic has reached 65%.Thus, he believed that if our concern is bandwidth consumption, we need to focus on the small number of requests to large objects. However, if our concern is to improve the overall user experience, then we must focus on the majority of requests for small files, despite their relatively small bandwidth demand. By analysis of users' log data, we found that the proportion of "small files" is not high in all requested files, less than 65%, and only 2.4% download traffic; while "large files" has 21.2% of all requested files, and its traffic accounts for 80.2% of the entire system traffic, as indicated in Figure 3.It can clearly be seen in Maze system, most of the traffic is generated by download larger files, and compared to Krishna P. Gummadi' conclusion, the request number and traffic of "big file" have higher proportion, and "small files" seems to have small contribution on requests number and download traffic.

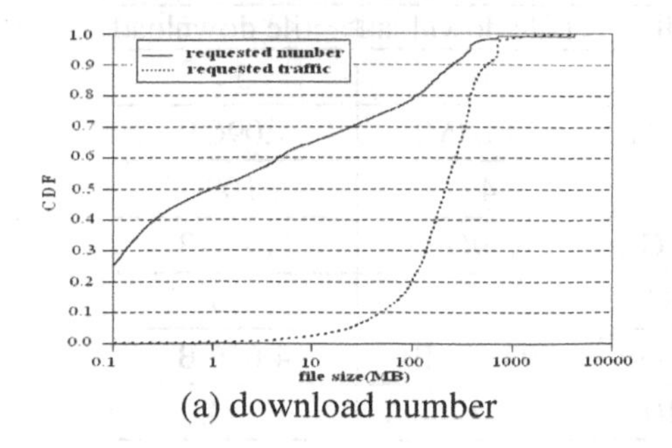

(a) download number

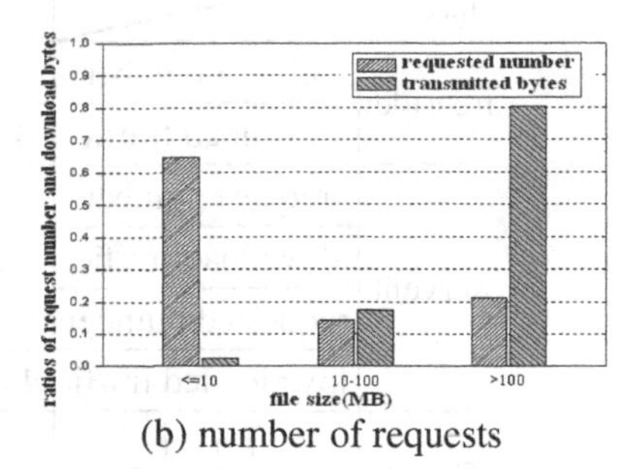

(b) number of requests

Fig. 3. Analysis of file length and load in Maze file-sharing system

According to different extension names, we divided all files into 10 types, namely, image files, video files, audio files, document files, executable files, web page files, compressed files, code files, flash files and other type ten categories to analyze load composition. In the light of file size, we found that video and compressed files are "large file" type, and apart from the executable file type belongs to "medium length file", the other file types are "small" file types. Eytan Adar [5] thought that Gnutella queries are mainly focused on some special or very popular files. We noticed that video files in requests number top 1% account for 29% of the total requests number, and 79% of the video file requests are focused on files requested former 20%. In addition, we also saw that users request to image, video and audio files have reached 75.87 percent ratio of the total requests number. So the traffic in Maze system is mainly composed by video files, while image, video and audio files as well as other multimedia files are the main download request sources of the system users.

4 User Behavior Characteristics

We observed that 45% of users produce upload traffic greater than 1GB in "servent" users. Upload traffic from these 45% "servent" users account for 97% of "servent" user upload traffic, 78% of total upload traffic in system; download traffic from above mentioned "servent" account for 60% of "servent" user download traffic, 22% of total download traffic, and in which 55% users download traffic is more than 1GB. 29% of "server" users upload traffic more than 1GB, accounting for 93% of the total "server" user upload traffic. "free rider" users have 27% users which download traffic more than 1GB, but these users account for 79% download traffic of "free rider" users, 50% of all download traffic.

User request number is download request number that one user put forward download requests to the other users, and the user requested number is the number that the other users request to this user. As table 2 shows, if one user can download more than one whole file, then its request times significantly more than the users never to download one intact file. Similarly, the requested number of users that downloaded more than one intact file is more than that of users never downloaded one intact file in a request.

Table 1. Behavior information of different types users'

class \ behavior		mean		
		all	<1file download	>=file download
freerider	request number	44	19	54
	download traffic	945.2MB	534.67MB	1.09GB
servent	request number	85	47	109
	download traffic	1.81GB	1.39GB	2.06GB
	requested number	184	9	211
	downloaded traffic	4.01GB	193.69MB	4.61GB
server	server	110	7	144
	servent	2.07GB	130.23MB	2.71GB

In P2P file sharing system, the active status means that users upload or download files when connect to the system. Because it's difficult to accurately grasp users' actual online situation [4], even though we can accurately describe its online situation, so studying the user's active status can accurately understand the user's transmission features and functions.

Table 2. Time length characteristics of one single session

Class \ Data		Mean(second)	
		All	≥1file transmit
session time	server	630	632
	free-rider	633	630
	servent	649	645
transmit time	server	157	156
	free-rider	188	162
	servent	182	179

Table 3. User activity time and transfer time characteristics

Class \ Data		mean(second)		
		all	≥1file transmit	0 file transmit
active time	server	16288	20776	1506
	free-rider	3851	4116	3309
	servent	20370	24529	5833
transmit time	server	16099	20538	1473
	free-rider	3788	4046	3252
	servent	18638	22030	5738

Define a session as the time period that between user puts forward download request and finishes download. Transmit time is defined as the duration that user start transmitting traffic and finish download in session. Firstly, we examine the time length of user session. From Table 2 shows that average session is not long, and several types of users' session time and average transmitting time are very similar.We can see from the Table 3 that "server" users and "servent" users have the longer average active time. Among the "server" and "servent" users, if the users can successfully transfer one whole file, they have more active time than that of the users that can not successfully transfer one whole file. And these users may be long-term online users, therefore those interesting users will search and download the shared directories of these users to find other interesting files, so will propose more download requests, which increases their active time.

5 Conclusion

In this paper, we confirmed that user requests and load transfer have time characteristics in P2P file-sharing system. We also observed that the main traffic load in Maze system is concentrated in a small number of video format files. And some "servent" users upload and download much traffic, which plays a major role to guarantee the normal running in system. In addition, activity time, upload traffic, download traffic, request number or requested number of users that once transmit the whole file content in a download request higher than that of users never done that.

References

[1] Satoshi, O., Konosuke, K.: A Study on Traffic Characteristics Evaluation for a Pure P2P Application. In: 16th Euromicro Conference on Parallel, Distributed and Network-Based Processing (PDP 2008), Toulouse, France, pp. 483–490 (2008)

[2] Vu, T.V., Kensuke, F., Nguyen, C.H., et al.: A study on P2P traffic characteristic evaluation in the WIDE backbone. In: Proceeding of The 6th International Conference on Information Technology and Applications (ICITA 2009), Hanoi, Vietnam, pp. 286–291 (2009)

[3] Han, Y., Park, H.: Distinctive Traffic Characteristics of Pure and Game P2P Applications. In: Proc. of 10th International Conference on Advanced Communication Technology, Phoenix Park, Gangwon-Do, Korea (South), pp. 405–408 (2008)

[4] Gummadi, K.P., Dunn, R.J., Saroiu, S., et al.: Measurement, modeling, and analysis of a peer-to-peer file-sharing workload. In: Proceedings of the 19th ACM Symposium on Operating Systems Principles, vol. 37(5), pp. 314–329. Association for Computing Machinery, New York (2003)

[5] Adar, E., Huberman, B.A.: FreeRiding on Gnutella, http://www.hpl.hp.com/research/idl/papers/gnutella/gnutella.pdf

Steady-State Value Computing Method to Improve Power System State Estimation Calculation Precision

Fang Liu and Hai Bao

North China Electric Power University, Beijing, 102206
liufangpaper@163.com, 512466314@qq.com

Abstract. With the respect to the inaccurate and divergence results shown in practical operation, the current researches focus on bad data and synchronous precision. Considering the precondition of state estimation, a new computing method was proposed for steady-state value calculation. The power system online measured data based on SCADA is analyzed through mathematical modeling. The theoretical analysis is simulated by using MATLAB/SIMULINK software, and it can provide an effective way to improve the precision of state estimation in practical operation.

Keywords: Power system, State Estimation, Calculation Precision, Steady-state Value.

1 Introduction

State Estimation is the basis of power system advanced application software such as load forecasting, online power flow calculating, etc. its main function is estimate the current operating status of power system according to online measured values. In addition, the function of power system state estimation also includes: distinguish and reserve bad data, improve data precision, estimate complete and accurate electrical values, correct the wrong switch state information according to remote measured data, process certain suspicious of unknown parameters as state variables, predict future tendency and possible state may occur according to present data, etc.

It is very important for state estimation to establish a reliable and accurate power grid database. Current researches of improving the precision of power system state estimation are mainly concentrated on the processing of bad data [1,2], but ignore the steady state electric parameters are the premise of state estimation. However, during the power grid operation, amplitude fluctuation of electric parameters caused by transient process and harmonics is far greater than the error of data itself. Therefore, processing power system online measured data is important and necessary.

2 On-Line Measured Electric Parameters Analyze

At present, the database of state estimation of power system is mainly built on SCADA system. According to related grid codes and industrial standards[3~5], power system

Z. Du (Ed.): Proceedings of the 2012 International Conference of MCSA, AISC 191, pp. 583–588.
springerlink.com

SCADA measures electric parameters of power grid at and above 35kV voltage level by using AC sampling method with the sampling frequency for at least 32 points/period. Calculation of data processing adopts effective value of electric parameters for each period, and then sends back the average of effective values every 3 seconds.

Effective value, also known as mean square root (RMS) value. If a AC current (or voltage) i (or u), effect on a resistance, the heat generated during the period of time equals to a steady current (or voltage), this value is called effective value, presented by I, namely:

$$I = \sqrt{\frac{1}{T}\int_0^T i^2 dt} \tag{1}$$

Discrete effective value also expressed as:

$$\begin{cases} U_{rms} = \sqrt{\frac{1}{N\cdot n}\sum_{i=1}^{N\cdot n} u_i^2}, I_{rms} = \sqrt{\frac{1}{N\cdot n}\sum_{j=1}^{N\cdot n} i_j^2} \\ \tilde{S}_{rms} = U_{rms}\cdot I_{rms} = \sqrt{\frac{1}{N\cdot n}\sum_{i=1}^{N\cdot n} u_i^2}\cdot\sqrt{\frac{1}{N\cdot n}\sum_{j=1}^{N\cdot n} i_j^2} = \frac{1}{N\cdot n}\cdot\sqrt{\sum_{i=1}^{N\cdot n}\sum_{j=1}^{N\cdot n} u_i^2\cdot i_j^2} \end{cases} \tag{2}$$

Where, U_{rms} is effective value of voltage, I_{rms} is effective value of current, S_{rms} is effective value of electric power, N is AC sampling points in each period, n is sampling periods. Remote electric parameters from SCADA system:

$$\begin{cases} U_r = \frac{1}{n}\sum_{i=1}^{n}\sqrt{\frac{1}{N}\sum_{k=1}^{N} u_{ik}^2}, I_r = \frac{1}{n}\sum_{j=1}^{n}\sqrt{\frac{1}{N}\sum_{l=1}^{N} i_{jl}^2} \\ \tilde{S}_r = U_r\cdot I_r = \frac{1}{n^2}\left(\sum_{i=1}^{n}\sqrt{\frac{1}{N}\sum_{k=1}^{N} u_{ik}^2}\right)\cdot\left(\sum_{j=1}^{n}\sqrt{\frac{1}{N}\sum_{l=1}^{N} i_{jl}^2}\right) = \frac{1}{N\cdot n^2}\cdot\sum_{i=1}^{n}\sum_{j=1}^{n}\sqrt{\sum_{k=1}^{N}\sum_{l=1}^{N} u_{ik}^2\cdot i_{jl}^2} \end{cases} \tag{3}$$

Where, U_r is remote value of voltage, I_r is remote value of current, S_r is remote value of electric power. Take electric power for example, error σ between on-line electric parameters and data from SCADA is present as:

$$\begin{aligned} \sigma = \left|\tilde{S}_r - \tilde{S}_{rms}\right| &= \left|\frac{1}{n^2}\left(\sum_{i=1}^{n}\sqrt{\frac{1}{N}\sum_{k=1}^{N} u_{ik}^2}\right)\left(\sum_{j=1}^{n}\sqrt{\frac{1}{N}\sum_{l=1}^{N} i_{jl}^2}\right) - \sqrt{\frac{1}{N\cdot n}\sum_{i=1}^{N\cdot n} u_i^2}\cdot\sqrt{\frac{1}{N\cdot n}\sum_{j=1}^{N\cdot n} i_j^2}\right| \\ &= \left|\frac{1}{N\cdot n^2}\cdot\sum_{i=1}^{n}\sum_{j=1}^{n}\sqrt{\sum_{k=1}^{N}\sum_{l=1}^{N} u_{ik}^2\cdot i_{jl}^2} - \frac{1}{N\cdot n}\cdot\sqrt{\sum_{i=1}^{N\cdot n}\sum_{j=1}^{N\cdot n} u_i^2\cdot i_j^2}\right| = \left|\frac{1}{N\cdot n}\cdot\left(\frac{1}{n}\sum_{i=1}^{n}\sum_{j=1}^{n}\sqrt{\sum_{k=1}^{N}\sum_{l=1}^{N} u_{ik}^2\cdot i_{jl}^2} - \sqrt{\sum_{i=1}^{N\cdot n}\sum_{j=1}^{N\cdot n} u_i^2\cdot i_j^2}\right)\right| \geq 0 \end{aligned} \tag{4}$$

Namely, when and only when the voltage and current during sampling time maintained as steady voltage and current, the error is zero. However, electric parameters seldom maintained as steady value in practical power system due to transient process and harmonic. Therefore it is necessary to process on-line electric parameters to meet the requirement of input data.

3 Mathematical Modeling

During grid fault or small disturbance in power system operation, electric parameters waveform is surge attenuate sine waves. In data processing, the attenuation envelope and oscillation function has the same characteristics. Therefore, oscillation attenuation envelope is adopted in analyzing transient response in under-damped system, the envelope equation:

$$y_{1,2}(t) = 1 \pm e^{\frac{-\xi w_n t}{\sqrt{1-\xi^2}}} \tag{5}$$

Regulating time constant t_s can reflect the speed of attenuation into steady-state:

$$t_s = -\frac{\ln\left(\sigma \cdot \sqrt{1-\xi^2}\right)}{\xi w_n} \tag{6}$$

Thus, under the circumstance of power grid keep steady state during a disturbance, power system oscillation time constant is similar to adjusting time constant. Also, the generator state equation and state value are:

$$\frac{d\delta}{dt} = (\omega - 1)\omega_0, \quad \frac{d\omega}{dt} = \frac{1}{2H}\left(P_T - \frac{E_q U}{x_{d\Sigma}}\sin\delta\right), \quad \delta = \delta_0 + \Delta\delta, \quad \omega = 1 + \Delta\omega \tag{7}$$

Where, δ refers to power angle, δ0 is initial power angle, ω is angular speed, ω0 is initial angular speed, △ω is angular speed variation, PT is mechanical power, Eq is no-load EMF, U is generator voltage, xd∑ is total reactance of the system, H is inertia time constant. Based on Taylor series expansion, electromagnetic power PE is obtained as:

$$\begin{aligned} P_E &= \frac{E_q U}{x_{d\Sigma}}\sin(\delta_0 + \Delta\delta) = \frac{E_q U}{x_{d\Sigma}}\sin\delta_0 + \left(\frac{dP_E}{d\delta}\right)_{\delta=\delta_0}\Delta\delta + \frac{1}{2!}\left(\frac{d^2P_E}{d\delta^2}\right)_{\delta=\delta_0}\Delta\delta^2 + \ldots \\ &\overset{\Delta\delta\to 0}{\approx} \frac{E_q U}{x_{d\Sigma}}\sin\delta_0 + \left(\frac{dP_E}{d\delta}\right)_{\delta=\delta_0}\Delta\delta = P_0 + \Delta P_E = P_T + \Delta P_E \end{aligned} \tag{8}$$

Its matrix form and characteristic roots are expressed as follows:

$$\begin{bmatrix}\Delta\dot{\delta} \\ \Delta\dot{\omega}\end{bmatrix} = \begin{bmatrix} 0 & \omega_0 \\ -\frac{1}{2H}\left(\frac{dP_E}{d\delta}\right)_{\delta=\delta_0} & 0 \end{bmatrix}\begin{bmatrix}\Delta\delta \\ \Delta\omega\end{bmatrix}, \quad p_{1,2} = \pm\sqrt{\frac{\omega_0}{2H}\left(\frac{dP_E}{d\delta}\right)_{\delta_0}} \tag{9}$$

It is obvious that, when p1,2 are a pair of virtual roots, thus △δ and △ω present persistent oscillation with the frequency fd :

$$f_d = \frac{1}{2\pi}\sqrt{\frac{\omega_0}{2H}\left(\frac{dP_E}{d\delta}\right)_{\delta_0}} \tag{10}$$

As for the on-line measured electric parameters, each period get an effective value, therefore, the length of time between effective value y(n) of the nth period and y(n-1) of the $(n-1)^{th}$ period is one period T. By analogy, the length of time between y(n) and y(n-x) is XT. Suppose n is infinite, then y(n) is considered as effective value of steady-state period, which can be draw numerical difference between all the effective values of sampling periods and steady-state period:

$$\Delta y^{(x)}(n)=\left|y(n)-y(n-1)\right|,\Delta y^{(x-1)}(n)=\left|y(n)-y(n-2)\right|,\ldots\ \Delta y^{(1)}(n)=\left|y(n)-y(n-x)\right| \quad (11)$$

Where, △y (1)(n), …, △y(x-1)(n), △y(x)(n) separately on behalf of the difference between 1st, …, (x-1)st, xst period and steady-state period effective values. The maximum value of this sequence is:

$$\Delta y_{\max} = \max\left\{\Delta y^{(1)}(n), \Delta y^{(2)}(n) \ldots \Delta y^{(x)}(n)\right\} \quad (12)$$

When △ymax ≤ σ (σ can be 5%, 2%), the electric parameters meet steady-state error conditions, and can be used for state estimation calculation, data cannot satisfy the requirements will be eliminated. The length of sequence x can be selected by oscillation frequency:

$$xT \geq T_d = \frac{1}{f_d} \quad (13)$$

$$x \geq \frac{1}{Tf_d} = \frac{2\pi}{T \cdot \sqrt{\frac{\omega_0}{2H}\left(\frac{dP_E}{d\delta}\right)_{\delta_0}}} = \frac{2\pi}{T \cdot \sqrt{\frac{\omega_0}{2H}\left(\frac{E_q \cdot U}{x_{d\Sigma}} \cos\delta_0\right)}} \quad (14)$$

4 Simulation and Results Analyze

According to the theory analyzed above, the simulation is applied to distribution model system. The parameters of generator, G, as followed: nominal power is 6*350MVA, line-to-line voltage is 13.8kV, frequency is 60Hz. Parameters of Transformer: winding 1 line-to-line voltage is 13.8 kV, winding 2 line-to-line voltage is 735 kV, and resistance is 0.004pu, the inductance is 0.16pu. The length of transmission line1 and line2 is 300km and 1,000km. Voltage levels of the three-wing transformer are 735kV/230kV/13.8kV. The capacity of Load1, Load2, Load4, Load5 and the Power System are marked on the figure. A disturbance occurs at the point of d in transmission line 2:

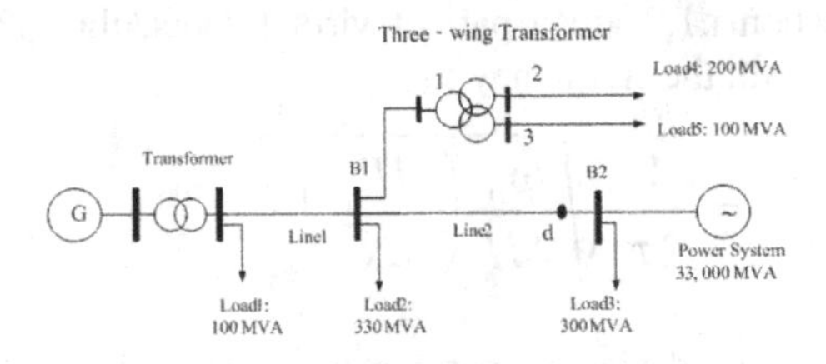

Fig. 1. Simulation model system

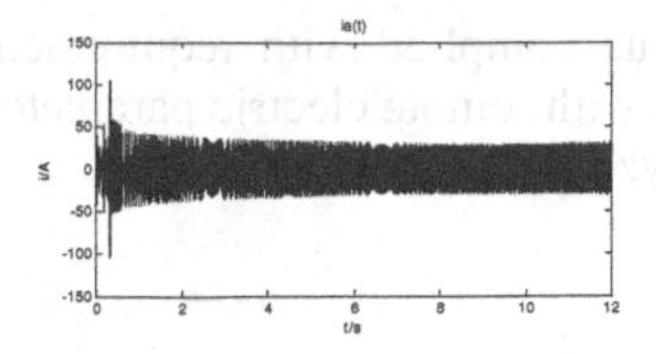

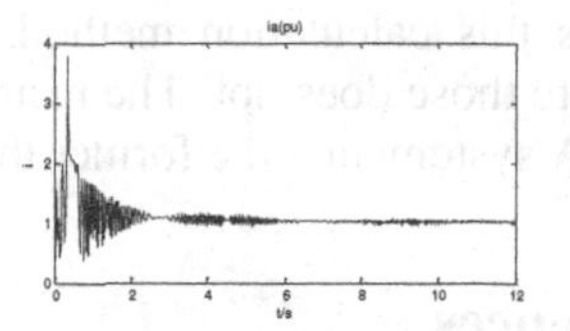

Fig. 2. Grid current continous value

Fig. 3. Grid current effective value

A disturbance occurs at location point d in Line2, continuous value of grid current is shown in Fig.2, and effective value of grid current is shown in Fig.3. As one can see from the figures, during 0s to 1.3s, after the disturbance occurred, the current reduces rapidly, and is relatively stable during 2.6s to 2.8s. If only one period effective value is compared with the steady-state effective value, these periods will be regard as steady-state period incorrectly. So it is very important to identify the system state by using the period length x. It could be worked out through calculation that, at the time of 5.8s, the system recovered into steady-state with the relative error of 5%, since the time of 9.5s, data relative error can reach 2%. The numerical difference between effective value of each period and steady-state period is shown in Table 1.

Table 1. Numerical difference of effective value between each period and steady-state

t(s)	0.183	0.35	0.58	1.43	1.93
σ	266.3%	175.6%	84.8%	38.2%	23.2%
t(s)	3.92	5.433	5.87	8.45	10.17
σ	12.3%	6.8%	4.2%	2.9%	1.8%

Consider the system recovery into steady-state when relative error is smaller than 5%. Table 1 shows that only after 5.87s, the data is acceptable, remote electric parameters from SCADA is shown in Table 2, during 12 seconds, 4 parameters will be sent back. According to the error requirement, the error of first two parameters, 0s~3s and 3s~6s is larger than 5%, which bring great impact on state estimation calculation result and cause the incorrect and divergence result in practical use.

Table 2. Remote electric parameters from SCADA system

Sampling period	0s~3s	3s~6s	6s~9s	9s~12s
SCADA Online measured data	1.2196	1.0592	1.0240	1.0073
σ	21.96%	5.92%	2.4%	0.73%

5 Summary

For the power system state estimation concerned, accurate and reliable electric parameters are very important. This paper is based on the premise of state estimation, puts forward a steady-state value calculation method, in order to improve the precision during practical operation. The simulation using MATLAB/SIMULINK software presents this calculation method, accept the data complied with requirements, and eliminate those does not. The result is compared with remote electric parameters from SCADA system, and the former theoretical analyze is validated.

References

[1] Huang, T., Lu, J., Zhang, H.: Measures to improve state estimation calculation accuracy in Guangdong provincial power dispatching center. Power System Technology 28(16), 78–81 (2004)
[2] Naka, S., Fukuyama, Y.: Practical distribution state estimation using hybrid particle swarm optimization. Power Engineering Society Winter Meeting 2, 815–820 (2001)
[3] Jin, Z., Ye, Z., Jiang, C., et al.: GB/T 13730-2002, Dispatching automation system for district power networks. China Electric Power Public, Beijing (2002)
[4] Liu, P., Tan, W., Xu, W., et al.: DL/T 630, Technical requirement for RTU with AC electrical quantities input, discrete sampling. China Electric Power Public, Beijing (1997)
[5] Yang, Y., Lin, J., Xu, W., et al.: GB/T13850, Electrical measuring transducers for converting AC electrical quantities into DC electrical quantities. China Electric Power Public, Beijing (1992)

Feasibility Study of P.E Contest Crowd Simulation

Bing Zhang

Institute of Physical Education, Huanggang Normal University, Huangzhou, Hubei, 438000, China
tiyuxi@qq.com

Abstract. As global physical education sports become more and more popular, multi-person P.E contest attracts more and more attention, such as Marathon, bicycle, orient cross-country, middle distance running, skate, etc. However, some problems come up during the contest, such as tumbling and trampling each other. As the digital human's sport simulation becomes increasingly mature, it has been a hot topic that whether simulation for a large group of people has a predictive and simulated effect. This paper, using method of historical documents, scientific analyzing method and inductive method, etc, studies worldwide works on crowd simulation and computer visualized crowd simulation, and applied situation and prospect of large-scale crowd simulation. The result shows, the field of P.E contest is single. The object of group is single, with obvious behavior characters. Therefore, P.E contest crowd simulation is feasible.

Keywords: Crowd simulation, P.E crowd, Contest visualization.

1 Introduction

By looking up historical documents, we realize large-scale crowd simulation has been applied widely, such as design of large transportation like boat, plane, etc, large public instruments like stadium and underground station. Take stadium as an example. According to stadium's scale, it can simulate ordinary situations such as leaving stadium, and assist in analyzing the rationality of width of veranda, and the number and positions of exits[1]. Besides, it can simulate crowd motions of special situations such as disturbance, fire and horrible attack, etc, help build emergency rescue plans, arrange extinguishers properly and allocate policemen. It can also be an assisted tool for security program rehearsal and training. Compared to traditional methods, it is more directive, flexible, effective, economic and out of risks, etc [2, 3].

Crowd and trampling often happen in the multi-person contests, such as going into bend track in bicycle contest, bend stage of skate and bend transcendence in long-distance running, etc, which always is the hot topic. As large-scale crowd simulation develops, it become a tool for generating vivid crowd animations, and realizes large-scale simulation. We can imagine that, if such technology can be applied to P.E contest, many benefits will be brought[4].

P.E contest crowd simulation is to study sports characters and rules of athletes under various environments, to build simulated models and present the process in three-dimensional way under computer visual environment. To realize crowd

Z. Du (Ed.): Proceedings of the 2012 International Conference of MCSA, AISC 191, pp. 589–593.
springerlink.com

simulation, two key problems have to be solved, one is to study and build simulated model and realizes its simulation, and the other is to study visualization of crowd sports.

P.E contest sports simulation mainly include real-time rendering of three-dimensional data. Global researchers have tried to solve such problem by Level of Detail-LOD, Point Sample Rending-PSR, etc. We will discuss its related technologies in the following.

2 Crowd Modeling

At present, existing crowd models basically in the general process. Several models will be introduced in the following.

The earliest particle system was brought into computer graphics in 1983, to model and visualize some natural fussy phenomena, such as cloud, water, gas and fire, etc. Frenchmen Eric Bouvier built a crowd model based on particle system [5].

In his model, each man is a simulated object-"particle". The whole crowd is considered as a particle system. The particles in the system can interactive. A particle collect equals to people with same behavioral modes. By setting various force fields in the system, using Newton mechanics and probability, solve the position, velocity, acceleration of each particle by simulated calculations, to realize crowd simulation. In the particle system, to make its motion to conform to human's characters and get constant velocity, friction must be added. Meanwhile cohesive coefficient can also be added to measure crowd's cohesion, so as to certain aggregation and division of crowd. Considering collision avoidance, Eric Bouvier uses relative velocity and relative mutual distance, and calculate collision probability in certain formulas, according to which regulates velocity to avoid collision.

Particle system builds a simple and useful crowd simulation. It is convenient for simulating some crowd behavior with obvious behavioral characters or crowd easy to be abstracted, such as crowd in underground station, etc. But to some complicate crowd behavior which need more decisive knowledge, they also require add lots of work, such as providing learning methods for crowd, and building decisive knowledge base, etc.

Dirk Helbing systematically studies crowd secession behavior in the shock state, and builds its simulated model. His model, only useful under the shock state, based on particle system, stress the force analysis. According to force situation, calculate individual's moving velocity, simulate crowd's typical behavioral characters, calculate velocities of crossing narrow aisle and exit and the number of people who might get hurt.

The relation between individual's force and velocity can be indicated in formula (1). mi is the mass of individual i. $v_i^0(t)$ is initial velocity of individual i. $e_i^0(t)$ is initial direction. $v_i(t)$ is expected velocity after Ti. $\sum_{i \neq j} f_{ij}$ is force generated by around crowd on individual i. $\sum_{w} f_{iw}$ is the force on individual i by around buildings such as wall, obstacles, etc. The result is shown as figure 1.

$$m_i \frac{dv_i}{dt} = m_i \frac{v_i^0(t)e_i^0(t) - v_i(t)}{\tau_i} + \sum_{j \neq i} f_{ij} + \sum_{w} f_{iw} \quad (1)$$

Although such model can only be useful in horrible state, it has high verisimilitude. It can simulate various typical phenomena, and estimate casualties, which has extreme high value.

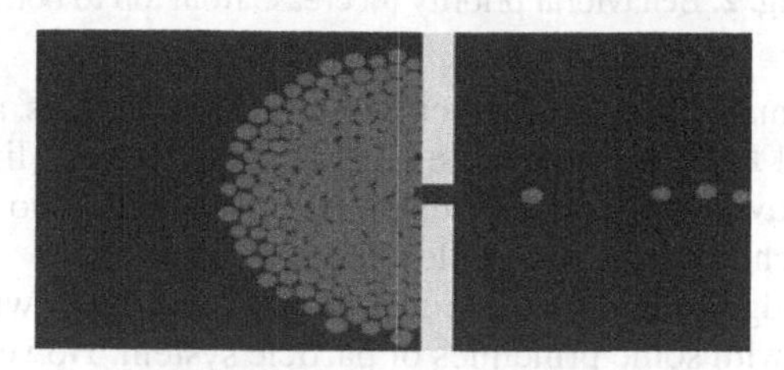

Fig. 1. Crowd evacuate through a narrow exit

ViCrowd model was proposed by Soraia, a man worked in computer graphics laboratory in Sweden Federal Technology Institute. ViCrowd definite a collection with three different levels: crowd, group, and agent. Besides it definite three important features: knowledge, beliefs and intentions.

Knowledge includes many kinds of information in virtual environment, such as obstacles, crowd action and behavior based on environment, or crowd, etc. Obstacle includes position, and areas which is accessible or not. Crowd action and behavior based on environment includes route plan in certain area, IP point (position people must go across) and AP point (position for taking certain action). Crowd knowledge is the varied information stored in the memory base of leaders.

Belief mainly includes action and behavior to be done. It can be classified into three categories:1) crow and its behavior: aggregation, following, changing objects, attraction, rejection, division, spatial occupation and safe check, etc. 2) emotional state: happiness, anger, sadness and laugh, etc. 3) individual belief: whether individual will be out of or into a crowd, and his position in the crowd, etc. Intention is crowd object.

Aside from such structures and characters, ViCrowd also definite three freedom levels, which are guided, programmed and autonomous. Guided behavior, also called as external control behavior, is generated by real-time interactive controlling. Programmed behavior, also called as scripted control behavior, is generated and controlled by programmable pre-setting. Autonomous behavior, which is innate behavior, is self-generated and controlled by virtual human according to self-knowledge and environment, etc. To crowd simulation, external control has highest priority, next is script control behavior and the last one is autonomous behavior, shown as figure 2.

External Control	Reaction
	Group Behavior
Scripted Control	Reaction
	Group Behavior
Innate Behavior	

Fig. 2. Behavioral priority (decrease from top to bottom)

ViCrowd systematically abstracts crowd behavior, divides and definite crowd's features, behavioral level and control level. ViCrowd, without limitation of situation, plot and crowd behavioral pattern, is a universal modeling method. But it is still only a skeleton, whose each level has to be filled and perfected.

Brogan and Hodgins model is mainly useful for crowd with obvious physical characters, mixing with some principles of particle system. However, in the process of modeling, more kinetic theories have been applied, with more considerations and restrictions on crowd's velocity, position and acceleration. Besides, it can also been analyzed in the aspect of physics. The result is shown as figure 3.

Fig. 3. Brogan and Hodgins

Such model is an expansion of particle system, which is applied to simulation of Marathon, bicycle contest and swimming.

3 Crowd Visualization

Most crowd simulation requires drawing of simulated results in three-dimensional way. Real-time rendering of large-scale three-dimensional data is a mature branch of computer graphical research, among which many technologies can be learned.

The distinguished LOD technology is progressive meshes, which was proposed by Microsoft Huges Hoppe based on mesh, in the late 1990s. In 1997, Garland & Heckbert in CMU used a quadric error metric for instructing modeling simplicity and optimization, under such thoughts. It has fine timing and spatial efficient, good simplicity effects and high value in research and practice.

To crowd simulation, crowd's position is flowing, in general. LOD technology not only reduces model's complication, but also provides smooth visual transition and dynamic control on model, which is suitable for three-dimensional visualization. Besides, LOD technology has been developed for many years. Many mature algorithms can be applied to crowd models after properly correcting. On current PC, LOD technology can be used to render the crowd with the scale of 100.

PSR breaks the pattern only rendering in polygons. In such technology, the model is demonstrated by series of sampling points of close planes. The points can be achieved from the mutually orthogonal direction, including color, depth, and normal line. Besides, it can compounded with Z and include Phong illumination and shadow, etc. The reason for such method is it is simple and fast. Summits are easy to be drawn, out of cutting polygons, scanning and transformation, texture mapping or bump mapping, so as to avoid large amounts of calculated spending. So such method not only uses large amounts of storages but also make the best of them.

PSR in reality, on current hardware instruments, its velocity is several numeric levels higher than polygonal rendering, and more vivid. Its visual effect is fine, and less pre-process. Under the support of current display card, the rendering scale can reach thousands.

4 Conclusion

The crowd simulation has been applied widely, among which the earliest and urgent area is public safeness, mainly used for crowd emergency evacuation in some accident. However, nowadays studies are lack of universal modeling methods and comprehensive skeletons. In the aspect of crowd visualization, the verisimilitude and rendering of visual effects have to be improved.

P.E contest crowd model is the abstracts of behavioral characters and mathematical description. Universal model is feasible. On one hand, although human is the most complicated biology, the analysis of factors is relatively simple. On the other hand, although each athlete is independent, they have common object. So the method to abstracting common characters is simple. All in all, the situation for P.E contest is single. Crowd has a single object with obvious behavioral characters. Therefore, the crowd model is valid.

References

[1] Bouvier, E., Cohen, E., Najman, L.: From crowd simulation to airbag deployment: particle systems, a new paradigm of simulation. Journal of Electronic Imaging 6(1), 94–107 (1997)
[2] Brodsky, D., Watson, B.: Model simplification through refinement. In: Proc. Graphics Inteface 2000, pp. 221–228 (2000)
[3] Brogan, D., Hodgins, J.: Group Behaviours for Systems with Significant Dynamics. Autonomous Robots 4, 137–153 (1997)
[4] Brogan, D.C., Metoyer, R.A., Hodgins, J.K.: Dynamically simulated characters in virtual environments. IEEE Computer Graphics and Applications 18(5), 58–69 (1998)

Design and Realization of Distributed Intelligent Monitoring Systems Using Power Plant

Zhemin Zhou

Hunan Chemical Vocation Technology College, China
13707419012@163.com

Abstract. In order to guarantee the smooth progress of the work in power plant, the production equipment and tools should be real-time monitored. The traditional artificial and statistical management methods have been unable to adapt to the present needs, therefore this paper presents a new distributed intelligent monitoring system based on the technology of intelligent terminal equipment and Ethernet communication. The system could realize real-time functions, such as the monitoring of equipment and tools, data statistics, system control and system alarm. Wherein, the communication between the different monitoring modules is completed used RS485 bus, and the communication between the monitoring module and control module is completed through Ethernet. Experimental results show that the system could run stably and real-time monitor the dynamic changes in power plant, and has good application prospect.

Keywords: Intelligent terminal equipment, Ethernet communication, RS485 bus, Distributed system.

1 Introduction

At present, the demand of power is increasing in various fields, which makes the power generation work facing greater challenges. So how to ensure the work is arranged in good order has great significance for the efficient and safe production of the plant, which is good for achieving the maximize use of the various equipment and tools. However, the traditional human management methods cannot meet current need, and it requires intelligent monitoring system for real-time control. The continuous development of network technology [1, 2], communication technology and intelligent terminal equipment provides good conditions for the design and implementation of intelligent monitoring system. On the basis of conditions mentioned previously, this paper introduces the overall structure of the distributed intelligent monitoring system, then describes the use of distributed communication technology in detail, and discusses the design of hardware and software at the same time. At last this paper has verified the feasibility of the whole system through experiments.

Z. Du (Ed.): Proceedings of the 2012 International Conference of MCSA, AISC 191, pp. 595–601.
springerlink.com

2 Overall Structure of Distributed Intelligent Monitoring System

The overall structure of distributed intelligent monitoring system is showed in Fig.1, the entire system consists of three parts, which are industrial computer control unit, dynamic monitoring unit and communication unit. Industrial computer control unit shows the running status of the operating conditions of the various production equipment and all kinds of tools. When the abnormal situation happens in the production equipment, such as the device is overheated, the industrial computer control unit should alarm in order to prompt staff, and send commands to control the corresponding devices to make respond. When some tools are in use, return time will be recorded and stored in a database. When the tool maintenance time arrives, the control unit will corresponding prompt. Through the control interface, the maintenance time of tools can be changed. As an intelligent node, dynamic monitoring unit with inputs and outputs for external analog and switching signal is mainly used to monitor temperature and humidity of production equipment and working environment, it could display the related data on the LCD screen and real-time displays the situation of various production tools. When the abnormal situation occurs, the monitoring equipment unit can send alarm information simultaneously, and make action through the control interface of the monitoring equipment. System communications contains Ethernet communication and RS485 bus communication, the Ethernet is used for communication between the industrial computer control unit and dynamic monitoring unit, which is achieved through a router. The method could reduce the communication connection and complete the distributed monitoring. RS485 bus is used for the communication between the different dynamic monitoring devices. This means of communication ensure real-time data aggregation, and it is convenient for function extension.

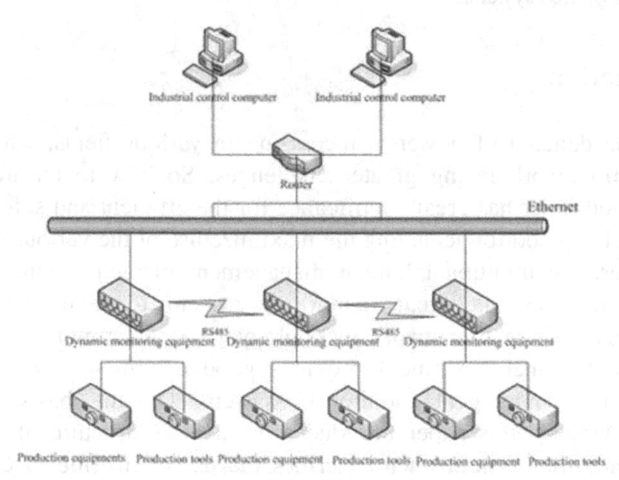

Fig. 1. The overall structure of distributed intelligent monitoring system

3 Data Communication Mode between Different Modules

The Ethernet technology was firstly created by the Xerox Corporation[3], which is jointly expanded by Xerox, Intel and DEC, and the three companies announced the specifications of the IEEE802.3 based on Ethernet. Ethernet uses CSMA and CD technology [4], and the communicate rate could be above 10Mpbs, when the signal is communicating on the various types of cable. Currently Ethernet technology includes not only the Ethernet specification in physical layer and data link layer, but also TCP and IP protocol suite, which contains an internet interconnection protocol in the network layer, transmission control protocol, user datagram protocol and so on. Therefore the Ethernet technology has many advantages, such as use widely, low-cost, high communicate speed, various hardware and software resources, sharing capabilities and potential for sustainable development[5-6]. The distributed intelligent monitoring system needs a special communication protocol for Ethernet data transfer. The protocol is the rules that manage how to communicate the system data. The protocol can guarantee the data transmitted from one node to another node correctly. Without the protocol, the equipment cannot resolve data signals sent by other device, neither the data can be transferred to any place. Simply speaking, the protocol defines the syntax and semantics of the communication between two parts. Combining the features of this system, this paper presents Ethernet communication protocol which suits for this system. Firstly it needs to design data frame. Considering the different achieve degree of the Ethernet transmit, send and receive, the Ethernet communication frame is set to three structures, which are 4 bytes, 5 bytes and 6 bytes, specific frame format as follows.

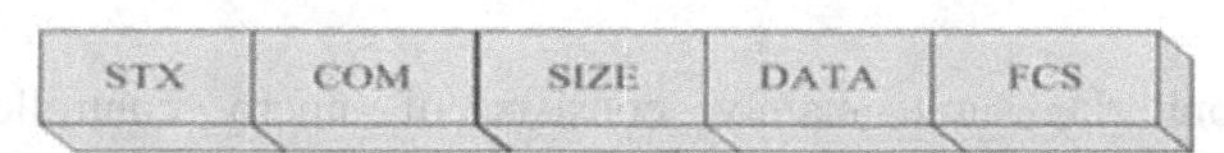

STX——the first frame sign (0XAAH);COM——Control byte;SIZE——The number of data bytes;DATA——Effective data segment;FCS——Frame checksum byte.

Fig. 2. The data format of frame

STX is used to determine the beginning of frame, which has a length of one byte; Control byte represents different control commands, that causes different perform commands or send different data. The data segment length has three kinds, which are given by the number of data bytes in the frame structure. The effective data segment is mainly used to store effective data in the communication process. The frame check is error detection encode mode, the cyclic redundancy check mode is used in this paper. According to the presented communication protocols and the actual need of communication process, this paper develops Ethernet communication command packet in the system. When industrial computer control equipment communicates with dynamic monitoring device, it will download two kinds of packet as follows.

When the system needs refresh the interface, it will send the data packet, whose format is "AA 11 00 **". "AA" means first data frame. "10" means control commands, which refresh the interface "00" means the number of bytes is 0. "**" indicates the cyclic redundancy check.

When the modified detection time data should be downloaded, the system sends the data packet, whose format is "AA 22 16 ×× ×× **". "AA" means first data frame. "22" means control commands, which send modified data; "16" means the number of bytes is 16; "×× ××" means all the data; "**" indicates the cyclic redundancy check. Dynamic monitoring equipment needs to upload data packet, it should regularly upload all the monitoring data, including temperature and humidity data of the working environment, and state data of tools. The format of the packet is "AA 33 16 ×× ×× **". "AA" means first data frame. "33" means control commands, which send state data "16" means the number of bytes is 16. "×× ××" means upload all the data; "**" indicates the cyclic redundancy check. According to the above formulated Communication protocol, the validity and accuracy of the command and data transmission will be guaranteed.

4 Hardware Design of Monitoring System

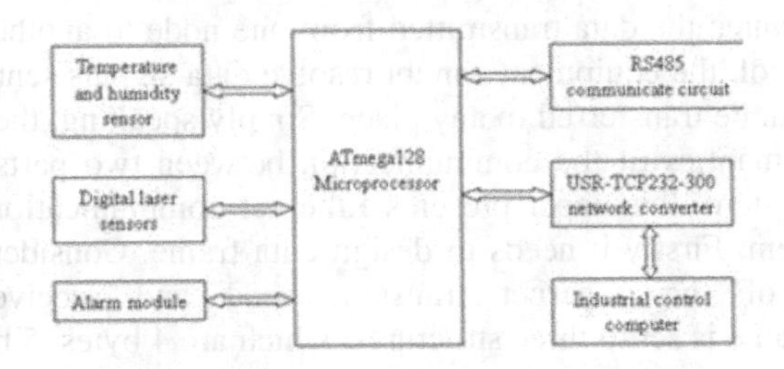

Fig. 3. The structure of the system hardware

Dynamic monitoring unit mainly consists of micro controller-ATmega128, temperature and humidity sensors-SHT10, digital laser sensors and RS485 communication module. ATmega128 is the eight series single chip microcomputer of ATMEL company, which is used widely. It has a 128K byte programmable flash, 53 programmable I/O port, thus it is convenient for function expansion. From Fig.3, it can be seen that SHT10 is used to collect the environmental temperature and humidity, and the relevant data is transferred to the microcontroller. If the temperature and humidity goes beyond a certain range, then the dynamic monitoring equipment is alarming. At the same time, the condition of power plant equipment and tools could be displayed in the main interface and condition data is uploaded to the computer control unit according to the fixed time, which is set to one second. Digital laser sensor is mainly used to monitor whether the various tools are used. If they are having been used, the output of sensor is high, otherwise the output is low. RS485 module mainly realizes the communication between the different dynamic monitoring devices, which is intended to summarize the condition of the equipment and tools. And when the abnormal situation is appeared, it can be better to guarantee the safety of the whole system. RS485 module needs the good anti-interference ability, so when the RS485 module is designed, it needs to do a special treatment, such as RS485 module should be far away from the radiation element.

5 Monitoring Interface Design

Monitoring system interface design consists of industrial control computer unit software design and dynamic monitoring unit design. The computer control unit is completed by Delphi software, and the dynamic monitoring unit is completed by the C language programming.

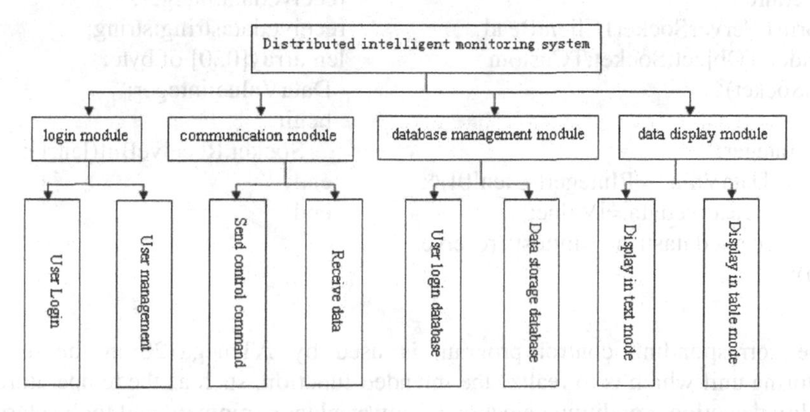

Fig. 4. Block diagram of industrial computer control unit software system

Block diagram of industrial computer control unit software system is showed in It is mainly divided into four modules, which are login module, communication module, database management module and data display module. The login module is mainly used to verify the identity of the operator in order to ensure the system is relatively safe. The structure diagram of user information table is showed in table 1. Database management module records the condition data and time data of the equipment and tools, ACCESS database is used in the module, which is flexible and convenient, and is suitable for the application environment of the system. The structure diagram of condition table is showed in table 2.

Table 1. The structure diagram of user information table

Field Name	Data Type	Note
User Name	Text	Primary Key
Password	Text	Display *

Table 2. The structure diagram of equipment condition table

Field Name	Data Type	Note
Serial Number	Data	Primary Key
System Time	Time	Null
Temperature and Humidity	Data	Null
Device Status Data	Data	Null
Tools Situation Data	Text	Null

The communication function of computer control unit is achieved using the Socket programming. Firstly, the computer receives all the characters according the communication protocol. Secondly, the concrete data is extracted from the characters. Thirdly, the concrete data is stored in the specified array and is displayed in the screen. Data receiving program is as follows.

```
procedure
TForm1.ServerSocket1ClientRead
(Sender:TObject;Socket:TCustom
WinSocket);
var
    i:integer;
        DataValue:=PInteger(@len[0])^;
          receivedata:=iValue;
      receivedatastring:=inttostr(receive
data);
receivedata:integer;
receivedatastring:string;
len:array[0..0] of byte;
  DataValue:integer;
  begin
     Socket.ReceiveBuf(len,1);
  end;
  end.
```

The corresponding control program is used by ATmega128 in the dynamic monitoring unit which is to realize the intended function, such as the temperature and humidity detection, condition detection of power plant equipment and tools, alarm for specific situations. The main program flow chart is shown in Fig.5.

6 Summary

Through the simulation experiment and field debugging, it is found that the distributed intelligent monitoring system is stable in operation, which can real-time monitor the dynamic changes of power plant. And the monitoring system realizes the data acquisition, data statistic, data storage, data display and abnormal alarm. Distributed Ethernet communication method makes the data exchange more accurate and convenient, which reduces the wires used in the system, and provides convenient condition for the system maintenance and repair. The construction and operation of distributed intelligent monitoring system will increase the lever of automation of power plant in a certain extent, and it relieves the contradiction between power grid development and management upgrade and realize the integration and share of the resources in power plant. Distributed intelligent monitoring system could ensure the different needs of dispatch and centralized control used effective technical measures, and it has a good application prospect.

References

[1] Dong, Q., Xie, J.: Fieldbus Network Implementation Based on RS-485. In: Proceeding of the 4th World Congress on Intelligent Control and Automation, pp. 2790–2793. IEEE (2001)

[2] Eidson, J., Kang, L.: IEEE 1588 standard for a precision clock synchronization protocol for networked measurement and control systems. In: Sensors for Industry Conference, pp. 98–105 (2002)

[3] Hung, M.H., Tsai, J., Cheng, F.T., et al.: Development of an Ethernet-based equipment integration framework for factory automation. In: Robotics and Computer-Integrated Manufacturing (2004)
[4] Lee, S.-H., Cho, K.-H.: Congestion control of high-speed Gigabit Ethernet Networks for industrial applications. Industrial Electronics 1(6), 260–265 (2001)
[5] Cushing. Process Control across the Internet. Chemical Engineering (5), 80-82 (2000)
[6] Kweon, S.K., Shin, K.G.: Statistics real-time communication over Ethernet. IEEE Transaction on Parallel and Distributed System 14(3), 322–335 (2003)

Output Feedback Control of Uncertain Chaotic Systems Using Hybrid Adaptive Wavelet Control

Chun-Sheng Chen

Department of Electronic Engineering, China University of Science and Technology, Taiwan

Abstract. The hybrid adaptive wavelet control scheme based on observer is proposed for a class of uncertain chaotic systems whose states are not all available. By applying "hybrid input" concept and combining adaptive control, robust control with wavelet neural networks, the output feedback control law and parameter adaptive law are derived. The proposed approach can guarantee the stability of the closed-loop, and achieve a good trajectory following performance without the knowledge of plant parameter.

Keywords: Direct adaptive control, Indirect adaptive control, Wavelet neural network, Chaotic system, Observer.

1 Introduction

Adaptive control techniques include the direct adaptive control (DAC) and the indirect adaptive control (IAC) algorithms [1]-[4], have the power of applying the system identification schemes to control uncertain systems. The wavelet neural networks (WNNs) inspired by both the feedforward neural networks and wavelet decompositions have been applied successfully to multi-scale analysis and synthesis, time-frequency signal analysis in signal processing [5], function approximation [6,7]. Inspired by the theory of multi-resolution analysis (MRA) of wavelet transforms and suitable adaptive control laws, an adaptive wavelet neural network is proposed for approximating arbitrary nonlinear functions. However, these control methods are based on the assumption that the system states are known or available for feedback. To treat this problem, an observer needs to estimate the unmeasured states and the Lyapunov design approach has been used to derive the adaptive and control laws [8,9], and the stability of the closed-loop system can be guaranteed.

2 Problem Formulation

Consider the nth-order chaotic system described in the following form:

$$\begin{aligned} x^{(n)} &= f(\boldsymbol{x}) + bu \\ y &= x \end{aligned} \tag{1}$$

where $f(\boldsymbol{x})$ is an unknown real continuous function, $u \in R$ and $y \in R$ are the control input and output of the system, respectively, b is a positive constant, and $\boldsymbol{x} =$

Z. Du (Ed.): Proceedings of the 2012 International Conference of MCSA, AISC 191, pp. 603–608.
springerlink.com

$[x, \dot{x}, \cdots, x^{(n-1)}]^T = [x_1, x_2, \cdots, x_n]^T$ is the state vector of system. In addition, only the system output y is assumed to be measurable. Rewriting (1) in the following form

$$\dot{\boldsymbol{x}} = A\boldsymbol{x} + B(f(\boldsymbol{x}) + bu)$$
$$y = C^T \boldsymbol{x} \tag{2}$$

where $A = \begin{bmatrix} \boldsymbol{0}_{(n-1)\times 1} & \boldsymbol{I}_{n-1} \\ 0 & \boldsymbol{0}_{1\times(n-1)} \end{bmatrix}$, $B = [0,0,\cdots,0,1]^T$, $C = [1,0,\cdots,0]^T$. The control objective is to force the output y for tracking a given bounded desired trajectory y_d in the presence of model uncertainties.

Define the output tracking error $e = y_d - y$, desired trajectory vector $\boldsymbol{y}_d = (y_d, \cdots, y_d^{(n-1)})^T$, output error vector $\boldsymbol{e} = \boldsymbol{y}_d - \boldsymbol{y}$. If the function $f(\boldsymbol{x})$ is known, the control law of the certainty equivalent controller can be chosen as

$$u^* = b^{-1}[y_d^{(n)} - f(\boldsymbol{x}) + K_c^T \boldsymbol{e}] \tag{3}$$

where $K_c = [k_1^c, k_2^c, \cdots, k_n^c]^T$ is the feedback gain vector. If K_c to be chosen such that the characteristic polynomial $h(p) = p^n + k_n^c p^{(n-1)} + \cdots + k_2^c p + k_1^c$ is Hurwitz, then it implies that the tracking error trajectory will converge to zero when time tends to infinity, i.e. $\lim_{t\to\infty} e(t) = 0$. However, $f(\boldsymbol{x})$ is unknown and only the system output is available for measurement, an observer will be designed to estimate the state vector and the wavelet neural network is utilized to approximate unknown function.

3 Wavelet Adaptive Output Feedback Control Design

In the standard form of WNN, the approximated wavelet-based representation for the nonlinear function $v(x)$ can be constructed as [1,5]

$$\hat{v}(\boldsymbol{x}, \boldsymbol{\theta}) = \sum_{j=M_1}^{M_2} \sum_{k=N_1}^{N_2} \theta_{jk} \psi_{jk}(\boldsymbol{x}) = \sum_{j=M_1}^{M_2} \sum_{k=N_1}^{N_2} \theta_{jk} \prod_{l=1}^{n} \psi_{jk}(x_l) = \boldsymbol{\theta}^T \Phi(\boldsymbol{x}) \tag{4}$$

where $\psi_{jk}(\boldsymbol{x}) = \psi_{jk}(x_1, x_2, \cdots, x_n) = \prod_{l=1}^{n} \psi_{jk}(x_l)$, $\boldsymbol{\theta} = (\theta_{M_1N_1} .. \theta_{M_1N_2} .. \theta_{M_2N_1} .. \theta_{M_2N_2})^T$ and $\Phi(\boldsymbol{x}) = (\psi_{M_1N_1}(\boldsymbol{x}) .. \psi_{M_1N_2}(\boldsymbol{x}) .. \psi_{M_2N_1}(\boldsymbol{x}) .. \psi_{M_2N_2}(\boldsymbol{x}))^T$ for some integers M_1, M_2, N_1, N_2. Let us use WNNs $\hat{f}(\hat{\boldsymbol{x}}) = \boldsymbol{\theta}_f^T \Phi(\hat{\boldsymbol{x}})$ and $u_D = \boldsymbol{\theta}_D^T \Phi(\hat{\boldsymbol{x}})$ to approximate $f(\boldsymbol{x})$ and direct adaptive wavelet controller, respectively, where $\hat{\boldsymbol{x}}$ is utilized to estimate $\boldsymbol{x}$. Thus, the resulting control law is

$$u = \alpha u_I + (1-\alpha) u_D + u_S \tag{5}$$

where $u_I = b^{-1}[y_d^{(n)} - \hat{f}(\hat{x}) + K_c^T \hat{e}]$ is the indirect adaptive wavelet controller, $\hat{e} = y_d - \hat{x}$, u_S is utilized to compensate the modeling error and $\alpha \in [0,1]$ is weighting factor. Design the adaptive observer

$$\begin{aligned} \dot{\hat{x}} &= A\hat{x} + BK_c^T \hat{e} + K_o(y - C^T \hat{x}) \\ \hat{y} &= C^T \hat{x} \end{aligned} \tag{6}$$

where $K_o = [k_1^o, k_2^o, \cdots, k_n^o]^T$ is the observer gain vector, which is selected such that the characteristic polynomial of $A - K_o C^T$ is strictly Hurwitz because (C, A) is observable. Define the observation error as $\tilde{e} = x - \hat{x}$ and $\tilde{y} = y - \hat{y}$, then subtracting (6) from (2), the observable error vector is given by

$$\begin{aligned} \dot{\tilde{e}} &= (A - K_o^T)\tilde{e} + B[\alpha(f(x) - \hat{f}(\hat{x})) + (1-\alpha)b(u_D(\hat{x}) - u^*) + y_d^{(n)} + bu_S \\ \tilde{y} &= C^T \tilde{e} \end{aligned} \tag{7}$$

Let x and $\hat{x}$ belong to compact sets U_x and $U_{\hat{x}}$, respectively, which are defined $U_x = \{x \in R^n : \|x\| \le \Lambda_1 < \infty\}$ and $U_{\hat{x}} = \{\hat{x} \in R^n : \|\hat{x}\| \le \Lambda_2 < \infty\}$, where $\Lambda_1, \Lambda_2 > 0$. Define the optimal parameter vector as $\theta_f^* = \arg \min_{\theta_f \in \Omega_f} [\sup_{x \in U_x, \hat{x} \in U_{\hat{x}}} | f(x) - \hat{f}(\hat{x} | \theta_f) |]$, $\theta_D^* = \arg \min_{\theta_g \in \Omega_g} [\sup_{x \in U_x, \hat{x} \in U_{\hat{x}}} | u^*(x) - u_D(\hat{x} | \theta_D) |]$, where θ_f and θ_D belong to compact sets Ω_f and Ω_D. Let the minimum approximation error be defined as

$$\omega = \alpha(f(x) - \hat{f}(\hat{x} | \theta_f^*)) + (1-\alpha)(u_D(\hat{x} | w_D^*) - u^*(x)) \tag{8}$$

Then the observable error (7) can be given by

$$\begin{aligned} \dot{\tilde{e}} &= (A - K_o C^T)\tilde{e} + B[-\alpha \tilde{\theta}_f \Phi(\hat{x}) + (1-\alpha)b\theta_D^T \Phi(\hat{x}) + y_d^{(n)} + bu_S + \omega] \\ \tilde{y} &= C^T \tilde{e} \end{aligned} \tag{9}$$

where $\tilde{\theta}_f = \theta_f - \theta_f^*$ and $\tilde{\theta}_D = \theta_D - \theta_D^*$.

Assumption 1. There exists positive definite matrix solution P for the equation (10)

$$\begin{aligned} &(A - K_o C^T)P + P(A - K_o C^T) + Q = 0 \\ &PB = C^T \end{aligned} \tag{10}$$

where $Q = Q^T$ is the given positive definite matrix..

Assumption 2. There exists a positive constant ε such that the wavelet neural approximation error satisfies $|\omega| \le \varepsilon$ for all $\boldsymbol{x} \in U_x$ ..

Theorem 1: The control law (5) stabilities the chaotic system (2) using the adaptation laws (12) and (13). Then the whole closed-loop system is stable, and the tracking errors converge to zero. The robust control and the adaptation laws are chosen as

$$u_S = -Kb^{-1}\operatorname{sgn}(\tilde{\boldsymbol{e}}^T PB), \quad K \ge \varepsilon \tag{11}$$

$$\dot{\boldsymbol{\theta}}_f = \gamma_1 \tilde{\boldsymbol{e}}^T PB\Phi(\hat{\boldsymbol{x}}) \tag{12}$$

$$\dot{\boldsymbol{\theta}}_D = -\gamma_2 \tilde{\boldsymbol{e}}^T PB\Phi(\hat{\boldsymbol{x}}) \tag{13}$$

Proof: Choose the Lyapunov-like function $V = \frac{1}{2}\tilde{\boldsymbol{e}}^T P\tilde{\boldsymbol{e}} + \frac{\alpha}{2\gamma_1}\tilde{\boldsymbol{\theta}}_f^T\tilde{\boldsymbol{\theta}}_f + \frac{1-\alpha}{2\gamma_2}\tilde{\boldsymbol{\theta}}_D^T\tilde{\boldsymbol{\theta}}_D$ with $P = P^T > 0$, $\gamma_1 > 0$, $\gamma_2 > 0$. Taking the time derivative of V yields

$$\begin{aligned}\dot{V} &= \tfrac{1}{2}\tilde{\boldsymbol{e}}^T((A-K_oC^T)^T P + P(A-K_oC^T))\tilde{\boldsymbol{e}} + \tfrac{\alpha}{\gamma_1}\tilde{\boldsymbol{\theta}}_f^T[-\gamma_1\hat{\boldsymbol{e}}^T PB\Phi(\hat{\boldsymbol{x}}) + \dot{\boldsymbol{\theta}}_f] \\ &\quad + \tfrac{1-\alpha}{\gamma_2}\tilde{\boldsymbol{\theta}}_D^T[\gamma_2\hat{\boldsymbol{e}}^T PB\Phi(\hat{\boldsymbol{x}}) + \dot{\boldsymbol{\theta}}_D] + \tilde{\boldsymbol{e}}^T PB(y_d^{(n)} + bu_S + \omega)\end{aligned} \tag{14}$$

where $\dot{\tilde{\boldsymbol{\theta}}}_f = \dot{\boldsymbol{\theta}}_f$, $\dot{\tilde{\boldsymbol{\theta}}}_D = \dot{\boldsymbol{\theta}}_D$. Using adaptation law (12)-(13), (10), and the definition of u_S, one has $\dot{V} \le -\frac{1}{2}\tilde{\boldsymbol{e}}^T Q\,\tilde{\boldsymbol{e}} + [\|\omega\| - K]\,|\,\tilde{\boldsymbol{e}}^T PB\,|$. Since $K \ge \varepsilon$, one has $\dot{V} \le -\frac{1}{2}\tilde{\boldsymbol{e}}^T Q\,\tilde{\boldsymbol{e}} < 0$. This means that the whole closed-loop system is stable.

4 Simulation Results

Consider the Duffing chaotic system whose dynamics can be described as follows:

$$\ddot{y} = -0.1x_2 - x_1^3 + 12\cos(t) + u(t) \tag{15}$$

where u and y are the control input and output of the system. The desired trajectory is $y_d = 0.1\sin(t)$. Select the Mexican Hat function $\psi(\hat{\boldsymbol{x}}) = (1-\hat{\boldsymbol{x}})\exp(-\hat{\boldsymbol{x}}^2/2)$ as our wavelet function and $[M_1, M_2, N_1, N_2] = [-15,15,-10,10]$ is the size of the wavelet network. Given $Q = diag[10\ 10]$ and choosing $K_c^T = [144,45]$, $K_o^T = [60,900]$, $\gamma_1 = 0.1$ and $\gamma_2 = 0.01$. The initial conditions are chosen as $\boldsymbol{x}(0) = [0.2, 0.2]^T$ and $\hat{\boldsymbol{x}}(0) = [1.5,1.5]^T$. Consider three cases of prescribed adaptation levels $\alpha = 1$, 0, 0.5, respectively, the simulation results are shown in Fig. 1 to Fig. 3.

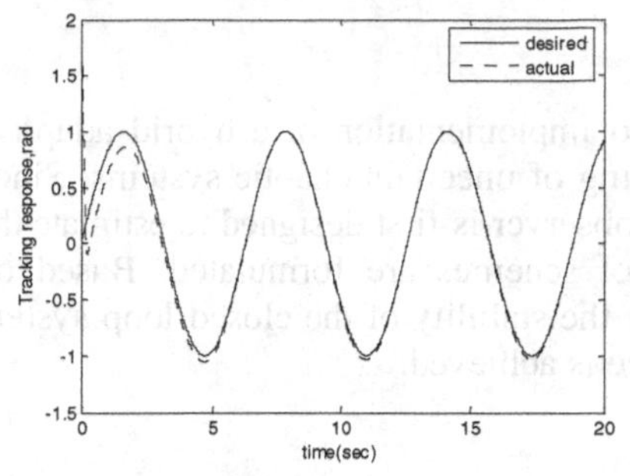

(a) Trajectories of the state x_1 and desired y_d

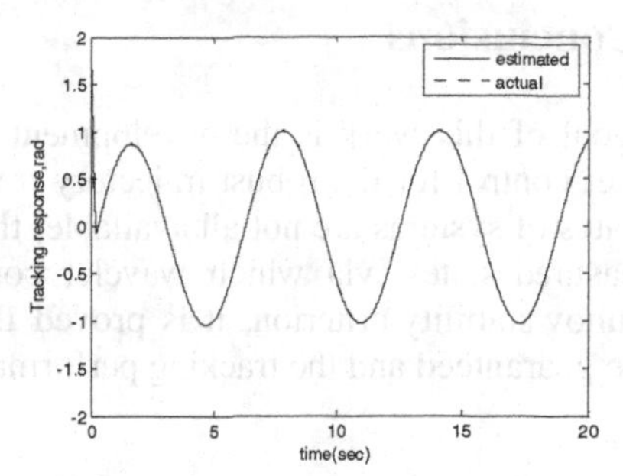

(b) Trajectories of the state x_1 and estimate $\hat{x}_1$

Fig. 1. The responses of Duffing forced oscillation system for $\alpha = 1$

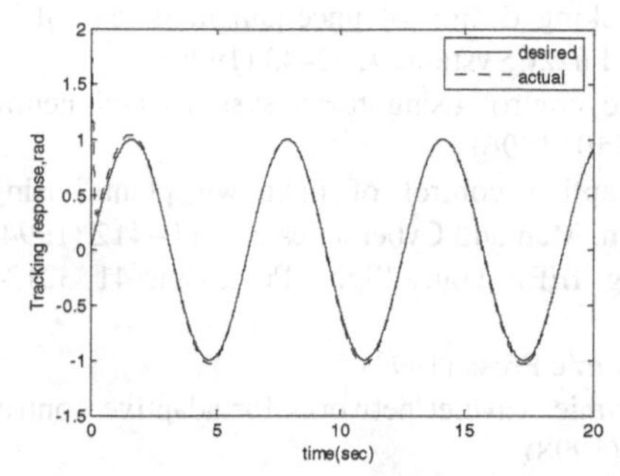

(a) Trajectories of the state x_1 and desired y_d

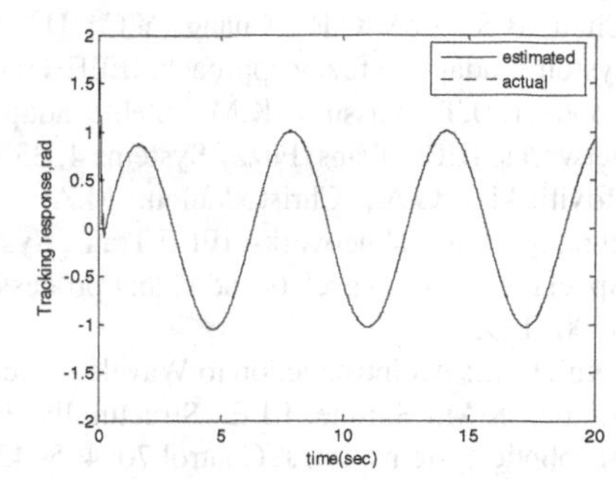

(b) Trajectories of the state x_1 and estimate $\hat{x}_1$

Fig. 2. The responses of Duffing forced oscillation system for $\alpha = 0.5$

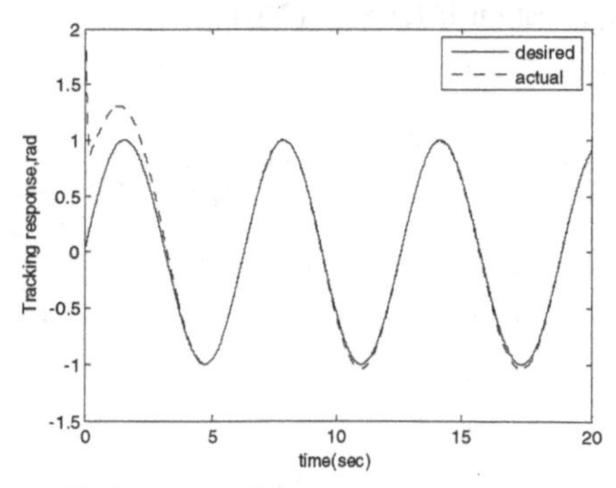

(a) Trajectories of the state x_1 and desired y_d

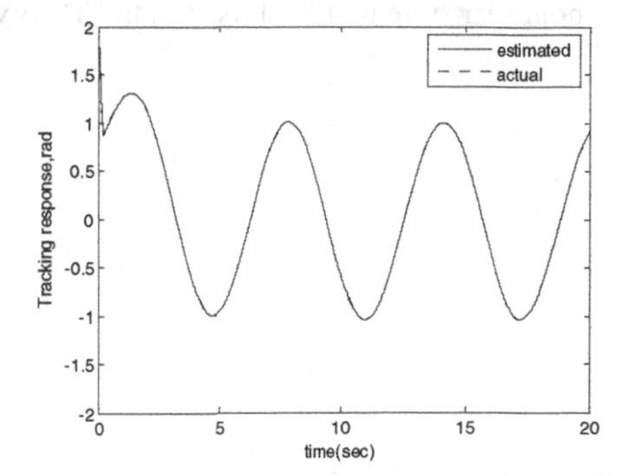

(b) Trajectories of the state x_1 and estimate $\hat{x}_1$

Fig. 3. The responses of Duffing forced oscillation system for $\alpha = 0$

5 Conclusions

The goal of this work is the development and implementation of a hybrid adaptive wavelet control for the robust trajectory tracking of uncertain chaotic systems. Since the states of systems are not all available, the observer is first designed to estimate the unmeasured states, via which wavelet control schemes are formulated. Based on Lyapunov stability criterion, it is proved that the stability of the closed-loop system can be guaranteed and the tracking performance is achieved.

References

[1] Wang, L.X.: Stable adaptive fuzzy control of nonlinear systems. IEEE Trans. Fuzzy Systems 1, 146–155 (1993)

[2] Chen, B.S., Lee, C.H., Chang, Y.C.: H^{∞} tracking design of uncertain nonlinear SISO systems: adaptive fuzzy approach. IEEE Trans. Fuzzy Systems 4, 32–43 (1996)

[3] Spooner, J.T., Passino, K.M.: Stable adaptive control using fuzzy systems and neural networks. IEEE Trans. Fuzzy Systems 4, 339–359 (1996)

[4] Rovithakis, G.A., Christodoulou, M.A.: Adaptive control of unknown plants using dynamical neural networks. IEEE Trans. System, Man and Cybernetics 24, 400–412 (1994)

[5] Special issue on wavelets and signal processing. IEEE Trans. Signal Processing 41, 3213–3600 (1993)

[6] Chui, C.K.: An Introduction to Wavelets. Academic Press (1992)

[7] Sanner, R.M., Slotine, J.J.E.: Structurally dynamic wavelet networks for adaptive control of robotic systems. Int. J. Control 70, 405–421 (1998)

[8] Leu, Y.G., Lee, T.T., Wang, W.Y.: Observer-based adaptive fuzzy-neural control for unknown nonlinear dynamical systems. IEEE Trans. Systems, Man, Cybernetics B 29, 583–591 (1999)

[9] Park, J.H., Kim, S.H.: Direct adaptive output-feedback fuzzy controller for a nonaffine nonlinear system. IEEE Proc. Control Theory Application 151, 65–72 (2004)

The Building of Oral Medical Diagnosis System Based on CSCW

Yonghua Xuan[1], Chun Li[1], Guoqing Cao[1], and Ying Zhang[2]

[1] Binzhou Polytechnic, Binzhou, Shandong, 256609 China
[2] Binzhou Medical University Affiliated Hospital, Binzhou, Shandong, 256603 China
xuanyonghua@cssci.info

Abstract. The CSCW based cooperative remote oral medical diagnosis information system has very important application value and practical prospect. The system has integrated the most advanced technologies of CSCW, multi-layers architecture, computer graph, videoconference, and the information processing to provide a cooperative environment for doctors and patients to communicate for dealing the medical consultations and oral heal care affairs. The system can be also be operated by number of doctors in different location to negotiate and diagnose the same patient oral medical disease through the oral image tools, videoconference tool and distributed software environment. The patient records can be shared between the different health professionals involved in the patient treatment.

Keywords: medical diagnosis system, oral medical system, CSCW.

1 Introduction

With the development of the network and the Internet, computer supported cooperative work (CSCW) technology also widely used in modern medical institutions. CSCW technology support groups of people engaged in a common task, provide an interface sharing environment [1, 2]. Based on CSCW cooperation distance medical diagnosis information system has very important application value and realistic prospect. It provides a dynamic environment, health experts visit patient information ability, improve patient consulting, interactive remote medical service, and the medical records management program for patient care.

2 CSCW for the Remote Medical Diagnosis System

CSCW and group of a rise in the 1980 s Shared interests from the product development personnel and researchers in various fields [1]. A computer is a technical support group and organizational work practice. It aims to develop computer-controllable network, virtual button to support synchronous collaboration. CSCW system support and cooperation work activities personal a set of common support service.

Z. Du (Ed.): Proceedings of the 2012 International Conference of MCSA, AISC 191, pp. 609–614.
springerlink.com

Research in the field of a medical information system is likely to be one of the most important of the application of CSCW technology [2]. There are many cooperation requirements in the medical information system in modern hospital for communication and cooperation to health professionals and department. Here is some cooperation on the medical work.

(1) The cooperative observations, interviews methods. The surgery observations and emulation with the instruction is the very important content in medical diagnosis and instruction. The activities of the surgery cannot carry out in the surgery room. With the CSCW based medical information system, the surgery picture can be transferred and observed to the conference room so that other doctors can observe the process in the videoconference. Therefore, the CSCW based medical information system can be widely used in the medical organization, the medicine colleges and universities.

(2) The collaboration and communication with patients. Doctors need to collaborate and communicate with other medical professional and patients in the process in the treatment of patients. The conferences between doctors, radiologists, and nurses are needed to communicate in the treatment.

(3) The patient records sharing. It is essential to sharing different records and material for several people to work together in patient treatment. Shared materials such as the medical records are used to coordinate the collaboration work between the different healths professionals involved in the patient treatment.

(4) The resources and tasks scheduling. It is essential to support the resources and task scheduling within different medical departments to collaborate one patient examination, tests, and operations within the hospital. The CSCW based medical information system in medical practice should include the electronic bulletin boards, electronic patient records, shared calendar systems, shared databases and videoconference. It would fully utilize the precious medicine resources through the network environment and the real-time multimedia broadcast technology.

3 System Design and Development of the Remote Cooperative Oral Medical Diagnosis System

The system arms to set up a cooperative distributed oral medical diagnosis environmental for hospital and dentists for providing the oral medical service to the remote districts located in different locations. It would be used as a fast channel for doctor to patient, hospital-to-hospital, hospital to clinic to provide the long-distance medical service, the long-distance medical consultation and the electronic medical records exchanging.

The overall framework of the cooperative oral medical diagnosis system is shown in figure 1. The main functions and principles for the system design are as the following:

(1) Provides the doctors consultation: It provides the patient's basic information to the long-distance medical consultation system, the doctor's advice and the examination, the inspection treatment report list, the medicine image material and diagnoses and treatment record and information.

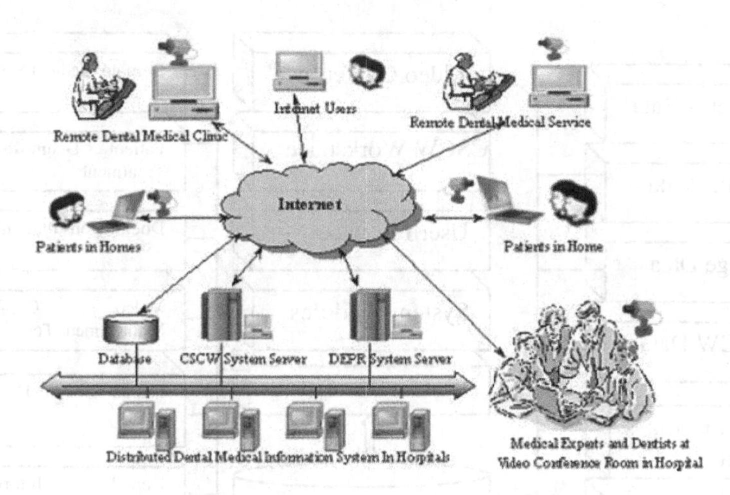

Fig. 1. System framework of the remote cooperative oral medical diagnosis system

(2) Information Sharing: The hospital information system can receive and stores the opposite part doctors' consultation information for patient's diagnosing and treatment, returns to original state and satisfies the precision request, which the clinical diagnosis needs. The local medical data, the oral picture, the medical record, the patient file can be transmitted to differently place. In addition, it can be used for long-distance education observation.

(3) Dynamic inquiry: The system would be designed to respond immediately for the patient request of the consultation from doctors at long-distance places. The system can receive the doctor's consultation result information; provide the medical record materials that preserved in remote hospitals.

(4) Dentistry graph data administration. The oral graph is needed to provide the exactly information about the patient oral disease information. All oral graphs of patient's OEPR database can be preserved in the system. It has enormously enriched the dentistry specialization electronic medical record.

(5) The video Conference Room: The videoconference room could provide a visual, bi-directional real-time video transmission for oral graphics that support the long-distance medical service and long-distance doctors' consultation.

The system architecture is shown in figure 2. The architecture is composed of three layers. The first the layer of the system is the database layer. The first system layer is the database layer. It saves as the information center of the system. The database of the system included the patients and the oral electronic patient records (OEPR) data, oral images data, multimedia data, CSCW data. The second layer is the system transaction layer, which is consisted of the system modules, such as the electronic medical records data management, user's management, oral images information collection, collaborative system functions, etc. The third layer is the user interface, which included the client side of the system, videoconference, camera, and CSCW workspaces, etc.

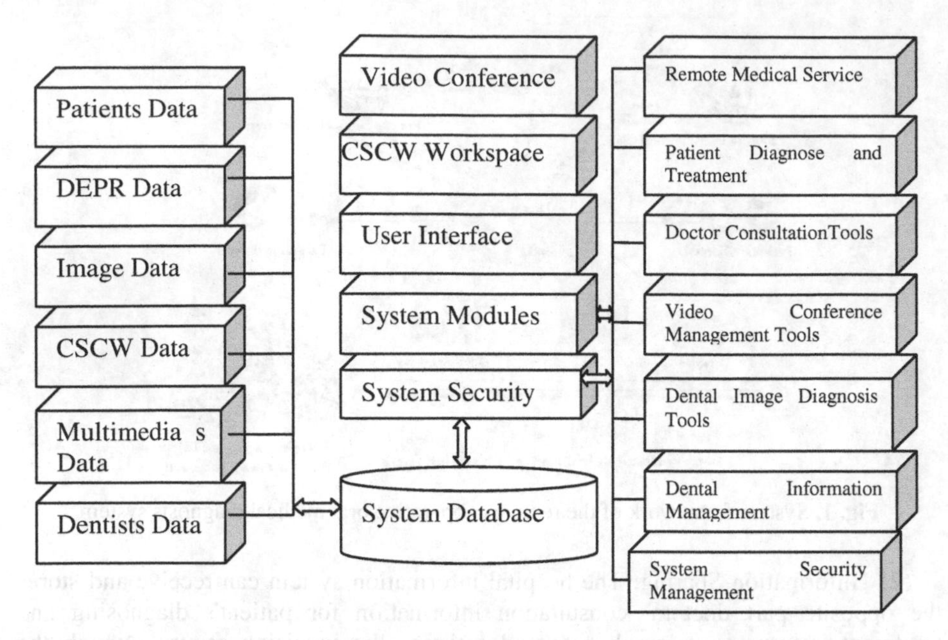

Fig. 2. System architecture of the multi-layers remote cooperative oral medical diagnosis system

Because there is many computers with different kinds of operating system will work in coordination in the system, the distributed, platform independent system architecture is needed to adapt the change in future use. To enable application system be accessed by lots of different computers in network, we choice the J2EE and EJB technology that offered a framework of service system to construct enterprise system structure.

The J2EE based system architecture is shown in figure 3. The middleware of CSCW, OEPR, oral images data, and multimedia EJB offers distributed transaction processing. Many host computers can join to offer the many services. Compared with other distributed technology, such as CORBA technology, the system structure of EJB has hidden the lower detail, such as distributed application, the events management, the target management, multi-thread management and connoting pool management etc. In addition, J2EE technology offers many kinds of different middleware to be applied to business logic. The data are stored and managed with EJB. Distributed computing enable users operate in any time, any place, and obtain business logic and data processing in remote server. The distributed systems enable the databases and services in the same or different computers. The database uses JBBC to communicate with EJB (back-end server). The system front-end uses the back-end server to provide service. The front-end client communicates with back end EJB.

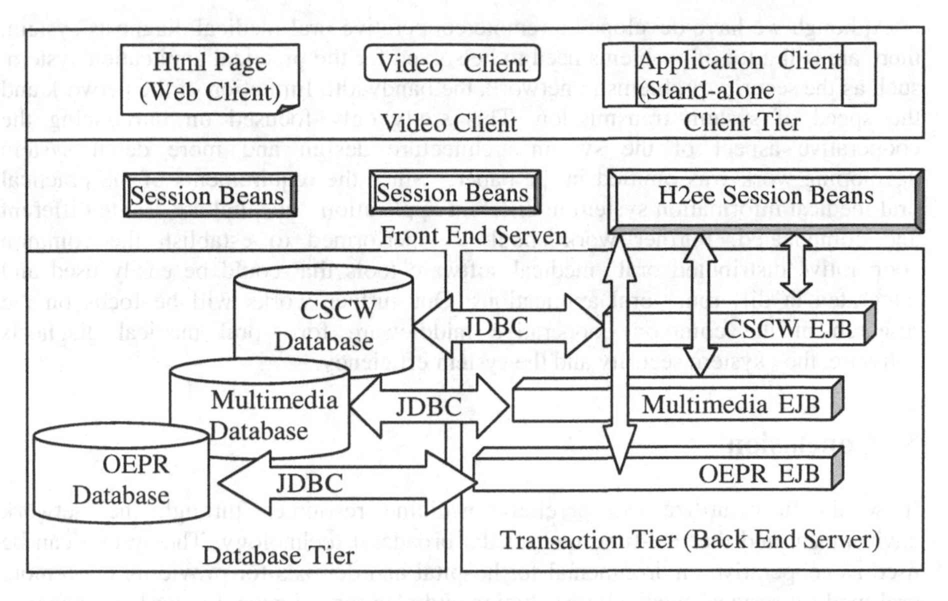

Fig. 3. Distributed system architecture for the remote cooperative oral medical diagnosis system

4 Dissuasions and Further Work

CSCW technology enables a group of collaborating individuals to perform the coordinated activities such as problem solving and communication. The CSCW technology has very important application prospect in the oral healthcare organization. The design and development of the CSCW based remote cooperative oral medical diagnosis and information system is a very interesting research field. According to the system framework, we have implemented the system of the cooperative oral medical diagnosis system and it has been used by more than ten oral hospitals and clinics in china. Figure 4 is the screenshot of the remote cooperative oral medical diagnosis system.

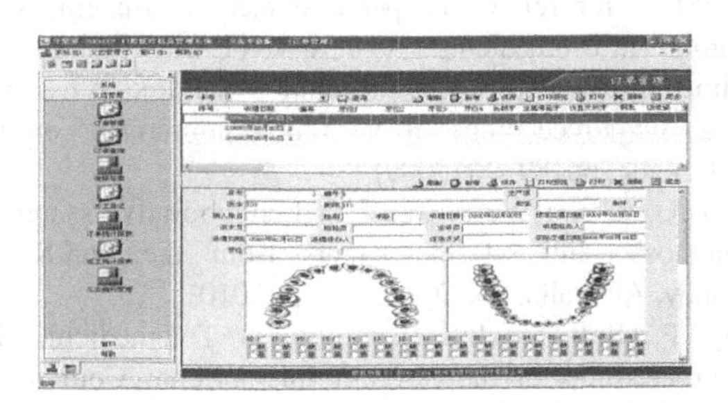

Fig. 4. A screenshot of the remote cooperative oral medical diagnosis system

Although we have developed a remote cooperative oral medical diagnosis system, there are still a lot of problems need to be solved for the practical application system, such as the security problems in network, the bandwidth limitation of the network and the speed of system transmission. This study only focused on introducing the cooperative aspect of the system architecture design and more detail system developing work was omitted in the paper. Since the requirements of the practical oral medical information system in different application hospital are quite different and complicated, Further work should be performed to establish the common cooperative distributed oral medical software tools that could be easily used and integrated in different oral applications. Our further works will be focus on the research of the common cooperative middleware for oral medical diagnosis software, the system security and the system efficiency.

5 Conclusion

It would fully utilize the precious medicine resources through the network environment and the real-time multimedia broadcast technology. The system can be used as cooperative environmental for hospital and dentists for providing the remote oral medical service, medical consultation and electronic medical records exchanging for the remote districts located in different locations.

References

[1] Grudin, J.: Computer-Supported Cooperative Work: History and Focus. Computer 27(5), 19–26 (1994)

[2] Wössner, U., Schulze, J.P., Walz, S.P., Lang, U.: Evaluation of a collaborative volume rendering application in a distributed virtual environment. In: Proceedings of the Workshop on Virtual Environments, Barcelona, Spain, pp. 113–118 (2010)

[3] Suebnukarn, S., Haddawy, P.: A collaborative intelligent tutoring system for medical problem-based learning. In: Proceedings of the 9th International Conference on Intelligent User Interface, Funchal, Madeira, Portugal, pp. 14–21 (2010)

[4] Gómez, E.J., del Pozo, F., Quiles, J.A., Arredondo, M.T., Rahms, H., Sanz, M., Cano, P.: A telemedicine system for remote cooperative medical imaging diagnosis. Computer Methods and Programs in Biomedicine 49(1), 37–48 (1996)

[5] Barakonyi, I., Fahmy, T., Schmalstieg, D.: Remote collaboration using Augmented Reality Videoconferencing. In: Proceedings of the 2004 conference on Graphics Interface, London, Ontario, Canada, pp. 89–96 (2009)

[6] Lim, Y.S., Feng, D.D., Cai, T.W.: A web-based collaborative system for medical image analysis and diagnosis. In: Selected papers from the Pan-Sydney workshop on Visualisation, Sydney, Australia, vol. 2, pp. 93–95 (2010)

[7] Ganguly, P., Ray, P.: Software Interoperability of Telemedicine Systems: A CSCW Perspective. In: Proceedings of the Seventh International Conference on Parallel and Distributed Systems (ICPADS 2000), July 2000 pp. 349–354 (July 2009)

[8] Greenshields, R., Yang, Z.: Architecture of a Distributed Collaborative Medical Imaging System. In: Proceedings of the Eleventh IEEE Symposium on Computer-Based Medical Systems, pp. 132–137 (1998)

Zhichan Soup on PD after Transplantation of Neural Stem Cells in Rat Brain Content of DA and Its Metabolites Related Data Mining Research

Shi Huifen[1] and Yang Xuming[2,*]

[1] Shanghai Hospital of Traditional Chinese Medicine, Shanghai 200071
[2] Shanghai University of TCM College of Acupuncture and Massage, Shanghai 201203

Abstract. Parkinson's disease is a high incidence of neurodegenerative diseases. This study was based on previous studies, using data mining methods, from dopamine (DA) Metabolic pathway, the use of artificial neural network algorithms and association algorithms to dig and forecast zhichan soup, only Chinese drug group, On the role of neural stem cells (NSC) transplantation Parkinson's disease (PD) in the rat substantia nigra DA and its metabolites in the meaning of content changes, to determine the soup ,only directed differentiation of NSC transplantation is non-linear effects of changes in real , and NSC transplantation for the treatment of PD to solve difficult problems in medicine to provide an effective solution to the experimental data.

Keywords: Parkinson's disease, dopamine, neural stem cell, dihydroxy phenyl acetic acid, homovanillic acid, association algorithm, Artificial Neural Networks.

1 Introduction

Parkinson's disease (PD) is second only to Alzheimer's disease (AD) in the second of neurodegenerative diseases [1]. Parkinson's disease in a variety of reasons (genetic, infection, drugs, oxidative stress) under the action of the brain dopaminergic (DA) neuron degeneration caused. Current treatment methods are drug therapy, surgery, cell transplantation and gene therapy. Stem cell transplantation and gene technology, with the stem cell research, and applied to PD treatment. By stem cell transplantation in patients with PD can rebuild the function of DA neurons. Moreover, PD is a stem cell transplantation is one of the best indications [2]. Main sources of stem cells come from embryonic stem cells, bone marrow stem cells, neural stem cells, in which neural stem cells(NSC) is considered the most suitable for cell replacement therapy and central nervous system cells in gene therapy. With the NSC in vitro and purification technology to improve, NSC transplantation as the most promising stem cell transplantation treatment of PD methods. The success of cell transplantation, including survival of transplanted cells to obtain a particular nerve cell phenotype and can be integrated into the recipient tissue function [3]. Therefore, how to promote the

* Corresponding author.

Z. Du (Ed.): Proceedings of the 2012 International Conference of MCSA, AISC 191, pp. 615–622.
springerlink.com

NSC transplanted into the brain to direct differentiation of DA neurons and the brain microenvironment change around the transplanted area and the signal [4] become a serious problem. Toxic side effects of Chinese medicine is less, it has the efforts of anti-free radical damage and neuroprotection, it let that PD and other neurodegenerative diseases have a certain effect [5-6].Zhichan soup is Professor Li Rukui many years experience in the treatment of PD summary, formerly Pingchan soup, in clinical studies to achieve better efficacy experiments [7-8]. In order to observe the effects of soup is just chatter to promote in vitro NSC and transplanted into the brain of PD rats NSC differentiation to DA neurons, and whether it can promote the migration of NSC, DA neurons in vivo proliferation and survival, and ultimately affect the dopamine synthesis and metabolism, traditional Chinese medicine to improve cell survival in the environment many factors that may have an advantage, is worth studying in the direction [9].

2 Experimental Design and the Significance of Data Mining

Data mining method used for animal experiments carried out analysis and forecasting.

2.1 Experimental Design

Experimental animals used 48 Wistar rats, both male, weighing 260 ± 20g, NSC transplantation into the normal group, model group and Zhichan soup group,each group has 12 rats, treated animals were in the 7th days, 14 days, 28 days after the last gavage behavioral test, the next morning anesthesia, decapitation, separating the brain, weighed, homogenized, centrifuged and the supernatant into the ultra-low temperature -80 °C refrigerator for use . Test samples, thaw at room temperature, each sample 10μl. High-performance liquid chromatography DA, dihydroxy phenyl acetic acid (DOPAC) and homovanillic acid (HVA) content analysis of the only soup on the micro-environment chatter of neural stem cells to dopaminergic neurons improve the role of the relevant factors [10]. In addition, it is observed only after transplantation of NSC PD rat brain content of DA and its metabolites effects of changes in [11].

2.2 The Significance of Data Mining

As traditional statistical methods and evidence-based medicine in the statistical methods, such as Meta-analysis, this information can not be a deeper level of relevance and regularity of treatment, because the traditional data analysis tools are based on proven methods, in these method, the user first data on the relationship between specific assumptions, and then analysis tools to determine or overturn these assumptions, of course, this information is obtained by floating on the surface, is the person's true feelings or feel that people really feel close; but data mining is a discovery-based approach, it is not explicitly assumed in the premise, the use of pattern matching and related algorithms to determine the important link between the data in order to discover the unknown, unpredictable, and even colleagues in the real feel exactly the opposite of knowledge.

3 Knowledge Discovery Association Rules

Association knowledge is associated with an event and other events between the dependent or associated knowledge. The important feature of association rules "associated with a natural combination," which attributes the discovery of a subset of all existing models are very useful, Exploration of traditional Chinese medicine only the soup, PD rat brain after NSC transplantation of DA and its metabolites associated with hidden changes in the impact of NSC transplantation for the treatment of PD to solve difficult problems in the experimental evidence for Chinese medicine. In this study, only the soup practical effect caused by the size of the drug should be: each drug group rat brain DA and its metabolites DOPAC and HVA content minus the change in the corresponding time points in all the model group of DA and its metabolites in rat brain changes in DOPAC and HVA content in the product difference.

Table 1. The ΔDA, ΔDOPAC and ΔHVA value after"Drug group - model group"

Time	group	ΔDA(nmol/g)	ΔDOPAC(nmol/g)	ΔHVA(nmol/g)
7D	1	54.9	53.7	5.7
	2	49.5	28.1	2.5
	3	62.4	54.5	-1.7
	4	46.2	31.5	-5.5
14D	1	54.5	25.3	3.3
	2	68.3	36.4	-2
	3	58.1	25.3	6.3
	4	20	18.5	0.5
28D	1	18.6	37.6	1
	2	14.4	26.7	2.5
	3	15.7	40.2	1.3
	4	9.7	23.9	3.3

Apriori algorithm [14] is associated with a classical algorithm, there are two main ideas: first, to find all frequent itemsets, by definition, These items set the frequency of emergence of at least a predefined minimum support as; second, the frequent itemsets generated by the strong association rules, by definition, these rules must satisfy minimum support and minimum confidence. In this study, starting from a combination of factors, the use of multi-dimensional association rules [15] for this issue is especially suitable for data mining. Can derive valuable experimental multi-dimensional association rules for clinical decision-makers to provide implicit in all aspects of the causal relationship between attributes.

In SQL Server 2008, in order to ΔDA as input, the causal relationship between predicted values ΔDOPAC computing the results obtained are as follows, we can see from Figure 1 from strong to weak dependencies are:

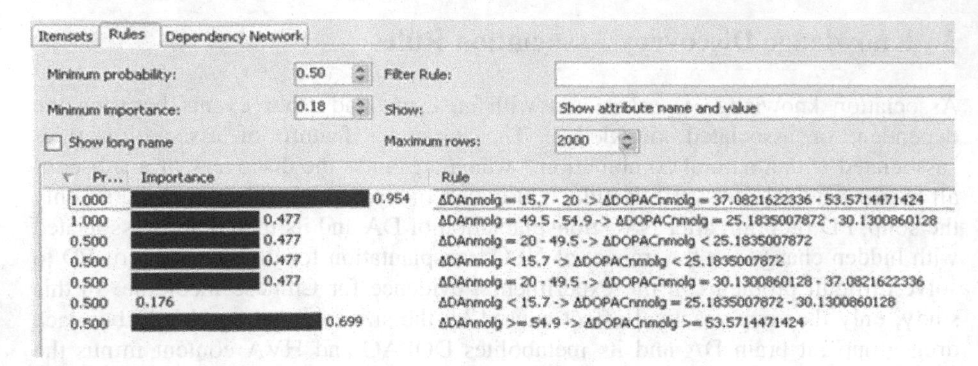

Fig. 1. ΔDA and ΔDOPAC association rules

From the known dependence of the strongest and the weakest, the strongest dependency is ΔDA = 20.00-49.50 within the range, resulting in ΔHVA <-2.19; and a relatively weak dependence of the resulting ΔHVA = 0.29-2.15; and, when ΔDA = 15.70-20.00 and ΔDA = 20.00-49.50, will lead to the same result ΔHVA = 0.29-2.15 pm. Thus, DA HVA caused by changes in relatively complex, the relationship is nonlinear, this situation is basically the same as the previous case study [11], so the specific ΔDA values to predict the value would be difficult to determine ΔHVA to grasp.

Similarly, NSC after transplant due to the size of the actual role should be: transplant group each rat brain DA and its metabolites DOPAC and HVA content changes in the model group minus the corresponding time points in all rat brain DA and its metabolites DOPAC changes in content and HVA difference.To ΔDA as input, predicted values of ΔDOPAC causal operations to get from strong to weak dependencies are:

Δ DA=6.1205-13.7176→ Δ DOPAC=10.1113-11.3562; Δ DA=18.2017-20.4998→ Δ DOPAC<4.8349;

Δ DA<6.1205→ Δ DOPAC<4.8349; Δ DA<6.1205→ Δ DOPAC=11.3562-15.3095; Δ DA>=20.4998→ Δ DOPAC=4.8349-10.1113; Δ DA=18.2017-20.4998→ Δ DOPAC=4.8349-10.1113; Δ DA=6.1206-13.7176→ Δ DOPAC=4.8349-10.1113.

And to ΔDA as input, predicted values of ΔHVA causal operations to get from strong to weak dependencies are:

Δ DA>=20.50→ Δ HVA=-0.6991-0.6973; Δ DA>=18.2017-20.4998→ Δ HVA>=1.4141;

Δ DA=6.1205-13.7176→ Δ HVA=0.6973-1.4141; Δ DA=18.2017-20.4998→ Δ HVA=-3.275—0.6691;

Δ DA=6.1206-13.7176→ Δ HVA=0.6973—1.4141.

From the above dependencies can be found ΔDA and ΔDOPAC, ΔHVA values are nonlinear relationship. And, because the same input value will be the results of different output values, and vice versa.

4 Artificial Neural Network Forecast

Artificial neural network developed since 1943 through the Enlightenment to the present, low and rehabilitation three times, before it has to type, feedback type, random type and competitive type of four self-organizing neural networks. BP algorithm using multi-layer feedforward network is by far the most widely used neural network can approximate any nonlinear curve, has a strong nonlinear mapping ability, and the middle layers of the network, the number of layers of processing units learning coefficient and network parameters can be set depending on the circumstances, very flexible, so it is in many application areas play an important role [16]. BP artificial neural network consists of input layer, hidden layer and output layer of three layers (hidden layer can contain multiple tiers), hidden layers commonly used sigmoid function. BP artificial neural network using a large number of samples for network training, the use of signal propagation from the input layer → hidden layer → output layer forward mode of transmission; error propagation using the comparison of the output layer to input layer error to reverse the direction of the right layer value adjustment (back propagation); weight regulation is generally used in a negative gradient descent, until the input sample to achieve convergence. BP artificial neural network mathematical description is as follows.

BP algorithm is based on the mathematical order derivative, it can reflect a function when the independent variable dependencies, can use the chain rule to determine the changes in the dependent variable and the relationship between the independent variables. The actual output data and calculated the error between the output data value is the size of neural network approximation of the actual system capacity. For a given set of training data, calculate the error signal for the derivative of any order parameter, then use the gradient descent method to the derivative of the order parameter towards the negative direction of the change, making the error signal drop.BP algorithm is described as follows [17]:

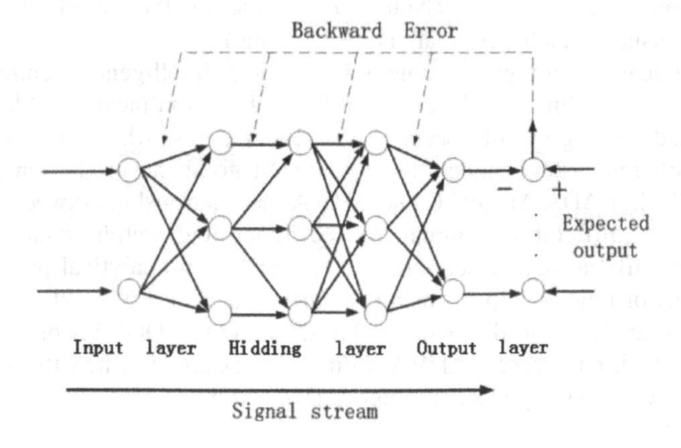

Fig. 2. BP artificial neural network mathematical description

Input: a given training set, each of which consists of a set of training samples are a set of input and output forms, all of the inputs and outputs are [0,1] between the floating-point data (if not, first through the data transformation to map them to [0,1]); neural network architecture: the number of hidden layer nodes; neural network each node, the parameters of the characteristic function.

Output: Each node neural network characteristic function of the parameters.

(1) In accordance with $\varepsilon_{l,i} = \frac{\partial^+ E_p}{\partial x_{l,i}} = \sum_{m=1}^{N_{l+1}} \frac{\partial^+ E_p}{\partial x_{l+1,i}} \bullet \frac{\partial f_{i+1,m}}{\partial x_{l,i}} = \sum_{m=1}^{N_{l+1}} \varepsilon_{l+1,m} \bullet \frac{\partial f_{l+1,m}}{\partial x_{l,i}}$

Calculate the overall error for each parameter order derivative formula (function).

(2) Choose a set of data as the initial parameters, the general choice (0,0, ..., 0) as the current parameters of this initial set of parameters.

(3) Based on the current parameters and type $E_p = \sum_{m-1}^{N_L} (y_{L,M} - x_{L,M})^2$

Calculate the overall error, if error is small enough, put the current parameters as output, exit; otherwise, continue the following steps.

(4) According to equation $\frac{\partial^+ E}{\partial \alpha} = \sum_{x \in S} \frac{\partial^+ E_p}{\partial x} \bullet \frac{\partial f}{\partial \partial}$

(S is the set of all nodes containing) and the current parameter values to calculate the overall error for each parameter value of the derivative order.

(5) According to equation $\Delta \alpha = -\eta \frac{\partial^+ E}{\partial \alpha}$ $(\eta = \frac{k}{\sqrt{\sum_{\partial} (\frac{\partial E}{\partial \alpha})^2}})$

η Serve as the learning rate, Increment of each parameter calculation, and calculate the size of the adjusted parameters. The adjusted parameters as the current parameters, return to step (3). (Note: η value of the selection is based on experience, is usually relatively small as a 0.01 value).

In recent years, as computer technology, artificial intelligence technology and the development of nonlinear science, multi-variable nonlinear problems for the prediction model has gradually become a research focus, while the use of artificial neural network forecasting models have achieved good the prediction [18]. If only through the Table 1 ΔDA,ΔDOPAC and ΔHVA the relationship between the observed concentration is difficult to summarize the input and output data relationships. However, the artificial neural network is to achieve non-analytical prediction of the primary means of relationship, so using BP artificial neural network data from these experiments to analyze and discover the laws, to ΔDA, ΔDOPAC prior assumptions for the input, to infer forecast ΔHVA value. For example, when the input ΔDA = 12.70, ΔDOPAC = 23.66, you can predict ΔHVA = 2.5.

5 Conclusion

This study was conducted in early pharmacodynamic and clinical studies, based on further observation only in vitro and transplantation soup PD rat brain content of DA and its metabolites associated with hidden impact of changes and prediction, to solve the NSC transplantation treatment of PD difficult issues in the Chinese medicine treatment of experimental data information mining and knowledge acquisition basis. Preliminary findings are: Zhichan soup can improve only after NSC transplantation in PD rat brain DA and its metabolites (DOPAC) levels, but the metabolites homovanillic acid (HVA) had no significant effect [11]. However, because these results are obtained by using SPSS statistical, while the traditional analysis tools and software limitations, not mining, hidden in the data analysis and prediction of internal diverse association, especially with the uncertainty associated with the noise of multiple powerless. Therefore, we must use methods of modern science and technology to accumulate and take the correct results and experimental methods, data mining techniques applied to a specific animal experiment, through specific algorithms to build data models to achieve zhichan soup to neural stem cell transplantation after the DA and its metabolites in PD rat brain content of the proper role of evaluation and prediction, in order to grasp its validity and accuracy, as the cause of human health services.

Acknowledgement. This research was supported by The National Natural Science Foundation of China (No. 30772870).

References

1. Beal, M.F.: Experimental models of parkinson's disease, pp. 325–334. Neurosci (2001)
2. Qian, Z., Li, L., Wangm, W., et al.: Stem cell transplantation for treatment of Parkinson's disease research. Chinese Journal of Neuroscience, 634–638 (2002)
3. Rossi, F., et al.: Opinion: neural stem cell therapy for neurological disease dream and reality, p. 401 (2002)
4. Englund, U., Fricker-Cales, R.A., Lundberg, C.: Transplantation of human neural progenitor cells into the neonatal rat brain:extensive migration and differentiation withlong distance axonal projection. Exp. Neural, 173 (2002)
5. Li, L., Sun, S., et al.: Protection of puerarin on Parkinson's disease in experimental research. Chinese Journal of Neuroscience, 7 (2002)
6. Cao, F., Shan, S., Wang, T., et al.: Ginkgo biloba extract inhibits experimental study of levodopa neurotoxicity. p.174. China University of Technology (2002)
7. Li, Y., Li, R.: Zhichan soup treating Parkinson's disease 31 cases observed. Shanxi Traditional Chinese Medicine, 16–17 (2002)
8. Li, R., Zhao, H., Tu, Y., et al.: Pingchan soup of Parkinson's disease model animal behavior and brain dopamine content affect. Chinese Medicine Research, 39 (2000)
9. Li, W., Shi, H., Wang, Y., et al.: Zhichan soup of the transplanted rat brain disease Rupajinsen proliferation and differentiation of neural stem cells in vitro. World Science and Technology (Modernization of Traditional Chinese Medicine), 371–373 (2009)
10. Liu, S.: Rat brain monoamine neurotransmitters and their metabolites in the detection method. Journal of Shandong University, 472–475 (2002)

11. Shi, H., Song, J., Li, W., et al.: Zhichan soup after transplantation of neural stem cells in rat brain PD DA and its metabolites impact of changes. Chinese Clinical Rehabilitation Tissue Engineering, 1165–1172 (2011)
12. Yang, X.: Acupuncture for treatment of juvenile myopia model of mining technology research. University of Shanghai (2011)
13. Zhang, Q., Liu, P., Zhang, W.: Data mining technology in TCM research in the application. Shanghai Traditional Chinese Medicine, 3–5 (2006)
14. Sun, Y., Sun, Y.: Apriori algorithm for association rule mining, pp. 82–84. Jilin Normal University (2009)
15. Tao, Z., Xie, K.: Audit data multidimensional association rule mining algorithm. Computer Application and Software, 160–168 (2008)
16. Yan, Y., Yang, J.: Investigation Of BP neural network algorithm. Scienice and Technology Development and Economic, 241–242 (2006)
17. Chan, P.: Data warehouse and data mining. Tsinghua University Press (2009)
18. Yang, L.: Study and Application on the Occurrence Prediction Model of paddy Stem Borer (Scirpophaga Incertulas) and Liriomya Huidobresis (Blanchard), University of Electronic Science and Technology of China (2009)

Remote Monitoring and Control of Agriculture

Xiao Chu, Xianbin Cui, and Dongdong Li

College of Control Science and Engineer, Shandong University, China

Abstract. This paper mainly describes the applying of Internet Of Things (IOT) technique in agriculture, both monitoring and control. In this system, the index values in the field such as temperature can be collected by wireless sensor system and uploaded to the internet, making it accessible by all the authorized visitors. In addition to the simply displacement of the data, we also make it possible for manger to realize remote control through system by simply typing in control index on the website. In this way, the network will be powerful because it combines monitoring and control together through internet. Remote control technique requires active sending message from the web server to host computer, which contradict the traditional way of communication. To implement this function, we designed a specific visiting method to give the control information from the website to the host computer.

Keywords: agriculture monitoring, agriculture remote control, TOT, wireless sensor.

1 Introduction

The tendency of using advanced wireless sensor system in agriculture has became predomination. With the movement of population from countryside to the cities and the rising of human costs put more emphasize on the intelligent system. The exploring of using wireless sensor system to keep an eye on the field is hot, during these years the complexity and the accuracy of the system have made great improvement.

However, one of the most important characters that mark the value of information is to some extent ignored. That is the timing data. How can the precise data collected by the sensors be more easily and timely get by the manager, even if they are not sitting in their control room looking exactly at their computer? If it is possible for the manager get control of the equipment in the field from a remote place simply through a computer?

The answer to these questions is IOT.

The Internet of Things (IOT) technology has been playing an important role in the modern information technology. With the help of internet, local physical data can be uploaded to a specific website. People who get permission can visit it. With these functions, people can easily make monitoring and management to remote objects.

When specific to this system, we designed a network, which can show the data collected by the sensor system and get simple control of the field equipment. It can technically be divided into two parts, one is monitoring, the other is control. In the monitoring part, we upload the physical index in the field (such as temperature,

Z. Du (Ed.): Proceedings of the 2012 International Conference of MCSA, AISC 191, pp. 623–627.
springerlink.com

moisture and sun light intensity) to the internet, and provide the function of drawing the tendency curve. Among all of these data, as long as there is one data overstep the threshold value or the slop of curve is large enough to indicate a coming overstep data, the system will send a message to the mobile phone of the manger to give a warm. In the management part, we set some function buttons on the internet. When manager received an warn, they can get access to the site through mobile phone, pad or remote PC, typing in some control index and clicking on the specific button to get control of the equipment in the field by sending the host computer a message.

2 System Display

2.1 Data Collection and Host Computer Parameter Setting

Data collected by the temperature probe and humidity sensor will be sent to the host computer by nrf2401 chip, at the same time will be displayed on the interface. What's more, we can set upper and lower bound alarm threshold values according to different requirements. So once the value exceeds the threshold value, some specific reactions will be triggered to remind the manager. All of these data and parameters will be packed and send to the servicer at the rate set by the manger (like once every 5 minutes).

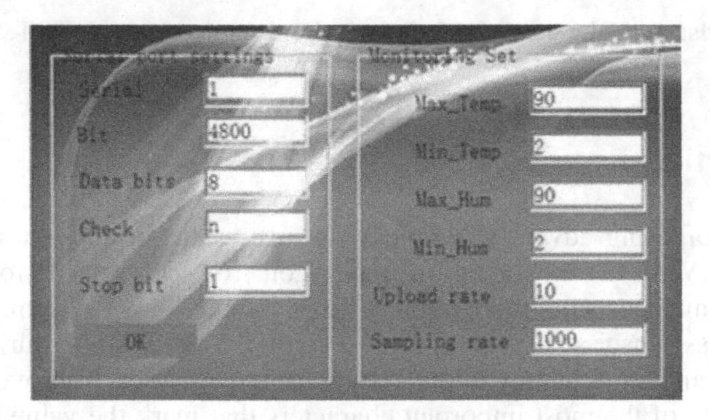

Fig. 1. Parameters set interface

2.2 Data Upload and Sharing

After parameter setting, the collected data is packed and sent to the server by HTTP protocol implemented with VC++ in our system. The data is post to the server in HTTP message form and received by corresponding PHP programs in the server side. Then, the received data is stored in MySQL, which is one of the most popular open source databases. Data stored in MySQL can be deal with by PHP program conveniently. The data can be visited by both PC and mobile phone.

People can visit the website through browser on a PC. In this condition, the transmission can be completed by HTML protocol. Except for the basic display of data, the page also provides alarming of the abnormal value of data. At the same time, we provide the function of drawing the curve which describes the changing tendency of temperature and humidity, to make it more clear and lucid for people to observe.

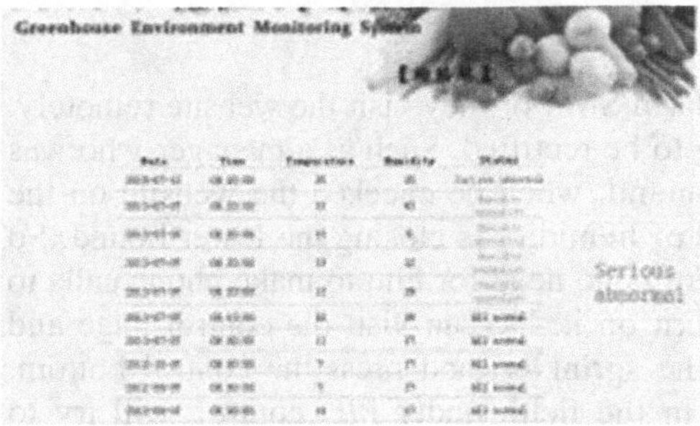

Fig. 2. Monitoring system

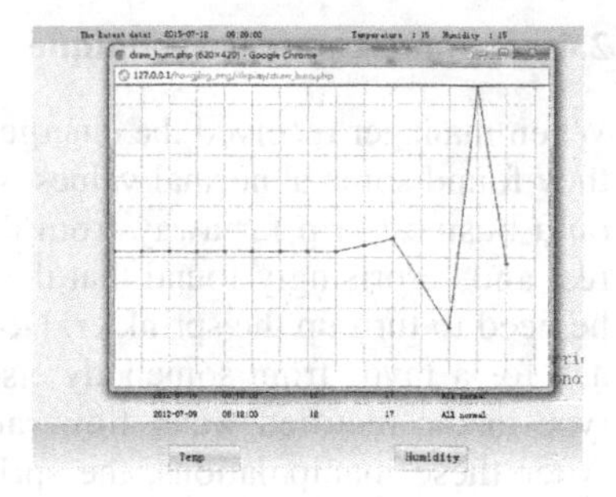

Fig. 3. Temperature changing curve

To ensure that people not sit at the computer can get access to these kinds of data collected from the field in time; we sent data to the wireless handheld device according to the WAP1.0 protocol.

2.3 Abnormal Data Emergency Alarm Send through SMS

When the value collected by the wireless sensor is abnormal, which means that it overstep the threshold value, the system will launch the sending SMS function. This function can be implement by sending a message to some specially electronic mail box, who provide the service of sending a reminding text message to bound mobile phone, and many of the prevailing mail boxes can offer this service.

Comparing to the tradition way of using GSM, this method will not only save the expense of hardware but also reduce the complexity of software.

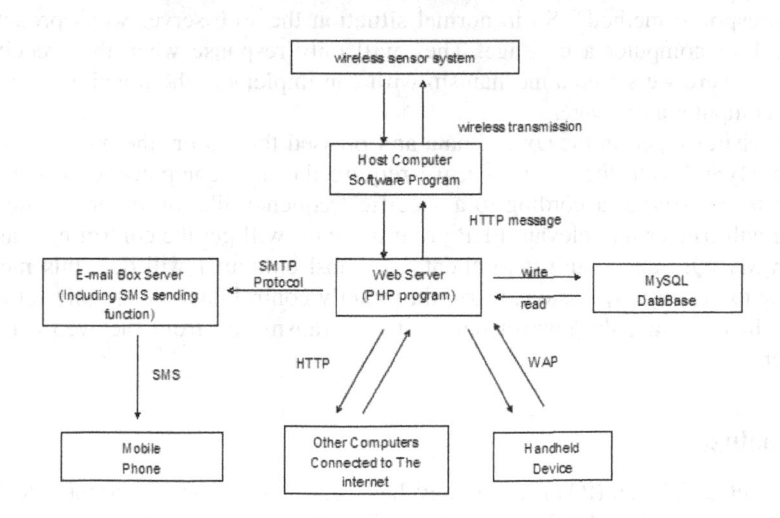

Fig. 4. System structure

2.4 Remote Control Technique

When manager received the emergency alarm SMS or they visit the website remotely, they found some abnormal values waiting to be rectified. Such as a manager who was on a business trip far away from the farmland, when he checked the website on the rest and surprisingly found that the value of humidity is closing the lower bound. So he need to turn up the sprinkler. Now, there is no need for him to make phone calls to ask for a favor from somebody else to turn on it. He can visit the control page and type in the rectified water flow rate of the sprinkler, and press the control bottom. After these manipulations, the sprinkler in the field, under PID control, will try to change itself to the flow rate you set.

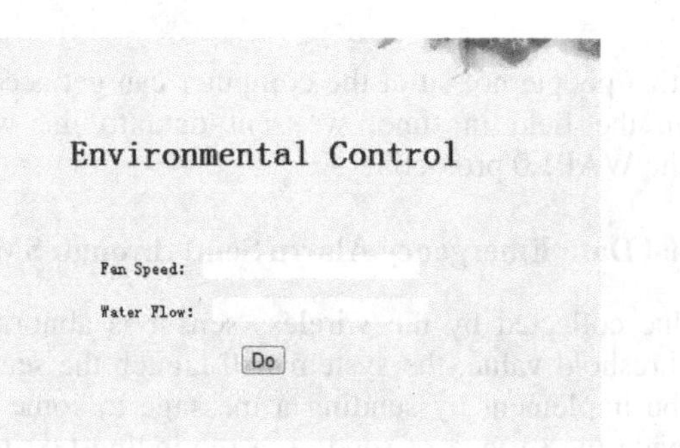

Fig. 5. Control index

So, here comes the problem. The normal communication mechanism we use is "request-response method". So in normal situation the web server won't proactively send the host computer a message. They will only response when they received a request. So, here we set up a mechanism with can implement the function of sending the host computer a message.

When visitor typed in the control data and pressed the button, the message will be saved in MySQL. On the other side, we require the host computer to send request message to web server according to a specific frequency like once per minute. This message will trigger the relevant PHP program, with will get the control message we saved in MySQL, and return it to client. The host computer will deal this message according to 'regular expression', to get the exactly control data we initially set on the website. In this way, the control data will be transmitted from the web server to computer.

3 Conclusion

The Internet of Things (IOT) technology has been playing an important role in the modern information technology. It is cool, because it provide the freedom of monitoring and managing the field without geographical restriction for the people who in charge of the field, it is a big emancipation of human. What's more, we also solve the communication problem even if WIFI is not accessible. To inform the manager, we send emergency message to them. To get control information, we design the website accessible through GPRS and other kinds of wireless used on mobile phone.

References

[1] Wang, N., Zhang, N., Wang, M.: Maohua Wang Wireless sensors in agriculture and food industry—Recent development and future perspective, vol. 50(1), pp. 1–14 (January 2006)
[2] Lemmon, H.: Comax: An Expert System for Cotton Crop Management. Science, New Series 233(4759), 29–33 (1986)
[3] Warle, T., Corke, P., Sikka, P., et al.: Transforming Agriculture through Pervasive Wireless Sensor Networks. IEEE Pervasive Computing 6(2), 50–57 (2007)
[4] Vitzthum, S., Konsynski, B.: CHEP: The Net of Things. Communications of the AIS 22, 485–500 (2008)
[5] Zhao, J.-C., Zhang, J.-F., Feng, Y., Guo, J.-X.: The study and application of the IOT technology inagriculture, 10.1109/ICCSIT.2010.5565120
[6] Information, http://www.hjp.at/doc/rfc/rfc2616.html
[7] Information, http://www.lavoisier.fr/livre/notice.asp?id=O3OWRLAROSSOWB

Research on Reform of Molding Materials and Technology

Pei Xuesheng

School of Art and Design, Henan University of Science and Technology

Abstract. At present, there is the teaching content in the teaching process in the design basic course in industrial design, modeling materials and technology "involves a wide range of knowledge points, and low student interest in learning, content and out of many other issues. The above problems, a series of Teaching Reform, had gained some experience in common with colleagues to questionable.

Keywords: Materials and Technology, Reform, Multi-discipline.

1 Introduction

"Modelling materials and technology" is the industrial design technology the basis of class, mainly about the product modeling design commonly used materials of organizations (metal, plastic, ceramic, glass, wood, paint, adhesives, etc.) performance and applications such as basic knowledge, at the same time familiar the general processing of some metals and non-metallic materials forming and surface treatment processes, and learn more about printing and the printing process. Students by learning to master the basic knowledge of commonly used materials in the product design, and learn how to choose the materials and processing technology.

2 Current Situation and Problems

Related to multi-discipline "shape, materials and processes of this course, the more content, and fewer hours, student learning is more difficult, so there are many institutions are carrying out the reform of the teaching, such as Xi'an Jiaotong University materials, laboratory sessions, the process of the Hong Kong school practical classes, etc. These explorations are experimental in nature. School teaching practice, through observation, feedback in the test and students after school and work and found there the following questions:

A molding material covers a wide range, the content is relatively trivial, contains a large number of definitions and concepts need to remember more, so the students feel the school is more boring, lack of interest. There is no interest there would be no motivation to learn, and the results of students do not master should have the knowledge. How to arouse students' interest, relatively boring things speak attractive is an urgent need to address the key issue.

Z. Du (Ed.): Proceedings of the 2012 International Conference of MCSA, AISC 191, pp. 629–632.
springerlink.com

Second, because most of the books is purely theoretical thing, the lack of appropriate production instance, when students encounter practical problems, or do not know how to apply these theoretical things. The simple theory rather than with the actual production combining a total feeling some paper. How will the combination of theoretical and practical most institutions are exploring a topic. Industrial design is a very practical professional to strengthen the abilities of students is a goal of our students.

In the process of designing is always faced with the choice of materials to meet the design requirements, you can choose the type of material is very much due to the lack of comprehensive consideration of the experience of students, in the end what kind of materials are often students feel I do not know from. For example, the selection of the chair, either using metal pipe can choose to use plastic, wood, etc., using which one can best show the shape of the design, which one best meets the needs of the market. Feel difficult to deal with these problem students.

Fourth, introduction of the material in the textbooks are a class of materials of general nature, the lack of specific material elements of composition, texture, performance and range of applications, there is no practical application example, the lack of practical significance, students are eager to know is how to flexibility in the application in product design.

Lacking knowledge of industrial design students in the mechanical design for the understanding of the machining process there are certain difficulties, in a limited time to allow students to better grasp the relationship between product structure and process, able to design new products to meet the processing requirements is one of the urgent need to address the problems of this class.

3 Objectives of Reform

For the issues raised above, we believe that the reform in teaching. Targets to be achieved are:

To increase the practical teaching to enhance students' interest in learning to enable students to understand the various properties of the materials and processing technology and methods as soon as possible.

To enable students to grasp the relationship between materials, processes and design, and learn how the availability of materials, aesthetics, craftsmanship and reasonable, economic, and design perfectly combine.

Apply the theory in practice, will analyze the characteristics of existing products, materials and processing, designed to meet the materials, products, according to the process requirements for the product to the appropriate choice of materials.

4 Implementation of Reform

Combined with the actual products to analyze this subject in the course of implementation, in order to achieve the above objectives, the talk about the theory, concepts and nature, analysis of the product material and the selection of the reason, can use other materials instead of active the advantages of the material and the lack of alternatives in there. Such an analysis, so that students can have a clear concept for the selection, no longer as said material is plastic or steel, but to grasp the characteristics of various materials and applications from the design and work in the future help.

First, we use multi-media teaching. Use as many pictures as possible, video material, so that students can have a real impression. Explain the ball timber, for example, we give dozens of photos of the ball, give students an appearance of feeling, let the students write before they can think of the ball material, and finally to summarize, look at some of the video image production process, so that the students have a ball with a plastic molding process with a comprehensive understanding unconsciously secondary schools to the materials, processing, molding, and other kinds of knowledge.

The second is the method of product analysis. To explain the basics of the relevant appliances, daily necessities, transport, IT classes, different uses of the products, analysis of the use of these products every part of the requirements, the materials used, the processing process, and difficulty surface treatment methods, materials, processing technology and surface treatment with the reality of products closely together to improve the students' interest, and learn how to use the knowledge learned to solve the issues related to product design.

Stressed the abilities of students. Material for a bicycle, we devoted one class time for students to dismantling the bike, draw some of its parts drawings, writing materials and processing methods used. Through this hands-on training so that students have a clear understanding for the different parts of the timber, and strengthen practical ability. We also specialize in one class time to explain the furniture, materials and processing methods, such as a chair, it is the object of the design has the same design; different materials have different effects and modeling. Shape to the selection, how to make better materials performance modeling, by allowing students to play the ability to innovate, design, different forms, different wood chairs, equipped with the knowledge in practice, get exercise.

Make full use of geographical advantages. Luoyang, Henan Province, advanced manufacturing technology base developed a manufacturing city, the Dachang more also more concentrated, machining, heat treatment, plastic processing, glass processing process is relatively complete, so to provide students with a very excellent visit internship conditions, with the actual production to explain the production process. For example, in explaining the metal processing technology can bring students to the various branches of a drag to visit, see a variety of production equipment and processing technology, know-how to play a multiplier effect.

Is a semi-open-book examination. The main solution for this class of materials processing and surface treatment, application of knowledge in product design, so in the examination, using a semi-open-book examination, part of a closed book, mainly to test the knowledge that must be mastered and memory points. Part of the open-book test the application of knowledge is a combination of a specific product, and allow students to choose the materials, the preparation process, select the appropriate surface treatment process. In the exam, allowing students to read textbooks, notes, and design manuals.

Set up a website on the Internet, molding materials and technology course curriculum goals, teaching plans and teaching content published to facilitate self-learning. Through the network Q, communication, homework, exams and other activities to consolidate learning and assessment of students.

The main features of this reform is to tell the story of the properties of the materials and processing, the actual product selection method, the material according to, and can achieve a multiplier effect. Each course combines a product and explains its use of materials and processing methods. Realities, to solve a single theory of boring speakers, can increase students' interest in learning.

5 Effect of Reform

This reform, our aim is to enable students to master how to properly select materials, how to ensure that the product complies with the design of the process in the design of the product. Through student reflection and student assignments and final examinations, the results were pretty good. Of course, the ultimate effect has to wait for students after graduation, to test in practical work. Through this reform, we also found some difficulties, such as how the product timber.

Characteristics about the process clear, and let the students according to the actual situation, learn the specific selection and application of the final design of the actual production products. Sometimes a product into a lot of content students is not easy to fully grasp, so it should be separate in terms of multiple products. Another is how to stimulate students' active learning ability of students coming from high school, accustomed to chalk and talk teaching methods often do not take the initiative to learn, and many of the materials is their own. The reform is a minor attempts to look forward to future teaching, be able to explore better ways to allow students more knowledge in a relaxed teaching middle school.

Acknowledgements. This work is supported by Education Reform Fund of Henan University of Science and Technology.

References

[1] Huang, X., Zhuang, J.: Case law in the teaching of construction materials. Fujian Building Materials (5), 127–129 (2008)
[2] Li, Y.: Materials and molding technology based curriculum reform of Mechanical Management and Development, vol. (5), pp. 166–169 (2008)
[3] Bi, F., Zhang, X.Y., Wang, L.: Engineering materials, teaching reform in practice, vol. 26(5), pp. 41–42. Daqing Normal University (October 2006)
[4] Du, G., Ding, H., Yang, J., Song, Y., Cui, Y.: Materials engineering based teaching experience. China Science and Technology 17, 293–295 (2007)
[5] Tang, S.: Polymer materials forming courses teaching reform and construction. Chemical Higher Education 25(1), 25–27 (2008)
[6] Zhao, C.H., Zhang, J., Zhong, X.Y., Chen, S.J., Liu, X.M.: Analysis of Tower Crane Monitoring and Life Prediction. IEIT Journal of Adaptive & Dynamic Computing (2), 12–16 (2012), doi: 10.5813/www.ieit-web.org/IJADC/2012.2.3
[7] Zhao, C.H., Chen, S.J., Liu, X.M., Zhang, J., Zeng, J.: Study on Modeling Methods of Flexible Body in ADAMS. IEIT Journal of Adaptive & Dynamic Computing (2), 17–22 (2012), doi: 10.5813/www.ieit-web.org/IJADC/2012.2.4
[8] Chen, G.Q., Jiang, Z.S., Wu, Y.Q.: A New Approach for Numerical Manifold Method. IEIT Journal of Adaptive & Dynamic Computing (2), 23–34 (2012), doi: 10.5813/www.ieit-web.org/IJADC/2012.2.5

Research and Implementation on Temperature and Humidity Measurement System Based on FBG

Luo Yingxiang

School of Electronics and Information Engineering, Chongqing Three Gorges University, Chongqing, China

Abstract. In the paper, we design a distributed fibber Bragg grating temperature and humidity measurement system, the clock pulse broadband light source, combined with time division and wavelength division multiplexing fiber Bragg grating sensor network technology, we will can solve complex problems with addressing to fiber Bragg grating, change of the PI film coating for the humidity, we propose a non-power temperature and humidity sensors. The use of arrayed waveguide grating sensor signal demodulation technology, through theoretical analysis confirmed the feasibility of the system. System for temperature and humidity measuring range (20 ~ 80) °C, (10 ~ 90) % RH measurement accuracy is achieved within ± 0.2 °C and ± 5% RH for real-time measurement.

Keywords: Distributed, Fiber, Measurements temperature and humidity.

1 Introduction

In agriculture, food storage is a very important parameter, the traditional method of measuring relative humidity and electromagnetic interference, and high humidity has the shortcomings difficult to measure; to play a non-power electromagnetic humidity sensor interference, fire, explosion-proof and other advantages to address the petrochemical, power, textiles and other areas of flammable and explosive environment temperature and humidity measurement and control problems, prompting people to study the new non-power humidity sensor [1]. In this paper, modified polyimide (PI) thin film humidity sensor for fiber grating (FBG) coated layer to form fiber grating-based temperature and humidity sensors. A sensor principle.

FBG-based temperature and humidity sensors mainly by the temperature and humidity sensitive FBG1 and temperature-sensitive FBG2 only [2-4] formed, as shown in Figure 1. Outsourced humidity film (PI) of FBG1 while the temperature-sensitive measurement of the change in the scene ΔT and relative humidity variation ΔH, without (PI) is only sensitive to humidity coating FBG2 temperature variation ΔT, and then into the center wavelength $\lambda B1$, $\lambda B2$ displacement $\Delta\lambda B1$, $\Delta\lambda B2$, then $\Delta\lambda B1$, $\Delta\lambda B2$ mathematically can be At the same time come to the specific values of temperature and humidity changes.

Z. Du (Ed.): Proceedings of the 2012 International Conference of MCSA, AISC 191, pp. 633–637.
springerlink.com

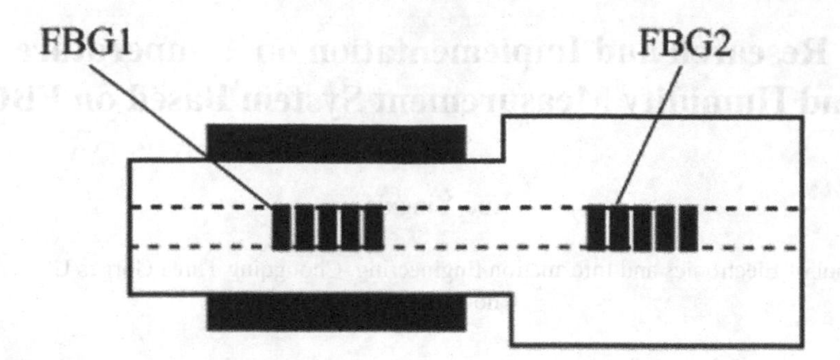

Fig. 1. Structure of the FBG temperature and relative humidity

According to the coupled mode theory, FBG reflection wavelength $\lambda_{Bi}(i=1,2)$ Satisfy the following formula:

$$\frac{\Delta\lambda_{Bi}}{\lambda_{Bi}}=\frac{\Delta n_{eff}}{n_{eff}}+\frac{\Delta\Lambda_i}{\Lambda_i} \tag{1}$$

The first right-hand side relative humidity of the ground caused by the change in ΔRH elastic-optic effect and temperature variation ΔT caused by the combined effect of thermo-optic effect of the results; ΔRH and the second is due to thermal expansion of axial strain and FBG.

As the humidity between the coating and fiber binding, ΔRH axial strain caused by FBG1 Free State for the axial strain and strain difference constraints, available from the elasticity theory:

$$\frac{\Delta\Lambda_i}{\Lambda_i}=C_1\beta\Delta\mathrm{RH} \tag{2}$$

$$C_1=\frac{E_{\mathrm{H}}(r_{\mathrm{H}}^2-r_{\mathrm{F}}^2)(1-2\mu_{\mathrm{F}})}{(1-2\mu_{\mathrm{H}})r_{\mathrm{F}}^2E_{\mathrm{F}}+(r_{\mathrm{H}}^2-r_{\mathrm{F}}^2)(1-2\mu_{\mathrm{F}})E_{\mathrm{H}}} \tag{3}$$

Where, respectively, the wet film humidity expansion coefficient, Poisson's ratio and Young's modulus; respectively, humidity and film cross-section of the fiber cladding radius. Consider the elastic-optic effect, thermo-optic effect and thermal expansion effects, FBG1 reflection wavelength can be expressed as relative change:

$$\frac{\Delta\lambda_{B1}}{\lambda_{B1}}=\mathrm{C}_1(1-\mathrm{p_e})\beta\Delta\mathrm{RH}+[\mathrm{C}_1(\alpha_H-\alpha_{\mathrm{F}})+\xi]\Delta T=K_{T1}\Delta T+K_{\mathrm{H1}}\Delta\mathrm{RH} \tag{4}$$

Where $\mathrm{p_e}$, ξ Optical fiber were effective elastic coefficient and thermal coefficient ; α_H , α_{F} Film and fiber, respectively humidity coefficient of linear expansion ; K_{T1} , K_{H1} Were FBG1 temperature and relative humidity sensitivity coefficient.

For the temperature-sensitive FBG2, since then $\beta = 0$, $K_{H2} = 0$ Value of the resonant wavelength can be expressed as the relative displacement

$$\frac{\Delta\lambda_{B2}}{\lambda_{B2}} = [(1 - p_e)\alpha_F + \xi]\Delta T = K_{T2}\Delta T \tag{5}$$

Where KT2 temperature sensitivity coefficient for the FBG2 1 solving the equation (4) and (5) the composition of the equations, you can also come to value the relative humidity and temperature changes ΔH value ΔT.

2 Measurements System of Temperature and Humidity of Distributed Fiber Grating

Distributed FBG strain and temperature measurement system at the same time, broadband light pulses emitted by a light pulse through the 3dB coupler, the arrival time delay switch [5]. It is the pulse of light into the sensing grating portal. Delay switch to turn it in turn connected to two channels gating to ensure that all the pulse arrival time delay of light in the switch, there will be a gated channel in the state, so that pulsed light can smoothly enter the sensing grating array , so that the whole system works. System, the time delay switch with the clock signal frequency remains the same. Two adjacent pulses of light that will enter different pathways to complete the different points of measurement. After a time delay switch gating, broadband light incident to the sensing FBG array. Satisfy the Bragg lattice wave length of light from the matching conditions reflected the various grating was demodulation receiver. Demodulation system using an array waveguide grating (AWG), each array from the AWG Reuters emitted light shines through each optical tube. At this point, the light signals into electrical signals. Signal input to the computer for processing, which can achieve wavelength demodulation of fiber Bragg grating, in order to achieve the simultaneous measurement of temperature and humidity.

Analysis of existing test results, it is easy to see:

In (20 ~ 80) °C, 10% ~ 90% RH range, the fiber grating humidity sensor output power and temperature and humidity changes is linear. FBG-type humidity sensors humidity hysteresis ≤ ± 1.5%, long-term stability is better than power humidity sensor; dynamic response time of less than 15s (depending on humidity-sensitive coating thickness PI). Mainly affected by the FBG demodulation system accuracy and PI humidity coating thickness uniformity of the limit, fiber grating sensor for relative humidity, humidity and temperature measurement accuracy was ± 5% RH and ± 0.2 °C.

FBG-based temperature and humidity sensor linearity thanks to the humidity, chemical stability of the PI film modified humidity, with design flexibility, long-term stability and interchangeability, fast response of the outstanding advantages of an addition, FBG sensors the inherent high accuracy, intrinsically safe, multiplexing ability, etc., so that the sensor in the petrochemical, power, textiles and other areas of high temperature, corrosion and other special circumstances in the multi-point, temperature and humidity measurement and control of distributed systems has a good prospect.

Will be calculated multimode fiber grating reflectivity spectrum and reported in the literature compare the experimental spectra: near 1550nm are presented in more than one peak reflectivity is relatively small, but the experimental spectra shows a peak a few more, mainly As the experiments measured the interaction between the main mode reflection is stronger than the theoretical calculation, resulting in larger number of each reflection peak, the peak number of total number; in dispersion shifted fiber, due to the excitation mode is less so than the 850nm peak number the calculated spectrum near a lot less, and the interaction between the peaks is almost no reflection, but the main mode of self-reflectivity is relatively large. This you can see, the use of incentive model number changes [1] and other means to reduce the number of patterns that can reduce or even eliminate the inter-reflection, self-reflection of the increased peak reflectivity, the multi-mode fiber grating to practical use.

3 Conclusions

In this paper, the concept of independent raster mode, type of gradient-index multimode fiber grating numerical simulation, has been a multi-mode fiber grating reflection spectrum. You can see, the axial dielectric grating perturbation, the grating length and operating wavelength of the grating reflection spectrum will have an impact. In general, the axial length of the dielectric perturbation changes led the General Assembly and the grating peak reflectivity increases, but the main mode of interaction between the different reflections in the axial perturbation larger dielectric constant in the case has been strengthened, while the grating length it has little effect. Meanwhile, the two have little impact on bandwidth. The working wavelength in the case of larger, will reduce the main mode, corresponding to the larger peak reflectivity and bandwidth is significantly increased. Calculated and experimental spectra in comparison, to get some of the similarities and differences: theoretical and experimental spectra are presented multi-peak structure, but the experiment of self-reflection theory of self-reflection than the small, 1550nm near the experimental spectrum, each reflection is much stronger than the theoretical calculations. In the dispersion-shifted fiber Bragg grating, because it is fewer models excitation, a significant reduction in the number of peaks, each reflecting little, but the main mode of self-reflectivity is relatively large.

References

[1] Kronenberg, P., Rastogi, P.K., Giaccari, P., et al.: Relativehumidity sensor with optical Bragg gratings. Optical Letters 27(16), 1385–1387 (2002)

[2] Konstantaki, M., Pissadakis, S., Pispas, S., et al.: Optical fiber long2period grating humidity sensorwith poly (ethylene oxide) / cobalt chloride coating. Applied Optics 45(19), 4567–4571 (2006)

[3] Kronenberg, P., Rastogi, P.K., Giaccari, P.M., et al.: Relative humidity sensor with optical Bragg

[4] Mizunami, T., Diambova, T.V., Niiho, T., Gupta, S.: Bragg Gratings in multimode and few-mode optical fibers. J. Lightwave Technol. 18(2), 230–235 (2000)

[5] Mizunami, T., Niiho, T., Djambova, T.V.: Multimode fiber Bragg grating for fiber optic bending sensors. In: Proc. SPIE, vol. 3746, pp. 216–219 (1999)

[6] Wanser, K.H., Voss, K.F., Kersey, A.D.: Novel fiber devices and sensors based on multimode fiber Bragg gratings. In: Proc. SPIE, vol. 2360, pp. 265–268 (1994)
[7] Szkopek, T., Pasupathy, V., Sipe, J.E., Smith, P.W.E.: Novel Multimode Fiber for Narrow-Band Bragg Gratings. IEEE J. Selected Topics In Quantum Electron 7(3) (May/June 2001)
[8] Fukushima, T., Yokota, T., Sakamoto, T.: Fabrication of 7×6 Multimoe Optical Fiber Grating Demulti-plexer-Star Coupler Using a Single GRIN-Rod Lens. J. Lightwave Technol. 15(10) (October 1997)
[9] Peral, E., Yariv, A.: Supermodes of Grating-Coupled Multimode Wavegrides and Aplication to Mode Conversion Between Copropagating Modes Mediated by Backward Bragg Scattering. J. Lightwave Technol. 17(5) (May 1999)
[10] Zhou, Y.Y.: Service Quality Measurement at University's Libraries by AHP Method. IEIT Journal of Adaptive & Dynamic Computing (2), 1–4 (2012), doi:10.5813/www.ieit-web.org/IJADC/2012.2.1
[11] Yu, J., Zhao, Y.: Weighted Approximation of Functions with Singularity by q-Baskakov Operators. IEIT Journal of Adaptive & Dynamic Computing (2), 5–11 (2012), doi: 10.5813/www.ieit-web.org/IJADC/2012.2.2

Clinical Analysis of Non-ST-Segment Elevation Acute Myocardial Infarction

Huang Zhaohe[1], Liang Limei[2,*], Pan Xingshou[1], Lan Jingsheng[1], He Jinlong[1], and Liu Yan[1]

[1] Department of Cardiology, Affiliated Hospital of Youjiang Medical University for Nationalites, Baise 533000

[2] School of Inspection, Youjiang Medical University for Nationalites, Baise 533000

Abstract. To investigate and analyze non-ST-segment elevation acute myocardial infarction clinical characteristics, we select hospital 130 patients with acute myocardial infarction from June 2009 to June 2011. According to body surface ECG ST-segment elevation, we divided them into ST-segment elevation group (n = 60) and non-ST-segment elevation group (n = 70), the clinical features of two groups of patients, coronary artery disease and hospitalization 1 the relevant results on a comparative analysis. The results of non-ST-segment elevation group rate history of diabetes, a history of recurrent angina was higher than ST-segment elevation group, by comparison were statistically significant (P <0.05). Group of non-ST-segment elevation myocardial infarction with ST-segment elevation group, P <0.05. Non-ST-segment elevation group, CK peak, TNI were lower than the peak ST-segment elevation group, by comparison were statistically significant (P <0.05). Non-ST-segment elevation group of three coronary artery disease was significantly higher than ST-segment elevation group, by comparison a statistically significant difference (P <0.05). Group of non-ST-segment elevation within 1 month of hospitalization with a trial fibrillation, cardiac pump failure, ventricular arrhythmia and death were significantly lower than the incidence of ST-segment elevation group, by comparison were statistically significant (P <0.05). Non-ST-segment elevation of CK peak, TNI peak lower than ST-segment elevation, coronary artery disease is higher than the ST-segment elevation, hospitalized a month prognosis is better than ST-segment elevation.

Keywords: Non-ST-segment elevation, ST-segment elevation, acute myocardial infarction, clinical features.

1 Materials and Methods

1.1 General Information

Choose from June 2009 to June 2011 our hospital 130 patients with acute myocardial infarction. Whether based on body surface ECG ST-segment elevation is divided into ST-segment elevation group and non-ST-segment elevation group. All patients were

* Corresponding author.

Z. Du (Ed.): Proceedings of the 2012 International Conference of MCSA, AISC 191, pp. 639–643.
springerlink.com

selected according to the Chinese Medical Society of Cardiology Council and other established diagnostic criteria for diagnosis [2]. (1) continuous chest pain or chest discomfort lasted ≥ 0.5h; (2) ST-segment depression ≥ 0.1mv, duration ≥ 0.5h (non-ST-segment elevation) or at least two adjacent ST-segment elevation and the evolution of a typical T wave (ST-segment elevation); (3) serum creatine kinase or troponin I were significantly increased. ST-segment elevation group (n = 60), including 31 males and 29 females; the youngest 37 years old, maximum 82 years, mean 60.18 ± 11.28 years old. Non-ST-segment elevation group (n = 70), including 37 males, 33 females; the youngest 39 years old, maximum 79 years, mean 62.07 ± 10.36 years old. Both groups were sex, age and other clinical data by comparison, were not statistically different (p> 0.05), differences between comparable.

1.2 Research Methods

(1) Records two groups of patients before the onset of past history (a history of recurrent angina, smoking history, diabetes, high cholesterol and hypertension incidence), to judge by the surface ECG myocardial infarction, detect and record the creatine kinase (CK) peak troponin I (TNI) peak.

(2) to determine coronary artery disease: coronary angiography through its judgments, the right coronary artery divided, the left anterior descending artery and left circumflex artery three, disease variables, divided into single-vessel disease, two diseases, three-vessel disease.

(3) Hospitalization within 1 month if there is re-infarction, atria fibrillation, cardiac pump failure, ventricular arrhythmias or deaths occurred.

1.3 Statistical Analysis

Using SPSS 16.0 statistical software for statistical analysis. All measurement data (mean ± standard deviation) indicated that the two sets of variables related to the line $\chi 2$ and t-test, univariate analysis of variance, taking P <0.05 for the difference was statistically significant.

2 Results

2.1 Comparison of Clinical Features

Non-ST-segment elevation group rate history of diabetes, a history of recurrent angina was higher than ST-segment elevation group, by comparison were statistically significant (P <0.05). Group of non-ST-segment elevation myocardial infarction with ST-segment elevation group, the difference was statistically significant (P <0.05). Non-ST-segment elevation group, CK peak, TNI were lower than the peak ST-segment elevation group, by comparison were statistically significant (P <0.05). Detailed in Table 1.

Table 1. Comparison of clinical characteristics of the two groups

	Non-ST-segment elevation group (n = 70)	ST-segment elevation group (n = 60)
Past history		
High cholesterol [n (%)]	20 (28.57)	16 (26.67)
Hypertension [n (%)]	26 (37.14)	20 (33.33)
Smoking history [n (%)]	38 (54.29)	37 (61.67)
Diabetes [n (%)]	19 (27.14)	6 (10.00) *
Recurrent angina [n (%)]	24 (34.29)	7 (11.67) *
Myocardial infarction		
Anterior [n (%)]	27 (38.57)	38 (63.33) *
Posterior wall of the [n %)]	43 (61.43)	22 (36.67) *
CK peak (IU / L)	845±468	2348±1063*
TNI peak (nag / ml)	0.83±0.49	1.67±0.71*

2.2 Comparison of Coronary Artery Disease

Non-ST-segment elevation group of single, two, three coronary artery disease were 9 cases (12.86%), 15 cases (21.43%), 20 patients (28.57%); ST-segment elevation group of single, two , three coronary artery disease were 10 cases (16.67%), 11 cases (18.33%), 8 cases (13.33%). Non-ST-segment elevation group of three coronary artery disease was significantly higher than ST-segment elevation group, by comparison ($\chi2 = 4.44$, $p <0.05$) difference was statistically significant.

2.3 Within 1 Month of Hospitalization Comparison

Group of non-ST-segment elevation within 1 month of hospitalization with atria fibrillation, cardiac pump failure, ventricular arrhythmia and death were significantly lower than the incidence of ST-segment elevation group, by comparison ($\chi2 = 8.41$, $\chi2 = 8.66$, $\chi2 = 12.88$, $\chi2 = 8.60$, both $p <0.05$) difference was statistically significant. Shown in table 2.

Table 2. Groups within 1 month of hospitalization comparison of the results

	Non-ST-segment elevation group (n = 70)	ST-segment elevation group (n = 60)
Re-infarction [n (%)]	8（11.43）	6（10.00）
Atria fibrillation [n (%)]	4（5.71）	14（23.33）*
Heart pump failure [n (%)]	6（8.57）	17（28.33）*
Ventricular arrhythmia [n (%)]	16（22.86）	32（53.33）*
Death [n (%)]	2（2.86）	11（18.33）*

Note: * p <0.05

3 Discussion

Non-ST-segment elevation acute myocardial infarction as the main pathogenesis of platelet thrombosis, platelet thrombus allows acute coronary subtotal occlusion, subendocardial injury, which showed no ST-segment elevation [3]. Whether the clinical ECG and elevation of ST-segment elevation can show the extent of myocardial injury [4]. Patients with single vessel coronary artery disease due to lack of myocardial protection of ischemic preconditioning, it is more sensitive to myocardial ischemia, severe injury [5]. Two, three due to myocardial infarction in patients with coronary artery disease before the existence of ischemic preconditioning, therefore making it less sensitive to myocardial ischemia, which is not easy to show ST segment elevation. Therefore, severe coronary artery disease to reduce the sensitivity of myocardial ischemia, myocardial necrosis, and reduced peak enzymes lower, clinical performance is not easy on the elevation of the ST segment.

In this paper, Africa ST-segment elevation group rate history of diabetes, a history of recurrent angina was higher than ST-segment elevation group; non-ST-segment elevation group of three coronary artery disease was significantly higher than ST-segment elevation group. Non-ST-segment elevation group, CK peak, TNI were lower than the peak ST-segment elevation group, by comparison were statistically significant (P <0.05). This group of researchers Africa ST-segment elevation acute myocardial infarction after myocardial infarction wall of the majority, and ST-segment elevation acute myocardial infarction myocardial infarction before the walls are mostly results of this study and Quan-Sheng Zhang et al study [6]

Non-ST-segment elevation acute myocardial infarction in patients with vasospasm due to self-discharge and the characteristics of thrombus autolysis and therefore making it relatively low hospital complications [7]. In this paper, Africa ST-segment elevation group hospital 1 month of atria fibrillation, the heart pump failure, the incidence of ventricular arrhythmia and death were significantly lower than the ST-segment elevation group, by comparison were statistically significant (P <0.05). But non-ST-segment elevation acute myocardial infarction in coronary artery disease is higher, the basis of more serious disease, it makes its long-term effect of the relatively poor prognosis.

In short, the clinical non-ST-segment elevation acute myocardial infarction mainly anticoagulant, ant platelet, lipid-lowering stations and other comprehensive treatment while actively maintaining cardiac function, reduce the incidence of complications, if necessary the use of coronary revascularization to improve their long-term prognosis and efficacy [8].

Acknowledgment. The paper is supported by Guangxi Science and Technology Agency Fund (Gui financial education [2010] No. 13).

References

[1] Wang style non-ST-segment elevation myocardial infarction in clinical analysis. PJCCPVD 17 (5), 404 (2009)
[2] Will be the Chinese Medical Society of Cardiology. Acute myocardial infarction diagnosis and treatment guidelines. Journal of Cardiology 29(12), 710–725 (2001)
[3] He, J., Lee: Such 15 cases of non-ST-segment elevation myocardial infarction diagnosis experience. Hainan Medical 16(1), 31–32 (2005)
[4] Shi-in: Non-ST-segment elevation myocardial infarction clinical analysis of 75 cases. China Modern Drug Application 3(11), 82–83 (2009)
[5] That open the constitution, Yu Ping should be the importance of non-ST-segment elevation acute myocardial infarction diagnosis and treatment. The world's Intensive Care Medicine 2(1), 533–536 (2005)
[6] Zhang, Q.-S., Zhilu, W., Yaqin, B.: ST-segment elevation and non-ST-segment elevation acute myocardial infarction comparative study. Chinese Critical Care Medicine 19(11), 697–698 (2007)
[7] Sun, Y.: Non-ST-segment elevation acute myocardial infarction diagnosis and treatment. Chinese Medical Innovation 6(9), 111–112 (2009)
[8] Nishi, M., Chow, Ray, J., et al.: Non-ST-segment elevation myocardial infarction Retrospective analysis of 216 cases. Articles 2(8), 892–899 (2008)

On the Batch Scheduling Problem in Steel Plants Based on Ant Colony Algorithm

Li Dawei[1], Zhang Ranran[2], and Wang Li[3]

[1] Journal Editorial Department, University of Science and Technology Liaoning, China
[2] School of Science, University of Science and Technology Liaoning, China
[3] School of Electronic and Information Engineering, University of Science and Technology Liaoning, China

Abstract. The continuous casting direct hot charge rolling (CC-DHCR) process is widely used by steel plants. This process is also put forward higher requirements for batch scheduling, especially in continuous casting-hot rolling stage. For this reason, an integrated production model is established about batch scheduling. With the objective of minimizing the production cost, the model satisfies three procedures constraints in steelmaking, continuous casting and hot-rolling. Then, coordination grouping ant colony is used to solve the model. Using practical production data in the experimental simulation, optimization effect and effective combination batches are obtained. It shows that the algorithm is feasibility and effectiveness for the batch scheduling problem.

Keywords: Continuous casting, Direct hot charge, Ant colony algorithm.

1 Introduction

As an advanced production process, CC-DHCR, with many advantages, such as saving energy consumption, increasing equipment utilization and production, has been applied in many steel plants [1, 2].

The CC-DHCR process needs that there are no intermittent domains among steelmaking, continuous casting and hot rolling process, so a variety of production constraints must be taken into account to optimize the production schedule, such that the CC-DHCR batch scheduling problem has become one of the hot research topics in iron and steel field [3-7].

This paper establishes a batch scheduling model about CC-DHCR process. Firstly, the model sorts a part of orders, and then develops production batch planning. Finally, an improved ant colony algorithm is applied to solve the model.

2 Scheduling Model Establishment about CC-DHCR Process

CC-DHCR process is a complex technology. The constraints considered in CC-DHCR process are as follows:

Z. Du (Ed.): Proceedings of the 2012 International Conference of MCSA, AISC 191, pp. 645–650.
springerlink.com

(1) steel making - continuous casting stage: steel kinds, slab width, furnace capacity, billet width changes, the rules of width changes (from wide to narrow, or, conversely), tundish life.
(2) hot rolling process: rolling hardness change, surface quality, rolling width and thickness, length limits with same or different width, total rolling length.

2.1 Model Description

According to the production capacity ratio of single continuous casting machine and single rolling line, the batch ratio between hot rolling unit and continuous casting unit has the following three modes [5], $1:1$, $1:n$, and $n:1$, which represents the number ratio between hot rolling unit and continuous casting unit in a batch. In this paper, the mode $1:2$ is adopted, which means that a hot rolling unit matches two continuous casting units in a batch.

2.2 Model Establishment

Firstly, the following decision variables may be defined:

$$X_{ijk} = \begin{cases} 1, & \text{order } j \text{ is scheduled after order } i \text{ in batch } k; \\ 0, & \text{otherwise.} \end{cases},$$

$$Y_{ik} = \begin{cases} 1, & \text{order } i \text{ in batch } k; \\ 0, & \text{otherwise.} \end{cases}, \text{ where } i, j = 1, 2, \cdots, n \text{, and } k = 1, 2, \cdots, m.$$

Then, the batch scheduling model can be established as follows:

$$\min \sum_{i=1}^{n} \sum_{j=1}^{n} \sum_{k=1}^{m} c_{ij} X_{ijk} \tag{1}$$

$$s.t. \quad x_i = \begin{cases} 1, & Q_i \geq T; \\ 0, & \text{otherwise.} \end{cases} \qquad i = 1, 2, \cdots, n. \tag{2}$$

$$\sum_{i=1}^{n} X_{ijk} \leq 1 \qquad j = 1, 2, \cdots, n; k = 1, 2, \cdots, m. \tag{3}$$

$$\sum_{j=1}^{n} X_{ijk} \leq 1 \qquad i = 1, 2, \cdots, n; k = 1, 2, \cdots, m. \tag{4}$$

$$\sum_{i=1}^{n} Y_{ik} \leq R \cdot L \qquad k = 1, 2, \cdots, m. \tag{5}$$

$$\sum_{i=1}^{n} Y_{ik} \cdot d_i \leq C_{\max} \qquad k = 1, 2, \cdots, m. \tag{6}$$

$$L_{\min} \leq \sum_{i=1}^{n} Y_{ik} \cdot r_i \leq L_{\max} \qquad k = 1, 2, \cdots, m. \tag{7}$$

$$\sum_{i=1}^{n} \sum_{j=1}^{n} X_{ijk} \leq n - 1 \qquad k = 1, 2, \cdots, m. \tag{8}$$

2.3 Model Explanation

In above model, the objective function (1) is to minimize the penalty values; constraints (2) denotes the timing control conditions; constraints (3) and (4) denote each order will at most be and can only be assigned to a batch; constraint (5) denotes the limitation number of continuous casting charge; constraint (6) is the length limitation of the slab with the same width; constraint (7) is the limitation for rolling length; constraint (8) avoids the emergence of sub-loop.

3 Model Solutions

The CC-DHCR batch scheduling problem can be attributed to the multiple traveling salesman problem, that is, a directed graph $G=(V,A,T)$, where V is the order set, the set of adjacent orders (i,j) corresponds to the set of edges in the graph arc $A=arc\{i,j\}$, the penalty value of the order j scheduled after orders i is the penalty function in each arc. The objective is to find the simplest sub-loop, such that the total penalty values are minimized.

Collaborative grouping ant colony algorithm based on the maximum and minimum of ant system (MMAS-CGACA) [8, 9] will be used for solving the CC-DHCR batch scheduling in the paper.

The difference between MMAS-CGACA and ant colony algorithm is that:

(1) Ant colony is divided into some groups, in which ant, respectively, starting from a different node in a particular place to meet, to collaborate to complete a search;
(2) The introduction of incentive factor;
(3) Path information to ensure that changes within the scope of $[\tau_{\min},\tau_{\max}]$, in order to avoid the algorithm by obstructing or into a local optimum; and
(4) Collaborative strategy to accelerate the search speed.

Ant groups are set a larger number, based on the search space and the distribution of the order properties.

$$k_{\max}=\text{int}\left(k_{\alpha}\cdot L/L_{\min}\right) \tag{9}$$

where " int " denotes taking integer, L is the total rolling length, $L_{\min}$ is the minimum rolling length of a batch, and $k_{\alpha}\le 1$ is a coefficient, set according to the distribution of the order properties.

4 Scheduling Simulation

4.1 Parameter Setting

Simulation is carried out in MATLAB 7.1 programming environment. Some other data in the model are: the minimum and maximum rolling length are 30km and 110km; the rolling length of the slab with the same width is 30km; the length limitation for inverse width-rolled is 10km, the limitation number of continuous casting charge is 10. Other data such as the change penalties of rolling width, thickness and hardness, please see references [4] and [10, 11].

MMAS-CGACA parameters are set as: pheromone inspiration factor $\alpha = 1$; expectations heuristic factor $\beta = 5$; pheromone evaporation coefficient $\rho = 0.1$, the total pheromone is 10000; $\tau_{\min} = 0.1$, and $\tau_{\max} = 10$.

4.2 MMAS-CGACA Solution Results

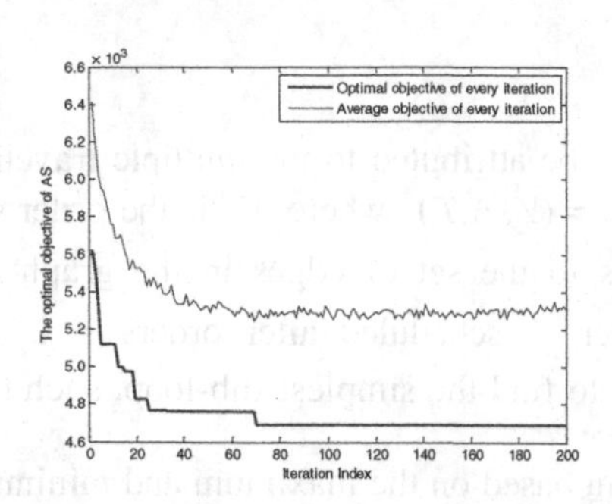

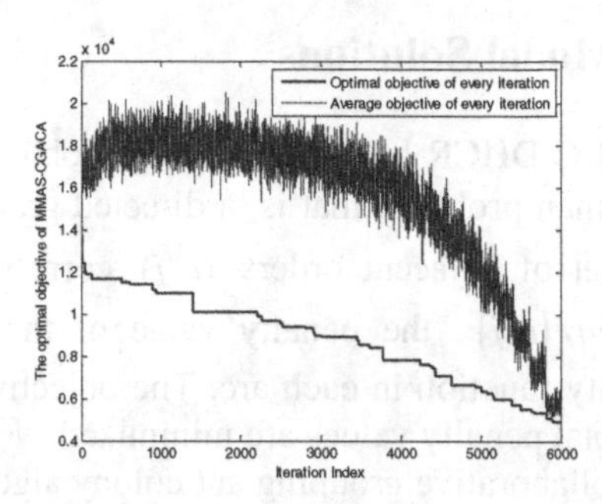

Fig. 1. The optimization results of AS and MMAS-CGACA

Figure 1 is the use of the simulation results of the AS iteration 200 to optimize performance and MMAS-CGACA algorithm for the collection of orders. From figure we can directly see that MMAS-CGACA algorithm is better than the AS algorithm.

Table 1. Scheduling results of using CC-DHCR process ($L_{\max}$=109.5km, $L_{\min}$=99.5km)

steel kind	length (km)	total charge	remainder charges	batches	batch ratio(%)	computing time(s)	Average time(s)
A	243.2	57	2	3	95.2	192	12
B	954.5	123	8	8	93.8		
C	472.3	70	9	5	88.8		
total	1506.2	250	19	16	92.6		

Table 2. A composition of a typical batch in the result

order	steel kind	casting width	rolling width	rolling thickness	surface quality	order priority	charg e	rolling length
1	B	1700	1700	7.67	1	3	5	12.9
2	B	1700	1680	7.65	1	3	2	10.5
3	B	1700	1650	6.50	1	3	5	11.1
4	B	1650	1600	5.38	2	2	5	29.7
5	B	1650	1550	6.54	2	2	6	25.8
6	B	1650	1520	5.86	2	2	3	11.7
7	B	1650	1500	5.94	2	2	2	7.9
total							28	109.6

It can be seen from table 2 that the rolling width, thickness and hardness are in the range of the jump coefficient. Moreover, continuous casting of different steel kinds and the width of the casting machine transitions also meet specification. Number of continuous casting in the tundish is within the scope of the economic times.

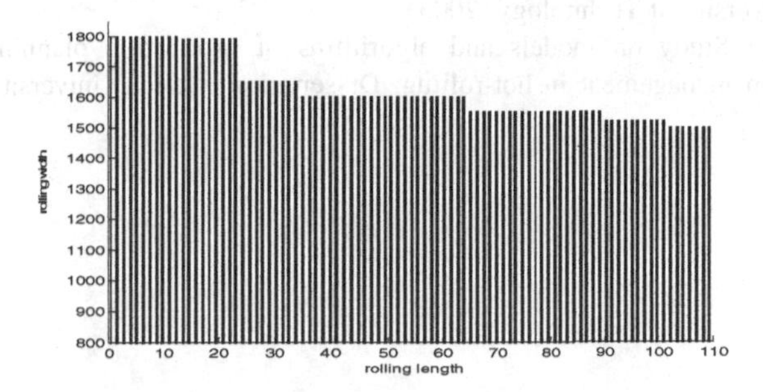

Fig. 2. Rolling width sequence in a batch

Figure 2 shows the rolling width sequence in a typical batch. The horizontal axis in the figure shows the total rolling length of a batch, while vertical axis shows the rolling width. The changes of slab width in a batch meet the specification, and the total rolling length is within the length limitation.

5 Summary

The CC-DHCR batch scheduling model established in this paper took into account many constraints which include rolling machines and slabs. The improved ant colony algorithm is introduced for solving the problem. The simulation results illustrate the feasibility and effectiveness of the algorithm in solving the problem.

References

[1] Sujian, L., Zhiming, C.: Logistics Management in Casting-rolling Production. Metallurgy Industry Press, Beijing (2001)
[2] Zensaku, Y.: Direct linking of steelmaking to hot rolling. Steel Times 4, 180–186 (1989)
[3] Cowling, P., Rezig, W.: Integration of continuous caster and hot strip mill planning for steel production. Journal of Scheduling 3, 185–208 (2000)
[4] Yaojia, Z.: Research on modeling and optimization algorithms for DHCR integrated scheduling problem. Dissertation, Shanghai Jiaotong University (2007)
[5] Shushi, N.: Research and application of integrated scheduling of steelmaking-continuous casting-hotrolling. Dissertation, Dalian University of Technology (2006)
[6] Lixin, T., Jiyin, L., Aiying, R., Zihou, Y.: A review of planning and scheduling system and methods for integrated steel production. European Journal of Operational Research 133, 1–20 (2001)
[7] Jian, X., Zhimin, L., Jinwu, X.: Model and algorithm of integrative batch planning based on parallel strategy for steelmaking-continuous casting-hot rolling. Control and Decision 21, 79–83 (2006)

[8] Shuo, M., Anping, W., Tianyu, Z.: Ant colony algorithm and its application to combinational optimization. Journal of Xianyang Normal University 23, 52–53 (2008)
[9] Jin, W., Shaomei, Z.: A grouped ACO algorithm based on MMAS with award and penalty strategy. Computer Applications and Software 26, 237–285 (2009)
[10] Jiewei, Z.: Converter automation control system design and implementation. Dissertation, Dalian University of Technology (2005)
[11] Yaohua, L.: Study on models and algorithms of production planning and slab-yard optimization management in hot-rolling. Dissertation, Dalian University of Technology (2005)

A Multi-channel Real-Time Telemetry Data Acquisition Circuit Design

Wang Yue, Zhang Xiaolin, and Li Huaizhou

Beijing University of Aeronautics and Astronautics, Beijing, China
Wang0011yue@126.com

Abstract. The paper presents a multi-channel real-time telemetry data acquisition circuit based on FPGA. The design achieves analog signals acquisition using the method of multi-access time-sharing acquisition. In the data acquisition circuit design, the transmission and collection of analog signals are easy to be disturbed. In order to solve the signal distortion occurred in the collection process, combining with the actual situation of the circuit, the paper analyzes and discusses the reason of the abnormal phenomena occurring in analog signal transmission, suggests improvements in the hardware analog signal conditioning circuit design, and optimizes the AD real-time sampling sequence. By practically using, analyzing and testing, the improving methods are presented in the system are realized to greatly simplify the hardware circuits and improve the precision, reliability and stability.

Keywords: real-time data acquisition systems, sampling accuracy, signal conditioning.

1 Introduction

The telemetry system is an important part of equipment in the missile (satellite) systems and other aircraft systems. For studying the performance of aircraft, the telemetry equipment can provide the flight parameters of the system working state data and environmental parameters, and also offers the basis for fault analysis [1]. With the continuous development of telemetry technology, more and more aircraft monitoring objects is needed for missile (satellite) and collection accuracy is increasingly demanding. The object which needs to be measured is usually various, the differences of signal frequency and amplitude are larger [2]. Therefore, how to collect the measured analog parameters accurately, simply, quickly and achieve high-precision sampling becomes one of the key points of telemetry systems. This paper carries out the research on this high-precision multi-channel telemetry system for real-time data acquisition. Paper describes the basic composition of the data acquisition circuit and proposes improvement measures to solve the analog signal distortion in collection process and it improves the accuracy of real-time data acquisition system sampling effectively.

Z. Du (Ed.): Proceedings of the 2012 International Conference of MCSA, AISC 191, pp. 651–655.
springerlink.com

2 Circuit Components and Working Principle

The diagram of multi-channel telemetry data acquisition device is shown in Fig. 1. It uses a modular design structure which mainly composed by the main control module, signal conditioning acquisition module, the internal bus.

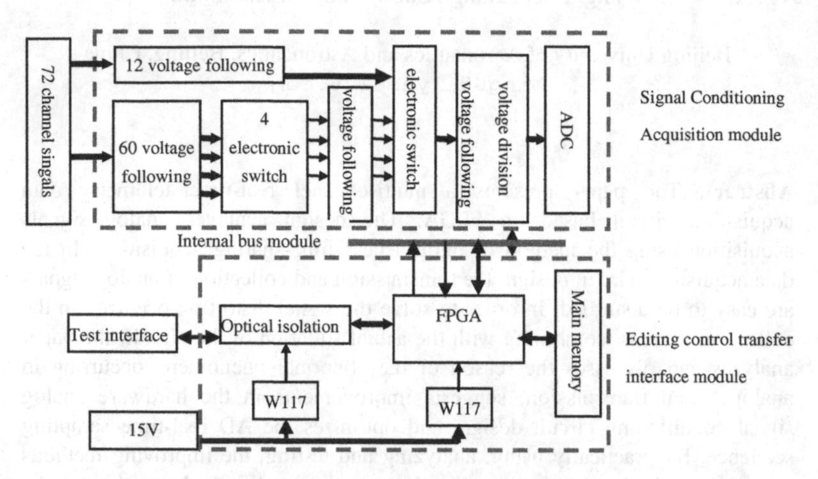

Fig. 1. Telemetry data acquisition circuit principle diagram

System uses FPGA (XC2S100E) as the main control chip and AD7667 as the ADC controller. The AD7667 is a 16-bit, 1 MSPS, charge redistribution SAR, analog-to-digital converter that operates from a single 5V power supply. The part contains a high-speed 16-bit sampling ADC, an internal conversion clock, internal reference, error correction circuits, and both serial and parallel system interface ports. AD7667's input range is 0 ~ 2.5V. FPGA control the analog electronic switch and the ADC achieve analog to digital conversion of multi-channel signals. According to the command of ground, the data is compiled in a format frame processing and then sent to the memory module or directly uploaded to the test equipment.

3 Circuit Design of Signal Conditioning Acquisition Module

In the telemetry system, often need to use the sensor to convert the physical quantity that under test to electrical signals, but the signal are often limited by the sensitive components and the characteristics of the detection circuit in the form, amplitude, etc. Generally it can not be directly collected by the test system. Signal conditioning is the processing of these signals and convert them into a suitable signal for AD conversion[3-4].

Analog signal conditioning circuit is channel of the sampled signal. To achieve high input impedance and signal isolation, the input analog signal must go through a voltage follower circuit firstly. As shown in Fig. 2, we use OPA4340 with rail-to-rail amplifier characteristics as the voltage follower in the design. OPA340 series rail-to-rail CMOS operational amplifiers are optimized for low voltage, single supply operation. Rail-to-rail input/output and high speed operation make them ideal for driving sampling analog-to-digital converters.

In practice, we supply +5.2V power can satisfy 0 ~ 5V signal following. At the same time signal transmission speed can satisfy the requirement of input signals and match subsequent analog switch speed and impedance. When the circuit open circle, the input through R1 will get down ensuring the circuit can work normally. Resistance R2 (10K) is current limiting resistor preventing to produce over-voltage damaged circuits.

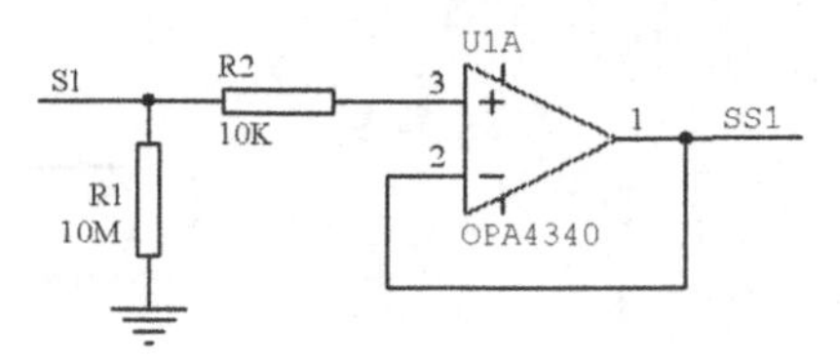

Fig. 2. Analog signal input follow circuit

After conditioning, multi-channel analog signals entered into multi-input electronic switches. Electronic switches devices required, in the choice, larger shutoff resistance, smaller conduction resistance, short switching time and adequate transmission accuracy. The ADG706 is low-voltage, CMOS analog multiplexers comprising 16 single channels and eight differential channels respectively. The ADG706 switches one of 16 inputs (S1–S16) to a common output, D, as determined by the 4-bit binary address lines A0, A1, A2, and A3. An EN input on both devices is used to enable or disable the device. The device is designed on an enhanced submicron process that provides low-power dissipation yet gives high switching speed, very low on resistance and leakage currents[5].We should add a voltage follower circuit after the electronic switch to ensure the transmission voltage signal is not distorted.

Because each channel collection of signal sampling rate is not identical, rapidly-varied signals sampling rate should be higher than slowly-varied signals. In the collection control, we need accord the minimum sampling rate to edit data frame. We should ensure the frame format in a complete data frame, minimum sampling rate of channel control choose only once, while other times the minimum sampling rate of N-channel gating is required N times.

4 System Optimization and Testing

Conducting circuit verification, input frequency of 4Hz, 0 ~ 5V amplitude square wave signal, and then start the acquisition. When the acquisition is to complete, read-back the data and redraw the waveform using the computer. But the waveform appeared seriously distortion, as shown in Fig. 3. With the oscilloscope amplifier testing output signal, the output signal show the overshoot and self-oscillation, shown in Fig. 4. Analysis of the circuit OPA4340 op amp voltage follower circuit in the role play, the voltage amplification factor close to 1, and in the circuit with a capacitive load, such as the analog switch is a capacitive load ADG706 (ADG706 features), rather in the OPA4340 output termination of a capacitor, it causes the op amp's output resistance of the output phase lag. When the input signal and output signal phase difference between the 180 °, the negative input and positive input exactly the same positive feedback circuit.

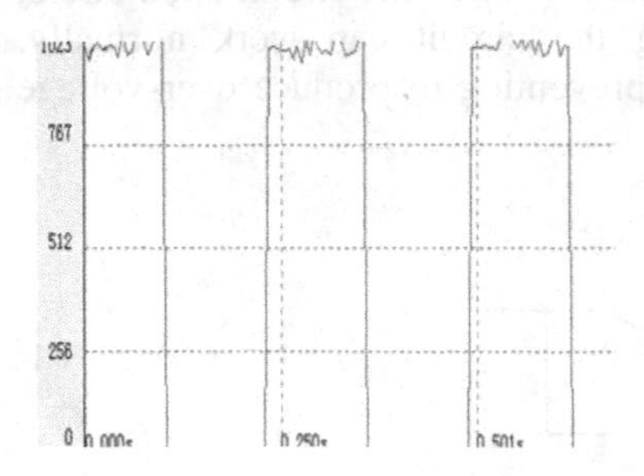

Fig. 3. The distortion waveform

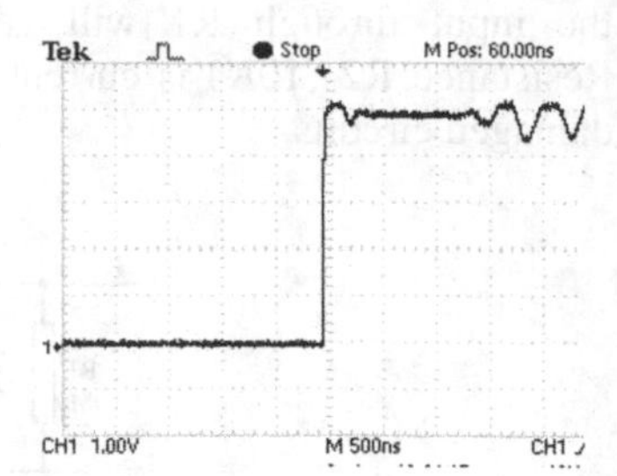

Fig. 4. Self-excited oscillation of the square wave

This condition of self-excited oscillation is consistent: the voltage amplification factor close to 1; additional phase shift reaches 180 °, the negative feedback into positive feedback. Meanwhile, in the process of signal acquisition, ADG706'EN in the use of the process could have been effective, when the analog switch gate of a channel, the channel has been in the conduction state, and the switch does not exist after strobe directional signal transmission, so ADG706 output signal to the next level will not only affect the output of op amp, while also gated by ADG706 return to this channel input, and the oscillation of the original input signal superposition, the output signal Interference enhancement; all the way to A/D converter inputs are subject to the impact of the oscillation signal, making the acquisition of the signal distortion, affecting acquisition accuracy.

In order to achieve the higher accuracy of the acquisition systems, we need to improve the stability of capacitive load for the op amp circuit. The paper proposes a optimized design for the analog signal interface circuit, as shown in Fig. 5, the op amp's two input termination resistor R173 (10K), the feedback terminal access resistor R174 (100K). This design is to improve the circuit's noise gain, which is the effective way to low-frequency circuit stability. That by increasing the closed-loop gain of the circuit (the noise gain) without changing the signal gain to make the circuit stable. The signal gain is 1, and the noise gain is: 1 + R147/R173 = 11, compared with the original design increase by 10 times. The disadvantage of this design makes the signal bandwidth is much more restricted. Because the gain-bandwidth product of the system is fixed, the system gain is increased, the bandwidth will be reduced. In the design, the new circuit cut down the system bandwidth, improve the system stability. Using the same square wave signal to verify the circuit after optimization, the op amp square wave output signal waveform shown in Fig. 6, the self-excited oscillation disappeared, but the overshoot still exists. In Fig. 7, computer collected data back to redraw the waveform showed that elimination of oscillation and the circuit stability has improved significantly.

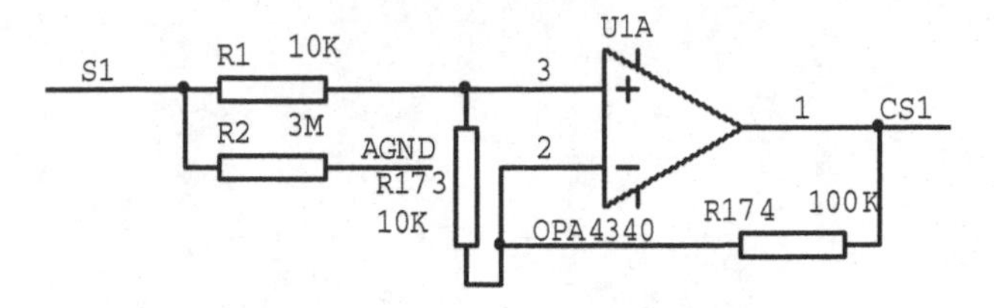

Fig. 5. Optimal Design of analog signal processing circuit

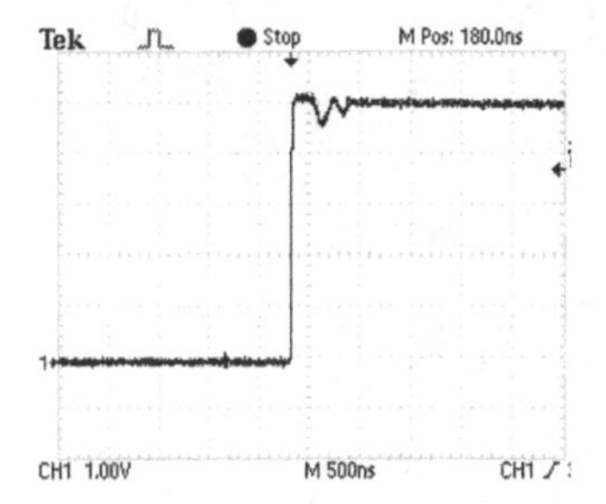

Fig. 6. Optimal design of the distortion waveform

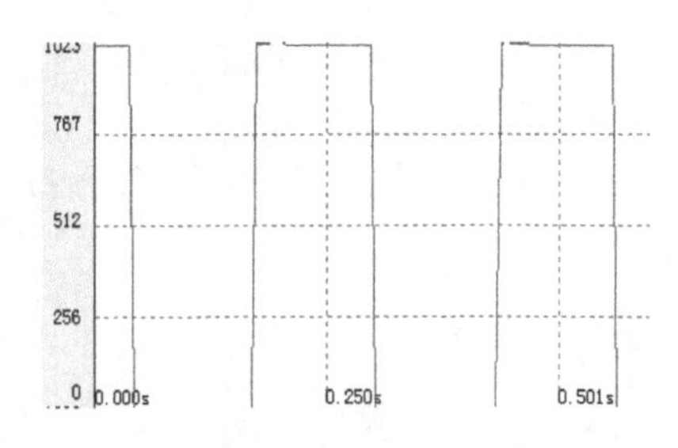

Fig. 7. Optimal design of square wave

5 Conclusions

Multi-channel real-time telemetry data acquisition equipment is a key component of the telemetry system. According to the design requirements, we design the circuit of multi-access time-sharing acquisition through real-time analog signal conditioning circuit and control the timing of the optimal sampling design, strengthen the stability of operational amplifier circuit in the capacitive load conditions, effectively improve the sampling accuracy. It's sampling accuracy is up to 0.1%. The hardware design is simple, practical to meet the requirements of the remote sensing analog signal acquisition. It has a certain value for application and promotion.

References

[1] Zhen, G., Sun, D., Zhao, H.: Mixed-signal data acquisition system. Instrument Technique and Sensor (12), 49–51 (2008)

[2] Operational Amplifiers design and Applications J.G.Graeme McGraw-Hill (1971)

[3] Yorozu, Y., Hirano, M., Oka, K., Tagawa, Y.: Electron spectroscopy studies on magneto-optical media and plastic substrate interface. IEEE Transl. J. Magn. Japan 2, 740–741 (1987); Digests 9th Annual Conf. Magnetics Japan, p. 301 (1982)

[4] Iyad, O., Nicolelis, M.A.L., Wolf, P.D.: A low power multichannel analog front end for portable neural signal recordings. Journal of Neuroscience Methods 133(1-2), 27–32 (2004)

[5] Analog Device Inc. 8/16-Channel Multiplexers ADG706/ADG707 (2000)

Research on Optical Fiber Inspection Technology Based on Radio Frequency Identification

Yu Feng[1], Liu Wei[1], and Liu Jifei[2]

[1] Network Center of Shenyang Jianzhu University, China
[2] Information and Control Engineering Faculty, Shenyang Jianzhu University, China

Abstract. We have studied the actual function demand for the optical cable patrol system and proposed an improved method of optical fiber inspection based on ARM9 processor and GPRS technology.RFID is used for malfunction maintenance and general management under online supervision.The system has the ability of real-time communication with inspection terminals by RFID and ARM technology.It can also make a combination with existing fiber optic monitoring system to deal with field diagnostic.Database programming and management are taken to improve the performance in real-time and two–way.

Keywords: inspeciont system, optical fiber, RFID, patrol.

1 Introduction

With the rapid development of communication and the increasing expansion of network capacity,the security and stability of optical fiber lines have attracted more attention. The intelligent cable patrol system provides a modern measure for quantifiable and dynamic management of communication lines inspection.It is also an important procedure in promoting the line maintenance departments to implement a scientific management nowdays. The existing optical fiber inspection systems consist mostly of the patrolling point,the portable instrument,and the management platform and the principle is reading and writing on the radiofrequency IC card.But the trajectories collected by line patrol members can not be transferred to the management platform real-timely and the management platform lacks communication with the line patrol members.Then, to meet the needs of lines maintenance for optical fiber communication, we propose a improved scheme with the combination of RFID,GARM and GPRS technology.It can intelligent manage the situation of regularly and temporary patrol and provide comprehensive information service for the managers in all classes.The experiments show our system has effectively solve the problem in intelligent patrol of optical fiber lines and has broad application prospects.

2 Research of Optical Fiber Inspect System

The safety and smooth flow of fiber optic cable line or not, in a certain sense, depending on the line daily to maintain the level of management and prevention of

Z. Du (Ed.): Proceedings of the 2012 International Conference of MCSA, AISC 191, pp. 657–661.
springerlink.com

barriers, rules of order transmission line maintenance staff should adhere to the fiber-optic cables regularly tour. Line inspection widely used traditional manual inspection, the use of paper-based manual records of work, the way there are many human factors, low efficiency, and can not supervise staff working conditions and other road defects.The apparent lack of timeliness of maintenance, management, lack of realism. Often result in delays in processing time, is difficult to clear up and down the responsibilities,relationships, conflict, inefficiency, and line obstacles and losses may increase. Therefore, we have before us a very pressing issue is how to use scientific and effective management methods, and promote the implementation of inspection system, the timely detection of problems in the inspection, the prevention of various types of blocking.The traditional inspection system is the manager of the inspection arrangements for inspection personnel schedule, inspection personnel in accordance with the time table prescribed time inspection, but they are only human on the line inspection, check the line status. In addition, the inspection staff routes to see which monuments, look carefully they are away from the manager or other work,so manager work of inspection staff can not do at a glance,or even knew anything about.Line intelligent inspection system is introduced to quantify the communication line inspection and dynamic management of modern methods, a new form of communication line maintenance departments to promote the implementation of scientific management is an important step. Intelligent data logging system to maintain the daily management of the main line departments to achieve the daily management of the maintenance department of computer dynamic management, strengthen the maintenance department management, scientific, institutionalized and improve the overall management of the maintenance department.

3 Research of Optical Fiber Inspect System

For routine inspection tasks cable line, designed based on RFID technology, integrated GPS technology and the current advanced ARM technology in one of the intelligent optical cable inspection system.The intelligent data logging system consists of data logging devices, inspection points and inspection management platform shown as figure 1.

The working principle is the implementation of inspection tasks, inspection personnel handheld data logging devices or vehicle inspection path detection apparatus according to the road, to reach inspection point, inspection point to read RFID tags on the information in the monuments, logging devices will read information and GPS data marked stones together, to create a record of inspection into the CF card. In turn checked all the inspection points, all inspection records have been recorded in the CF card, the task execution after the inspection with a CF card reader to transfer records to the management platform, the management system software analysis and processing, as required to generate Tour inspection re-ports for management staff to examine.

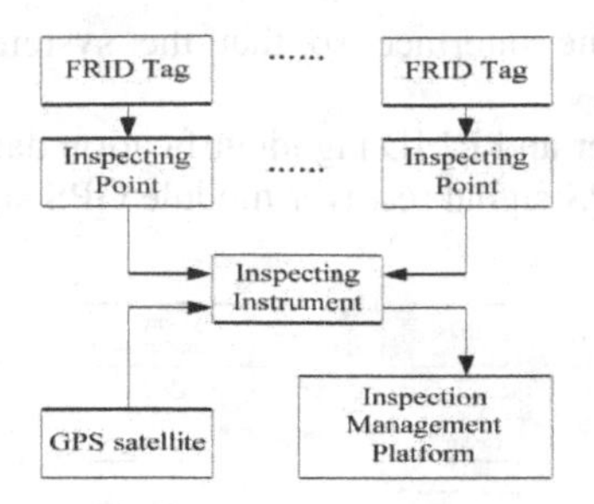

Fig. 1. Constitute of inspect system

3.1 Inspecting Point

Any inspection tasks should be pre-set inspection route,and strive to do the work of state inspection personnel managers well known, while increasing inspection personnel managers and transparency between. The inspection line is composed by a number of inspection points,and each point has a unique number. Inspection by the patrol route point number sequence is determined. The main role is to inspection point data logging devices with wireless communication, data logging devices to record the inspection point is the address, time and line status signal, indicating that this point has been inspection too.Inspection point circuit mainly by the RFID tag circuit.RFID tag information stored monuments installed in the standard rock, stone marked patrol officers arrived after the RFID tag read information, data logging devices in the CF card will record the number of the monuments, inspection location, inspection time and other information.

3.2 Inspecting Point

Detection apparatus is the key to the inspection system.It is not only with the Inspection Management Work-bench wired communications, but also with each inspection point for wireless communications, storage capacity required to ensure that the data does not overflow. Data logging devices overall structure of the ARM processor,memory, power, reset, system JTAG interface, network interface, manmachine interface, RFID reader modules,GPS signal receiver module circuit.The overall structure of the hardware shown in Figure 2.

1.Complete system reset circuit at power-on reset and the user when the system reset button. Power supply circuit is provided to LPC2210 I/O port operation 3.3V power supply and on-chip peripherals and core 1.8V power supply required.

2.FLASH memory to store user application has been debugged, embedded operating system or other systems the user needs to be saved after power-down data. SDRAM memory system is running as the main area, system and user data, the stack are located in SDRAM memory.

3.10M Ethernet interface for the system to provide Ethernet access to the physical channel, through the inter-face, the system can l0Mb/s Ethernet access rates.

4.Chip JTAG interface can access all the parts,through the interface of the system debugging, and programming; LCD display and keyboard scanning the main provider of human-machine interface so that the system has good man-machine interface.

5.Complete RFID reader and RFID tag identification data read.

6.Completion of the GPS signal receiver module GPS signal reception.

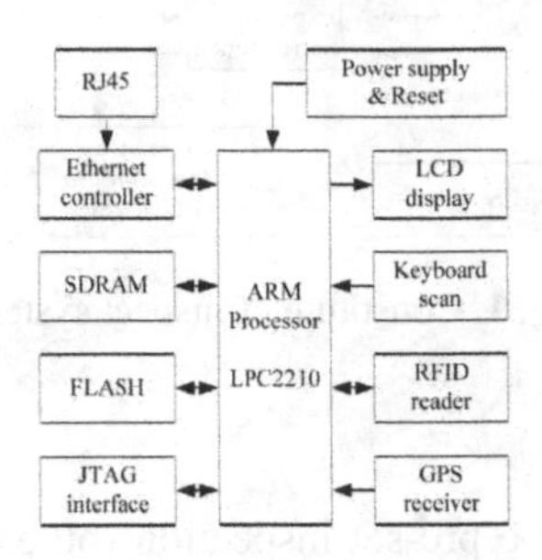

Fig. 2. Software structure of the system management

The design uses a CF card as storage units, which mechanical properties and stability, small size, large capacity.Inspection process, the processor to external data memory inspection point inspection data and inspection data to the file path stored in the CF card. In order to facilitate computer processing inspection instrument data management software, using the FAT16 file system to receive the data files stored on the CF card, CF card reader, or through the Ethernet port on the computer to read.Data logging devices in the boot after a software pro-gram running always in circulation, due to the operation of reading the RFID tag information is intermittent, so thereader with a button to control the operation.

3.3 Inspection Management Platform

Inspection Management Workbench mainly equipped with inspection by the management terminal management software components. Inspection Management software to complete the main features include: data communications, data query, system settings and data maintenance.Data communications, including: reading records, clear history, set patrol routes and other functions; data queries,including: query, statistics, printing and other functions;system settings, including: Patrol locale, set the data logging devices, set-up and inspection personnel inspection settings; data maintenance, including data backup and data recovery.

1.System management software is modular in design shown as figure 3.By the master interface to control the entire system operation to achieve the switching module, the module functions as follows:

2.Management system inspection personnel information management module, signal equipment, RFID tags, data logging devices and other information.

3.Tag to set the module configuration information stored in RFID tags, RFID tags to set passwords and other operations.

4.Inspection before the task, the task management module input inspection tasks, such as inspection personnel, inspection paths.

5.Check printing module queries inspection personnel, signal equipment, RFID tag information, and inspection of an officer, a patrol point inspection and so on, need to print out the information by category.

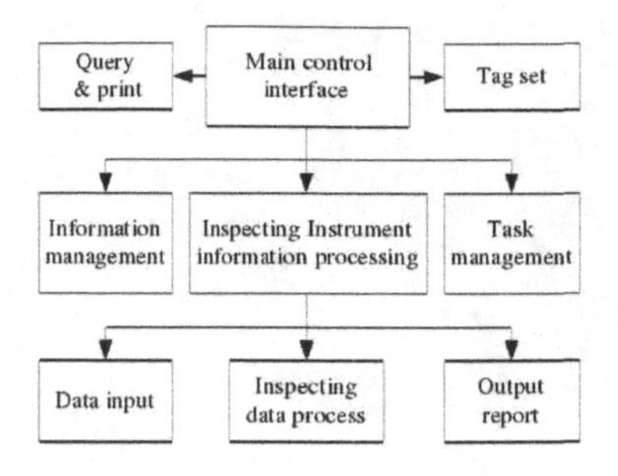

Fig. 3. Software structure of the system management

4 Conclusion

This design of intelligent optical inspection system consists of hand-held line inspection device, inspection points and inspection management software system. The system combines radio frequency identification (RFID),global positioning satellite (GPS), embedded systems technology, and to intelligently manage the regular cable line maintenance personnel traveling and temporary road conditions, while taking advantage of the system management platform built for all levels of maintenance and management personnel to provide intelligent information services.

System through simple improvements can be further extended to railway inspection, pipeline inspection, intelligent community patrol inspection and other key targets.This system not only can be used for line management,but also in the maintenance of special equipment, industrial pipelines, roads, railways, airways, urban transport,water conservancy fields, airports, and many have broad application prospects.

References

[1] Chartier, V.L.: Discussion of Evaluation of the potential for power line carrier to interfere with use of the nationwide differential GPS network. IEEE Transactions on Power Delivery 18(2), 649–650 (2003)

[2] Jerrel, L., Sung, C., Cheung, W.: Towards a framework for aligning RFID applications with supply chain strategies. In: Proceedings of the Ninth International Conference on Electronic Business (2009)

[3] Chu, C.H.: RFID Application Challenges and Risk Analysis. In: Proceedings of 2010 IEEE the 17th International Conference on Industrial Engineering and Engineering Management (2010)

[4] Feng, H., Fu, B., Shao, Z.: A safety patrolling model based on RFID. Journal of Shenyang University of Technology (2009)

[5] Mao, F., Li, Z., Lu, Y.: The Inspection Tour System for Intellegent Community Property Based on RFID Data Terminals. Microcomputer Information (2006)

Usability Analyses of Finger Motion in Direct Touch Technology

Xiaofei Li, Feng Wang*, Hui Deng, and Jibin Yin

Computer Technology Application Key Lab of Yunnan Province
Kunming University of Science and Technology, Kunming, China, 650500

Abstract. There is lots of research work on the direct touch technique. Among them, the study on gesture that is manipulated by the finger motions is an important part of current direct touch techniques. In this paper, we summarize the interactive gestures and the available properties of the finger motion. Single-touch, multi-touch and hand-touch interactive techniques are investigated in depth.

Keywords: HCI, Gesture recognition, Finger motion, Multi-touch.

1 Introduction

Due to its easy and intuitive, the application of the touch technique is becoming more popular in recent years. As an important part of direct touch interactions, a variety of action gestures are used in different touchable products. For example, Microsoft presented a series of finger actions such as tap, zoom, press and rotate and so on. All these actions make full use of single finger or multiple fingers motions. Most of the gestures are designed by changing the touch position. Because prior user interface widgets typically assume that the input device can only provide x-y position and binary button press information. Currently some devices support physical properties like pressure, orientation, shape of contact area and size of contact area and so on. So the gestures based on these physical properties become more and more popular. Although many interactive surfaces [5] sense the shape of the contact region, very few fully exploit the rich interaction information based on hand shape. In this paper, we make a finger motion investigation so as to help for future research.

2 Finger Interactive Properties

Many tasks implemented on direct-touch interactive surfaces rely on the finger motion. Prior literatures [1, 3] presented those two types of finger touch on interactive surfaces: vertical touch and oblique touch which distinguish with each other by the shape of contact area. A small rocking motion of the user's finger triggers the SimPress clicking technique [3]. SimPress clicking technique keeps the location fixed and changes the shape of contact area. Wu and Balakrishnan [4] have defined a

* Corresponding author.

Z. Du (Ed.): Proceedings of the 2012 International Conference of MCSA, AISC 191, pp. 663–667.
springerlink.com

gesture called Flicking which the manipulator can touch the screen and quickly slide single point away from itself, the gesture changes its location. All of these gestures mentioned above need manipulate by finger motion. Finger motion changes the value of interactive properties.

The mouse cursor acts as a digital proxy for a finger on graphical displays. But, one hand has ten fingers so that there are many degrees of freedom which are used to interact with the world. We posit that our fingers have more properties than mouse cursor, and then our fingers may perform richer and more fluid interaction. So far, the fingers are merely to position the cursor and click on currently available multi-touch interfaces, but in fact, the human hand is a complex mechanism. A total of 23 degrees of freedom (DOF) have been identified through medical and anatomical analysis. Wang and Ren [2] empirically investigated finger contact properties such as size, shape, width, length and orientation using a FTIR-based multi-touch surface. So we sort all the input properties of fingers into four aspects and illustrate them in Table 1. At one moment, if the finger's properties are fixed and keep unchanged we regard the finger is in a particular state. We say the finger state changes if more than one finger's properties change when we interact with physical world. To make it clear summarizing finger motion gestures, we define the changing of the finger state as finger motion in draft. For example, when we briefly touch surface with fingertip, the pressure, shape of contact area and size of contact area will change. When we move fingertip over surface without losing contact, the property of position will change. All these gestures are completed by finger motion.

Table 1. Classification of human finger properties

Input Property	Finger Property
Position property	Coordinate value (x, y)
Motion property	Velocity
	Acceleration
Physical property	Size of contact area
	Shape of contact area
	Orientation
	Pressure
Event property	Tap
	Flick

3 Interactive Techniques of Finger Motion

3.1 Single Touch and Its Application

In most cases, Touch-screen supports a single person operation. Sometimes user can touch the screen by only one single finger; we call the operation as single-touch motion. Wu and Balakrishnan [11] defined some single-touch motion gestures such as Tap, Double Tap, Flick and Catch. Shift and Escape selection techniques are also typical single-touch motion gestures. Touchpad on laptop PCs, a user can switch between moving the mouse and dragging with the mouse. It will trigger dragging state when user taps once and then quickly presses the finger down again. This gesture simulating mouse dragging operation is an obvious single-touch motion gesture.

SmartBoard [17] allows right clicking by pressing and holding until you see a right click menu, the gesture is also a kind of single-touch motion gesture. Computer-vision-based technologies are widely employed to enable direct-touch surfaces. Han et al. [14] introduced a multi-touch system based on Frustrated Total Internal Reflection (FTIR) which can detect not only the contact regions but also fingers hovering above the surface within a certain distance. So the surface can simply distinguish the hovering state and the dragging state. A user can put a finger above the surface within a certain distance to control the cursor movement and touch directly to drag an icon.

The finger touches the screen and keeps the position unchanged or changed very slightly except for changing the size of contact area, orientation or pressure. We regard this kind of finger motion as fixed-position finger motion. Fixed-position finger motion includes the gestures like changing finger orientation [2], changing the shape of the contact area or the size of contact area and changing the pressure. Simpress clicking technique keeps the location fixed and changes the shape of contact area. Some gestures directly control the finger angle (Roll, pitch, yaw) [6] between the finger and the touch surface to change the shape of contact area all belong to fixed-position finger motion. Shift [7] is a typical fixed-position finger motion application. When small targets are occluded by a user's finger, the proposed Shift technique reveals occluded screen content on the screen in a callout displayed above the finger. It allows users to fine tune with take-off selection technique [8, 9]. Shift belongs to fixed-position finger motion due to the finger motion is slight. Finger Sector Menu [2] based on pie menu, incorporation of the finger contact area size property improves the usability of the pie menu and makes the operation more natural. The finger sector menu is triggered by variations of finger contact area.

The finger can touch the surface and move to change the touch position to complete a gesture. We regard the kind of finger motion as unfixed-position finger motion. The Escape technique developed by Yatani et al. [10] uses a disambiguating gesture to select a target with a unique beak-like cue. A thumb tap followed by a gesture (without the release of the thumb) enables a user to select the target quickly and correctly even when it is small or occluded by other objects. The gesture sliding the thumb towards the icon direction is an obvious unfixed-position finger motion. Unfixed-position finger motion produces velocity, acceleration and changes coordinate value.

3.2 Multi-touch and Its Application

Nowadays more and more screens support Multi-touch operation, so we can complete some gestures by multiple fingers cooperating with each other. Fluid DTMouse solution [15] defines cursor hovering state. When two fingers (the thumb and middle finger in this case) are placed on the table at the same time; the cursor is moved, not dragged. When the third finger tap on the screen, now that the left mouse button has been engaged, dragging the thumb and index finger causes a drag operation centered between the fingers. When user taps the third finger again, the left mouse button will be released, cursor will restore in mouse move mode. The gesture is achieved by three fingers, so it is a kind of multi-touch motion gesture.

Microsoft defined a lot of multi-touch motion gestures. For example, to zoom out, a user should touch two points on the item and then move the fingers toward each other, as if you're pinching them together. To rotate, you should touch two points on the item and then move the item in the direction that you want to rotate it. Benko, Wilson and Baudisch [1] proposed Dual Finger Slider technique, which is a two fingers interaction that uses the distance between fingers to switch between cursor speed reduction modes. ShapeTouch [16] provides force-based interactions on 2D virtual objects that define three multi-fingers motion gestures, such as Pressing, Colliding and Friction. All the gestures above-mentioned achieved by more than one fingers are called multi-touch motion gestures.

3.2 Hand-Touch and Its Application

Some gestures are based on fingers motion, but some should be completed by the whole hand. RoomPlanner application [4] presents a variety of whole hand interaction techniques for those displays that leverage and extend the types of actions, which people perform when interacting on real physical tabletops. We call these gestures as hand-touch motion gestures. Some single hand and two hand gestures are realized on the displays where the shape of the user's hand can be sensed. Flat Hand is a hand-motion gesture that user can lay hand flat on surface. Vertical Hand gesture is triggered when a user touches surface with side of upright hand in a vertical manner. Horizontal Hand gesture is triggered when a user touches surface in a horizontal manner. Tilted Horizontal Hand is a hand motion gesture that a user can tilt top of horizontal hand away from self. Two Vertical Hands is a gesture completed by two hands that symmetrically slide two vertical hands together or apart. Two Corner-Shaped Hands is a two hand motion gesture that each hand makes a corner and touches the surface. RoomPlanner application uses these gestures to set the layout of a room, for example, a user can temporarily rotate the room layout by placing a hand flat on the table and translating that hand. When the side of a hand is placed on the surface of the table oriented that the contact surface is a vertical line, the user can use Vertical Hand gesture to sweep furniture pieces. All the gestures above-mentioned belong to hand-touch motion gestures.

4 Discussion and Future Works

Recent advances in sensing technology have enabled a new generation of tabletop displays that can sense multiple points of input from several users simultaneously. A technique for creating touch sensitive surfaces is proposed which allows multiple, simultaneous users to interact in an intuitive fashion. DiamondTouch [5] is a multi-user touch technology for tabletop front-projected displays. It enables several different people to use the same touch-surface simultaneously without interfering with each other. But now, there are few people research for more than one people working together to complete a task, this may be a future direction.

We summarize finger properties and some finger motion gestures. Meanwhile we classify finger motion gestures into single-touch motion, multi-touch motion, and hand-touch motion according to the number of the fingers. We summarize the finger motion gestures that previous have studied. Different people and companies have defined different gestures; some of the different gestures can complete the same task. In the future, we will do some research on the gestures that more than one people participate in and cooperate with each other to complete a task.

Acknowledgment. We would like to thank all the authors of the literatures which we have referred to. We appreciate the support from National Natural Science Foundation of China (61063027). We are also grateful to anonymous reviewers for your insights.

References

[1] Benko, H., Wilson, A.D., Baudisch, P.: Precise selection techniques for multi-touch screens. In: CHI 2006, pp. 1263–1272 (2006)

[2] Wang, F., Ren, X.: Empirical evaluation for finger input properties in multi-touch interaction. In: Proc. CHI 2009, pp. 1063–1072 (2009)

[3] Forlines, C., Wigdor, D., Shen, C., Balakrishnan, R.: Direct-touch vs. mouse input for tabletop displays. In: CHI 2007, pp. 647–656 (2007)

[4] Wu, M., Balakrishnan, R.: Multi-finger and whole hand gestural interaction techniques for multi-user tabletop displays. In: Proc. UIST 2003, pp. 193–202. ACM Press (2003)

[5] Dietz, P., Leigh, D.: DiamondTouch: a multi-user touch technology. In: UIST, pp. 219–226 (2001)

[6] Holz, C., Baudisch, P.: The generalized perceived input point model and how to double touch accuracy by extracting fingerprints. In: CHI 2010, pp. 581–559 (2010)

[7] Vogel, D., Baudisch, P.: Shift: a technique for operating pen-based interfaces using touch. In: CHI 2007: Proceeding of the Twenty-Fifth Annual SIGCHI Conference on Human Factors in Computing Systems, pp. 657–666. ACM, San Jose

[8] Potter, R., Weldon, L., Shneid Erman, B.: Improving the accuracy of touch screens: an experimental evaluation of three strategies. In: Proc. CHI 1988, pp. 27–32 (1988)

[9] Ren, X., Moriya, S.: Improving selection performance on pen-based systems: a study of pen-based interaction for se-lection tasks. ACM TOCHI 7(3), 384–416 (2000)

[10] Yatani, K., Partridge, K., Bern, M., Newman, M.W.: Escape: a target selection technique using visuallycued gestures. In: Proc. CHI 2008, pp. 285–294 (2008)

[11] Wu, M., Balakrishnan, R.: Multi-finger and whole hand gestural interaction techniques for multi-user tabletop displays. In: ACM UIST 2003, pp. 193–200 (2003)

[12] Synaptics Inc., http://www.synaptics.com/products/touchpad.cfm

[13] MacKenzie, C.L., Iberall, T.: The grasping hand. North Holland, Amsterdam (1994)

[14] Han, J.Y.: Low-cost multi-touch sensing through frustrated total internal reflection. UIST, 115–118 (2005)

[15] Esenther, A., Ryall, K.: Fluid DTMouse: better mouse support for touch-based interactions. In: Proc. AVI 2006, pp. 112–115 (2006)

[16] Cao, X., Wilson, A.D., Balakrishnan, R., Hinckley, K., Hudson, S.E.: Shapetouch: Leveraging contact shape on interactive surfaces. In: Proc. Tabletop, pp. 129–113. IEEE Computer Society (2008)

[17] Smart Technologies SMART Board, http://www.smarttech.com/SmartBoard

Design and Implementation of an Automatic Switching Chaotic System between Two Subsystems

Yao Sigai

Wuhan Polytechnic, 430074, China
yaosigai2012@163.com

Abstract. In order to generate complex attractor of chaos, an automatic switching system which is consists of two subsystems is constructed. Through the analog switch, the system can automatically switch between the three subsystems. With Lyapunov index and the bifurcation diagram analysis of the characteristics of the chaotic system. Analysis of the basic characteristics of the system, such as the equilibrium point, the fractal dimension, dissipation, and so on. Design of analog circuits to implement the switching function and switching chaotic system, through the circuit three chaotic systems automatically switch. Experimental results and computer simulation、Lyapunov exponent analysis and bifurcation diagram analysis are the same.

Keywords: Chaotic system, Switch function, Analog circuit.

1 Introduction

In 1963, Lorenz found a three-dimentioal autonomous chaotic system, which generated the well known Lorenz chaotic attractor[1]. The Lorenz system has been extensively studied in the fields of chaotic theory and dynamical systems[2-3]. However, the switched chaotic system has more complex dynamical behaviors than common chaotic system. It has been studied with increasing interest due to its theoretical and practical applications in technological fields, such as secure communication, laser, nonlinear circuits, neural networks, control and synchronization [4-10]. There are two main methods to generate the switched system. One is by manually operation[11-13], the other is switching automatically. The latter is better.

This paper proposes an auto-switched chaotic system, which is based on a chaotic system[4] and can change its behaviour automatically from one to another. It is also proved by the Lyapunov exponent spectrum, bifurcation diagram and experiments based on FPGA by EDA technology.

2 Construction of an Auto-Switched Chaotic System

Consider the following three-dimensional autonomous chaotic system based on a chaotic system[4]:

Z. Du (Ed.): Proceedings of the 2012 International Conference of MCSA, AISC 191, pp. 669–673.
springerlink.com

$$\begin{cases} \dot{x} = a(y - x) \\ \dot{y} = bx + cy - xz \\ \dot{z} = xf(x) - hz \end{cases} \tag{1}$$

where $f(x) = \begin{cases} x & x \geq 0 \\ y & x < 0 \end{cases}$, then an auto-switched chaotic system is obtained. When the state variable x of the auto-switched system (1) satisfies $x \geq 0$, the function $f(x)$ is x, thus it runs on one subsystem.

$$\begin{cases} \dot{x} = a(y - x) \\ \dot{y} = bx + cy - xz \\ \dot{z} = x^2 - hz \end{cases} \tag{2}$$

When $x < 0$ it runs on other subsystem

$$\begin{cases} \dot{x} = a(y - x) \\ \dot{y} = bx + cy - xz \\ \dot{z} = xy - hz \end{cases} \tag{3}$$

When a =20, b =14, c =10.6, h =2.8, subsystem (2) and subsystem (3) both have a chaotic attractor . The auto-switched chaotic system (1) transform its behavior randomly between subsystem (2) and (3) as the state x is varied. The chaotic attractor of auto-switched system (1) is shown in Fig 1, in which black parts express the orbits of subsystem (2) and red parts express the orbits of subsystem (3).

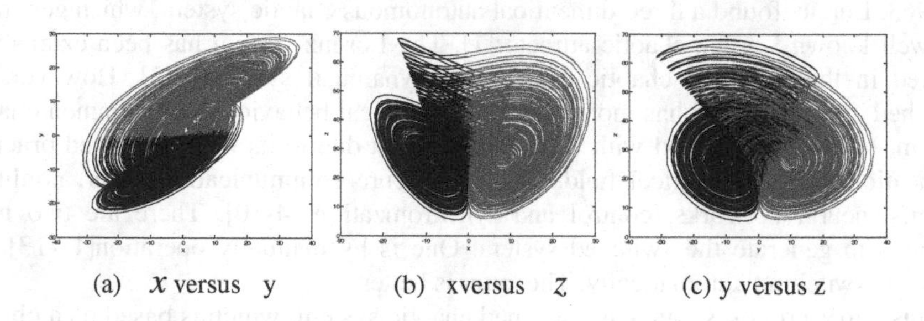

(a) x versus y (b) xversus z (c) y versus z

Fig. 1. The chaotic attractor of the auto-switched chaotic system (1) with a =20, b =14, c =10.6, h =2.8

3 The Basic Properties of the Auto-Switched Chaotic System

The new chaotic system (1) can be characterized with its Lyapunov exponents computed by Wolf method. For equilibrium points: $LE_3 < LE_2 < LE_1 < 0$;for periodic orbits: $LE_3 < LE_2 < 0, LE_1 = 0$; for quasi-periodic orbits: $LE_3 < 0$, $LE_2 = LE_1 = 0$; while for chaotic obits: $LE_3 < 0$, $LE_2 = 0$, $LE_1 > 0$, $LE_3 + LE_2 + LE_1 < 0$.

In order to observe the effect of the parameters on the dynamics of the chaotic system, we fix the parameters $a = 20$, $c = 28$, and let the parameter b vary in the interval [0 9].

The bifurcation diagram and the Lyapunov exponents are shown in Fig.2.

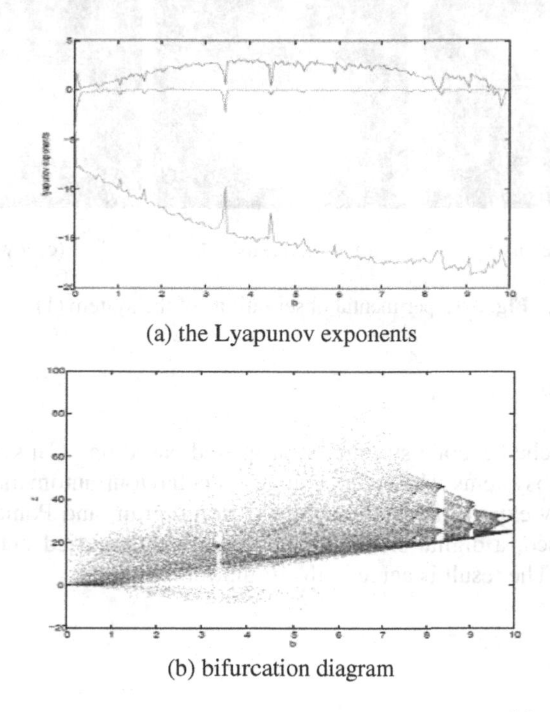

(a) the Lyapunov exponents

(b) bifurcation diagram

Fig. 2. Lyapunov exponents and bifurcation of the system (1) when $a = 20$, $c = 28$, and vary b

From Fig.2 (a) and (b), one can see that the Lyapunov exponents spectrum are completely consistent with the bifurcation diagram.

When $b \in [0, 9.45]$, the largest Lyapunov exponent is positive, implying that the system shows chaotic behaviour. Fig.1 depicts the chaotic attractor for $b = 2$.

When $b \in [9.45, 10]$, the three Lyapunov exponents are all less than zero, the system is equilibrium points, Fig.3 displays equilibrium points for $b = 9$.

4 Circuit Realization of the Switched Chaotic System Based on FPGA

The circuit of system (1) is designed by the DSP Builder tool from Altera Company in America. Adder, delay, multiplication, amplifier and data selector are selected from the DSP Builder component library, and the digital integrator is designed according to the relationship between first order differential coefficient and difference.

We can implement the system (1) model of the FPGA circuit in Matlab/Simulink and generate hardware description language automatically. The system (1) can be mapped into a rock-bottom hardware realized chaotic attractor based on FPGA directly. The attractors have been obtained by using D/A converter as shown in Fig.3. Experiment results are in accordance with simulation results.

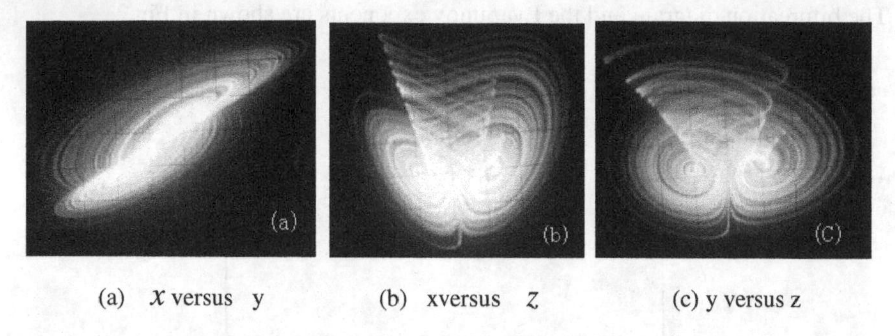

(a) x versus y (b) xversus z (c) y versus z

Fig. 3. Experimental observations of the system (1)

5 Conclusions

A novel auto-switched chaotic system is proposed based on Cai system, the system consists of two subsystems which can change its behaviour automatically from ont to another. Lyapunov exponent spectrum, bifurcation digram and Ponicare map of the system are analysed, a digital circuit experimentation is carried out upon FPGA by EDA technology. The result is agree with simulation.

References

[1] Lorenz, E.N.: Deterministic nonperiodic flow. Journal of the Atmosphere Science 141, 130–141 (1963)

[2] Chen, G.R., Ueta, T.: Yet another chaotic attractor. Int. J. Bifurc Chaos 9, 1465–1466 (1999)

[3] Lü, J.H., Chen, G.R., Cheng, D.Z., et al.: Bridge the gap between the Lorenz system and Chen system. Int. J. Bifur. Chaos 12(2), 2917–2926 (2002)

[4] Cai, G.L., Tan, Z.M., Zhou, W.H., et al.: Dynamical analysis of a new chaotic system and its chaotic control. Acta Physica Sinica 56(11), 6230–6237 (2007)

[5] Qi, G.Y., Chen, G.R., Du, S.Z., et al.: Analysis of a new chaotic system. Physica A 352(2-4), 295–308 (2005)

[6] Chen, Z.Q., Yang, Y., Yuan, Z.Z.: A single three-wing or four-wing chaotic attractor generated from a three-dimensional smooth quadratic autonomous system. Chaos, Solitons and Fractals 38, 1187–1196 (2008)

[7] Liu, C.X., Liu, T., Liu, L., et al.: A new chaotic attractor. Chaos Solitons & Fractals 22(5), 1031–1038 (2004)

[8] Li, Y.X., Chen, G.R., Tang, W.K.S.: Controlling a unified chaotic system to hyperchaotic. IEEE Trans. CAS 52, 204–207 (2005)
[9] Liu, Y.Z., Jiang, C.S.: Building and analysis of properties of a class of correlative and switchable hyperchaotic system. Acta Physica Sinica 58(2), 771–778 (2009)
[10] Liu, Y.Z., Jiang, C.S., Lin, C.S., et al.: A class of switchable 3D chaotic systems. Acta Physica Sinica 56(6), 3107–3112 (2007)
[11] Liu, Y.Z., Jiang, C.S., Lin, C.S., et al.: Four-dimensional switchable hyperchaotic system. Acta Physica Sinica 56(9), 5131–5135 (2007)
[12] Wang, G.Y., He, H.L.: A new Rosslor hyperchaotic system and its realization with systematic circuit parameter design. Chinese Physics B 17(11), 4014–4021 (2008)
[13] Wang, G.Y., Bao, X.L., Wang, Z.L.: Design and FPGA Implementation of a new hyperchaotic system. Chinese Physics B 17(10), 3596–3602 (2008)

Study on the Design and Application of Immersive Modeling System for Electronic Products with Multi-displayport Interface

Wang Xin-gong

Cangzhou Normal University, Cangzhou, 061001, China
wangxingong773@163.com

Abstract. In developing novel personal electronic products, the time has shortened to a few months due to high competition in the market. Now it has become hard to meet the time to market by evaluating products by making physical prototypes. To overcome the problem, immersive modeling system (IMMS) that enables us to interact with a digital product model using a multi-displayport interface is proposed. The IMMS allows the user to evaluate the product model, the architecture and main modules of the system are described in detail, and application examples to personal electronic products are also included.

Keywords: Multi-displayport interface, immersive modeling system, electronic products, application.

1 Introduction

In the electronics industry, due to the advances of communication technology and miniaturization of component parts, various kinds of personal electronic products are being developed year after year. It becomes harder and harder to meet the time to market by making physical prototypes of new product models. Thus, we need to develop a novel digital product model that can be evaluated and tested like the real physical counterpart. Till now, virtual prototyping systems have been developed to test features of a new product [1], these systems usually use CAD models and could resolve unforeseen problems that are only revealed after physical prototypes are developed. However, their roles are mainly limited to visual inspection of parts and interference checking between the component parts.

Generally speaking, two systems are used nowadays [2]: one system uses a head-mounted display to get a stereoscopic view of a product model, the other systems uses a haptic device to get force feedback or a touch feeling of the product model. These systems are designed to provide more realism for the users in viewing and handling the product model by providing a touch feeling, which also requires a lighter product model to achieve real-time interaction. Furthermore, most of these systems are designed to focus on one modality, and do not provide multi-displayport interactions.

In the paper, an immersive modeling system to provide multi-displayport interaction for personal electronic products is described. The immersive modeling

Z. Du (Ed.): Proceedings of the 2012 International Conference of MCSA, AISC 191, pp. 675–679.
springerlink.com

system (IMMS) aims to achieve a realistic interaction with a product model using multi-displayport senses that include visual, auditory and touch senses. In this immersive modeling system, the senses of taste and smell are not included since they do not contribute to the realism of the personal electronic products. The system is also portable in that it does not require big pieces of hardware and heavy computing power, and possible for a remote customer to evaluate a new product model by interacting with it in an immersive environment.

2 The Overview of Immersive Modeling System

The immersive modeling system (IMMS) consists of several hardware components and software modules as shown in figure 1. We use an ARvision-3D HMD [3] that gives "left" and "right" eye views, from which the user can see a stereoscopic image of the 3D model. The ARToolKit, a software library for building augmented reality applications, are used for a stereoscopic display interface. The force and tactile module increases the realism by providing the users with a natural and intuitive interface using the touch sense. In the IMMS, we use an open source haptic toolkit, the CHAI 3D SDK, to implement the haptic interface with the 3D model. Force and tactile realism for personal electronic products is presented to the user with kinesthetic feedback (force, motion) and tactile display (cutaneous touch). The auditory sense can be processed in parallel with the visual and touch senses. The control sound enhances the auditory realism by generating sounds when the user interacts with the 3D model when pressing a button on the model. The spatial sound adds more realism by enhancing user's perception to the position of sound sources. The effect of interaction with sound source is achieved by using audio processing techniques such as head-related transfer function [4]. The functional behavior is acquired by using a finite state machine that can represent the transitions between the states of the product.

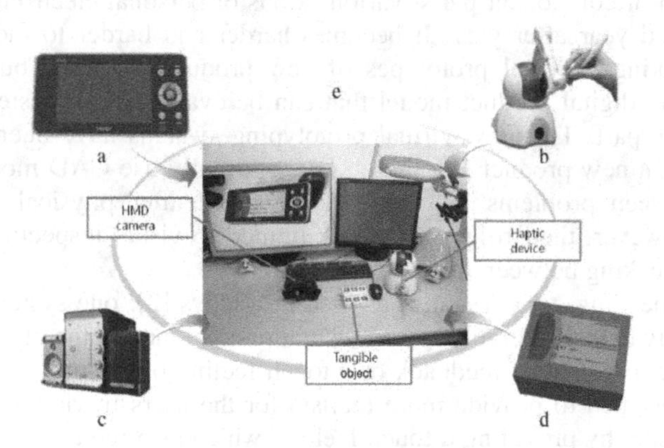

Fig. 1. The Overview of the Immersive Modeling System: (a) Visual realism, (b) Force tactile realism, (c) Auditory realism, (d) Functional realism, (e) IMMS

3 Design and Application of the System

In this study, we use multi-sensory feedback that includes visual, auditory and tactile senses to enhance immersion. Visual realism for personal electronic products is achieved by an immersive stereoscopic display of the model using an HMD. A post processing step is then required to filter the noise from the clouds of points that represent different views of the object and to merge them and quality. The QEM, a mesh simplification method, is used to maintain a high level of geometric detail while reducing the mesh size greatly. In an AR based product design environment, viewpoints between the virtual object and user interaction are integrated, so that the user wearing an HMD can directly touch or observe the virtual model by using their hands. This allows more intuitive and immersive interactions. The tangible object is designed as shown in figure 2(a), it can simulate different handheld personal electronic products that have a folder-type structure such as a game phone. The proposed hand segmentation works as shown in figure 3. At first, we reduce the processing time by processing only the hand area on the tangible object instead of processing an entire image. To process locally, the 3D outer points of the tangible object are projected to a 2D image, and then the convex hull points are extracted [5]. Using these points, contours (upper surface, bottom surface) are extracted and skin-regions are segmented. Then, we apply the statistical approach to segment the hand from the area of the tangible object. The hue and saturation components of the HSV color space are used to minimize the lighting effects. As a result, users interacting with augmented virtual objects by hand will experience better immersion and more natural images.

In order to realize the immersive experience of a personal electronic product, it is very important for the user to understand the functional behavior of the product with ease and clarity. Figure 4 shows the overall process of the HMI function simulation module that is used in this study. We capture the functional behavior of the product into an information model and utilize it using a finite state machine to simulate the functional behavior in an immersive environment.

The IMMS platform consists of many hardware components and software modules. The hardware components of our IMMS platform are shown in Figure 5. The PC-based IMMS platform uses AMD Opteron Dual Core175 (2.2 GHz) CPU and 4 GB PC2-3200 ECC DDR2 SDRAM for high speed computing. To accelerate the graphics rendering speed, we use Quadro FX4500 manufactured by nVIDIA. To give a stereoscopic display of 3D models, ARvision-3D [6] video see through head-mounted display is used. A PHANTOM Omni [7] haptic device from SensAble Technologies Inc. is used to provide force and tactile realism. A tangible object is also designed with an embedded Bluetooth module and two vibration motors. A universal computer-supported speaker Britz BR-5100D is used to implement auditory realism.

Application examples to personal electronic products are indicated in figure 6. It shows a state transition chart for the MP3 player where we consider only two main functional features: playing MP3 music and listening to an FM radio. The user can select any of them by pressing a button using a haptic interface, and Figure 6(b) shows the LCD window while playing MP3 music. To further demonstrate the applicability of the IMMS to personal electronic products, we also created 3D models of a game phone and a portable media player (PMP), and also prepared all the functions related to these products. These three personal electronic products can be fully operated using 3D models and the IMMS platform without making real physical models. Figure 6(c) and (d) show two games that can be played using the IMMS. The games are played by pressing buttons and also tilting the tangible object. The user can feel that he/she is operating a real game phone while he/she is playing these games in the IMMS. For example, the user can play a racing game displayed on an augmented virtual object by tilting its tangible object in all directions. These directions control the speed and direction of a car in the game. If the car collides with the wall or other cars, then 3D sound effects are played. To provide more realism, if a collision occurs during the game, it makes the tangible object vibrate.

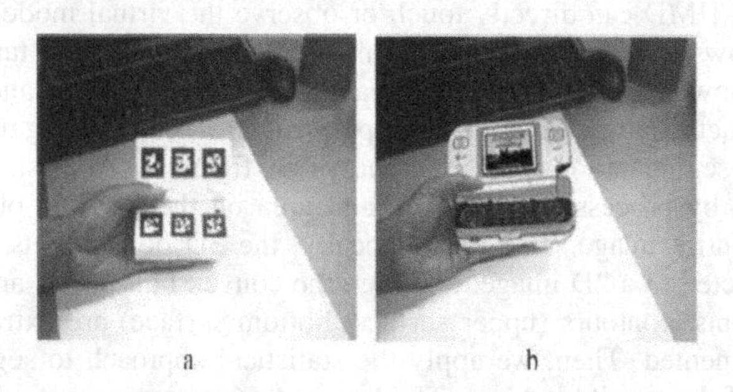

Fig. 2. (a) A folder-type tangible object; (b) Registration of a virtual object to the tangible object with corresponding scale

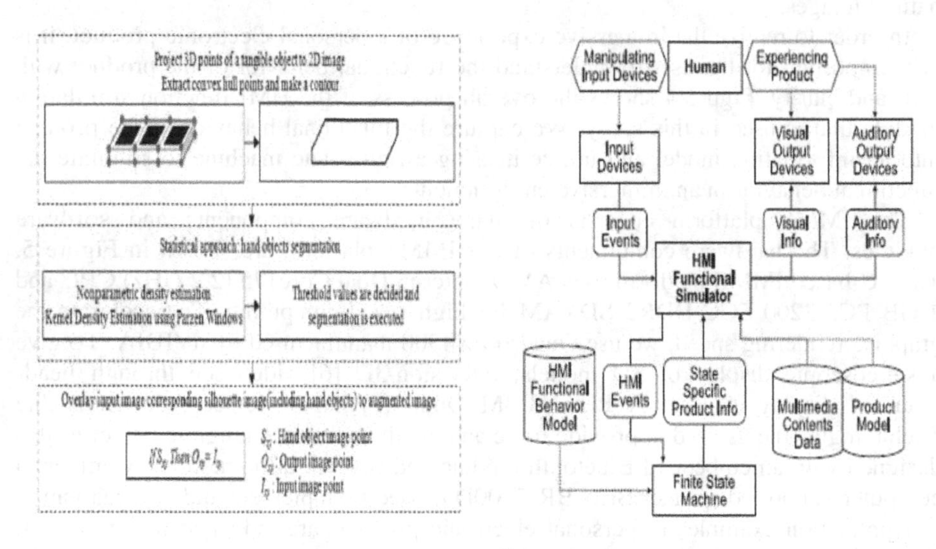

Fig. 3. Proposed Hand. Segmentation Process

Fig. 4. The Overall Process of HMI Functional Simulation

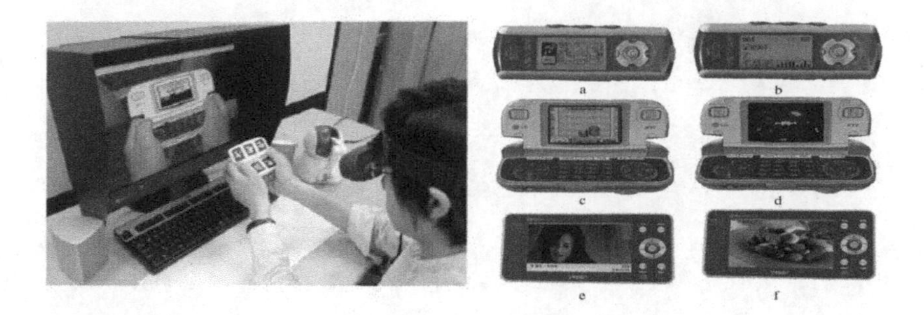

Fig. 5. The IMMS Hardware Configuration

Fig. 6. Application to MP3 Player, Game Phone and PMP

4 Conclusion

In this study, a novel multi-displayport interface called the IMMS has been developed. The system consists of five main modules: visual realism, auditory realism, force/tactile realism, functional realism and the core management modules. In the IMMS, the interfaces using different senses are combined into a single platform. It provides natural, intuitive and comfortable interactions for the user with a 3D product model in an immersive environment.

The result gave a reference for researching on wireless network security model. In order to visualize a 3D model with photorealistic quality, further research effort is required to build a multi-displayport interface system that runs in real-time with a 3D product model having a photorealistic quality. Moreover, the IMMS has not been implemented for a distant user. Further work in networking and data compression and decompression is needed to realize this remote operation.

References

[1] Gupta, R., Whitney, D., Zeltzer, D.: Prototyping and design for assembly analysis using multimodal virtual environments. Computer Aided Design 29(8), 585–597 (1997)

[2] Choi, S.H., Chan, A.M.M.: A virtual prototyping system for rapid product development. Computer Aided Design 36, 401–412 (2004)

[3] Bao, J.S., Ritchie, J.M., Gu, M.Q., Yan, J.Q., Ma, D.Z.: Immersive virtual product development. Journal of Materials Processing Technology 129(1-3), 592–596 (2002)

[4] Klinker, G., Dutoit, A.H., Bauer, M., Bayer, J., Novak, V., Matzke, D.: Fata Morgana-A presentation system for product design. In: ISMAR 2002, pp. 76–85 (2002)

[5] Harel, D.: Statecharts: A visual formalism for complex systems. Science of Computer Programming 8(3), 231–274 (1987)

[6] Rönkkö, J., Markkanen, J., Launonen, R., Ferrino, M., Gaia, E., Basso, V.: Multimodal astronaut virtual training prototype. International Journal of Human-Computer Studies 64(3), 182–191 (2006)

[7] Sarter Nadine, B.: Multimodal information presentation: Design guidance and research challenges. International Journal of Industrial Ergonomics 36(5), 245–257 (2006)

An Efficient Approach for Computer Vision Based Human Motion Capture

Wang Yong-sheng

Lanzhou Jiaotong University, Lanzhou 730070, China
wys_lz@126.com

Abstract. In this paper, we present a novel computer vision based human motion capture approach by human body reconstruction process and energy function minimizing. After analyzing 3D human model in detail, we conduct human motion capturing by four steps, which are 1) Capturing the video, 2) Recognizing human feature points, 3) Tracking the feature points, and 4) Representing the motion movement. To test the effectiveness of the proposed approach, we conduct experiments on HumanEva dataset under four metrics. Experimental results show that our approach can capture human motion precisely.

Keywords: Motion capture, Computer vision, Virtual human model, Energy minimization.

1 Introduction

Human motion capture refers to the process of recording movement of persons. It is widely used in many application fields, such as military, entertainment, sports, and medical applications and so on. In the field of filmmaking and computer games designing, human motion capture usually refers to recording actions of human actors, and utilizing this information to animate digital character models in two-dimensional or three-dimensional computer animation. If faces or fingers are included in human motion capture, it is usually denoted as performance capture.

Automatic human motion capture has been widely researched in recent years. Traditional human motion capture algorithms can be divided into two modes which are marker-based mode and markerless mode. For the first mode, researchers adopt a set of optically distinguishable markers placed in some landmarks of the body and are able to produce highly accurate results, mostly employed by the cinema and medical industries[1]. However, the hardware required to acquire these markers constrains the analysis to a setup scenario and the marker placement might be intrusive and/or uncomfortable for users[2].

For the second mode, markerless methods grant more freedom of movement and a more natural human motion capture scenario. These methods are based on obtaining a number of features derived from the video input such as edges and silhouettes[3,4] and then calculating the pose afterwards. In both the two modes, multi-camera acquisition systems are commonly used in order to obtain some different points of view of the same scene to solve occlusions and perspective issues[5,6].

Z. Du (Ed.): Proceedings of the 2012 International Conference of MCSA, AISC 191, pp. 681–685.
springerlink.com

The rest of the paper is organized as follows. Section 2 introduces overview of 3D human model. Section 3 presents computer vision based human motion capture algorithm. In section 4, we conduct experiments to show the effectiveness of the proposed algorithm. In Section 5, we conclude the whole paper.

2 Overview of 3D Human Model

3D human model is a local coordinate system, in which the bones obey the Parent-Child relationship as illustrated in Fig.1. The source node of human skeleton model is denoted as skeleton root, which connects with spine root. Afterwards, the spine root connects legs and spine. For each leg of human skeleton model, other parts are extended from it, such as thigh, shin, foot and toes. Spine is an important part in this model, which connects the upper body including shoulder, arm, hand, neck and head.

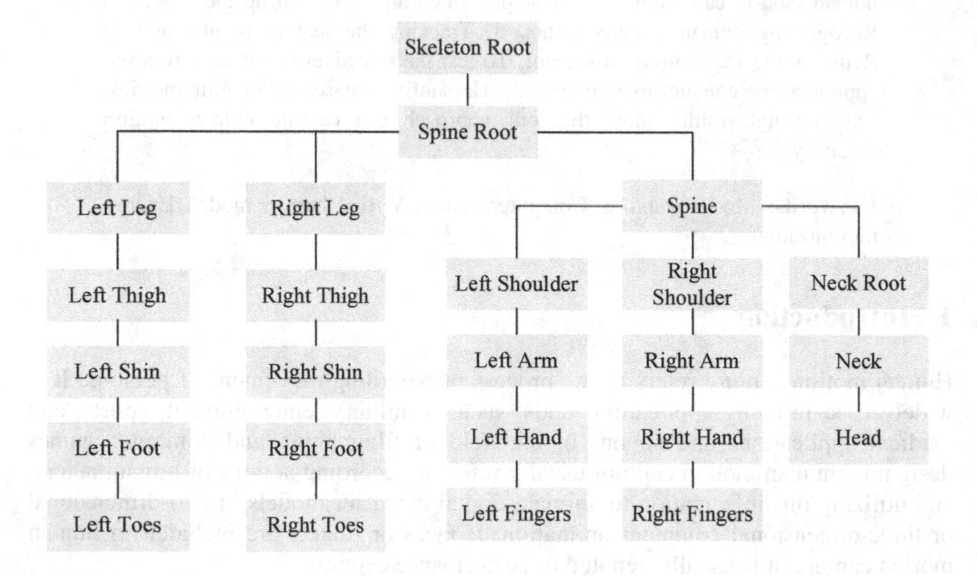

Fig. 1. Human skeleton model

Based on the design of human skeleton model, 3D virtual human model can be developed. As is shown in Fig.2, we give a 3D Virtual human model. Therefore, the motion capture problem can be reconstructed into a convex problem and simplified by using a hierarchical geometrical solver. In this model, parameters are set for the hierarchical procedure starting from the hip and all the way up to the head. Particularly, this model is scaled for model-based tracking, and, it combines the image data to overcome the lack of information about the model parameters.

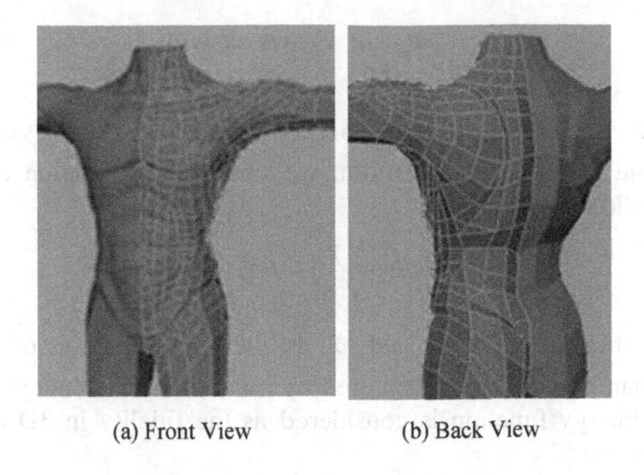

(a) Front View (b) Back View

Fig. 2. 3D Virtual human model

3 Computer Vision Based Human Motion Capture Algorithm

Our human motion capture is make up of four parts, which are 1) Capturing the video, 2) Recognizing human feature points, 3) Tracking the feature points, and 4) Representing the motion movement(shown in Fig.3). The main ideas of our algorithm lie in that our algorithm consists of two aspects, which are 1) Reconstructing by 3D active contours based on only the information of visual hull, 2) Obtaining the 3D prior reconstruction and human pose at the same time.

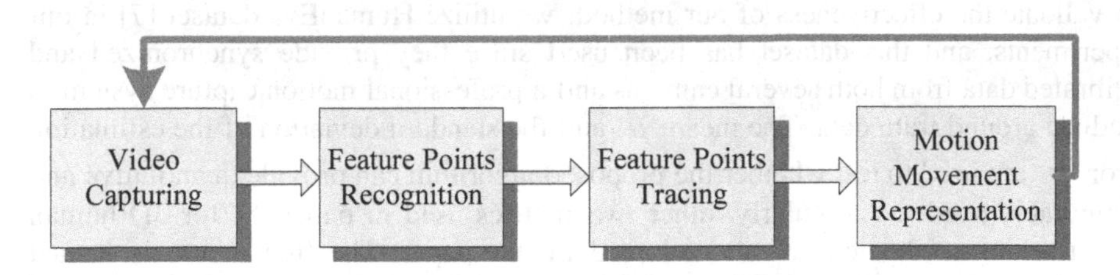

Fig. 3. Main steps of human motion capture

For the first step of our algorithm, 3D reconstruction of human body is constructed based on the visual hull V by 3D active contours. Afterwards, human body reconstruction process can be regarded as a process of distribute a label to each voxel. Hence, the energy of this voxel is defined as follows.

$$F(V,l)=\begin{cases} F_1(v)^2, & \text{if } l \text{ belongs to human body} \\ F_2(v)^2\,, & \text{otherwise} \end{cases} \tag{1}$$

Function $F_1(v)$ and $F_2(v)$ are defined as follows.

$$F_1(v)=\begin{cases} \alpha_1 & \text{if } v \in \text{visual hull} \\ \alpha_2 & \text{otherwise} \end{cases} \tag{2}$$

$$F_2(v)=\begin{cases}\alpha_2 & \text{if } v \in \text{visual hull} \\ \alpha_1 & \text{otherwise}\end{cases} \tag{3}$$

where parameter α_1 and α_2 are fixed coefficients and $\alpha_1 < \alpha_2$ is satisfied.

For the second step of our algorithm, we obtain the evolution equation of 3D contour C as following.

$$\frac{\partial \gamma}{\partial t} = \delta_\varepsilon \{u div(\frac{\nabla \gamma}{|\nabla \gamma|}) - F_1(V)^2 + F_2(V)^2\} \tag{4}$$

where γ is function of level set and δ_ε is the approximation of delta function. Afterwards, human motion can be captured by minimizing the energy function in the whole volumn. Energy function is considered as the fidelity in 3D active contours which is shown in Eq.5.

$$E(C) = \iiint_{\Omega_1} F_1(V)^2 dxdy + \iiint_{\Omega_2} F_2(V)^2 dxdy + \mu \oiint_C ds \tag{5}$$

Adopting the energy function, we can conduct inner optimization and outer optimization to obtain optimal result of pose p and curving surface C as follows.

$$Opt(C) = \arg\min_C E(C, p) \tag{6}$$

$$Opt(p) = \arg\min_p(\arg\min_C E(C, p)) \tag{7}$$

4 Experimental Results and Analysis

To validate the effectiveness of our method, we utilize HumanEva dataset [7] in our experiments, and this dataset has been used since they provide synchronized and calibrated data from both several cameras and a professional motion capture system to produce ground truth data. The mean μ and the standard deviation of the estimation error σ are used to test whether the proposed algorithm can provide quantitative and comparable results. Particularly, other two metrics used in paper [8] for 3D human pose tracking evaluation are also adopted in this paper. The first metric is named Multiple Marker Tracking Accuracy (MMTA), which is defined as the percentage of 3D body landmarks positions whose estimation error is below a predefined threshold. The second metric is called Multiple Marker Tracking Precision (MMTP), which is defined as the average of the estimation error of those landmarks considered by the MMTA. Using these metrics, experimental results is shown in Table.1 with the predefined threshold is set to 5cm.

Table 1. Experimental results of the proposed algorithm

Kind of Motion	μ	σ	MMTP	MMTA
Walk	45.27	23.36	34.9	88.42
Jog	42.56	24.56	42.13	82.28
Box	45.32	26.54	51.34	74.35
Gesture	55.19	26.31	47.54	78.12
Throw/Catch	43.12	27.43	41.89	82.31

As is shown in Table.1, we can see that the performance of our algorithm under HumanEva dataset is satisfactory.

5 Conclusion

This paper propose a new computer vision based human motion capture approach, which is made up of four main parts including 1) Capturing the video, 2) Recognizing human feature points, 3) Tracking the feature points, and 4) Representing the motion movement. Experimental results show that our approach can capture human motion precisely in HumanEva dataset.

References

[1] Cerveri, P., Pedotti, A., Ferrigno, G.: Robust recovery of human motion from video using Kalman filters and virtual humans. Human Movement Science 22, 377–404 (2003)

[2] Kirk, A., O'Brien, J., Forsyth, D.: Skeletal parameter estimation from optical motion capture data. In: Proceedings of IEEE International Conference on Computer Vision and Pattern Recognition, vol. 2, pp. 782–788 (2005)

[3] Deutscher, J., Reid, I.: Articulated body motion capture by stochastic search. International Journal of Computer Vision 61(2), 185–205 (2005)

[4] Mitchelson, J., Hilton, A.: Simultaneous pose estimation of multiple people using multiple-view cues with hierarchical sampling. In: Proceedings of British Machine Vision Conference (2003)

[5] Cheung, G., Kanade, T., Bouguet, J., Holler, M.: A real time system for robust 3D voxel reconstruction of human motions. In: Proceedings of IEEE Conference on Computer Vision and Pattern Recognition, vol. 2, pp. 714–720 (2000)

[6] Caillette, F., Galata, A., Howard, T.: Real-time 3D human body tracking using variable length Markov models. In: Proceedings of British Machine Vision Conference, vol. 1, pp. 469–478 (2005)

[7] Sigal, L., Balan, A., Black, M.: HumanEva: Synchronized video and motion capture dataset and baseline algorithm for evaluation of articulated human motion. International Journal Computer Vision 87(1-2), 4–27 (2010)

[8] Canton-Ferrer, C., Casas, J., Pardas, M., Monte, E.: Towards a fair evaluation of 3D human pose estimation algorithms, Tech. Rep., Technical University of Catalonia (2009)

Design and Implementation of Instant Communication Systems Based on Ajax

Liu Tian and Wu Jun

Faculty of Information Engineering
Jiangxi University of Science and Technology, Ganzhou 341000, China

Abstract. The instant communication tool is a hot topic in today's web applications. The method of interaction for polling refreshing every a period of time not only waste network traffics, but also led to the issue of white screen. To address these issues, in the paper, we take the chat system of college friend website as an example to introduce the design of instant communication systems based on Ajax in details.

Keywords: Chat, Ajax, Web, B/S construction.

1 Introduction

With the rapid development of Internet, people are increasingly inseparable from the Internet. Among the numerous network services, online chatting is very hot. Comparatively popular instant communication systems are the Tencent QQ, foreign ICQ, AIM, MSN and Yahoo Mes2senger etc [1]. These important communication systems used the traditional C/S construction, and the user must install the client software, which causes great inconvenience to the user; secondly, in the C/S construction, software upgrading and maintenance are rather difficult, because if the system version updates [2], the clients have to install the new version software. Therefore, currently more and more instant communication systems tend to use B/S construction. Generally, the traditional B/S interactive chat system is: Client-side page is refreshed timingly, reading the news of server database. During this period, each time when the server returns a new Html page to the client-side, the client needs to refresh the whole page once[3]. When the server processes request, the client can only wait idly, resulting in a tremendous waste of resources and the impact of real-time effects and page stable appearance, leaving users great inconvenience. Taking advantage of Ajax without refresh technology to exploit new chat system can solve this problem in a good way. Ajax (Asynchronous JavaScript and XML) technology is very popular in the field of Web development technology, which is like a bridge between a Web client script and the server language so that it can operate together, communicate with each other. The use of Ajax technology development with no page brush chat system is improved both in performance and user experience. This paper takes the friends-making network instant communication system as an example[4], designs and realizes instant communication system without refresh based on B/S structure, using Ajax technology.

Z. Du (Ed.): Proceedings of the 2012 International Conference of MCSA, AISC 191, pp. 687–693.
springerlink.com

2 Overview of AJAX Technology

In the traditional Web model, due to the adoption of synchronized information transmission mechanism, when the client sends a request, the client can only waiting for server to process the request in a static way, unable to conduct any other operation, only when the server finishes processing the request, before sending the data back to the client, the client cannot carry out the next operation. The traditional web request response mode client-side needs a long time, resulting in poor interaction in the systems, prolonged time for system evaluation response, system performance and other problems, not suitable for those chatting systems of strong instantaneity, strong interaction, limited transmission data but with large amount of information. In the traditional mode, even if little date is delivered, the server also needs to return to a complete Web page, which causes redundant transmission of resource and network bandwidth occupied.

AJAX(asynchronous JavaScript and XML Jesse JamesGarrett) is a Web development model proposed by Jesse JamesGarrett[5], which is not a new technology, but the combination of four techniques: CSS, DOM, JavaScript and XMLHttp. JavaScript is the core technology, defining workflow and business logic. Using JavaScript to operate DOM to change the user interface, reorganizing data displayed to the user, and processing the users' interaction based on the mouse and keyboard. CSS provides a reusable visualization style definition method for Web page element. The XMLHttpRequest object is used to have an asynchronous communication with the server.

3 System Structure

Instant messaging system generally has two modes: one is client/server mode, that is, the sending user and the receiving user must communicate through the server; the other is the user/user mode, which means the server establishes a TCP channel to each users to, and their communication is based on the TCP, without going through the server. For the HTTP protocol, the second model cannot be applied on top of it. Therefore, this paper uses the client/server mode. In this paper, the practical application of the system is instant communication system version V2.0 of making-friends network (hereinafter called the instant communication system). The system uses Microsoft NET technology, realizing one-to-many chatting mode. Different from the traditional C/S instant communication system, users can chat with a number of friends at the same time via this system.

3.1 Key Technology

The system adopts B/S mode, the traditional three layers construction: the presentation layer, business layer and data layer. The structure diagram is shown in figure 1. Wherein, presentation layer enables client-side to use the client browser and server to interact information, using Web services component to gain server information via Ajax engine, manipulating the DOM tree page update with the JavaScript scripting language. In the business layer, instant communication engine is a core component of the system, whose components are as follows:

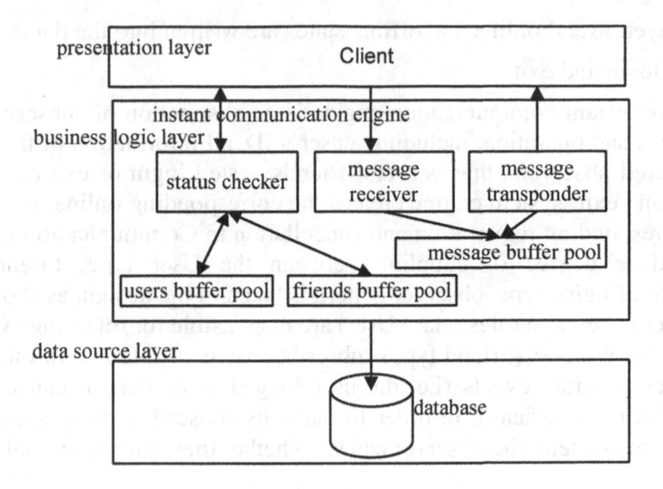

Fig. 1. System framework

(1)Status checker: Being responsible for detecting if user is online, whether to have new friend's request. The judgment is based on the user buffer pool and a buffer pool.

(2)Message receiver: All users need to send messages to the server, and the server message call distributor to distribute the message to the corresponding user. That is to say the server has the function to transmit messages. Message receiver puts the received message in the buffer pool.

(3)Message repeater: Taking out all of buddy chat log from users and their friends from the message buffer pool.

(4)Message buffer pool: Storing all users' chatting messages, managed centrally by the cache supervisor.

(5)Buffer pool: Caching all friends' information, managed centrally by the cache supervisor.

(6)User buffer pool: Caching all online user information, this is managed centrally by the cache supervisor.

All business logic operation of the layer makes all processed data packaged into XML format and returned to the Ajax engine. The Ajax engine analyzes the data transmitted by server, with the aid of DOM, DHTML to update page, display the data, and complete request response. Because all actions of the Ajax are performed in the background, which will not block the users' behavior, the users' waiting time can be saved. Meanwhile as the data transmitted and received by Ajax are relatively small, with high refresh speed, users will have the experience of real time response. From the system as a whole, all this shortens users' waiting time, and improves the system performance.

In the data layer, users' online and offline states are written into the database.

(1) The user login and exit

Users' type in instant communication engine is an abstraction of subscriber of users' basic attributes and operation, including a user's ID, all information of friends, as well as the registered observers, that is users' friends posted login or exit event. For each user's login and exit, system engine creates the corresponding online user objects and enter the corresponding registration and cancellation in Communication Engine[6]. In order to reduce degree of coupling between the User type, Friend type and CommunicationEngine type, observer pattern is taken in the design, as shown in figure 2. Among them, thematic roles (class User) are responsible for releasing events to their friends, and the observer (Friend type) subscribe events of interest. In this way, when thematic roles generate events (i.e. the user logged in or out), it can notice all the interested subscriber (Friend), in order to make its subscriber process corresponding operation, in this system, the observer can set whether their friends are online based on the event.

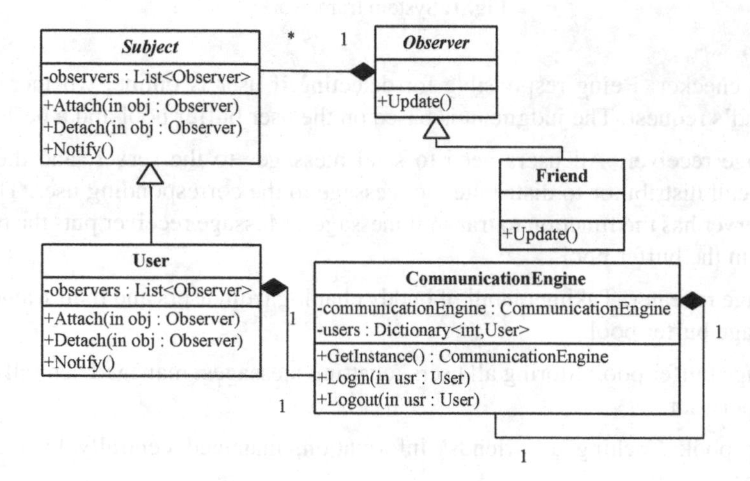

Fig. 2. Users' login and exit based on the observer design model

(2) online user management

After users log in, they need immediately access to current online users and user friend's online lists and other information. Therefore, the server needs to manage these information, and provide inquiry services. In order to realize fast search, the design in this paper uses data type Dictionary < int, User > to cache the online user object, this belongs to c# generic collection type[7]. In the dictionary, every element is a key value pair, time complexity of reading a value through a key is close to O. Because the user's ID is the unique one, system using ID as a key, and the value is the User object. Thus, according to the user ID, user object can be obtained from the user buffer pool. At the same time, in order to avoid security risks brought by the usage of global variables and reduce readability and maintainability of code, this paper adopts single instance mode to realize the online user management.

The CommunicationEngine type contains following properties:

(1) Buffer list for storing user information:
Private Dictionary < int, User > users
(2) Unique instance used to store the CommunicationEngine:
Private static CommunicationEngine communicationEngine
(3) Send and transmit messages

Sending and transmitting messages are managed by MsgStrategy, and this paper adopts the strategy pattern mainly because of the expansibility of the system, so that it can be easily extended to other algorithms for the management of message buffer pool to realize the message receiving and transmitting message. The model changes strategy of receiving and transmitting messages, having no influence on client-side. The process is illustrated in figure 3.

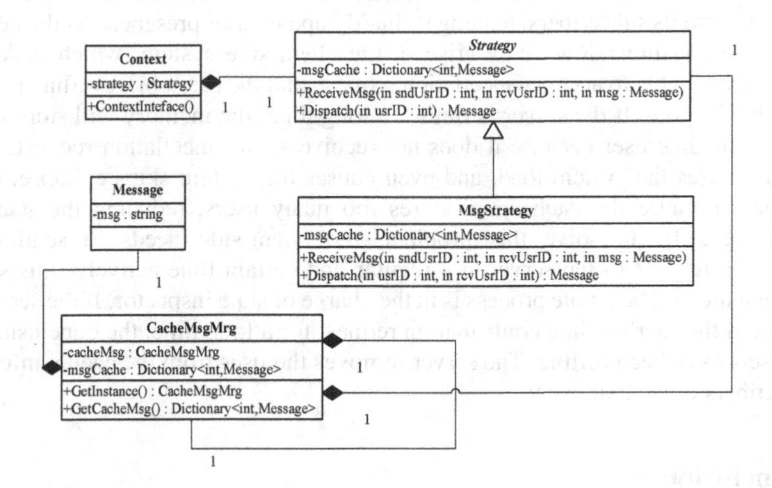

Fig. 3. Sending and transmitting messages based on strategy design model

Due to the fact that the buffer pool is a critical variable, which is mutually exclusive to write operation, this paper adopts the singleton pattern to prevent message buffer pool.

CacheMsgMrg contains properties:

(1) Buffer pool used to hold the chat log:
Private Dictionary < int, Message > msgCache
(2) Only instance used to hold the CacheMsgMrg:
Private static CacheMsgMrg msgCache

Strategy type includes the following operations:

(1) void ReceiveMsg (int sndUsrID, int rcvUsrID, Message MSG)
(2) Message DispatchMsg (int usrID, int friUsrID)

The two operations realize the message receiving and message forwarding, wherein, parameters sndUsrID and recUsrID in operation of ReceiveMsg are the user IDs of sending and receiving message, and the parameter MSG is the message sent, the operation of the user stores the the message sent to the friends into the buffer pool.

Parameters usrID and friUsrID in the operation of DispatchMsg are the user IDs of sending and receiving message, with a Message object returned.

The object contains a chat message. Because the HTTP protocol is a request-response protocol, the client-side cannot have real-time monitoring of server status changes, in order to realize the real-time message distribution, client-side sends update request actively every 5 seconds to the server using Ajax technology. If their friends send a new message, the server sends a new chat message to user for request, otherwise, return to "404" will have no new messages.

3.2 Processing Offline

When the user closes the chat window or exits the chat system, the client sends a cancellation action to the server. Chat engine releases object memory of user, and releases event to its subscribers, making its buddy update user presence. As the network is not stable and breakdown may arise in the client-side system, which makes the service server unable to accept cancellation request, and the user object offline is still in the user buffer pool. If the server is run for a long time, the memory will store a large number of off-line user because it does not receive such cancellation request, which greatly increases the system load, and even causes the system skips collapse, on the other hand, because the cache pool stores too many users, reducing the searching efficiency greatly, to solve this problem, the client-side needs to send online confirmation request to the server at a regular and certain time actively (this system uses 10 minutes). The whole process is in the charge of state inspector. If the server has not received the user's online confirmation request for a long time, the conclusion will be the users have been offline. The server removes the user from the cache, informing its subscribers of the exit event.

4 Conclusion

The realization of Instant Communication system is consisted of two parts: the server-side and client-side.The server requires a timely response to requests issued by the client-side. With the Ajax technology, the operation of chatting messages can be conducted hiding in the background, making up for the disadvantage of the traditional B/S architecture, which has low efficiency in refreshing frequently the entire page. This brings the user good experience of having no refresh, taking full advantages of the application of Ajax. This paper's next step is how to realize the functions of file transmission, voice and video.

References

[1] Yuanjie, L.: analysis of communication protocol in major instant messenger software. Application and Research of Computers, 243–250 (2005)
[2] Jiang, Y.: Chat system research based on multithreading technology. Computer Engineering and Design, 4046–4096 (2008)
[3] Crane, D., Pascarello, E., James, D.: Ajax in Action. Ajaxcn. Org., translated, pp. 25–50. People's post and Telecommunications Publishing House, Beijing (2006)

[4] http://www.gxqingyuan.com
[5] Garrett, J.J.: AJAX: A new approach J. to web applications. Adaptive Path (2005)
[6] Gamma, E., Helm, R., et al.: Design Patterns Elements al, of Reusable Object-Oriented Software, pp. 194–201. Publishing house of mechanical industry (2004); translated by Li Yingjun et
[7] Kauffman, J., Sussman, D.: ASP.NET 2.0 classic tutorial – C #. The posts and Telecommunications Press (2007); translated by Meng Xianrui, Yi Lei, et al.

Analysis of Characteristics in Japans Robot

Chen Beibei

Liaoning University of Technology, China

Abstract. Animation, as one kind of movie art with the most characteristics, has the characteristics of abstraction, exaggeration, imagination, deformation and collage different from real person film. Robot animation, as the most representative subject in Japanese animation, is more characterized with these art features of animation. And animation form most suitable to express robots is tried out. Generally speaking, there are the following characteristics of the art in animation of robots. First, great imagination space is set in the story setting. Overlength text can be easily produced to be suitable for TV broadcast, and generally the excellent works always have the style of epic. Second, the setting of characters is in accord with hero worship, which is mixed with inner factors of myth creation, added with modern technological elements, making this a kind of modern myth with certain thinking depth. Third, the style design is suitable for the market operation, adding the post of mechanical designer in the process labor division. The design is more important than the script itself. In the developing process of robot animation, there appears certain inner law, which conforms to the aesthetic principle of itself.

Keywords: Animation, setting, characters.

1 Introduction

Generally speaking, there is a great imaging space in the background setting of Japanese robot animation, which not only has the description of the mankind's future but also has the new era like what is constructed in Five Stars Stories. In general, most of the works welcomed worldwide link robots with the mankind's future. The assumed future is 30 to 100 year distant from today, combining the most advanced current technological development and creating the best actual sense.

2 Continuous Expansion of Imagination Space

Table 1. Imagination space in different periods

period	character	Plot space	Plot hot points
1960s	Astro Boy	Earth space	Robots with wisdom to save world
1970s-1980s	Gundam	Universe space	Humans fight with machines
1990s	Puppetmaster	Virtual space	Mixture of humans and machines

Z. Du (Ed.): Proceedings of the 2012 International Conference of MCSA, AISC 191, pp. 695–700.
springerlink.com

As we can see, both the design in robots' artificial intelligence and the handling of robots' goodness and badness indicate the appreciation of Japanese robots' animation for Euro-America science fiction literature. But Japanese animation is not short of its own originality but the most charming imagining part in the animation script, presenting a kind of unprecedented power. Beginning from Astro Boy, story is set in the setting like outer space. In the 1970s and 1980s when the Japanese animation is very splendid, more and more animations are set in galaxy and universe. In the 1990s, Japanese animation people expand the internet society into the virtual reality space. The space versatility in the work reflects Japanese animation peoples' good knowledge of space and their thinking of the developing trend towards era. In the previous Japanese animation, guarding the earth is a perpetual theme, both Astro Boy and Getta Robo, who are almost the mankind's hero, just like a savior. But when entering 1980s, the robot animation works represented by Gundam series are no longer in good favor of earth as mankind's home. With the advance of astronomical technology, human beings have known that earth is not more than a very common planet in the great universe, and people have more broad space to explore. Some lines like "only by destroying earth, can mankind walk out of cradle" appear in Gundam series animation. With continuous expansion of the background setting for robot animation, the animation people can created more and more overlength script and animation epic[1-3].

With the universe as the background, Japanese animation people begin to break the unreal feeling brought by many narrations from many angles, trying to bring audience more experience in reality. One kind of method is to modify the past technological civilization, adjusting the script of the past outdated technology in the 18th and 19th century, setting in the future space. Steam Boy made by Katsuhiro Otomo takes this kind of method, combining the sense of history together with future sense, getting a kind of real and virtual effect. Another method is starting from thinking of human's self knowledge towards space, overlapping virtual space with real space, expanding current virtual internet society, even changing human's thinking way towards space and redefine the background space, such as the future space expressed in Mamoru Oshii's Shell Tapping the Attacking Team, in which the future is expressed in thinking or ghosts and the meaning of body has disappeared. As we can see, in the expansion of background space, Gundam series denies the importance of earth, and Shell Tapping the Attacking Team discards body. This kind of breakthrough embodies creativity of Japanese robot animations.

What we should care is when Japanese animation people are dealing with the sense of alienation resulting from space change, the method they adopt is what is usually used in films, that is to enforce the reality of works through the emotions between people. In the 1940s, Big machines will stop its operation by Forster, Robot AL-76 Lost by Asimov, Sympathy circuit by Wyndham and so on, these works begin to think the problems of robots getting along with mankind after robots civilization comes. When Japanese animation becomes more and more mature, some famous animation producers including Hayao Miyazaki, Katsuhiro Otomo and so on are more and more willing to leave their personal traces onto the robots field which Japanese animation is good at, mixing animators' understanding of the future world into the works itself, constructing emotions of people and robots along with the scenes of people getting along with robots and trying to give an answer.

3 The Continuing of Modern Myth

In the animation of Japanese robots, a unique feature is the new myth produced in new era. We find the shadow of Son Goku in Chinese traditional culture obviously exists in Astro Boy. (Figure 4.2) we must see teenagers are the major audience of robots animation, which is a very natural thing for teenagers to admire power and speed. This kind of psychology roots in the lifestyle of living on catch and hunt in primitive society. Teenagers are longing for warm and lofty fighting scene and they continue to seek the repeatedly unbeaten spirit when defeating hardships. They can be satisfied in the field of robots fighting. This is because robots' fighting is less bloody than human's war, which can bring strong stimulus in sense organ without too more violent influence. The Chinese animators should also learn from this point.

Table 2. Comparison between ATongMu and Sun Wukong

Figure	power	speed	task	ending
ATongMu	Three big power lifts the building easily	Rockets spray	Protecting humans	Guard the peace of earth
Sun Wukong	golden cudgel is 13 thousand jins	One head turning can be very far away	Protecting Xuanzang	Succeed in pilgrimage for Buddhist scriptures

Table 3. Comparison between Gundam and Romance of the Three Kingdoms

Works	figures	plot
Gundam series	Many figures with complex relationship and different heroes appear in different periods	Humans fighting in the setting of universe
Romance of the Three Kingdoms	Many figures, and over a thousand people in the works, different depiction in different periods	The competition of different powers in the broad Chinese land

When the robot animation becomes mature, its creator also changes the script. The primary robot animation script mainly focuses on fairy tales. The ending that robots heroes defeat the wicked makes people feel the same as the happy reunion that the prince and princess live a happy life in the fairy tales, which has the suspicion of stylization. In the medium time, the script creation of robot animation is not only limited in similarity of appearance between theme character and myth character, and it also creates new myth script. Compared to myth story, the animation script obviously has longer story and more complex relationship between characters. This makes animation featured with multi-story structure which only exists in full-length novel, forming grand epic style. In the 1990s, based on predecessors' successful creation, the animators represented by Mamoru Oshii and Katsuhiro Otomo begin to stress the application of movie language, adding the formation of atmosphere to the originally careful moving plot of Japanese animation.

Although there are certain changes in the theme of Japanese robot animation in different periods, but generally speaking, when the robots' divinity is emphasized, the humanistic care has never been forgotten, which is embodied in the care for human's mind growth. In most of the robots animation, the idea of friendship, union, persistent in ideality and grittiness is delivered. Meanwhile, audience is told that the young heroes operating the great robots are also confused with growth and temporary confusion. This kind of care for youth's mind growth is not showed in homiletic form but closely connected with plot. In this way, personal mind growth is expressed in the detailed narration, adding the popularity among audiences. Both the Star Wild Iron man in Galaxy Express 999 and Amuro in Gundam 0079 are the successful main characters in robot animation. In this kind of animation, there is a young main character and even the wise robot Astro Boy is dressed like a child. Having teenagers go fighting is obviously not in accord with real life experience, but this can approach young audience more easily, making audience have the empathizing psychology, having the same experience with the main character, which is an important factor for establishing modern myth in Japanese robot animations.

4 Marketwise Style Design

The robots undergo a process from small to big and from big to small in style design, (Figure 4.4) but this process is absolutely not a simple return. Its process of changes shows that the creators of Japanese robot animation grasp the aesthetic psychology in the current times. In the style design of robot animation, we cannot deny the fact that market orientation dominates the mainstream. Although it is not sure to say all robot animations are produced for robots' derivation products sellers. But the factor of marketization is indeed the internal cause for many Japanese animation companies to produce great amount of robot animations.

Table 4. Comparison of style designs for animations in different periods

period	works	characters	type	Ratio of head and body	height
Period of Birth	Astro Boy	ATongMu	children	1: 5	120cm
Period of Development	Gundam series	Gundam	Machines and adults	1: 7	1800cm
Period of breakthrough	Shell tapping the attacking team	Suzi	adults	1: 7	170cm

In the 1970s, Japanese animation industry has entered a rapid growth period, greatly in need of bigger market, attracting more audience for animation, thus the market segmentation comes into being. The animation audience mainly have the following six types: first, younger age animation: Most of the audiences are preschoolers and lower graders, and the representative works are Clever Yixiu, Chibi Marukochan, Astro Boy and so on. Second, juvenile animation: the major audience are schoolboys, and the themes always focuses on motivation, friendship and pursuit for the true, good and beauty, represented by works such as Super Slams, Gundam series, Neon Genesis Evangelion and so on. Third, maiden animation: the major audience is schoolgirls, and the theme mainly focuses on romantic and careful maiden love, represented by Meteor Garden, Sailor Moon and so on. Four, teenager animation: most of the audience are university students and working adults, always adding scenes of violence and sex, and the main works are Pearly Gates, Shell Tapping the Attacking Team and so on. Five, female animation: most of the audience is family women or company female staff, with the theme of lengthy love story. Six, adult animation: adult animation is synonym of eroticism animation, produced catering for particular audience.

Basically, the coverage of animation for robot themes are in three aspects: Young – age animation, Juvenile animation and Youth animation and primarily focusing on Juvenile animation. Japanese robot theme animations in Gold Age such as "Galaxy Railway 999",the series of "Gundam" and "The Super Dimension Fortress" both have obvious characteristics of Juvenile animation. Reviewing the Japanese Animation development process, we can find that the start of market differentiation of Japanese Animation is basically consistent with the Gold Age of Japanese robot and both foster mutually. Differentiation market lets Japanese robot theme animations create with more fixed target, also forming basic characteristics grasping the market of juvenile animation. Just in this kind of environment, the style design of robot theme animations is closely linked to the derivatives sales, even the post for "machine settings" should be specially set in creating animations.

The commercial elements represented by derivatives sales are best inserted in in Japanese robot animation works. Astro Boy just creates a new mode of operation for Japanese animation industry. One third of the benefits are from domestic broadcasts, one third from foreign sell television rights and the rest one third from derivatives sales. Using Astro Boy as an opportunity, the Japanese Animation industry gradually formed a set of animation market operation mode and the Japanese government has also recognized economic and cultural significance of cartoon industry. The robot animations are easy to win the affirmation from overseas market for their distinct features. In addition to the two big traditional broadcast channels such as TV animation and Theatre animation outside, Japanese animation expands into OVA (original video animation)-a special mode, very suitable for Europe and the USA animation fans. In the 1970s, Japanese animation has gradually been accepted by the world. The cartoon "Space fort" made by adapting and clipping "Deja vu fortress" in the USA quickly popularizes in Europe and America. Later, the series of "Gundam" also created huge economic benefits.

As we can see, the Gold Age for robot animations is also the best-selling era of derivatives. "Gundam" model, a kind of derivative with playability, receives the most attention from consumers. When we summarize the artistic characteristics of Japanese robot animations, we must see that invasion of commercial elements into artistic works is inevitable. Especially from the angle of the industry development, the success of robots in style design makes great contributions for the Japanese animation industry to enter good circulation, although animators are the important factors for the Japanese animation to the world, but it is undeniable that the business is also the strong driving force for the development of Japanese animation. For this robot animation is the most representative.

5 Conclusion

When concluding the art characteristics of Japanese robot animation, we can see in the process that the Japanese robot animations form their own style, they learn a lot from the essence of art of sisters and to adapt to different aesthetic demand in the development of times. Continuous innovation is in step with times and even leads times. On the whole, Japanese robot animation is always breaking itself.

References

[1] Liu, J.: Legend of pictures—scanning of contemporary Japanese cartoon. Baihua Literature and Art Publishing House (2003)
[2] Ye, W., Tang, Y.: Sorrowful things and mystery. Guangxi Normal University Publishing House (2002)
[3] Boden, M. A., Liu, X., Wang, H.: Philosophy of artificial intelligence. Shanghai Translation Publishing House (2001)
[4] Bai, X.: Japanese cartoons. Chinese Tourism Publishing House (2006)
[5] Chen, X.: Screen Culture Research in Contemporary China. Beijing University Publishing House (2004)

The Development of Mobile Learning System Based on the Android Platform

Lingmei Kong

Tianhe College of Guangdong Polytechnic Normal University, GuangDong GuangZhou
konglingmei110@126.com

Abstract. Android is a completely open platform for third-party software developers, it has greater freedom in its development program, breaking the shackles of the traditional mobile phone. Mobile learning is a combination of computing power of mobile terminals and wireless network transmission capacity and the formation of a new mode of learning. The paper presents the development and design of mobile learning system based on the Android platform.

Keywords: Android platform, Mobile learning, web application.

1 Introduction

Android is open source mobile phone operating system developed by Google based on the Linux platform. It includes operating systems, user interface and applications - work required of the mobile phone all the software, but there is no exclusive rights of obstacles hinder the mobile industry, innovation. Google and Open Handset Alliance to develop Android, leader of the alliance by more than 30 technology and wireless applications, including China Mobile, Motorola, Qualcomm, HTC and T-Mobile [1]. We hope that with the establishment of standardized, open mobile phone software platform, the formation of an open ecosystem in the mobile industry through partnerships with operators, equipment manufacturers, developers and other interested parties formed a deep-seated, . We believe that this will promote better, faster innovation and unpredictable applications and services for mobile users.

Mobile learning is a combination of computing power of mobile terminals and wireless network transmission capacity and the formation of a new mode of learning [2]. Distinguish between types of mobile terminals and wireless networks; mobile learning can be divided into generalized mobile learning and narrow the mobile learning into two categories: general mobile learning users can use a variety of mobile terminals such as mobile phones, PDAs (Personal Digital Assistant) with even wireless card for laptops, learning through mobile communication networks, wireless local area network.

Mobile learning development of the terminal platform, according to the guiding ideology of the software engineering throughout the development process is divided into: needs analysis, structural design and function to achieve the three stages. The needs analysis is the first phase of the development of any system is to understand user requirements and software functionality with customer agreement, and ultimately the

Z. Du (Ed.): Proceedings of the 2012 International Conference of MCSA, AISC 191, pp. 701–706.
springerlink.com

formation of a complex process of the development plan. Needs analysis, the development of mobile learning terminal platform "what". must be answered the question, which is directly related to the success of software development.

2 Mobile Learning Needs Analysis of the Terminal Platform Based on Android

Android runtime library contains a set of core libraries (most of the functionality of the Java language core library), and the Dalvik virtual machine. Android provides a rich library support and the majority of open source code, such as 2D and 3D graphics library OpenGL ES, the database SQL.

The so-called "third screen" courseware is roughly divided into three areas: video player, select the directory area, presentation area. The video playback area to play the teachers teaching video or teaching scenarios; directory area is used to display the chapter directory, learners can play in the area to choose the relevant chapter, independent study: presentation area for video file synchronization, and constantly switch display related pictures, text, animation, program operation, and even teachers teaching video presentation and the process of mouth.

Extensive use of rapid prototyping in the development process of mobile learning terminal platform for mobile learning platform for the various modules, including online learning and communication exchange module can be used as a stand-alone application run separately, based on user feedback on the use of stand-alone applications. focus changes, such benefits on the one hand in the entire system prior to the completion of the various components by the end-user test, greatly improving the system of recognition, on the other hand this low coupling between the various functional modules to ensure that the entire system robustness.

Because the WAP-based mobile learning website by a lot of media to convert the information on the Internet using HTML language to describe the information described in WML and then displayed on the display of the mobile phone or other handheld devices, led to a page in the conversion process the lack of information, which greatly affect the integrity and vitality of the original content, is shown by figure1.

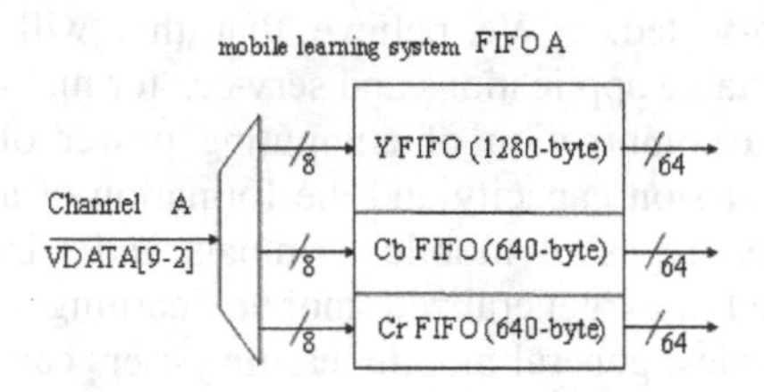

Fig. 1. WAP-based mobile learning Architecture

Third of screen courseware on learning in Internet-based network has the following advantage: the development of streaming media technology. Compared with the previous simple text and pictures, it has interactive, content-rich, self-control programs and progress; learners can see the live teaching by both, can also see the wonderful multimedia presentation, it can independently control the progress of the courseware.

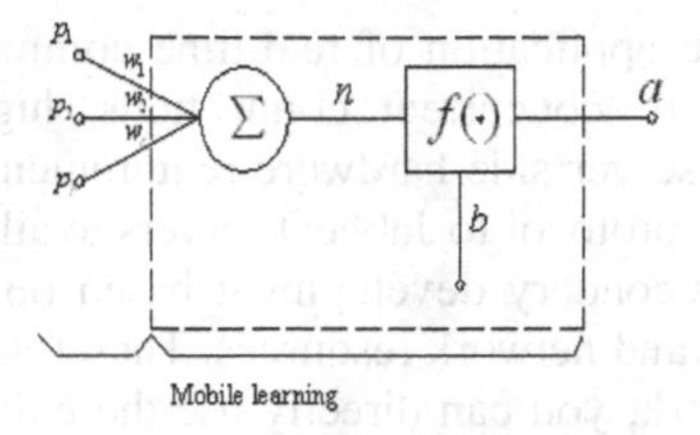

Fig. 2. Mobile learning Architecture

With the move has begun commercial operation of 3G, fully able to meet the complex mobile learning terminal platform for data transmission requirements, In addition, with the development of smart phones, a new generation of smart phone platform - Android for mobile learning Terminal Platform to provide high-performance terminal equipment and the development and testing tools can be seen that the platform has the ability to develop technically and conditions.

In order to maintain a large user base, the system uses a geographical form of the entire message space into different autonomous regions; each region is managed by a separate server, and set the application gateway and the other d II system interoperability. The partition structure is the autonomy with Taiwan [4]. Jabber address (also known as J II "contains the domain identifier (Domain), the node identifier (Node) and source identifier (Resource) --- part of route processing is also based on the J song address logical address way, the format is: node, as is shown by figure3.

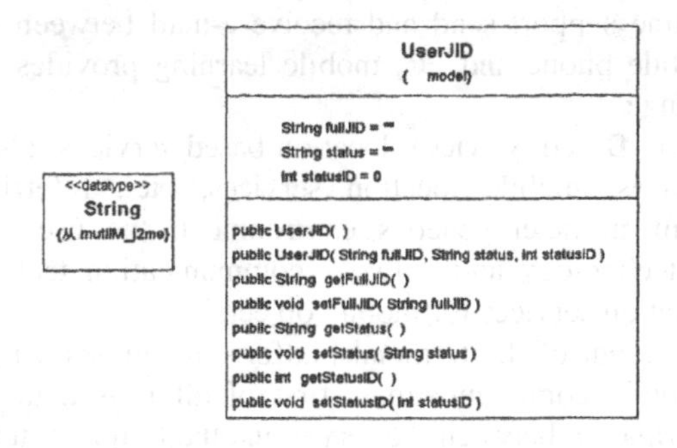

Fig. 3. Class diagram of the client code

MIME (Multipurpose Intermit Email Extension, multi-function Intermit Mail Extensions) is a Multipurpose Internet Mail Extensions, it is set to MIME (Multipurpose Intemet Email Extension, multi-functional Internet Mail Extensions) is a Multipurpose Internet Mail Extensions it is designed to support non-ASCII characters, binary format attachments in multiple formats mail messages. The meter is used to support the mail message ASCn character, binary format attachments in multiple formats.

A real investment in the application of real-time communication systems need to maintain a large number of concurrent connections, high processing power and network bandwidth on the server side hardware requirements. Google to comply with the Jabber framework open protocol to Jabber) servers available to other Jabber) stand with the free use, so the secondary development based on Jabber means free to use Google's existing hardware and network resources. Therefore, application development using open source framework, you can directly use the existing free server, to reduce the purpose of mobile learning applications operating costs. as is shown by figure4.

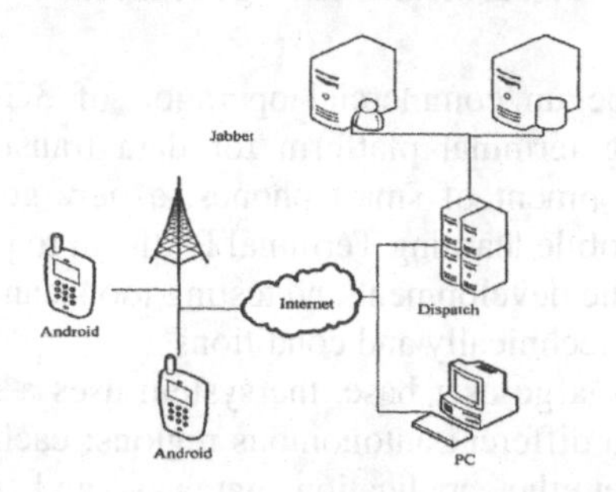

Fig. 4. Application framework based on the Jabber instant messaging system Architecture

Design an Android platform on the e-mail system, this system would directly support through the POP3 protocol to receive e-mail, send e-mail via SMTP, so you can overcome the inconvenience of SMS messages and WAP messages on the traditional mobile phone is slow, need protocol conversion, need continuing disadvantage of online support send and receive e-mail between mobile phones and mobile phones, mobile phone and PC, mobile learning provides the convenience of non-real-time exchange.

The LBS(Location. Based Service), location-based services, also known as spatial location-based services, mobile location services, etc., referring to the mobile computing environment, heterogeneous environment, the use of GIS technology, spatial positioning technology and network communication technology space-based geographical information services for mobile object.

Mobile learning design of the terminal platform integrates a number of advanced technologies for mobile communications, first of all it is a network terminal, the transmission of information between the server and the terminal, terminal and terminal by the mobile base stations connected to the Internet, followed by a location-aware and unique shared functions, built-in GPS module to obtain its own location information and instant communication and information passed between the different terminals of the location information, the last mobile terminal design with a user database in a non-networked environment for simple data processing.

3 The Development of Mobile Learning System Based on the Android Platform

Android application development software can be downloaded for free. We use the integrated development environment for Eclipse 3.3, complete the paper when Ann destroy the JSDK have JDK 1.6, the Android SDK, you need to install the integrated environment of Android development plug-in ADT, you need to choose a special version of Eclipse for Java development (Eclipseforjava), the other with the development of ordinary Java program similar to the SDK directory in the Windows system to the system environment variables.

RSS is a content publisher to publish information XI Yo According to the format, RSS feeds need to use a specific reader-friendly format to convert human-readable displayed to the user. Read z system we have developed is a reader on the Android platform [5].

LBS subsystem is to provide customized content for mobile learning user's channel. First mobile terminal needs to obtain accurate location information of the user's GPS receiver module, then the client application through the sharing function of the position location to provide location information for other clients..

Get your own latitude and longitude information, you first need to on in Goose printed label location-Googk Map divided into three categories, normal maps, satellite maps and map units into which the synthetic map satellite maps and general map of the transparent background superposition. These three maps are stitched together by a number of blocks of 256 × 256 ping image, each picture has only IJRI to distinguish the specialized function of latitude and longitude information into the URL for the location corresponding to the map, this processes the address resolution.

Android is a completely open platform for third-party software developers have greater freedom in its development program, breaking the shackles of the traditional mobile phone, etc. You can only add few fixed software; with Microsoft, Noki adifferent manufacturers, the Android operating system is free of charge to developers, which nearly cost savings. The Android platform is currently being mobile operators, handset manufacturers, developers and consumers to get strong support; we can see that there will be more innovative mobile applications based on the Android platform. The authors completed a terminal of the Android-based mobile learning platform, through the use of the platform, the mobile phone users access to rich online learning resources according to a summary of selectivity, due to the embedded WebKit browser engine, with a touch screen, advanced graphics display such as the Internet reading smart phone reader than improved significantly, as is shown by figure5.

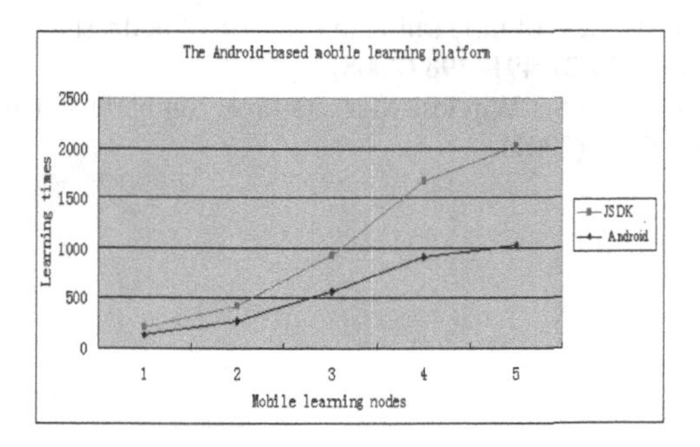

Fig. 5. Development terminal of the Android-based mobile learning platform compare pixture

This paper first describes the development environment to build, and then reading subsystem, communication subsystem, LBS subsystem and communication recorded a subsystem to achieve the key technologies described in detail, all subsystems can be successfully run on the Android emulator.

4 Summary

A comprehensive analysis of the terminal platform for mobile learning from the perspective of software development, including system requirements and technical analysis.In the requirements analysis phase, the demand function collected classification summarized, then the characteristics of mobile devices, mobile learning the difference between the terminal platform and learning platform; in the technical analysis phase, compared the commonly used mobile application solutionthe program's strengths and weaknesses, certainly have the technical feasibility of the development of new mobile learning terminal platform based on this platform by the next generation mobile application platform - Android.

The paper presents the development and design of mobile learning system based on the Android platform. Completed the design and realization of the end platform of the Android-based mobile learning and mobile development features of the system underlying data format, network communications, and embedded database, and other key issues were discussed. Given the key code of the platform section, and verify the availability of the mobile learning terminal platform in a simulated environment.

References

[1] Pasha, M.F., Supramaniam, S., Liang, K.K., Amran, M.A., Chandra, B.A., Rajeswari, M.: An Android-based Mobile Medical Image Viewer and Collaborative Annotation: Development Issues and Challenges. JDCTA 6(1), 208–217 (2012)

[2] Thompson, T.: The Android Mobile Phone Platform. The World of Software Development 33(9), 40–47 (2008)

[3] Norbayah, M.S., Norazah, M.S.: Mobile phone usage for m-learning: comparing heavy and light mobile phone users. Campus-Wide Information Systems 24(5), 355–365 (2007)

[4] Evans, C.: The effectiveness of m-learning in the form of podcast revision lectures in higher education. Computers 50(2), 491–498 (2008)

[5] Palenchar, J.: Android To Set Wireless Markets Free: Supporters. This Week in Consumer Electronics 22(24), 6–81 (2007)

Deep Study of Computer Simulation of Human's Thinking Way and Philosophical Value

Peilu Yang

Shandong University of Traditional Chinese Medicine, Jinan China
yangpeilu@yeah.net

Abstract. Advanced thinking of our humanity is divided into three categories: numerical thinking, logical thinking, visual thinking. Based on exploring the intrinsic relationship of these three types of thinking, this paper focused on the basic characteristic s of the human mind and the computer simulation problems. The author first proposes that the system 0, 1-code system is the ideological foundation of the birth of the computer, and reaches the basic thesis that ideal state of the calculator is computer. This paper argues the means about computer simulation of human three thinking ways and philosophical value, discusses basic intelligent expression of the 0,1 code system in detail and conducted in-depth. The proposed method has some academic value not only to computer science and technology, especially in scientific research of computer theory, but also to computer science and technology teaching and research.

Keywords: computer science and technology, code system, intelligent simulation, knowledge representation.

1 The Raise of the Issue

The dream of human uses machine operator to replace the artificial computing has been for thousands of years, but it has not been achieved until the coming of the computer in the half a century ago. How can the computer be produced in the end? What is the ideological foundation? After extensive research, author believe that the 0, 1-code system is the ideological foundation of the birth of the computer.

Because in the early days, people have formed a relatively fixed decimal mode of thinking, so our ancestors in order to study the computer endeavor to look for 10-state components corresponding to the decimal. It hasn't been found after thousands of years, also makes the study of computer standstill

Until now, people gradually discovered that the number of decimal system does not necessarily have to be fixed in decimal ,and binary, octal, hexadecimal, ... etc., are also possible. These different binary systems can be transformed into each other (see details in 2 (b) exposition), we apply the most simple binary.

Because it is very easy to find the two-state component which is corresponding to the binary code, it is no longer necessary to find the10-state components corresponding to the decimal. With the two states components to corresponding binary, we make conversion between binary and decimal. We use 0, 1 code software to correspond to a two-sate component hardware, and then the magic computer was born.

It is based on the above facts, the author first systematically proposed the thesis that 0,1code system is the ideological foundation of the birth of the computer.

Z. Du (Ed.): Proceedings of the 2012 International Conference of MCSA, AISC 191, pp. 707–712.
springerlink.com

2 Simulation Problems of Numerical Thinking

2.1 The Basic Concept

Binary counting system: using the binary count, referred to as hex.

Digital: a set of symbols used to represent a certain number system. Such as: 1, 2, 3, 4, A, B, C , I , II, III, IV, etc.

Base: The digital number used by the digital number system. Commonly used "R", said R hex. Such as the binary digital: 0, 1, then the base will be 2.

The right: refers to digital weight in different location In the binary counting system, in a different right of number the digital represent the different value For example, the decimal number 111,the single digits on a weight of $10 \wedge 0$, 10-digit on a weight of $10\wedge 1$, a right to the nearest hundred on a weight of $10 \wedge 2$. With this analogize the “nth” weights is $10 \wedge (n-1)$, if it is “mth” bits after the decimal point, then the weight is $10 \wedge (-m)$.

For the average number of system, some integer right is base(median -1), a decimal right is base - median.

2.2 Several Common Binary Counting Systems

Decimal: 0, 1, 2, 8, 9 ten digital composition, that means the base is 10. Features: dot and carry one, a when ten. It is represented by the letter D.

Binary: Formed by two digital of 0, 1, that the base is 2. Feature: meet two into borrow, a when two. It is represented by the letter B.

Octal: (abbreviated)

Hex (Hexadecimal): (abbreviated)

Corresponding relations between decimal binary octal, hexadecimal (abbreviated).

2.3 Decimal and Binary Operation Rules

1) Decimal addition rule table (two decimal numbers together, its rules of permutations are one hundred)

Addend + Augend	0	1	2	3	4	5	6	7	8	9
0	0	1	2	3	4	5	6	7	8	9
1	1	2	3	4	5	6	7	8	9	10
2	2	3	4	5	6	7	8	9	10	11
3	3	4	5	6	7	8	9	10	11	12
4	4	5	6	7	8	9	10	11	12	13
5	5	6	7	8	9	10	11	12	13	14
6	6	7	8	9	10	11	12	13	14	15
7	7	8	9	10	11	12	13	14	15	16
8	8	9	10	11	12	13	14	15	16	17
9	9	10	11	12	13	14	15	16	17	19

Fig. 1. Decimal addition rule

Picked up the red part of the upper-left corner rewrite decimal to binary 10(b) then get the binary addition rules table. As follows:

2)Rules table of binary addition (two binary numbers together, its rules' number of permutations and combinations is four)

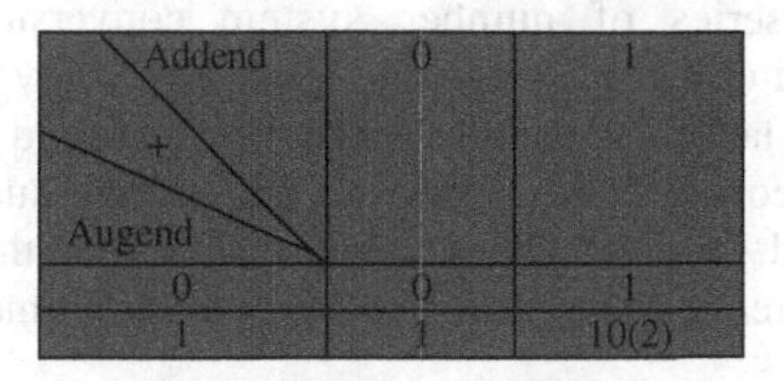

Addend + Augend	0	1
0	0	1
1	1	10(2)

Fig. 2. Binary addition rule

Similarly, we can get rules table of binary subtraction, multiplication, division as follows:

3)Rules table of binary subtraction

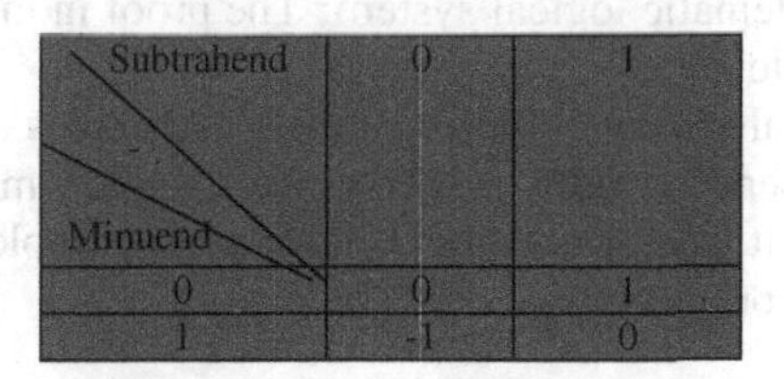

Subtrahend - Minuend	0	1
0	0	1
1	-1	0

Fig. 3. Binary subtraction rule

4) Binary multiplication rule table

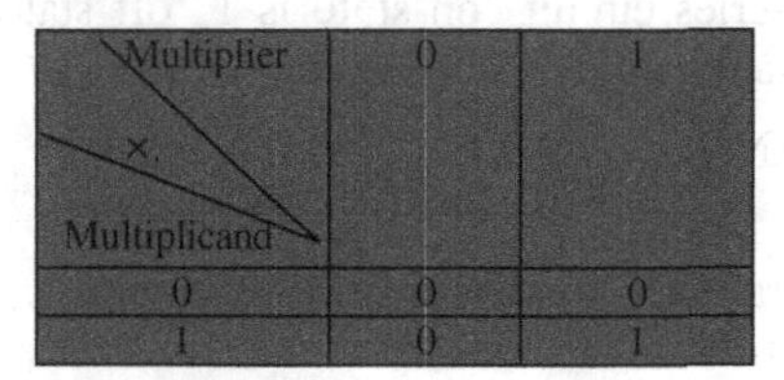

Multiplier × Multiplicand	0	1
0	0	0
1	0	1

Fig. 4. Binary multiplication rule

5) Binary division rule table

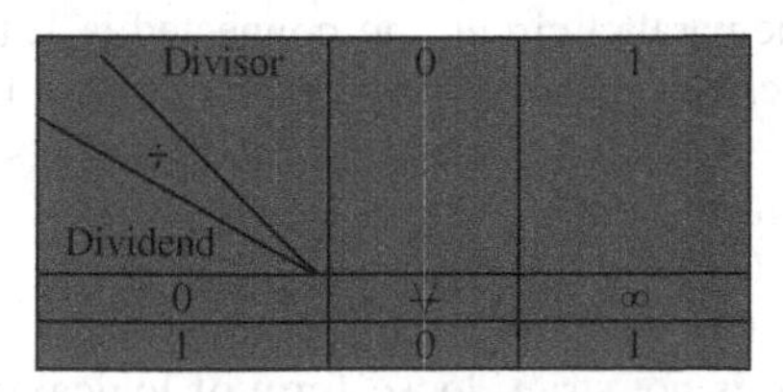

Divisor ÷ Dividend	0	1
0	$\forall$	∞
1	0	1

Fig. 5. Binary division rules

In binary using sixteen can fully demonstrated the 400 rules of the four operations in the decimal that makes the rules of computing greatly simplified

Due to the inherent relationship of the four operations: addition is the most basic operation, multiplication is the cumulative expression of the addition. Division is the inverse operation of multiplication (cumulative operation of the addition). So in the computer, through a series of number system conversion, the original code, complement code, anti-code, making our computer only counted addition can complete all computing tasks. As usually speaking, the speed of the computer is how many times means the computer can calculate per second: such as million times, ten millions times, hundred millions times. That means that this computer is able to calculate million times, ten millions times hundred million times per second.

3 The Simulation of Logical Thinking

Logical thinking is the very important thinking form of human mind. It is commonly exist in everyday life, such as: our judgment to question should be logical; when we speak, our argument, grounds of argument and their internal relations should be logical Like a typical chess game is a purely logical system Judges and lawyers' debate are also very systematic logical system; The proof in mathematics, and physics are also very logical system.

Logical reasoning is also very complex, but its basic rules are not complicated, the system has its basic operation rules just like the complex mathematical system has four operations, we call it, and or not rules. Other more complex logic problem can be completed by using and or, not rules.

3.1 AND Operator

AND operator, we can use a series circuit simulate:A and B represent two switches that connected in a series circuit Its connected state is 1, the disconnected state is 0.;C is the bulb on the series circuit on state is 1 off state is 0 $A \wedge B \rightarrow C$, also represent by 0,1 code, said a total of four states, as follows:

And the operator (AND): $0 \wedge 0 = 0$; $0 \wedge 1 = 0$;
$1 \wedge 0 = 0$; $1 \wedge 1 = 1$;

We can see that the operation is a strict form of computing, only all its premise is 1 can its conclusion be 1.

3.2 OR Operator

Or operation, we can use a parallel circuit simulate: A and B each represent the two switches connected in the parallel circuit the connected is 1, the disconnected ;is 0 C is a bulb of the parallel circuit and connected in series with the parallel body light on is 1, off is 0; $A \vee B \rightarrow C$ also represented by 0,1 code, said a total of four states as follows:

Or operator (OR): $0 \vee 0 = 0$; $0 \vee 1 = 1$;
$1 \vee 0 = 1$; $1 \vee 1 = 1$;

We can see that operator is not strict, loose form of logical operations, as long as at least one of its premise is 1, its conclusion might be 1

3.3 NOT Operator

Non-operator, also called the NOT operation, we can use a single switch ring circuit simulate: A represent a switch the connected state is 1, the disconnected; state is C is the bulb on this circular circuit, on is 1, off is 0.

NOT A → C (or written: A → C), also represent by 0,1code to said a total of two states, as follows:

Non-operator (NOT): 1=0 ; 0=1

4 Simulation Problem of Images Thinking

Human's image thinking is thinking activities based on image recognition. In daily life, we encounter a wide range of graphics, but the basic graphics is not too many, the following ten is enough:

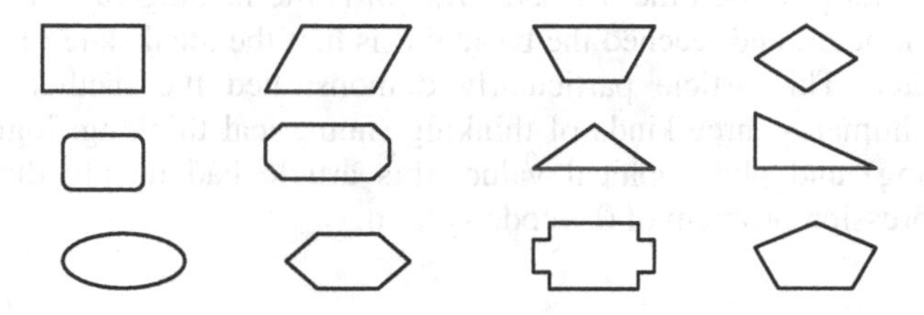

Fig. 6. Basic graphics

Of these basic graphics, right triangle is the most basic graphics, other graphics can be represented by it. Some people may ask how the graphics with arc be represented by triangle We can use the following methods: such as a circle, we can use a number of fan-shaped to represent it and then each fan-shaped can be represented by triangle; Of course, it is truth that the extensional part of curve cannot be expressed, but we can assume that when we use a sufficient number of fan-shaped, it can also means that when the triangle angle in the center of the circle is small enough, the circle represented by triangle also can be enough circular, as following.

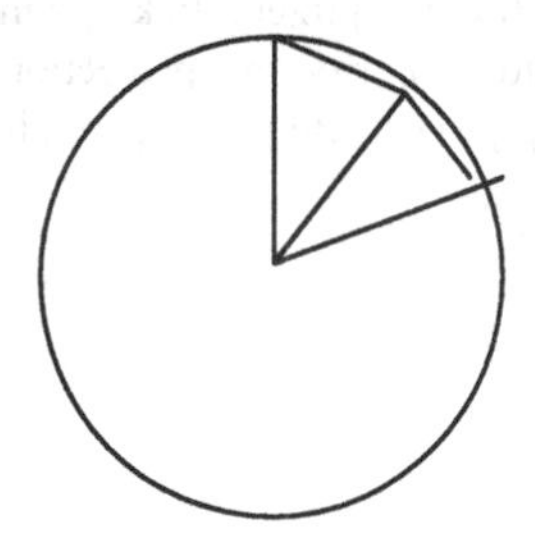

Fig. 7. Circle expressed by triangle

In our daily lives, there are voice recognition problems, we can first convert the voice into the audio image, in this way we convert voice recognition problem into the image recognition problem. Because in the front we have solved the problem of image recognition therefore the voice recognition problem will be solved. In other words the complex voice recognition problem can be converted into the very simple addition operation.

5 Conclusion

The author first affirmed the philosophical value of computer science, and divided advanced thinking of humanity into three categories: numerical thinking, logical thinking, visual thinking. On the basis of exploring the internal relations of these three types of thinking, this article focused on basic characteristics of the human mind and computer simulation problems.

The author first proposed the 0,1-code system is the ideological foundation of the birth of the computer, and reached the basic thesis that the ideal state of the calculator is the computer. This article particularly demonstrated the method of computer simulation of human's three kinds of thinking (numerical thinking, logical thinking, images thinking) and philosophical value, this article had deeply discussed basic intelligent expression problem of 0,1 code system.

References

[1] Yang, P.: Network application technology, designated materials for applied talents in the country training project. Beijing University of Aeronautics and Astronautics Press (September 2009)
[2] Yang, P.: Computer applications teaching and students' innovative ability cultivation of modern educational theory research. National Education Chinese Core Journals 9 (2009)
[3] Yang, P.: Information age university computer basis teaching reform and innovation. Chinese teachers, the Ministry of Education in Charge of Periodicals 9 (2009)
[4] Yang, P.: Computer teaching, management innovation and shaping the ecology of the educational environment. Modern Economy, China's Economic Core Journals (July 2009)
[5] Yang, P.: CATIA V5 R17 Chinese version of the Essentials, designated materials for applied talents in the country training project. Peking University Press (August 2009)
[6] Yang, P.: Network information security and protection, designated materials for applied talents in the country training project. Beijing University of Aeronautics and Astronautics Press (September 2009)

Water Quality Remote Monitoring System Based on Embedded Technology

Wang Xiaokai and Deng Xiuhua

Department of Physical Electronic Engineering,
Shanxi University, China

Abstract. Design a water quality monitoring system is important bescause water bloom break out without warning. Wireless communication is more convenient than wired communication network. While GPRS is wireless communication which is widly used. Therefore this paper Combine GPRS with embedded techology to design a water quality remote monitoring system. Using ARM9 CPU Micro2440 as main controller controls sensor and GPRS to work. 9 kinds of water quality parameters are collected and transported in time. It is realized that water quality is monititored autimatically.

Keywords: Embedded system, GPRS, Water quality, Monitor.

1 Introduction

In recent years, As the improvement of people's life quality, water pollution is more and more serious, especially in people activity area. Water bloom caused by blue-green algae is one of the most important factors which are the results of water pollution. It products something which can harm people's body[1]. Therefore, designing a water quality monitoring system is necessary. Then we can forecast the time of water bloom, and take effective measures to avoid water bloom in time.

It's studied and destroyed a Lake Water Quality Remote Monitoring System, which Combined with GPRS communication technology, Internet technology and data processing technology. The system transfer water quality data to monitoring center in fixed time, then monitoring center evaluates and forecasts water quality using wavelet BP Neural network. It offer reliable information to environmental protection department.

2 Main Structure Design of the System

Water Quality Remote Monitoring System is maked up of two parts, as the following figure1. Monitoring terminal is maked up of CPU, GPRS, GPS, power, data collect. And monitoring center is maked up of water quality evalute, water bloom forecast. Monitoring terminal collects many water parameters including temperature, specific conductance, total dissolved salt(TDS), salt, dissolved oxygen saturation(DOS), Dissolved oxygen(DO), pH, oxidation-reduction potential(Orp), chlorophyll(chl). First of all, water quality data is transformed form analog signals to digital signal.

Z. Du (Ed.): Proceedings of the 2012 International Conference of MCSA, AISC 191, pp. 713–717.
springerlink.com

Then, the result of transformation is transfered to monitoring center using GPRS. In monitoring center, many water quality evalute model are used to estimate water quality, and wavelet BP neural network is used to forecasts water quality.

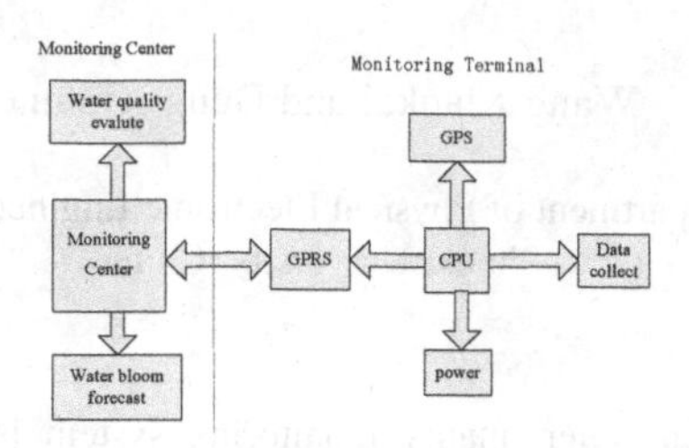

Fig. 1. System structure

3 Design of Monitoring Terminal

3.1 Hardware Design

ARM9 Micro2440 which is based on embedded is used for the CPU of monitoring terminal. Embedded system is divided to 4 levels,they are application software, operating system, various driver programs, embedded hardware[3]. choose Windows-CE as embedded operating system that is responsible for initialization of embedded hardware. Use C language to write driver programs and application programs, in order to realize user's tast. Monitoring terminal makes use of embedded system to realize data collected, managed, showed, transfer, etc.

Water Quality Remote Monitoring System establishs a water bloom forecast model which considers 9 species water quality parameters, they are temperature, specific conductance, total dissolved salt(TDS), salt, dissolved oxygen saturation(DOS), Dissolved oxygen(DO), pH, oxidation-reduction potential(Orp), chlorophyll(chl). Therefore, Data collected is required high for the Water Quality Remote Monitoring System. YSI6600 is a sensor for water quality monitor which can collect 17 kinds of water quality parameters at the same time. It has other advantages, for example long battery's life, small volume, strong function, etc. YSI6600 is appropriate for multi-point sample, Fixed-point collection of different water. It can monitor in water for long time. For these reasons we choose YSI6600 to collect data for monitoring terminal.

Data transfer is link of monitoring center and monitoring terminal, its stability and reliability affect Water Quality Remote Monitoring System Directly. MC55 is a Mature GSM/GPRS module, that has built-in TCP/IP protocol. And Up-link speed is 42.8 kB/s, down-link speed is 85.6 kB/s [4]. According to Design requirements, choose MC55 as GPRS in the system.

The three parties of monitoring terminal are connect though serial ports. The party of data collect uses YSI sensor that is supplied by Beijing Technology and Business University. At monitoring points, YSI collect water quality parameters, then make use of platform GPRS supplied by China Mobile to send data to monitoring center. Arm9 send AT command to GPRS though serial port to control GPRS module. Serial ports use RS232 serial port. We only use three pin to establish communication, SG(5), TXD(3), RXD(2). There are two wiring method as the following table 3.1. When data transfer, use a bit rather than a frame data to stimulate port event, therefor avoid data loss because of a frame data divided to many section in transmission process.

Table 1. Serial port wiring method

9 pin---------9 pin			
Wiring method	2	3	5
Extension cord	2	3	5
Cross line	3	2	5

3.2 Software Design

In order to make convenient for workers to check that water quality parameters is collected normal or not, not only need to collect data, but also need to show data in interactive interface. Analysed data, we found that spaces (20) during data between two symbols (0D0A) is regular. Every water quality parameters is separated by several spaces.While the number of spaces isn't equal, so we delete spaces, then using "," instead spaces . We can find "," to separate parameters.

Baud rate of serial port communication between monitoring terminal CUP ARM9 and other parties is 9600 bps. Data format is that one bit is parity checking,eight bits is data, one bit is stop. Whole process of serial port communication can be controlled in program though built a serial-port object . Figure 2 is a flow diagram of serial port communication for monitoring terminal.

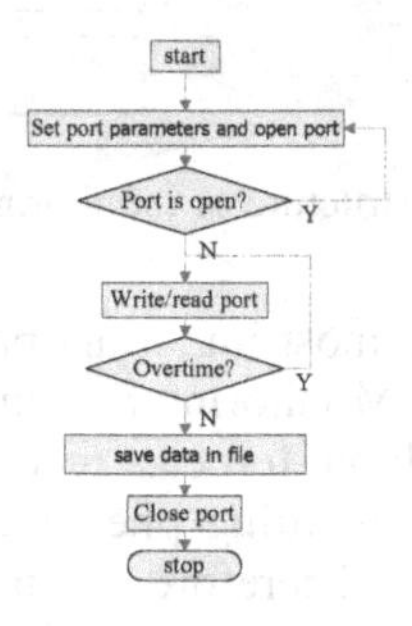

Fig. 2. Flow diagram of serial port communication

Some important property of serial-port object which are used when design the program of serial port communication are followed.

Portname: COM1,COM2,etc.
BaudRate: the speed of serial port communication
Parity: which can realize by choosing number in enumeration of parity.
DataBits
StopBits: which can realize by choosing number in enumeration of StopBits.

Arm9 control GPRS though AT command. AT command is standard of modem communication interface which can be recognized and executed[5]. The procedure of controlling GPRS by AT command is followed:

1. AT+CDNSORIP=0,set connecting way, '0' means using IP address of server,'1' means using domain of server. Monitoring center server has a fixed IP address, so we set that AT command value is '0'.
2. 2. AT+CIPSTART="TCP","117.79.130.124", "8080",Use this command to start TCP/UDP,"117.79.130.124" is a IP address which is used in experiment . "8080" is the port number.
3. 3. AT+CIPSEND,Send this command. Then after terminal receive '>',data can be send to monitoring center.

4 Design of Monitoring Center

Main function of monitoring center is following:receive data, inquire historical data, evaluate water quality, forecast water bloom, control monitoring terminal.When the value of chlorophyll exceed its threshold, monitoring center can control that monitoring terminal reduce the time of data transmission. Figure 3 is F-structure of monitoring center.

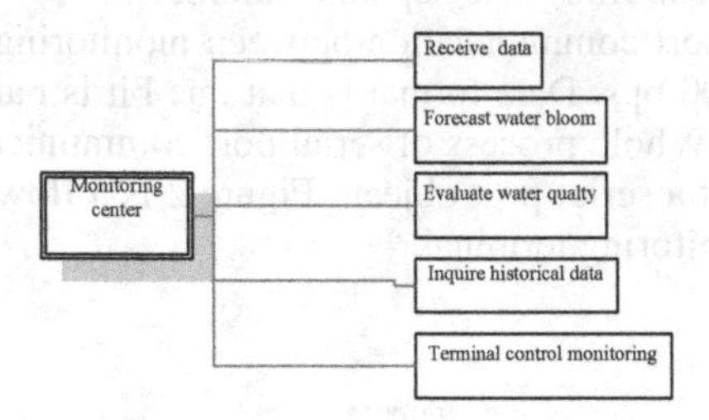

Fig. 3. F-structure of monitoring center

Forecasting water bloom is the most important part in monitoring center, it is the purpose of Water Quality Remote Monitoring System. Artificial neural network is the main method to establish water bloom forecast model. After do much experiment we found that it is easy to appear over-fitting phenomenon if only use artificial neural network to forecast water bloom. Therefore we need to optimize artificial neural network first, and then to forecast water bloom. There are two method that can be used to optimize artificial neural network, wavelet, fourier. Compared the results of these two method, we found that artificial neural network optimized by wavelet avoid blindness and local optimum which are caused by artificial neural network. Therefore, in this system, we choose artificial neural network optimized by wavelet to establish water bloom model.

Following is the procedure of establish water bloom model:

First, determine input and output of artificial neural network. The data we use to do experiment is supplied by Beijing Technology and Business University chemistry department.We need normalized data, X={x1, x2..., xN}is the input sample, Y= {y1, y2, ..., yN} is the output sample. Following is a formula of normalization.

$$x'_k = \frac{0.9-0.1}{x_{\max}-x_{\min}} x_k + \left(0.9 - \frac{0.9-0.1}{x_{\max}-x_{\min}} x_{\max}\right) \quad (4.1)$$

Second, use wavelet to denoise.

Third, establish neural network. It's the key step of the whole forecast model. Complexity of structure and value of parameter affect accuracy of forecast result. The number of input neuron is five,and the number output neuron is one. The number of hidden layer is decided by $\min(p,q) < n \le 2p+1$.

Fourth, train neural network which has been established.

Figure 4 is the fitting result of wavelet- neural network.

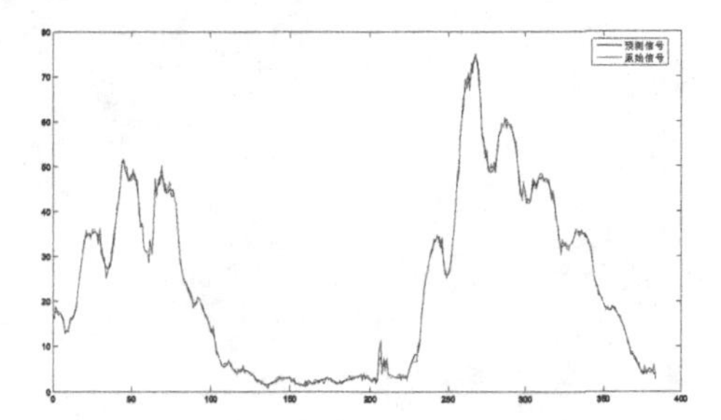

Fig. 4. Fitting result of wavelet- neural network

5 Conclusion

It's easy to realize Water Quality Remote Monitoring System with embedded system. Speed of this system is fast. The cast of data transfer can been saved, using GPRS wireless communication. And water bloom forecast in monitoring center can supply important condition for environmental protection department with that environmental protection department can take measures to avoid water bloom.

Acknowledgement. The work was supported by Shan Xi Province Science and Technology Project(20110321025-02).

References

[1] Yuhongman, L.: Talk About Harm and Prevention of Blue Algae. Journal of Beijing Fisheries 01, 20–30 (2004)
[2] Yuweihong: Environment Monitoring System and Designing Embedded Software Based on Linux Operating System. Tijian University (2008)
[3] Huangqiang, T.: A Remote Data Transmission Module Based on ARM and GPRS. Chinese Journal of Electron Devices 31(4), 1214–1218 (2008)
[4] Huangzhanhua, W.: Study of Realizing Multiple Message Transporting Based on GPRS and Socket Mechanism. Commuoications Technology 11(40), 376–378 (2007)

Electrical Automation Technology in the Thermal Power

Chen Yijun

Electronic Engineering, North China Electric Power University, Beijing, 102206, China

Abstract. Modern electrical automation system network highlights the advantages of strengthening the application of electrical information, can improve the efficiency of thermal power generation in the complex case, to improve the management level and economic efficiency of thermal power plants, promoting the stability of the electrical control,significant social and economic benefits.

Keywords: Thermal Power, Power System, Electric Automation.

1 Backgrounds

Thermal electrical automation system (referred to as ECS) is a new technology in the field of modern electrical automation, electrical automation in thermal power generation in a wide range of applications due to their own characteristics and network advantages, not only to better promotethe development of thermal power, but also improve the level of automation of power plants, enhanced electrical stability and advanced automation.

2 Obstacles Exist in the Traditional DCS Technology

Power Plant Electrical automation systems, referred to the EFCS or the ECS is a new hot spot for the rise in recent years in the field of power plant automation. Distributed control system (DCS) and the power plant focused on the monitoring of the thermal system corresponds to EFCS focused on the monitoring of power plant electrical system; focus on the part of the electrical control of power plants connected to the grid and power plant network monitoring system (NCS) is relatively ratio, EFCS focus on power generation Plant within the protection of low voltage electrical system auxiliary power, measure, measurement, control, analysis function.

EFCS system previously independently run the 6kV/10kV medium pressure system and 400V low pressure system type and number of protection devices, monitoring and control devices, automatic devices via the field bus or Ethernet links together to form the system, on the one hand, means of communication with the DCS system of information exchange, greatly reducing the cable investment in the investment and hard wiring of the DCS measurement points on the other hand, coordination and control of the electrical system through the network and back-office software, failure analysis and operation and management, to improve the level of automatic control of the entire power plant operation and management level.

Z. Du (Ed.): Proceedings of the 2012 International Conference of MCSA, AISC 191, pp. 719–724.
springerlink.com

(1) In electrical and automation systems, current and voltage of the electrical systems have long been the AC sampling, precision, high speed, digital; DCS such as voltage and current through the transmitter is converted to access the DCS, the second wiringcomplex, high cost, poor anti-interference performance;

(2) DCS limit of electrical measuring point, so many applications in the electrical system can not be achieved, such as fault diagnosis, fault analysis, economic analysis, valuation management. Thus unable to upgrade the electrical system operation and management level. In recent years, the successful application of Field bus, Industrial Ethernet network communication technology in the substation automation system, as well as the shortcomings of the hard wiring of DCS system gradually exposed, making the comprehensive plant electrical system automation level of the voice of increasingly higher. Since 2000, with some power at home and abroad automation equipment manufacturer and power planning, design and use, and test departments to actively explore a variety of EFCS program, and in some power plants were tested, and has accumulated valuable experience. The common feature of these programs are: plant electrical automation equipment through the fieldbus network; communication plus some hard-wired electrical system with the DCS to contact in order to reduce the number of cables; the establishment of electrical back-office systems, planning and the gradual development of various software applications.

3 The EFCS System Composition

(1) Spacer layer and the layer in order to promote the completion and improvement of the intelligent devices of a variety of specialized functions, including: pressure 6kV/10kV system series of protective and monitoring devices in the auxiliary power, auxiliary power low-voltage 400V System Series Intelligent Controland monitoring and control devices, auxiliary power fast switching devices, low voltage standby power source installed, automatic quasi-synchronization control device, a small current grounding line device, DC grounded feeder.Through the various functions of the software development for the operability of the system, the CPU, the A / D, RAM, EEPROM, fieldbus or Ethernet external communications interface.

(2) Communication management. This layer is the network and communications security management, to complete the variety of smart devices, DCS systems, electrical background monitoring system, power plants, other smart devices, power plants, other systems of communication. The main means of communication or industrial Ethernet and fieldbus such as PROFIBUS, CAN, modern communication management device has been achieved the conversion between the different field bus interface standard, interconnects, and communication.

(3) Station level. Changed to include the background monitoring computer hardware and a variety of professional applications, servers, workstations and other hardware, application software, including SCADA (data acquisition and monitoring), the auxiliary power meter, the recorded wave analysis, motor fault diagnosis, and other seniorcommunication interface between the software or infrastructure software, and back-office systems and power plant other management systems (eg, MIS) software.

4 Innovative Electrical Automation Technology in the Thermal Power

(1) Unified unit furnace unit. Automation technology should be in the thermal power of innovation to achieve the integration of mechanical, electrical, and gradually realize the thermal power plant machine, furnace, electrical integration, unified operation monitoring unit system. Distribution in the thermal control system (DCS) will be able to use the machine, furnace, electric unit system operating mode of the entire generator set operating parameters and status of mobile phones and analysis, to play the maximum potential of the thermal power generating units, while achieving their own with the control function, narrowing the control room, to simplify the monitoring system, to reduce the burden of management staff to improve the progress of work, while the maximum extent possible to reduce the cost; the same time, the unified unit furnace unit is also easy to thermal power generation in power plant information management system (MIS), information collection, strengthen the unified management of thermal power grid, finished in the tone of AGC instructions and requirements, improve the efficiency of power plant work, allows the plant to always maintain the best working conditions and to maximize the economic benefits. Therefore, the unified unit furnace unit will help improve the level of monitoring and automation level of thermal power units.

(2) Innovative control means of protection. System control and protection means in the traditional form of thermal power is mainly used for alarm and chain, and the grace of this way you limit alarm and interlock tripped volatility control and protection. And electrical automation technology by computer calculation and analysis, operations and fault diagnosis of electrical automation systems, improve the improvement of thermal power equipment, and reduce the failure occurs, change control and protection strategy and control system failure in prevention in the first place, to ensure the electrical and automation systems can continue to maintain the operating status has a positive meaning. In addition, preventive maintenance, electrical automation system can also passive and accident repair conversion predictable and equipment maintenance, preventive maintenance, and comprehensively improve the quality and efficiency of thermal power generation.

(3) To achieve electrical communication control. Practical point of view of China's thermal power generation, thermal power plant electrical automation system (ECS) is unable to meet the electrical automation systems, the distributed control system (DCS) to achieve electrical communication control, the speed and reliability of the communication problems that still exist. overall, relatively backward, electrical automation system (ECS) and the Distributed Control system (DCS) still have retained part of the hard-wired. Therefore, to achieve electrical communication, and good thermal process chain must be addressed to improve the level of practical application of the electrical system in the background in the basic stages of power generation to run real-time monitoring, substantially improve the control logic for electrical automation systems, control level, the level of automation, and operation and management level.

5 Modern Aautomation Mature Power Generation Technology

(1) Of the boiler drum level measurement technology to a major breakthrough.Qinhuangdao Huadian monitoring and control equipment Co., Ltd. and the Huai An Weixin Instrument Co., Ltd. After years of research and development, the introduction of the GJT-2000 high-precision, high reliability of the electrode drum level measurement devices, non-blind low deviation of two-color water level, drumbuilt-in electrode water level measuring device, multiple test holes take over the technology, and the steam drum internal soft power conditions

Gained extensive experience and sampling location configuration. These results, marking the boiler drum level measurement technology has achieved a major breakthrough. However, due to the current standards and regulations, including the 2001 State Power Corporation released a major accident to prevent electricity production 25 key requirements are established in the past drum level measurement devices and technologies on the basis of these procedures, therefore, to a certain extent, the norms seriously hindered the promotion and application of these new technologies. At present, according to the deviation between the above techniques, Water Level Indicator and the entire process can be controlled at less than 30mm, the drum level protection from point furnace reliable in operation. Tongliao Power Plant Drum built-in electrode type water level gauge test the long-running (more than two years), and through the identification, has been highly evaluated, and its success is expected to take the lead in the world so that the boiler drum level measurement site benchmark instrument breakthrough.

(2) SIS technology is mature. SIS concept first proposed by our experts a decade ago, after the great debate, the pilot and gradually promote the application has been over three hundred plants to promote and develop into an industry of dozens of companies in China. Although some power plants due to select developers improper or can not keep up the effect failed to appear, the majority of power plants are the obvious effect of varying degrees. SIS a power industry standard "Thermal Power Plant-level monitoring information system technical conditions" from 2004 has been awarded through a lot of practice, there has been considerable development in the SIS technology, network technology, including a one-way physical isolation devices to prevent the "loss" technology, etc. progress has been made, the application to constantly enrich and improve the optional expand the scope of real-time database systems, and some excellent home-made database in the market competition, many power plants and gradually formed a set of adaptation SIS system management power plant real-time system of the production process, methods and experience.

6 Electrical Automation Technology in Conventional Thermal Power Plant Control

(1) Local control. The development of smaller power plants, less the equipment used in specific operations, but still need to build an integrated control system, such as: boilers, turbines, generators and other equipment are integrated control.Automation technology can reasonably use the device to avoid the inconvenience due to equipment run separately.

(2) Centralized control. The scale of thermal power plants than any other plant, the number of power generation, power plant equipment is also more coordination between the perfect device operation is the management focus of the work, and this is one of the difficulties of the production planning. Automation technology can boilers, turbines, generators and other reasonable combinations, power plants, integrated control, and improved overall efficiency in the use of the device.

(3) Automatic control. Electric Automation will also promote the automation of electric power. Such as: the development of computer technology to reduce manpower management, promotion of automation not only improve the normative power plant management, and also maintain the safety of the personnel of production, reduce the difficulty of the electrical energy production, energy production, allows the plant to obtain more economic benefits.

(4) Failure to control. In the power generation process, the equipment is inevitable there will be a variety of fault and electrical automation technology can play a key role in control of the equipment failure. Technical staff through the computer building line monitoring system, power generation in a variety of exceptions to analyze and solve, but also according to the operation processing instruction to deal with glitches.

7 Conclusion

Electrical automation of the thermal power plants use the thermal power system management can improve automation and management level. The automation system through the computer, protection, measurement, hierarchical distributed control and communication technology, the operation of thermal power plant power system protection, control and fault handling, which undoubtedly is a comprehensive management system. Make full use of the information full advantage of the interconnection of the electrical system, successful electrical automation to complete the more complex electrical operation management.

References

1. Liu, H.: of the current situation and development direction of the electrical automation. Science and Technology Letter Interest Rates (6) (2010)
2. Wang, S.H., Li, G.: Analysis Application and Development Trend of electrical automation and control systems. Heilongjiang Science and Technology (20) (2011)
3. Fu, K.S.: Learning control systems and intelligent control systems. IEEE Trans. Auto. Contr. 16(1), 70–72 (1971)
4. Saridis, G.N.: Twards the realization of intelligentsia controls. Proc. of the IEEE 67(8), 1115–1133 (1979)
5. Astrom, K.J., Anton, J.J., Arzen, K.E.: Expert control. Automaton 22(3) (1986)
6. Levies, V., et al.: Challenges to control a collective view. IEEE Grans. Auto. Contr. 32(4), 275–285 (1987)
7. Recce, M., Plebe, A., Taylor, J., et al.: Video Grading of Oranges in Real Time. Artificial Intelligence Review 12, 117–136 (1998)

8. Boukouvalas, C., Kittler, J., Marik, R., et al.: Color Grading of Randomly Textured Ceramic Tiles Using Color Histograms. IEEE Trans. Industrial Electronics 46(1), 219–226 (1999)
9. Han, Y.J., Cho, Y.J., Lambert, W.E., et al.: Identification and Measurement of Convolutions in Cotton Fiber Using Image Analysis. Artificial Intelligence Review 12, 201–211 (1998)
10. Suen, P.H., Healey, G.: Modeling and Classifying Color Textures Using Random Fields in a Random Environment. Pattern Recognition 32, 1009–1017 (1999)
11. Van de Wounder, G., Saunderstown, P., Livens, S., et al.: Wavelet Correlation Signatures for Color Texture characterization. Pattern Recognition 32, 443–451 (1999)
12. Vishakhapatnam, S.T.S., et al.: Chaotic tailerons for online quality control in manufactural. Int. J. Adv. Nanotechnology 13, 95–100 (1997)
13. Rose, D.E.: Appropriate uses of hybrid system. In: Tourette, D.S. (ed.) Connection Models, Proc.1990 Summer School, pp. 277–286. Morgan Kaufmann, CA (1991)

The Research of Electrical Automation and Control Technology

Yu Xian

School of Electrical and Electronic Engineering,
North China Electric Power University, Beijing, 102206, China

Abstract. Described the course of development of electrical automation, influencing factors and the status quo to explore the features, functions and design philosophy of electrical automation and control systems, and electrical automation and control technology development trend of the outlook.

Keywords: Electrical Automation, Control Technology, Electrical Automation and Control Systems.

1 Backgrounds

Electrical automation is an important symbol of the industrial modernization and the core of modern advanced science technology, product operation, control and surveillance, in the absence of (or less) directly involved in the case, according to a predetermined plan or program automatically technology. Its role to improve the reliability of the work of running the economy, labor productivity, improvement of labor conditions, from heavy physical labor, part of the mental and poor, freed in a dangerous work environment, can enhance human understanding of the world the ability to transform the world[1].

2 The Development Process of Electrical Automation

2.1 The Development Process

Electrical automation technology is an electrical engineering technology disciplines closely linked with the electronics and information technology, electronic technology, information networks, intelligent control of the rapid development of electrical automation experience from scratch, from development to mature process.

In the 1950s, the term "automation" was proposed, gave birth to the electrical automation products such as electricity, motor, relays and contactors application allows the machine to complete the pre-arranged in accordance with the will and set good judgment and logic functions, prompting a change of electrical automation; in the 1960s, modern control theory and the application of computers to promote the integration process of automatic control and information processing, automation enter the stage of integrated automation can achieve production effective optimization of

Z. Du (Ed.): Proceedings of the 2012 International Conference of MCSA, AISC 191, pp. 725–729.
springerlink.com

process control and management, electrical, automation has been a qualitative leap; in the 1970s, with the rapid development of communications, IT, microelectronics and other technologies, automation objects and gradually extended to large, complex systems, many problems difficult through the use of modern control theory to be addressed through the study of these problems to automation theory and means to achieve innovation, resulting in a comprehensive utilization of systems engineering, computers, artificial intelligence, communication technology and other high-tech, high-level automation systems used in complex systems, and promote the electrical automated rapid development; from the 1980s, electrical automation, rapid development has been more mature, electrical automation is an important component part of the high-tech, has been widely used in industrial, defense, medicine, agriculture and other fields. greatly promoted the development of artificial intelligence, aerospace, transportation, and manufacturing technology in many fields, has played an important role in the development of the national economy.

2.2 Influencing Factors

The development of electronic automation by IT, the impact of the physical sciences, IT plays a decisive impact. Modern IT refers to the various means of development and use of information, computer and network technology, communication technology and other related technology, computer, optoelectronics, communications technology, including optoelectronics, microelectronics and components manufacturing information technology to collect, transport, processing, and use a variety of information technology and equipment, technology and application technology to achieve these functions. The development of information technology by the development of electrical automation, at the same time the development of information technology for the development of electrical automation to provide the necessary tools. In addition, the development of physical science plays a role in promoting the development of electrical automation, the development of the transistor, LSI technology has greatly promoted the progress of electrical automation electrical automation in close contact with the physical sciences gradually to expand the field of biological systems, micro-electromechanical systems.

2.3 Status Quo

First, electrical automation system information. IT in the vertical and horizontal electrical automation penetration, vertical, IT business data processing from the management face to infiltrate the use of information technology can effectively access financial and other management data, the dynamics of the production process monitoring, real-time production information and to ensure comprehensive, complete and accurate; horizontal, IT penetration, microelectronics and other technology equipment, systems and other applications of information control system, PLC and other equipment boundaries gradually become blurred from the well-defined, the software structure, group the state of the environment, communications capabilities, such as the increasingly prominent role, such as networking, multimedia technology has been widely applied.

Second, electrical automation systems, maintenance and repair facilitation. Windows NT has become the standard of electrical automation and control platform, norms and language Windows-based man-machine interface has become the mainstream of electrical automation, and based on the Windows control system has a flexible, easy integration advantages, has also been widely used . Windows operating platform makes use of electrical automation systems, maintenance and repair more simple and convenient.

Finally, the distributed control applications. Electrical automation systems via serial cable to connect the central control room, the PLC, the scene, industrial computer, the CPU of the PLC, remote I / O station, smart instrumentation, low voltage circuit breakers, inverter, motor starters, etc. to connect field devices The information collected from the central controller. Distributed control applications through the branch structure of the digital serial connection automation systems and smart devices, two-way transmission communication bus, the PLC and field devices with the corresponding I / O devices connected together, so that input and output modules play the on-site inspection and enforcement role.

3 Features, Functions and Design Philosophy of the Electrical Automation and Control Systems

3.1 Feature

Compared with the heat engine equipment, electrical control system of the control object, a small amount of information, the operating frequency is low, but with a fast, accurate and advantage. Electrical equipment requires a higher level of protection automatic device reliability and rapid response capability and high anti-jamming capability, electrical control system with more chain protection, to meet the requirements of effective control.

3.2 Function

Based on the characteristics of electrical control, electrical automation and control systems to achieve effective control of the electrical system of the generator - transformer unit must meet the following basic functions: the effective control and operation of the generator - transformer unit export isolation switch and circuit breakers; power generation machine - transformer unit, excitation transformer, high transformer protection control; excitation generator excitation system operation, de-excitation operation, changes in the magnetic operation, stable investment and retirement, control mode switch; switch automatically, manually over the same period and the net; high voltage power supply monitoringand operation and monitoring of the switching device, start, such as investment and retirement; monitoring and operation of low voltage power supply and self-investment and device control; high-voltage transformer control and operation; control and operation of the generator set; LPS, DC system monitoring.

3.3 Design Concept

Electrical automation and control systems have centralized monitoring, remote monitoring, fieldbus monitor in ways designed to The characteristics of centralized monitoring and centralized system functions by a processor for processing, the advantages of simple design, low protection requirements, easy operation and maintenance. Processor workload is too heavy, resulting in low processing speed of all electrical equipment to monitor the host redundancy will lead to lower increase in the number of cables, causing investment to increase, and long-distance cable interference will affect the system, Isolating Switch the circuit breaker uses a hard-wired and easy to produce auxiliary contacts are not in place, check the line is not convenient, increases the chance of misuse. The RMON way with flexible configuration and save cable and installation costs, materials, and high reliability due to electrical equipment communication than on works, CAN, and other field bus communication speed is not high, the way only for small monitoring of the system can not meet the requirements of large-scale electrical automation systems. The universal application of the field bus, Ethernet technologies and the corresponding operation of the accumulation of experience, intelligent electrical equipment has been a rapid development, and network control systems gradually applied to the electrical system, the specific circumstances of the field bus monitoring methods for electrical systems design, not only has all the advantages of remote monitoring, but also save the analog transmitter, isolation devices, I / O card. In addition, the smart device easy to install, you can save the control cable and the corresponding investment and installation.

4 Electrical Automation and Control Technology Development Trend

With the OPC (OIJE for Process Control) the emergence of the technology and the promulgation of IEC61131 and Microsoft's Windows platform, widely used in the future combination of electrical technology, computers play an increasingly irreplaceable role. IEC61131 has become an international standard of electrical automation and control technology, being the major control system vendors widely adopted. PC client / server architecture, Ethernet and Internet technologies has led to the electrical automation revolution again and again, the needs of the market-driven automation and integration of IT platforms, and the popularity of e-commerce will accelerate this process. Internet / Intranet technology and multimedia technology has wide application prospects in the field of automation and enterprise management using a standard browser to access the corporate management of all aspects of financial and other data, can also monitor the dynamic picture of the current production process, the first time to understand the most comprehensive and accurate production information. The application of virtual reality technology and video processing technology, the future of automation products will have a direct impact, such as the design of man-machine interface, and equipment maintenance system, corresponding to the software architecture, communications capabilities and ease of use and unified configuration environment variable important, the importance of software has been improving, this trend is towards integrated systems from a single device.

5 Conclusion

In summary, with the rapid development of intelligence, information technology, electrical automation technology will continue to the trend of science and technology, information technology, open development in the areas of electrical automation will continue to increase, technology updates will continue to accelerate. electrical automation and control technology will also be the rapid development and continuous improvement.

References

1. Liu, H.: of the current situation and development direction of the electrical automation. Science and Technology Letter Interest Rates (06) (2010)
2. Wang, S.H., Li, G.: Analysis Application and Development Trend of electrical automation and control systems. Heilongjiang Science and Technology (20) (2011)
3. Fu, K.S.: Learning control systems and intelligent control systems. IEEE Trans.Auto. Contr. 16(1), 70–72 (1971)
4. Saridis, G.N.: Towards the realization of intelligentsia controls. Proc. of the IEEE 67(8), 1115–1133 (1979)
5. Astrom, K.J., Anton, J.J., Arzen, K.E.: Expert control. Automaton 22(3) (1986)
6. Levies, V., et al.: Challenges to control a collective view. IEEE Grans. Auto. Contr. 32(4), 275–285 (1987)
7. Recce, M., Plebe, A., Taylor, J., et al.: Video Grading of Oranges in Real Time. Artificial Intelligence Review 12, 117–136 (1998)
8. Boukouvalas, C., Kittler, J., Marik, R., et al.: Color Grading of Randomly Textured Ceramic Tiles Using Color Histograms. IEEE Trans. Industrial Electronics 46(1), 219–226 (1999)
9. Han, Y.J., Cho, Y.J., Lambert, W.E., et al.: Identification and Measurement of Convolutions in Cotton Fiber Using Image Analysis. Artificial Intelligence Review 12, 201–211 (1998)
10. Suen, P.H., Healey, G.: Modeling and Classifying Color Textures Using Random Fields in a Random Environment. Pattern Recognition 32, 1009–1017 (1999)
11. Van de Wounder, G., Saunderstown, P., Livens, S., et al.: Wavelet Correlation Signatures for Color Texture characterization. Pattern Recognition 32, 443–451 (1999)
12. Vishakhapatnam, S.T.S., et al.: Chaotic tailerons for online quality control in manufactural. Int. J. Adv. Nanotechnology 13, 95–100 (1997)
13. Rose, D.E.: Appropriate uses of hybrid system. In: Tourette, D.S. (ed.) Connection Models, Proc.1990 Summer School, pp. 277–286. Morgan Kaufman, CA (1991)

The Military Strategy Threat of Cognitive System Based on Agent

Tiejian Yang, Minle Wang, and Maoyan Fang

Xi'an Hi-tech Institute,
China Xi'an 710025

Abstract. Human subjective consciousness and objective world is connected by Agent in order to enhance the cognitive abilities on the threat of military strategy. The research is based on the threat theory and the complexity of the military struggle, architecture, working principle, structural model and implementation of military strategy threat of cognitive system were studied based on Agent. The results can improve the cognitive abilities of military strategy threat for military field.

Keywords: military strategy threat of cognitive system (MSTCS), cognitive system, Agent.

1 Introduction

The cognitive is the process of processing information for individual, cognitive psychology regard cognitive processes as information acquisition, encoding, storage, extraction and use of a series of consecutive cognitive operations stage according to certain information processing system[1]. Threat cognitive theory suggests that: the threat is not perceived by the parties, even in the face of the obvious objective evidence, making it nearly impossible to mobilize the defensive resources[2]. On the contrary, when the threat is perceived, even if the opponent is assumed to be non-malicious, will take the appropriate anti-threat measures[3,4].

In recent years, Agent technology continues to mature[5,6], Agent technology from the field of artificial intelligence developed unique mental properties in the study of complex systems has been widely used, the cognitive system based on Agent is mainly carried out[7]: 1) human subjective perception of consciousness and the objective world through the Agent will closely integrate and enhance the capacity of people's awareness of the threat of military strategy[8]; 2) make the appropriate intentions and strategies for problem-solving response by Agent perception of environmental information and awareness.

Z. Du (Ed.): Proceedings of the 2012 International Conference of MCSA, AISC 191, pp. 731–735.
springerlink.com

2 System Architecture

Military strategic threat to the cognitive system includes three layers: resource layer, agent system layer and interactive layer. Structure of the system framework is shown in Figure 1.

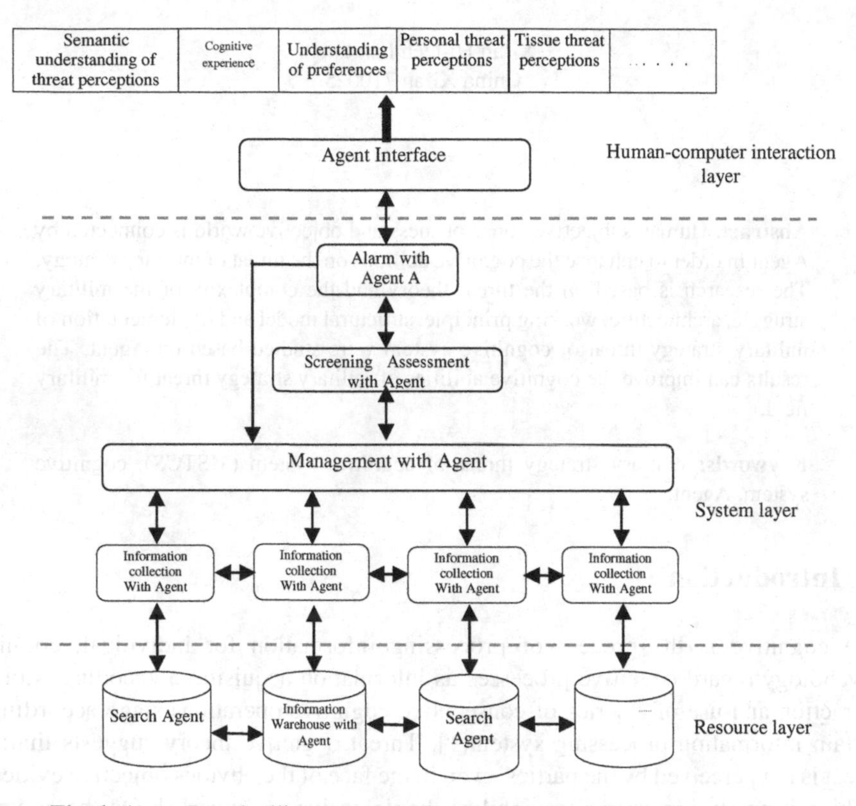

Fig. 1. Agent-Based military strategic threat to the cognitive system structure

3 System Principles

The system works in two ways: 1) bottom-up cognitive process, and the Agent through its own characteristics, take the initiative to analyze the threat of military strategy, and implementation of alarm through the critical point of information to remind; 2) Another agent to respond to people, according to the requirements of the threat of cognitive support. This article focuses on the study and exploration of the first approach, as Figure 2.

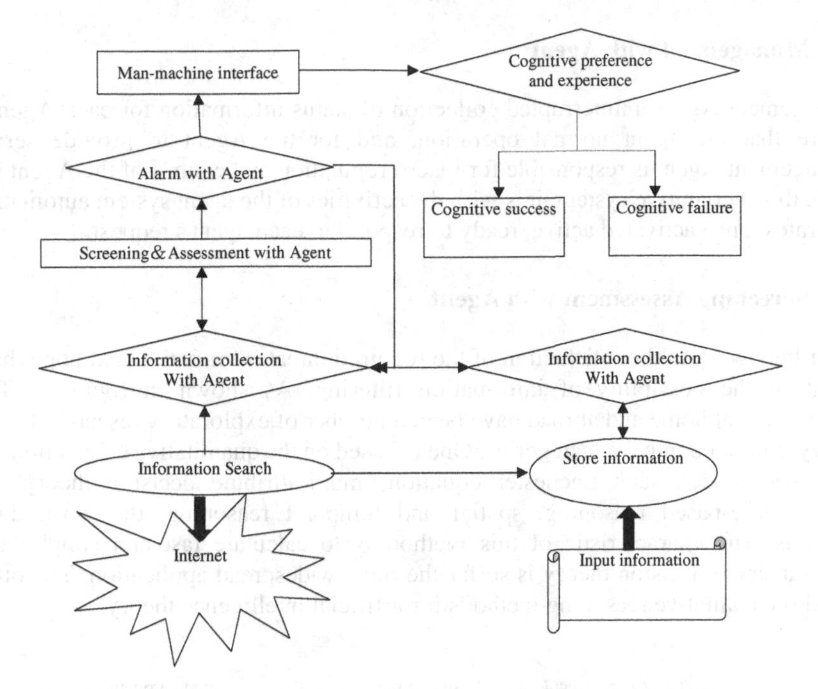

Fig. 2. Agent military strategic threat to the cognitive system works diagram

4 System Model

Prototype based on the kernel of the Intelligent Agent, in accordance with the agent's design needs, easily in the prototype design, free to add functional modules, and then construct the different functions of the Agent.

4.1 Search Agent

The main components of the search Agent is dynamically identify and search engine of the external environment.

4.2 Information Warehouse with Agent

Agent warehouse consists of intelligent memory. Its function first, save through the Interface Agent the information entered, the Search Agent part of the information is stored.

4.3 Collection of information with Agent

The information collection agent structure consists of the following side: information collection interface module including a communication control module and the information exchange module.

4.4 Management with Agent

Management Agent uninterrupted collection of status information for each Agent and ensure that the agent normal operation, and for the Agent to provide services. Management Agent is responsible for macro regulation and control of the Agent in the entire threat cognitive system, it is with the activities of the agent system automatically generates, once activated active, ready to respond to each agent's request.

4.5 Screening Assessment with Agent

The filters work to the satisfaction of the required threat information and hope that the threat of the possibility of information filtering. As shown in figure 3. Threat Assessment at home and abroad have been a number of exploratory research, the main theory is divided into two categories: One is based on the quantitative calculation of the analytical model, such Lachester equation, multi-attribute decision theory, game theory, case-based reasoning, spatial and temporal reasoning, the work domain methods. The characteristic of this method is to calculate fast and simple, which multi-attribute decision theory is so far the most widespread application. The other is based on qualitative reasoning methods for artificial intelligence theory.

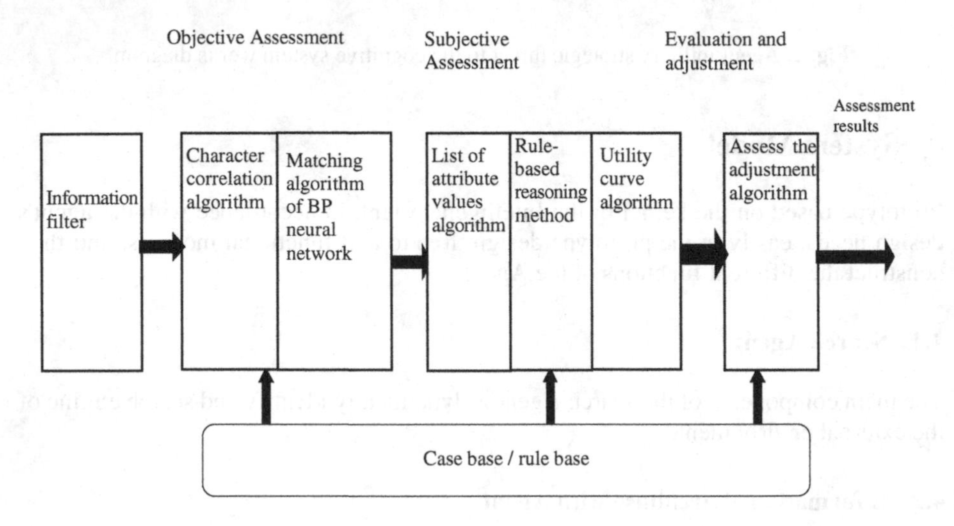

Fig. 3. Screening assessment agent assessment process diagram

4.6 Alarm with Agent

Alarm Agent core components of the threat of quantizer and threats standards library, its main function: First, alarm, alarm premise to assess the results quantify the threat identification, probability and event price or the product of assessment of the severity of events.

4.7 Agent Interface

The Interface Agent is an intelligent Agent to act as the role of man-machine interface in the system, it is an intelligent man-machine interface, or it could be considered an agent of the user. Its functions:1) It can provide services according to user requirements. 2) It can take the initiative to prompt the user with detailed information and the assessment of the threat information for the use of user awareness.

5 Conclusion

Combination of military strategy, cognitive process and artificial intelligence, agent technology research, build the model in order to find the best way to military strategy, information cognition, which is a preliminary attempt, which also exist system structure should be further improved, Agent function richer, this study should be applied to many problems in practice to be further addressed.

Acknowledgement. The work was supported by Xi'an hi-tech institute Research Foundation (EPGC2012).

References

[1] Knorr, K.: Threat Perception Historical Dimensions of National Security Problems. Lawrence Press, Kansas (1976)
[2] Quoted by Raymond Cohen. Threat Perception in International Crisis. University of Madison, Wisconsin Press (1979)
[3] Wang, S.-X.: Research on Agent-based Modeling and Simulation of Military Counter System. Fire Control Command Control (2009)
[4] Quc, H.: A Method of Threat Assessment Using Multiple Attribute Decision Making. In: 2002 6th International Conference on Signal Processing, pp. 1091–1095 (2002)
[5] Looneycg, L.R.: Cognitive Situationand Threat Assessment of Ground Battle spaces. Information Fusion 4(2), 297–308 (2003)
[6] Petterson, G., Axelsson, L., Jensen, T.: Multi source Integration and Temporal Situation Assessment in Air Combat. Information Decision and Control, 371–375 (1999)
[7] Department of Defense. Network Centric Warfare (2001)
[8] Darbyshire, P., Abbass, H., Barlowve, M., McKay, B.: A Prototype Design for Studying Emergent Battlefield Behavior through Multi-Agent Simulation. School of Information Systems, VUT, Australia (2000)

Based on Zigbee Technology Greenhouse Monitoring System

Mingtao Ma and Gang Feng

Jilin Agricultural Science and Technology College, Jilin, 132101 China
mmt800@126.com

Abstract. This paper account of the features of farmland greenhouse designs the corresponding monitor system.based on the self-designed test platform, the paper introduces the key issues of farmland greenhouse sensor network, analysis of the mechanic structure and network topology of wireless network protocol,designs the hardware structure of the system, and complete the software of the system.The experiment proved that the system is easy to use, stable and reliable,has a certain practical value.

Keywords: Zigbee, Network, Monitor.

1 Introduction

The weather is cold in winter in the northern part of China and it doesn't fit for the growth of various crops. However with the development of people's living standard and the need to enlarge the vegetable market in winter, the greenhouse provides a good artificial environment for the growing vegetables. While most of the current greenhouses are controled by artificial methods, it is difficult to collect and adjust various environmental parameters on time and accutately. Part of the greenhouses adopt the indicator parameters of the wired communication technology monitoring system, it still exists the shortcomings of the high cost of installation and maintenance, system power consumption , not flexibility. ZigBee is a low complexity, low power, low-cost, full-duplex wireless communication technology, this paper features this technology to the greenhouse environment monitoring.

2 The Introduction of Zigbee

ZigBee is a wireless personal area network technology, which fits for the occasion of small amount of data communication, the low rate of data transter but with a certain requirements for security and reliability of data. It develops the physical layer and data link layer standard according to IEEE 802.15.4 and ZigBee Alliance developed standards of the network layer and the application layer according to the wireless communication network . ZigBee supports three network topologies[1]: star-shaped network, the tree network and mesh , it is difficult to achieve high-density expansion with relatively poor flexibility and limitd coverage. Tree structure keeps the advantage

Z. Du (Ed.): Proceedings of the 2012 International Conference of MCSA, AISC 191, pp. 737–740.
springerlink.com

of simplicity of the star topology and needs only small upper routing information low memory requirements and low costs. Network structure has greater flexibility, scalability, reliability, but with complex structure and routing information , high cost and difficulty to maintain.

There are three logical network devices: coordinator, router and end devices. Coordinator: launching a new network, setting the network parameters, managing network nodes and storage nodes information in the network, network, only one coordinator, the other nodes as routers and; router: responsible for the routing information within the network; end devices: the node to achieve sense, among these coordinator and router have the functions to let devices join or leave the network.

3 System Components and the Working Process

The system uses a tree topology[2], and is composed of PC, coordinator, routers and end monitoring nodes ,and communicates with the monitoring computer with 485 interfaceof the structure of the system is shown in Figure 1.

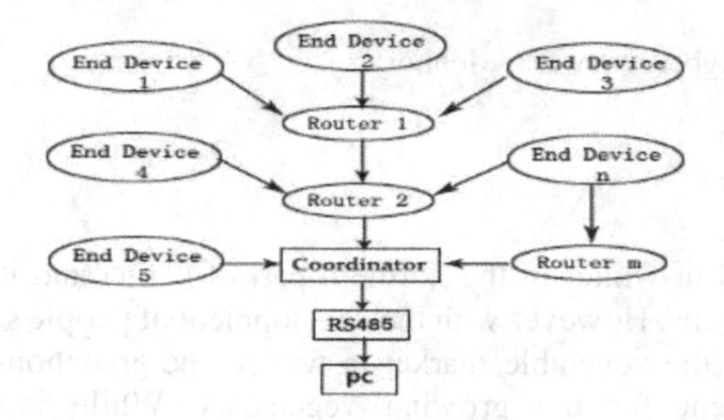

Fig. 1. System structure

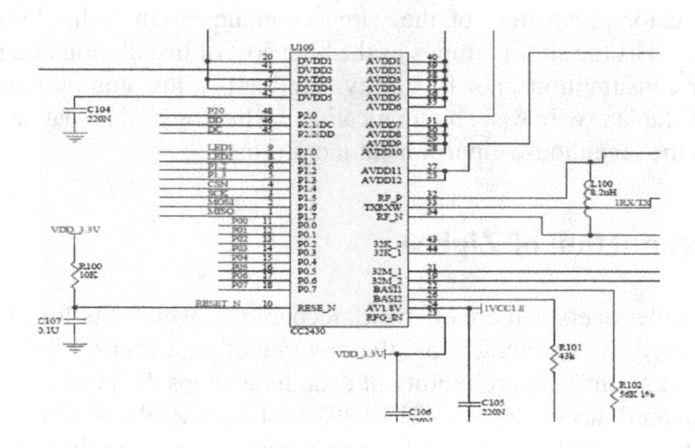

Fig. 2. Node of electric circuit

Coordinator starts the whole network, receives commands from PC and sends data or control commands from PC to the nodes directly or through router. Router is used as a transit point for data transmission, the node can communicate to the coordinator through multi-level routing; terminal nodes collect a variety of environmental information, send them to the coordinator through a router or directly and control temperature and humidity of the greenhouse through the intelligent control algorithm according to the instructions received from the coordinator.

4 System Hardware Design

System hardware includes the following parts: node design, power supply circuit, the serial circuit. Node circuit is shown in Figure 2, using TI's CC2430 as a core device.

In order to ensure system reliability and low power consumption, the coordinator is supplied by the main power directly. Routers and end monitoring nodes are supplied by using 3.3V battery-power. System power supplying circuit is shown in Figure 3.

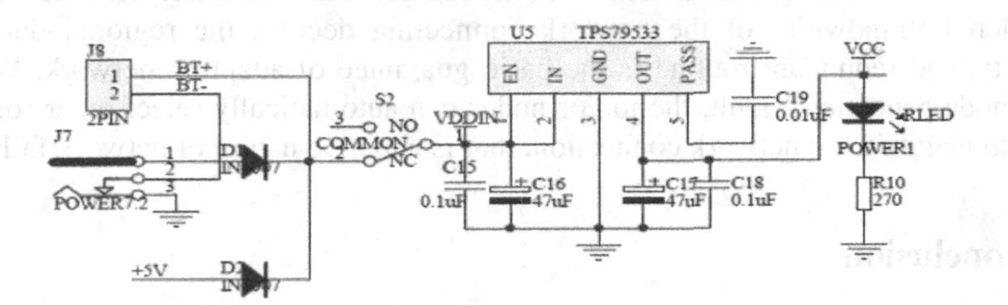

Fig. 3. System power supply circuit

5 Software Design of System

As the network coordinator, it can be divided into two parts according to the functions : network creation and management ; data transfer. The former is mainly responsible for the formation of a ZigBee network distribution network address and maintenance binding table. Coordinator creates a new network by scanning an empty channel to maintain a current list of connecting devices and support independent scanners to ensure that previously connection device can be able to rejoin the network. The latter is used as a gateway between Zigbee and RS485 and connect the two networks using different protocols together and converts the data with each other. Controller software flow is shown in Figure 4:

Coordinator is responsible for starting the network and initializing the hardware and the network, set up a ZigBee wireless sensor networks. Then the coordinator node is entering the wait state, waiting for a command sent by the PC or the information sent to the coordinator node .

6 Technical Difficulties

Ultra low power consumption can extend the life of the network nodes and the node energy consumption is in three areas: data acquisition of sensor components, data storage and data processing of micro-processing unit and wireless module receiver / transmitter. Among these the largest energy consumption is in the RF signal transmission process. Setting the node sleep and wake-up mechanism can minimize energy consumption. There is only 1.6μA current and 0dbm output power when the node is in deep sleep, the average current is about 250μA to wake up period of 1s.

ZigBee routing nodes[3] can participate in route discovery, data transmission, routing maintenance and expansion of the scope of the network, controlled by the ZigBee coordinator and it must be FFD. ZigBee end-nodes can only participate in routing mechanism through its parent node , which can be FFD or reduced functionality devices (RFD).

Fixed network nodes wireless sensor are the main carrier of connecting greenhouse wireless sensor nodes , with relative stability. In the process of setting up fixed network nodes , the connecting bandwidth of the nodes and the redundancy should be fully considered. Bandwidth of the network connecting decides the regional data load capacity, and redundant routing is the basic guarantee of adapting network. When a fixed node generates a fault, the junior nodes can automatically select other routes in order to maintain the network connection, that is adaptive nature of network failure.

7 Conclusion

Testing and actual operation proves that the use of the characteristics of ZigBee wireless technology realizes wireless automatic monitoring and controlling of the environment of greenhouses, and solve the problems of existing transmission such as the high cost, wiring complexity, maintenance problems, poor flexibility and scalability, save human resources and it is good for greenhouses's intelligent and unified management. At the same time, some minor changes to the system can be applied to other wireless sensor networks, which has great application value.

References

[1] Song, W., Wang, B., Zhou, Y.B.: Wireless Sensor Network Technology and Applications. Electronic Industry Press (March 2007)

[2] Li, W., Duan, C., Zig, B.: Wireless Networking Technology and Real Practice. Beijing Aviation and Aerospace University Press (April 2007)

[3] Lvzhi, A., Zig, B.: Networking Principles and Application Development. Beijing Aviation and Aerospace University Press, Beijing (February 2008)

Author Index

Bao, Hai 583
Bao, Qingpeng 129
Bao, Xue 421
Bian, Zhicheng 413, 421

Cao, Guoqing 609
Chen, Baogang 117, 577
Chen, Beibei 695
Chen, Bin 449
Chen, Chun-Sheng 603
Chen, Jingsong 19
Chen, Peng 197
Chen, Yijun 719
Chenggeng, Jinyan 363
Chi, ZhenFeng 259
Chu, Xiao 623
Cui, Xianbin 623
Cui, Xingyu 461

Dai, Qiwei 145
Deng, ChenHao 137
Deng, Hui 663
Deng, Weiming 301
Deng, Xiuhua 713
Dong, Junhua 523

Fang, Maoyan 731
Fang, Yuan 63
Feng, Aifen 203, 209
Feng, Gang 737
Feng, Lizhu 95

Gao, Ying 31
Gao, Yongming 1
Geng, Shu 129
Genjian, Yu 39
Gong, Qinhui 83
Gu, Rui 203, 209
Guo, Hongyan 295
Guo, Jun 495
Guo, Xiuli 283

Han, Shuhua 289
Han, Shuying 283
Hao, Chuanhui 197
He, Jinlong 639
He, Junliang 413
He, Li 381
Hong, Zhiling 77
Hou, Hailong 203, 209
Hu, Guangzhou 129
Hu, Jinlong 117, 577
Hu, Liang 167, 173, 179
Hu, Xiaona 413
Hu, Xiaowei 461
Hu, Xiaoya 95
Hu, Xiao-ying 479
Huang, Han-Chen 433, 441
Huang, Linna 229
Huang, Zhaohe 639
Huo, Liang 57

Jiang, Chunmin 507
Jiang, Yan 159
Jie, Ligui 455
Jin, Qiwen 393

Kong, Lingmei 701

Lai, Han 271
Lai, Wen 271

Lan, Jingsheng 639
Li, Baichao 111
Li, Baiqing 529
Li, Changcun 393
Li, Chun 609
Li, Chunmei 253
Li, Dawei 645
Li, Dongdong 623
Li, Haiyan 25
Li, Haoru 129
Li, Huaizhou 651
Li, Jing 47
Li, Juan 449
Li, Na 485
Li, Taowei 145
Li, Tinggui 83
Li, Wang 455
Li, Xiaofei 663
Li, Xiaohui 295
Li, Yan-xia 53
Li, Yuhong 381
Li, Zhijun 405
Li, Zhuguo 95
Liang, Limei 639
Liang, Zhihong 111
Lin, Li 253
Liu, Chunli 229
Liu, Fang 583
Liu, Gang 47
Liu, Hong 25
Liu, Jianjun 547
Liu, Jifei 657
Liu, Jin-sheng 223
Liu, Ruifang 393
Liu, Tian 687
Liu, Wei 657
Liu, Xi 427
Liu, Yan 639
Liu, Yanxia 547
Liu-Suqi 215
Lu, Houjun 427
Luo, Da-yong 151
Luo, Yingxiang 633

Ma, Jianfeng 559
Ma, Mingtao 737
Ma, Tao 101
Ma, Zongli 7
Mi, Weijian 427
Min, He 517
Mu, Ruihui 501

Pan, Xingshou 639
Pei, Xuesheng 629
Pei, Yan 235

Qi, Yang 387
Qin, Yihui 101

Ren, Ailian 511
Ren, Haoli 1
Ru, Cun-guang 491
Ru, Qian 553

Shen, Weihua 351
Shi, Huifen 615
Su, Xuemei 289
Sun, Zengyou 197

Tan, Hanlin 145
Tang, Lvhua 535
Tian, Bin 313
Tu, Chunhua 321

Wang, Bingwen 95
Wang, Bingzhang 259
Wang, Feng 663
Wang, Fenglei 511
Wang, Haiyin 405
Wang, Hongfeng 405
Wang, Hongxia 295
Wang, Jinghui 467
Wang, Lei 461
Wang, Li 645
Wang, Lifen 345
Wang, Minle 731
Wang, Qiang 53
Wang, Tao 57
Wang, Xiaojie 71
Wang, Xiaokai 713
Wang, Xin-gong 675
Wang, Yong-sheng 681
Wang, Yue 651
Wang, Zhen 399
Wei, Jingjing 461
Wei, Shaoqian 101
Wei, Ya-li 223
Weiyan, Lu 215
Wenyan, Wang 215
Wu, Baozhu 101

Wu, Gang 363
Wu, Hao 495
Wu, Jian 63
Wu, Jun 687
Wu, Lu 517
Wu, Meihong 77
Wu, Zexun 105

Xia, Guilin 13
Xia, Yanhua 375
Xie, Jing 159
Xie, Qiang 369
Xie, Xiaojie 63
Xiong, Xiaohong 307
Xiong, Yan 467
Xu, Jianchao 47
Xuan, Yonghua 609

Yan, Wei 413, 421
Yan, Xiong 455
Yang, Li 507
Yang, Liping 541
Yang, Liu 57
Yang, Peilu 707
Yang, Shen 25
Yang, Tiejian 731
Yang, Xin 381
Yang, Xing-dong 357
Yang, Xuming 615
Yang, Yang 495
Yang, Yunfeng 185
Yao, Sigai 669
Ye, Yanqing 145
Yi, Xiazi 399
Yin, Jibin 663
You, Yangming 259
Yu, Chen 473
Yu, Feng 657
Yu, Jin-tao 495
Yu, Xian 565, 725
Yu, Xue 571
Yue, Sufang 7

Zeng, Zhao-xian 223
Zhai, Ran 191
Zhang, Baixin 241
Zhang, Bichuan 71
Zhang, Bing 589
Zhang, Chunzhi 129
Zhang, Cuiping 327, 333, 339
Zhang, Guoqing 259
Zhang, Jianru 265
Zhang, Jun 151
Zhang, Qiansheng 167, 173, 179
Zhang, Ranran 645
Zhang, Wei 145
Zhang, Xiaolin 651
Zhang, Ying 609
Zhang, Ying-jian 53
Zhang, Zhi 89
Zhao, BaoBin 259
Zhao, Ning 421
Zhao, Wanzhe 247
Zheng, Jianlin 31
Zheng, Liqun 467
Zheng, Shao-peng 123
Zhou, Chang-xian 123
Zhou, Xueguang 449
Zhou, Zhemin 595
Zhu, Dan 511
Zhu, Hui-ling 159
Zhu, Wenjie 277
Zhu, Xin-yin 159